Recarbonization of the Biosphere

Rattan Lal • Klaus Lorenz • Reinhard F. Hüttl
Bernd Uwe Schneider • Joachim von Braun
Editors

Recarbonization of the Biosphere

Ecosystems and the Global Carbon Cycle

IASS
POTSDAM

Springer

Editors
Rattan Lal
Carbon Management and
Sequestration Center
The Ohio State University
Coffey Road 2021
Columbus, OH, USA

Reinhard F. Hüttl
Deutsches GeoForschungsZentrum GFZ
Helmholtz-Zentrum Potsdam
Telegrafenberg, Potsdam
Germany

Joachim von Braun
Zentrum für Entwicklungsforschung,
Universität Bonn,Walter-Flex-Str. 3 Bonn,
Germany

Klaus Lorenz
Global Soil Forum
IASS Institute for Advanced Sustainability
Studies
Berliner Str. 130
Potsdam, Brandenburg
Germany

Bernd Uwe Schneider
Deutsches GeoForschungsZentrum GFZ
Helmholtz-Zentrum Potsdam
Telegrafenberg 309G
Potsdam, Brandenburg
Germany

ISBN 978-94-007-4158-4 ISBN 978-94-007-4159-1 (eBook)
DOI 10.1007/978-94-007-4159-1
Springer Dordrecht Heidelberg New York London

Library of Congress Control Number: 2012935781

Printed on acid-free paper

Springer is part of Springer Science+Business Media (www.springer.com)

Foreword

Recognizing that human activities are a significant geologic force through, for example, land-use change and fossil fuel combustion, Paul Crutzen and Eugene Stoermer (2000) introduced the term "Anthropocene". Through expansion of agricultural and urban ecosystems along with combustion of biomass and fossil fuel, humans have drastically changed the "Face of the Earth". Increase in atmospheric concentration of greenhouse gases (GHGs) and the attendant global climate change are among the principal ramifications of the Anthropocene. Whereas emissions from fossil fuel combustion began with the on-set of industrial revolution ~1800 A.D., those from land-use conversion and deforestation commenced ~10–14 millennia ago with the dawn of settled agriculture. The cumulative emissions from land-use conversion over millennia are estimated by some at about 480 Pg carbon (C), compared with 275 Pg C from fossil fuel combustion and cement production between 1850 and 2000. World soils may have lost 60–100 Pg C through drainage, excessive tillage and the widespread use of extractive farming practices, which does not only affect C emissions but the development potential of people depending on soil resources. The magnitude of loss of the soil C pool and interrelated negative consequences for humans are exacerbated by degradation processes such as accelerated erosion, salinization, acidification, nutrient depletion, and elemental toxicity. Decarbonization of the terrestrial C pool has adversely affected numerous ecosystem services such as the provision of food and fresh water, supporting services such as elemental (C, N, P, S etc.) cycling, biodiversity, among others. The severe decline in agronomic productivity may be the most prominent negative effect of management practices depleting the soil C pool.

Recarbonization of the biosphere is an essential step to reduce the net anthropogenic C emissions through sequestration of carbon dioxide (CO_2) in terrestrial sinks comprising of forests, soils, and wetlands. The drawdown of atmospheric CO_2 through C sequestration in the terrestrial biosphere is estimated at ~50 ppm over a century. In addition to being a cost-effective and a natural process based on enhanced net primary production through adoption of restorative land use and management, the strategy of C sequestration in the terrestrial biosphere has numerous co-benefits such as increasing agronomic productivity and advancing global food security.

Terrestrial biosphere, with a high priority for an immediate action, and those with a high C sink capacity include: drained peat lands, soils degraded by erosion and salinization, agriculturally marginal lands, tropical rainforests and acid savanna soils, and urban ecosystems. In addition to afforestation and reforestation, avoidance of tropical deforestation, conversion of degraded ecosystems to restorative land use, and adoption of recommended soil and crop management practices are important options for recarbonization of the biosphere.

A wide spread adoption of recommended management practices, in particular, by the resource-poor farmers and small land holders of the developing countries requires appropriate, just, and fair process to incentivize them. Land managers could be incentivized, if implemented effectively, through payments for generating ecosystem services such as sequestering C to mitigate climate change, improving quality of natural waters, enhancing biodiversity, and improving environmental quality.

The workshop entitled "Recarbonization of the Biosphere" was held at the IASS Institute for Advanced Sustainability Studies Potsdam, from 20 to 22 March 2011. It was attended by about 30 participants from Germany, USA, Australia, Brazil, Italy, France, Spain, China representing soil science, forestry, climatology, sociology, and political sciences along with policy makers and media representatives. Specific recommendations emerging from the deliberations include the following:

Land-Based Carbon Sinks: Priority ecosystems with large C stocks which must be protected and sustainably managed are: permafrost, wetlands, peatlands, tropical rainforests, tropical savannas, urban lands, degraded or desertified lands, and agricultural lands.

Carbon Sink Capacity Management: The C sink capacity of land-based sinks are continuous, and require long-term management and protection strategy because it grows with a progressive improvement in ecosystem health.

Ecosystem Services and Co-Benefits of Land-Based Carbon Sinks: Protecting and enhancing land-based sinks generate numerous ecosystem services. Important among these are:

- Food security,
- Quality and quantity of renewable fresh water resources,
- Adaptation and mitigation to climate change, and
- Biodiversity

Global Soil Forum: The key roles in advancing food security and providing numerous co-benefits and ecosystems services of vital significance to humanity necessitate identification, nurturing and support of a transdisciplinary process with primary focus on sustainable management and protection of world soils.

This volume is based on the papers presented at the workshop. The organization of the workshop and publication of the volume were made possible by the staff of IASS in cooperation with The Ohio State University. The volume is edited by an inter-disciplinary team of scientists comprising of Drs. R. Lal, K. Lorenz, R. F. Hüttl, B. U. Schneider, and J. von Braun. Special thanks are due to all authors for their contributions and willingness to share the knowledge and expertise with others. The efforts of all others who contributed to publishing this volume in a timely manner are greatly appreciated.

Prof. Dr. Klaus Töpfer
Executive Director
IASS, Potsdam, Germany

Preface

With increase in world population from seven billion in 2011 to the projected 9.2 billion in 2050, several challenges facing humanity during the Anthropocene include the followings: (i) increasing atmospheric concentration of CO_2 and other greenhouse gases with the attendant risks of global warming, (ii) the need to increase food production by at least 70%, and even more in densely populated developed countries, and the severe decline in per capita productive agricultural soil/land, (iii) high risks of soil degradation and land desertification because of anthropogenic perturbations and those exacerbated by the projected climate change, (iv) pollution and contamination of natural waters, and the severe projected decline in per capita availability of renewable freshwater supply, and (v) a rapid and widespread decline in both above- and below-ground biodiversity.

Some relevant strategies of addressing these challenges include: (i) sequestering carbon (C) in the biosphere (i.e., soils, trees), (ii) making agriculture climate resilient and, as much as possible, C-neutral, (iii) improving quality of soils of agroecosystems, thereby increasing agronomic and net primary production (NPP), (iv) adopting technologies of sustainable intensification and saving land for nature conservancy, and (v) making soils/land integral to any strategy of mitigating the climate change and improving environmental quality.

Thus, the workshop entitled "Recarbonization of the Biosphere" was held at the IASS Institute for Advanced Sustainable Studies in Potsdam, Germany, from 20 to 22 March 2011. About 30 researchers and policy makers, from about ten countries, were invited to present the-state-of-the-knowledge review on thematic issues related to the topic. Two Nobel Prize Laureates, Prof. Dr. Paul Crutzen and Prof. Dr. Carlo Rubbia, also attended the workshop and presented articles on specific topics. The invited participants were specifically chosen to promote and enhance trans-disciplinary and inter-institutional cooperation and discussion. Thus, participants represented soil science, forestry, physics, chemistry, climatology, geology, hydrology, bio-geochemistry, sociology, and political science.

The workshop deliberations made the following conclusions:

1. Sustainable soil and land management is essential to achieve food security, sustainable development, to adapt to and mitigate climate change, and to maintain environmental quality.
2. Global problems of food insecurity, soil degradation, water scarcity and pollution, and energy insecurity are exacerbated by the changing climate, increasing population, and increasing standards of living.
3. For these reasons, IASS was encouraged to pursue the idea of creating a "Global Soil Forum" as an organizational mechanism with focus on sustainable management and protection of world soils.

Based on these conclusions, the workshop made the following recommendations:

1. Invest much more in soil resources, assessment and management in order to advance food security, adapt to and mitigate climate change, and enhance ecosystem services.
2. Identify key technological options to enhance and sustain soil-based ecosystem services, especially those that serve food security and nutrition.
3. Strengthen related institutional and human resource capacity, especially by providing long- and short-term training opportunities and research capacities.
4. Support and coordinate related activities, including program targets, monitoring and evaluation systems, network development and policy implementation at local, regional and global levels, and
5. Create a globally coordinated state-of-the-soils data bank that will facilitate identification and analysis of their properties for specific ecosystem services.

This 25-chapter volume is based on the papers presented at the workshop, and some additional chapters to strengthen the thematic discussion on all biomes. The volume addresses the potential and challenges of recarboninzation of the natural and managed biomes. Important among these are peatlands, forests, savannas/steppes, degraded and desertified lands, croplands, grazing lands and coastal ecosystems.

The organization of the workshop and publication of the volume were made possible by the visionary leadership and support of Prof. Dr. Klaus Töpfer, the Executive Director of IASS. The editors also thank all the authors for their outstanding efforts to document and present the state-of-the-knowledge scientific information on themes of interest to them. Their efforts have contributed to enhancing the overall understanding of the science, practices and policies of this theme of global significance. These articles have enhanced the frontiers of knowledge.

Special thanks are also be given to the staff of IASS and the Carbon Management and Sequestration Center (C-MASC) for their help and support in organizing the Workshop. Valuable contributions were made by Ms. Corinna Bobzien, Ms. Thando Tilmann, Dr. Falk Schmidt, Ms. Astrid Lehmann and other staff of IASS. We thank Ms. Theresa Colson of C-MASC, OSU for production support in preparing the volume for the final submission to the publisher. Thanks are also due to the staff of

Springer Verlaag, Dordrecht, The Netherlands, (Ms. Maryse Walsh and Ms. Melanie van Overbeek) in publishing the volume and making it available to the scientific community, policy makers, NGOs, and land managers. The efforts of many others, who supported this project in one way or another, are gratefully acknowledged.

Editors
Rattan Lal
Klaus Lorenz
Reinhard F. Hüttl
Bernd Uwe Schneider
Joachim von Braun

Editors Personal Profiles

Rattan Lal is a Distinguished University Professor of soil physics in the School of Environment and Natural Resources, and Director of the Carbon Management and Sequestration Center, College of Food, Agricultural, and Environmental Sciences/ Ohio Agriculture Research and Development Center, at The Ohio State University. He has worked at the University of Sydney, Australia (1968–1969), the International Institute of Tropical Agriculture (IITA), Ibadan, Nigeria (1970–1987) and at The Ohio State University (1987–). He has been affiliated with the Institute for Advanced Sustainability Studies (IASS), Potsdam as a member of the Scientific Advisory Board (2010) and with the Global Soil Forum (2011–). Professor Lal is a fellow of the Soil Science Society of America, American Society of Agronomy, Third World Academy of Sciences, American Association for the Advancement of Sciences, Soil and Waste Conservation Society, and Indian Academy of Agricultural Sciences. He is a recipient of the International Soil Science Award of the Soil Science Society of America, the Hugh Hammond Bennett Award of the Soil and Water Conservation Society, the 2005 Borlaug Award and 2009 Swaminathan Award. He also received an honorary degree of Doctor of Science from Punjab Agricultural University, India, from the Norwegian University of Life Sciences,

Aas, Norway, and the Alecu Russo Balti State University in Moldova. He is the past president of the World Association of the Soil and Water Conservation, the International Soil Tillage Research Organization and Soil Science Society of America. He was a member of the U.S. National Committee on Soil Science of the National Academy of Sciences (1998–2002, 2007–2010). He has served on the Panel of Sustainable Agriculture and the Environment in the Humid Tropics of the National Academy of Sciences. He has authored and coauthored about 1,500 research papers. He has also written 15 and edited or co-edited 50 books. Prof. Lal has been academic advisor and mentor of ~100 graduate students, 50 postdoctoral researchers and 85 visiting scholars from around the world.

Klaus Lorenz is Research Fellow at the IASS Institute for Advanced Sustainability Studies in Potsdam, Germany. He studied biology at University of Freiburg, Germany, and obtained his Ph.D. in Agricultural Sciences from University of Hohenheim, Germany. His research focuses on agricultural, forest and urban soil use and management to enhance soil organic carbon sequestration for climate change mitigation and adaptation. From 2004 to 2011, he was employed as Research Scientist at the Carbon Management and Sequestration Center, The Ohio State University, USA. He has written the book 'Carbon Sequestration in Forests Ecosystems' (Co-author Rattan Lal). At IASS, he is organizing workshops addressing the sustainable use of soils to meet increasing anthropogenic demands and mitigating global change. Together with colleagues in the IASS Cluster Global Contract for Sustainability, he is implementing the Global Soil Forum and organizing the first Global Soil Week to be held 2012.

Prof. Dr. Dr. h.c. Reinhard Hüttl is the Scientific Executive Director and Chairman of the Board at the Helmholtz Centre Potsdam – GFZ German Research Centre for Geosciences, the National Research Centre for geosciences in Germany. Reinhard Hüttl is also President of the German National Academy of Science and Engineering (acatech) as well as Vice-president of the Helmholtz Association and Coordinator of the Research Field "Earth and Environment".

Reinhard Hüttl was born on January 1, 1957. He studied Forest and Soil Sciences at the Albert-Ludwigs-University (ALU), Freiburg, Germany and at the Oregon State University (USA) from 1978 to 1983 and was awarded a Ph.D. at the ALU Freiburg in April 1986. From 1986 to 1992 he was Head of the International Research Department at the mining company Kali & Salz AG/BASF-group, Kassel, Germany. He worked as an Assistant Professor for Geobotany at the University of Hawaii, Honolulu, USA in 1990/1991 and was appointed to the Chair of Soil Protection and Recultivation at the Brandenburg University of Technology (BTU) in Cottbus, Germany in 1992 where he is Director of the collaborative research centre "Artificial Watershed Catchment" which is funded by the German Science Foundation.

He was member of the Council of Experts for Environmental Questions of the German Government, and both member and chairman of the Scientific Commission of the Scientific Council of the German Government as well as member of the Ethics Commission on the Safe and Secure Provision of Energy of the German Government. He is also member of a number of national and international Academies and was conferred with the Cross of Merit, First Class of the Federal Republic of Germany in July 2008.

Dr. Bernd Uwe Schneider is Head of Staff of the Scientific Executive Board at the Helmholtz Centre Potsdam – GFZ German Research Centre for Geosciences, the National Research Centre for geosciences in Germany.

Uwe Schneider completed his degree in Forest Engineering at the Faculty of Forest Sciences in Göttingen in 1984 and was awarded a Ph.D. in Soil Sciences and Plant Nutrition at the University of Bayreuth, Faculty of Geosciences in 1989.

He spent 5 years in Guatemala and Honduras as a Forest Expert for the German Technical Mission (GTZ). From 1995 to 2006 he was employed at the Brandenburg University of Technology (BTU) in Cottbus where he headed different working groups on soil biology, soil-root-interactions and spatial structure of root systems, forest ecology, renewable energy production and alternative land-use systems.

Uwe Schneider has coordinated several EU and German joint research projects in addition to a number of basic research projects funded by the German Research Foundation (Deutsche Forschungsgemeinschaft, DFG) and the BMBF. He is member of the German Soil Science Society (DBG) and has lectured at both German and international study courses. He was member of the Faculty Council, the Scientific Commission and the Senate at the BTU in Cottbus. Uwe Schneider is co-founder of the Institute for Advanced Sustainability Studies (IASS) in Potsdam, and the Lead Science Coordinator on behalf of the Helmholtz Association for the Helmholtz Alberta Research Initiative (HAI).

Joachim von Braun is Director of the Center for Development Research (ZEF), and Professor for Economic and Technological Change at University of Bonn, Germany. His research addresses international development economics topics, economics of natural resources, poverty; science and technology; food policy, and trade. von Braun was Director General of the International Food Policy Research Institute (IFPRI) based in Washington, DC, U.S.A. 2002–2009. 2000–2003 he was President of the International Association of Agricultural Economists (IAAE).He serves on various international, UN, and European advisory councils, and is member of the International Policy Council (IPC). von Braun is member of the German National Academy of Science and Engineering, the Academy of Science of North-Rhine Westphalia, Fellow of the American Association for the Advancement of Sciences; received a honorary Doctoral degree from University of Hohenheim, and the Bertebos Prize of the Swedish Academy of Agricultural Sciences 2009 and the Justus von Liebig Prize of the Fiat Panis foundation 2011 for his research on world food security.

Contents

Contributors

Tim Beringer Earth System Analysis, PIK, Potsdam Institute for Climate Impact Research, Potsdam, Germany

Martial Bernoux IRD, UMR Eco&Sols (Montpellier-SupAgro, Cirad, Inra, IRD), Montpellier Cedex 2, France

Jagtar Bhatti Northern Forestry Centre, Canadian Forest Service, Edmonton, AB, Canada

Michael F. Bird School of Earth and Environmental Sciences, James Cook University, Cairns, QLD, Australia

T. Andy Black Biometeorology and Soil Physics Group, University of British Columbia, Vancouver, BC, Canada

Julia Boike Alfred Wegener Institute for Polar and Marine Research, Potsdam, Germany

H.-R. Bork Institute for Ecosystem Research, Christian-Albrechts-University, Kiel, Germany

Klaus Butterbach-Bahl Karlsruhe Institute of Technology, Institute for Meteorology and Climate Research, Atmospheric Environmental Research (IMK-IFU), Garmisch-Partenkirchen, Germany

Department Atmosphere/Biosphere, Karlsruhe Institute of Technology, Garmisch-Partenkirchen, Germany

Xiaojing Cao UNEP-Tongji Institute of Environment for Sustainable Development, Tongji University, Shanghai, China

College of Environmental Science and Engineering, Tongji University, Shanghai, China

Carols Clemente Cerri Laboratório de Biogeoquímica Ambiental, CENA Centro de Energia Nuclear na Agricultura, USP Universidade de São Paulo, Piracicaba, SP, Brazil

Carlos Eduardo Pellegrino Cerri Dept. Ciência do Solo, ESALQ Escola Superior de Agricultura Luiz de Queiroz, USP Universidade de São Paulo, Piracicaba, SP, Brazil

Abad Chabbi Laboratoire de Biogéochimie et Ecologie des Milieux Continentaux (BIOEMCO), Centre INRA Versailles-Grignon, Bâtiment EGER, CNRS, Thiverval-Grignon, France

Unité de Recherche Pluridisciplinaire Prairies et Plantes Fourragères UR P3F, Centre de Recherche Poitou- Charentes, INRA, Lusignan, France

Galina Churkina Geography Department, Humboldt Universität zu Berlin, Berlin, Germany

Independent Research Group, Leibniz-Centre for Agricultural Landscape Research, Müncheberg, Germany

Richard Theodore Conant Natural Resource Ecology Laboratory, Colorado State University, Fort Collins, CO, USA

Institute for Sustainable Resources, Queensland University of Technology, Brisbane, QLD, Australia

Stephen Crooks Climate Change Services, ESA PWA | Environmental Hydrology, San Francisco, CA, USA

P.J. Crutzen Department of Atmospheric Chemistry, Max-Planck-Institute for Chemistry, Mainz, Germany

Ch. Dahlke DFG-Graduiertenkolleg Interdisziplinäre Umweltgeschichte, Georg-August-Universität Göttingen, Göttingen, Germany

Department of Geography, University of Hildesheim, Hildesheim, Germany

Michael Dannenmann Tree Physiology, Institute of Forest Botany and Tree Physiology, University of Freiburg, Freiburg, Germany

James Edwards The Belle Baruch Institute for Marine & Coastal Sciences, University of South Carolina, Columbia, SC, USA

Marine Sciences Program, University of South Carolina, Columbia, SC, USA

Stefan Erasmi Cartography, GIS and Remote Sensing Section, Faculty of Geoscience and Geography, Georg August-University, Göttingen, Germany

Brigitte Josefine Feigl Laboratório de Biogeoquímica Ambiental, CENA Centro de Energia Nuclear na Agricultura, USP Universidade de São Paulo, Piracicaba, SP, Brazil

Sabine Fiedler Forschungszentrum Jülich GmbH, IBG-3, ICG-4, Agrosphere, Jülich, Germany

Nicolas Gerber Center for Development Research, University of Bonn, Bonn, Germany

Stephan Glatzel Landscape Ecology and Site Evaluation, University of Rostock, Rostock, Germany

Ru Guo UNEP-Tongji Institute of Environment for Sustainable Development, Tongji University, Shanghai, China

College of Environmental Science and Engineering, Tongji University, Shanghai, China

Stefanie Heinze Department of Soil Science & Soil Ecology, Geographical Institute Bochum, Bochum, Germany

Felix Heitkamp Landscape Ecology, Faculty of Geoscience and Geography, Georg August-University, Göttingen, Germany

R. Hill The University of Adelaide, Adelaide, SA, Australia

R.A. Houghton Woods Hole Research Center, Falmouth, MA, USA

Reinhard F. Hüttl Deutsches GeoForschungszentrum GFZ Helmholtz Center Potsdam, Telegrafenberg, Potsdam, Germany

Anna Jacobs Department of Agronomy, Institute for Sugar Beet Research, Göttingen, Germany

Rachhpal Jassal Biometeorology and Soil Physics Group, University of British Columbia, Vancouver, BC, Canada

Hermann F. Jungkunst Landscape Ecology, Faculty of Geoscience and Geography, Georg August-University, Göttingen, Germany

J. Paul Krüger Landscape Ecology, Faculty of Geoscience and Geography, Georg August-University, Göttingen, Germany

Yakov Kuzyakov Department of Soil Science of Temperate Ecosystems, Büsgen Institute, Georg August-University, Göttingen, Germany

Rattan Lal Carbon Management and Sequestraton Center, The Ohio State University, Columbus, OH, USA

Moritz Langer Alfred Wegener Institute for Polar and Marine Research, Potsdam, Germany

Hugues Lantuit Alfred Wegener Institute for Polar and Marine Research, Potsdam, Germany

Susan Laurance School of Marine and Tropical Biology, James Cook University, Cairns, Australia

Fengting Li UNEP-Tongji Institute of Environment for Sustainable Development, Tongji University, Shanghai, China

College of Environmental Science and Engineering, Tongji University, Shanghai, China

Klaus Lorenz Global Soil Forum, IASS Institute for Advanced Sustainability Studies e.V., Potsdam, Germany

Fei Lu State Key Lab of Urban and Regional Ecology, Research Center for Eco-Environmental Sciences, Chinese Academy of Science, Haidian District, Beijing, China

Wolfgang Lucht Potsdam Institute for Climate Impact Research, Potsdam, Germany

Department of Geography, Humboldt Universität zu Berlin, Berlin, Germany

Bernd Marschner Geographisches Institut Bodenkunde/Bodenoekologie, Ruhr-Universitaet Bochum, Bochum, Germany

A. David McGuire Department of Biology and Wildlife, U.S. Geological Survey, Institute of Arctic Biology, University of Alaska Fairbanks, Fairbanks, AK, USA

James T. Morris Department of Biological Sciences, University of South Carolina, Columbia, SC, USA

The Belle Baruch Institute for Marine & Coastal Sciences, University of South Carolina, Columbia, SC, USA

Sina Muster Alfred Wegener Institute for Polar and Marine Research, Potsdam, Germany

Paul Overduin Alfred Wegener Institute for Polar and Marine Research, Potsdam, Germany

Cecilio Oyonarte Deptamento de Edafología & Química Agrícola, Universidad de Almería, Almería, Spain

Enrique Reyes Department of Biology, East Carolina University, Greenville, NC, USA

Kurt Roth Institute of Environmental Physics, University of Heidelberg, Heidelberg, Germany

Carlo Rubbia IASS Institute for Advanced Sustainability Studies e.V., Potsdam, Germany

Cornelia Rumpel Laboratoire de Biogéochimie et Ecologie des Milieux Continentaux (BIOEMCO) Centre INRA Versailles-Grignon, Bâtiment EGER, CNRS, Thiverval-Grignon, France

Torsten Sachs Inorganic and Isotope Geochemistry, GFZ German Research Centre for Geosciences, Potsdam, Germany

Enrique P. Sánchez-Cañete Estación Experimental de Zonas Àridas (EEZA, CSIC), Almería, Spain

Bernd Uwe Schneider Deutsches GeoForschungszentrum GFZ Helmholtz Center Potsdam, Telegrafenberg, Potsdam, Germany

Penélope Serrano-Ortiz Estación Experimental de Zonas Àridas (EEZA, CSIC), Almeria, Spain

Department of Applied Physics, Universidad de Granada, Granada, Spain

Klaus Töpfer IASS Institute for Advanced Sustainability Studies e.V., Potsdam, Germany

Michelle Venter School of Earth and Environmental Sciences, James Cook University, Cairns, Australia

Oscar Venter School of Marine and Tropical Biology, James Cook University, Cairns, Australia

Joachim von Braun Zentrum für Entwicklungsforschung, Universität Bonn, Bonn, Germany

Hongtao Wang UNEP-Tongji Institute of Environment for Sustainable Development, Tongji University, Shanghai, China

College of Environmental Science and Engineering, Tongji University, Shanghai, China

Xiaoke Wang State Key Laboratory of Urban and Regional Ecology, Research Center for Eco-Environmental Sciences, Chinese Academy of Sciences, Haidian District, Beijing, China

Matthias Wendland Institut für Agrarökologie, ökologischen Landbau und Bodenschutz, Bayrische Landesanstalt für Landwirtschaft, Freising, Germany

Sebastian Westermann Department of Geosciences, University of Oslo, Oslo, Norway

Le Yang State Key Lab of Urban and Regional Ecology, Research Center for Eco-Environmental Sciences, Chinese Academy of Sciences, Haidian District, Beijing, China

Jing Zhang UNEP-Tongji Institute of Environment for Sustainable Development, Tongji University, Shanghai, China

College of Environmental Science and Engineering, Tongji University, Shanghai, China

Chapter 1
Terrestrial Biosphere as a Source and Sink of Atmospheric Carbon Dioxide

Rattan Lal, K. Lorenz, R.F. Hüttl, B.U. Schneider, and Joachim von Braun

Abstract The terrestrial biosphere has lost a considerable amount of its antecedent carbon (C) pool because of anthropogenic activities since the dawn of settled agriculture about 12–14 millennia ago. Deforestation and land use conversion has presumably caused cumulative emission of 476 Pg C (1 Pg = 10^{15} g). Of this, 78 ± 12 Pg C may have been depleted from world's soils. Globally, about 2,300 Pg C are stored to 3-m depth in the soil organic carbon (SOC) pool, 1,700 Pg C in permafrost, 600 Pg C in peatlands, and up to 1,700 Pg C in the soil inorganic carbon (SIC) pool. While a large fraction of C emissions may have been absorbed by the ocean and land-based sinks, the knowledge about the historic loss provides a reference point about the technical C sink capacity of the terrestrial biosphere. The later may be as much as a draw-down of 50 ppm of atmospheric carbon dioxide (CO_2) over century or more, which in view of the already accumulated levels of atmospheric CO_2 of 390 ppm is significant. Priority soils and ecosystems for recarbonization of the

R. Lal (✉)
Carbon Management and Sequestration Center, The Ohio State University, 2021 Coffey Road, Kottmann Hall 422b, Columbus, OH 43210, USA
e-mail: lal.1@osu.edu

K. Lorenz
Global Soil Forum, IASS Institute for Advanced Sustainability Studies e.V., Berliner Strasse 130, 14467 Potsdam, Germany
e-mail: klaus.lorenz@iass-potsdam.de

R.F. Hüttl • B.U. Schneider
Deutsches GeoForschungszentrum GFZ Helmholtz Center Potsdam, Telegrafenberg, 14473 Potsdam, Germany
e-mail: reinhard.huettl@gfz-potsdam.de; bernd_uwe.schneider@gfz-potsdam.de

J. von Braun
Zentrum für Entwicklungsforschung, Universität Bonn, Walter-Flex-Str. 3, 53113 Bonn, Germany
e-mail: jvonbraun@uni-bonn.de; a.vdveen@uni-bonn.de

R. Lal et al. (eds.), *Recarbonization of the Biosphere: Ecosystems and the Global Carbon Cycle*, DOI 10.1007/978-94-007-4159-1_1,

biosphere include degraded soils (eroded, salinized, depleted, polluted and drained peatland soils), and desertified ecosystems. Whereas the generic technologies for sustainable intensification exist for croplands, grazing lands, forest lands and restoration of degraded soils, these technologies must be validated and fine-tuned for soil-and-site specific conditions. The adverse externalities of land use change for both, climate and soils requires policy actions for corrective incentives. However, there is no panacea, and a wide range of technological options need to be carefully and prudently evaluated under site-specific situations. Policy interventions must incentivize land managers for implementing sustainable land use-, soil- and crop management practices that are avoiding the adverse effects for climate and soils. Incentives that foster the natural process of recarbonization of the biosphere can be a cost-effective strategy, and would have numerous co-benefits.

Keywords Abrupt climate change • Terrestrial biosphere • Land use • Land cover change • Urbanization • Fossil fuel combustion • Ecosystem C pool • Anthromes • Afforestation • Deforestation • Soil restoration • Soil erosion • Carbon transported by erosion • Mineral-associated carbon • Rubisco • C_4 plants • C_3 plants • Policy implications • Ecosystem services • Payments for ecosystem services • Zero emission technology • Aerosols • Co-benefits • Payments for ecosystem services

Abbreviations

ACC	Abrupt climate change
C	Carbon
CCS	Carbon capture and storage
CDM	Clean Development Mechanism
Gha	Gigahectare
GCC	Global carbon cycle
GHGs	Greenhouse gases
LULCC	Land use land cover change
Mha	Million hectare
NPP	Net primary productivity
ppm	Parts per million
Pg	Petagram
SIC	Soil inorganic carbon
SOC	Soil organic carbon

1.1 Introduction

The terrestrial biosphere impacts and is impacted by the climate. The abrupt climate change (ACC) can affect the ecosystem carbon (C) pools through losses by erosion and mineralization, and decline in biodiversity along with the attendant changes and

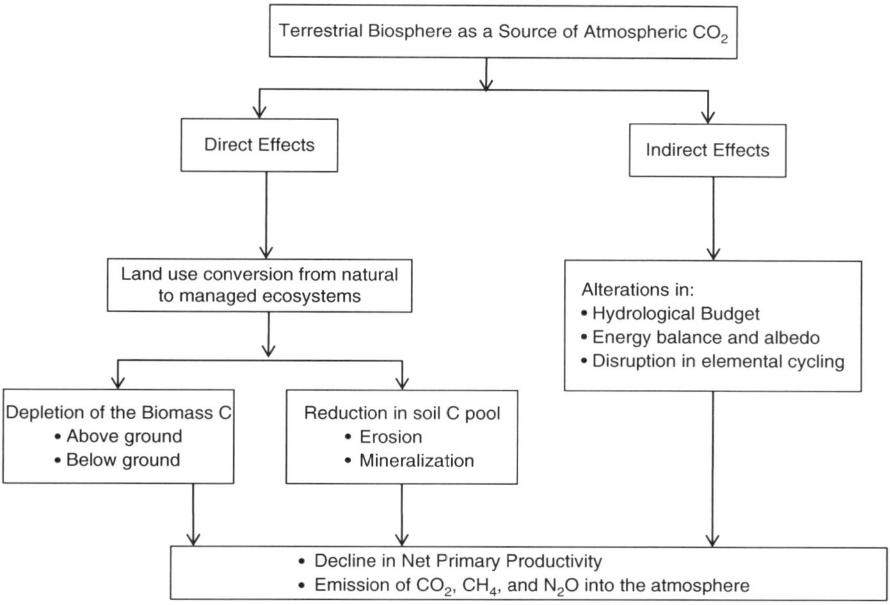

Fig. 1.1 Human-induced alterations in the terrestrial biosphere

spatial shifts in major biomes (Fig. 1.1). Land use change, including conversion from natural to managed ecosystems and specific management practices to raise plants and animals, has strong influence on the C pools in the above and below-ground biomass and in the soil profile. Humans have drastically transformed the terrestrial biosphere ever since the dawn of settled agriculture about 12–14 millennia ago (Ruddiman 2003, 2005). Principal biomes of the world (Table 1.1) are major reservoirs of C, stored in biomass and soil. Anthropogenic perturbations have disturbed the ecosystem C pools of the natural biomes through conversion and land use change to cropland, grazing/pasture land, forest plantations, urban settlements and rural communities, and industrial and infra-structure development. In addition, drainage of wetlands has also reduced the C pool in these ecosystems through oxidation of peat. It has been estimated that prior to the industrial revolution in 1700, nearly half of the terrestrial biosphere was wild, and most of the remainder (~45%) was in a semi-natural state (Ellis et al. 2010). By 2000, only 25% was wild, <20% in semi-natural state, and the remainder had been converted to agricultural and settled anthropogenic biomes (anthromes) or biomes drastically altered by human. Majority of the natural biomes have been completely transformed into croplands, rangelands, plantations, and urban/rural settlements. Agricultural expansion and intensification, in particular, are the main driver of global land use/land cover change (LULCC, Pielke et al. 2011). The global area of cropland dramatically increased to about 11% of earth's total ice-free land surface as did the global area used for grazing livestock to about 25% of earth's total ice-free land surface. Further, about 0.5% of the total global land area is urban (Schneider et al. 2009). At the

Table 1.1 Principal global biomes and climates

Climates	Biome
I. Low latitude climates	
1. Tropical rainforest (Af)	Annual precipitation >250 cm, temperature 27°C, humidity 77–88%, 10° N and S of equator
2. Wet-dry tropical Savanna (Aw)	Annual precipitation 75–150 cm, temperature range of 15°
3. Dry tropical climate (Bw)	Annual precipitation 25–50 cm, temperature range 15°C, latitude 15–25°N and S
II. Mid latitude climates	
4. Temperate Savanna/Steppe (Bs)	Annual precipitation of <10 cm, temperature range 24°C, latitude 35–55°N
5. Mediterranean climate/Chaparral (Cs)	Wet winter/dry summer, annual precipitation of 25–45 cm, temperature range 7°C, latitude 30–50°N and S
6. Dry middle climates (grasslands (Bs))	Annual precipitation of 75–80 cm, temperature range 30°C, latitude range of 30–55°N and S
7. Moist deciduous forest (Cf)	Annual precipitation 80 cm, temperature range 30°C, latitude 30–55°N and S
III. High latitude climates	
8. Boreal forest/Taiga (Dfc)	Annual precipitation 30 cm, temperature range 40°C, latitude 50–70°N and S
9. Tundra (E)	Annual precipitation 20 cm, temperature range −22°C to 6°C, latitude 60–75°N
10. Alpine/highland (H)	Annual precipitation of 20–25 cm, temperature range −18°C to 10°C, Latitude (depends on altitude)

beginning of the twenty-first century, 39% of the earth's total ice-free surface had been converted into agricultural land and settlements, and an additional 37% has been embedded within managed biomes (Ellis et al. 2010), and was vulnerable to human activities. Consequently, these anthromes have been drastically decarbonized, leading to increase in atmospheric concentration of CO_2 and other greenhouse gases (GHGs), and increased risks of ACC.

Land use conversion and creation of anthromes have drastically influenced the global C cycle (GCC). It is estimated that since 1850 about 35% of anthropogenic emission resulted from land use conversion (Foley et al. 2005). Perhaps as late as 1950s relatively more anthropogenic emissions were generated from LULCC than from fossil fuel combustion. In addition to perturbing the GCC, creation of anthromes also affects energy and hydrologic balance, elemental cycling, biodiversity and natural habitat, and the attendant degradation of the quality of soil and water resources (Fig. 1.1). There has been a large-scale deforestation, including that in North America and Europe, since the middle of seventeenth century (Table 1.2). The rate of deforestation accelerated in the humid tropics (e.g., Amazon Basin, Congo Basin, West Africa and Sumatra) during the middle of twentieth century. By 1990, global forest cover decreased by about 1,100 Mha, and only a few desert regions, parts of

Table 1.2 Estimates of global deforestation between 1650 and 1978 (Adapted from Williams 1994)

Year	Land area deforested (10^6 ha)	
	High estimate	Low estimate
Pre-1650	152.2	98.6
1650–1749	75.8	59.8
1750–1849	159.2	168.0
1850–1978	418.5	418.5
Total area	805.7	744.9

Table 1.3 Increase in population, industrial output, energy, use of irrigation, fertilizers and pesticides between 1900 and 2000 (Ponting 2007)

Parameter	Increase factor between 1900 and 2000
Population	3.8
Urban population	12.8
Industrial output	35
Energy use	12.5
Oil production	300
Water use	9
Irrigated area	6.8
Fertilizer use	342
Organic chemicals	1,000
Urban water use	20

Table 1.4 Agricultural intensification, and energy-related emissions between 1890 and 1990 (Adapted from Crutzen 2002)

Parameter	Increased factor between 1980 and 1990
I. Agriculture	
Cattle population	4
Pig population	9
Cropland area	2
Forest area	0.8
Use and fish catch	35
II. Energy related emissions	
Coal production	4
CO_2 emissions	17
SO_2 emissions	13
Lead emissions	8

the central Amazon and Congo Basins, and the Arctic and Antarctic had not been affected by LULCC by 2000 (Pielke et al. 2011). In recent decades, the global deforestation rate decreased but is still at about 13 Mha year^{-1} for the period 2000–2007, with deforestation rate for tropical forests alone of 12 Mha year^{-1} (Pan et al. 2011). Further, agricultural lands are being used intensively to increase agronomic/food production through input of water for irrigation, chemical fertilizers as plant nutrients, and pesticides to control pests and pathogens. Increase in population during the twentieth century by a factor of four accentuated the demand for production of food and other basic necessities. Consequently, there has been a drastic increase in the area under cropland and especially irrigated croplands, fertilizers and pesticide use, and the cattle population (Tables 1.3 and 1.4). The era since the industrial revolution is appropriately termed the "Anthropocene" (Crutzen 2002). Over the 40 year period ending in 2006, agricultural intensification doubled crop production even though the area under arable land use increased by only 12%, fertilizer use increased by 700%, and mechanization increased drastically (Foley et al. 2007). As much as 24% of the Earth's net primary productivity (NPP) is appropriated by humans (Haberl et al. 2007). Thus, the objective of this chapter is to describe the principles, practices and policies for recarbonization of the biosphere to mitigate and adapt to ACC.

1.2 Loss of Carbon from the Terrestrial Biosphere

Anthropogenic CO_2-C emission from fossil fuel combustion and cement production increased from about 0.5 Pg C year^{-1} in 1900 to 6.6 Pg C year^{-1} in 2000 (Sterman 2008), 8.7 Pg C year^{-1} in 2008 (Canadell et al. 2007a,b; Le Quéré et al. 2009), and 9.1 Pg C year^{-1} in 2010 (Peters et al. 2011). The emissions from land-use change in 2010 were about 0.9 Pg C year^{-1}, and ~40% of the total emission of 10 Pg C year^{-1} remained in the atmosphere (Peters et al. 2011). Thus, in conjunction with emission from land use conversion, total anthropogenic emissions have increased atmospheric concentration of CO_2 from 280 ppm during the pre-industrial era to 390 ppm in 2010 (WMO 2010). Increase in atmospheric CO_2-C is influenced both by land use conversion, and fossil fuel combustion and cement production. In contrast to the onset of Anthropocene around 1,800 (Crutzen and Stoermer 2002), Ruddiman (2003) hypothesized that the Anthropocene began thousands of years ago with the dawn of settled agriculture. Increase in atmospheric CO_2 concentration may have begun about 8,000 years ago and that of methane (CH_4) about 5,000 years ago corresponding with start of forest clearance and cultivation of irrigated rice paddies, respectively (Ruddiman 2003). Emissions of CO_2-C over 8,000 years are estimated by Ruddiman (2003) as follows:

(a) Pre-industrial Era (7,800 years) @ 0.04 Pg C year^{-1} = 312 Pg C
(b) Industrial Era (200 years) @ 0.8 Pg C year^{-1} = 160 Pg C

Thus, total emission from the terrestrial biosphere equal 472 Pg. While most of these emissions may have been absorbed by ocean and by the forest and land sinks, the cumulative emission equals ~120 ppm (~4 Pg = 1 ppm) (Broecker 2007). Of this cumulative emission, 78 ± 12 Pg C may have been depleted from the pedosphere, especially from the soils of agroecosystems and those which have been degraded or desertified (Lal 1999). Crude and tentative as these estimates may be, these data provide a reference point or base line towards judging the technical C sink capacity of the terrestrial biosphere.

1.3 Recarbonization of the Terrestrial Biosphere

There is a strong need to identify a wide range of feasible options to reduce and sequester anthropogenic CO_2 emissions. The cost of not acting may be much higher than that of an appropriate and timely action by the world community. It is precisely in this context that the importance of the strategy of recarbonization of the biosphere cannot be over-emphasized. The philosophy of recarbonization through restoration and rehabilitation of degraded soils and desertified ecosystems and judicious/prudent management of the biosphere is also in accord with numerous cultural believes and societal values such as the one expressed by Voltaire "Il faut cultivar notre jardin" (Read 2006).

There has been a drastic reduction in the land area under biomes with a high ecosystem C pool. The data in Table 1.5 show the decrease in the land area under forest and savanna biomes between 1700 and 2000. In contrast, there has been a strong increase in anthromes, especially the cropland and grazing land. However, even before 1500 humans had immense impacts on forests in the northern, temperate and the tropical regions (Williams 2000). Also, the distribution of historical LULCC over time is highly regionalized (Pielke et al. 2011). For example, by 1500 large areas of Western Europe had been partially cleared, and LULCC intensified, in particular, in this region through 1800 while significant LULCC also occurred over much of Asia including India and China. By 1750, only Western Europe and perhaps parts of Northern China were strongly affected by LULCC. However, by 1990 intensive LULCC had impacted parts of the United States, much of Western Europe, India, Northern China, and elsewhere. Further, large areas of the Southern Hemisphere underwent LULCC throughout the nineteenth century (Pielke et al. 2011). Globally, up to 800 Mha of closed canopy forest and up to 300 Mha of open woodland and shrubland were cleared (Williams 2000). Such LULCC and, in particular, deforestation have depleted the ecosystem C pool, and created C sink capacity which can potentially be refilled to some extent through conversion to a restorative land use, and adaption of recommended practices of soil and vegetation management.

Technical options for restoration of croplands, forest lands, wetlands and degraded/desertified lands are outlined in Fig. 1.2. These are generic technologies, but there is no single solution that is universally applicable. These technologies must be validated and adapted under soil/site-specific situation, with due consideration to social, cultural, economic and political factors which involve the human dimensions of land use and management.

Maintaining/preserving natural ecosystems is essential to enhancing/sustaining ecosystem services and functions. Assuming that 2.4 Gha of land can be used for enhancing the biosphere C pool, it has a technical potential of sequestering at least 3.5 Pg C $year^{-1}$, even at a modest rate of 1.5 Mg C ha^{-1} $year^{-1}$.

Table 1.5 Global area (10^9 ha) under different biomes between 1700 and 2000

Year	Forests/ Woodland	Savanna/Grass /Steppe	Abandoned forests/Woodland	Abandoned Savanna/ Grass/Steppe	Cropland	Pasture land
1700	5.28	3.23	0	0	0.41	0.32
1750	5.19	3.20	0	0	0.54	–
1800	5.09	3.17	0	0	0.68	0.51
1850	4.99	3.14	0	0	0.82	–
1900	4.80	3.02	0.01	0.002	1.14	1.29
1950	4.60	2.83	0.05	0.02	1.53	2.47
1990	4.40	2.67	0.15	0.06	1.79	3.34
2000	4.09	2.03	–	–	1.53	3.43

Loveland et al. (2000), Ramankutty and Foley (1999), FAOSTAT (2010), and Ruddiman and Ellis (2009)

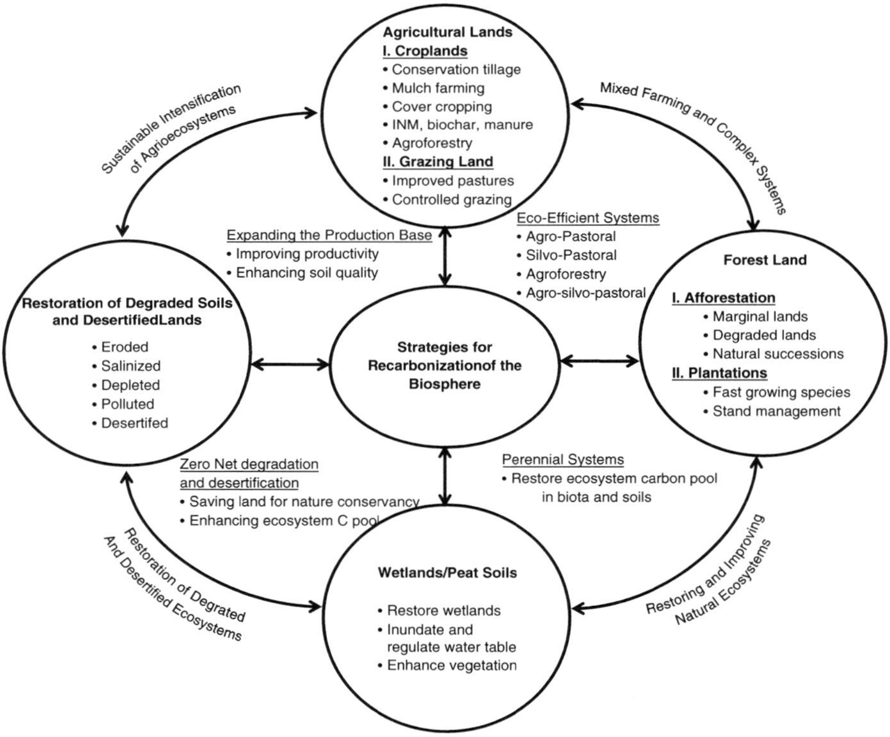

Fig. 1.2 Strategies towards recarbonization of the biosphere

Management of the terrestrial C pool may also imply leaving the natural ecosystem intact to the extent possible. Strategies to manage the C pool in these lands have been proposed by Read (2007) Read and Parshotam (2007), Fischer and Schrattenholzer (2001) among others. These strategies include the following:

1. Afforestation on marginal lands and degraded soils to enhance the NPP and increase the input of C into the soil,
2. Restoration and enhancement of soil quality and biomass productivity of degraded/depleted agricultural soils,
3. Restoration of wetlands and peatlands, and
4. Adoption of sustainable-intensive land use in soil/crop management systems.

Most of the available land for recarbonization exists in developing countries where the future incremental demand for food, feed and fibers is also the highest. Thus, it is strategically important to involve all countries in the decision-making process, especially those where surplus land exists for the recarbonization process. Establishment of biofuel plantation by deforestation of tropical rainforest or clearance of peatlands (such as in Indonesia or Malaysia) is counter-productive because of a large C-debt created by this land use conversion (Righelato and Spracklen 2007; Fargione et al. 2008).

Recarbonization of biosphere C must particularly focus on restoring soil C pools, comprising the soil inorganic carbon (SIC) and soil organic carbon (SOC) pools, as recarbonized soils contribute to climate change adaptation and mitigation. Further, restoring the SOC pool enhances soil quality, ecosystem services and food security. LULCC has and continues to deplete soil C pools. For example, SOC pools to 100-cm depth decreased by 42% when native forest was converted to cropland but no significant changes occurred below 60-cm depth (Guo and Gifford 2002). However, the variability in subsoil C changes is high but not many studies report data on subsoil SOC (Prechtel et al. 2009; Poeplau et al. 2011). In temperate regions, in particular, deforestation and conversion to cropland caused a rapid SOC loss of 32 ± 20% to 29 ± 14 cm depth with a new SOC equilibrium being reached after 23 years (Poeplau et al. 2011). The change rate of SOC increased with temperature but decreased with increase in clay content and with decrease in soil texture. Further, changes in the subsoil (>20 cm depth) were not different from changes in the topsoil. However, studies on organic soils and wetlands soils were not included (Poeplau et al. 2011). When primary forests in the tropics are converted, SOC losses were 12 ± 2% to 36 ± 3 cm depth when converted to grassland, 25 ± 3% to 36 ± 4 cm depth when converted to cropland, and 30 ± 3% to 48 ± 8 cm when converted to perennial crops, respectively (Don et al. 2011). The relative SOC loss in the subsoil was similar on grassland but not significant for croplands. The SOC losses increased with increasing temperature, and for conversion into grassland also with increasing precipitation. However, current hot spots of LULCC in South East Asia and Africa, and those in C-rich tropical wetland forests were not taken into account (Don et al. 2011). In general, about 25–30% of the SOC stored in the top meter of soil is released by cultivation of native soils, whether under forest or prairie vegetation (Houghton 2010).

Improved vegetation and soil management practices can recarbonize the biosphere by recovering some of the SOC released in the past. For example, in temperate regions cropland conversion into forest increased SOC by 117±54% to 28±15 cm depth when the forest floor was included, and the SOC increase was negatively correlated with temperature and soil depth but positively with precipitation (Poeplau et al. 2011). When only the mineral soil was considered, SOC changed by 83±39% to 40±25 cm depth, and SOC increases decreased with higher precipitation. However, no SOC equilibrium may be reached more than 120 years after converting cropland into forest (Poeplau et al. 2011). When grassland was converted into forest and the forest floor included, SOC increased 28±11% to 39±12 cm depth and this increase was negatively correlated with precipitation but may probably not be in equilibrium after 200 years. However, when only the mineral soil was considered SOC did not change to 25±16 cm depth. This may continue for at least 150 years after conversion and, in particular, not change with increase in temperature and soil depth, and decrease in clay content (Poeplau et al. 2011). Thus, experiments to recarbonize the soil must study changes in soil profile C and be accompanied by modeling as reaching a new equilibrium is a long-term process (Wutzler and Reichstein 2007). The persistence of SOC also depends on soil type as it determines reactive mineral surfaces, water availability, soil acidity and soil redox state (Schmidt et al. 2011).

The SOC losses after deforestation in tropical regions are also partly reversible. If cropland is afforested, SOC increased by 50±12% to 44±6 cm depth (Don et al. 2011). Further, afforestation of grassland increased the SOC pool by 18±8% to 35±6 cm depth. The average time period was only 33 years since afforestation but it is not known whether SOC equilibrium was already reached (Don et al. 2011). In summary, wherever land use change decreases soil C, the reverse process usually increases soil C (Guo and Gifford 2002). However, the simplifying assumption in many models that SOC pools can reach equilibrium has been challenged by the observation that some old forest soils do still accumulate C (Wutzler and Reichstein 2007). Similar, agricultural soils may never reach a theoretical equilibrium SOC level because of changing conditions and partial resets by disturbances such as erosion (Polyakov and Lal 2004). However, the chemically and biochemically protected SOC pools may be influenced by C-saturation behavior (Stewart et al. 2009). Once the chemically protected SOC pool is filled, added C may accumulate in the physically and in the non-protected fractions. Mineral-associated SOC pools, in particular, eventually saturate. In topsoils, reactive mineral surface area is a finite resource, and C-saturation may occur in SOC associated with the mineral phase in topsoils (Séquaris et al. 2010). Thus, arbitrarily defined soil fractions may have different C-saturation dynamics (Stewart et al. 2009).

Aside replacing SOC-depleting with SOC-accreting land use and management, recarbonization of the soil can be achieved by SOC sequestration implying an additional net transfer of C from the atmosphere to the soil via biomass (Powlson et al. 2011). Through enhancing the SOC pool the capacity of soil to produce food, feed, fiber and fuel can be restored for sustainable development. Thus, SOC-accreting soil and land use management practices must be implemented for the recarbonization of the soil.

Among management strategies to enhance SOC is phyto-engineering, i.e., the breeding and cultivating of plants for SOC sequestration. Strategies include improving the carboxylation efficiency of the enzyme Rubisco which catalyzes the first major step of C fixation during photosynthesis (Spreitzer and Salvucci 2002). Another approach includes increasing the proportion of C_4 plants in warmer climates as plants using the C_4 photosynthetic pathway are more efficient in converting solar radiation into biomass under these climatic conditions than those using the C_3 photosynthetic pathway (Zhu et al. 2010). However, as the overall efficiency of C_3 plants may be higher at lower temperatures (Jansson et al. 2010), the C_3/C_4 species mixture needs to be optimized for SOC sequestration in colder climates. As plant roots and associated microorganisms are the major soil C input (Lorenz and Lal 2005), breeding of plants with deeper and bushy root ecosystems and their cultivation may contribute to SOC sequestration (Kell 2011). Further, replacing annual crops with perennial crop relatives may contribute to the recarbonization of the soil by an additional net transfer of C from the atmosphere into soil (Glover et al. 2010). However, it must be evaluated whether (i) production of plants with deeper and bushy root ecosystems will be at the expense of aboveground biomass yields and (ii) whether soil resources (e.g., nutrients, water) are sufficient to support the cultivation of perennial crop relatives.

Engineering towards recarbonized soils can also be achieved through the construction of reclaimed mine soils and urban soils with high C contents in the stabilized SOC fraction (Hüttl and Gerwin 2005; Macías and Arbestain 2010). The soil addition of black C compounds (e.g., char, charcoal, biochar) may also contribute to a net transfer of C from the atmosphere but C sequestration strategies based on adding recalcitrant material to soil must be critically evaluated (Schmidt et al. 2011). For example, whether C can be sequestered by soil application of biochar together with organic wastes in land uses, soil types and climates other than those associated with Terra Preta (*do Indio*) in the Brazilian Amazon is not known (Glaser 2007; Liang et al. 2010; Sohi et al. 2010).

1.4 Policy Implications

There are important policy implications to implement programs effective in recarbonization of the biosphere. In view of the potentials in countries of all hemispheres, involvement of all countries is called for. Furthermore, adoption of 100% zero-emission technologies is relevant (Read 2006). This obligates the desired rate of technology adoption and facilitates positive approaches to market share and competitive edge (rather than setting one company/country against another as is the case in the zero emission caps) (Read 2006).

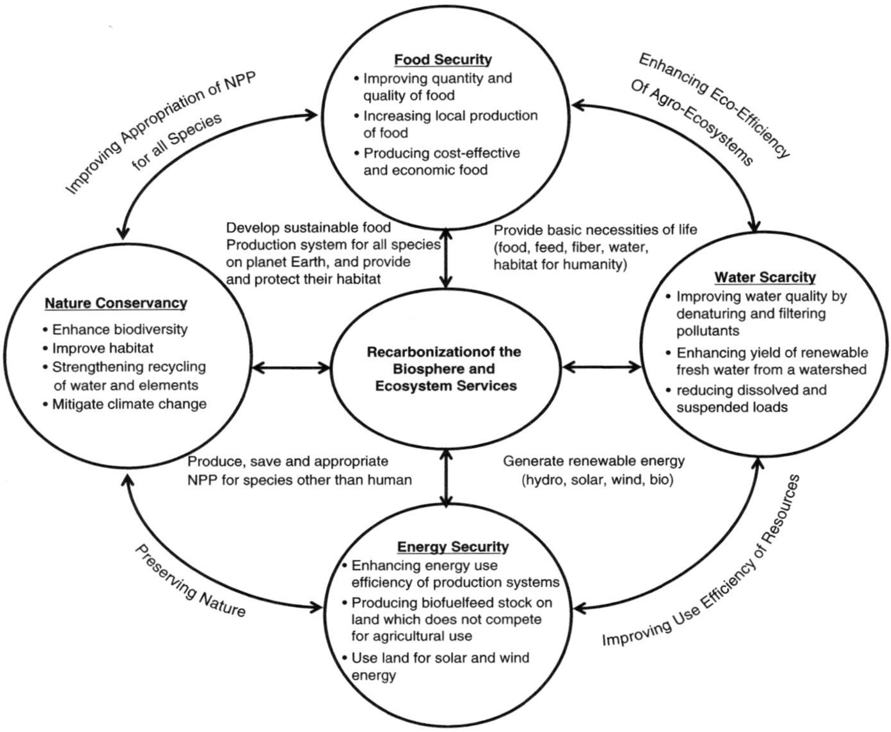

Fig. 1.3 Enhancement and provision of ecosystems services created by recarbonization of the terrestrial biosphere

While the ideas of carbon trade and of the Clean Development Mechanism (CDM) are not as attractive now as these were a decade ago, the concept of compensating land managers for generating/strengthening ecosystem services, and for buying and selling pollution (i.e., Australia's carbon tax, Petherick 2012) is valid and being vigorously pursued. The strategy of recarbonization of the biosphere would complement these new and emerging innovations being adopted by Australia, Japan, New Zealand, Switzerland and the European Union. The carbon trade could be linked with the concept of "zero emission technology". While the growth of C market is desirable, the price of C sequestered in the biosphere should be based on just, fair and transparent criteria.

Rather than subsidies, payment to farmers for enhancing ecosystem services is another positive strategy. Recarbonization of the biosphere strengthens numerous ecosystem services (Fig. 1.3). Important among these are increase in C sequestration, improvement in quantity and quality of water resources, and enhancement of biodiversity leading to food security, water security and energy security while also saving the land for nature conservancy. The adoption of technologies outlined in Fig. 1.2 can be vastly increased by incentivizing the land manager through payments for ecosystem services.

1.5 Conclusions

While the control of the elements has been in the prayers and thoughts of ancient civilizations for millennia (i.e., God Indra of the Indo-Aryans) and of the writers and philosophers (i.e., Homer's "Odyssey" and Shakespeare's "The Tempest") (Schneider 2008), recarbonization of the terrestrial (and aquatic) biosphere does provide some control of the runaway element of the ACC to the so called "Carbon Civilization" (Lal 2007). In strong contrast to the geoengineering techniques (CCS, sulfur aerosols, space reflectors), the strategy of recarbonization of the biosphere seems cost-effective (McKinsey & Co. 2009), and would have numerous co-benefits. Restoration of degraded soils and desertified ecosystems is a natural process, and is strongly linked to numerous ecosystem services including food security, water scarcity, and energy security.

There is no single technology for sustainable management of C pool in world's soils and biota. Most of the known/proven technologies must be validated, specified and adapted under soil-site specific conditions. Over and above the biophysical factors, the human dimensions involving social, cultural, ethnic, economic, and political factors must also be considered.

Incentivizing the land managers is important to adoption of the recommended soil/forest/land management options. Payments for ecosystems services, rather than subsidies, are positive approaches. It would promote innovation and stewardship of land resources, while adapting to and mitigating the ACC.

References

Broecker WS (2007) CO_2 arithmatic. Science 315:1371

Canadell JG, Le Quere C, Raupach MR et al (2007a) Contributions to accelerated atmospheric CO_2 growth from economic activity, carbon intensity, and efficiency of natural sinks. Proc Natl Acad Sci USA. doi:10.1073/pnas.0702737104.104718866-18870

Canadell JG, Pataki DE, Gifford R et al (2007b) Saturation of the terrestrial carbon sink. Chapter 6. In: Canadell JG, Pataki D, Pitelka L (eds) Terrestrial ecosystems in a changing world, The IGBP series. Springer, Berlin/Heidelberg

Crutzen PJ (2002) The "anthropocene". J Phys IV Fr. doi:10.1051/jp4:20020447

Crutzen PJ, Stoermer EF (2002) The "anthropocene". IGPP Newsl 41:12

Don A, Schumacher J, Freibauer A (2011) Impact of tropical land-use change on soil organic carbon stocks – a meta-analysis. Glob Change Biol 17:1658–1670

Ellis EC, Goldewiijk KK, Siebert S et al (2010) Anthropogenic transformation of the biomes, 1700 to 2000. Global Ecol Biogeogr. doi:10.111/j.1466-8238.2010.00540.x

FAOSTAT (2010) Global land use. http://faostat.fao.org/site/567/DesktopDefault.aspx?PageID=567#ancor

Fargione J, Hill J, Tilman D et al (2008) Land clearing and the biofuel carbon dept. Science 319:1235–1238

Fischer G, Schrattenholzer L (2001) Global bioenergy potentials through 2050. Biomass Bioenergy 20:151–159

Foley JA, DeFries R, Asner GP et al (2005) Global consequences of land use. Science 309:570–574

Foley JA, Monfreda C, Ramankutty N et al (2007) Out share of the planetary pie. Proc Natl Acad Sci USA 104(31):12585–12586
Glaser B (2007) Prehistorically modified soils of central Amazonia: a model for sustainable agriculture in the twenty-first century. Philos Trans R Soc B 362:187–196
Glover JD, Reganold JP, Bell LW, Borevitz J, Brummer EC, Buckler ES, Cox CM, Cox TS, Crews TE, Culman SW, DeHaan LR, Eriksson D, Gill BS, Holland J, Hu F, Hulke BS, Ibrahim AMH, Jackson W, Jones SS, Murray SC, Paterson AH, Ploschuk E, Sacks EJ, Snapp S, Tao D, Van Tassel DL, Wade LJ, Wyse DL, Xu Y (2010) Increased food and ecosystem security via perennial grains. Science 328:1638–1639
Guo LB, Gifford RM (2002) Soil carbon stocks and land use change: a meta analysis. Glob Change Biol 8:345–360
Haberl H, Erb K-H, Krausmann F, Gaube V, Bondeau A, Plutzar C, Gingrich S, Lucht W, Fischer-Kowalski M (2007) Quantifying and mapping the human appropriation of net primary production in earth's terrestrial ecosystems. Proc Natl Acad Sci USA 104:12942–12947
Houghton RA (2010) How well do we know the flux of CO_2 from land-use change? Tellus 62B:337–351
Hüttl RF, Gerwin W (2005) Disturbed landscapes – development of ecosystems. Geoderma 129:1–3
Jansson C, Wullschleger SD, Kalluri UC, Tuskan GA (2010) Phytosequestration: carbon biosequestration by plants and the prospects of genetic engineering. Bioscience 60:685–696
Kell DB (2011) Breeding crop plants with deep roots: their role in sustainable carbon, nutrient and water sequestration. Ann Bot. doi:10.1093/aob/mcr175
Lal R (1999) Soil management and restoration for C sequestration to mitigate the greenhouse effect. Prog Environ Sci 1:307–326
Lal R (2007) Soil science and the carbon civilization. Soil Sci Soc Am J 71:1095–1108
Le Quéré C, Raupach MR, Canadell JG, Marland G, Bopp L, Ciais P, Conway TJ, Doney SC, Feely RA, Foster P, Friedlingstein P, Gurney K, Houghton RA, House JI, Huntingford C, Levy PE, Lomas MR, Majkut J, Metzl N, Ometto JP, Peters GP, Prentice IC, Randerson JT, Running SW, Sarmiento JL, Schuster U, Sitch S, Takahashi T, Viovy N, Van Der Werf GR, Woodward FI (2009) Trends in the sources and sinks of carbon dioxide. Nat Geosci 2:831–836
Liang B, Lehmann J, Sohi SP, Thies JE, O'Neill B, Trujillo L, Gaunt J, Solomon D, Grossman J, Neves EG, Luizão FJ (2010) Black carbon affects the cycling of non-black carbon in soil. Org Geochem 41:206–213
Lorenz K, Lal R (2005) The depth distribution of soil organic carbon in relation to land use and management and the potential of carbon sequestration in subsoil horizons. Adv Agron 88:35–66
Loveland TR, Reed BC, Do B et al (2000) Development of a global land cover characteristics database and IGBP DISCover from 1 km AVHRR data. Int J Remote Sens 21(6/7): 1303–1330
Macías F, Arbestain MC (2010) Soil carbon sequestration in a changing global environment. Mitig Adapt Strateg Glob Change 15:511–529
McKinsey & Co (2009) Pathways to a low-carbon economy. Version 2 of the global greenhouse gas abatement cost curve, London
Pan Y, Birdsey RA, Fang J, Houghton R, Kauppi PE, Kurz WA, Phillips OL, Shvidenko A, Lewis SL, Canadell JG, Ciais P, Jackson RB, Pacala SW, McGuire AD, Piao S, Rautiainen A, Sitch S, Hayes D (2011) A large and persistent carbon sink in the world's forests. Science 333:988–993
Peters GP, Marland G, Le Quéré C, Boden T, Canadell JG, Raupach MR (2011) Rapid growth in CO_2 emissions after the 2008–2009 global financial crisis. Nat Clim Change. doi:10.1038/nclimate1332
Petherick A (2012) Duty down under. Nat Clim Change 2:20–21
Pielke RA, Pitman A, Niyogi D, Mahmood R, McAlpine C, Hossain F, Goldewijk KK, Nair U, Betts R, Fall S, Reichstein M, Kabat P, de Noblet N (2011) Land use/land cover changes and climate: modeling analysis and observational evidence. WIREs Clim Change 2:828–850

Poeplau C, Don A, Vesterdal L, Leifeld J, Van Wesemael B, Schumacher J, Gensior A (2011) Temporal dynamics of soil organic carbon after land-use change in the temperate zone – carbon response functions as a model approach. Glob Change Biol 17:2415–2427

Polyakov V, Lal R (2004) Modeling soil organic matter dynamics as affected by soil water erosion. Environ Int 30:547–556

Ponting C (2007) A new green history of the world. Penguin Books

Powlson DS, Whitmore AP, Goulding KWT (2011) Soil carbon sequestration to mitigate climate change: a critical re-examination to identify the true and the false. Eur J Soil Sci 62:42–55

Prechtel A, von Lützow M, Schneider BU, Bens O, Bannick CG, Kögel-Knabner I, Hüttl RF (2009) Organic carbon in soils of Germany: status quo and the need for new data to evaluate potentials and trends of soil carbon sequestration. J Plant Nutr Soil Sci 172:601–614

Ramankutty N, Foley JA (1999) Estimating historical changes in global land cover: croplands from 1700 to 1992. Global Biogeochem Cycles 13(4):997–1027

Read P (2006) Biosphere carbon stock management: addressing the threat of abrupt climate change in the next few decades: an editorial essay. Clim Chang. doi:10.1007/s10584-007-9356-y

Read P, Parshotam A (2007) Holistic greenhouse gas management strategy (with reviews comments and authors' rejoinders). Institute of Policy Studies working paper 07/1. Victoria University of Wellington. http://ips.ac.nz.publications/files/f659c9661de.pdf

Righelato R, Spracklen DV (2007) Carbon mitigation by biofuels or by saving and restoring forests. Science 317:902

Ruddiman WF (2003) The anthropogenic greenhouse era began thousands of years ago. Clim Chang 61:261–293

Ruddiman WF (2005) How humans took control of the climate. Princeton University Press, Princeton

Ruddiman WF, Ellis EC (2009) Effect of per-capita land use changes on Holocene forest clearance and CO_2 emissions. Elsevier Q Sci Rev. doi:10.1016/j.quascirev.2009.05.022

Schmidt MWI, Torn MS, Abiven S, Dittmar T, Guggenberger G, Janssens IA, Kleber M, Kögel-Knabner I, Lehmann J, Manning DAC, Nannipieri P, Rasse DP, Weiner S, Trumbore SE (2011) Persistence of soil organic matter as an ecosystem property. Nature 478:49–56

Schneider SH (2008) Geoengineering: could we or should we make it work? Philos Trans R Soc A 366:3843–3862

Schneider A, Friedl MA, Potere D (2009) A new map of global urban extent from MODIS satellite data. Environ Res Lett 4, 044003. doi:10.1088/1748-9326/4/4/044003

Séquaris J-M, Guisado G, Magarinos M, Moreno C, Burauel P, Narres H-D, Vereecken H (2010) Organic-carbon fractions in an agricultural topsoil assessed by the determination of the soil mineral surface area. J Plant Nutr Soil Sci 173:699–705

Sohi S, Krull E, Lopez-Capel E, Bol R (2010) A review of biochar and its use and function in soil. Adv Agron 105:47–82

Spreitzer RJ, Salvucci ME (2002) Rubisco: structure, regulatory interactions, and possibilities for a better enzyme. Annu Rev Plant Biol 53:449–475

Sterman JD (2008) Risk communication on climate: mental models and mass balance. Science 322:532–533

Stewart CE, Paustian K, Conant RT, Plante AF, Six J (2009) Soil carbon saturation: implications for measurable carbon pool dynamics in long-term incubations. Soil Biol Biochem 41:357–366

Williams M (1994) Forests and tree cover. In Meyer WB and BL Turner II (eds) Changes in land use and land cover: a global perspective. Cambridge University Press, Cambridge, U.K. 97–124

Williams M (2000) Dark ages and dark areas: global deforestation in the deep past. J Hist Geogr 26:28–46

WMO. (2010). Greenhouse bulletin: the state of the greenhouse gases in the atmosphere based on observations through 2009, Geneva, Switzerland

Wutzler T, Reichstein M (2007) Soils apart from equilibrium – consequences for soil carbon balance modelling. Biogeosciences 4:125–136

Zhu X-G, Long SP, Ort DR (2010) Improving photosynthetic efficiency for greater yield. Annu Rev Plant Biol 61:235–261

Chapter 2
Climate Change Mitigation by Managing the Terrestrial Biosphere

Rattan Lal

Abstract The terrestrial C pool, 3,900 Pg comprising of C in soil and the vegetation, is the third largest after the oceanic and the geological pools. Rather than with the on-set of the Industrial Revolution around 1800 AD, the Anthropocene may have commenced 10–12 millennia ago with the dawn of settled agriculture. Given this hypothesis, the terrestrial biosphere may have lost as much as 480 Pg C, of which 70–100 Pg may have been depleted from the world's soils. Whereas most of the C emitted from the terrestrial biosphere may have been reabsorbed by the ocean and the land-based sinks, the depleted C pool created a C sink capacity that can be refilled through conversion to judicious land use(s) and adoption of recommended management practices of soil, vegetation and domesticated animals. Principal biomes for C sequestration, those with high capacity for C sequestration/emission avoidance and with numerous co-benefits, are peatlands and wetlands, degraded soils and desertified ecosystems, soils of agroecosystems and urban lands. Effective erosion control can also avoid emissions exacerbated by mineralization of C transported in runoff and eroded sediments and redistributed over the landscape. In addition to mitigating climate change, other co-benefits of C sequestration in the terrestrial biosphere are advancing food security, improving quality and quantity of water resources, and increasing the habitat and total biodiversity. Being an engine of economic development, improvements in productivity of agricultural, forestry and other managed ecosystems can advance the Millennium Development Goals, promote political and social stability, and improve standards of living. This is a win-win option in the context of climatic, environmental, political, economic and social issues facing the Carbon Civilization of the twenty-first century.

R. Lal (✉)
Carbon Management and Sequestraton Center, The Ohio State University, 2021 Coffey Rd., 210 Kottman Hall 422b, Columbus, OH 43210, USA
e-mail: lal.1@osu.edu

R. Lal et al. (eds.), *Recarbonization of the Biosphere: Ecosystems and the Global Carbon Cycle*, DOI 10.1007/978-94-007-4159-1_2,

Keywords Anthropocene • Terrestrial biosphere • Global C pools • Terrestrial C pool • Soil C pool • Biotic C pool • Land-based sinks • Oceanic sink • Co-benefits • Emission avoidance • Food security • Millennium development goals • The carbon civilization • Charcoal • Primary carbonates • Lithogenic carbonates • Pedogenic carbonates • Secondary carbonates • Adapting and mitigating climate change • Recarbonization • Biome • Net primary productivity • Forests • Savannas • Deserts • Boreal region • Tundra • Cryosols • Celisols • Ocean acidification • Urban growth • Peatlands • Wetlands • Croplands • Pasture lands • Megacities • Urbanization in India • Greenbelts • The human dimensions • Global soil forum

Abbreviations

ACC	abrupt climate change
ADAM	adapting and mitigating
CCS	carbon capture and storage
CRP	conservation reserve program
Eg 10^{18} g	exagram
GCC	global carbon cycle
GSF	Global Soil Forum
GHGs	greenhouse gases
HCC	hidden C cost
MRT	mean residence time
NPP	net primary productivity
NIMBY	not in my backyard
SIC	soil inorganic C
SOC	soil organic C
SOM	soil organic matter

2.1 Introduction

Earth has a large reservoir of C. Expressed in exagram (Eg, 10^{18} g), total C pool is 22×10^4 Eg in the Earth mantle, 60 Eg as carbonate in rocks, 15 Eg as organic C (fossil fuel) in rocks, 42 Eg in oceans, 4 Eg in soils, 0.8 Eg in atmosphere and 0.6 Eg in the biosphere. Together, the C pool in the soils and biota is called the terrestrial C pool, or C pool in the terrestrial biosphere.

The terrestrial C pool, with an important impact on the global carbon cycle (GCC), is the third largest pool, after the oceanic and geologic (fossil fuel). It comprises of two components: biotic pool with total reserve of ~620 Pg, and soil pool with a total reserve of ~3,294 Pg to 3-m depth (Fig. 2.1). The biotic C pool has three components: the aboveground biomass, the below-ground biomass, and the detritus material (Fig. 2.1a). The soil C pool has two distinct components: soil organic

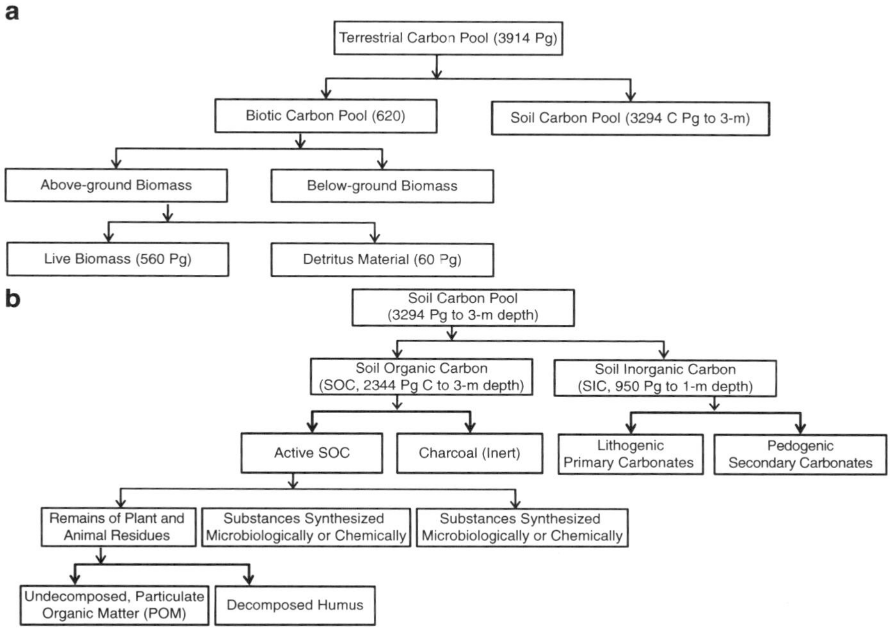

Fig. 2.1 Components of the terrestrial C pool (The data on SOC pool from Batjes 1996; Jobbágy and Jackson 2000, SIC pool from Eswaran et al. 2000, biotic pool from Lal 2004)

C (SOC) and soil inorganic C (SIC) (Fig. 2.1b). The SOC comprises of active humus and relatively inert charcoal. The humus is part of the total soil organic matter (SOM). It is a dark brown or black amorphous material, has large surface area, high charge density, and high reactivity with clay and other colloids. Humus is also an important determinant of soil quality, and its depletion leads to degradation of soil physical, chemical and biological quality. The SIC pool includes elemental C and primary minerals (e.g., calcite, dolomite and gypsum). There are two types of carbonates in soils: lithogenic or primary carbonates, and pedogenic or secondary carbonates. Primary carbonates are derived from the parent material, and secondary carbonates are formed through dissolution of CO_2 in the soil and reaction of the dilute carbonic acid with Ca^{2+} or Mg^{2+} brought in from outside the ecosystem. Thus, soil C sequestration implies transfer of atmospheric CO_2 into soil either as recalcitrant humus or as secondary carbonates so that it is not returned into the atmosphere for a long period of decades to centuries to millennia. The long mean residence time (MRT) may be attained through enhancement of physical, chemical or biological/ecological mechanisms of protecting of humus and SOM against microbial and physico-chemical processes.

Climate change is a major global issue of the twenty-first century. Yet, the science behind the climate change remains to be a debatable issue (Sherwood 2011;

Somerville and Hassol 2011; Pierrhumbert 2011). Because of its global significance, and effect on lives of all population, the debate about climate change has also spread to classrooms in secondary and high schools (Reardon 2011). Regardless, there is a strong need to identify win-win options of adapting and mitigating (ADAM) climate change. Therefore, the objective of this chapter is to deliberate the role of terrestrial biosphere as a source and sink of atmospheric CO_2 through recarbonization of different biomes by a judicious and planned management of their diverse C pools.

2.2 Principal World Biomes

The term "biome" refers to a grouping of terrestrial ecosystems with similar vegetation structure, rainfall, temperature, seasonality and physiognomy. Principal world biomes can be classified into three groups, which have been described at length by Whittaker (1975), Olson et al. (2001) and Strahler and Strahler (1984), among others.

2.2.1 Low-Latitude Biomes

The climate is influenced by equatorial region and consists of:

2.2.1.1 Tropical Forests

This biome is characterized by characteristics of the dense, diverse, and evergreen vegetation, and comprising of numerous vines and epiphytes. The biome has a high constant temperature (~25–27°C), high annual rainfall (~250 cm), and high relative humidity (75–90%). The biome has a high net primary productivity (NPP), and soils of acidic pH and low inherent fertility.

2.2.1.2 Tropical Savannas and Grasslands

This biome is characterized by wet and dry seasons, and the dry season is somewhat cooler than the wet season. This ecoregion occurs in the latitude range of 15–25° N and S, and has an annual temperature range of ~15°C.

2.2.1.3 Deserts and Semi-deserts

This biome occurs within the latitude of 18–28° N and S around the tropics of Cancer and Capricorn. These regions have high pressure, so these are seldom penetrated by air masses which produce rain. Thus, this biome is characterized by low rainfall (<25 cm/annum). Dominant plant species include cacti, sagebrush, mesquite, and some halophytes. These regions have low NPP.

2.3 Mid-latitude Biomes

The climate is influenced by two air masses: the tropical air mass moving towards the poles and the polar air mass moving towards the Equator. The relative influence of these two air masses determines the specific climate.

2.3.1 Temperate Grasslands and Shrublands

These regions are called grasslands, prairies, plains, pampas, savannas, velds, or steppe. Some regions are also called "Chaparral" such as those in California. The biome has an annual temperature range of 25°C and precipitation range of 10 cm (dry) to 50 cm (moist steppe). This biome occurs within the latitude of 30–50° N and S of Equator. The vegetation is fire-dependent, which is an important factor for the dominance of grasses rather than the trees. Soils of these biomes, have a higher SOC concentration than those of temperate forests.

2.3.2 Temperate Forests

The regions is characterized by large differences in summer and winter temperatures (~30°C), and has an average precipitation of ~80 cm. This biome occurs within the latitude of 30–55° N and S of the Equator. The vegetation is characterized by hardwood trees (beech, maple, oak, hickory) which are deciduous. The NPP of managed and natural ecosystems is high because of fertile soils and favorable climate with high annual rainfall and a temperate climate.

2.4 High Latitude Biomes

The climate is dominated by arctic and polar air masses. Thus, these biomes are characterized by cold climates including the permafrost.

2.4.1 Boreal Forests

The biome is characterized by long and cold winters, and short and cool summers. The land is dominated by conifers, and consists of numerous lakes, marshes and

Table 2.1 Land area and the carbon stock in world biomes (Adapted from WBGU 1988; IGBPC 1999)

Biomes	Present area (10^9 ha)	Global carbon stock (Pg C)		
		Plant	Soil (1-m)	Total
Tropical forests	1.75	340	213	553
Temperate forests	1.04	139	153	292
Boreal forests	1.37	57	338	395
Tropical savannas and grasslands	2.25	79	247	326
Temperature grasslands and shrublands	1.25	23	176	199
Deserts and semi-deserts	3.00	10	159	169
Tundra	0.95	2	115	117
Croplands	1.42	4	165	169
Wetlands	0.35	15	225	240
Ice	1.55	–	–	–
Total	14.93	669	1,791	2,683

peatlands The annual range is 40°C for temperature, and 30 cm for annual precipitation. It occurs within the latitude 50–70° N. This biome is also called the Taiga region, with specific reference to the biome in Russia.

2.4.2 *Tundra*

Polar and arctic air masses dominate the climate. The latter is characterized by long and severe winter and no summer. The temperature ranges from –25°C to 5°C, annual precipitation is ~20 cm, and the latitude is 60–75° N. The vegetation of the boreal region is highly stunted. The predominant soils of the region are Cryosols/Gelisols, and are characterized by permafrost.

2.4.3 *Alpine Biome*

The high altitude climate is characterized by cool and cold biome (i.e., Tibetan Plateau, the Andean region, Alps). Increase in altitude by 300 m has a similar effect on the vegetation as is the northward/poleward shift of ~1,000 km. The seasons in highlands are determined by the biome in which highlands are located. The mean temperature range from –20°C to 10°C, and the annual precipitation from 20 to 25 cm.

Estimates of land area of these biomes are shown in Table 2.1. Undisturbed, under native vegetation cover, these biomes together are a major C pool, and have played an important role in the GCC. Total C pool in world's biomes consists of ~620 Pg in the plant biomass and 1,800 Pg in the soil to (1-m depth) and 4,000 Pg to 3-m depth. Thus, the terrestrial biosphere is the third largest reservoir of C, after

Table 2.2 Ratio of C density (Mg C/ha) in soil:plant (Recalculated from Table 2.1)

Biome	Ratio of C density (soil:plant)
Tropical forests	1.0
Temperate forests	1.7
Boreal forests	5.4
Tundra	21.2
Tropical savannas/grasslands	4.03
Temperate grasslands/savannas	33.7
Deserts/semi deserts	21.0
Croplands	40.0
Wetlands	15.0

ocean and the geologic or fossil fuel (Table 2.1). The ratio of C pool in the soil: plant is a characteristic of the biome. The data in Table 2.2 show the ratio of C density (Mg C/ha) in soil:plant as ~1 for tropical forests, ~2 for temperate forests. ~5 for boreal forests, and >20 for tundras. Because of their high SOC concentration, the ratio is also high for temperate grasslands and steppes.

2.5 Principal Soils and Their Carbon Pools

In accord with a wide diversity of C pools in different biomes (Tables 2.1 and 2.2), there also exists a large variation in C pools among principal soil orders (Table 2.3). Total magnitude of the SOC pool to 1-m depth is ~1,590 Pg. The SOC density (Mg C/ha) ranges from 38 in Aridisols to 1,193 in Histosols. Based on the variations in land area under specific soil orders, the magnitude of SOC pool ranges from 42 Pg in Vertisols to 296 Pg in Gelisols. There is also a large variation in the SIC pool, which has a global magnitude of 965 Pg (Table 2.3). The SIC density (Mg C/ha) ranges from 0 in soils of the humid regions (Andisols, Histosols, Spodosols, Ultisols) to 290 in Aridisols. Based on the land area under different orders, the magnitude of SIC pool also ranges from 0 in soils of the humid climates to 456 Pg in Aridsols and 263 Pg in Entisols (Table 2.3).

The C pool also varies widely among soils in different biomes and moisture regimes (Table 2.4). Relatively high SOC pool occurs in soils of Tundra and Boreal (Taiga) biomes. In contrast, high magnitude of SIC pool occurs in soils of temperate climates (Table 2.4). In terms of moisture regime, expectedly high SOC pools exist in soils of humid and permafrost regimes. In contrast, high SIC pools occur in soils of arid and semi-arid climates (Table 2.4).

Peatland soils have an exceptionally high SOC pool (Table 2.5). Of the total area of 442 Mha, peatlands in the northern latitude cover 400 Mha or 90% of the global peatland area. Accordingly, the total magnitude of SOC pool is in peatlands of the northern latitude is 547 Pg or 89% of the total SOC pool of 612 Pg in peatlands. The

Table 2.3 Area and carbon pool in world soils (Adapted and recalculated from Eswaran et al. 1995, 2000)

Soil order	Area in ice-free land (10^9 ha)	SOC pool to 1-m depth		SIC pool	
		Total (Pg)	Density (Mg/ha)	Total (Pg)	Density (Mg/ha)
Alfisol	1.26	158	101	43	34
Andisols	0.09	69	767	0	0
Aridisols	1.57	59	38	456	290
Entisols	2.11	90	43	263	125
Gelisols	1.13	296	262	11	10
Histosols	0.15	179	1,193	0	0
Inceptisols	1.29	190	143	34	26
Mollisols	0.90	121	134	116	129
Oxisols	0.98	150	153	21	21
Spodosols	0.34	98	288	0	0
Ultisols	1.11	137	125	0	0
Vertisols	0.32	42	131	21	65
Shifting sands	0.53	–	–	–	–
Rocky land	1.31	–	–	–	–
Total	13.09	1,589	–	965	–

Table 2.4 Soil carbon pool in different biomes and moisture regimes (Adapted and updated from Eswaran et al. 2000)

Biome/moisture regime	Soil C pool (Pg)	
	SOC	SIC
I. Biome		
Boreal	376	256
Temperate	355	518
Tropical	408	149
Tundra	387	18
Total	1,526	940
II. Moisture regime		
Arid	145	732
Semi-arid	337	134
Mediterranean	40	50
Humid	535	4
Per humid	83	2
Permafrost	387	18
Total	1,527	940

Table 2.5 Global peatlands and carbon pool (Yu et al. 2011)

Region	Area (Mha)	SOC Pool (Pg)	Holocene Rate (kg C/ha/year)
Northern	400	547	186
Tropical	37	50	128
Southern	5	15	220
Total	442	612	

mean rate of SOC accumulation in global peatlands over the Holocene has ranged from 128 to 220 kg C/ha/year. Of the total SOC pool of 1,589 Pg (Table 2.3), peatland soils contain about 38.5% (612 Pg) of the total pedological pool to 1-m depth. With reference to anthropogenic activities, it is important to preserve the SOC pool in peatland soils, and to safeguard the thaw or melting and decomposition of SOC pool in the permafrost (Cryosols, Gelisols) in the Tundra region.

2.6 Anthromes

The human-induced transformation of the terrestrial biosphere led to creation of anthropogenic biomes or Anthromes (Ellis et al. 2010). The conventional biomes outlined in Sect. 2.2 have been drastically altered by human activities through change in land use for agriculture, forestry, human settlements, urbanization and development of infrastructure. Thus, management of anthromes is essential to sustainable land use, adaption and mitigation (ADAM) of climate change, and to recarbonization of the biosphere. Important among these anthromes are: croplands, rangelands and pastures, urban settlements, semi-natural regions and parks, etc. Most of the anthromes have been created since ~1,700. Only 45% of Earth's ice-free land was disturbed to some level by 1,700, and rapid intensification occurred during the twentieth century. In addition to cropland, large area were converted into rangelands and pastures. There was a near sixfold increase in land area under pasture between 1,800 and 2,000. Together, croplands and grazing land cover nearly 40% of the ice-free land on earth. An additional 37% of ice-free land (presently not being used) is in close proximity to or embedded within anthromes, and is prone to land use change (Ellis et al. 2010). Thus, there is hardly 23% of undisturbed natural land, most of which exits in cold and arid biomes. Therefore, the era since the industrial revolution is called "Anthopocene".

2.7 Terrestrial Biosphere as a Source of Carbon

The Anthropocene, industrial era activities which have altered the atmospheric concentration of greenhouse gases or GHGs (Crutzen and Stoermer 2000), supposedly began ~1,800. The era is characterized by combustion of the fossil fuel and emission of CO_2 and other GHGs into the atmosphere. In contrast, Ruddiman (2003, 2007) argued that the anthropocene began about 12–14 millennia ago with the dawn of settled agriculture. Ruddiman argued that a slow increase in atmospheric concentration of GHGs began ~8,000 years ago for CO_2 corresponding with large scale deforestation, and ~5,000 years ago for CH_4 corresponding with cultivation of rice paddies and domestication of cattle. Ruddiman estimated that warming caused by emissions from the terrestrial sources in recent millennia may be as much as 0.8°C globally and as much as 2°C in high latitudes. Further, these warming had been

Table 2.6 Global carbon budget since 1850 (Houghton 2007)

I.	**Total emissions (Pg)**	500
	Fossil fuel	375
	Land use change	100
	Cement production	25
II.	**Sinks (Pg)**	
	Oceans	150
	Terrestrial biosphere	125
	Atmosphere	225

substantial to affect the Little Ice Age (1300–1900 AD). The cumulative C loss from the terrestrial biosphere was estimated at 480 Pg C as follows (Ruddiman 2003):

$$7{,}800 \text{ years} \times 0.04 \text{ Pg C / year} = 312 \text{ Pg} \tag{2.1}$$

$$200 \text{ years} \times 0.8 \text{ Pg C / year} = 160 \text{ Pg} \tag{2.2}$$

Similar to Ruddiman, Olofsson and Hickler (2008) estimated C release from land use change between 4000 BC to 2000 AD at 262 Pg, of which 114 Pg occurred during the pre-industrial era and 148 Pg during the industrial period from 1850 to 1990.

Over and above the pre-historic depletion of the terrestrial C pool since the dawn of settled agriculture (320 Pg, Ruddiman 2003, 2007; Ruddiman and Ellis 2009), there has been drastic perturbation of natural biomes since the onset of industrial revolution, and particularly since 1850. Total anthropogenic emission since 1850 are estimated at 500 Pg comprising of 375 Pg (75%) from the fossil fuel combustion, 100 Pg (20%) from land use change, and 25 Pg (5%) from cement production (Table 2.6, Houghton 2007). Of the anthropogenic emissions, 150 Pg (30%) are supposedly absorbed by the ocean, 125 Pg (25%) by the terrestrial biosphere, and the remainder 225 Pg (45%) by the atmosphere (Houghton 2007). Between 2010 and 2030, additional emissions are estimated at 190 Pg from fossil fuel combustion and 106 Pg from land use change and deforestation (Holdren 2008).

The contemporary global C budget on a decadal scale shown in Table 2.7 indicate an increase in anthropogenic emissions from 6.5 Pg C/year in 1980s to 9.5 Pg C/year in 2008. With emissions from land use change remaining the same (~1.5 Pg C/year), the increase in anthropogenic emissions is primarily caused by increase in fossil fuel combustion. The data in Table 2.7 show a progressive increase in the annual uptake of C by the atmosphere from 3.1 Pg C/year in 1980s to 4.3 Pg C/year n 2008. An updated version of Table 2.7, also on a decadal scale from 1960s to 2000s, shown in Table 2.8, indicate an annual increase in uptake by the atmosphere at 1.8 Pg C/year in 1960s to 4.1 Pg C/year in 2000s. Taking into account the unknown C sink, presumably in the temperate and boreal forest biomes, the capacity of natural sinks (terrestrial and oceanic combined) has ranged from 61% in the 1960s and 1990s to 51% in 1980s, and 55% in 1970s and 2000s. It is feared that capacity of natural sinks is finite and may decline over time, because of ocean acidification, soil C saturation, and soil quality degradation.

The short-term GCC, outlined in Fig. 2.2, show principal C pools and annual fluxes among them. The strong inter-dependence and connectivity among the

Table 2.7 Contemporary global C budget (Canadell et al. 2007)

Parameter	1980s	1990s	2000s
1. Sources			
Fossil fuel	5.4	6.5	8.0
Land use	1.5	1.6	1.5
Total	6.9	8.1	9.5
2. Sinks			
Atmospheric	3.1	3.2	4.2
Ocean	2.0	2.2	2.3
Total	5.3	6.1	7.1
3. Unknown sink	1.6	2.0	2.5
Natural sink (%)	55.0	60.0	57.0

Table 2.8 Global carbon budget (Pg C/year) for the period 1960–2008 (Recalculated from Le Quéré et al. 2010)

Sources/sinks	1960s	1970s	1980s	1990s	2000s	2008
Sources						
1. Fossil fuel & cement	3.1	4.7	5.5	6.4	7.7	8.7
2. Land Use	1.5	1.3	1.5	1.6	1.4	1.2
Subtotal	4.6	6.0	7.0	8.0	9.1	9.9
Sinks						
1. Atmosphere	1.8	2.7	3.4	3.1	4.1	3.9
2. Ocean	1.5	1.7	2.0	2.2	2.3	2.3
3. Land	1.2	2.6	1.8	2.3	3.0	4.7
Subtotal	4.5	7.0	7.2	7.9	9.4	10.9
Unaccounted land sinks	0.1	−1.0	−0.2	0.1	−0.3	−1.1
% Uptake by natural sinks	60.9	55.0	51.4	61.3	54.9	59.6

terrestrial and atmospheric pools necessitate an objective and a prudential management of the terrestrial biosphere for C sequestration.

2.8 Carbon Sequestration

Carbon sequestration implies transfer of atmospheric CO_2 into other pools (e.g., geologic, oceanic, pedologic, biotic) through chemical, biological or engineering processes, to ensure that it is not re-emitted into the atmosphere over decades, centuries or millennia. Among numerous options outlined in Fig. 2.3, the choice of an appropriate strategy depends on numerous criteria : (i) cost/Mg C or the economic factor, (ii) risks of re-emission, (iii) sink capacity, (iv) permanence or MRT, (v) safety and social acceptance (free from "not in my backyard" or NIMBY syndrome), and (vi) co-benefits. Each of the strategy outlined in Fig. 2.3 has its own pros and cons. The sink capacity is high for geologic and oceanic sequestration

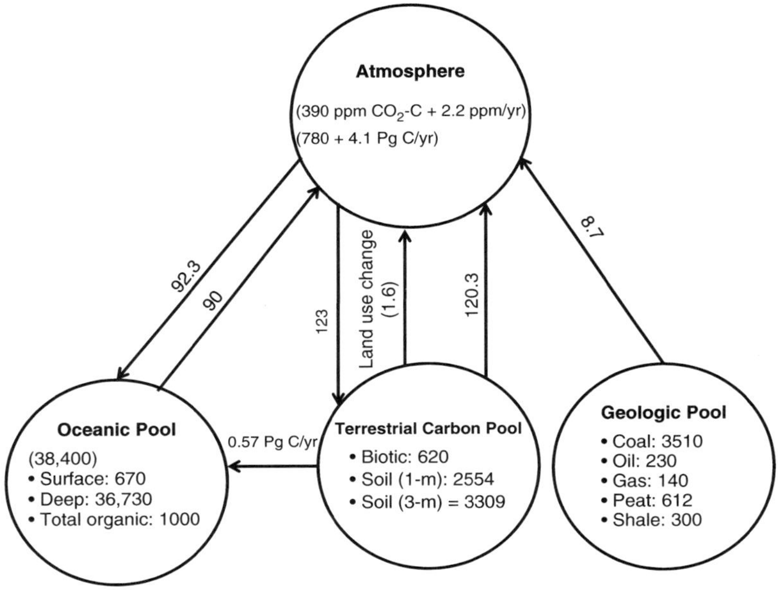

Fig. 2.2 The short-term global C cycle showing pools (Pg C) and annual fluxes (Pg C/year). Flux values, (Adopted from Le Quéré 2010), refer to the period 2000–2008. The pool values are from Lal (2004, 2008). Loss of C from soil to ocean by erosion is from Lal (2003)

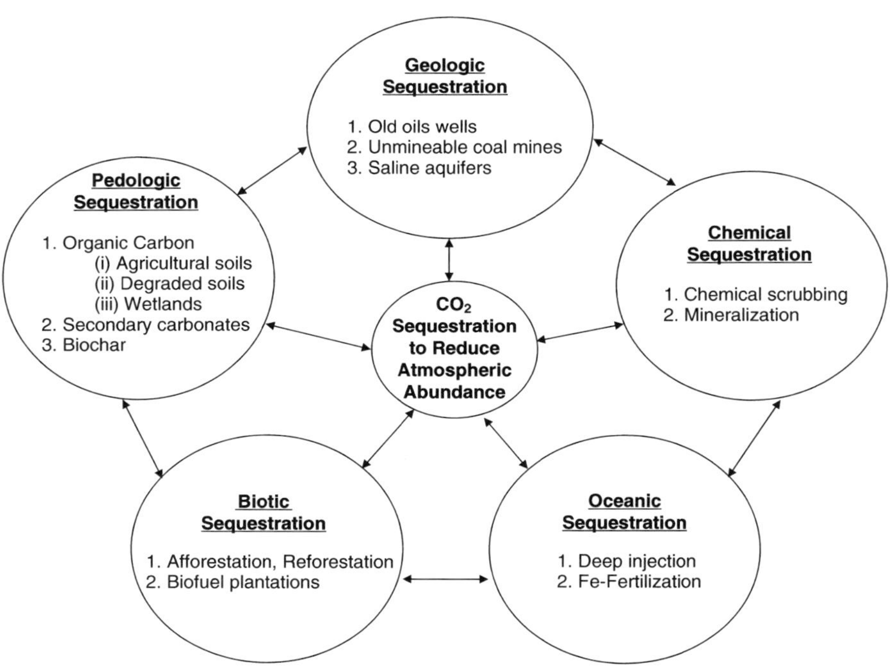

Fig. 2.3 Technological options for sequestration of atmospheric CO_2 in geologic, oceanic, and terrestrial pools

Table 2.9 Estimates of land area converted from natural vegetation to cropland (Recalculated from Ramankutty and Foley 1999)

Biome	Prehistoric (10^9 ha)	Area in 1992 (10^9 ha)	Change (10^9 ha)
Total forests/woodland	5.53	4.39	1.14
Total savanna/grassland	3.34	2.67	0.67
Total shrubland	1.79	1.59	0.20
Total tundra	2.36	2.33	0.03
Deserts	1.53	1.52	0.01
Total	13.09	10.98	2.03

Table 2.10 Strategies of avoiding emissions from and sequestering emissions into the terrestrial biosphere

Avoiding emissions	Sequestering emissions
1. Preserve existing forests	1. Restore degrades soils and desertified ecosystems by afforestation and reforestation, and re-inundation of peatlands and wetlands
2. Maintain peatlands, marshes, bogs and wetlands	2. Adopt technologies in managed ecosystems which create a positive C budget (e.g., conservation agriculture, mulch farming, integrated nutrient management, soil and water conservation, drip irrigation, precision farming, biochar)
3. Adopt strategies towards sustainable intensification of agroecosystems by using land saving technologies	3. Convert marginal agricultural soils to natural ecosystems (e.g., CRP, set aside, land retirement)
4. Optimize the use of inputs based on the combustion of fossil fuel (e.g., fertilizers, pesticides, irrigation, farm machinery)	4. Use eco-efficient production systems characterized by high use efficiently of inputs
5. Reduce human appropriation of natural resources and NPP	5. Enhance productivity pre unit land area, save land for nature conservancy
6. Educate general public about the benefits of plant-based diet	6. Strengthen science-policy nexus

compared with those of biotic and pedologic sequestration. Yet, the cost-effectiveness and co-benefits of the terrestrial sequestration in world soils (especially in soils of croplands and grazing lands) is the improvement in soil quality, the attendant increase in agronomic productivity, and the scope of advancing the global food security. Furthermore, C sequestration in the biosphere has also important co-benefits in terms of water security, improvements in biodiversity, and overall increase in ecosystem services and functions. The co-benefits of C sequestration are especially relevant in large areas converted from natural to managed ecosystems by deforestation (Table 2.9), especially the agriculturally marginal lands. Sequestration of C in agricultural soils must consider both SOC and SIC components. Strategies of avoiding emissions and sequestering emissions in the terrestrial biosphere are outlined in Table 2.10. The goal is to minimize deforestation, reduce drainage and conversion of wetlands and peatlands, restore degraded and desertifed ecosystems, and create a positive ecosystems C budget in all managed ecosystems (Table 2.10).

Table 2.11 Merits of using soil C pool as an indicator of climate change and environment quality

1. It is a familiar property,
2. It involves direct measurement,
3. It can be measured in 4 dimensions (length, width, depth, time),
4. It lends itself to repeated measurements over the same site,
5. It is linked to ecosystem performance and services,
6. It is a key driver of soil formation,
7. It is important to soil fertility, and global food security,
8. It has memory,
9. It has well defined properties,
10. It can be used in synergism with other indicators,
11. Its uncertainty can be quantified,
12. Its pathways across the landscape can be followed,
13. It is an important archive of paleo-environmental conditions
14. It can be traded as a farm commodity, and
13. It is the essence of all terrestrial life as an energy source of microbial processes and transformations.

There are several advantages and co-benefits of C sequestration in world soils, provided that the recommended land use and management options are adopted on a long-term basis to prolong the MRT of the C thus sequestered. The SOC pool can also be used as an indicator or proxy of climate change. There are several merits of using SOC pool as an indicator (Table 2.11). Based on the estimates of historic loss of the SOC pool from soils of the managed ecosystems (76 ± 12 Pg; Lal 1999), the soil C sink capacity can be filled within 25–50 years. It has a drawdown capacity of reducing atmospheric CO_2 concentration by 15–20 ppmv. Together with the sink capacity of world biota, recarbonization of the terrestrial biosphere has a drawdown capacity of ~50 ppm of CO_2 over about 100 years (Hansen et al. 2008).

2.9 Priority Land Uses and Biomes for Recarbonization of the Biosphere

Recarbonization is a high priority for those biomes which either have a significant capacity as a C sink or have substantial benefits of emission avoidance. The priority biomes must also generate numerous ecosystem services and co-benefits, upon recarbonization and conversion to a restorative land use.

2.9.1 Peatlands

In terms of the C sink capacity, preservation and restoration of peatlands and mangroves, inundation of wetlands, and sustainable management of riparian forests (gallery forests) are a high priority (Fig. 2.4). Both in the context of

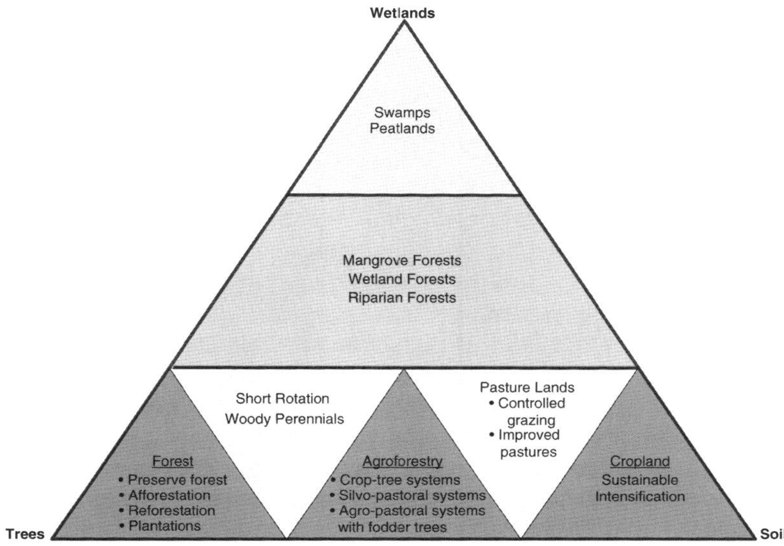

Fig. 2.4 Priority biomes for C sequestration presented as a pyramid to enhance the terrestrial C pool/bank

reducing and sequestering emissions, the importance of peatlands cannot be over-emphasized.

Peatlands are organic soils formed under wetland conditions, and cover a total area of ~400 Mha (Kaat and Joosten 2008). Countries with large peatland area include Russia (38%), Canada (31%), Indonesia (7%), and USA (6%) (Wetland Intl 2009; Yu et al. 2011). Being formed under wetland conditions, there exists a strong correspondence between the land area under and that under peatlands (Aselmann and Crutzen 1989). Large areas of peatlands also occur in the permafrost zone, which is vulnerable to positive feedback caused by global warming (Laine et al. 2009; Tarnocai et al. 2009). However, not all wetlands lead to peat formation (e.g., rice paddies). Further drainage of wetlands for agricultural and forestry land uses can lead to rapid oxidation of SOM in peat. Thus, high emissions of GHGs from degrading peatlands are a major concern. Emission rates of CO_2-C by 575 kg/ha/year have been observed (Tanneberger and Wichtmann 2011).

Peatlands are normally formed under boreal/temperate/arctic climates. However, peatlands also occur in tropical lowlands such as in Southeast Asia (e.g., Indonesia, Malaysia) and in the Amazonian regions including Brazil and Peru. Field observations form 17 wetland sites in the Peruvian lowland Amazonia indicated that 16 sites contained peat deposits ranging in thickness from 0 to 5.9 m (Lähteenoja et al. 2009). The historic yearly peat C accumulation rates in the Amazonia peat ranged from 0.94 ± 0.99 to 4.88 ± 1.65 mm or from 26 ± 3 to 195 ± 70 g C/m^2. The long-term rates, similar to those of the Indonesian peat, ranged from 1.69 ± 0.03 to 2.56 ± 0.12 mm/year or 39 ± 10 to 85 ± 30 g C/m^2/year. Global distribution of natural

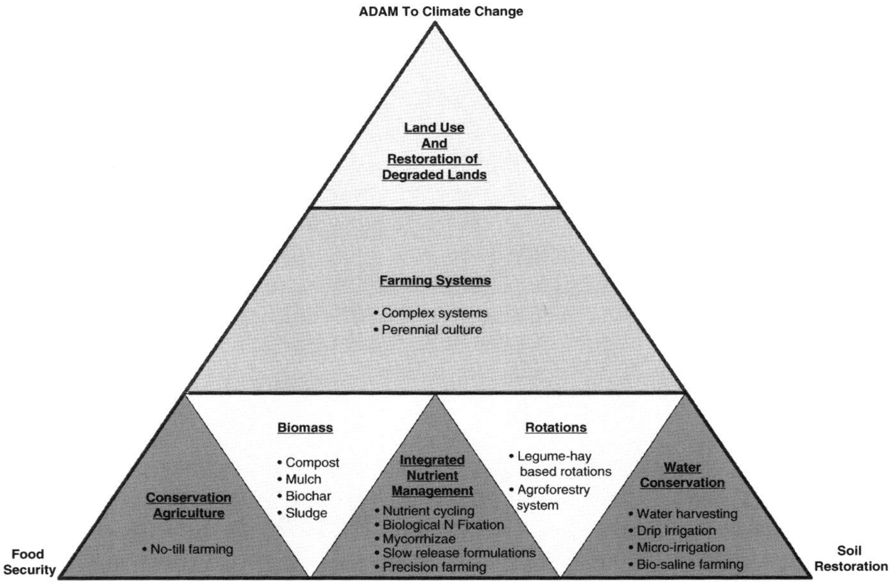

Fig. 2.5 Linking adaptation and mitigation of climate change with the need for soil restoration and achieving food security

freshwater wetlands, their net primary productivity, and other characteristics are detailed by Aselmann and Crutzen (1989), among others.

Whereas the estimates of peatland area are credible (~442 Mha), those of SOC pool of peatlands are preliminary and work in progress. The SOC pool in peatlands is estimated at 462 Pg by Maltby and Immirzi (1993), 357 Pg by Eswaran et al. (1993), 350–535 Pg by Gorham (1991, 1995), 550 Pg by Wetland Intl. (2009), and 612 Pg by Yu et al. (2011).

These highly productive areas and large reservoirs of terrestrial C pool are prone to anthropogenic perturbations including drainage and cultivation, urbanization, and climate change. Thus preservation and restoration of peatlands and the related ecosystems are on the top of the pyramid as indicative of high priority(Fig. 2.4).

Land use, and soil management (tillage, drainage, fertilizer input) remain to be an important source of anthropogenic emissions from peatlands (Tables 2.7, 2.8, 2.9, Fig. 2.2). Thus, judicious land use and management are important to both reducing and sequestering emissions. The schematic in Fig. 2.5 show the strong link between the ADAM strategies of climate change with the need to advance food security via soil restoration through choice of appropriate land use and sustainable intensification of agroecosystems. There being no silver bullet or panacea, it is critically important to validate, fine-tune and adapt different options outlined in Fig. 2.5 under soil/site specific conditions with due consideration to the biophysical and other factors related to the human dimensions.

2.9.2 Degraded Soils and Desertified Ecosystems

There are estimated 3.5 Bha of degraded soils and desertified ecosystems (Bai et al. 2008). Conversion to a restorative land use (afforestation, set side, etc.) can restore the ecosystem C pool, and off-set anthropogenic emissions. The technical potential of C sequestration through restoration of degraded lands is estimated at 0.5–1.4 Pg C/year, of which that through the restoration of world's saline soils is 0.3–0.7 Pg C/year (Lal 2010b). Accelerated soil erosion, through an increase in decomposition of C being translocated and redistributed over the landscape, can be a source of atmospheric CO_2 (Lal 2003). Although a small part of the C displaced by erosion is carried into the depositional sites/aquatic ecosystems and is buried (Van Oost et al. 2007), most of it is emitted into the atmosphere. Anaerobosis and inundation also cause methanogenesis and accentuate nitrification and denitrification. Over and above the benefits of mitigating the climate change, restoration of degraded lands can generate numerous ecosystem services. Important among these are increase in biodiversity, improvement in water quality, etc.

2.9.3 Agricultural Soils

Soils of agro-ecosystems are depleted of their SOC pool, especially those managed by the resource-poor farmers through extractive farming practices and also those which are prone to degradation by erosion, salinization, and nutrient/C depletion. Land areas of agricultural soils include those of croplands covering about 1.5 Bha and grazing land covering about 3.5 Bha. Additional croplands may be needed to feed the growing population, of which >1 billion are already food-insecure (Lele 2010), and changing dietary preferences (Wild 2003). Because of the scarcity of good quality arable land, it will become increasingly difficult to produce all what the burgeoning and affluent world population may demand. Thus, it is prudent to choose the essentials and decide how to produce these indispensible commodities efficiently and sustainably. In addition to the economics, the choice of production technologies must also be based on the hidden C cost (HCC) and the overall C foot print. The strategy is to produce the essentials through sustainable intensification. Accordingly, the goal is to grow more produce from less land, more crop per drop of water, more yield per unit input of fertilizers and pesticides, more food per units of energy, and more biomass per unit of C and environmental foot print (Lal 2011).

In comparison, grazing lands occupy much larger area (~3.5 Bha) and are managed less intensively. The technical potential of C sequestration in cropland soils is estimated at 0.4–1.2 Pg C/year (Lal 2004). In contrast, technical potential of C sequestration through improved management of grazing land is 0.3–0.5 Pg C/year (Lal 2010a). Technological options for improved management of grazing lands include: controlled grazing at low stocking rate and rotational grazing, choice of growing appropriate species adapted to specific ecoregions, fire management,

Table 2.12 The population growth of 15 Indian cities between 1950 and 2025 (Adapted and redrawn from Kazmin 2011)

	Population (10^6 inhabitants)			
City	1950	1980	2010	2025
1. Ahmedabad	0.9	2.5	5.7	7.6
2. Banglore	0.7	2.8	7.2	9.5
3. Calcutta	4.5	9.0	15.6	20.1
4. Hyderbad	1.1	2.5	6.8	8.9
5. Indore	0.3	0.8	2.6	2.9
6. Jaipur	0.3	1.0	3.1	4.2
7. Kanpui	0.7	1.6	3.4	4.5
8. Lucknow	0.5	1.0	2.9	3.9
9. Madras	1.5	4.2	7.5	9.9
10. Mumbai	2.9	8.7	20.0	25.8
11. Nagpur	0.5	1.3	2.6	3.5
12. New Delhi	1.4	5.6	22.2	28.6
13. Patna	0.3	0.9	2.3	3.1
14. Pune	0.6	1.6	5.0	6.6
15. Surat	0.2	2.5	4.2	5.6
Total of 15 cities	16.4	46.0	111.1	144.7
% of population living in cities of >1 million	3.1	5.8	13.0	15.6

nutrient management and soil and water conservation. In comparison with the cost of carbon capture and storage (CCS) through geoengineering, C sequestration in agroecosystems is the most cost effective option (McKinsey & Co. 2009).

2.9.4 Urban Ecosystems

More than 50% of the world's population lives in urban ecosystems, and the proportion will reach 70% by 2050. World's megacities cover large areas such as 0.87 Mha by New York, 0.70 Mha by Tokyo, 0.55 Mha by Chicago and 0.47 Mha by Philadelphia. Urban land area in the U.S. is estimated at ~27 Mha (Lal and Augustin 2012). The global rate of urban encroachment of 2 Mha/year comprises of 1.6 Mha on cropland, 0.2 Mha on forest land, and 0.2 Mha on woodland/grassland. It is estimated that 40,000 ha of land is needed for providing basic living needs of 1 million people. With annual population increase of 70–80 million, 3 Mha/year additional land is converted to urban space. The urban population is rapidly increasing, especially in densely populated countries and emerging economics. The data in Table 2.12 show very rapid population in major cities in India. Population of 15 major Indian cities was 16.4 million in 1950, 46.0 million in 1980 (+180%), 111.1 million in 2010 (+577%), and is projected to be 144.7 million in 2025 (+782%). Percent of total population living in cities of >1 million was 3.1 in 1950, 5.8 in

1980, 13.0 in 2010 and 15.6 in 2025 (Table 2.12). Cities with major growth in population between 1950 and 2025 are Bangalore (from 0.7 to 7.6 million by a factor of 10.9), Calcutta (from 4.5 to 20.1 million by a factor of 8.1), Madras (from 1.5 to 9.9 million by a factor of 6.6), Mumbai (from 2.9 to 25.8 million by a factor of 8.9), and New Delhi (from 1.4 to 28.6 million by a factor of 20.4) (Kazmin 2011). Urban lands, globally covering ~2% of the land area, are source of ~70% of all anthropogenic emissions, and up to 60% of those are from transportation and building sectors. Indeed urban ecosystems transform the structure of local C flows over larger areas than they occupy. Furthermore, cities have more intensive metabolism per unit area which require inflow of artificial energy. Megacities consume a large amount of materials (e.g., food, water, energy, wood, metal), and also emit a large amount of GHGs and pollutants. Thus, urban ecosystems can be judiciously managed to minimize emissions and also sequester C. Accordingly, greenbelts must be established in cities to sequester C in lawns, trees, shrubs, ponds, lakes and green roofs. Indeed, NPP of managed urban ecosystems can be more than those of natural ecosystems (4 Mg C/ha/year vs. 5 MgC/ha/year). The concept of "free city space" can be advantageously used to recarbonize the urban ecosystems.

2.10 Conclusions and Priorities

Depletion of the C pool from the terrestrial biosphere, including 70–100 Pg from world's soils, has exacerbated global warming and the abrupt climate change (ACC), degraded quality of soils, polluted and contaminated water, and destabilized economic, social and political conditions globally. The trilemma of the downward spiral described as the effects and consequences of anthropogenic factors (expressed as rates/min) is as follows (Lal 2011):

I. Causes		
Population increase	:	150 people (250 births, 100 deaths)
Energy use	:	1 PJ
CO_2 carbon increase	:	6,150 Mg
Tropical deforestation	:	25 ha
Urban encroachment	:	5.5 ha
II. Effects		
Desertification	:	23 ha
Soil Degradation	:	10 ha
Deaths from hunger	:	16 people (including 12 children)
III. Consequences		
Political instability		
Civil strife		

Thus, recarbonization of the biosphere is an important strategy to reverse the degradion trends, and improve the environment. Priority biomes with a high potential to sequester C and avoid emissions are peatlands and wetlands, degraded soils

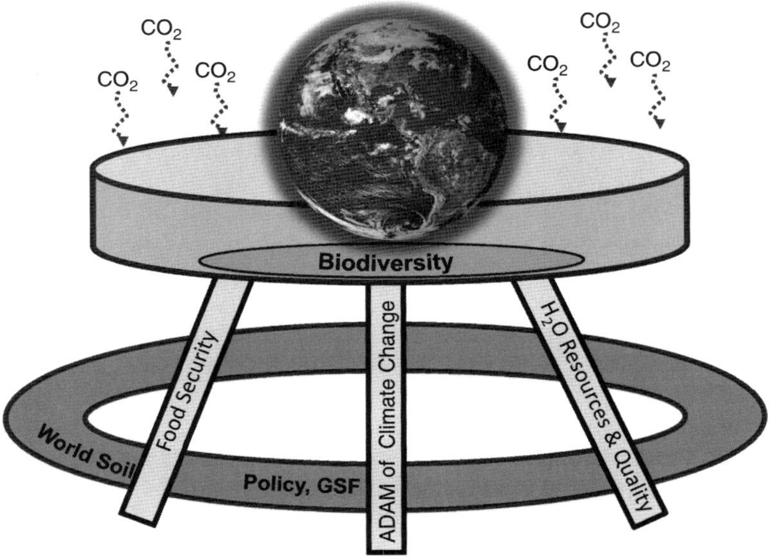

Fig. 2.6 Carbon sequestration in the terrestrial biosphere to achieve food security, improve water resources, enhance biodiversity, and adapt to and mitigate climate change. Soil carbon sequestration leads to healthy soils, healthy people and stable climate, there is a need to establish Global Soil Forum to implement appropriate practices and polices

and desertified ecosystems, soils of agroecosystems, and urban lands. Enhancing soil C pool has numerous co-benefits, and soil C pool can also be used as an indicator of climate change. Over and above adaptation to and mitigating the climate change, sequestration of atmospheric CO_2 into the terrestrial biosphere has numerous co-benefits (Fig. 2.6). Important among these are food security, water security, biodiversity including habitat and species. Furthermore, healthy soils (through C sequestration to a level above the threshold of 1.5–2.0% in the root zone) are also essential to healthy people and stable climate—politically, socially, economically, and environmentally. It is in this context that establishment of a Global Soil Forum (GSF) is essential to: create awareness in policy makers and the general public about the need to recarbonize the biosphere, and provide information to facilitate implementation of the recommended land use and soil/vegetation management practices. The GSF, depicted as a pyramid (Fig. 2.7), is aimed at creating synergisms and complimentarily among national, regional and international organizations addressing relevant issues including food security (FAO), climate change (IPCC), desertification, land degradation and drought (UNCCD) and biodiversity (UNFCBD). While the base of the pyramid is sustainable management of world soils, the apex is strengthened by policy intervention and by initiatives which effectively address issues of the human dimension.

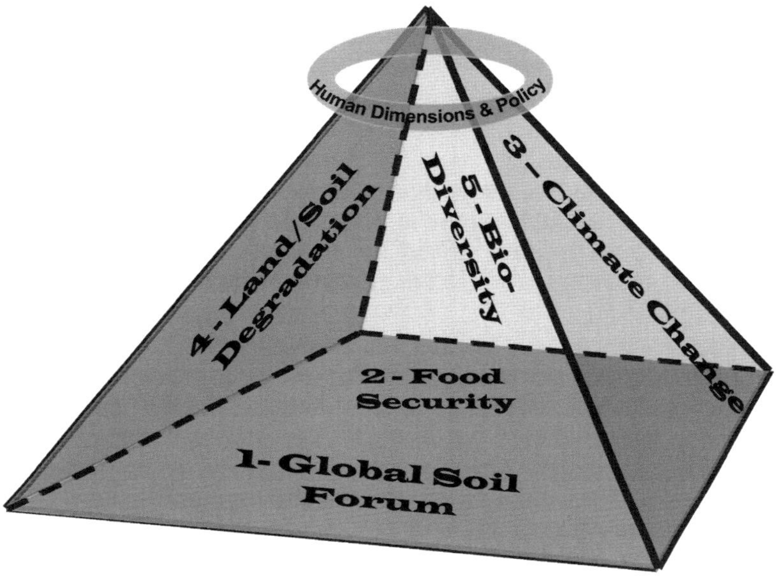

Fig. 2.7 Principal components of Global Soil Forum: (*1*) global soil forum (*base*), (2) food security (*front side/wall*)

References

Aselmann I, Crutzen PJ (1989) Global distribution of natural freshwater wetlands and rice paddies, their net primary productivity, seasonality and possible methane emissions. J Atmos Chem 8:307–358. doi:10.1007/BF00052709

Bai ZG, Dent DL, Olsson L et al (2008) Proxy global assessment of land degradation. Soil Use Manage 24:223–234

Batjes NH (1996) Total carbon and nitrogen in soils of the world. Eur J Soil Sci 47:151–163

Canadell JG, Pataki D, Gifford R et al (2007) Saturation of the terrestrial C sinks; Chapter 6. In: Canadell JG, Pataki D, Pitelka L (eds) Terrestrial ecosystems in a changing world, The IGBP series. Springer, Berlin

Crutzen PI, Stoermer EF (2000) The "anthropocene". IGBP Newsl 41:12

Ellis EC, Goldewijk KK, Siebert S et al (2010) Anthropogenic transformation of the biomes: 1700 to 2000. Global Ecol Biogeogr 19:589–606

Eswaran H, Van den Berg E, Reich P (1993) Organic carbon in soils of the world. J Soil Sci Soc Am 57:192–194

Eswaran H, Van den Berg E, Reich P et al (1995) Global soil carbon resources. In: Lal R, Kimble JM, Levine ES, Stewart BA (eds) Soil and global change. CRC/Lewis Publishers, Boca Ratan, pp 27–43

Eswaran H, Reich PF, Kimble JM et al (2000) Global soil carbon stocks. In: Lal R, Kimble JM, Eswaran H, Stewart BA (eds) Global climate change and pedogenic carbonates. Lewis Publishers, Boca Raton, pp 15–25

Gorham E (1991) Northern peatlands: role in the carbon cycle and probable responses to climatic warming. Ecol Appl 1:182–195. doi:10.2307/1941811

Gorham E (1995) The biogeo chemistry of northern peatlands and its possible responses to global warming. In: Woodwell GM, MacKenzie FT (eds) Biotic feedbacks in the global climatic system. Oxford University Press, New York, pp 169–187

Hansen J, Sato M, Kharecha P et al (2008) Target atmospheric CO_2: where should humanity aim? Open Atmos Sci 2:217–231
Holdren JP (2008) Meeting the climate change challenge. 8th Annual John H. Chaffe Memorial Lecture, 17 January 2008, Ronald Regan Bld, Washington, D.C.
Houghton RA (2007) Balancing the global carbon budget. Annu Rev Earth Planet Sci 35:313–347
IGBPC (International Geosphere-Biosphere Program – Data Information Service) (1999) Soil carbon layer. Royal Swedish Academy of Sciences, Stockholm, Sweden
Jobbagy EG, Jackson RB (2000) The vertical distribution of soil organic carbon and its relation to climate and vegetation. Ecol Appl 10:423–436
Kaat A, Joosten H (2008) Fact book for UNFCCC policies on peat carbon emissions. Wetlands Intl., Ede, the Netherlands
Kazmin A (2011) India's megacities: a case of rise and sprawl. Financial Times 8 November 2011, World News, p 8
Lähteenoja O, Roukolainen K, Schulman L et al (2009) Amazonian peatlands: an ignored C sink and potential source. Global Change Biol 15:2311–2320. doi:10.1111/j.1365-2486.2009.01920.x
Laine J, Minkkinen K, Trettin C (2009) Direct human impacts on the peatland carbon sink. Geophys Mono Series 184, Am Geophys Union. doi: 10.1029/2008GM000808
Lal R (1999) Soil management and restoration for C sequestration to mitigate the greenhouse effect. Prog Environ Sci 1:3070326
Lal R (2003) Soil erosion and the global carbon budget. Environ Int 29:437–450
Lal R (2004) Soil carbon sequestration impacts on global climate change and food security. Science 304:1623–1627
Lal R (2010a) Managing soils and ecosystems for mitigating anthropogenic carbon emissions and advancing global food security. Bioscience 60:708–721
Lal R (2010b) Beyond Copenhagen: mitigating climate change and achieving food security through soil C sequestration. Food Secur 2:169–177
Lal R (2011) Harnessing science knowledge for combating desertification, land degradation and drought. Key note lecture, 10th session of COP to UNCCD, 17–18 October 2011, Changwan, Korea
Lal R, Augustin B (eds) (2012) Carbon sequestration in urban ecosystems. Springer, Dordrecht Holland, p 385
Le Quéré C (2010) Trends in the land and ocean carbon uptake. Curr Opin Environ Sustain 2:219–224
Le Quéré C, Canadell JG, Ciais P et al (2010) An international carbon office to assist policy-based science. Curr Opion Environ Sustain 2:297–300
Lele U (2010) Food security for a billion poor. Science 326:1554
Maltby E, Immirzi CP (1993) Carbon dynamics in peatlands and other wetlands soils: regional and global perspectives. Chemosphere 27:999–1023
McKinsey & Co (2009) Pathways to a low carbon economy. Version 2. Global greenhouse gas abatement cost curve. McKinsey & Co, New York
Olofsson J, Hickler T (2008) Effects of human land-use on the global carbon cycle during the last 6000 years. Veg Hist Archaeobot 17:605–615
Olson DM, Dinerstein E, Wikramanayake ED et al (2001) Terrestrial ecosystems of the world. A new map of life on earth. Bioscience 51:933–938
Pierrhumbert R (2011) Infra red radiation and planetary temperatures. Physics Today, January 2011, pp 48–53
Ramankutty N, Foley JA (1999) Estimating historical changes in global land cover: croplands from 1700 to 1992. Global Biogeochem Cycles 13:997–1027
Reardon S (2011) Climate change sparks battle in classrooms. Science 333:688–689
Ruddiman WF (2003) The anthropogenic greenhouse era began thousands of years ago. Clim Chang 61:261–293
Ruddiman WF (2007) The early anthropogenic hypothesis: challenges and responses. Rev Geophys 45:2006 RG000207R

Ruddiman WF, Ellis EC (2009) Effect of per capita land use changes on Holocene forest clearance and CO_2 emissions. Quaternary Sci Rev 28:3011–3015
Sherwood S (2011) Science controversies: past and present. Physics Today, October 2011, pp 39–56
Somerville RC, Hassol SJ (2011) Communicating the science of climate change. Physics Today, October 2011, pp 48–53
Strahler AN, Strahler AH (1984) Elements of physical geography. Wiley, New York
Tanneberger F, Wichtmann W (eds) (2011) Carbon credits from peatland rewetting, climate, diversity, land use, science, policy, implementation, and recommendations of a pilot project in Belarus. Schweizerbart Science Publishers, Stuttgart
Tarnocai C, Canadell JG, Schuur EAG et al (2009) Soil organic carbon pools in the northern circumpolar permafrost regions. Global Biogeochem Cycles 23. doi:10.1029/2008GB0033277
Van Oost K, Quine TA, Glovers G et al (2007) The impact of agricultural soil erosion in the global carbon cycle. Science 318:626–629
WBGU (*Wissenschaftlicher* Beirat der Bundesregierung Globale Umweltveränderungen) (1988) Die Anrechnung biologischer Quellen und Senken im Kyoto-Protokoll: Fortschritt oder Rückschlang für den globalen Umweltschutz. Sondergutachten 1988. Bremerhaven, Germany
Wetland Intl (2009) Wetlands, for water and life: peatlands and CO_2 emission. Wageningen, the Netherlands
Whittaker RH (1975) Communities and ecosystems. Macmillan, New York
Wild A (2003) Soils, land and food, managing the land during 21st century. Cambridge University Press, Cambridge
Yu Z, Beilman DW, Frolking S et al (2011) Peatlands and their role in the global carbon cycle. EOS 92(12):97–98

Chapter 3
Atmospheric Chemistry and Climate in the Anthropocene

K. Lorenz, P.J. Crutzen, Rattan Lal, and K. Töpfer

Abstract The effects of human activities are increasingly overwhelming the geologic, biological and chemical processes that drive changes in the abundance of trace and greenhouse gases (GHGs) in the atmosphere and Earth System behavior. Thus, mankind has opened a new geological epoch or age – the Anthropocene. This development is driven by the strong population increase which may result in ten billion people by 2100. With this population increase, human demands for food and animal protein in particular, clean water, natural resources and nutrients such as fixed nitrogen (N) and phosphorus (P), land and energy will continue to increase strongly. It is also hypothesized that human enterprise is responsible for the sixth mass species extinction. Further, releases of gases such as sulfur dioxide (SO_2), nitric oxide (NO) and chlorofluorocarbons (CFC) into the atmosphere are several times higher than natural emissions. CFCs give rise to highly active radicals in the stratosphere which destroy ozone (O_3) by catalytic reactions. However, CFC emissions have been drastically reduced and stratospheric O_3 concentrations are increasing.

K. Lorenz (✉)
Global Soil Forum, IASS Institute for Advanced Sustainability Studies e.V., Berliner Strasse 130, 14467 Potsdam, Germany
e-mail: klaus.lorenz@iass-potsdam.de

P.J. Crutzen
Department of Atmospheric Chemistry, Max-Planck-Institute for Chemistry, P.O. Box 3060, 55020 Mainz, Germany
e-mail: paul.crutzen@mpic.de

R. Lal
Carbon Management and Sequestration Center, The Ohio State University, 2021 Coffey Road, 210 Kottman Hall 422b, Columbus, OH, USA
e-mail: lal.1@osu.edu

K. Töpfer
IASS Institute for Advanced Sustainability Studies e.V., Berliner Straße 130, 14467 Potsdam, Germany
e-mail: klaus.toepfer@iass-potsdam.de

R. Lal et al. (eds.), *Recarbonization of the Biosphere: Ecosystems and the Global Carbon Cycle*, DOI 10.1007/978-94-007-4159-1_3,

In contrast, concentrations of GHGs such as carbon dioxide (CO_2), methane (CH_4) and nitrous oxide (N_2O) continue to increase and are now well above the pre-industrial levels as increasing human demand for energy is met by burning fossil fuels. As a consequence, land and ocean temperatures are increasing, glaciers are melting and sea levels are rising. Strong reductions, in particular, of the anthropogenic CO_2 emissions are needed to minimize risks of future warming and its consequences as CO_2 is the major GHG with part of it having a long atmospheric residence time. However, CO_2 emissions continue to increase even during the periods of global economic crises and reduced industrial activities. Thus, geoengineering or climate engineering techniques are discussed to cool Earth indirectly by removing CO_2 from the atmosphere or directly by increasing backscattering of solar radiation into space. For example, injecting sulfur (S) into the stratosphere has been proposed as sulfate particles reflect sunlight but many issues remain unresolved. Thus, a strong reduction in anthropogenic CO_2 emissions is needed to mitigate climate change.

Keywords Anthropocene • Atmospheric chemistry • Mitigating climate change • Greenhouse gases • Climate engineering • Energy production • Ozone layer • Black carbon • Chlorofluorocarbons • Methane • Nitrous oxide • Global biogeochemical cycles • Renewable energy • Nuclear energy • Carbon capture and storage • Solar radiation management

Abbreviations

BNF	biological N fixation
BC	black carbon
C	carbon
CCS	carbon capture and storage
CO_2	carbon dioxide
CDR	carbon dioxide removal
CO	carbon monoxide
CFC	chlorofluorocarbon
GHGs	greenhouse gases
CH_4	methane
EJ	exajoules
NPP	net primary production
NO_3^-	nitrate
N	nitrogen
N_2O	nitrous oxide
OC	organic carbon
O_3	ozone
ppm	parts per million
P	phosphorus
Pg	petagrams
S	sulfur

SRM	solar radiation management
SO_2	sulfur dioxide
Tg	teragrams

3.1 Introduction

The presence of life on Earth determines the chemical conditions of the atmosphere, the oceans, and Earth's crust (Schlesinger et al. 2011). In particular, Earths' radiative balance and atmospheric chemistry critically depend on atmospheric abundance of the trace gases carbon dioxide (CO_2), ozone (O_3), methane (CH_4), nitrous oxide (N_2O) and halogen-containing compounds (Seinfeld and Pandis 2006). Over the past two centuries, geologic, biological and chemical processes that drive changes in abundance of trace gases and Earth System behavior have been overwhelmed by the rapidly increasing effects of human activities (Crutzen and Stoermer 2000; Steffen 2010). As the global population and economy are growing at an unprecedented rate, climate and atmospheric chemistry are being altered directly and indirectly (Crutzen 2002; Victor 2010). The central feature is the enormous increase in the use of fossil fuels since the onset of industrialization (Steffen et al. 2007). Thus, the recent increases in atmospheric trace gases have occurred at an extraordinary pace, and the composition of the atmosphere is changing rapidly on the global scale (Seinfeld and Pandis 2006).

During the past three centuries the global population has increased tenfold to six billion, and quadrupled during the twentieth century (Potts 2009). With this exponential increase human demand for food is also increasing strongly. For example, from 1995 to 2005 cereal production increased from 1897 to 2200 million tons (20%), and meat production increased from 207 to 260 million tons (26%, FAO 2006a). Further, the number of cattle increased by a factor of four during the past century totaling now about 1.4 billion (FAO 2006b). There are about 20 billion farm animals worldwide. Global marine fish catch increased strongly from less than 20 million tons in the mid-twentieth century to about 80 million tons by 2000 (Watson and Pauly 2001). However, overfishing is one of the most serious conservation concerns in marine ecosystems but understanding which fish species are most at risk remains a challenge (Worm et al. 2009; Pinsky et al. 2011). The direct and indirect human pressures on the biosphere are strongly increasing along with the human demand for energy. For example, global energy production doubled over the past half-century while petroleum consumption increased 3.5-fold since 1960 (Steffen et al. 2007; Steffen 2010).

3.2 Changes in the Biosphere

The increasing demands of the growing population result in widespread land use and land use changes within the biosphere. The biosphere extends about 10 km above and 10 km below Earth's surface (Schlesinger et al. 2011). More than half of the terrestrial biosphere has been transformed from mostly wild land to anthropogenic land by human action (Ellis et al. 2010). In addition, the global population has become increasingly urban as urbanization (i.e., the expansion of urban land uses, including commercial, industrial, and residential uses) increased more than tenfold in the past century (UNFPA 2007). About 0.5% of the global ice-free land area was urban land in 2002 but only 0.01% in 1700 (Ellis et al. 2010; Schneider et al. 2009). For the first time in history, more than 50% of the global population lives in urban centers, towns and settlements (UNFPA 2007).

The human alteration of Earth System behavior includes also drastic changes in biodiversity and evolutionary processes. The evolution of new species typically takes at least hundreds of thousands of years. However, technology and human population growth also affect evolutionary trajectories and dramatically accelerates evolutionary change especially in commercially important, pest and disease organism (Palumbi 2001). This is apparent in antibiotic resistance, plant and insect resistance to pesticides, rapid changes in invasive species, life-history change in commercial fisheries, and pest adaptation to biological engineering products. Species extinction rates are also dramatically accelerated. Before humans existed, the average fossil species extinction rate was about 1.8 extinctions of species per million species-years (Barnosky et al. 2011). This is distinctively lower compared to the maximum observed rates during the last thousand years which are even above the late Pleistocene megafaunal diversity crash. Recent average species extinction rates are also too high compared to pre-anthropogenic averages. However, whether currently Earth's sixth mass extinction (i.e., loss of three-quarters of species in geological time interval) is under way is not known (Barnosky et al. 2011).

3.3 Human Alterations of Global Biogeochemical Cycles

Life has left its imprint on the chemistry of the planet resulting in characteristic biogeochemical cycling of elements driven mainly by microbial engines (Falkowski et al. 2008; Schlesinger et al. 2011). However, humans are increasingly altering global biogeochemical cycles. For example, prior to human intervention the Earth's nitrogen (N) cycle was almost entirely controlled by microbes (Falkowski et al. 2008). Atmospheric reactions and slow geological processes controlled Earth's earliest N cycle, and by about 2.7 billion years ago, a linked suite of microbial processes evolved to form the modern N cycle with robust natural feedbacks and controls (Canfield et al. 2010). However, the creation of reactive N (i.e., all N forms except non-reactive N_2) by humans for fertilizers is increasing every year, with drastic transformation of the global N cycle (Galloway et al. 2008). All biological systems need reactive N. As plant growth is usually constrained by soil N availability in most terrestrial ecosystems, increases in N fertilization and deposition may stimulate plant growth (LeBauer and Treseder 2008; Lu et al. 2011).

All the N used in food production is added to the environment, as is the N emitted to the atmosphere during fossil-fuel combustion (Gruber and Galloway 2008). In the 1990s, these two sources of anthropogenic N to the environment amounted to about 160 teragrams (Tg) N per year. Globally, this is more than that supplied by natural biological N fixation (BNF) on land (110 Tg N per year) or in the ocean (140 Tg N per year). Further, humans are likely to be responsible for doubling the turnover rates not only of the terrestrial N cycle but also of the N cycle of the entire Earth (Gruber and Galloway 2008). However, the fate of the human-enhanced N inputs to the land surface is little understood (Schlesinger 2009). For example, increases in terrestrial ecosystem N effluxes caused by N addition were much greater than those in plant and soil pools except soil nitrate (NO_3^-), suggesting a leaky terrestrial N system (Lu et al. 2011). Anthropogenic transformations of global N cycles interact also with C sequestration processes. However, the net effect of increase in N availability on the forest ecosystem carbon (C) balance, for example, is not well understood (Lorenz and Lal 2010).

Similar to N, phosphorus (P) is an essential macronutrient for plant growth, and considered important in determining the biodiversity and biomass of natural ecosystems (Cramer 2010). However, while P does not limit global primary production in the oceans, long-term degradation of soil and terrestrial ecosystems occurs over millennia in the absence of soil-resetting disturbance as P is lost to groundwater and by occlusion in strongly weathered soils (Walker and Syers 1976; Falkowski et al. 2000). The terrestrial ecosystem mass balance of P, in particular, is controlled by depletion, soil barriers and low P-parent material (Vitousek et al. 2010). Thus, as it is removed with harvest P is often a limiting nutrient in agriculture (Sánchez 2010). External P inputs may be required to sustain both primary productivity in terrestrial ecosystems and other biological processes (Vitousek et al. 2010). To overcome the P limitations, P inputs to the biosphere have increased about fourfold primarily due to mining of P compounds for fertilizer (Falkowski et al. 2000). Thus, human activities profoundly alter the global P cycle. Human release of P to the environment is causing widespread eutrophication of surface freshwaters (Carpenter and Bennett 2011). Biodiversity in many natural ecosystems such as Mediterranean terrestrial ecosystems is threatened by super-abundance of the formerly limiting resource P (Tilman et al. 2001). However, the global distribution of P is uneven, and soils of many regions remain P-deficient and those of others are P-saturated (MacDonald et al. 2011). This heterogeneity complicates major challenges for environmental management in the twenty-first century, i.e., the provision of food and high quality freshwater (Carpenter and Bennett 2011). Similar to N, whether the anthropogenic amplification of the P cycle enhances net primary production (NPP) of natural ecosystems and terrestrial C sequestration is not clearly understood (Lorenz and Lal 2010).

The enhanced global flows of P to the biosphere are the result of the growing consumption of inorganic P fertilizers derived from mining of nonrenewable phosphate rock (Smil 2000). Fertilizers account for about 80% of global use of phosphate rock (Van Vuuren et al. 2010). To support agricultural production and meet growing demand for food there will be potentially not enough P in the future as global P demand may exceed global P supply (Cordell et al. 2009). However, when this 'peak P' is reached is difficult to predict as peaks in mineral resources are generally difficult to forecast (van Kauwenbergh 2010). Further, only four countries (i.e., Morocco and Western Sahara, China, Jordan, South Africa) control 80% of the world's global phosphate rock reserves that can be economically mined using current technologies (USGS 2010). Global implications are geopolitical 'shortages' in phosphate supply as have been observed in the past for crude oil and rare minerals. However, the nonrenewable resource oil can be replaced by renewable energy sources but there is no replacement for nonrenewable rock phosphates and no substitutes for P as plant nutrient in agriculture (Van Vuuren et al. 2010). Thus, sustainable P use and management is needed which includes maximizing the efficient conversion of phosphate rock into fertilizer (Cordell et al. 2009). Further options include reducing the large flows of P to surface waters by recycling and recovering from municipal and other waste products and the efficient P use in agriculture including recycling of animal and human excreta (Cordell et al. 2009). Resolving agronomic P imbalances is particularly possible with more efficient use of P fertilizers (MacDonald et al. 2011). In summary, some of the biggest gains in the finite P resource can probably be made from the recovery and recycling of phosphates (Gilbert 2009).

3.4 Atmospheric Chemistry

The increasing anthropogenic pressures within the Earth System affect also atmospheric chemistry. The atmospheric concentrations of greenhouse gases (GHGs) and those of conventional air pollutant such as sulfur dioxide (SO_2), nitrogen oxides ($NO_x = NO + NO_2$), carbon monoxide (CO), primary carbonaceous particles of black carbon (BC), organic carbon (OC) and CH_4 are increasing (Cofala et al. 2007). For example, sulfur (S) is ubiquitous in the biosphere and often occurs in relatively high concentrations in fossil fuels such as coal and crude oil deposits (Smith et al. 2011). The widespread combustion of fossil fuels for energy production has greatly increased emissions of SO_2 into the atmosphere, with the anthropogenic component now substantially greater than natural emissions on a global basis. Global SO_2 emissions peaked in the early 1970s and decreased until 2000. However, in recent years emissions increased further due to increased emissions in China, India, the Middle-East, Brazil, and by international shipping, aviation and developing countries in general (RIVM 2005). About 120 Tg SO_2 have been emitted in 2005 by fossil fuel combustion and industrial processes but only about 4 Tg SO_2 in the same year by forest and grassland burning, and by agricultural waste burning (Smith et al. 2011). Emissions of SO_2 can result in sulfuric acid deposition that can be detrimental to ecosystems, harming aquatic animals and plants, and damaging to a wide range of terrestrial plant life. Further, SO_2 forms sulfate aerosols that have a significant effect on global and regional climate. Sulfate aerosols reflect sunlight into space and also act as condensation nuclei, which tend to make clouds more reflective and change their lifetimes, causing a net cooling (Smith et al. 2011).

Similar to the emissions of SO_2, global anthropogenic emissions of NO_x decreased in the early 1990s but increased afterwards to about 117 Tg NO_2 in 2005 (RIVM 2005). The most important anthropogenic NO_x sources are road transport (41% of anthropogenic NO_x emissions in 2000) followed by power plants, industry and non-road vehicles (21%, 16% and 13% respectively, Cofala et al. 2007). In addition to the 83 Tg NO_2 emitted in 2000, NO_x emissions from biomass burning, international shipping and aviation were responsible for 24.1, 10.0 and 2.5 Tg NO_2, respectively. Releases of NO_x may cause regional high surface (tropospheric) O_3 levels by the reaction of sunlight with air containing hydrocarbons and NO_x (Seinfeld and Pandis 2006). The O_3 in the lower atmosphere can have adverse effects on human health and plants.

The abundance of several important GHGs has substantially increased. For example, atmospheric CO_2 concentrations fluctuated between 180 and 300 parts per million (ppm) by volume over the glacial-interglacial cycles during the past 650,000 years (Jansen et al. 2007). However, about 100 years ago the levels of CO_2 began to increase markedly to about 390 ppm, and continue to rise at annual growth rates >2 ppm (Seinfeld and Pandis 2006; Tans 2011). Similar, atmospheric CH_4 concentrations fluctuated between 0.36 and 0.70 ppm during the last 420,000 years (Spahni et al. 2005). However, since the start of the industrial revolution CH_4 concentrations more than doubled to 1.78 ppm (Forster et al. 2007).

Among the GHGs are the almost inert chlorofluorocarbon (CFC) gases (Forster et al. 2007). Their photochemical breakdown in the stratosphere including those of halons gives rise to highly reactive chlorine and bromine gases which destroy O_3 by catalytic reactions (Crutzen 2002). This anthropogenic O_3 destruction caused the Antarctic 'ozone hole' but CFCs are now regulated (Mäder et al. 2010). Gradually towards the end of the twenty-first century the amount of the CFCs will be diminished so much that N_2O will become the main source of NO_x, which in turn depletes O_3 by catalytic reactions (Ravishankara et al. 2009). Nitrogen oxides are known to catalytically destroy O_3 (Crutzen 1970). Reduction of stratospheric O_3 results in increased UV-B radiation from the sun which may lead to an enhanced risk of skin cancer.

3.5 Climate in the Anthropocene

With the dawn of settled agriculture with attendant deforestation, soil cultivation, spread of rice paddies and raising cattle, it is hypothesized that a trend of increase in atmospheric CO_2 concentration began 8,000 years ago (Ruddiman 2003). Similarly, the rise in atmospheric CH_4 may have begun 5,000 years ago but causes for the late Holocene rise in atmospheric CH_4 concentration are discussed controversially (Singarayer et al. 2011). However, since the beginning of the nineteenth century human activities have major and growing impacts on atmosphere and Earth. Thus, the current geological epoch has been appropriately named Anthropocene to emphasize the central role of mankind in geology and ecology within the current interglacial, the Holocene (the last 11,600 years; Crutzen and Stoermer 2000). Humans are now clearly affecting climate and can deliberately do so.

Climate forcings are imposed radiative perturbations of the Earth's energy balance, and can be of natural and anthropogenic origin (Newman et al. 2010). Primary forcings such as the orbital solar insulation changes are externally imposed on the climate systems. Secondary forcings are feedbacks within the climate system. Examples for secondary forcings are mineral dust, GHGs, land cover, sea ice, and continental ice and sea level which all impact the radiative balance of the atmosphere. The GHG concentrations for past millennia are well known from ice core analysis. The atmospheric concentrations of CO_2, CH_4 and N_2O have increased by 36%, 148% and 18%, respectively, since 1750 (IPCC 2007). Primarily, fossil fuel use and to a lower extent land use change cause the increase in CO_2 whereas agriculture is primarily responsible for the increase in CH_4 and N_2O. The CO_2 and CH_4 levels are much higher than at any time during the last 650,000 years. Further, for about 11,500 years before the industrial period, the concentration of N_2O varied only slightly but increased relatively rapidly toward the end of the twentieth century. The combined radiative forcing due to the increases in CO_2, CH_4 and N_2O is +2.3 W m^{-2} (IPCC 2007). Further, radiative forcing of +0.35 W m^{-2} is caused by tropospheric O_3 changes due to emissions of O_3-forming chemicals (NO_x, CO, and hydrocarbons). The direct radiative forcing due to changes in halocarbons is +0.34 W m^{-2}. Changes in surface albedo due to deposition of BC aerosols on snow cause a radiative forcing of +0.1 W m^{-2}. Changes in solar irradiance since 1750 are estimated to cause a radiative forcing of +0.12 W m^{-2}. In contrast, a cooling effect results from negative radiative forcing by anthropogenic contributions to aerosols (primarily SO_4^{2-}, organic C, BC, NO_3^- and dust), with a total direct radiative forcing of −0.5 W m^{-2} and an indirect cloud albedo forcing of −0.7 W m^{-2}. Further, changes in surface albedo due to land cover changes have also a cooling effect with a radiative forcing of −0.2 W m^{-2}. In summary, the global average net effect of human activities since 1750 has been one of warming, with a radiative forcing of +1.6 W m^{-2} (IPCC 2007).

3.6 The Evidence of Climate Change

Warming of the climate system is unequivocal (IPCC 2007). It is evident from observations of increases in global average air and ocean temperatures, widespread melting of snow and ice, and rising global average sea level. Specifically, global surface temperature has risen by 0.76°C from 1850–1899 to 2001–2005. Since the late 1970s, the warming trend of the global surface temperature was 0.15–0.20°C per decade (Hansen et al. 2010). Also increasing is the average temperature of the global ocean as shown by observations since 1961. Specifically, the ocean has been absorbing more than 80% of the heat added to the climate system. Thus, ocean water is expanding and this contributes to sea level rise (IPCC 2007).

The current warming is unusual when viewed from the millennial perspective provided by multiple lines of proxy evidence and the 160-year record of direct temperature measurements (Thompson 2010). Some of the strongest evidence that a large-scale, pervasive, and, in some cases, rapid change in Earth's climate system is underway is provided by the ongoing widespread melting of high-elevation glaciers and ice caps, particularly in low to middle latitudes. For example, observations of the twentieth and twenty-first century show that glaciers in the Andes, the Himalayas, and on Mount Kilimanjaro are shrinking. Ice cores retrieved from shrinking glaciers around the world confirm their continuous existence for periods ranging from hundreds of years to multiple millennia, suggesting that climatological conditions that dominate those regions today are different from those under which these ice fields originally accumulated and have been sustained (Thompson 2010). Further, summertime melting of Arctic sea-ice has accelerated far beyond the expectations of climate models (Allison et al. 2009).

The mean sea level remained nearly stable since the end of the last deglaciation about 3,000 years ago (Lambeck et al. 2002). However, the sea level is now rising due to heat-induced ocean water expansion and due to widespread decreases in glaciers, ice caps and the ice sheets of Antarctica and Greenland (IPCC 2007). Specifically, since 1950 the sea level has risen by an average of 1.7 mm $year^{-1}$, and from 1993 to 2009 by 3.3 mm $year^{-1}$ (Church and White 2006; Ablain et al. 2009). Contributions of ocean temperature change to the global mean sea level and those of glaciers are estimated to be both ~30% each for 1993–2009 (Cazenave and Llovel 2010; IPCC 2007). These sea level rises will almost certainly accelerate through the twenty-first century and beyond because of climate change (Nicholls and Cazenave 2010).

Further evidence of climate change is found in cold regions. Permafrost (perennially frozen ground) is a unique characteristic of polar regions and high mountains, and is fundamental to geomorphic processes and ecological development in tundra and boreal forests (Jorgenson et al. 2010). Thawing of permafrost affects surface hydrology and changes in soil drainage alter the degradation and accumulation of soil C and emissions of CO_2 and CH_4 (Christensen et al. 2004; Schuur et al. 2008; Turetsky et al. 2007). Moderate warming irreversibly thaws and decays permafrost C and initiate a positive permafrost C feedback on climate (Schaefer et al. 2011).

3.7 Mitigating Climate Change

3.7.1 Reductions in Anthropogenic Greenhouse Gas Emissions

Fossil fuel CO_2 emissions increased by 29% between 2000 and 2008 whereas emissions from land-use changes were nearly constant (Le Quéré et al. 2009). The fraction of CO_2 emissions that remained in the atmosphere has also increased to 45% in the past 50 years. Further, the contribution of CO_2 to the CO_2 equivalent concentration sum of all GHGs is projected to increase from 55% in 2005 to 75–85% by the end of this century (NRC 2011). However, efforts to mitigate climate change require in particular the stabilization of atmospheric CO_2 concentrations as CO_2 is the single most important climate-relevant GHG in Earth's atmosphere (Lacis et al. 2010; Montzka et al. 2011). Thus, CO_2 emissions must be cut strongly, i.e., by more than 80%

to stabilize atmospheric CO_2 concentrations for a century or so (House et al. 2008). Stabilization of atmospheric CO_2 at any level requires in particular anthropogenic CO_2 emissions to go eventually to zero (Weaver 2011).

After CO_2, CH_4 is the second most important anthropogenic GHG in the atmosphere. Since CH_4 has a relatively short lifetime of about 9 years and it is very close to a steady state, reductions in its emissions would quickly benefit climate (Dlugokencky et al. 2011; Montzka et al. 2011). However, in contrast to CO_2 the sinks and, in particular, the sources for CH_4 are poorly quantified. The CH_4 emission rate for the 2000–2004 period has been estimated to be twice the preindustrial period, and 60–70% of the emissions during this period were of anthropogenic origin many of which related to agriculture (Denman et al. 2007; NRC 2011). Present-day CH_4 emissions are estimated to be composed of 64–76% of biogenic, 19–30% of fossil, and 4–6% of pyrogenic sources (Neef et al. 2010). After three decades with little change, atmospheric CH_4 concentrations are increasing since 2007 but the causes are debatable (Rigby et al. 2008; O'Connor et al. 2010; Heimann 2011). While Aydin et al. (2011) explained the slow-down of global atmospheric CH_4 growth during the past two decades by a decline in fossil-fuel emissions, Kai et al. (2011) explained it by a reduction in microbial CH_4 sources in the Northern Hemisphere (i.e., drying northern wetlands, decreasing emissions from rice agriculture in China; Heimann 2011). Further, the main reasons discussed for recent renewed growth are enhanced natural wetland emissions during 2007 and 2008 as a result of higher precipitation in the tropics and anomalously warmer temperatures in the Arctic (Bousquet et al. 2011). However, it is unclear why atmospheric CH_4 concentrations continued to increase in 2009 and 2010 (Montzka et al. 2011). Thus, the reduction in anthropogenic CH_4 emissions needed to stabilize atmospheric concentrations at the current level is unknown. Relative large emission reductions are possible for landfills, and the production of coal, oil and gas (van Vuuren et al. 2007). However, it will be difficult to bring anthropogenic CH_4 emissions to zero in the long term given the continuing need for agriculture to feed the world's population (NRC 2011). Further, significant increases in CH_4 emissions are likely in a future climate from wetland emissions, permafrost thaw, and destabilization of marine hydrates (O'Connor et al. 2010).

N_2O is the fourth largest contributor to radiative forcing in the atmosphere, and second to CH_4 in radiative forcing among non-CO_2 GHGs (Denman et al. 2007). Sources of N_2O to the atmosphere from human activities are approximately equal to those from natural systems. Specifically, strong emissions occur in the Tropics with high temporal variability (Kort et al. 2011). Human activities that emit N_2O include transformation of fertilizer-N into N_2O and its subsequent emission from agricultural soils, biomass burning, raising cattle and some industrial activities but the release of N_2O is poorly understood (Manning et al. 2011). However, atmospheric concentrations of N_2O continue to rise linearly as most emissions are associated with feeding the world's growing population (Forster et al. 2007; Montzka et al. 2011).

The chlorofluorocarbons CFC-11 and CFC-12 are long-lived GHGs and were extensively used in the past as refrigeration agents and in other industrial processes (Forster et al. 2007). However, after their presence in the atmosphere was found to cause stratospheric O_3 depletion, an efficient reduction in global anthropogenic emissions was reached by the Montreal Protocol in 1987.

3.7.2 *Reductions in Greenhouse Gas Emissions from Energy Production*

The primary source of the increased atmospheric concentration of CO_2 since the pre-industrial period is fossil fuel use for energy production (IPCC 2007). In 2004, CO_2 from fossil fuels contributed 56.6% to the total anthropogenic GHG emissions (Rogner et al. 2007). Further, emissions increased from 15,627 petagrams (Pg) CO_2 in 1973 to 29,236 Pg CO_2 in 2008 (IEA 2010). The recent growth in CO_2 emissions parallels a shift in the largest fuel emission source from oil to coal (Le Quéré et al. 2009). In addition to CO_2, fossil fuel use also contributes to the observed increase in atmospheric CH_4 concentration. Thus, energy savings must focus on a strong reduction in fossil fuel use as other energy sources were only responsible for 0.1–0.4% of total CO_2 emissions from energy use (IEA 2010). Improving efficiency in fossil fuel energy use will also contribute to reduced GHG emissions but behavioral interventions are required to improve energy efficiency (Allcott and Mullainathan 2010).

Renewable energy is an option for lowering GHG emissions from the energy system while still satisfying the global demand for energy services (Edenhofer et al. 2011). The most important renewable energy sources to mitigate climate change are bioenergy, direct solar energy, geothermal energy, hydropower, ocean and wind energy. Of the 12.9% of the total 492 Exajoules (EJ) of primary energy supply in 2008 provided by renewable energy, biomass contributed 10.2%, with the majority being traditional biomass used in cooking and heating applications in developing countries but with rapidly increasing use of modern biomass (i.e., all other biomass except traditional biomass). Hydropower represented 2.3% of renewable energy whereas other sources accounted for 0.4%. In particular, GHG emissions from renewable energy technologies for electricity generation are lower than those associated with fossil fuel options (Edenhofer et al. 2011). Although most current bioenergy systems result in GHG emission reductions, most biofuels produced through new processes (i.e., advanced or next generation biofuels) could provide higher GHG mitigation. However, land use changes and corresponding emissions and removals may also affect the GHG balance of biofuels. In total, renewable energy has a large potential to mitigate GHG emissions by 2050 with the largest contributions from modern biomass, wind and direct solar energy use (Edenhofer et al. 2011).

Nuclear energy has been viewed for a long time as playing an important role to meet increasing electricity demand while at the same time decreasing CO_2 emissions from energy production (Widder 2010). However, safety is a major concern and the secure disposal of spent nuclear fuel in a repository is unresolved. Further, the nuclear fuel cycle in many countries such as the United States is unsustainable. Thus, mere absence of GHG emissions may not be sufficient to assess nuclear power as mitigation strategy for climate change (Kopytko and Perkins 2011).

In particular, inland locations of nuclear power plants encounter great problems with interrupted operations while safety is of primary concern at coastal locations. The latter was recently emphasized by the March 11, 2011 disaster at the Fukushima Daiichi nuclear power facility located at the Pacific coast in Japan after a strong earthquake off-coast triggered a devastating tsunami. In the following days, massive explosions at the power plant scattered nuclear fuel around the site. The health effects of the release of large amounts of radioactive material especially with respect to contamination of food and water are uncertain (Butler 2011). Also, the clean-up process at and around Fukushima could last for many decades or even a century (Brumfiel 2011). As a consequence, Germany is looking at options to shut down all its nuclear power plants in the near future as risks for the human society appear to be unpredictable independent of any possible GHG savings by using nuclear power for electricity generation (Ethics Commission for a Safe Energy Supply 2011).

With the recent skepticism towards nuclear power, fossil fuels will account for even more than half of the projected increase in global energy consumption by 2035 with coal remaining the dominant fuel for electricity generation (IEA 2010). Thus, carbon capture and storage (CCS) technologies will be required to reduce increasing CO_2 emissions from coal-fired power plants. CCS refers to a range of technologies that aim to capture the CO_2 in fossil fuels either before or after combustion, and store it for very long time in underground formations such as depleted oil and gas reservoirs, deep saline formations and un-mineable coal seams (Global CSS Institute 2011). Currently, it is uncertain how much and when CCS can contribute to CO_2 sequestration as governments and industry are still in the early stages of implementing large scale international programs to shorten the timeframe for the commercial deployment of CCS. These programs are focused on the demonstration phase for developing and improving capture technologies in new industrial applications and proving the safe and secure long-term storage of CO_2. However, the demonstration phase is likely to last for over a decade and CO_2 emissions from existing and new coal-fired power plants will continue to strongly increase. For example, the world's first large-scale CCS project in the power sector, the Southern Company Integrated Gasification Combined Cycle project in the United States, is still under construction (Global CSS Institute 2011).

3.8 Climate Engineering

Global efforts to reduce GHG emissions have not yet been sufficiently successful to provide confidence that the reductions needed to avoid dangerous climate change will be achieved (The Royal Society 2009). For example, the CO_2 emission growth rate is increasing from increasing global economic activity and from increasing C intensity of the global economy while at the same time land and ocean based CO_2 sink activity appear to weaken (Canadell et al. 2007). Thus, additional actions involving climate engineering or geoengineering may become necessary to cool the Earth. Geoengineering can be defined as the deliberate large-scale intervention in the Earth's climate system in order to moderate global warming (The Royal Society 2009).

Geoengineering methods include carbon dioxide removal (CDR) techniques which remove CO_2 from the atmosphere and solar radiation management (SRM) techniques which reflect a small percentage of the sun's light and heat back into space. CDR techniques include (i) land use management to protect or enhance land C sinks, (ii) the use of biomass for C sequestration as well as a C neutral energy source, (iii) enhancement of natural weathering processes to remove CO_2 from the atmosphere, (iv) direct engineered capture of CO_2 from ambient air, and (v) the enhancement of oceanic uptake of CO_2, for example by fertilization of the oceans with naturally scarce nutrients, or by increasing upwelling processes (The Royal Society 2009). However, only permanent sequestration has the potential to decrease total cumulative CO_2 emissions over time and, thus, decrease the amount of climate change that occurs (Matthews 2010). By contrast, nonpermanent sequestration will generally only delay emissions as very slow leakage of C stored in ocean or geological reservoirs occurs. Thus, total cumulative emissions over time are not decreased and long-term temperature changes would be similar to a scenario without CDR.

SRM techniques include (i) increasing the surface reflectivity of the planet, by brightening human structures (e.g., by painting them white), planting of crops with a high reflectivity, or covering deserts with reflective material, (ii) enhancement of marine cloud reflectivity, and (iii) placing shields or deflectors in space to reduce the amount of solar energy reaching the Earth. However, little research has yet been done on most of the geoengineering methods and there have been no major directed programs of research on the subject although CDR appears to be less risky than SRM. Otherwise, CDR may have a much slower effect on reducing global temperature, and SRM techniques may be ineffective in offsetting changes in rainfall patterns and storms (Irvine et al. 2010). In summary, there are major uncertainties regarding effectiveness, costs and environmental impacts of geoengineering methods (The Royal Society 2009). The most appropriate response to human-induced climate change is to decrease the level of overall human intervention in the climate system by decreasing GHG emissions (Matthews 2010).

Among the SRM techniques discussed is mimicking the cooling effects of volcanic eruptions by injecting sulfate aerosols into the lower stratosphere (Crutzen 2006). Sulfate particles can act as cloud condensation nuclei and thereby influence the microphysical and optical properties of clouds, affecting regional precipitation patterns, and increasing cloud albedo. The deposition of 1–2 Tg S annually in the stratosphere would be sufficient to cool the climate similar to the cooling that occurred after release of 10 Tg S by the eruption of Mount Pinatubo in June 1991. However, the residence time of S in the stratosphere is only about 2 years and the required annual S inputs would be large and need to continue for more than a century. Further, unresolved issues are particle size, effects on long wave radiation, and possible O_3 loss through ClO/Cl reactions. Thus, the sulfate albedo scheme should only be used if climate changes drastically and no serious side effects occur. The reduction in anthropogenic GHG emissions must be the focus in climate change mitigation (Crutzen 2006).

3.9 Summary

Human population is drastically increasing with severe consequences for atmospheric chemistry and the Earth System behavior. Ecosystem services such as the provision of food and water are increasingly exhausted by the ever increasing population. Atmospheric levels of trace gases are higher than during the pre-industrial period. Fortunately, emissions of chlorofluorocarbons (CFFs) are now regulated and stratospheric ozone (O_3) levels are recovering. However, increasing energy demand is met by burning fossil fuels and atmospheric concentrations of greenhouse gases (GHGs) are increasing. Thus, climate is warming and sea levels are rising as oceans are expanding and glaciers are melting. Without drastic reduction in CO_2 emissions, Earth is committed to long term warming with unpredictable consequences for human society. Deliberately reducing atmospheric CO_2 by geoengineering techniques has been proposed but actively reducing solar radiation reaching Earth's atmosphere may be a fast action method to cool Earth. However, environmental issues associated with geoengineering techniques such as the release of large amounts of S for more than a century in the stratosphere remain unresolved. Thus, efforts to mitigate climate change must focus on the reduction in anthropogenic CO_2 emissions.

References

Ablain M, Cazenave A, Valladeau G et al (2009) A new assessment of the error budget of global mean sea level rate estimated by satellite altimetry over 1993–2008. Ocean Sci 5:193–201

Allcott H, Mullainathan S (2010) Behavior and energy policy. Science 327:1204–1205

Allison I, Bindoff NL, Bindschadler RA, et al (2009) The Copenhagen diagnosis, 2009: updating the World on the latest climate science. The University of New South Wales Climate Change Research Centre (CCRC), Sydney, Australia

Aydin M, Verhulst KR, Saltzman ES et al (2011) Recent decreases in fossil-fuel emissions of ethane and methane derived from firn air. Nature 476:198–201

Barnosky AD, Matzke N, Tomiya S et al (2011) Has the Earth's sixth mass extinction already arrived? Nature 471:51–57

Bousquet P, Ringeval B, Pison I et al (2011) Source attribution of the changes in atmospheric methane for 2006–2008. Atmos Chem Phys 11:3689–3700

Brumfiel G (2011) Fukushima set for epic clean-up. Nature 474:135–136

Butler D (2011) Fukushima health risks scrutinized. Nature 472:13–14

Canadell JG, Le Quéré C, Raupach MR et al (2007) Contributions to accelerating atmospheric CO_2 growth from economic activity, carbon intensity, and efficiency of natural sinks. Proc Natl Acad Sci USA 104:18866–18870

Canfield DE, Glazer AN, Falkowski PG (2010) The evolution and future of Earth's nitrogen cycle. Science 323:192–196

Carpenter SR, Bennett EM (2011) Reconsideration of the planetary boundary for phosphorus. Environ Res Lett 6:014009

Cazenave A, Llovel W (2010) Contemporary sea level rise. Annu Rev Mar Sci 2:145–173

Christensen TR, Johansson T, Malmer N et al (2004) Thawing sub-arctic permafrost: effects on vegetation and methane emissions. Geophys Res Lett 31:L04501

Church JA, White NJ (2006) A 20th century acceleration in global sea-level rise. Geophys Res Lett 33:L01602
Cofala J, Amann M, Klimont Z et al (2007) Scenarios of global anthropogenic emissions of air pollutants and methane until 2030. Atmos Environ 41:8486–8499
Cordell D, Drangert J-O, White S (2009) The story of phosphorus: global food security and food for thought. Glob Environ Chang 19:292–305
Cramer MD (2010) Phosphate as a limiting resource: introduction. Plant Soil 334:1–10
Crutzen PJ (1970) The influence of nitrogen oxides on the atmospheric ozone content. Q J R Meteorol Soc 96:320–325
Crutzen PJ (2002) Geology of mankind. Nature 415:23
Crutzen PJ (2006) Albedo enhancement by stratospheric sulfur injections: a contribution to resolve a policy dilemma? Clim Chang 77:211–219
Crutzen PJ, Stoermer EF (2000) The "anthropocene". IGBP Newsl 41:17–18
Denman KL, Brasseur G, Chidthaisong A et al (2007) Couplings between changes in the climate system and biogeochemistry; Chapter 7. In: Intergovernmental Panel on Climate Change (ed) Climate change 2007: the physical science basis. Cambridge University Press, Cambridge
Dlugokencky EJ, Nisbet EG, Fisher R et al (2011) Global atmospheric methane: budget, changes and dangers. Philos Trans R Soc A 369:2058–2072
Edenhofer O, Pichs-Madruga R, Sokona Y et al (2011) Summary for policymakers. In: Edenhofer O, Pichs-Madruga R, Sokona Y et al (eds) IPCC special report on renewable energy sources and climate change mitigation. Cambridge University Press, Cambridge/New York
Ellis EC, Goldewijk KK, Siebert S et al (2010) Anthropogenic transformation of the biomes, 1700 to 2000. Global Ecol Biogeogr 19:589–606
Ethics Commission for a Safe Energy Supply (2011) Germany's energy transition – a collective project for the future. http://www.bundesregierung.de/Content/DE/__Anlagen/2011/05/2011-05-30-abschlussbericht-ethikkommission__en,property=publicationFile.pdf
Falkowski PG, Fenchel T, Delong EF (2008) The microbial engines that drive Earth's biogeochemical cycles. Science 320:1034–1039
Falkowski P, Scholes RJ, Boyle E et al (2000) The global carbon cycle: a test of our knowledge of earth as a system. Science 290:291–296
Food and Agriculture Organization of the United Nations (FAO) (2006a) FAO statistical databases, Rome. Available at http://faostat.fao.org/default.aspx
Food and Agriculture Organization of the United Nations (FAO) (2006b) Livestock's long shadow: environmental issues and options. Food and Agriculture Organization, Rome
Forster P, Ramaswamy V, Artaxo P et al (2007) Changes in atmospheric constituents and in radiative forcing. In: Solomon S, Qin D, Manning M et al (eds) Climate change 2007: the physical science basis. Contribution of working group I to the fourth assessment report of the Intergovernmental Panel on Climate Change. Cambridge University Press, Cambridge/New York
Galloway JN, Townsend AR, Erisman JW et al (2008) Transformation of the nitrogen cycle: recent trends, questions, and potential solutions. Science 320:889–892
Gilbert N (2009) The disappearing nutrient. Nature 461:716–718
Global CCS Institute (2011) The global status of CCS: 2010. Canberra, Australia
Gruber N, Galloway JN (2008) An earth-system perspective of the global nitrogen cycle. Nature 451:293–296
Hansen J, Ruedy R, Sato M et al (2010) Global surface temperature change. Rev Geophys 48:RG4004. doi:10.1029/2010RG000345
Heimann M (2011) Enigma of the recent methane budget. Nature 476:157–158
House JI, Huntingford C, Knorr W et al (2008) What do recent advances in quantifying climate and carbon cycle uncertainties mean for climate policy? Environ Res Lett 3:044002
IEA (2010) Key world energy statistics. Organization for Economic Co-operation and Development/International Energy Agency, Paris, France
IPCC (2007) Summary for policymakers. In: Solomon S, Qin D, Manning M, Chen Z, Marquis M, Averyt KB, Tignor M, Miller HL (eds) Climate change 2007: the physical science basis.

Contribution of working group I to the fourth assessment report of the intergovernmental panel on climate change. Cambridge University Press, Cambridge
Irvine PJ, Ridgwell A, Lunt DJ (2010) Assessing the regional disparities in geoengineering impacts. Geophys Res Lett 37:L18702. doi:10.1029/2010GL044447
Jansen E, Overpeck J, Briffa KR et al (2007) Palaeoclimate. In: Solomon S, Qin D, Manning M et al (eds) Climate change 2007: the physical science basis. Contribution of working group I to the fourth assessment report of the Intergovernmental Panel on Climate Change. Cambridge University Press, Cambridge/New York
Jorgenson MT, Romanovsky V, Harden J et al (2010) Resilience and vulnerability of permafrost to climate change. Can J For Res 40:1219–1236
Kai FM, Tyler SC, Randerson JT et al (2011) Reduced methane growth rate explained by decreased Northern Hemisphere microbial sources. Nature 476:194–197
Kopytko N, Perkins J (2011) Climate change, nuclear power, and the adaptation–mitigation dilemma. Energy Policy 39:318–333
Kort EA, Patra PK, Ishijima K et al (2011) Tropospheric distribution and variability of N_2O: evidence for strong tropical emissions. Geophys Res Lett 38:L15806. doi:10.1029/2011GL047612
Lacis AA, Schmidt GA, Rind D et al (2010) Atmospheric CO_2: principal control knob governing Earth's temperature. Science 330:356–359
Lambeck K, Yokoyama Y, Purcell T (2002) Into and out of the last glacial maximum: sea-level change during oxygen isotope stages 3 and 2. Quaternary Sci Rev 21:343–360
LeBauer DS, Treseder KK (2008) Nitrogen limitation of net primary productivity in terrestrial ecosystems is globally distributed. Ecology 89:371–379
Le Quéré C, Raupach MR, Canadell JG et al (2009) Trends in the sources and sinks of carbon dioxide. Nat Geosci 2:831–836
Lorenz K, Lal R (2010) Carbon sequestration in forest ecosystems. Springer, Dordrecht
Lu M, Yang Y, Luo Y et al (2011) Responses of ecosystem nitrogen cycle to nitrogen addition: a meta-analysis. New Phytol 189:1040–1050
MacDonald GK, Bennett EM, Potter PA et al (2011) Agronomic phosphorus imbalances across the world's croplands. Proc Natl Acad Sci USA 108:3086–3091
Mäder JA, Staehelin J, Peter T et al (2010) Evidence for the effectiveness of the Montreal Protocol to protect the ozone layer. Atmos Chem Phys 10:12161–12171
Manning AC, Nisbet EG, Keeling RF et al (2011) Greenhouse gases in the earth system: setting the agenda to 2030. Philos Trans R Soc A 369:1885–1890
Matthews HD (2010) Can carbon cycle geoengineering be a useful complement to ambitious climate mitigation? Carbon Manage 1:135–144
Montzka SA, Dlugokencky EJ, Butler JH (2011) Non-CO_2 greenhouse gases and climate change. Nature 476:43–50
Neef L, van Weele M, van Velthoven P (2010) Optimal estimation of the present day global methane budget. Global Biogeochem Cycles 24:GB4024. doi:10.1029/2009GB003661
Newman L, Kiefer T, Otto-Bliesner B et al (2010) The science and strategy of the past global changes (PAGES) project. Curr Opin Environ Sustain 2:193–201
Nicholls RJ, Cazenave A (2010) Sea-level rise and its impact on coastal zones. Science 328: 1517–1520
NRC (2011) Climate stabilization targets: emissions, concentrations, and impacts over decades to millennia. Committee on stabilization targets for atmospheric greenhouse gas concentrations, National Research Council. The National Academies Press, Washington, DC
O'Connor FM, Boucher O, Gedney N et al (2010) Possible role of wetlands, permafrost, and methane hydrates in the methane cycle under future climate change: a review. Rev Geophys 48:RG4005. doi:10.1029/2010RG000326
Palumbi SR (2001) Humans as the world's greatest evolutionary force. Science 293:1786–1790
Pinsky ML, Jensen OP, Ricard D et al (2011) Unexpected patterns of fisheries collapse in the world's oceans. Proc Natl Acad Sci USA 108:8317–8322
Potts M (2009) What's next? Philos Trans R Soc B 364:3115–3124

Ravishankara AR, Daniel JS, Portmann RW (2009) Nitrous oxide (N_2O): the dominant ozone-depleting substance emitted in the 21st century. Science 326:123–125

Rigby M, Prinn RG, Fraser PJ et al (2008) Renewed growth of atmospheric methane. Geophys Res Lett 35:L22805. doi:10.1029/2008GL036037

RIVM (2005) Emission database for global atmospheric research-EDGAR. RIVM, Bilthoven, The Netherlands

Rogner H-H, Zhou D, Bradley R et al (2007) Introduction. In: Metz B, Davidson OR, Bosch PR, Dave R, Meyer LA (eds) Climate change 2007: mitigation. Contribution of working group III to the fourth assessment report of the Intergovernmental Panel on Climate Change. Cambridge University Press, Cambridge/New York

Ruddiman WF (2003) The anthropogenic greenhouse area began thousands of years ago. Clim Chang 61:261–293

Sánchez PA (2010) Tripling crop yields in tropical Africa. Nat Geosci 3:299–300

Schaefer K, Zhang T, Bruhwiler L et al (2011) Amount and timing of permafrost carbon release in response to climate warming. Tellus 63B:165–180

Schlesinger WH (2009) On the fate of anthropogenic nitrogen. Proc Natl Acad Sci USA 106:203–208

Schlesinger WH, Cole JJ, Finzi AC et al (2011) Introduction to coupled biogeochemical cycles. Front Ecol Environ 9:5–8

Schneider A, Friedl MA, Potere D (2009) A new map of global urban extent from MODIS satellite data. Environ Res Lett 4:044003. doi:10.1088/1748-9326/4/4/044003

Schuur EAG, Bockheim J, Canadell JG et al (2008) Vulnerability of permafrost carbon to climate change: implications for the global carbon cycle. Bioscience 58:701–714

Seinfeld JH, Pandis SN (2006) Atmospheric chemistry and physics: from air pollution to climate change, 2nd edn. Wiley, Hoboken

Singarayer JS, Valdes PJ, Friedlingstein P et al (2011) Late Holocene methane rise caused by orbitally controlled increase in tropical sources. Nature 470:82–86

Smil V (2000) Phosphorus in the environment: natural flows and human interferences. Annu Rev Energy Environ 25:53–88

Smith SJ, van Aardenne J, Klimont Z et al (2011) Anthropogenic sulfur dioxide emissions: 1850–2005. Atmos Chem Phys 11:1101–1116

Spahni R, Chapellaz J, Stocker TF et al (2005) Atmospheric methane and nitrous oxide of the late Pleistocene from Antarctic ice cores. Science 310:1317–1321

Steffen W (2010) Observed trends in earth system behavior. WIREs Clim Chang 1:428–449

Steffen W, Crutzen PJ, McNeill JR (2007) The Anthropocene: are humans now overwhelming the great forces of nature? Ambio 36:614–621

Tans P (2011) Trends in atmospheric carbon dioxide – global. http://www.esrl.noaa.gov/gmd/ccgg/trends/

The Royal Society (2009) Geoengineering the climate: science, governance and uncertainty. The Royal Society, Science Policy, London

Thompson LG (2010) Climate change: the evidence and our options. Behav Anal 33:153–170

Tilman D, Fargione J, Wolff B et al (2001) Forecasting agriculturally driven global environmental change. Science 292:281–284

Turetsky MR, Wieder RK, Vitt DH et al (2007) The disappearance of relict permafrost in boreal North America: effects on peatland carbon storage and fluxes. Glob Change Biol 13:1922–1934

United Nations Populations Fund (UNFPA) (2007) State of world population 2007: unleashing the potential of urban growth. United Nations Population Fund, New York

U.S. Geological Survey (USGS) (2010) Mineral commodity summaries 2010: U.S. Geological Survey

van Kauwenbergh SJ (2010) World phosphate rock reserves and resources. International Fertilizer Development Center, Muscle Shoals

van Vuuren DP, den Elzen MGJ, Lucas PL et al (2007) Stabilizing greenhouse gas concentrations at low levels: an assessment of reduction strategies and costs. Clim Chang 81:119–159

van Vuuren DP, Bouwman AF, Beusen AHW (2010) Phosphorus demand for the 1970–2100 period: a scenario analysis of resource depletion. Glob Environ Chang 20:428–439
Victor P (2010) Questioning economic growth. Nature 468:370–371
Vitousek PM, Porder S, Houlton BZ et al (2010) Terrestrial phosphorus limitation: mechanisms, implications, and nitrogen–phosphorus interactions. Ecol Appl 20:5–15
Walker TW, Syers JK (1976) The fate of phosphorus during pedogenesis. Geoderma 15:1–19
Watson R, Pauly D (2001) Systematic distortions in world fisheries catch trends. Nature 414:534–536
Weaver AJ (2011) Toward the second commitment period of the Kyoto Protocol. Science 332:395–396
Widder S (2010) Benefits and concerns of a closed nuclear fuel cycle. J Renew Sustain Energy 2:062801
Worm B, Hilborn R, Baum JK et al (2009) Rebuilding global fisheries. Science 325:578–585

Chapter 4
Historic Changes in Terrestrial Carbon Storage

R.A. Houghton

Abstract Human use of land has reduced the amount of carbon (C) in terrestrial ecosystems, probably since the first use of fire as a tool for clearing land thousands of years ago. Because variations in climate have also affected C storage over this period, it is difficult to attribute long-term changes in terrestrial C to direct human activity. Over the last 150–300 years, however, reconstructions of land use and land-use change suggest that between ~100 and ~200 Pg (1 Pg = 10^{15} g) C were lost from land, largely from the conversion of forests to agricultural lands. This loss of C over the past century or so is greater than the loss attributable to human activity for all of time before 1850. Most of the loss since 1850 has been from forest biomass, while the loss of C from soil organic matter (SOM) as a result of cultivation is estimated to have contributed ~25% of the net loss. The restoration of forests on cleared lands could, in theory, re-carbonize the biosphere with 100–200 Pg C; but most of these lands are currently in use and unlikely to be returned to forests. Management practices would have to reverse the centuries-long loss of C.

For most of the last 300 years, the net annual loss of C from land use seems to explain (i.e., is roughly equivalent to) the net terrestrial flux of C to the atmosphere. Starting near the middle of the twentieth century, however, the annual net emissions of C from land use appear to have been offset by a terrestrial C sink *not* directly related to land use. The explanations for this residual terrestrial sink include carbon dioxide (CO_2) fertilization, nitrogen (N) deposition, variations in climate, and, possibly, a centuries-long reduction of natural disturbances. Much of the offsetting C sink is thought to be in forests. The residual C sink indicates that terrestrial ecosystems, despite land use, have removed C from the atmosphere over the last decades. The magnitude of this sink is large relative to the effect human management could have, but recent evidence suggests that the sink may be beginning to saturate.

R.A. Houghton (✉)
Woods Hole Research Center, 149 Woods Hole Road, Falmouth, MA 02540, USA
e-mail: rhoughton@whrc.org

R. Lal et al. (eds.), *Recarbonization of the Biosphere: Ecosystems and the Global Carbon Cycle*, DOI 10.1007/978-94-007-4159-1_4,

If the residual terrestrial sink were to disappear or become an additional source of C as a result of climate change, managing the global C cycle would be much more difficult than envisioned today.

Keywords Carbon management • Deforestation • Global carbon budget • Land use • Residual carbon sink • Terrestrial carbon pool • Biotic carbon pool • Soil carbon pool • Residual C sink • Missing C • Historic C loss • Deforestation • Soil degradation • Global C budget • Direct human effects • Croplands • Pastures • Settled lands • Wood harvest • Reforestation • Fire management

Abbreviations

C	carbon
SOC	soil organic carbon
SOM	soil organic matter

4.1 Introduction

The global carbon (C) cycle is not in equilibrium. The C accumulated in fossil reservoirs over millions of years is being mobilized into the atmosphere over a few centuries and is being redistributed into the oceans and land (Fig. 4.1). Human management of land has offset to a large extent the terrestrial uptake of fossil C, but the net terrestrial sink of recent decades may be changing. This chapter reviews historical changes in the global C cycle, with an emphasis on changes in terrestrial C storage 1850–2005.

The three major drivers of changes in terrestrial C storage are: (1) natural processes, (2) direct effects of human management, and (3) indirect effects of human activity. Regarding the natural processes of de- and re-carbonization, the highest C densities on land appear in forests, wetlands, tundra, and prairies, as a result of natural processes. Two aspects of these processes are important here. First, most of these high C densities accumulated over centuries or millennia, and the accumulation rates today are generally low. Nevertheless, large areas of low accumulation rates can add up to be globally significant.

Second, these C-rich ecosystems have been sinks in the past and may be sinks today where they still exist (Luyssaert et al. 2008; Dommain et al. 2011; Donato et al. 2011). However, the direct effects of management, such as draining and mining of wetlands for crops, forestry, and fuel, have released to the atmosphere much of the C built-up in these ecosystems over centuries. More than that, the loss of these ecosystems has eliminated the potential C sink that might otherwise have persisted (Gitz and Ciais 2003; Bridgham et al. 2006; Strassmann et al. 2008).

The third driver of changes in terrestrial C storage is the indirect effect of human activity, such as increasing concentrations of carbon dioxide (CO_2) in the atmosphere,

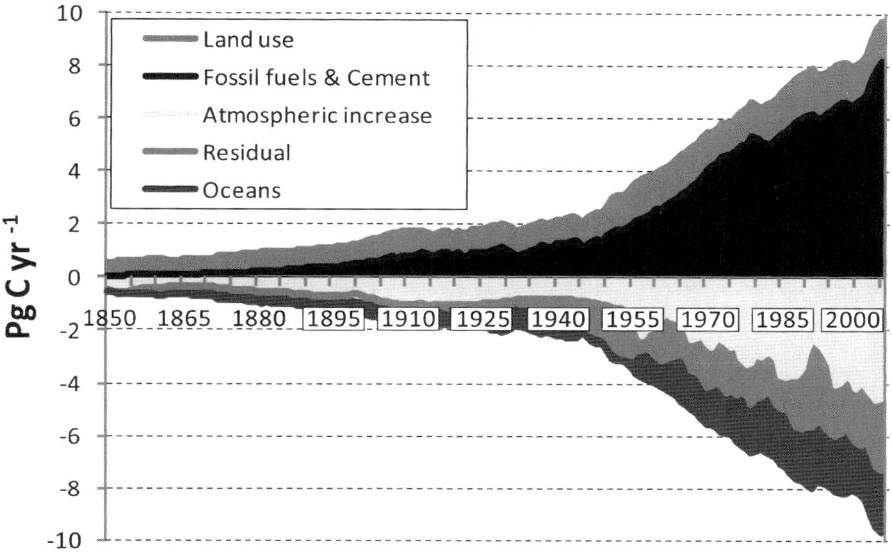

Fig. 4.1 Annual sources and sinks in the global C balance, 1850–2008 (From Houghton, this study; Canadell et al. 2007; Le Quéré et al. 2009). Total annual emissions from fossil fuels and land-use change equal the sum of the annual increase in the atmosphere plus annual uptake by oceans and land (residual)

increased rates of deposition of reactive nitrogen (N), and changes in climate. These environmental factors may affect the rates of photosynthesis and respiration, thereby potentially changing the storage of C on land. The factors are the indirect effects of human activity, in contrast to the direct effects of land management.

This chapter considers all three effects, but focuses on the direct effects of human management; i.e., land use and land-use change. In contrast to most of the following chapters, which consider the potential for re-carbonization of the biosphere, this chapter deals largely with de-carbonization because the net effect of land management over millennia has been to reduce the amount of C in terrestrial ecosystems (Kaplan et al. 2009; Ellis 2011; Pinter et al. 2011). Furthermore, the rate of loss has accelerated over the last ~150 years, although not equally in all regions (Fig. 4.2). On the contrary, the gradual loss, globally, is composed of recent, accelerated losses in tropical countries and little change or re-carbonization in developed countries of the temperate zone and boreal regions.

4.1.1 The Global Carbon Budget 1850–2005

With the start of the industrial revolution in the eighteenth century, fossil C began being increasingly added to the atmosphere. Carbon was already being emitted from the expansion of agricultural lands, and not until half way through the twentieth century did the annual emissions of C from fossil fuels exceed the annual net

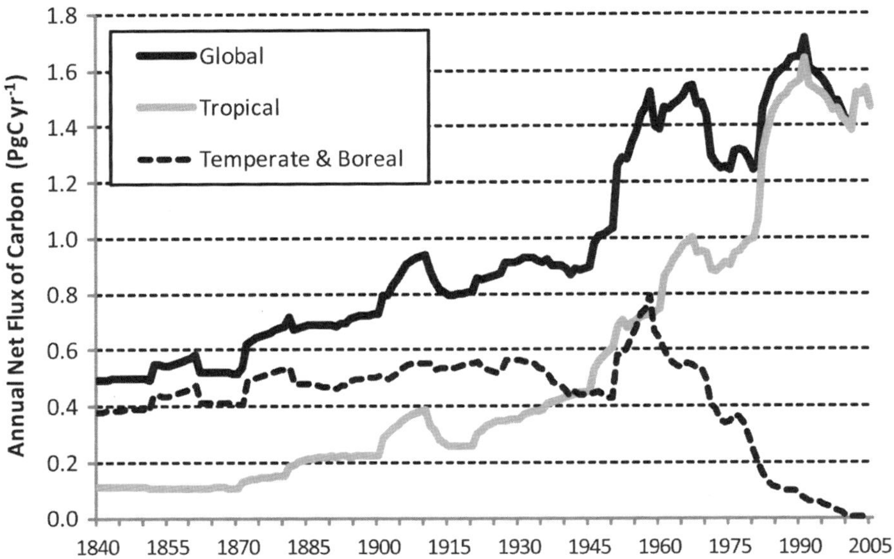

Fig. 4.2 Annual net loss of carbon from land use in global, tropical, and temperate & boreal lands (Houghton, this study)

Table 4.1 Sources (+) and sinks (–) of carbon over the period 1850–2005 (Updated from Houghton 2007) and 2000–2008 (From Le Quéré et al. 2009)

Carbon sources/sinks	1850 Pg C	2000–2008 Pg/year
Emissions from fossil fuels	320	7.7
Emissions from land-use change	156	1.4
Atmospheric increase	–198	–4.1
Oceanic uptake	–155	–2.3
Residual terrestrial sink	–123	–2.7

emissions from land-use change (Fig. 4.1). Between 1850 and 2005 the emissions from fossil fuels (and cement production) were 320 Pg C, while the emissions from land-use change were 156 Pg C. Total anthropogenic emissions of C were, thus, 476 Pg C. The amount of C in the atmosphere increased by 198 Pg C (~33%) during this 155-year period, from ~280 ppm (595 Pg C) to ~374 ppm (793 Pg C) (Table 4.1). The accumulation of C in the atmosphere was ~42% of total emissions; land and oceans accumulated 28% and 30%, respectively. The fraction remaining in the atmosphere has increased slightly, averaging 37% between 1850 and 1955, and 45% between 1956 and 2005.

One interesting feature of Fig. 4.1 is that the terrestrial component of the budget appears twice. It appears as a net source of C from land use and land-use change. Before ~1950 the net emissions from land management were larger than global emissions of C from fossil fuels. Obviously, the proportion of emissions from fossil fuels has grown dramatically since then, and the proportion of total emissions from land use has declined from ~30% over the period 1850–2005 to 20% for the 1990s to 15% in 2008.

But the land has also been a net sink for C, at least for the last several decades, for reasons that are not well understood. This 'residual terrestrial sink', unlike the other terms, is not independently determined. Rather, it is determined from the other terms in the global C budget and the requirement that sources and sinks must balance (Table 4.1 and Fig. 4.1). The residual terrestrial sink does *not* include those C sinks that result from management; e.g., regrowth following wood harvest or agricultural abandonment. The latter sinks are the result of direct human activity and are included in the net land use term. The residual terrestrial sink is the result of indirect and/or natural effects. It is worth noting that before ~1900 the annual residual terrestrial sink was close to zero. This implies that the net loss of C from direct human management was nearly equivalent to the net terrestrial flux. A natural or indirect effect is not apparent until after 1900.

Possible explanations for the residual terrestrial C sink include errors or omissions in the estimated net flux from land-use change or errors in the other terms of the global C balance (fossil fuels, atmosphere, oceans).

On the other hand, the residual terrestrial C sink may be explained by terrestrial processes not included in analyses of land use and land-use change. There are three general processes potentially responsible. First, many management practices are not included in analyses of land use and land-use change, although the major practices are described in the following chapters. Second, natural disturbance regimes (including recovery) may act independently of land use and land-use change, and may themselves be changing, causing C to accumulate (Marlon et al. 2008; Wang et al. 2010). It must be noted, however, that in many regions the effects of climate change (droughts and fires) appear over the last decades to be causing additional C to be lost rather than accumulated (Gillett et al. 2004; Westerling et al. 2006; Kurz et al. 2008). Finally, C stocks may be influenced by environmental changes in climate, CO_2, or biologically available N. For the last few decades the net effects of these three processes has been to sequester C on land, but their relative contributions are unknown.

The important questions for the residual terrestrial sink, globally or locally, are:

When did it become significant in the net balance?
How large is it now?
Where is it?
What are its causes?
Will it persist?

A recent analysis of forest inventories suggests that at least a portion of the residual sink is in forests, worldwide (Pan et al. 2011). Regarding the sink's persistence, the thawing of permafrost in response to the current and committed global warming is likely to expose labile organic C to decomposition and may increase the loss of C from land. This positive feedback to the warming is an example of a process that would reduce the residual sink.

The interactions of direct human effects, indirect effects, and natural effects on terrestrial C emissions are complex. The component measured most directly seems to be the direct effects; i.e., the changes in C due to land use and land-use change. Estimates of this change are important for both political and scientific reasons. The political reasons relate to assigning C credits and debits as incentives to reduce emissions of C through management (e.g., REDD, Reduced Emissions from Deforestation and forest Degradation). The scientific reasons relate to understanding the global C cycle and, in particular, the rate and extent of feedbacks between climate change and the C cycle. Arguably the most important feedback is the effect of climate change on terrestrial and oceanic C sinks. Over the last several decades, those sinks have been responsible for removing about ~55% of anthropogenic C emissions (Le Quéré et al. 2009) (Table 4.1). Remarkably, these sinks have increased in proportion to emissions. The single best indicator of whether the sinks are continuing to grow is the airborne fraction, the ratio of growth in atmospheric C to total emissions (land use and fossil fuel). Of the three terms needed to evaluate the airborne fraction, the net emissions from land use and land-use change are the most uncertain. The trend of annual emissions over the last three decades is known barely well enough to suggest that the airborne fraction is increasing (Canadell et al. 2007; Le Quéré et al. 2009), but that conclusion has been challenged (Knorr 2009; Gloor et al. 2010). Reducing the uncertainty in the land-use flux would enable a more precise measure of trend in the airborne fraction.

The calculation of a residual terrestrial sink for the globe suggests that residual sinks (and sources) exist at other scales as well; that is, the changes in C contributed by land use and land-use change in individual regions or sites may be different from the sources and sinks actually measured with forest inventories or fluxes of CO_2. Measured changes include natural or indirect effects as well as the direct effects of land management. For example, the net sink of C in Europe's terrestrial ecosystems (Janssens et al. 2003) is not necessarily attributable to management.

Most ecosystems where C is accumulating are recovering from an earlier disturbance, either short term (e.g., fire) or long term (climatic variations). "Disturbance" is perhaps not the appropriate term to describe climatic variations, but terrestrial C storage varies significantly over glacial cycles. The warming that followed the last glacial maximum increased terrestrial C storage by 550–694 Pg C (Prentice et al. 2011). This increased storage associated with warming is larger than total anthropogenic emissions (fossil and land use) since ~1850 (476 Pg C) (Table 4.1). The magnitude of this change, attributable to natural processes, illustrates the difficulty of separating climatic from land-use effects in pre-historic times (Pinter et al. 2011; Reick et al. 2011). It also reminds us that management effects may be overwhelmed by natural processes. Today natural processes on land and in the oceans are removing about 55% of anthropogenic emissions, but in past warming periods land and ocean have been additional sources of C.

4.2 Direct Human Effects on De- and Re-carbonization

Direct human effects on terrestrial C storage from land use, land-use change, or management, are divided here into changes before 1850 and changes since 1850.

Table 4.2 Net loss of carbon (Pg C) from global changes in land use (From Pongratz et al. 2009; Houghton 2010)

Reference	Preindustrial	Industrial	Until present
DeFries et al. (1999)	48–57 (until 1850)	125–151 (1850–1990)	182–199 (until 1987)
Houghton (2003)	–	155 (1850–2000)	–
Houghton this study	–	156 (1850–2005)	–
Strassmann et al. (2008)	45 (until 1700)	188 (1700–1999)	233 (until 1999)
Olofsson and Hickler (2008)	114 (4000 BC to 1850)	148 (1850–1990)	262 (4000 BC to 1990)
Pongratz et al. (2009)	63 (until 1850)	108 (1850–2000)	171 (until 2000)
Shevliakova et al. (2009)	–	164–188 (1850–2000)	–

4.2.1 Losses Before 1850

Estimates of the amount of terrestrial C lost before 1850 (Table 4.2) are based on a combination of three sets of data: estimates of natural vegetation cover, the associated C densities (in vegetation and soil), and the conversion of those natural covers to croplands and pastures, the latter based on a combination of agricultural maps and population density. Four estimates of pre-historic C loss range between 45 and 114 Pg C (Table 4.2).

4.2.2 Losses Between 1850 and 2005

Two approaches have been used to calculate the more recent losses of C from land use and land-use change. One approach uses process-based ecosystem models to estimate biomass density (e.g., Le Quéré et al. 2009 for some contemporary models). These models simulate spatial and temporal variations in ecosystem structure and physiology. They differ in detail with respect to number of plant functional types and number of C pools. To simulate the correct biomass density, however, the models account for disturbances and recovery, whether natural or anthropogenic.

A second approach uses empirically-based response curves to define annual per hectare changes in C density (Mg C ha^{-1} $year^{-1}$) as a result of management (Houghton 1999, 2003). Carbon density declines with management as a result of burning and decay; it increases as a result of forest growth (recovery) and some forms of land management. Rates of decay and regrowth vary by geographic region, ecosystem type, and type of land use. Changes are defined for living vegetation (above- and belowground), coarse woody debris, wood products, and soil organic matter (SOM). The per hectare changes are initiated each year by the areas affected by land use or land-use change; that is, by the areas logged or the areas converted from one ecosystem to another. A C-tracking model sums the changes per hectare over all of

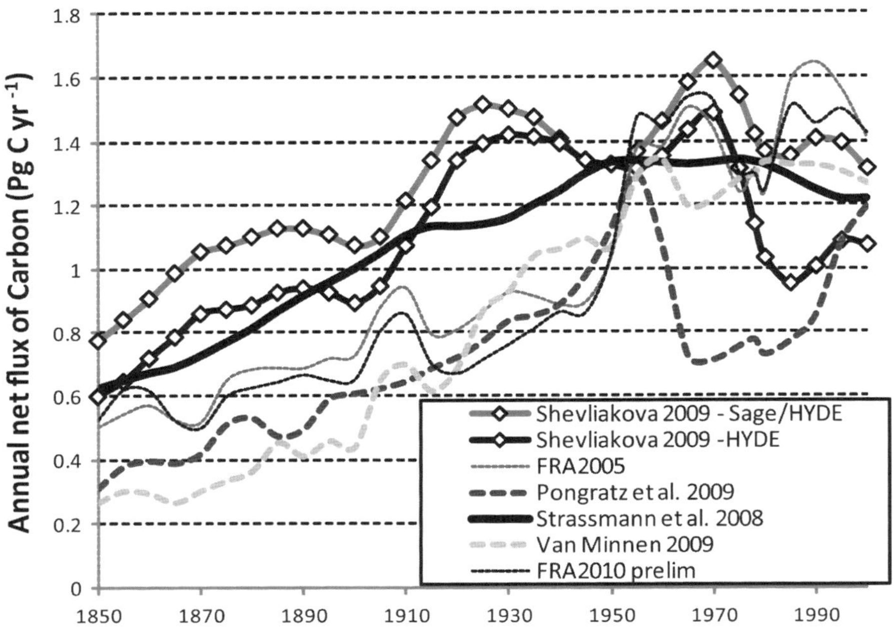

Fig. 4.3 Recent estimates of the annual net loss of carbon from land, globally, as a direct result of land use. FRA2005 refers to the Houghton's analysis in Le Quéré et al. (2009) (Also Houghton, this study). FRA2010 prelim refers to Houghton's analysis in Friedlingstein et al. (2010)

the hectares affected by management. Hectares unaffected by management are assumed to be unchanged with respect to C density.

This second approach was used in the analyses reported by Canadell et al. (2007) and Le Quéré et al. (2009), which extended Houghton's (2003) earlier work to 2005 using rates of deforestation from the U.N. Food and Agriculture Organization (FAO) (FAO 2006) and rates of wood harvest from FAOSTAT (FAO 2009). Many of the results presented here have not previously been published and are cited here as Houghton, this study. Preliminary estimates of C emissions based on the most recent Forest Resources Assessment (FRA) from FAO extended Houghton's analysis to 2010 (Friedlingstein et al. 2010), but those preliminary estimates are not included here. They have only a small effect on the long-term fluxes (1850–2005).

The total net loss of C from land as a result of land use and land-use change over the period 1850–2005 is estimated, using both methods, to have been 108–188 Pg C (Table 4.2). Annual rates of net C loss increased from ~0.5 Pg C year^{-1} in 1850 to ~1.5 Pg C year^{-1} in 2005 (Fig. 4.3), but the regional contributions to this global trend have varied. Until the mid-twentieth century, the major losses of C were from non-tropical regions (North America, Europe, Russia, Australia) as a result of wood harvests and agricultural expansion. After ~1950 deforestation for agricultural expansion was predominantly in tropical regions. The emphasis in this chapter on global,

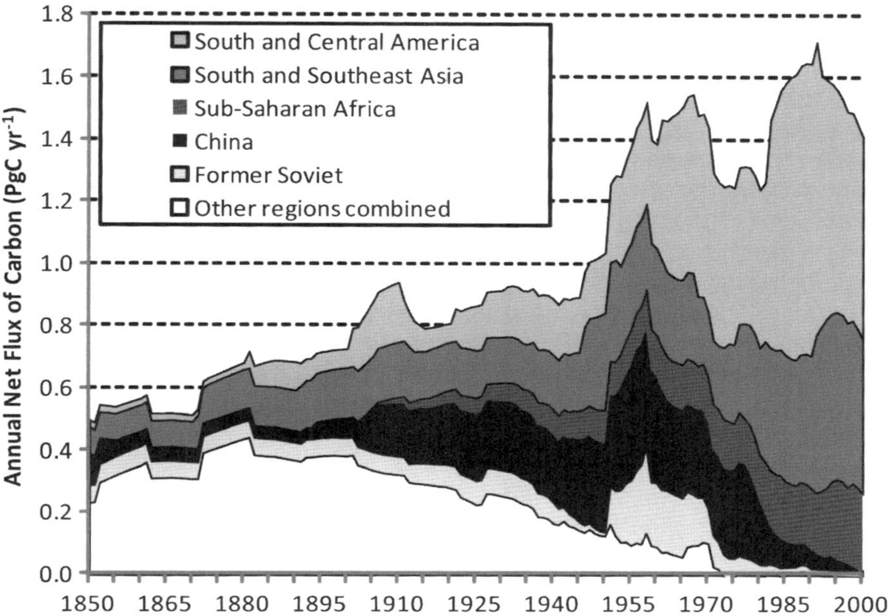

Fig. 4.4 Annual net losses of carbon from world regions (Houghton, this study)

tropical, and non-tropical regions hides the fact that many sub-regions may have been sinks for C as a result of forest growth (Kauppi et al. 2006; Pan et al. 2011). In fact, most regions include both sources and sinks of C as a result of past and present management practices. Figure 4.4 shows the net source/sink history of individual regions.

4.2.2.1 Deforestation

The amount of C lost from land or sequestered as a result of land-use change and management practices depends on two factors: the total areas affected by any particular land use or land-use change and the per hectare changes in C density associated with a management practice. These two factors are described below for the human activities believed to be responsible for the largest changes in C storage: deforestation, degradation, reforestation, and other management activities.

Croplands. Globally, the conversion of lands to croplands has been responsible for the greatest loss of C from land-use change (91 Pg C over the period 1850–2005) (Houghton, this study). The emissions are large because the global area of croplands has grown substantially in the last one and a half centuries and because the changes in C stocks per hectare are large when lands, especially forests, are converted to croplands. The estimated net release may be an overestimate, however, because it is based on an estimate of change that is larger than other estimates (Table 4.3).

Table 4.3 Estimates of global cropland and pasture areas (10^6 ha)

Land use/cover	Area in 1850	Area in 2000	Net change	Reference
Croplands	376	1,414	1,035	This study
	515	1,390	875	Hyde
	824	1,750	926	Sage
	–	1,550	–	Ramankutty et al. (2008)
Pastures	1,559	2,755	1,196	This study
	1,550	2,650	1,500	Hyde
	1,250	3,025	1,775	Sage
	–	2,800	–	Ramankutty et al. (2008)

Permanent croplands in 2000 are estimated to have covered 1,414–1,750 × 10^6 ha (Table 4.3), or between 10% and 12% of the Earth's total land surface (12–15% of the productive land surface – rock, sand, snow, ice, and water bodies excluded). Estimates of the increase over the last 150 years (1850–2000) vary from 875 to 1,035 × 10^6 ha, a doubling or tripling, depending on the estimates. The three different estimates of global cropland area are based on data initially published by Ramankutty and Foley (1999) (SAGE), Klein Goldewijk (2001) (HYDE), and Houghton (this study). Ramankutty and Foley (1999) estimated changes in croplands from 1700 to 1992 from historical cropland inventories. Klein Goldewijk (2001) used population density to estimate the spatial and temporal distribution of croplands from 1700 to 1990. The original SAGE and HYDE data sets have been compared (Klein Goldewijk and Ramankutty 2004), revised (Klein Goldewijk and van Drecht 2006), and used by others (Hurtt et al. 2006; Pongratz et al. 2008; Strassmann et al. 2008).

The changes in vegetation and soil, at least in the top meter, that result from clearing and cultivation are among the changes in terrestrial C stocks best documented. Essentially all of the initial vegetation is replaced by crops, so if the initial vegetation and its biomass are known, it is, in principle, straightforward to calculate the net loss of C associated with clearing. Because forests hold so much more C per unit area than grasslands, the loss of C associated with cropland expansion depends primarily on whether the croplands were claimed from forests or open lands. The variation in C stocks of different crop types is relatively small as long as tree (permanent) crops are differentiated from herbaceous crops. Some uncertainty results from estimating the time it takes for the release or uptake of C to occur. How much of the biomass is burned at the time of clearing? How much woody material is removed from site (wood products) and not decayed immediately? Answers vary across regions and through time (e.g., Morton et al. 2008). Estimates of annual sources and sinks depend on the answers, yet site-specific data are generally lacking. A few case studies usually provide the values used in calculation of C emissions and uptake over large regions.

On average, soil organic carbon (SOC) in the upper meter of soil is reduced by 25–30% as a result of cultivation, and this average has been documented in a large number of reviews (Mann 1985, 1986; Detwiler 1986; Schlesinger 1986; Johnson 1992; Davidson and Ackerman 1993; Post and Kwon 2000; Guo and Gifford 2002; Murty et al. 2002). There is some variation about this average, but the loss is broadly robust across all ecosystems, despite the variety of soil types, cultivation practices, and decomposition processes. Below a meter SOC seems to be more dynamic than previously thought (Schipper et al. 2007), and these dynamics add uncertainty to the estimated change.

The remaining uncertainty with respect to changes in SOC in response to cultivation concerns the fate of C lost from soil. Is all of it, in fact, released to the atmosphere, as most analyses assume, or is some of it eroded and moved to a different location, perhaps buried in anoxic environments and thereby sequestered? Comparison of erosion rates with the amount of organic C in freshwater sediments suggests that some of the C lost through erosion may accumulate in riverbeds, lakes, and reservoirs (Stallard 1998; Smith et al. 2001; Berhe et al. 2007). To the extent that this is so, the calculated loss of C from cropland expansion are overestimated.

When croplands are abandoned, C re-accumulates in vegetation as the land reverts to the natural ecosystem. The greater the biomass of the returning ecosystem, the greater the long-term C sink associated with recovery. In the short term, however, the magnitude of the annual sink for a particular parcel of land will vary with rate of recovery, which may be affected by the intensity of previous land use or by biophysical factors, such as distance from seed source, herbivory, soil fertility, or climatology (Uhl et al. 1988; Kozlowski 2002). The rate of recovery of vegetation can also depend on both climate conditions (growing season length) and soil type (Johnson et al. 2000). Soil organic C may also re-accumulate after abandonment of cultivation, although the rates of C accumulation in mineral soil are generally modest, albeit less well known (Post and Kwon 2000), especially when compared to the much faster rates of C accumulation in vegetation, surface litter, or woody debris (e.g., Harrison et al. 1995; Huntington 1995; Barford et al. 2001; Hooker and Compton 2003). Globally, C accumulation in mineral soils recovering from past cultivation is likely to amount to less than 0.1 Pg C year^{-1} (Post and Kwon 2000), although this estimate could change as deeper soil profiles are investigated further.

Pastures. The global expansion of pastures over the last 150 years is estimated to have caused a de-carbonization of 24 Pg C, the second largest loss from land-use change (Houghton, this study). The estimate is probably low because it is based on a low estimate of pasture expansion (67% and 80% lower than the estimates from the HYDE and SAGE data sets). The original SAGE data set (Ramankutty and Foley 1999) did not consider pastures. Changes in pastures attributed to SAGE were generated by merging the HYDE data on pastures with the cropland data from Ramankutty and Foley (1999) (Hurtt et al. 2006; Shevliakova et al. 2009).

In the year 2000, pastures covered between 2,650 and 3,378 × 10^{6} ha, or 18–23% of the Earth's land surface (Table 4.3). Pastures and croplands, together, covered 4,040–4,775 × 10^{6} ha, or 28–33% of total land area.

The net loss of C from changes in pasture area has been less than the loss from cropland expansion, despite the larger growth in pasture area, because many pastures expanded into natural grasslands rather than forests (thus changing aboveground C stocks little) and because pastures are generally not cultivated, and thus lose little C from soils. The primary exception to this grassland origin of pastures has been in Latin America where cattle pasture is still the main driver of deforestation.

The changes in SOC resulting from the conversion of forests to pastures are highly variable, however, with both increases and decreases observed (Post and Kwon 2000; Guo and Gifford 2002; Osher et al. 2003; Parfitt et al. 2003). For example, pasture soils cleared from forests in the Brazilian Amazon have been shown to lose C in some cases and gain it in others (Neill and Davidson 2000). The direction of change may be related to rainfall, site fertility, fertilizer practices, species of grass planted, or other factors that govern site productivity. In a meta-analysis of 170 studies, Guo and Gifford (2002) observed a modest mean increase in SOC (about 10%) in upper soil layers (<100 cm) when forests were converted to pastures; however, some sites had large C gains and others had large losses. When pastures are converted to croplands, SOC is lost as a result of cultivation. It should also be noted that overgrazing, particularly in dry lands, often leads to a loss of C from soil (Lal 2001).

Shifting cultivation. Shifting cultivation is a rotational form of agriculture, where crops alternate with periods of forest recovery (fallow). The initial clearing for shifting cultivation requires land conversion, most often forest or savanna. In terms of historical de-carbonization, shifting cultivation is fifth in importance, releasing ~9 Pg C over the last 150 years.

The areas in shifting cultivation and fallow are not well documented. The FAO includes the cropping portion of shifting cultivation in "arable and permanent crops" and excludes the fallow areas if they are older than 5 years, but these definitions may not be applied consistently by all countries reporting.

Houghton and Hackler (2006) estimated the areas annually deforested for shifting cultivation in Africa when the loss of forests reported by the FAO (2006) was greater than the increase in permanent croplands and pastures (FAOSTAT 2009). The assumption that the difference was explained by shifting cultivation seemed consistent with the definitions used by the two sources within the FAO. Subsequent revisions to FAOSTAT (2009) suggest the differences were more likely due to uncertainties than to real changes in agricultural area. However, errors in assigning deforestation to shifting cultivation, as opposed to permanent cropland, affect the attribution of C fluxes from one land use to another, but they do not have a large effect on the calculated net flux of C, which is determined more by the rate of deforestation than by the end land use.

According to the data and assumptions used to characterize the areas and changes in shifting cultivation, the practice is estimated to have occupied 411×10^6 ha in 2000, about 3% of the Earth's land surface. The increase in shifting cultivation (1850–2005) was 161×10^6 ha, only 10% of the increase in croplands or pastures. Neither Ramankutty and Foley (1999) nor Klein Goldewijk (2001) included fallow areas in their estimates of croplands and pastures, but shifting cultivation is included in the data set assembled by Hurtt et al. (2006). Fallow periods can be long or short, and generally the stocks of C in fallow forests re-cleared for cultivation are less than the stocks in undisturbed forests. Because the cultivation does not involve tillage, the loss of C from soil is less than the loss under cultivation of "permanent" croplands. Thus, the net per hectare changes in C stocks (both biomass and soil) are smaller under shifting cultivation than under permanent cultivation.

In many areas of tropical Asia and Africa, the fallow periods are being reduced as land becomes scarce (Myers 1980; Uhlig et al. 1994). Often the shortened fallow does not allow the recovery of nutrients necessary for crop production, and this intensification may lead to an increase in degraded lands that support neither crops nor forests, and a gradual reduction in C stocks. Some of these degraded lands gradually return to forest, but the changes are not systematically documented either on the ground or with satellites (Grainger 2008, 2009).

Settled lands. Because the area of urban ecosystems is small, globally, <0.5% (Schneider et al. 2009) to 2.4% (Potere and Schneider 2007) of the land surface, urban areas have been ignored in most estimates of C emissions from land use and land-use change. However, exurban areas were nearly 15 times greater than urban areas in the U.S. in 2000 (Brown et al. 2005). Furthermore, much of the deforestation in developed countries and China is currently for residential, industrial, and commercial use rather than for agriculture (Jeon 2011). The magnitude of net C emissions from the expansion of settled lands is uncertain. Newly established areas may be net sources initially but may become net sinks as trees are re-established. In arid areas settlements may become sinks if irrigation is used to expand the areas vegetated or to increase the C density of existing vegetation and soil.

4.2.2.2 Degradation

Degradation is defined here as a reduction in C density within a land cover.

Wood harvest. The harvest of wood from forests can increase the amount of C on land if the wood products have a long residence time and, thereby, accumulate enough mass to offset the reductions in forest biomass. In practice, however, little of the C initially held in the forest ends up in long-term products, especially because half of global harvests are for fuelwood (FAOSTAT 2009). Furthermore, although a constant rate of logging would eventually yield a net flux nearly zero, as decay and regrowth offset each other, as long as rates of harvest are increasing, the losses in biomass will be greater than the accumulations in wood products, yielding a net overall loss of C from land. Such has been the case globally over the last 150 years or more.

Estimates of the effects of harvest on vegetation C stocks are based on information on preharvest biomass and the fractions of this biomass harvested, damaged, and left living. Wood removed from the forest enters the forest products stream, whereas wood left behind enters the harvest residue pool. Woody debris provides a large source of C to the atmosphere as the dead wood decomposes, with the rate and duration of this source dependent on the amount and condition of wood left on-site. The amount of C in the dead wood pool is large during the years after harvest, decreases as the slash pool decomposes, and then increases again later in succession as dead wood accumulates (e.g., Harmon et al. 1990; Idol et al. 2001). The rate of C accumulation in vegetation during forest recovery after harvest, as after other disturbances, varies with climate and soil conditions (Johnson et al. 2000).

Table 4.4 Terrestrial carbon storage in 1850 and 2005 (From Houghton, this study)

C storage	1850	2005	Change
Living vegetation	623	482	−141
Soils	1,738	1,697	−41
Slash	7.7	14.4	7
Wood products	7.4	26.5	19
Total	2,376	2,220	−156

At the global level, approximately the same volumes of wood are harvested for industrial wood (timber, pulp) and fuelwood, although most of the wood harvested in developed countries is for industrial wood, while most harvests in developing countries are for fuelwood (FAOSTAT 2009). Fuelwood use is probably minor in affecting C stocks except in those regions where the supply of wood is less than the demand. Particularly around urban centers, demand often exceeds supply, and C stocks are reduced (Ahrends et al. 2010). Such degradation, and its effects on C stocks, are not well documented, globally.

Illegal logging makes estimates of industrial logging uncertain, and fuelwood use is also difficult to assess. Nevertheless, the cumulative area harvested for industrial wood (~$1{,}500 \times 10^6$ ha according to the analysis by Houghton (this study)) is larger than the cumulative area of forest cleared for croplands ($875–1{,}035 \times 10^6$ ha) (Table 4.3). The total area harvested is also nearly twice the area computed to be in secondary forest, indicating that many forests were harvested more than once or converted to other uses. Despite the large area logged, even the gross emissions of C from wood harvest are smaller than from deforestation for crops because harvests may have little effect on the C content of mineral soil (Johnson 1992; Johnson and Curtis 2001; Nave et al. 2010) (e.g., Zummo and Friedland (2011) for an exception). The forest floor often loses C after harvest, due largely to reduced C inputs and to the mechanical transfer of forest floor material to deeper soil layers (Currie et al. 2002; Yanai et al. 2003; Nave et al. 2010). Because forest harvest usually causes little loss of SOM, little additional accumulation of SOC occurs with forest recovery.

The net loss of C from industrial wood harvest (including losses from wood products and gains in regrowing forests) over the last 155 years was 17 Pg C according to the analysis by Houghton (this study). At the same time, 19 Pg C accumulated in wood products (Table 4.4). That accumulation accounts for ~40% of the industrial wood harvested over the period. The fraction is consistent with efficiencies reported for harvests in the U.S., taking into account that the fraction of harvested products going to long-term storage (>5 years) has increased over recent decades from ~20% to 30–40% (Harmon et al. 1990; Smith et al. 2006). On the other hand, the fraction of the original forest biomass held in wood products 100 years after harvest may be only 1%, with another 13% in landfills (Ingerson 2010).

Other. Most forms of management other than wood harvest have received little attention in global estimates of C flux from land use and land-use change. An exception is the net release of C estimated to have occurred in China between 1900 and 1980 (Houghton and Hackler 2003). During this interval, the net loss of forest area was more than three times greater than the net increase in croplands and pastures. The loss may have resulted from unsustainable harvests, from deliberate removal of forest cover (for protection from tigers or bandits), and from the deleterious effects of long-term intensive agriculture on soil fertility. Unlike croplands, pastures, and forests, the area in degraded lands is rarely enumerated (Oldeman 1994), yet the losses of C may be equivalent to the losses resulting from cultivation, especially if the degradation results from worn-out cultivated lands, abandoned but not returning to forest.

4.2.2.3 Reforestation and Management

Reforestation, afforestation, and other types of land management have run counter to the global trend of de-carbonization. Some of these activities are described below.

At the national level, forest area may be related to the country's stage of development – the "forest transition" (Mather 1992). Frontier lands may have large forests areas, which are used to fuel economic growth. The result is a loss of forest area. With development, however, forest areas expand again. Examples include China, India, Europe, and the northeastern and mid-western regions of the U.S. The trends are broadly consistent with the emissions from tropical and non-tropical regions (Fig. 4.2). The developed countries lost C in the past, but many are now sequestering C. In contrast, the losses of C from tropical countries are still high. There are exceptions, however. Forests are being lost, again, in New England (US) – this time for residential and commercial uses rather than for agriculture (Jeon 2011). As many of these residences are second homes, they suggest a new wrinkle to the forest transition paradigm.

Plantations. Plantations, globally, have not been a large net sink for C, as might be expected, because plantations are often established on forest lands, and the accumulation of C in growing plantations is offset by the emissions of C from deforestation. Further, many plantations are timber or fuelwood plantations, periodically harvested and thus have an average biomass less than the forests from which they were derived. The conversion of native forests to plantations normally leads to a net reduction in C stocks.

Although the rate of plantation establishment has recently increased, especially in China, Russia, and Vietnam, the area annually planted globally was ~25% of the area deforested (FAO 2006). This estimate does not include changes in the areas of orchards or oil palm plantations, which are "permanent crops" according to the FAO. The rate of accumulation of C aboveground is well documented for plantations, but the spatial heterogeneity (which types of plantation, planted where?) is not readily available for large regions. For example, plantations may be established for timber, shelter belts, or fuel, and the stocks of C in biomass vary considerably. Whether plantations are established on non-forest lands or on recently cleared forests also affects the net changes in biomass and SOC that result. Reviewing more than 100 observations, Guo and Gifford (2002) found that the establishment of plantations on forest lands or pastures generally decreased SOC stocks, while establishment on croplands increased them. This finding is consistent with the observations, above, that cultivation causes a 25–30% decline in the top meter, while pastures, often not cultivated, lose considerably less or even gain C. In another review Paul et al. (2002) found that plantations established on agricultural lands (both croplands and pastures) lost SOC during the first 5–10 years but gained it over periods longer than 30 years. The length of time land has been cultivated before being converted to a plantation may explain some of this variability. Overall, the changes in SOC were small relative to the gains in biomass.

Agricultural management. The changes in SOC that result from the conversion of natural ecosystems to croplands and their subsequent cultivation are addressed above (Sect. 4.2.1), but changes in C stocks result from cropland management, including cropping practices, irrigation, use of fertilizers, different types of tillage, changes in crop density, and changes in crop varieties. Many studies have addressed the potential for management to sequester C. Fewer studies have tried to estimate past or current C sinks. Recent analyses for the U.S. suggest a current sink of 0.015 Pg C $year^{-1}$ in croplands (Eve et al. 2002), while a recent assessment for Europe suggests a net source of 0.300 Pg C $year^{-1}$, perhaps because of reduced application of manure to cropland (Janssens et al. 2003). In Canada, the flux of C from cropland management is thought to be changing from a net source to a net sink, with a current flux near zero (Smith et al. 2000). Globally, the current flux is uncertain but probably not far from zero.

Aside from the losses of C resulting from cultivation of native soils and the re-accumulation of C in abandoned croplands soils, changes in agricultural management have not generally been included in global analyses of land use and land-use change although they have been included in regional analyses at high spatial resolution (e.g., Kutsch et al. 2010; West et al. 2010). The effects of erosion and redeposition of organic C, discussed above under croplands, pertain here as well.

Fire management. The emissions of C from fires associated with the conversion of forests to croplands and pastures are included in analyses of land-use change, but fire management has largely been ignored despite the fact that fire exclusion, fire suppression, and controlled burning are practiced in many parts of the world. In many regions, fire management may cause a terrestrial sink (Houghton et al. 1999; Marlon et al. 2008). In other regions it increases the net source. In particular, the draining and burning of peatlands in Southeast Asia are thought to add another 0.3 Pg C $year^{-1}$ to the net emissions from land-use change (not included in the estimates reported here) (Hooijer et al. 2009).

Woody encroachment. The expansion of trees and woody shrubs into herbaceous lands, although it cannot be attributed definitively to natural, indirect (climate, CO_2), or direct effects (fire suppression, grazing), is, nevertheless, re-carbonizing land in many regions. Scaling it up to a global estimate is problematical, however (Scholes and Archer 1997; Archer et al. 2001), in part because the areal extent of woody encroachment is unknown and difficult to measure (e.g., Asner et al. 2003). Also, the increase in vegetation C stocks observed with woody encroachment is in some cases offset by losses of soil C (Jackson et al. 2002). In other cases the soils may gain C (e.g., Hibbard et al. 2001) or show no discernable change (Smith and Johnson 2003). Finally, woody encroachment may be offset by its reverse process, woody elimination, an example of which is the fire-induced spread of cheatgrass (*Bromus tectorum*) into the native woody shrublands of the Great Basin in the western U.S. (Bradley et al. 2006).

The net effect of woody encroachment and woody elimination is, thus, uncertain, not only with respect to net change in C storage, but also with respect to attribution. It may be an unintended effect of management, or it may be a response to indirect or natural effects of environmental change.

4.3 Summary and Conclusions

4.3.1 The Past

The amount of C lost from terrestrial ecosystems over the last ~150 years is 1–3 times greater than the losses before 1850 (Table 4.2). Since 1850, most of the loss has been from biomass; soils contributed only about 25% of this net release (Table 4.4), although the proportion varies by ecosystem type. The amount of C in slash and wood products is small, but the increases over 155 years are significant relative to the changes in living biomass and soil (Table 4.4)

More than one third of the total land surface of the Earth has been altered directly by human activity (nearly half of the productive land surface – rock, sand, snow, ice, and water bodies excluded). Croplands and pastures, together, occupy a quarter to a third of the land surface, and secondary forests recovering from some form of management account for ~30% of the world's forests. All of the lands that are known to have been managed in 2000 account for 37% of total land area, or 46% of the productive land surface. Hurtt et al. (2006) estimated that 42–68% of the land surface had been affected by land-use activities, but even those estimates are probably conservative. It seems likely that most of the productive land surface has been used in one way or another over the last 150 years, but many uses are never reported or recorded.

Direct human activity has caused a total net decarbonization of 170–260 Pg C from the world's terrestrial ecosystems (Table 4.2). As discussed in Sect. 4.1, this decarbonization in recent years has been offset to a large extent by natural or indirect human effects that have led to an accumulation of C on land. Nevertheless, the net effect of management, to date, has been to decarbonize rather than recarbonize the land surface. Most (60–70%) of this long-term decarbonization occurred over the last 155 years (108–188 Pg C). According to the analysis reported here, about 90% of this more recent decarbonization has resulted from the clearing and management of forests; cultivation of prairie soils account for the other 10%. Estimates are that 30–40% of the world's forests have been lost (Pongratz et al.

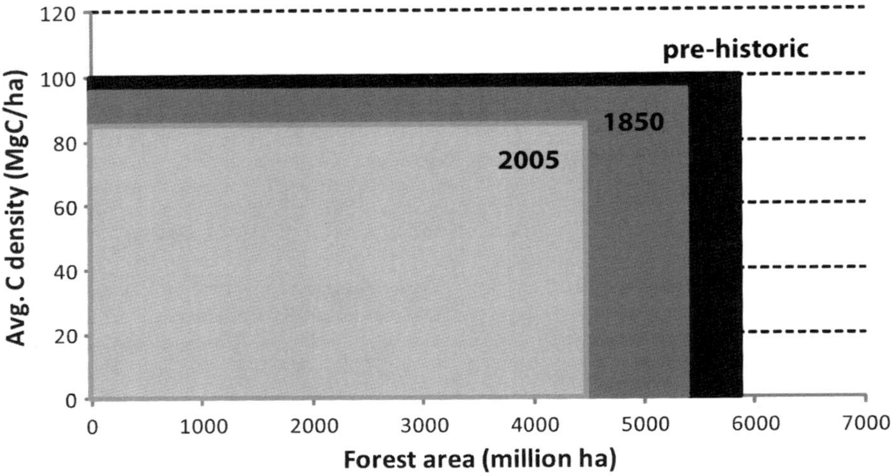

Fig. 4.5 The area and average carbon density of the world's forests before human activity, in 1850, and in 2005 (Houghton, this study)

2008; World Commission on Forests and Sustainable Development 1999). The loss of ~1,074 × 10^6 ha of forest over the last 155 years has had the greatest effect on decarbonization, ~110 Pg C, but widespread harvest of wood and other processes degrading forests have contributed ~40 Pg C (~25%). This degradation has reduced the carbon density of forest biomass from a global average of 103 Mg C/ha to 101 Mg C/ha by 1850 to 91 Mg C/ha by 2005 (Fig. 4.5).

The rate of these transformations over the surface of the Earth has been accelerating, globally, although not equally in all regions. In the northern mid-latitudes (largely developed countries) the area in agriculture has been nearly constant over the last decades, while in the tropics agricultural lands have been expanding.

4.3.2 *The Future*

This short review has sketched broad global trends. It has not done justice to the numerous management practices that counter these trends in decarbonization. Indeed, the larger the decarbonization in the past, the greater the opportunity for recarbonization in the future. Most of the activities that have released C to the atmosphere might, with appropriate incentives, store it again on land. Examples include the management of forests and use of wood (Kohlmaier et al. 1998) and the management of agriculture (Smith et al. 2007).

Two activities have the potential to affect terrestrial C sinks at scales large and rapid enough to stabilize the concentration of CO_2 quickly: massive reforestation and a halt to deforestation. On the order of 200–300 × 10^6 ha of new forest would be required to remove 1–1.5 Pg C year^{-1} from the atmosphere, and that uptake would decline after several decades. The area is large, but not in comparison to current areas in croplands and pastures. The magnitude of 1–1.5 Pg C $year^{-1}$ assumes, optimistically, an average sequestration rate in wood and soils of 5 Mg (1 Mg = 10^6 g) C $ha^{-1}year^{-1}$, while the lands available for such afforestation are unlikely to be the most fertile and productive.

Deforestation is responsible at present for annual emissions of 1–1.5 Pg C. A halt to deforestation, combined with a massive program of reforestation, could thus reduce C emissions by 2.5–3 Pg C $year^{-1}$. The reduction is more than half the rate at which C is accumulating in the atmosphere at present (~4 Pg C $year^{-1}$) (Table 4.1). Similar reductions of 1–1.5 Pg C $year^{-1}$ in the emissions of C from fossil fuels, which are now nearly 9 Pg C $year^{-1}$ (Friedlingstein et al. 2010), would stabilize the concentration of CO_2 in the atmosphere immediately. Additional reductions would be required over time to bring the concentration of CO_2 in the atmosphere back to 350 ppm, but the management of C on land offers a short-term solution to stabilization of atmospheric CO_2 concentrations. It has the advantage of being technically achievable and cheap relative to other emissions reductions.

The emphasis in this chapter on net losses of C from land hides the much larger *gross* sources and sinks of C associated with land use and land-use change. The gross loss of C from land management, for example, was more than 4 Pg C $year^{-1}$, while the net loss was only 1–1.5 Pg C $year^{-1}$ (Richter and Houghton 2011). These gross fluxes indicate a much larger potential than the global net flux for reducing current losses and for re-carbonizing terrestrial ecosystems through management.

Stopping deforestation and establishing 200–300 × 10^6 ha of new forest would require a reversal of the global trends in land use and land-use change that have accelerated over the last 300+ years. Such reversals have already happened in many nations, as suggested by the forest transition (see Sect. 4.2.2.3). Can they become global? And can they become global when the demands for food are growing, as well as the demands for meat, and bioenergy? The competition for land looms as a particularly important issue over the next century.

References

Ahrends A et al (2010) Predicable waves of sequential forest degradation and biodiversity loss spreading from an African city. Proc Natl Acad Sci USA 107:14556–14561

Archer S, Boutton TW, Hibbard KA (2001) Trees in grasslands: biogeochemical consequences of woody plant expansion. In: Schulze E-D, Harrison SP, Heimann M, Holland EA, Lloyd J, Prentice IC, Schimel D (eds) Global biogeochemical cycles in the climate system. Academic, San Diego, pp 115–1337

Asner GP, Archer S, Hughes RF, James R, Wessman CA (2003) Net changes in regional woody vegetation cover and carbon storage in Texas drylands, 1937–1999. Glob Chang Biol 9:316–335

Barford CC et al (2001) Factors controlling long- and short-term sequestration of atmospheric CO_2 in a mid-latitude forest. Science 294:1688–1691

Berhe AA, Harte J, Harden JW, Torn MS (2007) The significance of the erosion-induced terrestrial carbon sink. Bioscience 57:337–347

Bradley BA, Houghton RA, Mustard JF, Hamburg SP (2006) Invasive grass reduces aboveground carbon stocks in shrublands of the Western US. Glob Chang Biol 12:1815–1822

Bridgham SD, Megonigal JP, Keller JK, Bliss NB, Trettin C (2006) The carbon balance of North American wetlands. Wetlands 26:889–916

Brown DG, Johnson KM, Loveland TR, Theobald DM (2005) Rural land-use trends in the conterminous United States, 1950–2000. Ecol Appl 15:1851–1863

Canadell JG et al (2007) Contributions to accelerating atmospheric CO_2 growth from economic activity, carbon intensity, and efficiency of natural sinks. Proc Natl Acad Sci USA 104:18866–18870

Currie WS, Yanai RD, Piatek KB, Prescott CE, Goodale CL et al (2002) Processes affecting carbon storage in the forest floor and in downed woody debris. In: Kimble JM et al (eds) The potential for U.S. forest soils to sequester carbon and mitigate the greenhouse effect. CRC Press, Boca Raton, pp 135–157

Davidson EA, Ackerman IL (1993) Changes in soil carbon inventories following cultivation of previously untilled soils. Biogeochemistry 20:161–193

DeFries RS, Field CB, Fung I, Collatz GJ, Bounoua L (1999) Combining satellite data and biogeochemical models to estimate global effects of human-induced land cover change on carbon emissions and primary productivity. Global Biogeochem Cycles 13:803–815

Detwiler RP (1986) Land use change and the global carbon cycle: the role of tropical soils. Biogeochemistry 2:67–93

Dommain R, Couwenberg J, Joosten H (2011) Development and carbon sequestration of tropical peat domes in south-east Asia: links to post-glacial sea-level changes and Holocene climate variability. Quaternary Sci Rev 30:999–1010

Donato DC, Kauffman JB, Murdiyarso D, Kurnianto S, Stidham M, Kanninen M (2011) Mangroves among the most carbon-rich forests in the tropics. Nat Geosci 4(5):293–297

Ellis EC (2011) Anthropogenic transformation of the terrestrial biosphere. Philos Trans A Math Phys Eng Sci 369:1010–1035

Eve MD, Sperow M, Paustian K, Follett RF (2002) National-scale estimation of changes in soil carbon stocks on agricultural lands. Environ Pollut 116:431–438

FAO (2006) Global forest resources assessment 2005. FAO forestry paper 147, Rome

FAO (2009) http://faostat.fao.org/site/377/default.aspx#ancor (11/09)

FAOSTAT (2009) http://faostat.fao.org/site/377/default.aspx#ancor (11/09)

Friedlingstein P, Houghton RA, Marland G et al (2010) Update on CO_2 emissions. Nat Geosci 3:811–812

Gillett NP, Weaver AJ, Zwiers FW, Flannigan MD (2004) Detecting the effect of climate change on Canadian forest fires. Geophys Res Lett 31:L18211

Gitz V, Ciais P (2003) Amplifying effects of land-use change on future atmospheric CO_2 levels. Global Biogeochem Cycles 17:1024. doi:10.1029/2002GB001963

Gloor M, Sarmiento JL, Gruber N (2010) What can be learned about the carbon cycle climate feedbacks from the CO_2 airborne fraction? Atmos Chem Phys 10:7739–7751

Grainger A (2008) Difficulties in tracking the long-term global trend in tropical forest area. Proc Natl Acad Sci USA 105:818–823

Grainger A (2009) Measuring the planet to fill terrestrial data gaps. Proc Natl Acad Sci USA 106:20557–20558

Guo LB, Gifford RM (2002) Soil carbon stocks and land use change: a meta analysis. Glob Chang Biol 8:345–360

Harmon ME, Ferrell WK, Franklin JF (1990) Effects on carbon storage of conversion of old-growth forests to young forests. Science 247:699–702

Harrison KG, Post WM, Richter DD (1995) Soil carbon turnover in a recovering temperate forest. Global Biogeochem Cycles 9:449–454

Hibbard KA, Archer S, Schimel DS, Valentine DW (2001) Biogeochemical changes accompanying woody plant encroachment in a subtropical savanna. Ecology 82:1999–2011

Hooijer A et al (2009) Current and future CO_2 emissions from drained peatlands in Southeast Asia. Biogeosci Discuss 6:7207–7230

Hooker TD, Compton JE (2003) Forest ecosystem carbon and nitrogen accumulation during the first century after agricultural abandonment. Ecol Appl 13:299–313
Houghton RA (1999) The annual net flux of carbon to the atmosphere from changes in land use 1850–1990. Tellus 51B:298–313
Houghton RA (2003) Revised estimates of the annual net flux of carbon to the atmosphere from changes in land use and land management 1850–2000. Tellus 55B:378–390
Houghton RA (2007) Balancing the global carbon budget. Annu Rev Earth Planet Sci 35:313–347
Houghton RA (2010) How well do we know the flux of CO_2 from land-use change? Tellus B62:337–351. doi:10.1111/j.1600-0889.2010.00473.x
Houghton RA, Hackler JL (2003) Sources and sinks of carbon from land-use change in China. Global Biogeochem Cycles 17:1034. doi:10.1029/2002GB001970
Houghton RA, Hackler JL (2006) Emissions of carbon from land use change in sub-Saharan Africa. J Geophys Res 111:G02003. doi:10.1029/2005JG000076
Houghton RA, Hackler JL, Lawrence KT (1999) The U.S. carbon budget: contributions from land-use change. Science 285:574–578
Huntington TG (1995) Carbon sequestration in an aggrading forest ecosystem in the southeastern USA. Soil Sci Soc Am J 59:1459–1467
Hurtt GC et al (2006) The underpinnings of land-use history: three centuries of global gridded land-use transitions, wood-harvest activity, and resulting secondary lands. Glob Chang Biol 12:1–22
Idol TW, Filder RA, Pope PE, Ponder F (2001) Characterization of coarse woody debris across a 100 year chronosequence of upland oak-hickory forests. For Ecol Manage 149:153–161
Ingerson A (2010) Carbon storage potential of harvested wood: summary and policy implications. Mitig Adapt Strat Glob Chang. doi:10.1007/s11027-010-9267-5
Jackson RB, Banner JL, Jobbágy EG, Pockman WT, Wall DH (2002) Ecosystem carbon loss with woody plant invasion of grasslands. Nature 418:623–626
Janssens IA et al (2003) Europe's terrestrial biosphere absorbs 7–12% of European anthropogenic CO_2 emissions. Science 300:1538–1542
Jeon SB (2011) The effect of land-use change on the terrestrial carbon budget of New England. PhD dissertation, Boston University
Johnson DW (1992) Effects of forest management on soil carbon storage. Water Air Soil Pollut 64:83–120
Johnson DW, Curtis PS (2001) Effects of forest management on soil C and N storage: meta analysis. For Ecol Manage 140:227–238
Johnson CM, Zarin DJ, Johnson AH (2000) Post-disturbance above-ground biomass accumulation in global secondary forests. Ecology 81:1395–1401
Kaplan JO, Krumhardt KM, Zimmermann N (2009) The prehistoric and preindustrial deforestation of Europe. Quaternary Sci Rev 28:3016–3034
Kauppi PE, Ausubel JH, Fang J, Mather AS, Sedjo RA, Waggoner PE (2006) Returning forests analyzed with the forest identity. Proc Natl Acad Sci USA 103:17574–17579
Klein Goldewijk K (2001) Estimating global land use change over the past 300 years: the HYDE database. Global Biogeochem Cycles 15:417–433
Klein Goldewijk K, Ramankutty N (2004) Land cover change over the last three centuries due to human activities: the availability of new global data sets. GeoJournal 61:335–344
Klein Goldewijk K, van Drecht G (2006) HYDE3: current and historical population and land cover. In: Bouwman AF, Kram T, Klein Goldewijk K (eds) Integrated modeling of global environmental change. An overview of IMAGE 2.4. Netherlands Environmental Assessment Agency (MNP), Bilthoven, The Netherlands
Knorr W (2009) Is the airborne fraction of anthropogenic CO_2 emissions increasing? Geophys Res Lett 36:L21710. doi:10.1029/2009GL040613
Kohlmaier GH, Weber M, Houghton RA (eds) (1998) Carbon dioxide mitigation in forestry and wood industry. Springer, Berlin
Kozlowski TT (2002) Physiological ecology of natural regeneration of harvested and disturbed forest stands: implications for forest management. For Ecol Manage 158:195–221

Kurz WA et al (2008) Mountain pine beetle and forest carbon feedback to climate change. Nature 452:987–990
Kutsch WL et al (2010) The net biome production of full crop rotations in Europe. Agric Ecosyst Environ 139:336–345
Lal R (2001) Potential of desertification control to sequester carbon and mitigate the greenhouse effect. Clim Chang 51:35–72
Le Quéré C, Raupach MR, Canadell JG et al (2009) Trends in the sources and sinks of carbon dioxide. Nat Geosci 2:831–836
Luyssaert S, Schulze E-D, Börner A et al (2008) Old-growth forests as global carbon sinks. Nature 455:213–215. doi:10.1038/nature07276
Mann LK (1985) A regional comparison of carbon in cultivated and uncultivated alfisols and mollisols in the central United States. Geoderma 36:241–253
Mann LK (1986) Changes in soil carbon storage after cultivation. Soil Sci 142:279–288
Marlon JR, Bartlein PJ, Carcaillet C et al (2008) Climate and human influences on global biomass burning over the past two millennia. Nat Geosci 1:697–701
Mather AS (1992) The forest transition. Area 24:367–379
Morton DC, DeFries RS, Randerson JT et al (2008) Agricultural intensification increases deforestation fire activity in Amazonia. Glob Chang Biol 14:2262–2275
Murty D, Kirschbaum MF, McMurtrie RE, McGilvray H (2002) Does conversion of forest to agricultural land change soil carbon and nitrogen? A review of the literature. Glob Chang Biol 8:105–123
Myers N (1980) Conversion of tropical moist forests. National Academy of Sciences Press, Washington, DC
Nave LE, Vance ED, Swanston CW, Curtis PS (2010) Harvest impacts on soil carbon storage in temperate forests. For Ecol Manage 259:857–866
Neill C, Davidson EA (2000) Soil carbon accumulation or loss following deforestation for pasture in the Brazilian Amazon. In: Lal R, Kimble JM, Stewart BA (eds) Global climate change and tropical ecosystems. CRC Press, Boca Raton, pp 197–211
Oldeman LR (1994) The global extent of soil degradation. In: Greenland DJ, Szaboles I (eds) Soil resilience and sustainable land Use. CAB International, New York, pp 99–118
Olofsson J, Hickler T (2008) Effects of human land use on the global carbon cycle during the last 6,000 years. Veg Hist Archeobot 17:605–615. doi:10.1007/s00334-007-0126-6
Osher LJ, Matson PA, Amundson R (2003) Effect of land use change on soil carbon in Hawaii. Biogeochemistry 65:213–232
Pan Y, Birdsey RA, Fang J et al (2011) A large and persistent carbon sink in the world's forests, 1990–2007. Science 333:988–993
Parfitt RL, Scott NA, Ross DJ, Salt GJ, Tate KR (2003) Landuse change effects on soil C and N transformations in soils of high N status: comparisons under indigenous forest, pasture and pine plantation. Biogeochemistry 66:203–221
Paul KI, Polglase PJ, Nyakuengama JG, Khanna PK (2002) Change in soil carbon following afforestation. For Ecol Manage 168:241–257
Pinter N, Fiedel S, Keeley JE (2011) Fire and vegetation shifts in the Americas at the vanguard of Paleoindian migration. Quaternary Sci Rev 30:269–272
Pongratz J, Reick C, Raddatz T, Claussen M (2008) A reconstruction of global agricultural areas and land cover for the last millennium. Global Biogeochem Cycles 22:GB3018. doi:10.1029/2007GB003153
Pongratz J, Reick CH, Raddatz T, Claussen M (2009) Effects of anthropogenic land cover change n the carbon cycle of the last millennium. Global Biogeochem Cycles 23:GB4001. doi:10.1029/2009GB003488
Post WM, Kwon KC (2000) Soil carbon sequestration and land-use change: processes and potential. Glob Chang Biol 6:317–327
Potere D, Schneider A (2007) A critical look at representations of urban areas in global maps. GeoJournal 69:55–80

Prentice IC, Harrison SP, Bartlein PJ (2011) Global vegetation and terrestrial carbon cycle changes after the last ice age. New Phytol 189:988–998
Ramankutty N, Foley JA (1999) Estimating historical changes in global land cover: croplands from 1700 to 1992. Global Biogeochem Cycles 13:997–1027
Ramankutty N, Evan AT, Monfreda C, Foley JA (2008) Farming the planet: 1. Geographic distribution of global agricultural lands in the year 2000. Global Biogeochem Cycles 22:GB1003. doi:10.1029/2007GB002952
Reick CH, Raddatz T, Pongratz JM, Claussen M (2011) Contribution of anthropogenic land cover change emissions to preindustrial atmospheric CO_2. Tellus 62B:329–336
Richter DdeB, Houghton RA (2011) Gross CO_2 fluxes from land-use change: implications for reducing global emissions and increasing sinks. Carbon Manag 2:41–47
Schipper LA, Baisden WT, Parfitt RL, Ross W, Claydon JJ, Arnold G (2007) Large losses of soil C and N from soil profiles under pasture in New Zealand during the past 20 years. Glob Chang Biol 13:1138–1144
Schlesinger WH (1986) Changes in soil carbon storage and associated properties with disturbance and recovery. In: Trabalka JR, Reichle DE (eds) The changing carbon cycle: a global analysis. Springer, New York, pp 194–220
Schneider A, Friedl MA, Potere D (2009) A new map of global urban extent from MODIS satellite data. Environ Res Lett 4. doi:1088/1748-9326/4/4/044003
Scholes RJ, Archer SR (1997) Tree-grass interactions in savannas. Annu Rev Ecol System 28:517–544
Shevliakova E et al (2009) Carbon cycling under 300 years of land use change: importance of the secondary vegetation sink. Global Biogeochem Cycles 23:GB2022. doi:10.1029/2007GB003176
Smith DL, Johnson LC (2003) Expansion of *Juniperus virginiana* L. in the Great Plains: changes in soil organic carbon dynamics. Global Biogeochem Cycles 17:1062. doi:10.1029/2002GB001990
Smith WN, Desjardins RL, Pattey E (2000) The net flux of carbon from agricultural soils in Canada 1970–2010. Glob Chang Biol 6:557–568
Smith SV, Renwick WH, Buddemeier RW, Crossland CJ (2001) Budgets of soil erosion and deposition for sediments and sedimentary organic carbon across the conterminous United States. Global Biogeochem Cycles 15:697–707
Smith JE, Heath LS, Skog KE, Birdsey RA (2006) Methods for calculating forest ecosystem and harvested carbon with standard estimates for forest types of the United States. Gen. Tech. Rep. NE-343. U.S. Department of Agriculture, Forest Service, Northeastern Research Station, Newtown Square, PA, 216 pp
Smith P, Martino D, Cai Z et al (2007) Agriculture. In: Metz B, Davidson OR, Bosch PR, Dave R, Meyer LA (eds) Climate change 2007: mitigation. Contribution of Working Group III to the fourth assessment report of the Intergovernmental Panel on Climate Change. Cambridge University Press, Cambridge/New York
Stallard RF (1998) Terrestrial sedimentation and the carbon cycle: coupling weathering and erosion to carbon burial. Global Biogeochem Cycles 12:231–257
Strassmann KM, Joos F, Fischer G (2008) Simulating effects of land use changes on carbon fluxes: past contributions to atmospheric CO_2 increases and future commitments due to losses of terrestrial sink capacity. Tellus 60B:583–603
Uhl C, Buschbacher R, Serrao EAS (1988) Abandoned pastures in eastern Amazonia. I. Patterns of plant succession. J Ecol 76:663–681
Uhlig J, Hall CAS, Nyo T (1994) Changing patterns of shifting cultivation in selected countries in Southeast Asia and their effect on the global carbon cycle. In: Dale V (ed) Effects of land-use change on atmospheric CO_2 concentrations. South and Southeast Asia as a case study. Springer, New York, pp 145–200
Wang Z, Chappellaz J, Park K, Mak JE (2010) Large variations in southern hemisphere biomass burning during the last 650 years. Science 330:1663–1666
West TO et al (2010) Cropland carbon fluxes in the United States: increasing geospatial resolution of inventory-based carbon accounting. Ecol Appl 20:1074–1086

Westerling AL, Hidalgo HG, Cayan DR, Swetnam TW (2006) Warming and earlier spring increase western US forest wildfire activity. Science 313:940–943

World Commission on Forests and Sustainable Development (1999) Our forests, our future. Cambridge University Press, Cambridge

Yanai RD, Currie WS, Goodale CL (2003) Oil carbon dynamics after forest harvest: an ecosystem paradigm revisited. Ecosystems 6:97–212

Zummo LM, Friedland AJ (2011) Soil carbon release along a gradient of physical disturbance in a harvested northern hardwood forest. For Ecol Manage 261:1016–1026

Chapter 5
Soil Erosion and Soil Organic Carbon Storage on the Chinese Loess Plateau

Ch. Dahlke and H.-R. Bork

Abstract Agricultural soil erosion and its importance to the global carbon (C) cycle is a debatable issue. China, the largest global emitter of anthropogenic carbon dioxide (CO_2), and the Chinese Loess Plateau, with extremely high soil erosion rates, are typical regions for studying soil erosion and soil organic carbon (SOC) storage. Soil erosion dynamics are reconstructed on the basis of the sediment deposits in check-dam reservoirs, soil-sediment sequence analysis, differential global positioning system measurements (DGPS), map analysis and expert interviews. The reservoirs encompass 50 years of soil erosion history in Yangjuangou near Yan'an City. The Yangjuangou catchment (2.02 km^2) is a subcatchment of the Yellow River. In Yangjuangou at least 250,000 Mg of soil were eroded between 1957 and 2003. The data also indicate that on average 2.48 Mg SOC ha^{-1} were stored in the reservoir between 1957 and 2003. These studies in Yangjuangou provide a better understanding of the water erosion history and the recent SOC budget in the watershed over 46 years.

Keywords Soil erosion • Soil organic carbon storage • SOC budget • Loess Plateau • China

---

C. Dahlke (✉)
Department of Geography, University of Hildesheim, Marienburger Platz 22, 31141 Hildesheim, Germany
e-mail: dahlkec@uni-hildesheim.de

H.-R. Bork
Institute for Ecosystem Research, Christian-Albrechts-University, Olshausenstr. 75, 24118 Kiel, Germany
e-mail: hrbork@ecology.uni-kiel.de

R. Lal et al. (eds.), *Recarbonization of the Biosphere: Ecosystems and the Global Carbon Cycle*, DOI 10.1007/978-94-007-4159-1_5,

Abbreviations

C	carbon
CO_2	carbon dioxide
DGPS	differential global positioning system
GHG	greenhouse gases
IPCC	Intergovernmental Panel on Climate Change
CH_4	methane
N	nitrogen
N_2O	nitrous oxide
OM	organic matter
P	phosphorus
SOC	soil organic carbon

5.1 Introduction

In the context of climate change and the global warming debate, there is a growing concern about effects of the ever increasing atmospheric concentrations of greenhouse gases (GHG). The reports and the assessment by the Intergovernmental Panel on Climate Change (IPCC) have contributed to the discussion on climate change mitigation strategies and carbon (C) storage technologies. Several measures for sequestration of atmospheric carbon dioxide (CO_2) have been discussed such as engineering techniques for geological and oceanic storage of C, and its sequestration in the terrestrial biosphere (Lal 2009).

Beside the reduction of GHG emissions and the conservation of existing C pools in contemporary ecosystems such as forests or wetlands, issues of land-use change and their influence on the climate have been addressed (Watson et al. 2000). A large fraction (40–50%) of the Earth's ice-free surface is used for agriculture, which is responsible for significant emissions of carbon dioxide (CO_2), methane (CH_4) and nitrous oxide (N_2O) (Smith et al. 2007).

A range of mitigation strategies through agricultural management have been proposed for reduction of GHG emissions into the atmosphere. For example, Lal (2010) described two general paths for mitigation of climate change through agricultural management which include: (i) avoiding or reducing GHG emissions in the agricultural sector, and (ii) sequestering CO_2 in the terrestrial biosphere such as that in the biomass and soil.

Agricultural soil erosion and its importance to the global C cycle is a debateable topic (Kuhn et al. 2009; Berhe et al. 2007). The contributions of agricultural soil erosion to the atmospheric CO_2 concentration and the question whether soil erosion is a net source or a sink of soil organic carbon (SOC) are not clearly understood (Berhe et al. 2007). On the one hand, it is argued that erosion of soil C and its mineralization during the transport process and in the depositional areas in the landscape can be an important sources of atmospheric CO_2 (Lal 2003, 2004; Lal and Pimentel 2008). On the other hand, Van Oost et al. (2007, 2008) propose an opposite hypothesis. The fate of the transported soil C through erosion and the deep-burial of SOC-organic rich sediment are the key questions in the debate. Kuhn et al. (2009) reported the importance of different environments for soil C conservation due to geomorphic and ecological factors. Lal (2003) described the difficulties in generalizing the series of complex interacting processes for calculating a SOC budget on a watershed scale. Lal (2003) observed that accelerated erosion exacerbates the depletion of SOC by increase in its oxidation on site, during the transport and redistribution over land, in the depositional sites, and in river and other aquatic ecosystems. Yet, some fraction of the transported SOC is stored by deep burial and in depressional sites, or transported and eventually buried in aquatic ecosystems.

China provides several research sites for studying the soil erosion dynamics on a long-term perspective, and the related SOC storage. China is the largest global emitter of anthropogenic CO_2 (United Nation Statistic Division 2011), and has a rapidly increasing economy. The Chinese Loess Plateau is well known for extremely high soil erosion rates (Fu and Gulinck 1994; Cai 2001). Further, China may have a high potential for soil C sequestration through the restoration of degraded soils. Lal (2002) estimated that about 20% of the annual industrial C emissions in China can be off-set by the recarbonization of the biosphere, and specifically through soil C sequestration. Yan et al. (2007) reported large differences in the potential for SOC sequestration in China's soils because of large diversity in climate, soil properties and crop productivity. Song et al. (2005) assessed the impact of cultivation and soil management on SOC storage in China. Specifically, land use history is a relevant factor for understanding current soil quality and its dynamics. Pan (2009) summarized studies on China's SOC and recommended additional research on this topic.

This chapter presents a case study about a site on the Chinese Loess Plateau near Yan'an City in Shaanxi Province. The study was conducted to assess water erosion dynamics of a small subcatchment of the Yellow River during the few last decades. Existing documents (days to century) and geoarchives (days to millennia) can be used for reconstructing historical rates of soil erosion (years to decade). Despite the large potential to reconstruct historical erosion on a catchment scale with a high resolution in time and space, little attention has been given to colluvial deposits and check-dam sediments on the Chinese Loess Plateau. In contrast to calculations based on sediment storage in the middle part, the lower part or the delta area of the Yellow River, investigations of sediment storage in small catchments near the sediment source area have only a small cascade-effect (Lang and Hönscheidt 1999).

This chapter first explains how soil erosion dynamics are reconstructed on the basis of the sediment deposits in check-dam reservoirs, soil-sediment sequence analysis, differential global positioning system measurements (DGPS), map analysis and expert interviews. Subsequent description include procedure of calculations of sediment budgets for reservoirs and soil erosion rates in the study area. Finally, the data generated by these calculations are used to develop a scenario to describe the effects of soil erosion and SOC storage in the Yangjuangou catchment. The present investigation provides an improved understanding of the recent SOC budget in the region.

5.2 Study Area

The Yangjuangou catchment (36°42′ 6″ N, 109°31′17″ E) is situated in Northern China, in the Shaanxi Province, approximately 300 km north of Xi'an (Fig. 5.1a, b). The catchment was chosen because of the comparable geological uniformity and ideal conditions for quantifying erosion and deposition during the second half of the twentieth century. The Yangjuangou catchment covers a total area of 2.02 km², and is sited at an elevation of 1,056–1,273 m above sea level (Li and Lindström 2001). The study area is a second order tributary of the Yan River, which drains into the Yellow River. The accumulation of loess in Central China began during Early Pleistocene about 2.5 million years ago (An et al. 1991), with different loess formations on the Chinese Loess Plateau. The geology of the study area can be divided into two main stratigraphic units. The Wucheng and Lishi Loess are exposed (mostly decalcified 'red loess') on the lower slope areas. However, Malan Loess (calcareous 'yellow loess') predominates on the upper slope sections of the dry valleys.

The two main dry valleys of the study area are separated by a ridge with steep slopes. Presently, the upper and middle segments of the slopes are characterized by intensive gullying. Calcareous Malan Loess is exposed on the upper and middle segment of the slopes. The dominant soil type is Calcaric Regosol (WRB, Li and Lindström 2001; Schindler et al. 2004). Grain size analysis of loess soil samples taken in the Yangjuangou catchment under different vegetation covers and from different slope positions show a uniform grain size distribution with contents of >80% for silt, 10–12% for clay, 5–7% for sand by weight, and that of calcium carbonate of 10–15% by weight (Schindler et al. 2004). The SOC concentration ranges between 1.6 and 3.7 gC kg^{-1}, and is significantly higher in the surface layer ($\alpha = 10\%$) than in the subsoil (1.6 gC kg^{-1}) (Schindler et al. 2004). Land use and management related changes in SOC sequestration also differ at patch, hill slope transect, and small watershed scale (Wang et al. 2011).

The present climate in the Yangjuangou region is semiarid and continental. Detailed data on the climate is available for Yan'an City 14 km northeast of

Fig. 5.1 (**a**) Location of the study area Yangjuangou, Shaanxi Province (Wang et al. 2011) and (**b**) Loess landscape in Yangjuangou with steep slopes

Yangjuangou. The average annual temperature is 10.1°C, with a maximum of 40°C in summer and a minimum of −30°C in winter (Schindler et al. 2004). The average annual precipitation is about 564 mm. The rainfall distribution on the Loess Plateau shows a characteristic seasonal pattern because of the East Asian monsoon, and about 70% of the annual rain falls between June and September. Rainfall events strongly influence the development of gullies and the destruction of earth check-dams inside the study area. Extreme rainfall events were observed in 1968, 1974 (pers. communication Mr. Ma), 1977, 1983 and in July 1996 (Li et al. 2003) for the study area Yangjuangou.

This study area has been under agricultural land use since the last millennium (Bork and Li 2002). Thus, the vegetation of the area has been altered by human activities. Expert interviews indicate that significant land use changes in the Yangjuangou catchment also occurred during the Great Leap Forward era between late 1950s and early 1960s. New crops and crop rotation systems were introduced during this era, the arable land was expanded, and poor agricultural management was prevalent. Li et al. (2003) reported an increase in grassland area including fallow land from 44% in 1984 to 48% in 1996. New forests were expanded from 11% of the land area in 1984 to 15% in 1996. The conversion of sloping croplands to forests and orchards was an important change in land use. Thus, the area under orchard increased from 4% in 1984 to 15% in 1996, and that under cropland decreased from 41% in 1984 to 22% in 1996. A detailed vegetation cover map from the Yangjuangou catchment shows land use types in 1998 and in 2006 (Fig. 5.2a, b). These maps also document

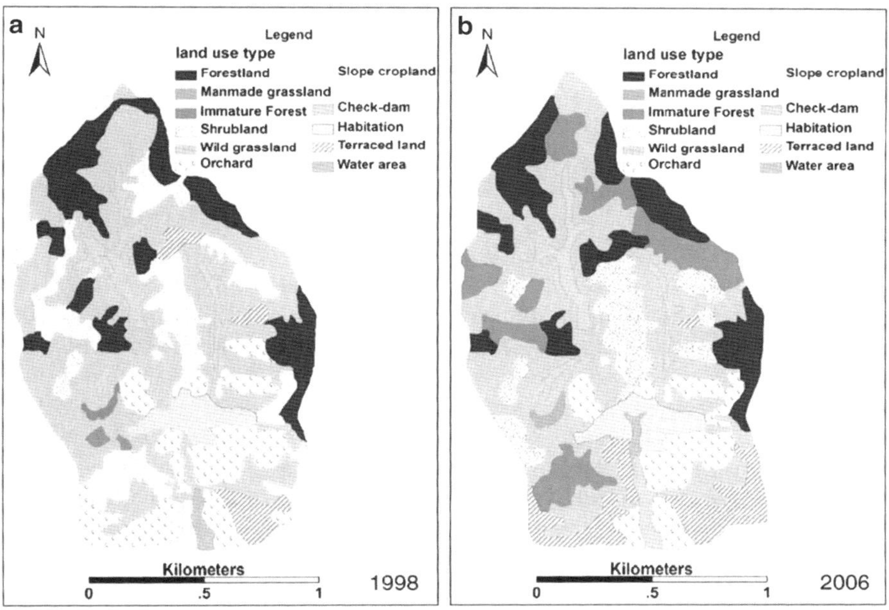

Fig. 5.2 (**a**) Land use types at Yangjuangou catchment in 1998 (Wang et al. 2011); (**b**) and in 2006 (Wang et al. 2011)

an ongoing change in vegetation cover pattern in the Yangjuangou watershed from 1998 to 2006, with additional vegetation restoration on sloping croplands (Wang et al. 2011). The conversion of cropland to forests and orchards was implemented under the programs sponsored by the Chinese government (Du 2005). Consequently, farmers earned higher incomes by selling fruits (Fu et al. 2000) and through financial compensation for afforestation (Du 2005). The non-fruit trees in the watershed of Yangjuangou are predominantly pine (*Pinus tabulae formis*) and robinia (*Robinia pseudoacacia* L.) (Schindler et al. 2004). On the sloping farmland, Millet (*Pennisetum glaucum* (L.) R. Br.), potatoes (*Solanum tuberosum* L.), soy beans (*Glycine max* (L.) Merr.) and vegetables are cultivated (after Li et al. 2006). Maize (*Zea mays* L.) is cultivated on farmland with check-dams (Li et al. 2006).

5.3 Material and Methods

5.3.1 Soil-Sediment Sequence Analysis

The exposures on the backside of the check-dam reservoirs and dams were opened in the western main valley of the Yangjuangou catchment. These exposures provided information about the texture of alluvial sediments of the reservoirs, and the attendant effects of construction and destruction of the dams.

5.3.2 Differential Global Positioning System Measurements

In order to quantify long-term dynamics of water erosion during the last 50 years in the western main valley in Yangjuangou (reservoirs II, III, V, VI, VIII, IX, X), its neighbour valleys (reservoirs IV, VII, XI) and the final reservoir of the reservoir cascade near the village Zhongzuimao (reservoir I) were studied (Figs. 5.3, 5.4). The extension of the 11 reservoir surfaces was determined by using a DGPS. Reservoir II was assessed by using a measuring tape because of the topography the DGPS signal was not accurate. Reservoir II is the only reservoir without an earthen dam constructed perpendicular to the main water flow direction.

The mean thickness of each reservoir fill was also analysed to calculate the sediment budgets of the small subcatchments. These calculations are based on a reconstruction of the valley morphology before the check-dams were built. The slopes were reconstructed by using cross sections and length profiles of the valleys. The estimation of the intercepted sediment mass of the check-dam system is based on sediment mass values of each reservoir and the average dry bulk density of the sediment estimated at ~1.5 Mg m^{-3}.

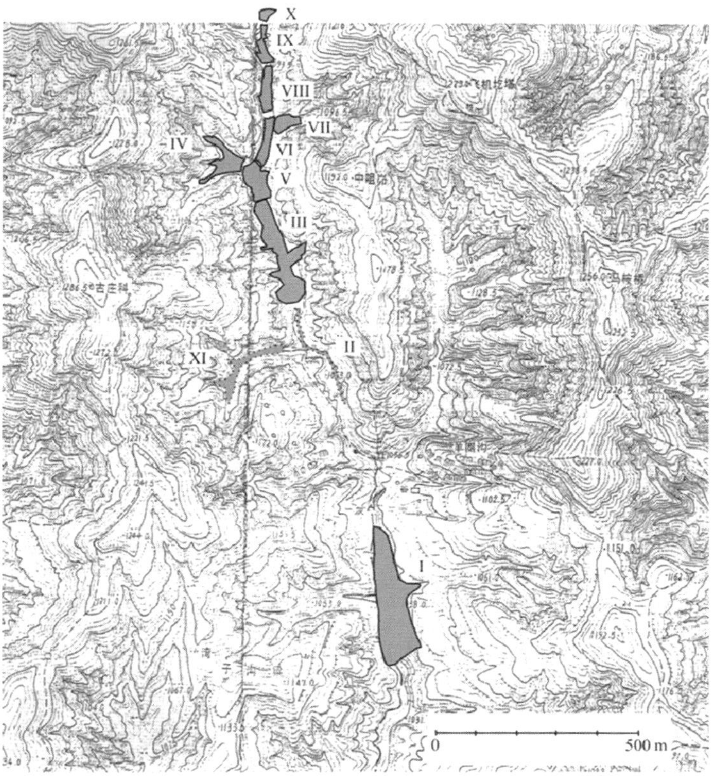

Fig. 5.3 Map of the reservoirs *I–XI* in Yangjuangou. (Source of the map: topographical map no J-49-112-(57), Nianzhuang. Year of measurement 1977, printed 1979, orig. scale 1:10,000. Suveyor Bureau of Revolution Committee in Shaanxi Province). The *broken line* around *II* and *XI* indicates reconstructed reservoir shapes

5.3.3 Map Analysis

The extension of the Yangjuangou catchment and the size of the subcatchment of the reservoir XI were measured by the DGPS. The sizes of all other subcatchments were determined by analysing a topographical map (No.: J-49-112-(57) provided by the Surveyor Bureau of Revolution Committee of the Shaanxi Province 1977). The map has an original scale of 1:10,000, which was enlarged to a scale of 1:5,000. The size of the subcatchment reservoir X was determined by aerial photograph interpretation. A topographical map covering the total relevant subcatchment was not available. The reservoirs of the side valleys (XI, VII, IV) and reservoir X are situated at the beginning of a check-dam cascade, and define the sediment delivery areas. All other reservoirs contain material transported through reservoirs XI, VII, IV, and X

Fig. 5.4 Part of the cascade of check-dam reservoirs in the western main valley in Yangjuangou

or temporary intercepted sediment, remobilised by broken check-dams, or eroded material from the steep slopes.

5.3.4 Expert Interviews

The expert interviews in Yangjuangou village were conducted for calculating water erosion rates in Yangjuangou of the last 50 years. Villages were questioned about the date when the check-dams were constructed and when these were totally filled with sediments. Mr. Ma, a local agricultural expert, provided additional information about the land use history of the subcatchment XI. This information was supplemented by additional interviews with local farmers.

5.3.5 Quantification of Water Erosion and Mass Balances

The down slope areas or dells were studied to quantify historical soil erosion accumulation areas. Colluvial and alluvial deposits (i.e., sediment intercepted above check-dams) provide the minimum values of historical water erosion for the documented erosion events (Bork et al. 1998). The balances give average annual minimum water erosion rates per surface area of the catchment. Secondary or tertiary remobilisation of sediment rarely occurs in small catchments. Therefore, these catchments are especially suitable for mass balances by documenting the primary erosion (Lang and Hönscheidt 1999). The sediment budgets of this study area are based on stratigraphies of reservoir sediments in the main western valley on the basis of their sediment source areas. The alluvial sediment budgets show the absolute mass of the sediment. The soil erosion rates are documented as the soil loss per unit area.

Steps of quantification:

(a) Calculation of sediment volumes [m^3]: Profiles show the average thickness of the sediment bodies [m] of each reservoir. The spatial distribution of the sediment layers [m^2] were measured by the DGPS.
(b) Calculation of sediment masses of each sediment body [Mg]: The volumes of each sediment body [m^3] were multiplied with the specific weight of the sediment [g cm^{-3}] in order to get the sediment masses [Mg]. The average dry density of the material was measured by laboratory analysis.
(c) Calculation of the average annual soil erosion rate due to the sediment body volumes [m^3 $year^{-1}$]: The time of deposition [year] of each sediment body of a reservoir was defined on the basis of expert interviews. The volume of each sediment body was divided by the maximum time of sedimentation, to compute an average value of the annual deposit for each sediment body.
(d) Calculation of the average annual soil erosion rate due to the sediment body mass [Mg $year^{-1}$]: The results of step b were divided by the deposition time.
(e) Calculation of the average annual soil erosion rate due to sediment body volumes [Mg $year^{-1}$] and source areas of the sediment [m^3 ha^{-1} $year^{-1}$]: Source areas [ha] of the sediment layers were reconstructed by topographical maps and additional DGPS measurements. The results of step c were divided by the size of the respective sediment source areas.
(f) Calculation of the average annual soil erosion rate due to sediment mass and deposition time [Mg ha^{-1} $year^{-1}$]: The results of step e were multiplied by the average specific weight of the sediments.

5.4 Results

5.4.1 Soil-Sediment Sequence Analysis

The results of the profile analysis can be summarized in the three aspects of the soil erosion history in Yangjuangou: (i) the check-dam construction, (ii) the destruction of check-dams by soil erosion, and (iii) sediment storage inside the reservoirs.

The exposure of check-dams in the western valleys of the study site shows that the dams are built on in situ red loess. The compacted lower parts of the dams are characterized by laminated texture. Cross sections of the dams document that the earth dams have had no spillways for extreme runoff events. Inside the profiles of

Table 5.1 Results of the differential global positioning system measurements of the check-dam reservoirs in the main western valley of Yangjuangou, Shaanxi Province

		Sediment		
Reservoir	Surface area [10^3 m^2]	Average thickness [m]	Volume [10^3 m^3]	Mass [10^3 Mg]
I	22	4.4	99	148
II	2	0.6	1	2
III	12	2.4	30	45
IV	4	1.2	5	8
V	4	1.4	6	9
VI	3	0.9	2	4
VII	2	0.9	2	3
VIII	3	1.2	3	5
IX	2	0.8	1	2
X	1	0.7	1	1
XI	4	4.5	16	24

the Guzhuangzi dam (reservoir XI), relics of several older dams and gullies were identified which dissected the former dams. The first dam of reservoir XI with a height of about 7 m was built in 1958 according to the interviews of the villagers. Four years later the reservoir was filled, the dam was flooded, eroded and rebuilt again. A more detailed description of the land use changes and the soil erosion history of the subcatchment XI was provided by Dahlke and Bork (2006a). Check-dams in Yangjuangou are still being damaged by surface and subsurface erosion. The fillings of the reservoirs consist of reworked yellow loess which is exposed on the upper and middle slopes of the dry valleys.

5.4.2 *Results of the DGPS Measurements, Expert Interviews and Map Analysis*

The data in Table 5.1 show the results of the reservoir measurements which form the basis of the calculation of the sediment budgets of the check-dam reservoirs. The reservoirs of the main western valley in Yangjuangou and reservoir I (Fig. 5.5) cover a total area of 59,700 m^2. The mean sediment thickness of the reservoir fillings is 0.5–4.5 m. The reservoirs I to XI (Fig. 5.3) are intercepting more than 160,000 m^3 of sediment which accumulated between 1957 and 2003. Taking the average dry bulk density of the sediment into account, more than 250,000 Mg of sediment were deposited. The mean sediment storage in a reservoir exceeds 15,000 m^3. Sizes of the subcatchments range between about 3 and 75 ha, the total catchment area of Yangjuangou is 2.02 km^2 (202 ha).

The data in Table 5.2 show the results of the mass balance. The average annual minimum erosion rates ranged from 0.6 Mg ha^{-1} $year^{-1}$ (reservoir II) to 228 Mg ha^{-1} $year^{-1}$ (reservoir XI). Reservoir XI Guzhuangzi was established in 1958, and was filled with sediments by 1962. In 1966, a second dam was constructed on the same place and the new reservoir was again silted up by 1968.

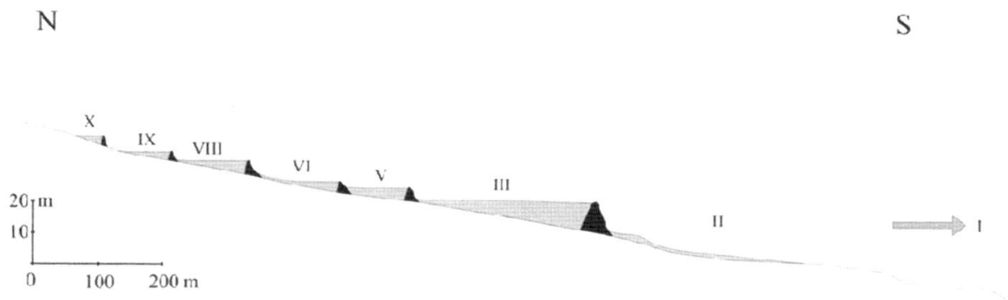

Fig. 5.5 Cascade of check-dam reservoirs in the western main valley of Yangjuangou (2003), north-south profile (*black colour*=check-dams, *gray colour*=check-dam sediment). The *arrow* points towards the final reservoir of the cascade (Fig. 5.3)

Table 5.2 Mass balance of the western main valley of the Yangjuangou catchment

Reservoir	Deposition years	Period of sedimentation [year]	Size of subcatchment [ha]	Avg. annual min. erosion rate [10^3 Mg year^{-1}]	Avg. annual min. erosion rate/area [Mg ha^{-1}year^{-1}]
I	1979–2000	21	202	7	35
II	1958–2003	45	75	0.05	0.6
III	1957–1964	7	55	6	117
IV	1967–1970	3	15	3	177
V	1965–1966	1	43	9	216
VI	1967–1970	3	26	1	47
VII	1972–1975	3	11	1	90
VIII	1963–1965	2	14	3	181
IX	1972–1975	3	9	1	80
X	1972–1975	3	(3)*	0.03	78
XI	1958–1962 and 1966–1968	6	17	4	228

*Reservoir without a check-dam in the southern direction, therefore "only" sediment storage in the original relief position

5.5 Discussion

5.5.1 *Case Study Results*

5.5.1.1 Soil-Sediment Sequence Analysis

Numerous check-dams were established on the middle region of the Loess Plateau during the late 1950s and the early 1960s (Xu et al. 2004). The first check-dam in the Nianzhuang catchment, containing the Yangjuangou catchment and its subcatchment Guzhuangzi, was built in September 1956. The first check-dam in Yangjuangou was constructed in 1957 at reservoir III. One year later, farmers established a dam with 7 m height in the Guzhuangzi during the campaign of the Great Leap Forward. The most recent dam in Yangjuangou was built in 1979 at the south end of reservoir I.

Study of the reservoir fillings document a detailed history of water erosion for the last 50 years. In contrast to the down slope areas in Yangjuangou dominated by red loess, fillings of the reservoirs consist of yellow loess exposed in the agricultural land on the upper slopes. The sediment was eroded on the upper parts of the catchment. However, new terraces are being created in the dry valleys through the accumulation of the alluvial material. About 3% of the catchment was covered by "new" reservoir fields by 2003. Step by step, the reservoir basins have lost their functions of water and sediment retention areas and flood protection for the villages (König 1987; Xu et al. 2004). The new terraces of alluvial loess inside the valley are valuable to agricultural production. In comparison to sloping farmland, the check-dam fields are characterized by abundant soil moisture (Li 1995 cit in Xu et al. 2004) and high concentrations of nitrogen (N), phosphorus (P) and organic matter (OM) (Li and Bai 2003). Field studies in Shanxi Province by Fang (1999 cit. in Xu et al. 2004) and Wang et al. (2000) indicate higher crop yields on fields with check-dams compared to those from sloping farmland in the same areas, except in years check-dam fields are inundated.

5.5.1.2 DGPS Measurements, Map Analysis and Expert Interviews

Runoff during heavy rainfall events with high sediment loads caused intensive accumulation in the reservoirs above the earth dams. The check-dams in the western main valley are 3.5–20 m high (Li et al. 2003a). Thus, the sediment volume differed in each reservoir. The highest deposition and the erosion rates were calculated for reservoir XI (228 Mg ha^{-1} $year^{-1}$) and the lowest for reservoir II (0.6 Mg ha^{-1} $year^{-1}$). The extremely high soil erosion rate of reservoir XI is the result of the extremely steep slopes and the lack of temporary accumulation areas. Reservoir II is the only artificial basin without a dam at the south end. This is the reason for the very low deposition rates for reservoir II. Further research is needed for a detailed interpretation of each water erosion rate of the subcatchments in Yangjuangou.

Study of the reservoirs and the intercepted alluvial sediment in Yangjuangou, and further investigation of colluvial deposits in this area (unpublished) indicate a strong increase of the average annual minimum soil erosion rate in the second half of the twentieth century since 1958 AD compared to a long period between about 1200 BC and 1958 AD (Bork and Li 2002). Reasons for this development were pointed out by the survey of agricultural experts. The agrarian reforms in 1958 during the campaign of the Great Leap Forward had a large impact (Dahlke and Bork 2006b). Land use changes, new crops and crop rotation systems as well as altered field structure accelerated the soil erosion rates. The expert interviews also indicated a psychological phenomenon as probably the most important indirect reason for the accelerated soil erosion primarily because of the decrease in individual and collective attention to the cultivation of the fields in the study area.

Are the qualitative results of the Yangjuangou case study representative for the development of erosion in the early 1960s on the Chinese Loess Plateau? There are no data available about the development of the water erosion during the last 50 years for areas in close proximity of the study site. Therefore, only "soft" data based on literature are available. In Yangjuangou, a strong increase of water erosion was indicated by the use of ^{137}Cs dating (unpublished), and by expert interviews. There occurred a drastic increase in erosion following the implementation of the Great Leap Forward. The increase in erosion during the early 1960s is apparently not an isolated case on the Chinese Loess Plateau. Betke (1987) described the following situation for Guyuan in Ningxia Province: "…clear-cuttings and the agricultural use of grassland [remark authors: often conversion to cropland] have changed the drainage system; soil erosion have increased: In 1970 in parts of the county soil erosion increased 25-times compared to the period before 1966." Wein (1986) mentioned an even stronger link between erosion and the Great Leap Forward. Sediment delivery data of the Yellow River also indicated an increase in water erosion on the entire Chinese Loess Plateau during the 1960s (He et al. 2004; Milliman et al. 1987).

The quantitative assessment of erosion in Yangjuangou show that at least 250,000 Mg of soil were eroded between 1957 and 2003 from an area of only 2.02 km^2. Following a long period of almost no soil erosion, rates reached a height of 200 Mg ha^{-1} $year^{-1}$ after 1958. These rates were computed on the basis of the reservoirs maintained for long time period. Is Yangjuangou a representative case for the quantitative development of the water erosion on the Chinese Loess Plateau during the second half of the twentieth century? The average water erosion rate for the whole Chinese Loess Plateau may be lower as in the study area, because the slopes in the study area are very steep compared to those in some other parts of the Loess Plateau. For a region north of Yan'an City, Chen et al. (1989) calculated average annual water erosion rates of 100–150 Mg ha^{-1} $year^{-1}$ based on the average annual sediment load of the Huang He before 1983. Only 11% of the Loess Plateau Region is characterized by average annual erosion rates of 100–150 Mg ha^{-1} $year^{-1}$ (Chen et al. 1989).

5.6 Soil Erosion Rates and the Soil Carbon Balance on the Chinese Loess Plateau

The reconstruction of soil erosion rates in Yangjuangou based on field data indicated that at least 250,000 Mg of material accumulated in the check-dam reservoir cascade of the main western valley between 1957 and 2003. An analysis of the soil erosion processes in the upper and middle segments of the steep slopes showed that 80% of the material, which deposited in the reservoirs, resulted from sheet erosion processes and consists of relocated topsoil. About 20% of the sediment eroded in gullies, and predominantly comprised of the parent calcareous loess. For the scenario calculation, the SOC values were measured in 2002 and were 3.7 g C kg^{-1} for the surface layer and 1.6 g C kg^{-1} for the sub-soil (Schindler et al. 2004). Based on the sediment analysis, the average SOC concentration of the reservoirs estimated to ~2 gkg^{-1}. Thus, about 500 Mg of SOC were stored in the sediment load of 250,000 Mg between 1957 and 2003, and amounted to 2.48 Mg SOC ha^{-1}.

Wang et al. (2011) analyzed the effects of vegetation restoration on soil C sequestration at multiple scales (patch, hill slope, watershed) in Yangjuangou over 8 years. They observed positive effects of vegetation restoration at all scales, and reported that the SOC stock increased by 19% in the top 20 cm between 1998 and 2006. This increase is equivalent to 19.92 Mg C_{org} km^{-2} $year^{-1}$ (Wang et al. 2011). The "Grain for Green" project initiated in 1999 has been implemented on the semi-arid Loess Plateau, and is the largest vegetation restoration program in China (Wang et al. 2011). In addition to measurements of SOC, the Chinese research group has also sampled soil for ^{137}Cs inventory. The ^{137}Cs measurements were widely used to estimate redistribution of soil on the Loess Plateau. Using this method, Wang et al. (2011) developed a relationship between increase in SOC and the corresponding decrease in soil erosion at the watershed scale.

5.7 Conclusions

The alluvial sediments intercepted in the reservoirs provide information about 50 years of soil erosion by water in the Yangjuangou catchment. Measurements of these sediment demonstrate that soil erosion in the research area increased drastically during the second half of the twentieth century. The restoration of vegetation and soil conservation activities in Yangjuangou today reduced the rate of soil erosion. Wang et al. (2011) demonstrated that the restoration of vegetation also had positive effects on SOC sequestration at all spatial scales. The following conclusions can be drawn in view of Lal's (2010) two general paths for mitigation of climate change through agricultural management: The present study indicates that on average in the Yangjuangou region 2.48 Mg SOC ha^{-1} was stored between 1957 and 2003 in reservoir sediments. Wang and colleagues also showed that the ecological restoration with the increase in biomass aboveground and the accumulation of SOC is a sink for atmospheric CO_2. Wang et al. (2011) also observed that soil erosion by water can modify effects of SOC sequestration in semiarid and topographical complex landscapes. The study by Wang and colleagues in Yangjuangou indicated a strong correlation between ^{137}Cs concentration and the SOC stock, but concluded that the net effect of soil erosion on SOC, the question of sink or source, remains unresolved. Thus, additional long-term monitoring and interdisciplinary research is necessary to answer this question.

Acknowledgments The research was supported by the Evangelisches Studienwerk Villigst, Germany. We thank Prof. Dr. Yong Li and Dr. Ling-yu Bai from the Institute for Agricultural Environment and Sustainable Development, Chinese Academy of Agricultural Science in Beijing for their support during the field work in China. We thank Prof. Dr. Bojie Fu and Dr. Yafeng Wang from the State Key Laboratory of Urban and Regional Ecology, Research Center for Eco-Environmental Sciences, Chinese Academy of Sciences in Beijing for providing two Figs. 5.1a, and 5.2a, b and Elsevier for the reprinting permission of these figures.

References

An Z, Kukla G, Porter St C et al (1991) Magnetic susceptibility evidence of monsoon variation on the Loess Plateau of Central China during the last 130,000 years. Quaternary Res 36:29–36

Berhe AA, Harte J, Harden JW et al (2007) The significance of the erosion-induced terrestrial carbon sink. Bioscience 57:337–346

Betke D (1987) Geschichte der Landentwicklungsprogramme im Wuding-Gebiet und am Mittellauf des Huang He seit Ende des 19. Jahrhunderts. In: Betke D, Küchler J, Obenauf KP (eds) Wuding und Manas: Ökologische und sozio-ökonomische Aspekte von Boden- und Wasserschutz in den Trockengebieten der VR China, Urbs et Regio. Kasseler Schriften zur Geographie und Planung 43. Gesamthochschulbibliothek, Kassel, pp 44–59

Bork H-R, Li Y (2002) 3200 Jahre Reliefentwicklung im Lößplateau Nordchinas. Das Fallbeispiel Zhongzuimao. Petermanns Geogr Mitt 146(2):80–85

Bork H-R, Bork H, Dalchow C et al (1998) Landschaftsentwicklung in Mitteleuropa. Klett-Perthes, Gotha

Cai Q (2001) Soil erosion and management on the Loess Plateau. J Geogr Sci 11(1):53–70

Chen Y, Jing K, Lu J et al (1989) The temporal and spatial variation of erosion and sediment yield on the Loess Plateau. In: Brush LM, Wolman MG, Huang B-W (eds) Taming the Yellow River: silt and floods, proceedings of a bilateral seminar on problems in the lower reaches of the Yellow River, China. Kluwer Academic Publishers, Dordrecht, pp 275–288

Dahlke C, Bork H-R (2006a) Der Erddamm von Guzhuangzi (Shaanxi, Nordchina). In: Bork H-R (ed) Landschaften der Erde unter dem Einfluss des Menschen. WBG, Darmstadt, pp 29–32

Dahlke C, Bork H-R (2006b) Nicht vom Himmel abhängig: Chinas Großer Sprung nach vorn? In: Bork H-R (ed) Landschaften der Erde unter dem Einfluss des Menschen. WBG, Darmstadt, pp 34–40

Du Q (2005) Six priority programs and legal issues on conservative forestry development in the PRC. In: Strategies, science and law for the conservation of the world resources. International Workshop Selfoss, Iceland, 14–18 Sept 2005. AUI Publication 4, pp 115–124

Fu B, Gulinck H (1994) Land evaluation in an area of severe erosion: the Loess Plateau of China. Land Degrad Rehabil 5:33–40

Fu B, Chen L, Ma K et al (2000) The relationships between land use and soil conditions in the hilly area of the loess plateau in northern Shaanxi, China. Catena 39:69–78

He X, Tang K, Zhang X (2004) Soil erosion dynamics on the Chinese Loess Plateau in the last 10,000 years. Mt Res Dev 24(4):342–347

König A (1987) Landwirtschaftlicher Wasserbau. In: Betke D, Küchler J, Obenauf KP (eds) Wuding und Manas: Ökologische und sozio-ökonomische Aspekte von Boden- und Wasserschutz in den Trockengebieten der VR China, Urbs et Regio. Kasseler Schriften zur Geographie und Planung 43. Gesamthochschulbibliothek, Kassel, pp 37–43

Kuhn NJ, Hoffman TH, Schwanghart W et al (2009) Agricultural soil erosion and global carbon cycle: controversy over? Earth Surf Process Landforms 34:1033–1038. doi:10.1002/esp1796

Lal R (2002) Soil carbon sequestration in China through agricultural intensification, restoration of degraded and desertified ecosystems. Land Degrad Dev 13:469–478. doi:10.1002/ldr.531

Lal R (2003) Soil erosion and the global carbon budget. Environ Int 29:437–450. doi:10.1016/S0160-4120(02)00192-7

Lal R (2004) Soil carbon sequestration impacts on global climate change and food security. Science 304:1623–1627

Lal R (2009) Sequestering atmospheric carbon dioxide. Crit Rev Plant Sci 28:90–96. doi:10.1080/07352680902782711

Lal R (2010) Managing soils and ecosystems for mitigating anthropogenic carbon emission and advancing global food security. Bioscience 60(9):708–721. doi:10.1525/bio2010.60.9.8

Lal R, Pimentel D (2008) Soil erosion: a carbon sink or source? Science 319:1040–1041

Lang A, Hönscheidt S (1999) Age and source of colluvial sediments at Vaihingen-Enz, Germany. Catena 38:89–107

Li Y, Bai L (2003) Variations of sediment and organic carbon storage by check-dams of the Chinese Loess Plateau. J Soil Water Conserv 17(2):1–4 (Chinese)

Li Y, Lindström MJ (2001) Evaluating soil quality – soil redistribution relationship on terraces and steep hillslope. SSSA J 265(5):1500–1508

Li Y, Poesen J, Yang YC et al (2003) Evaluating gully erosion using ^{137}Cs and $^{210}Pb/^{137}Cs$ ratio in a reservoir catchment. Soil Tillage Res 69:107–115

Li Y, Zhang QW, Reicosky DC et al (2006) Using ^{137}Cs and $^{210}Pbex$ for quantifying soil organic carbon redistribution affected by intensive tillage on steep slopes. Soil Tillage Res 86(2):176–184

Milliman JD, Qin Y-Sh, Ren M et al (1987) Man's influence on the erosion and transport of sediment by Asian rivers: The Yellow River (Huanghe) example. J Geol 95:751–762

Pan G (2009) Stock, dynamics of soil organic carbon of China and the role in climate change mitigation. Adv Clim Change Res 5(Suppl):11–18

Schindler U, Li Y, Funk R (2004) Soil properties and soil water conditions in the Yangjuangou catchment of the Chinese Loess Plateau. Arch Agron Soil Sci 50:467–476

Smith P, Martino D, Cai Z et al (2007) Agriculture. In: Metz B, Davidson OR, Bosch PR et al (eds) Climate change: mitigation. Contribution of working group III to the fourth assessment report of the Intergovernmental Panel on Climate Change. Cambridge University Press, Cambridge

Song G, Li L, Pan G et al (2005) Topsoil organic carbon storage of China and its loss by cultivation. Biogeochemistry 74:47–62. doi:10.1007/s10533-004-2222-3

United Nation Statistic Division (2011) Millennium development goals indicators. Carbon dioxide emissions. http://mdgs.un.org/unsd/mdg/SeriesDetail.aspx?srid=749&crid. Accessed 12 May 2011

Van Oost K, Quine TA, Govers G et al (2007) The impact of agricultural soil erosion on the global carbon cycle. Science 318:626–629

Van Oost K, Six J, Govers J et al (2008) Reply to letter on 'Soil erosion: a carbon sink or source?' by R. Lal and D. Pimental. Science 319:1040–1042

Wang ZH, Hu ZH, Wei ZH et al (2000) Soil erosion control and damland system agriculture in Weijiayu watershed. In: Laflen JM, Tian J, Huang C-H (eds) Soil erosion and dryland farming. CRC Press, Boca Raton, pp 81–91

Wang Y, Fu B, Lü Y et al (2011) Effects of vegetation restoration on soil organic carbon sequestration at multiple scales in semi-arid Loess Plateau, China. Catena 85:58–66. doi:10.1016/j.catena.2010.12.003

Watson RT, Noble IR, Bolin B et al (eds) (2000) Land use, land-use change and forestry IPCC special report. Cambridge University Press, Cambridge

Wein N (1986) Agriculture and forestry in the Loess Plateau region of China. The example of the Chung-hua District. GeoJournal 12(1):57–64

Xu X, Zhang H, Zhang O (2004) Development of check-dam systems in gullies on the Loess Plateau, China. Environ Sci Policy 7(2):79–86

Yan H, Cao M, Liu J et al (2007) Potential and sustainability for carbon sequestration with improved soil management in agricultural soils of China. Agric Ecosyst Environ 121:325–335

Chapter 6
Methane Emissions from China's Natural Wetlands: Measurements, Temporal Variations and Influencing Factors

Xiaoke Wang, Fei Lu, and Le Yang

Abstract Natural wetlands received increased attention for their ecosystem services and high methane (CH_4) emissions. In China, the total area of wetlands is about 38 million ha (M ha), accounting for about 4% of the land. Natural wetlands include peatlands (35.6%), coastal wetlands (15.4%), rivers (21.3%), lakes (21.7%) and reservoirs (6.7%). Human activities and settlement development have drastically reduced wetland areas worldwide. A recent estimate showed that 33% were lost between 1978 and 2008, and land reclamation accounted for more than 70% of the total loss in China. Effects of human activities on wetlands in China were studied at 16 sites across the country, i.e., 6 peatlands, 3 coastal wetlands, 5 lakes and 2 reservoirs. The mean CH_4 emissions were 6.0 (range 1.0–15.6) mg CH_4 m^{-2} h^{-1} for peatlands, 1.6 (0.5–2.4) mg CH_4 m^{-2} h^{-1} for coastal wetlands, 3.1 (0.9–9.7) mg CH_4 m^{-2} h^{-1} for lakes and 0.2 (0.1–0.3) mg CH_4 m^{-2} h^{-1} for reservoir. The annual CH_4 emissions from natural wetlands in China was estimated to be 10.5 Tg CH_4 $year^{-1}$ (1 Tg = 10^{12} g), which accounted for 7.3% (145 Tg CH_4 $year^{-1}$) of the global emissions from wetlands and 171% (6.147 Tg CH_4 $year^{-1}$) of the CH_4 emission from rice paddies in China. The significant contribution of CH_4 emission from natural wetlands should be taken into account in national greenhouse gas inventory.

Keywords Methanogensis • Peatlands • El Nino • Labile organic carbon • Landsat thematic mapper • Greenhouse gases • Atmospheric carbon pool • Aquatic systems • Ethane emission • Natural wetland • Lake • Reservoir • China • Northeast Asia • Tibetan Plateau • Photosynthesis • Hydrophytes • Coastal wetlands • Meadows • Methane emissions • Temporal variations in CH_4 emission • Solar radiation • Hydrology • Water table • Vegetation

X. Wang (✉) • F. Lu • L. Yang
State Key Lab of Urban and Regional Ecology, Research Center for Eco-Environmental Sciences, Chinese Academy of Sciences, No.18 Shuangqing Road, Haidian District, Beijing 100085, China
e-mail: wangxk@rcees.ac.cn; feilu@rcees.ac.cn; yangle3012@163.com

R. Lal et al. (eds.), *Recarbonization of the Biosphere: Ecosystems and the Global Carbon Cycle*, DOI 10.1007/978-94-007-4159-1_6,

Abbreviations

ATCM	Atmospheric Transport and Chemical Model
CCICED	China Council for International Cooperation on Environment and Development
ENSO	El Nino Southern Oscillation
GHG	greenhouse gas
LOC	labile organic organic
ETM+	Landsat Enhanced Thematic Mapper plus
TM	Landsat Thematic Mapper
NPP	net primary production
SFA	State Forestry Administration

6.1 Introduction

Methane (CH_4) is the second most important anthropogenic greenhouse gas (GHG) in terms of radiative forcing to global warming after carbon dioxide (CO_2). Although its residence time in the atmosphere is short (about 12 years), the ability of CH_4 to absorb infrared radiation makes it 20–30 times more efficient than CO_2 in trapping energy (Weaver 2011). CH_4 is also an important atmospheric constituent by influencing the cleansing capacity [concentration of hydroxyl radicals (OH)], and the formation of ozone (IPCC 2007). The global atmospheric CH_4 concentration has increased from a pre-industrial level of about 715–1,732 ppb in the early 1990s, and further to 1,774 ppb in 2005 (Zhuang et al. 2009). Growth rates have declined since the early 1990s, consistent with trends in CH_4 emissions (sum of anthropogenic and natural sources) being nearly constant during this period. It is very likely that the observed changes in CH_4 concentration are due to anthropogenic activities, predominantly agriculture and fossil fuel use, but relative contributions from different source types are not well determined (IPCC 2007). A recent estimate of annual global CH_4 emission was 582 Tg CH_4 $year^{-1}$ ($1\ Tg=10^{12}$ g, IPCC 2007), of which 70% was released from biogenic sources. These sources include wetlands, rice agriculture, livestock, landfills, forests, oceans and termites. Natural wetland has been proposed to be the single largest CH_4 source based on recent estimates combining bottom-up and top-down fluxes, and global observations of atmospheric CH_4 concentrations in a three dimensional Atmospheric Transport and Chemical Model (ATCM) simulation (Chen and Prinn 2005, 2006). Bastviken et al. (2011) compiled CH_4 emissions from 474 freshwater ecosystems including lakes, impoundments, and rivers and estimated CH_4 emissions to be 103 Tg CH_4 $year^{-1}$ based on recent data on area and distribution of inland water bodies. Expressed as CO_2 equivalents (eq), this emission corresponds to 0.65 Pg CO_{2eq} $year^{-1}$ ($1\ Pg=10^{15}$ g) or 25% of the estimated land greenhouse gas (GHG) sink, assuming that 1 kg of CH_4 corresponds to 25 kg of CO_2 over a 100-year period (Bastviken et al. 2011).

Although most wetlands accumulate organic carbon (C) and are important sinks for atmospheric C, the high global warming potential of CH_4 makes wetlands net sources of GHGs. The mean CH_4 emission rates for wetlands are about 200 kg CH_4 ha^{-1} $year^{-1}$ (Mitra et al. 2005), which would mitigate a C sequestration of 1.5 Mg C ha^{-1} $year^{-1}$ (1 Mg = 10^6 g). This value is slightly higher, but in the same order of magnitude of what can be derived as average C sequestration (0.2–1.4 Mg C ha^{-1} $year^{-1}$ based on a global estimate by Wojick 1999). With the increasing number of field measurements and significant improvement of biogeochemical models, global and national CH_4 budgets have been improved significantly. However, the range is still large from 100 Tg CH_4 $year^{-1}$ (Wuebbles and Hayhoe 2002) to 231 Tg CH_4 $year^{-1}$ (Fletcher et al. 2004). A median was reported by Chen and Prinn (2006) with an estimate of 145 Tg CH_4 $year^{-1}$ emitted 1996–2001. With the increasing availability of regional and national CH_4 budgets, the uncertainties can be reduced significantly. For instance, CH_4 emissions were 9 Tg CH_4 $year^{-1}$ from wetlands in North America with an uncertainty greater than 100% (Bridgham et al. 2006), 5.2 Tg CH_4 $year^{-1}$ from European wetlands and water bodies (Saarnio et al. 2009), and 1.8 Tg CH_4 $year^{-1}$ from natural wetlands in China based on field measurements during 1995–2004 (Ding et al. 2004a).

With more field measurements conducted, the processes and factors controlling CH_4 emissions were investigated in detail. It has been recognized that the key processes are fermentation, methanogenesis, and sulfate, iron, and nitrate reduction in the anaerobic zone (Kayranli et al. 2010). The main factors controlling CH_4 emissions from wetlands are soil temperature (Christensen et al. 2003), water table depth (Moore et al. 1998), and the amount and quality of decomposable substrates (Christensen et al. 2003). However, the diverse combination of environmental factors and processes specific to sites make it difficult to predict CH_4 emission accurately. Although CH_4 emissions from wetlands can be estimated based on their areas and emission rates, in fact, wetland area and CH_4 emission rates are not easy to be determined for a country because of significant temporal and spatial variability.

Although available reviews (e.g., Kayranli et al. 2010) have analyzed the processes and factors influencing CH_4 emissions worldwide, few investigations carried out in China have been included because of the difficulty to access to those reports written in Chinese. In recent years, field measurements of CH_4 emissions from wetlands have been conducted covering different types and locations of wetlands. Considering that China is the largest global CO_2 emitter and most populous country with rapid economic development and significant changes in land use, it is required to review CH_4 emissions from wetlands in China. In this study, the area and CH_4 emission rates of wetlands in China were reviewed, and a national estimate of CH_4 emissions from wetlands was carried out based on the best data available.

6.2 Wetland Area and Changes in China

Wetland can be defined as an area of land whose soil is saturated with moisture either permanently or seasonally at the interface between terrestrial and aquatic systems (http://en.wikipedia.org/wiki/Wetland). Wetland areas may also be covered partially or completely by shallow pools of water. A boarder definition was given under the Ramsar International Wetland Conservation Treaty, i.e., wetlands are areas of marsh, fen, peatland or water, whether natural or artificial, permanent or temporary, with water that is static or flowing, fresh, brackish or salty, including areas of marine water whose depth at low tide does not exceed 6 m. In China, more people prefer to the definition of wetlands made by the Ramsar Treaty, which includes swamps, peatlands as well as lakes and reservoirs (Niu et al. 2009). The rice paddy or other agricultural used wetlands are excluded.

There are different estimations of wetland area in China, ranging from about 26 to 66 million ha (M ha) (Zhang et al. 2008). In 2004, the State Forestry Administration (SFA) issued a report stating that the total area of wetlands individually larger than 100 ha is about 38 M ha which is about 4% of total land area. These estimations include natural and constructed wetlands. The later occupy about 3 M ha or 7.8% of total wetland area. Natural wetlands include coastal (5.9 M ha, 15.4%), riverine (8.2 M ha, 21.3%), lakes (8.4 M ha, 21.7%) and peatlands (13.7 M ha, 35.6%). The area of reservoirs is estimated to be 2.56 M ha, about 6.7% of total wetland area. A recent mapping of the China's wetlands using Landsat Enhanced Thematic Mapper plus (ETM+) data indicated that a total of 35.9 M ha wetlands are of non-agricultural use, of which 33.9 M ha are inland wetland, 0.28 M ha are non-agricultural artificial wetland, and 1.76 M ha are coastal wetland (Niu et al. 2011). The late estimate is similar to that by the SFA with a difference of only 13%. However, the area of coastal wetland surveyed by SFA is three times the area reported by Niu et al. (2009), because single-date remote sensing data cannot capture the land in the intertidal zone and its underwater part (Niu et al. 2009).

In China, peatlands are concentrated in the northeastern and southwestern regions. Sanjiang Plains and Zoige are the two largest peatlands. Other small peatlands are widely distribute in alpine regions such as Qiang-Tibet Plateau and in boreal regions such as Daxing'an and Xiaoxing'an Mountains. Coastal wetlands occur in eastern China along the Pacific Ocean with sandy and muddy beds, and rich vegetation in the northern part and rocky beds in the southern part. Lakes can be divided into five regions according to the differences in climate and topography, i.e., Eastern Plains, Inner Mongolia and Xinjiang Plateau, Yun-Gui Plateau, Qing-Tibet Plateau and Northeastern Plains (Liang et al. 1999). Reservoirs are generally smaller except some big dams, i.e., Three Gorge Reservoir, and are located mainly in southwestern China.

Human activity and settlement development have drastically reduced wetland area not only worldwide but also in China. It was reported that there were abundant wetlands in ancient China reaching 65.7 M ha in coverage. With the increasing demands for grains production, and industrial and residential uses, wetlands have been vanishing for a long period of time. In China, the most intensive wetland reclamation occurred in Ming and Qing Dynasties (1368–1911), and 1950–1990. After the foundation of the People's Republic China in 1949, wetland area in was drastically reduced. In Sanjiang Plains, northeastern China, where the largest peatlands are located, croplands have increased from about 78 M ha in 1949 to 366 M ha by a factor of 4.2, mostly conversion from peatlands (Zhao and Gao 2007).

Another report indicated that 1.36 Mha of wetlands disappeared on the Sanjiang Plain between 1950 and 2000 with a decrease in area from about 52% to 16% of the total area (Hou et al. 2006). In the middle and low reaches of the Yangtze River, the cover of lakes larger than 15 ha had decreased by 43.5% from 1950s to 1980s. Coastal wetland area decreased by 50% or 2 Mha (Zhao and Gao 2007). A recent report by the China Council for International Cooperation on Environment and Development (CCICED), a joint Chinese and international advisory board to the government, indicated that 57% of the country's coastal wetlands have disappeared since the 1950s, largely due to land reclamation (Qiu 2011).

A recent estimate indicated that 33% of wetland area was lost between 1978 and 2008, and land reclamation accounted for more than 70% of the loss (Niu et al. 2011). Based on Landsat Thematic Mapper (TM) images from 1987 to 1992, and (ETM+) images from 1999 to 2002, the wetland distribution in China was mapped between 1990 and 2000, respectively (Gong et al. 2010). In 1990, the total wetland area was 35.5 Mha whereas in 2000 it decreased to 30.5 Mha with a net loss of 5.0 Mha. During 10-year period, inland wetlands were reduced in cover from 31.8 to 25.8 Mha, coastal wetland area decreased from 1.4 to 1.2 Mha, but artificial wetland area increased from 2.3 to 3.5 Mha. The greatest natural wetland loss occurred in Heilongjiang, Inner Mongolia, and Jilin with a total loss of over 5.7 Mha. In western China, over 1.3 Mha of wetlands were created in Xinjiang, Tibet, and Qinghai. About 1.2 Mha of artificial wetlands were also created for fish farms and reservoir constructions.

6.3 Methane Emissions from China's Wetlands

Because of the large area covered by rice paddies in China, CH_4 emissions have been measured since the 1980s while measurements of emissions from natural wetlands or reservoirs began later. since the 1990s, CH_4 emissions from natural wetlands were measured in Sanjiang Plain in 1995 (Cui 1997) and in Tibet-Qinghai Plateau in 1996–1997 (Jin et al. 1999). Since then, additional measurements of CH_4 emissions were made. Review of the published literature showed that the measurements were carried out in at least 16 sites across the country, i.e., 6 peatlands, 3 coastal wetlands, 5 lakes and 2 reservoirs (Fig. 6.1). All measurements for at least one growing season were compiled in Table 6.1, except some measurement data measured for a short period that were designed for specific studies. For example, Duan et al. (2006) reported CH_4 emissions for 3 days to investigate their responses to hydrophyte photosynthesis. Hirota et al. (2005) measured CH_4 emission for two clear days to investigate their response to grazing in an alpine wetland on the Qinghai-Tibetan Plateau. Although there are large variations in CH_4 emission rate across sites (Fig. 6.2) depending on vegetation, and environmental conditions, it can be summarized that the mean CH_4 emissions were 6.0 (range 1.0–15.6) mg CH_4 m^{-2} h^{-1} for peatlands, 1.6 (0.5–2.4) mg CH_4 m^{-2} h^{-1} for coastal wetlands, 3.1 (0.9–9.7) mg CH_4 m^{-2} h^{-1} for lakes, and 0.2 (0.1–0.3) mg CH_4 m^{-2} h^{-1} for reservoirs (Table 6.1).

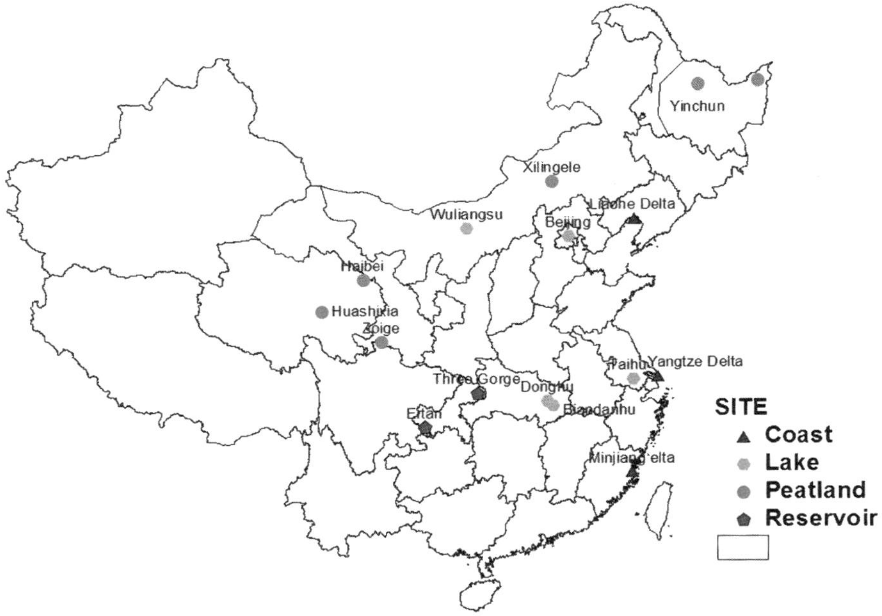

Fig. 6.1 The wetland CH_4 measurement sites in China

6.3.1 Peatlands

Although there is wide distribution of peatlands across China, large area of peatlands are concentrated in Sanjiang Plain, northeastern China and Qinghai-Tibetan Plateau, western China. In both regions, the climate is characterized by cold, long winters and short, cool summers with relatively high precipitation The annual mean air temperature and precipitation were 2.52°C and 558 mm in Sanjiang Plain (Song et al. 2009) and 1.7°C and 650 mm in Qinghai-Tibet Plateau (Chen et al. 2009), respectively. In summer, the relatively humid climate supports high plant productivity, which induces a large input of organic C to the soil. As the low temperatures and high soil moisture regimes are not favorable to decomposition, organic C accumulates in soils, and builds up large quantities of peat, which are prone to anaerobic CH_4 production. Since 1990s, the measurements of CH_4 emissions have begun in these two regions.

The Sanjiang Plain was probably the largest wetland in China. It covered 5.3 million ha in the early 1950s (Liu and Ma 2000). Unfortunately, wetland area shrunk very quickly with development of agriculture. CH_4 emissions were measured during the plant growing season typically from May through October over 1 or 2 years period (Cui 1997; Ding et al. 2004b; Hao et al. 2011; Yang et al. 2004, 2006a; Wang et al. 2006a; Zhang et al. 2007), and 4 years period (Song et al. 2009). High CH_4 emissions were reported during a short summer period (Cui 1997). In this study, the

Table 6.1 The field measurements of CH_4 from wetlands in China

Type	Location	Treatment	Period	Frequency	Mean	Reference
Peatland	Tongjiang	*Carex lasiocarpa*	July and Sept., 1995, May 1996	Twice a month	20.67	Cui (1997)
		Glyceria spiculosa			28.92	
		Carex schmidtii			13.19	
		Calamagrostis angustifolia			18.81	
		Phragmites communis			39.28	
		Carex lasiocarpa	2 years	Twice a week	19.60	Ding et al. (2004b).
		Carex lasiocarpa	May–Aug., 2001	Twice a week	17.30	
		Carex lasiocarpa	May–Aug., 2002	Twice a week	22.00	
		Carex lasiocarpa	Jun–Aug	3 consecutive days/month	20.06	Ding et al. (2004b)
		C. meyeriana	June–Aug.		17.60	
		Forest patch	June–Sept., 2003	1–weeks	0.05	Yang et al. (2004)
		Carex lasiocarpa	June–Sept., 2003	Once a week	10.80	Yang et al. (2006a)
		Carex pseudocuraica			11.40	
		Deyeuxia angustifolia			1.59	
		Carex lasiocarpa	2002–2005	2×/week, early May-late September	4.50	Song et al. (2009)
		Deyeuxia angustifolia			0.50	
		Shrub		1×/month, non-growing season	0.02	
		Corex lasiocarpa	Non-growing	Twice per week	2.43	Song et al. (2009)
			Growing		23.46	
			2002–2003		25.89	
			Non-growing		1.70	
			Growing		47.57	
			2003–2004		49.27	
			Non-growing		–0.17	
			Growing		43.20	
			2004–2005		43.03	
			Growing (2005–2006)		32.30	
			Annual mean		39.40	

(continued)

Table 6.1 (continued)

Type	Location	Treatment	Period	Frequency	Mean	Reference
		Deyeuxia angustifolia	Non-growing		0.39	
			Growing		6.09	
			2002–2003		6.45	
			Non-growing		0.18	
			Growing		5.65	
			2003–2004		5.82	
			Non-growing		–1.00	
			Growing		0.90	
			2004–2005		0.81	
			Growing (2005–2006)		0.00	
			Annual mean		4.36	
		Shrub	Non-growing		0.06	
			Growing		0.21	
			2002–2003		0.27	
			Non-growing		0.20	
			Growing		0.14	
			2003–2004		0.34	
			Non-growing		–0.04	
			Growing		0.06	
			2004–2005		0.02	
			Growing (2005–2006)		0.00	
			Annual mean		0.21	
		Corex Lasiocarpa	June–Oct., 2002		11.90	Hao et al. (2004)
		Deyeuxia angustifolia			8.50	
		Shrub			0.75	

Peatland	Haibei	*Kobresia tibetica*	June-Sept., 2003	Twice a week	1.10	Hu et al. (2005)
		Corex pamirensis			6.92	
		Corex Lasiocarpa	June-Oct., 2001		17.29	Wang et al. (2003)
	Haibei, Qinghai	*Potamogeton pectinatus*	July-Sept., 2002	2 weeks	1.38	
		Hippuris vulgaris			8.92	
		Scripus distigmaticus			4.57	
		Carex allivescers			8.19	Hirota et al. (2004)
	Zoige	*Carex muliensis*	May–Sept., 2001	Twice a week	2.87	Wang et al. 2002
		Carex meyeriana			4.51	
		Open fen on hills	June–Sept., 2006	Monthly	2.21	Chen et al. (2010a)
		Smooth littoral wetland			0.65	
		Steep riparian zone			0.00	
		Natural meadow			0.00	
		Steep littoral wetland			11.95	
		Mead			0.00	
	Huashixia	*Kobresia humulis*	July–Aug., 1996	Once a month	1.80	Jin et al. (1999)
		Batrachium trichophyllum			0.54	
		Hippuris vulgaris			–0.01	
		Kobresia tibetica			1.91	
		Kobresia humulis	Apr.–Sept., 1997		0.56	
		Batrachium trichophyllum			0.28	
		Hippuris vulgaris			2.47	
		Carex atrofusa			2.91	
	Yichun	*Alnus sibirica swamp*	June-Oct., 2007	10 days	0.64	Song et al. (2009)
		Betula platyphylla			0.03	
		Larix gmelinii-Carex schmidtii			–0.04	
		Larix gmelinii-moss			–0.04	
		Larix gmelinii-Sphagnum			2.34	

(continued)

Table 6.1 (continued)

Type	Location	Treatment	Period	Frequency	Mean	Reference
		Corex schmidtii	June–Oct., 2007		*1.85*	Song et al. (2009)
		Shrub			0.09	
		*Larix gmelinii*_10yr_R			0.01	
		*Larix gmelinii*_20yr			–0.07	
		*Larix gmelinii*_10yr_D			0.10	
		*Larix gmelinii*_20yr			0.06	
	Yichun	*Carex schmidtii*	June–Oct., 2007	10 days	1.88	Sun et al. (2009b)
	Xilingele	Hummock_upper site	July–Aug., 2003	4 days	5.51	Wang et al. (2005)
		Hollow_upper site			10.18	
		Hummock_middle site			4.53	
		Hollow_middle site			20.32	
		Hummock_lower site			1.18	
		Hollow_lower site			16.85	
		Drained			–0.08	
Coast	Minjiangkou	*Spartina alterniflora*			2.35	Tong et al. (2008)
	Liaohe Delta	*Phragmites australis*	Apr.–Nov., 1997	Monthly	0.52	Huang et al. (2001b)
	Changjiangkou	*Scirpus mattfeldianus*_M	May 2004–Apr. 2005	Monthly	2.06	Yang et al. (2007)
		*Scirpus mattfeldianus*_L			0.04	
Lakes	Wuliangsu	*Phragmites australis*-H	2003	2× month	17.93	Duan et al. (2007)
			2004		19.13	
		Phragmites australis-L	2003		12.67	
			2004		14.87	
		Potamogeton petinatus –H	2003		2.09	
			2004		4.26	
		Potamogeton petinatus-L	2003		2.32	
			2004		4.20	

	Donghu	I	Apr. 2003–May 2004	Monthly	1.12	Xing et al. (2005)
		II			0.88	
		III			1.14	
	Biandanhu	I	May 2003–Apr. 2004	Monthly	0.62	Xing et al. (2006)
		II			0.70	
		III			1.31	
	Taihu	Eulittoral	Aug. 2003–Aug. 2004	Once a week	2.60	Wang et al. (2006a)
		Pelagic			0.50	
		Infralittoral			0.40	
		Supralittoral			0.10	
	Beijing	Urban	2009	2× month	2.61	Ai et al. (2009)
Reservoir	Ertan	Open water	May 2008–Apr. 2009		0.12	Zheng et al. (2011)
	Three Gorge	Sandouping	2010	2× month	0.28	Yang et al. (2012)
		Zigui			0.12	
		Wushan			0.26	
		Yunyang			0.59	

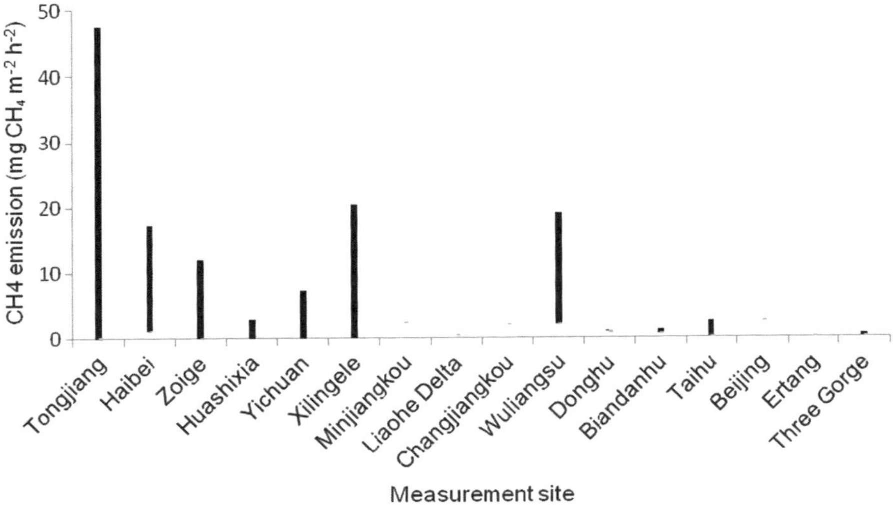

Fig. 6.2 The ranges of CH_4 emission from different wetland sites in China

effects of vegetation, cutting and water depth on CH_4 emission were assessed with a limited data. The 2-year mean emission was 19.6 ± 12.8 mg CH_4 m^{-2} h^{-1} from *Corex lasiocarpa* EHRH from May to August 2001 and 2002 (Ding et al. 2004b). The effects of plant species and cutting on CH_4 emission were also assessed (Ding et al. 2004c, 2005). Since 2002, Song et al. (2008, 2009) has monitored CH_4 emissions from three wetlands [permanently inundated wetland (PI), seasonally inundated wetland (SI) and shrub swamp (SS)] for a period of 4 years. The mean CH_4 emissions were 6.0 ± 1.0 mg CH_4 m^{-2} h^{-1} for PI, 0.7 ± 0.3 mg CH_4 m^{-2} h^{-1} for SI, and 0.03 ± 0.02 mg CH_4 m^{-2} h^{-1} for SS, respectively (Song et al. 2009). Substantial interannual variation of CH_4 fluxes were reported due to significant climatic variability. There was another study of CH_4 emissions carried out by Yang et al. (2004) from June to September 2003 in similar types of wetlands located 10.9 km from the measurement site of Song et al. (2009). The mean CH_4 emissions were 10.8, 11.4 and 1.6 mg CH_4 m^{-2} h^{-1} for *Corex lasiocarpa*, *C. pseudocuraica* and *Deyeuxia angustifolia* peatlands, respectively (Yang et al. 2006a). It is important to note that CH_4 fluxes from a forest patch within a peatland were also measured from June 1 to September 28 of 2003 in Sanjiang Plain, and were in the range of –0.03 to 0.04 mg CH_4 m^{-2} h^{-1} depending on the water level (Yang et al. 2004).

Large areas of peatland are located in Qinghai-Tibetan Plateau with an estimated total area of 13.3 Mha. Previous measurements conducted in a transect across four types of peatlands in Huashixia Permafrost region in 1996 and 1997 (Jin et al. 1999) indicated that CH_4 emissions were in the range of –0.01 to 1.91 mg CH_4 m^{-2} h^{-1} from July to August 1996, and 0.27–3.00 mg CH_4 m^{-2} h^{-1} from April to September 1997, respectively. In Haibei, CH_4 emissions were measured from June 30 to September 4, 2003, along a moisture gradient from 38.5% to 100% (v/v), and at

Potentilla fruticosa scrub meadow, *Kobresia humilis* meadow, *Koresia tibetica* meadow to seasonal wetland. The mean values varied from −0.03, −0.03, 1.10–6.92 mg CH_4 m^{-2} h^{-1}, respectively (Hu et al. 2005). In the same region, CH_4 emissions were measured in three emergent-plant zones, dominated by *Carex allivescers* (ZCar), *Scirpus distigmaticus* (ZSci), or *Hippuris vulgaris* (ZHip), and one submerged plant zone dominated by *Potamogeton pectinatus* (ZPot) along a gentle gradient of shallow to deep water. The smallest CH_4 fluxes were 0.5–2.8 mg CH_4 m^{-2} h^{-1} at ZPot. The highest were 4.4–12.3 mg CH_4 m^{-2} h^{-1} at ZHip. ZSci and ZCar had intermediate values from 1.1 to 6.5 mg CH_4 m^{-2} h^{-1} and 1.9 to 10.6 mg CH_4 m^{-2} h^{-1}, respectively (Hirota et al. 2004).

Zoige peatland, located at the eastern edge of Qinghai-Tibetan Plateau, is one of the largest peatlands in China with an area of about 0.4 Mha (Wang et al. 2002). From May to September 2001, CH_4 emissions were measured at two types of peatlands. One is a *Carex muliejsis* marsh and the other a *C. meyeriana* marsh. CH_4 emissions were 0.5–8.2 mg CH_4 m^{-2} h^{-1} with an average of 2.9 mg CH_4 m^{-2} h^{-1} from the *C. muliejsis* marsh and 0.4–10.0 mg CH_4 m^{-2} h^{-1} with an average of 4.5 mg CH_4 m^{-2} h^{-1} from *C. meyeriana* marsh, respectively (Wang et al. 2002). In 2005, a short-term (from mid-June to mid-August and mid-June to mid-September) measurement of CH_4 emissions at Zoige Plateau was conducted to assess the effects of vegetation on CH_4 emission (Chen et al. 2010a). The mean CH_4 emission was 2.5 mg CH_4 m^{-2} h^{-1}, with the highest emission rate of 12.0 mg CH_4 m^{-2} h^{-1} from steep littoral wetlands, and the lowest flux rate of −0.007 mg CH_4 m^{-2} h^{-1} in steep Riparian zones. Significant difference may occur among microsites even under the same vegetation, and water tables which are important factors controlling CH_4 emissions from peatlands (Chen et al. 2010a).

Although forest swamps are widely distributed in northeastern China including Daxing'an and Xiaoxing'an mountainous regions, CH_4 emissions were not measured until very recently. During June to October 2007, Sun et al. (2009a) measured CH_4 emission from different swamps in Yichun of Xiaoxing'an Mountains at 10 days interval. The CH_4 emissions during the growing season were 2.3 ± 8.4, 0.6 ± 0.6, 0.03 ± 0.04, −0.04 ± 0.07 and −0.04 ± 0.1 mg CH_4 m^{-2} h^{-1} for *Alnus sibirica* swamp, *Betula platyphylla* swamp, *Larix-Carex schmidtii* swamp, *L. gmelinii*-moss swamp and *L. -Sphagnum* spp. swamp, respectively. In their another report, Song et al. (2009) reported CH_4 emissions during the growing season from a marsh, a swamp, and 10- and 20-years—old plantations were 1.9 ± 2.3 mg CH_4 m^{-2} h^{-1}, 0.09 ± 0.09 g CH_4 m^{-2} h^{-1}, 0.01 ± 0.11 g CH_4 m^{-2} h^{-1} and −0.07 ± 0.06 g CH_4 m^{-2} h^{-1}, respectively. Sun et al. (2009b) reported a mean CH_4 emission of 1.9 mg CH_4 m^{-2} h^{-1} in a *Carex schmidtii* marsh from June to October 2007 measured in the same region.

In semiarid grasslands, plenty of small patches of wetlands in areas of low relief and poor drainage regions are scattered which may be CH_4 sources (Wang et al. 2005). Field measurements of CH_4 emissions were carried out in riparian mires and adjacent uplands on the Xilin River basin in 2004. Mean emission was 9.8 mg CH_4 m^{-2} h^{-1} during the growing season, but varied by 2–3 orders of magnitude (Wang et al. 2005).

6.3.2 Coastal Wetlands

In the eastern part of China, wetlands are widely distributed along the coast of the Pacific Ocean. CH_4 emissions from coastal wetlands have been measured in Liaohe Delta (northeastern China), Yangtze Delta (Municipality of Shanghai) and Minjiang Delta (Fujian Province). CH_4 emissions were −1.0 to 2.7 mg CH_4 m^{-2} h^{-1} from reed wetlands in Liaohe Delta (Huang et al. 2001a), 2.1 mg CH_4 m^{-2} h^{-1} in middle flat and 0.04 mg CH_4 m^{-2} h^{-1} in low flat wetlands, respectively, in the Chongming east intertidal flat (Yang et al. 2007), and 13.1 and 12.9 mg CH_4 m^{-2} h^{-1} from *Spartina alterniflor* before flood and after ebb in Minjiang Delta, respectively (Tong et al. 2008).

6.3.3 Lakes

In China, CH_4 emissions from lakes have been measured in field since 2003. A 2-year measurement during two growing seasons from April to October 2003 and 2004 indicated that the mean CH_4 emission rate from *Potamogeton pectinatus* (submerged macrophyte) growing zones was 3.4 ± 1.6 mg CH_4 m^{-2} h^{-1}, 78% lower than that from *Phragmites australis* (emergent macrophyte) in Wuliangsu Lake, Inner Mongolia (Duan et al. 2007). The average CH_4 emission was 23.3 ± 18.6 mg CH_4 m^{-2} h^{-1} with strong seasonal variation in Donghu lake, Hubei Province (Xing et al. 2005), and 0.6 ± 0.4, 0.7 ± 0.4, and 1.3 ± 0.6 mg CH_4 m^{-2} h^{-1} from three replicates, respectively, in Biandantang lake, Hubei Province (Xing et al. 2006). In hypereutrophic Meiliang Bay of Taihu Lake, measurements showed that the macrophyte-covered infralittoral zones were the "hotspots" of CH_4 emission with large temporal variations for CH_4 fluxes, ranging from about −2 to 131 mg CH_4 m^{-2} h^{-1} in the macrophyte-covered littoral zone (Wang et al. 2006a).

6.3.4 Reservoirs

The debate about the effect of energy generation by hydropower on climate is related to emissions of CH_4 from reservoir constructed for electricity production. In 2008, the first reported all-year measurement of CH_4 emissions from the newly established reservoir in Ertan showed that the surface of the reservoir was a net source of CH_4 during the sampling period with a mean CH_4 flux of 0.12 ± 0.06 mg CH_4 m^{-2} h^{-1} (Zheng et al. 2011). Measurements conducted at the Three Gorge Reservoir, one of the largest reservoirs in the world, showed that CH_4 emissions from the water surface were 0.3 ± 0.1 mg CH_4 m^{-2} h^{-1} during summer (Lu et al. 2011).

6.3.5 Geographical Variation in Methane Emissions

It is difficult to depict a clear geographical distribution of CH_4 emission from wetland because the available data are relatively limited, and many factors (e.g., vegetation, water table) also influence CH_4 emissions. The two largest peatlands in China are located in Zoige Plateau and Sanjiang plain. Although there are remarkable variations in CH_4 emission in a region because of vegetation and measurement sites, available results show that CH_4 emissions in Sanjiang peatland are higher than those in Zoige peatland (Table 6.1). The simultaneous measurements showed that the average value of CH_4 emissions from Sanjiang peatland was 4.7 times more than that of Zoige peatlands due to different water and temperature regimes (Wang et al. 2003). In coastal wetlands, CH_4 emissions were the lowest in the northern region (e.g., Liaohe Delta) and higher in the southern region (e.g., Mingjiang Delta), which may be related to temperature differences. The annual air temperature were 8°C in Liaohe Delta (Chen et al. 2011) and 19.7°C in Minjiang Delta (Li et al. 2009), respectively. In lakes, higher CH_4 emissions were reported in northwestern China (e.g., Wuliangsu Lake) than in other regions which may be mainly driven by higher vegetation biomass (18.6–25.5 Mg ha^{-1} for reed, Duan et al. 2007) in the Wuliangsu Lake.

6.4 Temporal Variation in Methane Emissions

6.4.1 Diel Variation

Understanding the diel variation in CH_4 emissions is of importance to designing the temporal sampling strategy for estimating the amount of CH_4 emissions and to investigating environmental variables controlling CH_4 emission. Although diel variations of CH_4 emissions have been observed at various sites, no consistent pattern has emerged. CH_4 emissions from *Corex lasiocarpa* and *Deyeuxia angustifolia* freshwater marsh in Sanjiang Plains exhibited a unique peak at 9:00 in the morning, lagging about 4 h behind sunrise and the lowest CH_4 emission at 0:00 midnight about 5 h after sunset. Both plant species showed higher daytime emissions than during nighttime. However, the diel variation in CH_4 emission was much smaller at the *D. angustifolia* than the *C. lasiocarpa* site (Ding et al. 2004c). Correlation analysis showed that CH_4 emissions were not significantly related to air temperature and soil pore water temperature at 0–35 cm depth (Ding et al. 2004c).

In peatland of Zoige, there was an apparent diurnal variation pattern in CH_4 emission with one minor peak at 06:00 in the morning and a major one at 15:00 in the afternoon. The sunrise peak was consistent with a two-way transport mechanism for plants (i.e., convective at daytime and diffusive at nighttime). The afternoon peak could not be explained by diurnal variations in soil temperature, but may be attributable to changes in CH_4 oxidation and production driven by gas transport mechanisms within the plants (Chen et al. 2010b).

Field measurements have been conducted in the Inner Mongolia marshes to compare the diurnal CH_4 fluxes between summer and winter, between a sandy site and an organic site, and between a wet meadow and a waterlogged habitat at each site (Wang and Han 2005). The results indicated that during summer apparent diurnal patterns in CH_4 emissions with unique peaks at sandy sites and plant photosynthesis greatly affected the processes of CH_4 production, oxidation and transport. This resulted in a diurnal variation of CH_4 emission with a peak in the late afternoon and the lowest value immediately prior to next day's sunrise. At the OM-rich site, plant-mediated CH_4 transport together with the absence of a significant relationship between CH_4 flux and environmental variables indicated that diurnal CH_4 flux was determined by vascular plants (Wang and Han 2005).

Distinct trends were observed based on measurements of CH_4 emissions from Wuliangsu Lake as in July and August. However, significant differences in CH_4 emissions between day and night were observed which may be related to irradiation during the growing season (Duan et al. 2005). The CH_4 emissions from Pondweed-dominated plots were low during the nighttime, increased in the morning and reached a maximum primarily in the afternoon between 14:00 and 18:00 when the sediment temperature was the highest (Duan et al. 2005).

The diurnal CH_4 fluxes followed the same trends during the wet and dry seasons reported at the Ertan Reservoir (Zheng et al. 2011). The fluxes peaked during early afternoon and then gradually declined. The maximum CH_4 flux occurs during the afternoon at 14:00 in the wet season and at 16:00 during the dry season (Zheng et al. 2011).

6.4.2 Seasonal Variation

In general, CH_4 emissions are higher during summer or the growing season than those during winter or the non-growing season. The apparent seasonal pattern appears in a sinusoidal shape with the vertex around August and nadir around January for CH_4 emission at the Sanjiang Plain (Song et al. 2009). Actually, the monthly CH_4 emissions from three kinds of wetland start increasing from March or April when plant growth begins, and peaks when plant biomass reaches its maximum. CH_4 emissions during the entire growing season (April to October) account for more than 90% of the annual flux (Song et al. 2009). Although there were measurable CH_4 emissions (0.1–2.3 mg CH_4 m^{-2} h^{-1}) in the freshwater marsh during winter (November through March), it was only about 3.8%, 5.5%, and 2.2% of the annual emissions 2003, 2004, and 2005, respectively (Song et al. 2009). During the freeze–thaw period (April–June), CH_4 emissions increased significantly, and were about 31%, 21%, and 216% of the annual emissions 2003, 2004, and 2005, respectively (Yang et al. 2006b).

Similar seasonal variations in CH_4 fluxes were also apparent at reed-dominated lakes. The CH_4 flux increased from early April when the shoots began to emerge from the water, reached its maximum in summer, and then declined. The CH_4 fluxes were positively correlated with soil temperature in reed-dominated lakes with an exponential correlation. In contrast, there was only a small seasonal variation in CH_4 release from the pondweed dominated plots but the highest emission rate was observed in August (Duan et al. 2005).

In Donghu Lake, Hubei, the mean CH_4 emission rates for all stations remained low during spring, autumn and winter, while they markedly increased during summer. The CH_4 emissions were positively correlated to net primary production (NPP), chlorophyll a concentrations, and water and sediment surface temperatures in shallow lakes (Xing et al. 2005).

The seasonal variation in CH_4 emissions during wet and dry seasons were of minor importance at Ertan Reservoir as emission rates were low all year round (Zheng et al. 2011).

6.4.3 Inter-annual Variation

As most measurements of CH_4 emissions are conducted during only 1 year, the available data on multiple-year variation are inconclusive. For example, 2-year measurements of CH_4 emissions from Wuliangsu Lake showed that emissions from reed dominated wetlands was larger in 2004 than in 2003 because of higher reed biomass 2004. The flux from pondweed dominated wetlands in 2004 was lower than that 2003 due to blooming of blue algae which limited pondweed growth. Song et al. (2009) conducted measurements of CH_4 emissions in Sanjiang Plains from 2002 to 2005. The CH_4 emission varied significantly year by year. The biggest change was more than one order of magnitude. For example, the CH_4 flux from shrub swamp wetland ecosystems in 2004 was nearly 17 times that of 2005. The CH_4 flux from seasonally inundated wetland ecosystems in 2003 was about eight times that of 2005 (Song et al. 2009), which may have resulted from the lower precipitation in northern China caused by the relatively strong El Nino Southern Oscillation (ENSO) (Song et al. 2009).

6.5 Environmental Variables and Their Effects on Methane Emissions

The CH_4 emissions are depending on three processes namely production, oxidation and transport. All processes depend on variety of climate, hydrology, soil and vegetation. In order to estimate and predict CH_4 emissions from wetlands, studies about the effects of varying environmental factors are important. During the measurements of CH_4 emission from various wetlands in China, factors influencing emissions were also monitored and analyzed simultaneously. The main results are discussed in the following section.

6.5.1 Solar Radiation

Solar radiation provides nearly all energy to all living organisms. As increasing photosynthesis stimulates roots to exude more organic compounds providing substrates for methanogenic bacteria to produce CH_4, there is a close relationship between photosynthesis and solar radiation. Thus, the increase in solar radiation, especially photosynthetically active radiation (PAR), results in increasing CH_4 production. Duan et al. (2005) reported that diurnal variations in CH_4 emissions were related to irradiation, and during growing stage, higher emissions occurred during sunny days in reed dominated plots at Wuliangsu Lake. When PAR was blocked by covering (e.g., chamber covered with black clothe), CH_4 emissions declined significantly (Duan et al. 2006) assuming the chamber effect on air temperature and humidity were minor.

Diurnal changes in CH_4 flux in three types of peatlands in Haibei, Qinghai-Tibet Plateau, dominated by *Scirpus distigmaticus*, *Hippuris vulgaris* and *Potamogeton pectinatus*, respectively, were significantly correlated with changes in PAR, especially in the late season (Hirota et al. 2004). The chambers were covered with sheets of aluminum foil for light–dark experiments which suggested the presence of stomatal control of CH_4 fluxes (Hirota et al. 2004).

6.5.2 Temperature

Temperature is an important factor influencing microbial activities which may produce CH_4 in anaerobic environments. Thus, CH_4 emissions were positively and exponentially related to soil temperature in reed-dominated wetlands in Wuliangsu Lake (Duan et al. 2005), to water surface and sediment temperature in a shallow hypereutrophic subtropical lake (Xing et al. 2005), and to soil temperature in 5-cm depth during the growing season in freshwater marsh in northeast China (Song et al. 2008). To accurately quantify the sensitivity of gas fluxes to air temperature, Song et al. (2009) used Q_{10} as the index to monitor the increases in gas fluxes when the air temperature increased from 10°C to 20°C, and reported that Q_{10} for CH_4 emission from peatlands in Sangjiang Plain was 1.9–2.7, indicating global warming may stimulate CH_4 emissions from wetlands.

6.5.3 Hydrology

Methane is produced under anaerobic environment controlled by water depth or soil moisture regime. Duan et al. (2005) compared CH_4 emission rates from wetlands with different depth and found that a relatively high mean emission rate was observed in reed-dominated plots with the deepest standing water depth., Ding et al. (2002) reported that CH_4 emissions from *Carex lasiocarpa*, *Carex meyeriana* and *Deyeuxia angustifolia* marshes in the Sanjiang Plain in Northeastern China increased as standing water depth increases from 5 to 20 cm. In Yichun of Xiaoxing'anling Mountains, average CH_4 emission rates were higher with deeper water table among the forested swamps, except for *Larix gmelinii—Sphagnum* spp. Swamp (Song et al. 2009). A critical point of water table for atmospheric CH_4 source or sink in Xiaoxing'an mountains was reported to be at 34.5–30.8 cm, below which the peatlands turn into atmospheric CH_4 sinks (Song et al. 2009). In Ertan hydroelectric reservoir, shallow-water areas emitted more CH_4 than deep-water regions. The reasons are (1) higher CH_4 fluxes in shallow areas probably resulting from higher CH_4 production in the sediments due to OM inputs in the littoral zone and (2) the nutrients in runoff from the catchment apparently accelerate OM accumulation and plant productivity in the littoral zone, thereby boosting CH_4 emissions (Zheng et al. 2011).

Water table influences CH_4 emissions not only directly by changing the anaerobic environment where CH_4 is produced but also indirectly by determining the distribution of hydrophytes. For example, emergent macrophytes, *Scirpus acutus* and *Typha latifolia* grow in the shallow zone of Wuliangsu Lake while *Phragmite australis* appears in the zones where the water depth is under 2 m with higher CH_4 emissions (Duan et al. 2005). Ding et al. (2002) also reported that standing water depth determined the type of marsh plants, which governed CH_4 transport, and the amount of plant litter, which resulted in the difference in labile organic organic (LOC) for methanogenesis among marshes in Sanjiang Plains.

6.5.4 Vegetation

In recent years, vegetation has been recognized as key factor influencing spatial variation in CH_4 emissions. Vegetation properties, such as density, life form, and species composition, affect three processes—production, consumption, and transport of CH_4—and, thus, interfere with emissions from wetlands. Through litter production and root exudates, plants can provide substrates for methanogenesis. Aquatic plants provide gas conduits which transport both CH_4 from waterlogged soils to the atmosphere and oxygen from the atmosphere to the soil. The measurements in Wuliangsu Lake indicated that mean CH_4 emissions from submerged plant (*Potamogeton pectinatus*) growing zones were 2.2 mg CH_4 m^{-2} h^{-1} which was 86% lower than that from emergent macrophyte (*Phragmites australis*) growing zones (Duan et al. 2005). In Liaohe Delta, CH_4 emissions from the wetland dominated with reeds were 15 times that without reed (Huang et al. 2001a). The emergent plants can transport CH_4 to the atmosphere directly whereas CH_4 emission from the submerged plants was only distributed to the water column, and its emission was controlled by ebullition and diffusion from plants and sediment surfaces.

In Sanjiang plains, CH_4 emissions from *Carex lasiocarpa* dominated wetlands was significantly higher than those of *Deyeuxia angustifolia* dominated and shrub wetlands (Song et al. 2009). Measurements of CH_4 emissions in six different littoral zones of Huahu Lake on the Qinghai-Tibetan Plateau in the peak growing season 2006 and 2007 indicated that emergent plant zones (*Hippuris vulgaris* and *Glyceria maxima* stands) had the highest CH_4 flux rates. The CH_4 emission in the floating mat zone of *Carex muliensis* was significantly lower than those of emergent plant zones. CH_4 fluxes in the floating leaved zone of *Polygonum amphibium* and bare lakeshore showed no significant difference and was low, only higher than that of a littoral meadow (*Kobresia tibetica*) (Chen et al. 2009). In Haibei of Qinghai-Tibet Plateau, there were three emergent-plant zones (*Hippuris*-dominated; Scirpus-dominated; and *Carex*-dominated) and one submerged-plant zone (*Potamogeton*-dominated). The lowest CH_4 emission (seasonal mean 1.4 mg CH_4 m^{-2} h^{-1}) was observed in the *Potamogeton* dominated zone, which occupied about 74% of the total wetland area . The highest CH_4 flux (seasonal mean 8.9 mg CH_4 m^{-2} h^{-1}) was observed in the *Hippuris*-dominated zone, also the second deepest water area (Hirota et al. 2004). In Yichun of Xiaoxing'an mountain, varying CH_4 emission rates were reported among vegetation from CH_4 sink to source during the growing season (Song et al. 2009)

6.5.5 Other Factors

Human activities influence CH_4 emissions from wetlands in many ways. The most common is to drain wetlands or convert them into cropland which may significantly reduce CH_4 emissions and even turn the converted land into a CH_4 sink. Other practices also influence CH_4 emissions. For example, Hirota et al. (2005) investigated the effect of grazing on CH_4 emissions in Haibei at the Qinghai-Tibet Plateau. After grazing for about 3 months, ecosystem CH_4 emissions were remarkably greater than under non-grazing conditions. This increased CH_4 emissions resulted mainly from the flux increase within the gas-transport system of grazed plants (Hirota et al. 2005). Zhang et al. (2007) reported that exogenous applied N significantly enhanced the mean seasonal CH_4 emission rates by 181% (application rate 6 gN m^{-2}), 254% (12 gN m^{-2}) and 155% (24 gN m^{-2}), respectively (Zhang et al. 2007).

It is evident from this study that processes controlling CH_4 emissions from wetlands are highly complex and vary with climate, vegetation and human activities. Thus, no single environmental variable can explained variations in CH_4 emissions entirely. Process-based biogeochemical models are therefore helpful tools. Huang et al. (2010) developed a CH_4 MOD model which had been validated by CH_4 emissions data from peatlands in Sanjiang Plain and Zoige. The model consists of 20 basic functions. Methanogenic substrates derived from plants via root exudates, from the decomposition of above- and below-ground plant litter and from SOM are simulated. Plant growth and senescence, CH_4 production and emission, and the influence of environmental factors on CH_4 production and emission are simulated. However, this model has not yet been used to estimate CH_4 emissions from wetlands on a national scale.

6.6 Regional and National Estimates of Methane Emission

The regional and national CH_4 emissions are generally estimated based on limited numbers of field measurements by summing-up the product of area and its CH_4 emission for each habitat. Up to now, the CH_4 emissions have been estimated for Qinghai-Tibet Plateau, Zoige peatland, Sanjiang peatland, Wuliangsu Lake, and Donghu Lake (Table 6.2). Due to significant variations in CH_4 emissions among different habitats, regional CH_4 emissions are not attributed proportionally to land area of each source. For example, 85% of the CH_4 emissions were from a *Phragmites australis* community dominated area with less than 50% of total area cover of Wuliangsu Lake (Duan et al. 2007). Although the riparian mires cover only 0.4% of the Xilin River basin, their CH_4 emissions were about half of the amount of CH_4 consumed by the upland grassland which covered 89.7% of the land (Wang et al. 2006b).

The annual CH_4 emissions from peatlands in China were estimated to be 1.8 Tg CH_4 year^{-1} (Ding et al. 2005) and 1.7 Tg CH_4 year^{-1} (Ding and Cai 2007) based on measurements at peatlands in Qinghai-Tibet Highland and the freshwater marsh in Sanjiang Plain 2001 and 2002, and measurements 1995–2004, respectively. Jin et al. (1999) reported CH_4 emissions of 2.0 Tg CH_4 year^{-1} in Qinghai-Tibet Plateau based on field measurement at Huashixia, Maduo, Qinghai. Based on the data in this study, the annual CH_4 emission is 10.5 Tg CH_4 year^{-1} with a range between 2.5 and 33.1 Tg CH_4 year^{-1} in China (Table 6.3), of which 69% occurs in peatlands, 8% in coast wetlands, 22% in lakes, 0.4% in reservoirs and 1.4% in rivers, respectively.

Table 6.2 Regional and national estimates of CH_4 emission from wetland in China

Wetland	Area (km^2)	CH_4 emission (Mg year^{-1})	Reference
Sanjiang	11192.9	0.9607	Cui (1997)
Northeastern China	2.4×10^4	1.36	Huang et al. (2010)
Zoige Peatlands	4,038	0.052	Wang et al. (2002)
Qinghai-Tibet Plateau	13.3×10^4	0.79	Jin et al. (1999)
Wuliangsu Lake	293	0.0012	Duan et al. (2005)
Donghu Lake	19.03	0.00016	Xing et al. (2005)
Country (peatlands)	9.4×10^4	1.76	Ding et al. (2004b)
	9.4×10^4	1.7	Ding and Cai (2007)
Country	25×10^4	2.0	Jin et al. (1999)
	38.4×10^4	9.97	This study

Table 6.3 CH_4 emission from different wetlands in China

Type	CH_4 emission ($mg\ m^{-2}\ h^{-1}$) Mean	Max	Min	Area ($10^3\ km^2$)	CH_4 emission ($Tg\ year^{-1}$) Mean	Max	Min
Peatland	6.00	15.63	1.02	137.00	7.20	18.76	1.22
Coast wetland	1.64	2.35	0.52	59.40	0.86	1.22	0.27
Lake	3.09	9.68	0.88	83.52	2.26	7.08	0.64
Reservoir	0.21	0.31	0.12	22.56	0.04	0.06	0.02
River	0.21			82.02	0.15	6.00	0.32
Sum				384.50	10.51	33.13	2.48

Note: The CH_4 emission for river is deployed the same figure as that for reservoir because of lack of field measurement data

6.7 Conclusions and Outlook

Natural wetlands have received increased attention because of their importance for ecosystem services and CH_4 emissions. In China, the total area of wetlands is about 38 Mha accounting for about 4% of the total land area. Natural wetlands include peatlands (36%), coastal wetlands (15%), river (21%), lakes (22%) and reservoirs (7%). Human activities and the development of settlements have drastically reduced the wetland area. A recent estimate indicated that 33% of wetlands were lost between 1978 and 2008, and land reclamation accounted for more than 70% of this loss.

CH_4 emissions from natural wetlands are measured in Sanjiang Plain since 1995 and in Tibet-Qinghai Plateau since 1996–1997. Based on previous studies, measurements of CH_4 emissions were carried out at 16 sites across the country, i.e., six peatlands, three coastal wetlands, five lakes and two reservoirs. The mean CH_4 emission rates were 6.0 (range 1.0–15.6) mg CH_4 $m^{-2}\ h^{-1}$ for peatlands, 1.6 (0.5–2.4) mg CH_4 $m^{-2}\ h^{-1}$ for coastal wetlands, 3.1 (0.9–9.7) mg CH_4 $m^{-2}\ h^{-1}$ for lakes, and 0.2 (0.1–0.3) mg CH_4 $m^{-2}\ h^{-1}$ for reservoirs.

Although diel variations of CH_4 emissions have been observed at many sites, there is no consistent pattern. In general, CH_4 emissions are higher in summer or during the growing season than that in winter or non-growing seasons. Because of the difference in site characteristics, seasonal patterns of CH_4 emissions change with vegetation, freeze-thaw periods, and climate types. Inter-annual variations of CH_4 emissions are significant because of changes in vegetation growth and precipitation.

CH_4 emission are generally determined by three processes, i.e., CH_4 production, oxidation and transport. The processes depend on changes in climate, hydrology, soil properties and vegetation type and production. CH_4 emissions increase with increase in solar radiation, especially at diurnal scales. Further, CH_4 emissions were positively and exponentially correlated with air or soil temperature in many wetlands. Water table influences CH_4 emissions not only directly by changing the anaerobic environment where CH_4 is produced but also indirectly by determining the distribution of hydrophytes. Vegetation has been recognized as a key factor affecting spatial variation in CH_4 flux by providing methanogenic substrates and conduits for CH_4 transport. In addition, human activities influence CH_4 emissions from wetlands in many ways, such as reductions in emission by wetland drainage and intensification of grazing.

The regional or national CH_4 emissions are estimated based on a limited number of field measurements which indicate that regional CH_4 emissions are not attributed proportionally to land area of each source. The annual CH_4 emission rates from natural wetlands in China are 10.5 Tg CH_4 year^{-1}, accounting for 7% (145 Tg CH_4 year^{-1}) of global emissions from wetlands, and is 171% (6.1 Tg CH_4 year^{-1}) of the CH_4 emissions from rice paddies in China.

Field measurements are essential for more accurate estimations of CH_4 emissions from natural wetlands of China. To date, there are three measurement methods available, i.e., the chamber method, the air-water interface method and the eddy covariance method. Up to now, only chamber methods have been used in China. However, chamber effects may influence temperature, photosynthesis and transpiration. Although black chambers coated by isolative material can prevent the temperature rise within the chamber, the reduction in photosynthesis will reduce root exudation which supplies important substrates to methanogenic bacteria. In addition, CH_4 emission is monitored discontinuously, and CH_4 fluxes at night or during rapid changes in water levels are generally not monitored. Also, very few measurements of CH_4 emissions have been carried out during the winter season.

To reduce the uncertainty in estimating CH_4 emissions from wetlands, process-based biogeochemical models are valuable tools. The models can capture the main processes controlling CH_4 emissions with regard to high temporal and spatial resolution of environmental factors. There are some biogeochemical models employed to estimate CH_4 emissions from wetlands, e.g., Wetland-DNDC (Zhang et al. 2002), PEATLAND (van Huissteden et al. 2006). With the increasing number of field experiments conducted in China, more critical parameters embedded in biogeochemical models will be available, which will promote the development of biogeochemical models specific to China's wetland. CH_4MOD developed by Huang et al. (2010) has been verified in China and will soon be used to estimate CH_4 emissions.

It is crucial to quantify CH_4 emissions from different types of wetlands and to determine the driving factors. CH_4 emission into the atmosphere is the result of production, oxidation and transportation driven by temperature, hydrology, and vegetation. These driving factors vary by location. The resolution of vegetation distribution is critical to improve accurate estimations of CH_4 emissions because there are significant differences in CH_4 emissions among vegetation covers and their distribution depending on hydrology, which is also important for CH_4 emissions. The relationship between CH_4 emission and NPP is important for estimating CH_4 emissions from remote sensing data which are used for large scale NPP estimations.

It is well known that water depth is a very important factor controlling CH_4 emission, and CH_4 source and sinks greatly depend on water level which varies seasonally, and even hourly. Unfortunately, the information of water depth is generally not available because of its considerable temporal and spatial variation. Continuous monitoring and remote sensing data would be valuable for accurately measuring water depth.

Climate change is one of the most threatening forces that will affect future CH_4 emissions because climate influences temperature, evapotranspiration, water level, distribution of aqueous vegetation, and growing season length. The recent drought affecting wetlands and lakes on the Qinghai-Tibetan Plateau caused a change in vegetation zones from submerged plants to emerged plants as result of a rapid decrease in water depth. This change in plant community composition may decrease CH_4 fluxes from alpine wetlands in the future. The prolonged drying of many wetland regions in the world as a result of drainage and climate change may have resulted in a reduction in CH_4 emissions . Thus, it is also necessary to investigate CH_4 emissions under a future climate change.

Although occupying only about 3% of the world's land area, wetlands provide many important ecosystem services. Costanza et al. (1997) estimated the total global value of services provided by coastal areas and wetland ecosystems to be 15.5 trillion US$ per year or 46% of the total value of services global ecosystems provide. The major functions of wetlands are water storage and groundwater recharge, flood control, shoreline stabilization, water quality control, moderating climate and community structure, biodiversity and wildlife support. Unfortunately, wetlands have historically been the in the center of large-scale drainage efforts for agricultural and real estate developments, or flooding for creating recreational lakes. Drainage to cropland will turn wetlands from a CH_4 source into sink. The important issue is how to deal with tradeoffs between ecosystem service and mitigation of wetland CH_4 in decision making.

Reducing CH_4 emissions from natural wetland is one of the important measures to mitigate climate change. Based on the available investigations, the following measures may be considered: (1) wetlands can be flooded by intermittent management which will mitigate CH_4 emission as well (Huang et al. 2001b). (2) Invasion by vascular plants such as weed (for lowering water table) and eutrophication should be prevented because they stimulate CH_4 emission. (3) Climate change should be mitigated because the rise of temperature will stimulate CH_4 emissions.

Acknowledgements This study is partly supported by the Chinese Academy of Sciences for Strategic Priority Research Program (XDA005050602 and XDA05060102). Thank for support from Institute for Advanced Sustainability Studies e.V. for participating the Recarbonization of Biosphere Workshop in Potsdam, Germany. Two anonymous reviewers should be appreciated for their very helpful improvement of this manuscript.

References

Ai Y, Wang X, Gong X et al (2009) Evolutions and seasonal characteristics of methane concentrations on urban lakes in Beijing. Urban Environ Urban Ecol 22(6):1–5

Bastviken D, Tranvik LJ, Downing JA et al (2011) Freshwater methane emissions offset the continental carbon sink. Science 311(6013):50

Bridgham SD, Megonigal JP, Keller JK et al (2006) The carbon balance of North American wetlands. Wetlands 26(4):889–916

Chen Y, Prinn R (2005) Atmospheric modeling of high- and low-frequency methane observations: Importance of interannually varying transport. J Geophys Res 110:D10303. doi:10.1029/2004JD005542

Chen Y, Prinn R (2006) Estimation of atmospheric methane emissions between 1996 and 2001 using a three-dimensional global chemical transport model. J Geophys Res 111:D10307. doi:10.1029/2005JD006058

Chen H, Wu N, Yao S et al (2009) High methane emissions from a littoral zone on the Qinghai-Tibetan Plateau. Atmos Environ 43:4995–5000

Chen H, Wu N, Wang Y et al (2010a) Methane fluxes from Alpine wetlands of Zoige Plateau in relation to water regime and vegetation under two scales. Water Air Soil Pollut 217(1–4):173–183

Chen H, Wu N, Yao S et al (2010b) Diurnal variation of methane emissions from an alpine wetland on the eastern edge of Qinghai-Tibetan plateau. Environ Monit Assess 164(1–4):21–28

Chen S, Ma A, Li Z (2011) Landscape pattern changes and its driving mechanism of the wetland in Liaohe Delta. Periodical Ocean Univ China 41:081–087

Christensen TR, Ekberg A, Ström L et al (2003) Factors controlling large scale variations in methane emissions from wetlands. Geophys Res Lett 30(7):1414. doi:10.1029/2002GL016848

Costanza R, D'Arge R, Groot R et al (1997) The value of the world's ecosystem services and natural capital. Nature 387:253–260. doi:10.1038/387253a0

Cui B (1997) The rule and estimation of CH_4 emission on the marshes of Sanjiang Plain. Sci Geogr Sin 17:93–95

Ding W, Cai Z (2007) Methane emission from Natural wetlands in China: summary of years 1995–2004 studies. Pedosphere 17(4):475–486

Ding W, Cai Z, Tsuruta H et al (2002) Effect of standing water depth on methane emissions from freshwater marshes in Northeast China. Atmos Environ 36:5149–5157

Ding W, Cai Z, Wang D (2004a) Preliminary budget of methane emissions from natural wetlands in China. AE Int-Asia 38:751–759

Ding W, Cai Z, Tsuruta H (2004b) Summertime variation of methane oxidation in the rhizophere of a *Carex* dominated freshwater marsh. Atmos Environ 38:4165–4173

Ding W, Cai Z, Tsuruta H (2004c) Diel variation in methane emissions from the stands of *Carex lasiocarpa* and Deyeuxia angustifolia in a cool temperate freshwater marsh. Atmos Environ 38:181–188

Ding W, Cai Z, Tsuruta H (2005) Plant species effects on methane emissions from freshwater marshes. Atmos Environ 39:3199–3207

Duan X, Wang X, Mu Y et al (2005) Seasonal and diurnal variations in methane emissions from Wuliangsu Lake in arid regions of China. Atmos Environ 39:4479–4487

Duan X, Wang X, Ouyang Z (2006) Plant-mediated CH_4 emission from a *Phragmltes*-dominated wetland in a arid region, China. J Freshwater Ecol 21(1):139–145

Duan X, Wang X, Chen L (2007) Methane emission from aquatic vegetation zones of Wuliangsu lake, Inner Mongolia. Environ Sci 28(3):455–459

Fletcher SEM, Tans PP, Bruhwiler LM et al (2004) CH_4 sources estimated from atmospheric observations of CH_4 and its $^{13}C/^{12}C$ isotopic ratios: 2. Inverse modeling of CH_4 fluxes from geographical regions. Global Biogeochem Cycles 18. doi:10.1029/2004GB002224

Gong P, Niu Z, Chen X et al (2010) China's wetland change (1990–2000) determined by remote sensing. Sci China Earth Sci 53:1–7

Hao Q, Wang Y, Song C et al (2004) Study of CH_4 emission from wetlands in Sanjiang Plain. J Soil Water Conserv 18:194–199

Hao Y, Cui X, Wang Y et al (2011) Predominance of precipitation and temperature controls on ecosystem CO_2 exchange in Zoige Alpine wetlands of southwest China. Wetlands 31(2):413–422. doi:10.1007/s13157-011-151-1

Hirota M, Tang Y, Hu Q et al (2004) Methane emissions from different vegetation zones in a Qinghai-Tibetan Plateau wetland. Soil Biol Biochem 36:737–748

Hirota M, Tang Y, Hu Q et al (2005) The potential importance of grazing to the fluxes of carbon dioxide and methane in an alpine wetland on the Qinghai-Tibetan Plateau. Atmos Environ 39:5255–5259

Hou W, Kuang W, Zhang S et al (2006) Analysis of cultivated land reclamation process and the ecological effects in north of Sanjiang Plain from 1950s. Ecol Environ 15(4):752–756

Hu Q, Wu Q, Li D et al (2005) Comparative study on methane emissions from alpine grasslands with different soil water content. Chin J Ecol 24(2):118–122

Huang G, Xiao D, Li Y et al (2001a) CH_4 emissions from the reed wetland. Acta Ecol Sin 21(9):1494–1498

Huang G, Li Y, Chen G et al (2001b) Influence of environmental factors on CH_4 emission from reed wetland. Environ Sci 22(1):1–5

Huang Y, Sun W, Zhang W et al (2010) Marshland conversion to cropland in northeast China from 1950 to 2000 reduced the greenhouse effect. Global Change Biol 16:680–695

IPCC (2007) Climate change 2007: the physical science basis. In: Solomon S, Manning M, Chen Z et al (eds) Contribution of working group I to the fourth assessment report of the Intergovernmental Panel on Climate Change. Cambridge University Press, Cambridge/New York

Jin H, Wu J, Cheng GD et al (1999) Methane emissions from wetlands on the Qinghai-Tibet Plateau. Chin Sci Bull 44(24):2282–2286. doi:10.1007/BF02885940

Kayranli B, Scholz M, Mustafa A et al (2010) Carbon storage and fluxes within freshwater wetlands: a critical review. Wetlands 30(1):111–124

Li L, Chen J, Chen H et al (2009) Climate change of Minjiang River Estuary Wetland and its impact on eco-environment. Chin Agric Sci Bull 25:245–249

Liang H, Zhao G, Chen K (1999) Wetland vegetation in China. Science Press, Beijing

Liu X, Ma X (2000) Influence of large-scale reclamation on natural environment and regional environmental protection in the Sanjiang Plain. Sci Geogr Sin 1:14–19

Lu F, Yang L, Wang X et al (2011) Preliminary report on methane emissions from the Three Gorges Reservoir in the summer drainage period. J Environ Sci 23(12):2029–2033

Mitra S, Wassmann R, Vlek PLG (2005) An appraisal of global wetland area and its organic carbon stock. Curr Sci 88:25–35

Moore TR, Roulet NT, Waddington JM (1998) Uncertainty in predicting the effect of climatic change on the carbon cycling of Canadian peatlands. Clim Chang 40:229–245

Niu Z, Gong P, Cheng X et al (2009) Geographical characteristics of China's wetlands derived from remotely sensed data. Sci China Ser D 52(6):723–738

Niu Z, Zhang H, Gong P (2011) More protection for China's wetlands. Nature 471:305

Qiu J (2011) China faces up to 'terrible' state of its ecosystems. Nature 471:19

Saarnio S, Winiwarter W, Leitao J (2009) Methane release from wetlands and watercourse in Europe. Atmos Environ 43:1421–1429

Song C, Zhang J, Wang Y (2008) Emission of CO_2, CH_4 and N_2O from freshwater marsh in northeast of China. Atmos Environ 88:428–436

Song C, Xu X, Tian H et al (2009) Ecosystem-atmosphere exchange of CH_4 and N_2O and ecosystem respiration in wetlands in the Sanjiang Plain, Northwastern China. Global Change Biol 15:672–705

Sun X, Mu C, Feng D et al (2009a) Effects of wetland draining for forestation on methane emissions in Xiaoxing'an mountain, Northeast China. Acta Ecol Sin 29(8):4251–4259

Sun X, Mu C, Shi L et al (2009b) Methane emission from forested swamps in Xiaoxing'an mountain, Northeastern China. Chin J Plant Ecol 33(3):535–545

Tong C, Yan Z, Wang W et al (2008) Methane flux from invasive species (*Spartina alterniflora*) and influencing factors in the Min River Estuary. Sci Geogr Sin 28(6):826–832

van Huissteden J, van den Bos R, Alvarez IM (2006) Modeling the effect of water-table management on CO_2 and CH_4 fluxes from peat soils. Neth J Geosci 85:3–18

Wang Z, Han X (2005) Diurnal variation in methane emissions in relation to plants and environmental variables in the Inner Mongolia marshes. Atmos Environ 39:6295–6305

Wang D, Lu X, Ding W et al (2002) Methane emission from marshes in Zoige Plateau. Adv Earth Sci 17(6):877–880

Wang D, Ding W, Wang Y (2003) Influence of major environmental factors on difference of methane emission from Zoige Plateau and Sanjiang Plain wetlands. Wetland Sci 1(1):63–67

Wang Z, Han X, Li L et al (2005) Methane emission from small wetlands and implications for semiarid region budgets. J Geophys Res 110:D13304
Wang H, Lu J, Wand W et al (2006a) Methane fluxes from the littoral zone of hypertrophic Taihu Lake. China J Geophys Res 111:D17109
Wang Z, Han X, Li L (2006b) Methane emission patches in riparian marshes of the Inner Mongolia. Atmos Environ 40(28):5528–5532
Weaver A (2011) Toward the second commitment period of the Kyoto Protocol. Science 332:795–796
Wojick DE (1999) Carbon storage in soils: The ultimate no-regrets policy? A report to greening earth society. http://www.greeningearthsociety.org/pdf/carbon.pdf.
Wuebbles DJ, Hayhoe K (2002). Atmospheric methane and global change. Ear-Sci Rev 57(3–4):177–210
Xing Y, Xie P, Yang H et al (2005) Methane and carbon dioxide fluxes from a shallow hypereutrophic subtropical lake in China. Atmos Environ 39:5532–5540
Xing Y, Xie P, Yang H et al (2006) The change of gaseous carbon fluxes following the switch of dominant producers from macrophytes to algae in a shallow subtropical lake of China. Atmos Environ 40:8034–8043
Yang J, Yu J, Liu J et al (2004) NO_2 and CH_4 fluxes in an island forest in wetland, Sanjiang Plain. Ecol Environ 13(4):476–479
Yang W, Song C, Zhang J (2006a) Dynamics of methane emissions from a freshwater marsh of northeast China. Sci Total Environ 371:286–292
Yang W, Song C, Luan Z et al (2006b) Methane emission from freshwater marshes in Sanjiang Plain and its affecting factors in arid year. Chin J Ecol 25(4):423–427
Yang H, Wang D, Chen Z et al (2007) Seasonal variation of CH_4 emission from Chongming east intertidal flat of Yangtze River Estuary. Sci Geogr Sin 27(3):408–413
Yang L, Lu F, Wang X et al. (2012), Methane emissions from the drawdown area of the Three Gorges Reservoir, J Geophys Res (submitted)
Zhang Y, Li C, Trttin CC et al (2002) An integrated model of soil, hydrology, and vegetation for carbon dynamics in wetland ecosystems. Glob Biogeochem Cycles 16:1061
Zhang L, Song C, Wang D et al (2007) The variation of methane emission from freshwater marshes and response to the exogenous N in Sanjiang plain northeast China. Atmos Environ 41:4063–4072
Zhang X, Li D, Pan G et al (2008) Issues on conservation of wetland soil C stock and global climate change of China. Adv Climate Change Res 4:202–208
Zhao Q, Gao J (2007) Ecosystem services of wetlands and their delineation in China. Chin J Eco-Agric 15:1–4
Zheng H, Zhao X, Zhao T et al (2011) Spatial-temporal variations of methane emissions from the Ertan hydroelectric reservoir in southwest China. Hydrol Process 25(9):1391–1396
Zhuang Q, Melack J, Walter K et al (2009) Global methane emissions from wetlands, rice paddies, and lakes. EOS 90(5):37–44

Chapter 7
Accounting More Precisely for Peat and Other Soil Carbon Resources

Hermann F. Jungkunst, Jan Paul Krüger, Felix Heitkamp, Stefan Erasmi, Sabine Fiedler, Stephan Glatzel, and Rattan Lal

Abstract In the context of "recarbonization", it is important to know where the soil C stocks are located and how much of these are prone to emission to the atmosphere. While it may appear to be a trivial question considering available global estimates and maps, yet there is a strong need to emphasize that erroneous estimates are made in assessing the global soil C stocks. Without doubt, peatlands hold the single most important soil C stock at the global scale, and these soils are mostly located in the northern latitudes between 50°N and 70°N. However, there are additional wetlands or other ecosystems which also hold potentially relevant amounts of soil C stocks. From the soil science perspective, it implies that there

H.F. Jungkunst (✉) • J.P. Krüger • F. Heitkamp
Landscape Ecology, Faculty of Geoscience and Geography, Georg August-University, Goldschmidtstr. 5, 37077 Göttingen, Germany
e-mail: hjungku@uni-goettingen.de; janpaul.krueger@gmx.de; fheitka@uni-goettingen.de

S. Erasmi
Cartography, GIS and Remote Sensing Section, Faculty of Geoscience and Geography, Georg August-University, Goldschmidtstr. 5, 37077 Göttingen, Germany
e-mail: serasmi@gwdg.de

S. Fiedler
Institute for Geography Soil Science, Johannes Gutenberg-Universität Mainz, Johann-Joachim-Becher-Weg, 2155099 Mainz
e-mail: s.fiedler@geo.uni-manz.de

S. Glatzel
Landscape Ecology and Site Evaluation, University of Rostock, Justus von Liebig Weg 6, 18059 Rostock, Germany
e-mail: stephan.glatzel@uni-rostock.de

R. Lal
Carbon Sequestration and Management Center, School of Environment and Natural Resources, The Ohio State University, 2021 Coffey Road, Kottmann Hall 422b, Cloumbus, OH 43210, USA
e-mail: lal.1@osu.edu

R. Lal et al. (eds.), *Recarbonization of the Biosphere: Ecosystems and the Global Carbon Cycle*, DOI 10.1007/978-94-007-4159-1_7,

are other hydromorphic soils, besides Histosols and potentially other soil types, also containing relevant amounts of soil C stock. Differences in scientific approaches, which include terminology, definitions, depth to which soil C is considered, and bulk density, etc., lead to different estimates of soil C stocks. Recent estimates indicate that peatlands cover only 3% of the global land surface but contain 40% of the soil C stocks to 1-m depth. Consequently, only small differences in the estimate of the land coverage lead to great differences in the soil C stock estimates. Typically peatlands, wetlands and other ecosystems rich in soil C, cover only small parts of the landscapes, and yet are not easily accounted for by any inventory or mapping attempts. With estimates presented in this chapter, hydromorphic soils, aside Histosols, add 10% soil C stock to the estimates of peatland's Histosols. Additionally, non hydromorphic Podzols add another 10% to the soil C stock. Above all, soils from the steppe biome must also be considered. The soil C stock of Cryosols (frozen soil C not separated from peatlands) contain as much as 1,500 Pg C, which is as much C as the total stock estimated in world soils to 1-m depth. Thus, coordinated and substantial efforts are needed to improve the mapping of ecosystems, particularly of those which are rich in soil C stocks. One option is to improve remote sensing techniques for wetlands. These efforts must be undertaken quickly because soil C stocks are being depleted not only by the positive feedback with the climate system but also directly by land use change. The conversion of peatlands to agricultural and forestry uses is not sustainable because of the depletion C stocks, and especially not for conversion of peatlands for "biofuels" production.

Keywords Hydromorphic soils • Histosols • Fluvisols • Gleysols • Planosols • Chernozems • Phaeozem • Kastanozem • Greyzem • Podzols • Podzols • Aerobic decomposition • Anaerobic decomposition • Turnover rate • Wetlands • Methane • Soil carbon budget • Seasonally inundated • Mire • Marsh • Swamp • Fen • Bog • Temperate peatland • Tropical peatland • Biofuel • Oil palm • Biodiesel • Sphagnum • Peatland conversion • Ethanol • Prairie • Methane • Water table • Permafrost • CO_2-equivalent • Peatland distribution • Land cover • Land use change • Abiotic • Anoxic sites • Carbon sequestration

Abbreviations

C	carbon
DOC	dissolved organic C
GCC	global carbon cycle
GLS	Global land cover
GLCC	Global land cover characteristics
GHGs	greenhouse gases
LCCS	land cover classification system
Mha	million ha

OM	organic matter
SOC	soil organic carbon
SOM	soil organic matter
GWP	global warming potential

7.1 Introduction

There are myriads of articles and book chapters on the importance of peat for global and local ecosystem carbon (C) budgeting (see review by Limpens et al. 2008). Both large C pools and biogeochemical processes in peatland are of utmost importance for the climate system. Feedbacks of releasing or storing atmospheric C are possible (Davidson and Janssens 2006). The rationale for writing this chapter lies in the fact that there are some important aspects that still need to be highlighted more intensively. The focus on solely peat and may be even the permafrost does not account for all important soil C recourses. There may be other organic rich soils which have not yet been studied more thoroughly e.g. other hydromorphic (or wetland) soils, Podzols and Chernozems or similar soils from the steppe. Last but not the least a book on "Recarbonization of the Biosphere – Ecosystems and the Global Carbon Cycle" would be incomplete without a chapter on peat.

Peat is an organic material formed *in situ* from the remains of plants (and animals) under anaerobic conditions, and with a minimal thickness of 30 cm (Joosten and Clarke 2002) to 40 cm (National Wetlands Working Group 1997) above the mineral horizon. Thus, peatland refers to a peat covered landscape (Rydin and Jeglum 2006). All peats are formed in wetlands, but not all wetlands are associated with peat. Wetlands are characterized by saturated soil conditions or standing water (inundation), hydromorphic soils of a high organic matter (OM) content, and hydrophytic (water loving) vegetation. A water body without hydrophytic vegetation growth is a pond or a lake, but not a wetland, because not all wetlands have the conditions to form peat (Rydin and Jeglum 2006).

There are many **wetland definitions**, which can differ largely from one to another even though it appears to be a rather straight forward expression. A wetland, as suggested by the nature of the name, consists of two interacting natural media: water and soil (Richardson et al. 2001). Generally, wetlands experience water saturation for a sufficient duration of time during the year and usually are defined from "botanical" aspects. Wetlands are permanently or seasonally inundated or water saturated areas with a vegetation which is adapted to water saturated soil conditions (Joosten and Clarke 2002). Aselmann and Crutzen (1989) divided wetlands into six categories: bogs, fens, swamps, marshes, floodplains and shallow lakes. According to the Ramsar Convention, wetlands are areas of marsh, fen, peatland, or water, whether natural or artificial, permanent or temporary, with water that is static or flowing, fresh, brackish, or salt (Ramsar Information Bureau 1971). Mitsch and Gosselink (2007) divided wetlands into two main groups, coastal and inland wetlands.

Table 7.1 Types of wetlands related to peat (Adapted from Tiner 1998, 1999; Rydin and Jeglum 2006; Mitsch and Gosselink 2007)

Type	Description
Mire	A wetland or wet spongy earth, and dominated by living peat-forming plants
Marsh	A wetland subject to continuous or frequent floods, with standing or flowing water, and with or without peat
Swamp	A wetland flooded under a shallow depth of water, with swamp-woody forests and shrub forests (transitional), and comprising fresh, brackish or sea water
Fen	A wetland with either shallow depth of standing water or water table just beneath the soil surface. Fens are associated with peat depth often >40 cm, and are two types basin fen (topogenous) and sloping fen (soligenous).
Bog	A wetland (also called moor or quagmire) associated with acidic peat derived from mosses and (lichens under arctic climates), and either receiving natural precipitation (rain-fed or ombrotrophic) or acidic groundwater.

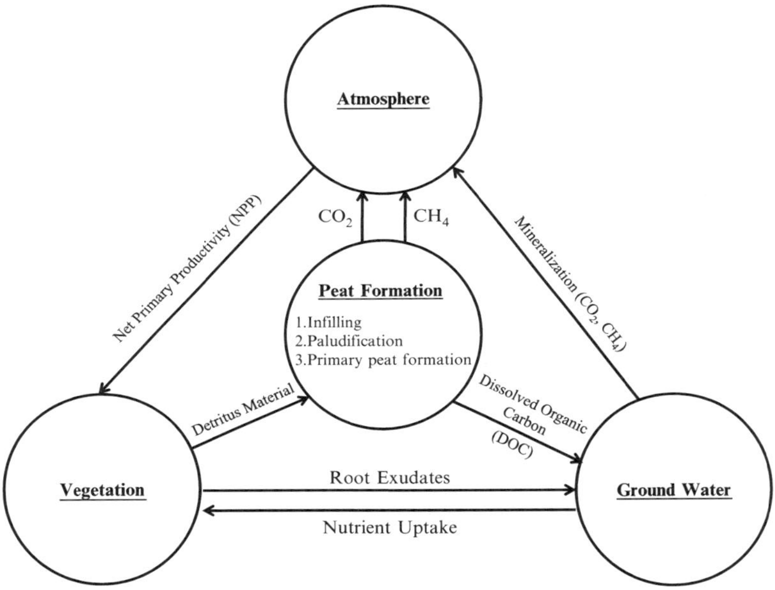

Fig. 7.1 Schematic of carbon cycling in peatland

In this context, peatlands belong to the group of inland wetlands and represent the largest area (3.5×10^6 km^2) of wetlands worldwide (5.8×10^6 km^2) (Mitsch et al. 2009). Apparently wet forest ecosystems are not particularly mentioned, but at least from mangroves, floodplain forests and forested peatlands it is clear that forests and wetlands can coincide. The most important consequence of water saturation is the oxygen depletion leading to conditions of low redox potential to which plants have to adapt. Therefore, more soil oriented definitions could be useful but not easy to map either with or without means of remote sensing. Different types of wetlands are described in Table 7.1. A schematic of C cycling in peat land indicates the C flow between peat on the one hand and atmosphere, vegetation and the groundwater on the other. Thus, world's peatlands are a major reservoir of soil C pool, and play an important role in the global carbon cycle (GCC) through interaction with vegetation or the biosphere, the hydrosphere and the atmosphere (Fig. 7.1).

7.2 Peat Formation

Undisturbed peatlands are a net sink for atmospheric CO_2. The C sink in peatlands is attributed to a high productivity. The magnitude of net C sequestration in natural peatlands ranges from 20 to 1,120 kg C ha^{-1} $year^{-1}$ (Strack 2008). Sequestration of C in peatlands happens through photosynthesis by hydrophytic vegetation and some of it being retained as the net gain in the ecosystem. The latter is estimated in C ha^{-1} $year^{-1}$ at 160–2,440 for herbs and graminoids, 500–1,800 for shrubs, and 1,220–1,880 for bryophytes (Rydin and Jeglum 2006). Because of the low rate of decomposition under anaerobic/inundated conditions, OM accumulates as partially decomposed peat. An anaerobic environment created by waterlogging/inundation reduces the rate of decomposition in peatlands, and is the principal determinant of the rate of net C sequestration. In addition to anaerobiosis, the rate of decomposition is also determined by the composition of the biomass, pH, redox potential, temperature, and can be aerobic and/or anaerobic. Composition of the biomass varies among plant species. The aerobic decomposition increases with drop in the water table, and the anaerobic with the rise in the water table and inundation. It is the anaerobic decomposition which leads to methanogenesis. Hence, natural peatlands are also net sources of CH_4. The anaerobic conditions in peatlands enhance methanogenesis with CH_4 flux of 20–330 kg C ha^{-1} $year^{-1}$ (Strack 2008). With global warming potential (GWP) of ~21, CH_4 emission is an important factor affecting the magnitude of radiative forcing by trace gases. In addition, the loss of C from wetlands also occurs in the form of dissolved organic C (DOC). The global efflux of DOC is estimated at 30–210 kg C ha^{-1} $year^{-1}$ (Strack 2008). Despite the emission of CH_4 and loss of DOC, natural peatlands are net C sink.

Under complete anaerobic conditions, decomposition is slower than the rate of C input as biomass. Thus, peat formation by accumulation of biomass through net ecosystem C gain occurs by three principle processes (Rydin and Jeglum 2006; Wieder and Vitt 2006): (i) **Infilling or terrestrialization** is the process by which new peat formation occurs on the margins of the wetland and in regions with shallow waters of a pond or a slow-flowing streams/rivers, (ii) **Paludifucation** is the process of peat formation over the ground which was previously not flooded or was less wet, and thus does not have any prior sediment deposition. Thus, these types of peat often have woody mass and stump underneath. Paludification can either happen by the rise of water table and new waterlogging of an adjacent upland, or by changes in pedogenesis. The latter is set-in-motion by alterations in hydrologic processes within the soil profile (e.g., decrease in permeability, runon, pan formation). Paludification is promoted by natural or anthropogenic factors like deforestation, fire, etc., (iii) **Primary** peat formation happens when new peat is deposited on top of the mineral soil such as on land exposed by sea level change, fresh volcanic deposit, glacial moraine, mined lands etc. The basic conditions of inundation and growth of hydrophytic vegetation are important pre-requisites.

Other processes of peat formation include **dischargelspring** and **ombrotrophication**. Dischargelspring implies creation of continuous wetland along a hillside due to the lateral flow or interflow (spring). In comparison, the term ombrotrophication implies transformation of mineral soil to peat formation by an upward development of peat surface (Rydin and Jeglum 2006). This transition can be triggered by any climatic shift.

On the basis of formation, there are two principal types of peat. **Ombrotropic bogs** involve peatlands domed above the surrounding landscape and the input of water and nutrients occurs through the atmosphere. Thus, the peat principally involves the remains of *Sphagnum* mosses (Gorham 1991). In contrast, **minerotrophic fens** are formed where water enriched in bases and nutrients percolates into the peat from the surrounding soil. In case of poor, generally sandy substrate, minerotrophic fens may be poor in nutrients, favoring vegetation generally found in ombrotrophic bogs. Minerotrophic fens primarily contain reed, cattail and sedges, and in case of poor fens, *Sphagnum* mosses and shrubs.

Humification (chemical and structural alteration of OM) is another on-going and a continuous process in peat formation. The degree of humification can be assessed by visual examination or by laboratory-based analyses (Blackford and Chambers 1993). Laboratory analysis is based on different extractability of OM with strong acid or bases and classification and fulvo acid, humic acid and humin (Waksman 1936). However, the methodological procedure can create a strong artifact in assessment of the humification (Caseldine et al. 2000), and the concept of different extractibility is currently questioned (Schmidt et al. 2011).

Boreal and sub-arctic peatlands, comprising a total C pool of 455 Pg accumulated since the post-glacial period, have on average accumulated C at the rate of 0.096 Pg $year^{-1}$ (Gorham 1991). The present rate of C sequestered in peatlands is ~0.076 Pg $year^{-1}$ (Clymo 1984; Gorham 1991). The long-term drainage, on the other hand, can exacerbate oxidation and emission of 0.0085 Pg $year^{-1}$, to which should be added the emission from peat fuel at 0.026 Pg $year^{-1}$ (Gorham 1991). In addition CH_4 emission from world peatlands is estimated at 0.046 Pg $year^{-1}$ (Gorham 1991).

Peatlands are also located in the tropics. However, tropical peat lands differ from those in temperate regions in vegetation and other physiographic characteristics (Table 7.2). With ~90% of the global area of peatlands located in northern latitudes (MacDonald et al. 2006), those in Europe and North America are among the most widely studied.

With a large C pool (~550 Pg), peatlands are an important component of the GCC (Yu et al. 2010). There is a strong and growing interest in peatlands because of the projected risks of positive feedback to climate change (Friedlingstein et al. 2006), by which peatlands may become a major source (rather than being a sink) of atmospheric CO_2 and other greenhouse gases (GHGs). Large areas of peatlands

Table 7.2 Differences in temperate vs. tropical peatlands (adapted from Rydin and Jeglum 2006)

Characteristics	Tropics	Temperate
1. Vegetation	Tropical rainforest	Sphagnum and herbaceous
2. Material	Partly decomposed, woody, and covered with leaf litter, with a high hydraulic conductivity	Non-woody and relatively more decomposed, with a distinct *catotelm* layer of low hydraulic conductivity
3. Gradient	Gentle (<0.5 m/km)	Steeper gradient
4. Depth	>25 m	<10 m
5. Duration	Since the last 5,000 years, but some as old as 40,000 years	Glacial melting
6. Uses	Source of timber	Source of peat

have been drained for conversion to agricultural and forestry lands uses. Drained peat decompose rapidly (Fargione et al. 2008), and drained peatlands are prone to fire (Strack 2008). Together with other wetlands, they also form the largest natural source of atmospheric CH_4.

Thus, the objective of this chapter is to describe the difficulties of accounting for soil C resources, offer management strategies for restoring the wetlands, and minimize the risks of emissions of GHGs from peatlands.

7.3 Ecological Characteristics of Peatlands and Other Ecosystems Rich in Soil C

Peatlands are wetland with accumulated (partially) decomposed OM (peat, i.e., material with high fibre content). According to most definitions peat has at least 30% (dry mass) of OM. The definition of peatland varies between the different classifications from 20 up to 70 cm thickness of accumulated peat layer (Joosten and Clarke 2002). Histosols, the typical peat soils, are defined by WRB (2007) by having a histic horizon (>30% OM and >10 cm thickness) of 40 cm thickness within the upper 80 cm. National definitions differ from this. For example in Germany peat soils are defined as soils with >30% dry mass of OM and >30 cm thickness (AG-Boden 2005). One of the most wide-ranging studies of northern peatlands was conducted by Gorham (1991), who used 30 cm of peat as minimum to distinguish between peat and non-peat.

Different definition of peatland or wetland affects directly estimates of organic C stored in the soils of these ecosystems. For example, 'organic soils' and peatlands are commonly found in both North Russia and North America. According to the Russian definition these contain ≥30 cm thick organic horizons (e.g., histic horizon). In contrast, in North America 'organic soils' and peatlands, by definition, have to be thicker than 40 cm (Tarnocai et al. 2009). Consequently,

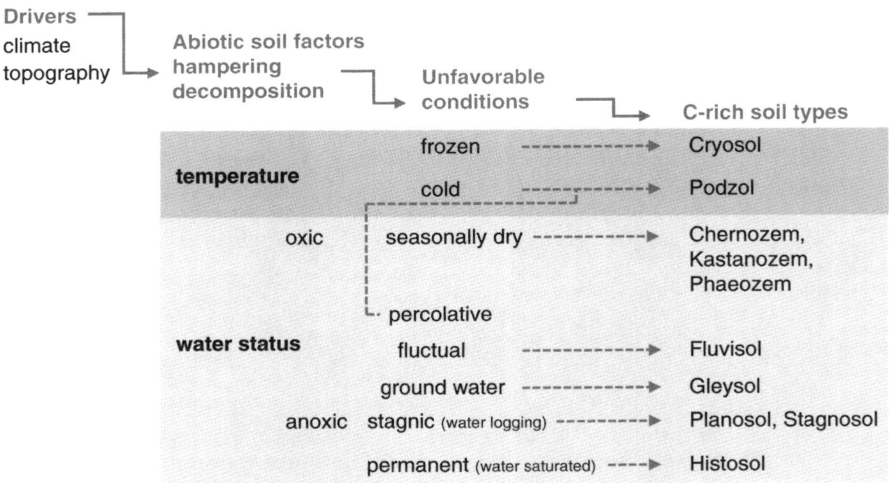

Fig. 7.2 Main pathways of organic matter preservation in soils

areas (1,162–2,730 × 10^3 km^2) and C stocks (94–215 Pg) of peatlands in published estimates vary widely (Tarnocai et al. 2009) (see also Fig. 7.2).

The story of feedbacks between soils and the climate system (Fig. 7.1) is predominately a C story. Plants assimilate C by performing photosynthesis and building up the OM. The term soil organic matter (SOM) refers to the total sum of all substances that occur in soils and contain elements in organic form (Collins and Kuehl 2001). The SOM is a complex mixture of organic residues of plant and animal origin (Kögel-Knabner 2002) that accumulate in the soil and undergo continuous transformation.

Compared to "regular" oxic sites a larger part of this C is stored below-ground at anoxic sites. The reason is that decomposition is hampered by anoxic or hypoxic conditions. Other conditions hampering decomposition are acidic, frozen or cold conditions (Fig. 7.2).

The SOM transformation predominately depends on **temperature and water status** which are the main controls on soil microbial activity. Transformation consists of a series of steps which can be associated by gaseous by-products such as nitrous oxide (N_2O) and end-products CO_2 (aerobic and anaerobic decomposition) and CH_4 (anaerobic decomposition). These gaseous C and nitrogen (N) products of SOM decomposition are the main feedbacks to the climate system. All individual steps are governed by a complex interplay of (a) soil organisms, (b) the quantity and quality of OM (Paul and Clark 1996), and (c) prevailing environmental conditions (e.g., temperature, water supply, Fig. 7.2).

In general, aerobic decomposition of OM is more efficient and more rapid than anaerobic decomposition. D'Angelo and Reddy (1999) postulated that aerobic C

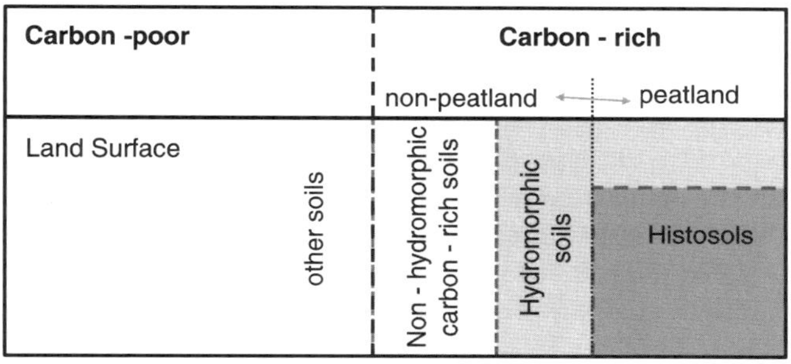

Fig. 7.3 Soil C characteristics of peatlands, wetlands and other ecosystems

transformation is about 3 times faster than an anaerobic transformation. Therefore, in **hydromorphic soils**, the accumulation rate is much higher and hence overrides the rate of decomposition. For that reason, the quantity of OM tends to increase in hydromorphic soils (e.g., undecomposed plant tissues) to a greater extent than in aerated soils. However, both thickness and darkness of H and Ah horizons are positively correlated with the exposure time of the soils to reducing conditions (Thompson and Bell 1996).

The OM undergoes a much slower transformation in acidic soils in cold and humid climates. In strongly acid soils, the growing conditions for microorganisms are poor, resulting in low levels of biological oxidation of OM (Primavesi 1984). As a result, Podzols usually have much thicker organic horizons than their less acidic counterparts (like Cambisols). The pH-values affect humus formation in two ways: hampered decomposition, and biomass production. However, decomposition is usually more affected than biomass production, resulting in accumulation of OM.

Extremes in climate trigger the turnover rates of SOM. A prominent example is the formation of soils under steppe vegetation (Chernozems, Phaeozems and Kastanozems). These biomes are dominated by cold winters and dry, hot summers. Plant productivity by these conditions is less reduced than decomposition.

Besides the lack of monitoring networks with adequate temporal and spatial resolution (Prechtel et al. 2009), one of the major obstacles to correctly account for soil C stocks and identifying the spatial hot spots are differing definitions from diverse scientific disciplines and cultures (countries, administration). The latter are the cause why maps on soil C, peatland etc. may show abrupt changes at administrative borders which are not explainable by the natural causes. This is among numerous reasons for very different regional estimates that have to rely on mapping which applies to any attempt to identify soil C hotspots and account for soil C stocks. One illustrating example is the fact that forests are frequently and solely considered to be upland and therefore non-wetland ecosystems. This simplification leads to serious bias for landscape biogeochemical budgets, e.g., methane (Fiedler et al. 2005) and soil C budgets.

7.4 Predominant Soils of Peatlands and Other Ecosystems Rich in Soil C

At the pedological perspective, wetlands build **hydromorphic soils**. These include biologically active organic as well as mineral soils in which the water contents are permanently or temporally large enough to inhibit oxygen diffusion into soil and stimulate anaerobic processes. In turn, these are characterised by accumulation of SOM. In total, hydromorphic soils cover 17.4×10^6 km^2 (based on FAO 1995) or 12.9% of the land surface, which is more than other wetland estimates. Diagnostic horizons of soils are outlined in Table 7.3.

The most prominent hydromorphic soils are **Histosols** (2.5×10^6 km^2 and 1.8%) which are formed in response to permanent water saturation. FAO (1995) estimates the global area of Histosols at about 3.3×10^6 km$^{2.}$ In theory, Histosols and peatlands should be approximately the same, however, these estimate are lower than most peatland estimates.

Histosols, by definition according to both WRB (2007) and the US soil taxonomy (2006), are soils consisting primarily of organic materials. Soil type is extremely poorly drained because the OM has a high water retention capacity.

Estimated average C contents of Histosols range between 720 and 1,250 Mg C ha^{-1} (0–100 cm depth) and 1,230–2,640 Mg C ha^{-1} (0–200 cm) (Batjes 1996). The majority of Histosols are located in the boreal, subarctic and low arctic regions of the Northern hemisphere. During the revision of soil classifications (US Soil Taxonomy 2006; WRB 2007), a new soil type defined was 'Cryosols' (WRB 2007)/ Gelisols (US Soil Taxonomy 2006) which includes some soils formerly classified as Histosols. Therefore, it is difficult to precisely differentiate between Histosols and Cryosols. By definition, **Cryosols** are soils at or below the freezing point of water for two or more years. The extent of permafrost (should be Crysols) is estimated by

Table 7.3 Some typical diagnostic horizons (WRB 2007) of carbon-rich soils with some properties important for SOC storage

Cause for OM accumulation	Diagnostic horizons (WRB 2007)	Properties
Too wet	Folic	10 cm, organic, <30 days water saturation
	Histic	10 cm, organic, >30 days water saturation
Too dry	Mollic, umbric	20 cm, >0.6[a]/2.5% SOC
	Voronic	35 cm, >1.5[a]/6% SOC
Too acidic	Sombric	Illuvial humus
	Spodic	Illuvial humus, >0.5% SOC, pH <5.9
	(Fulvic, melanic)	30 cm, >4% SOC, andic soils
Anthropogenic	Anthric	Wide range
	Hortic	20 cm, >1% SOC,
	Plaggic	20 cm, >0.6% SOC, commonly acidic

[a]If Munsell chroma <3

Tarnocai et al. (2009) to be 18.8×10^6 km^2 (equal to 16% of the total soil area) but these estimates vary and the actual extension will change with the projected climate changes.

Cryosols can store extremely high amounts of C. In a recent study of Tarnocai et al. (2009) the C storage in Cryosols is estimated between 1,400–1,700 Pg worldwide. This estimate include: (a) C stock in circumpolar soils at 496 Pg to 1-m depth, (b) 1,024 Pg to 3-m depth, and (c) the estimated C stock of 1,672 Pg in all Cryosols and peatland including thus far unconsidered stocks of 648 Pg contained in the deltaic deposits, and the Sibirian yedoma sediments below 3-m depth. Nonetheless, these estimates are large because only 1,500 Pg is the estimate C stock to 1-m depth in all world soils (Batjes 1996).

OM decomposition is generally faster under periodic than continuous water saturation. Therefore, Histosols contain the highest SOM content among all other wetland soils. All other hydromorphic soils (i.e. Stagnosols, Planosols, Gleysols, and Fluvisols) (WRB 2007), are mineral soils and formed in response to temporary water saturation. The diffusion of oxygen into hydromorphic soils is drastically curtailed, at least episodically, forcing biological and chemical processes to change synchronously (or slightly delayed) with water saturation. The water regimes transform soils from an aerobic to an anaerobic environment and vice versa leading to predominantly redoximorphic features (e.g., mottled soil matrix). These changing redox conditions are associated with highly dynamic biogeochemical cycling.

Predominant mineral hydromorphic soils are **Gleysols** (8.3×10^6 km^2, 6%) which are saturated by ground water near the surface (for long time). These soils occur at foot slopes or at landscape depressions. The prevalent water table creates a typical redox gradient in soil profiles. Soil horizons above ground water table are mostly oxic and characterized by Fe accumulation which explains their brown colour. Grey coloured horizons below the water table are predominately anoxic. The highest C accumulation (up to 150 Mg C ha^{-1} in 0–30 cm) is observed in the topsoil which increases with increase in the mean of annual water table.

Fluvisols (3.2×10^6 km^2, 2.4%) are young soils in alluvial deposits, and are widely distributed in floodplains or terraces including alluvial plains, river fans, valleys and tidal marshes which are inundated or flooded frequently. The hydrology of Fluvisols is dictated by the periodically presence of ground water and flood water. The accumulation of C in Fluvisols is influenced by import as well as export of fresh OM due to alluvial transport and wet conditions. The lateral and horizontal changes in sediment deposits results in a large variability in the C content of Fluvisols. Additionally, Fluvisols are commonly associated with Gleysols in great river deltas and Histosols in the meandering river areas. Such a physiographic combination makes soil survey maps of floodplains difficult to interpret regarding the C stocks, because they cannot adequately delineate all characteristics required to fully characterize these soils.

The most prominent mineral hydromophic soils formed on concave or flat topography and plateaus are **Planosol** and **Stagnosols.** These soils are formed by the influence of stagnant water caused by dense soil layers, which impede water percolation. These soil types (3.4×10^6 km^2 equal to 2.5% of the total land) are distributed among all continents and climatic zones with maxima in the zone between 10° and 50°N covering 35% of the total area. Global estimation of C stocks (1-m depth) of these soils range from 57 to 138 Mg C ha^{-1}.

The largest anthropogenic wetlands are paddy soils, classified as hydromorphic **Anthrosols** (WRB 2007). These soils occupy 1.55×10^6 km^2 of the total land surface (http://beta.irri.org/index.php/). Paddy soils can originate from different soils (e.g., Andosols, Cambisols, Vertisols), however, specific management leads to alternating redox conditions and therefore to soil properties and morphology independent of the antecedent soil unit. The formation of these Anthrosols is induced by tilling the wet soil (puddling), and the flooding and drainage regime associated with the development of a plough pan (Kögel-Knabner et al. 2010). Anthropogenic activities (large inputs of plant residues, organic fertilizers, temporal flooding) enhance accumulation of SOC in these soils (Lal 2002, 2004; Sahrawat 2004).

Podzols occur mainly in humid areas in the boreal (3.2×10^6 km^2) and temperate zones (1.3×10^6 km^2), and locally in the tropics (1×10^5 km^2). On the global scale, Podzols store a significant amount of C (275 Pg in 0–200 cm depths), especially in the 'spodic horizon'. Humus accumulation is enhanced due to strongly acidic soil conditions, but the mechanisms are not well understood (Sauer et al. 2007).

Chernozems (equivalent to Mollisols of the US Soil Taxonomy) and the associated soils (**Kastanozem, Phaeozem and Greyzem**) are mineral soils which usually develop ubiquitously under steppe vegetation. Chernozems constitute 4–6% of the total global land area. By definition, Chernozems must have a mollic epipedon which is a thick (≥25 cm), dark coloured surface horizon containing C content ≥6 g kg^{-1} and a base saturation of ≥50% (WRB 2007). Factors which trigger genesis of Chernozems include semiarid to subhumid climates, grassland vegetation, and a calcium-rich parent material. The combination of these factors control the amount of SOM by favouring increased below-ground biomass production, deposition of lignin-rich residues and development of stabilizing bonds with Ca that slow the rates of OM decomposition (Thompson and Bell 2001). Chernozems are highly productive soils, but have a vulnerability to decomposition.

7.5 Distribution of Peatland and Hydromorphic Soils

Peatlands cover only approximately 3% of the land surface (Yu et al. 2011), or approximately 4×10^6 km^2, but contain 40% of the SOC to 1-m depth (612 Pg out of 1,550 Pg). Regarding the huge C stocks in peatlands it is important to realize that (a) the area covered by peatland is not the only factor for assessing the importance of these ecosystems to global biogeochemical cycle, and (b) erroneous estimates are made by different assumptions about the area under peatlands. Estimating the mean or actual depth and bulk density of peatland are also very challenging. Hence global and regional (continental) estimates of C stored in peatlands vary with approach and the time when the estimate was made (Fig. 7.4).

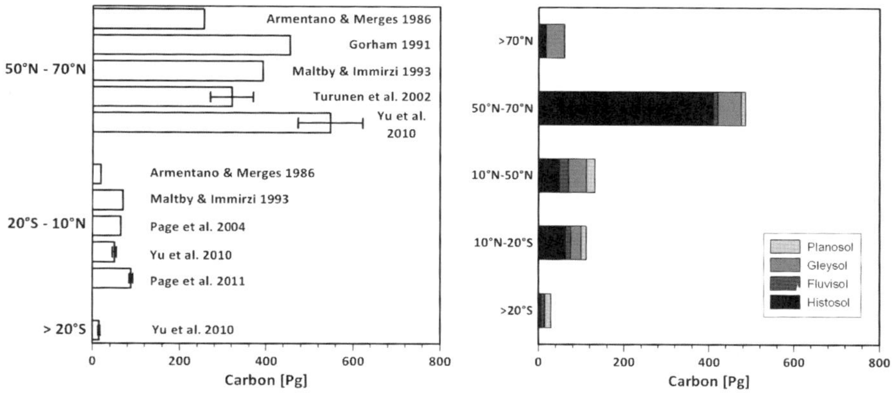

Fig. 7.4 *Left*: Zonal distribution of soil carbon of peatlands by various authors. *Right*: Zonal distribution of soil carbon of hydromorphic soils (2 m depth) (FAO 1995; Batjes 1996)

Therefore, studies are inherently imprecise and coordinated efforts are needed to assess the C stocks of peatland and hydromorphic soils. The qualities of those estimates depend on the transparency of the chosen procedure, and providing the degree of uncertainty. One uncertainty that is particularly difficult to capture is the spatial extent, because wetlands and peatlands are not clearly defined. It is also important to know the location and area of other C-rich but nonpeat and partially wet soils which are even more difficult to delineate. The comparison between the low land area and high importance as C stock provides an important reason to improve the scientific understanding of peat, peat-like and wetland systems. Credible data are needed about the land area and associated processes for more precisely predicting the future soil to atmosphere C feedbacks.

A large proportion of the global peatland area is located in the northern latitudes (50–70° north), especially in boreal and subarctic regions of Canada and Russia (Aselmann and Crutzen 1989). The humid and cold climatic conditions in these regions are suitable for a high rate of peat accumulation. In addition to the largest area of peatlands at these latitudes, there soils are also characterized by the highest C densities (Yu et al. 2011). Northern peatlands accumulated about 550 Pg of C since the last glacial maximum (Yu et al. 2010) with a mean net rate of about 0.1 Pg C year^{-1} (Gorham 1991). The formation of peatlands in the high northern latitudes started mainly after the last glacial maximum with the peak accumulation in the early Holocene (Yu et al. 2010). These peatlands store about 15–30% of the world's soil C stock (Limpens et al. 2008) which is lower than the estimate stated above. All hydromorphic soils, which include Histosols, store approximately 41% of the global soil C stock. Podzols store an additional 10%.

The second largest land area covered by peatland is found between 20° south and 10° north (Aselmann and Crutzen 1989) with approximately 11% of the global peatland area. Moreover the second largest concentration of C with about 85 Pg or

15–19% of the global peatland C stock is estimated in tropical peatlands. Most of the tropical peatlands occur in Southeast Asia (with about 5% of the global peatland area), and with the largest C stock in Indonesia (Page et al. 2011). Tropical peatlands have the lowest C accumulation rates during the Holocene, but these peatlands started to form much earlier than northern peatlands (Yu et al. 2010).

Peatlands in the southern hemisphere, mainly in Patagonia, South America, have accumulated approximately 15 Pg of C since the last glacial maximum (Yu et al. 2010). These peatlands are characterized by the highest accumulation rate of C during the Holocene with 220 kg C ha^{-1} $year^{-1}$ compared to northern and tropical peatlands (Yu et al. 2010). Understanding the mechanisms of these high accumulation rates is of utmost importance particularly in the context of the need for "recarbonization" of the biosphere. Additionally, the spatial distribution of peat and peat-like ecosystems may be underestimated for Africa (a case study is presented later in this chapter). Besides evapotranspiration and precipitation, topography also plays an important role in peat formation. Peatland formation can also occur in even in wetlands of the semiarid regions, which are abundant in Africa (Tooth and McCarthy 2007). In the tropics, countries with known and mapped peat lands include Indonesia (27 Mha), Brazil (5.5 Mha), Peru (5.0 Mha), Papua New Guinea (2.0 Mha), Malaysia (2.5 Mha), Congo (1.5 Mha), Uganda (1.4 Mha), Colombia (1.0 Mha), Venezuela (1.0 Mha) and Zambia (1.0 Mha).

The huge amount of C stored in peatlands has mainly been accumulated since the last glacial maximum with a net sink rate of more than 5 Pg of C per century on average (Yu et al. 2010). Mainly due to water-saturated (hypoxic = oxygen low) soil conditions, build-up of SOM is faster than decay and these wet and mostly peaty soils act as prolonged atmospheric C sinks. The fate of this locked C in a changing world (See Chap. 18) under changing conditions is not understood, but is important to predicting any feedback mechanisms with climate system (Davidson and Janssens 2006).

7.6 Differences Between Wetland and Non Wetland Soils

7.6.1 A Case Study South Africa

Only very preliminary work has been done in African and South American peatlands, and at this stage even their areal extent is not well quantified, much less their C content, vegetation composition, or C cycle characteristics (Frolking et al. 2011). Therefore, soil C content of four different soils was determined in an investigation in the Drakensberg, Republic of South Africa. Two wetlands and two non wetlands were chosen for the study in this region. The differences between wetland and dryland were obvious in the vegetation and in the geomorphology, but only slight differences were measured in the top soil C content (see Fig. 7.5). Wetland soils have a depth of >0.5 m and up to 1.5 m, whereas the dryland soils were only about 20–30 cm thick. However, a thickness of 30 cm would qualify these soils as peatland

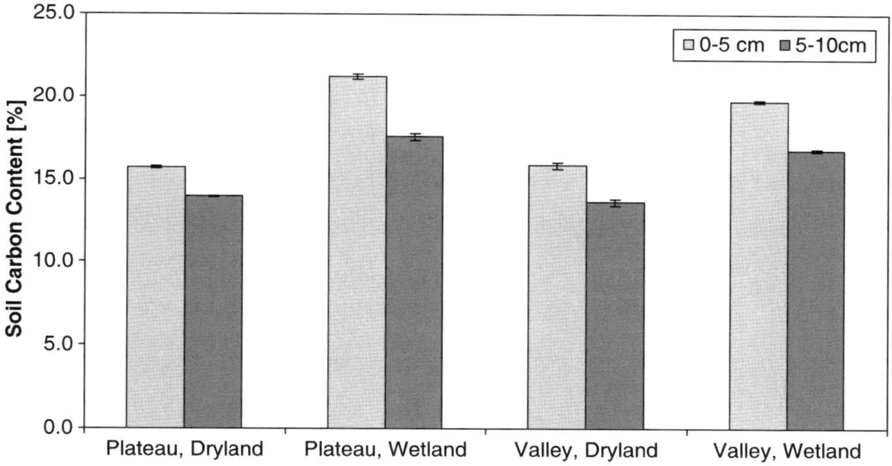

Fig. 7.5 Mean soil organic carbon content [%] with standard deviation of four different locations [n=3] with depths of 0–5 cm and 5–10 cm, Drakensberg region, Republic of South Africa

(for example Gorham 1991). This case study shows that even non wetland soils may have high C contents, and C content and stocks are underestimated. Therefore, the simple differentiation of soils into high C-rich wetlands and C-poor drylands is inadequate. Based on the large areal extent of the dryland compared to the wetland soils, credible assessment of the C content is an important component in estimating regional and global C stocks.

7.7 Global Soil Carbon Hot Spots: Potential Sources for Atmospheric CO_2

In the context of "recarbonization", it is important to know where the soil C hot spots are. While it may appear to be a trivial question considering available global estimates and maps (FAO 1995), yet there is a strong need to stress that erroneous estimates are made in precisely accounting for the global soil C resources. The difficulty lies not only in insufficient means to measure soil C resources at the global scale – because of the insufficient data – but also due to other factors such as consideration of different depths, including peat and other wetland soils, and different definitions or perceptions of soil types, all of which hamper obtaining the reliable estimates of global and regional soil C stocks. Therefore, there exists a large uncertainty not only about the magnitude of soil C stocks but also which part is prone to being emitted into the atmosphere. World soils most likely store more than the "official" estimate of 1,500 Pg C (Fig. 7.6), which is already twice the amount contained in the atmosphere. The available estimates of soil C stocks, for example, do not include the latest estimates for peatland C stock of Yu et al. (2010). Furthermore,

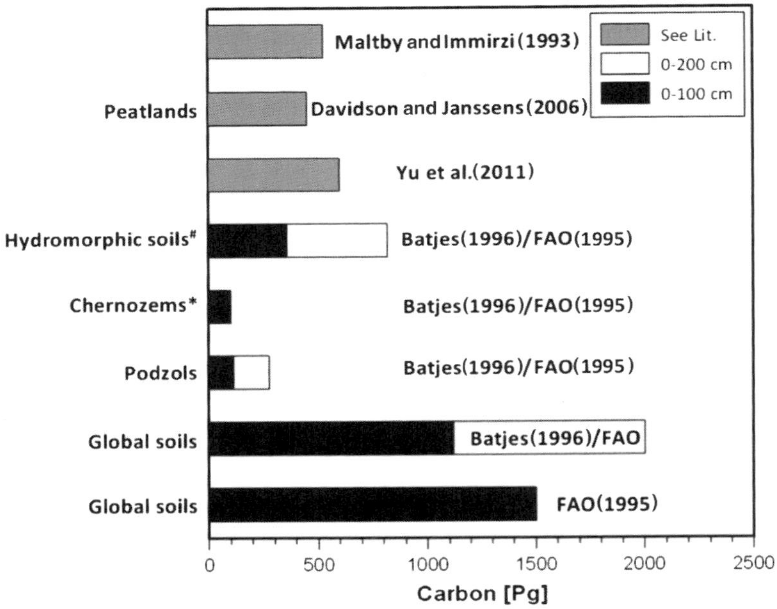

Fig. 7.6 Global soil carbon estimations. Hydromorphic soils# (Histosols, Fluvisols, Gleysols, Planosols), Chernozems* (Chernozem, Phaeozem, Kastanozem, Greyzem), (FAO 1995; Batjes 1996)

the present estimates also underestimate the deep soil C storage. Nonetheless, the deep soil C is not prone to being emitted into the atmosphere in the near future. Therefore, a credible inventory of vulnerable and less vulnerable soil C stocks would be advantageous for predicting feedbacks to the climate system. In this context, the first simplest step is to differentiate between wetland and upland ecosystems. Hydromorphic soils store a greater part of soil C (Fig. 7.6) which is, at the same time, undeniably highly vulnerable to changes in both temperature and precipitation.

Grouping the world into wet- and dryland or low and high C-rich soils is not as simple as it may seem. Addressing the exchange processes between the atmosphere and the biosphere by disregarding the complexities can lead to erroneous estimates. Soils characterized by transient water logging conditions (semi-wet) are very common and react, at least partially like wetlands. Therefore, all wetlands must be included, as has been done for the present study (Fig. 7.6). All hydromorphic soils (0–200 cm) potentially store 818 Pg C which is more than the latest estimates for peatland alone, even though the former conservative estimates by Batjes (1996) were used and multiplied by the land area given by the FAO (1995).

There are more conditions that hamper OM decomposition leading to higher soil C resources. When summers are too dry for decomposition, typically C-rich Chernozems are formed in the steppe. These Chernozems (includes Chernozem, Phaeozem, Kastanozem, Greyzem) store at least another 100 Pg C (Fig. 7.6). Potentially these steppe soils store about another ~25% in 1–2 m depth as compared to the top 1 m (Mikhailova and Post 2006). Podzols have both C rich topsoils due to hampered decomposition and high subsoil C due to percolation transport. The rough estimate of the mean values given by Batjes (1996) and multiplied by the area derived from the FAO soil type maps is 275 Pg C (for 0–2 m, Fig. 7.6). All C-rich soils and particularly any in-between states need to be included in biogeochemical feedbacks estimates between soils and the climate system. And even less C-rich soils are important because they cover larger part of the terrestrial surface. The global estimates derived from Batjes (1996) and area given by the FAO is 1,120 Pg C (0–1 m) and 2,003 Pg C (0–2 m). Using the soil C data provided by the FAO (1995), the mean value is 1,501 Pg C (also reported by Schlesinger 1997; Glatzel 2011), with a range of 963–2,057 Pg C.

7.8 Peatland Conversion to Agricultural Use

Drainage, deforestation and tillage lead to rapid decomposition. In the temperate zone, rate of oxidation of drained peatlands, used for seasonal/annual crops, can be 1–2 cm $year^{-1}$. The rate (Mg Cha^{-1} $year^{-1}$) of CO_2-C emission from drained and cultivated peatlands have been measured at 2.8–6.7 for Germany, 4.9–11.2 for Poland, 3.5–17 for Sweden, 2.2–8.2 for the Netherlands, 5.5 for Russia, 1.9–2.3 for Canada, 3.3–8.3 for Finland (Strack 2008). The site-specific rate of C emission depends on local conditions related to temperature and moisture regimes, pH and the soil and crop management practices.

Area of non-tropical peatlands converted to managed ecosystems are estimated at 25 Mha for agriculture and 15 Mha for forestry (Joosten and Clarke 2002). Total area of peatlands drained for agricultural and forestry uses in Europe is estimated at 10–30 Mha (Joosten and Clarke 2002; Rydin and Jeglum 2006), with large land use conversion in Russia, Finland, Sweden, U.K., Netherlands, etc. As much as 14% of the peatland area in Europe has been converted to managed ecosystems. Large areas of peatlands have also been converted to agricultural and forestry land uses in Thailand, Malaysia and Indonesia, and additional areas are being deforested and drained in southeast Asia for establishment of biofuel plantations. More than 50% of the original peatlands in south-east Asia have already been converted. Deforestation for timber is a serious threat. Drainage ditches for log transport alter the hydrology, and exacerbate the risks of fire.

Rather than emission, drained and cultivated peatlands may be a small sink for CH_4. The sink capacity also depends on site-specific conditions and varies widely among locations. Further, there is a large spatial and temporal variability within each site. The CH_4 uptake on cultivated peatlands has been measured at rate (kg CH_4 ha^{-1} $year^{-1}$) of –0.2 to –4.9 (Strack 2008). Depending on the moisture regime, cultivated peatlands can also emit CH_4 at rates of 0.2–3.6 kg CH_4 ha^{-1} $year^{-1}$ (Strack 2008).

In addition to gaseous emissions, other environmental impacts of drainage and cultivation of peat include loss of dissolved organic carbon (DOC), emission of N_2O

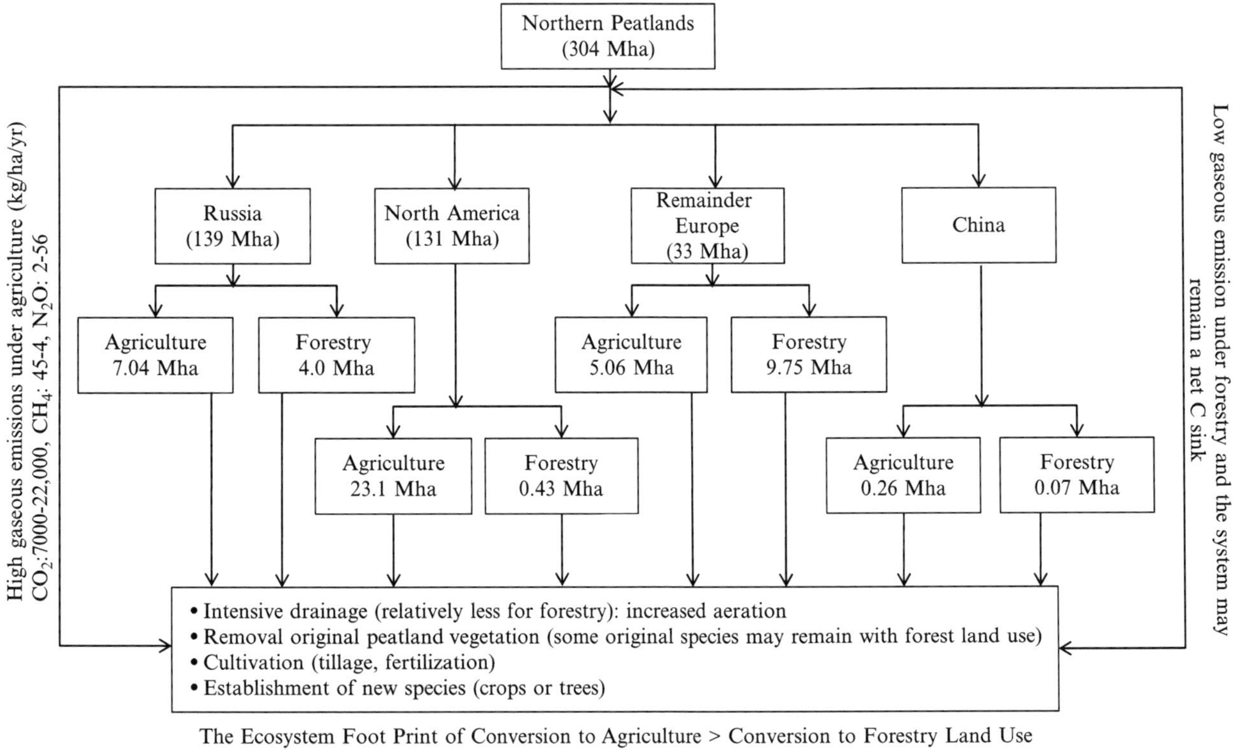

Fig. 7.7 Anthropogenic perturbations of northern peatlands for agriculture and forestry (Redrawn from the data by Lane et al. 2009). C density agriculture, forestry and other uses

Table 7.4 Land clearing and biofuel debt (Recalculated from Fargione et al. 2008)

Biofuel	Former land use	Location	C debt ($Mg CO_2/ha$)	Time to repay biofuel C debt (year)
Palm biodiesel	Tropical forest	Indonesia/Malaysia	702 (175)	86
Palm biodiesel	Peatland forest	Indonesia/Malaysia	3,452 (2,952)	423
Soybean biodiesel	Tropical forest	Brazil	737 (200)	319
Sugarcane ethanol	Cerrado (woods)	Brazil	165 (100)	17
Soybean biodiesel	Cerrado (grasslands)	Brazil	85 (85)	37
Corn ethanol	Central grassland	USA	134 (10)	93
Corn ethanol	Abandoned cropland	USA	69 (5)	48
Prairie biomass ethanol	Abandoned cropland	USA	6	1
Prairie biomass ethanol	Marginal cropland	USA	0	No debt

The number in parenthesis is soil carbon as a part of total carbon debt

through the nitrification and denitrification processes, and leaching of nitrates. Drainage and lowering of water table can lead to emission of N_2O at 2–56 kg N_2O ha^{-1} $year^{-1}$ (Strack 2008).

Increase in demand for production of biofuel has accentuated the rate of deforestation of peatlands in Southeast Asia. Deforestation, drainage and cultivation create a large ecosystem C debt (Fargione et al. 2008). Deforestation and drainage adversely impact both the biomass and soil C stocks. In addition, burning exacerbates the ecosystem C loss. Fargione et al. (2008) estimated the C debt computed as CO_2 released during the first 50 years of land use conversion. The data in Table 7.4 show CO_2 debt and the time needed to repay the debt through the production of biofuel. The highest C debts (Mg CO_2 ha^{-1}) for biomass, and soil, respectively, was 3,452 and 2,952 with 423 years to repay it were estimated for conversion of tropical peatlands (Indonesia, Malaysia) to oil palm plantation for the biodiesel production. In comparison, establishment of soybean-based biodiesel plantation in Brazil created CO_2 debt of 737 and 200 with 319 years to repay it for the tropical rainforest ecosystem, compared with 165 and 100 for 17 years to repay it for soybean-based biodiesel from the cerrado/savanna ecosystem. Thus, conversion of peatlands to biodiesel production is not sustainable. Conversion of virgin peat swamp forest to oil palm (including biomass, fire), created a total C-loss of 17 Mg ha^{-1} $year^{-1}$ over 25 years (Hergoualc'h and Verchot 2011).

7.9 Interaction with the Climate System

Increasing temperatures lead to increasing microbiological activity and, therefore, enhanced decomposition (Arrhenius equation, Vicca et al. 2009). Thus, a direct feedback of climate warming and more CO_2 emission from soils may appear convincing. However, plant growth might also be enhanced and sensitivity of decomposition

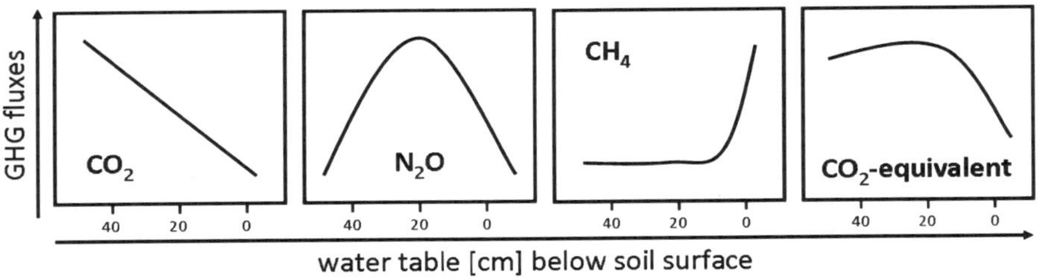

Fig. 7.8 Schematic relationship between greenhouse gas fluxes and water table in wetland soils (According to Jungkunst et al. 2008)

processes is not completely understood (Davidson and Janssens 2006). A more comprehensive discussion can be found in Chap. 18. On the other hand, higher temperatures lead to higher evapotranspiration which eventually leads to dryer conditions that may hamper decomposition despite the better temperature conditions. From a wet and peatland perspective, drying usually leads to a shift from anoxic conditions (low redox values) that hamper decompositions to more favorable oxygen richer conditions (higher redox values). Therefore, increasing temperature and higher redox values will both enhance soil C decomposition from hydromorphic soils and increase the feedback loop. In some cases predicted increase in precipitation can promote low redox condition associated with soil C lockage. Increasing variability of precipitations is the most difficult feature of predicted climate change conditions because (a) both agriculture and ecosystem cannot easily adapt to ever changing and "unreliable" conditions, and (b) feedback reactions are most difficult to predict because every wetness state is associated to a different composition of GHGs releases from soils (Fig. 7.8).

For example the latest observations that atmospheric CH_4 is increasing at lower rates is explained by: (a) drying of northern wetlands, or (b) a change in the agriculture practice in rice fields mainly in Southeast Asia (Heimann 2011). The change in the rice agriculture in Asia in the last three decades has contributed to the reduction in CH_4 emission (Kai et al. 2011). A greater application of fertilizer and shorter inundation period can reduce the CH_4 emissions from these soils (Heimann 2011; Kai et al. 2011). However, both less anoxic conditions and more available fertilizer (mainly N) promotes nitrous oxide (N_2O) emissions. New agriculture practices likely reduce CH_4 emissions, but potentially increases N_2O emissions (Frolking et al. 2004). For a complete atmospheric – soil feedback assessment, all three important GHGs (i.e. CO_2, CH_4 and N_2O) must be considered (Jungkunst and Fiedler 2007).

The net release of CO_2, as the common "end-product" of OM decomposition, is substantially lower under anoxic than oxic conditions (Fig. 7.8). However, under strictly anoxic conditions the release of CH_4 is high. The debate, what is worse for the climate system: (a) high CO_2 release from dried peat, or (b) low CO_2 and associated high CH_4 under wet peat conditions, is common. Details are even more complicated. Both N and C cycles are closely coupled and cause the conjoined feedbacks to the climate systems. In this context, N_2O is highly important, which is preferably released under intermediate wet conditions that fall in the hypoxic category (0–5% O_2) (Fig.7.8). Further, N_2O is particularly important for managed soils and vegetation free peat e.g. in the artic (Repo et al. 2009). Generally, N_2O has to be considered for estimation of the impact of climate change in the northern peatlands. A warming of this region would probably increase the N_2O emissions from these ecosystems and which will be a positive feedback to climate change (Elberling et al. 2010; Jungkunst 2010). Besides being large CH_4 sources, several northern peatlands (drained, low C:N ratio and the absence of vegetation) are considered as hot spots for N_2O emissions (Marushchack et al. 2011). Therefore, despite the available knowledge, there is a need for more process-based understanding on peat, wetland and soil biogeochemistry. Theses alarming findings from higher latitudes are unlikely to be restricted to boreal and arctic hydromorphic soils.

7.10 Climate Change and the C Cycle in Peatlands

The climate-induced uncertainties in the fate of C in peatlands are due to the changes in water table and the temperature regimes. These uncertainties may be especially large in regions where melting of the permafrost, thermokarst erosion, and formation of melt/thaw lakes may change the surface and sub-surface hydrology. Yet, another uncertainty with a strong impact may be caused by the frequency and intensity of fire. The climate change and other anthropogenic factors have pertubated the natural balance.

The projected climate change may also affect the C cycle in peatlands through changes in temperature and moisture regimes, and the attendant alterations in the water table (Strack 2008). For example, Roulet et al. (1992) predicted the increase in temperature by 3°C and rainfall by 36.5 cm year^{-1} may lower the water table in peatlands by 14–22 cm. These factors may lead to alterations in species compositions, and thus affect the quantity and quality of biomass addition (Weltzin et al. 2000). An increase in atmospheric CO_2 concentration may have no net effect on ecosystem C budget, because any gains by the CO_2 fertilization effect on the biomass production may be negated by increase in respiration at elevated temperature and the attendant decomposition (Strack 2008). Increase in the length of the growing season may alter the total biomass production in northern latitude but not so in the tropical or equatorial regions. Therefore, the SOC pool in the northern peatlands may be vulnerable to decomposition and to creating a positive feedback. Bridgeham et al. (2008) studied the effects of changing water and temperature regimes over 8 years in 54 peat monoliths from a bog and fen in northern Minnesota, USA. Increase in water availability increased C sequestration in bogs but not in fens. These trends were primarily due to water table effects on Sphagnum moss production in bogs, and to decomposition in fens. The controlled study indicated that peatlands can gain or lose large amounts of soil C until the new equilibrium with the water table depth is attained. The projected climate change may also lead to other C cycle-hydrological feedbacks at a larger/global scale. The projected climate change may lead to new peat accumulation in sub-arctic climate where the prevailing cool climate led to the development of permafrost (Zoltai and Tarnocai 1975) due to increased biomass production and water levels.

7.11 Distribution of Soil Carbon Resources

The distribution of soil carbon (0–1 m) is similar to the distribution of area covered with soils (Fig. 7.9). The exceptions are the "boreal" latitudes (50–70°N) that have proportionally more C than soil area. For the other for zones area and soil carbon stocks correlete well ($r^2 = 0.926$) and the average million square kilometer hold 12 Pg soil C (0–1 m). This number for the "boreal" latitudes is 20 Pg soil C (0–1 m) per 10^6 km^2. This elucidates the relative importance of the "boreal" latitudes for soil C resources, but the absolute importance of the "mid-latitudes" (10–50°N) is at least as important (Fig. 7.9). If the soil type specific mean values given by Batjes (1996) are used and multiplied with the area of the individual soil types by the FAO (1995), the mid-latitudes appear even more important in absolute numbers than the boreal regions. For many estimates, these common mid-latitudes are not considered, whereas tropical zones are. In the context of these estimates, the "tropical" latitudes (20°S–10°N) appear less important. Nonetheless, all zones must be appropriately considered.

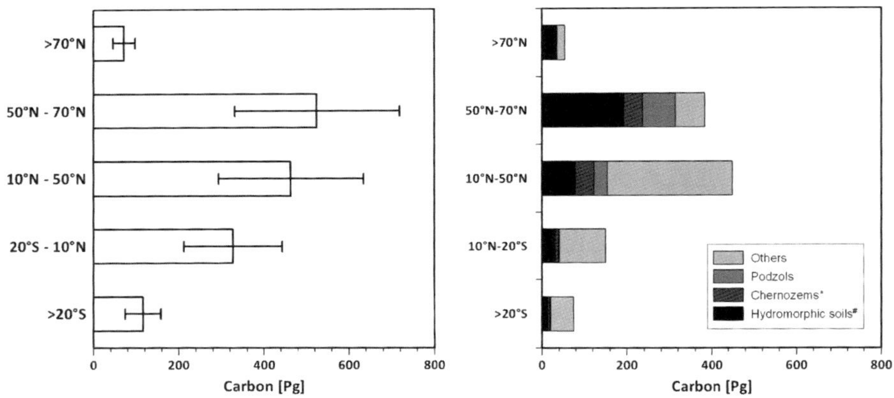

Fig. 7.9 Soil carbon resources referring to 1 m soil depth. *Left*: Zonal distribution of soil carbon (FAO 1995). *Right*: Zonal distribution of soil carbon of different soil types (by FAO 1995; Batjes 1996), Hydromorphic soils (Histosols, Fluvisols, Gleysols, Planosols), Chernozem* (Chernozem, Phaeozem, Kastanozem, Greyzem)

7.12 Peat Extraction

In regions rich in peat, peat extraction has been a part of traditional land use for millennia (Charman 2002). Peat has been used, among others, as surgical dressing, diaper absorbent and in traditional Chinese medicine (Harris 2008), as a prerequisite for manufacturing of whisky, for horticultural purposes and last but not the least as energy source. Presently, 8,105 km^2 of peat area is under active peat extraction for energy purposes using industrial techniques. Most of the affected area (6,590 km^2) is located in the boreal zone, and an additional 1,609 km^2 are in the temperate zone, but only 6 km^2 in the tropics (IPS 2011). These estimates do not include peat extraction for other purposes, such as that for horticultural use, which is widespread in Canada and Baltic countries, and for many small-scale operations in Latin America, Africa and elsewhere.

This quick glance at the diversity of peat use shows that, during the last century, only the scale of peat extraction but not the land use type itself has changed. On the global scale, considering the large peat resources, peat use for energy purposes is small, amounting to 17 Gg $year^{-1}$ (World Energy Council 2007). Since, 80% of the peatlands, most of them located in high latitudes, are still pristine, the question of the sustainability of peat extraction depends on the scale: Globally, more peat is formed in the world's intact peatlands than is extracted for energy purposes. However, considering the smaller scales, it becomes clear that peat extraction contributes to species loss as well as to jeopardizing the last remaining peatlands in many countries in the temperate zone.

7.13 Peat Restoration

As has been already discussed, using peatlands for forestry, agriculture, or extracting peat is a C-consuming process, leading to a net rise in GHG emissions. Recognizing this, numerous measures for restoring peatlands have been adopted since 1980s.

The aim of peatland restoration is to regenerate a self-sustaining naturally functioning mire ecosystem that accumulates carbon and retains nutrients from through-flowing waters (Wheeler and Shaw 1995). There are numerous strategies of peatland restoration ranging from just raising the water table while continuing agricultural use less intensively to flooding the peatland and converting it, for some time, into a shallow lake. The choice of the best restoration method depends on environmental conditions, the specific goals, and the available infrastructure. Until recently, the major goal of peatland restoration has mostly been motivated by nature conservancy. More recently, following the concept of "ecosystem services" (SCEP 1970), the "sponge" function of peatlands (e.g., protection from flooding and, especially their ability to store C) have become the main motivation of peatland restoration.

As restoration of peatlands focusing on the mitigation of GHG emissions is a very recent concept, there is no universally accepted "best practice" suited to all situations. In general, the local situation determines the best practice. However, as a general guideline, it is well known that a permanent water table at −10 to 0 cm below ground minimizes the emissions of GHGs (Couwenberg et al. 2011). Nonetheless, it is difficult to maintain these conditions throughout the year.

Flooding is much easier and also economical in most locations. In Central Europe, some studies have shown extremely high CH_4 release following flooding, which results in elevated net GHG emission compared to the prior drained situation (Augustin and Chojnicki 2008; Glatzel et al. 2011). It is not yet clear, however, whether these emission peaks are representative of the flooding situations, and they are probably transient. Regardless of this, duration of the transient phase is not yet known, and is related to the trophic status of the site. In any case, a transient period of increased CH_4 release may contribute to GHG mitigation when followed by an extended period of C storage and low emissions. Thus, it is necessary to ensure decades of high water table in order to maximize GHG sequestration following flooding.

7.14 Feedbacks to Climate Change

The ability of peatlands to maintain the C-sink function under changing environmental conditions will depend on the balance between C-inputs and outputs (i.e., the net ecosystem C balance, NECB). Furthermore, fluxes of CH_4 and N_2O must also be taken into account (Schulze et al. 2009), and numerous uncertainties exist regarding how the NECB of terrestrial soils will be influenced by increasing concentrations of atmospheric CO_2, warming, and increases in drying and rewetting or freezing and thawing cycles (see Chap. 18). For example, warming and CO_2 fertilization increase biomass production, but also the rate of mineralization. As a result, the NECB may be unaffected if effects of increased biomass production are negated by increase in the rate of SOC-mineralization. The situation is even more complex in hydromorphic soils. Despite processes relevant to terrestrial soils, water level and hydraulic conductivity are dominant controls on biomass production, species composition, decomposition, gas fluxes and production of dissolved organic carbon (DOC) (Limpens et al. 2008).

In pristine peatlands, vegetation influence the C-cycle in several ways: (a) species composition influences the C-input into the ecosystem; (b) litter chemistry determines the rate of decomposition (Limpens and Berendse 2003); (c) diffusion, hydraulic conductivity and redox conditions are influenced by the structure of the peat, which is determined by the peat-forming genus (Limpens et al. 2008); and (d) an increase in graminoid species can increase CH_4 emissions by transport through aerenchymatic tissue (Nilsson et al. 2001). Shifts in vegetation are driven by water level, temperature and N-deposition. Decreasing water level as well as increasing temperature and N-availability stimulate growth of vascular plants at the expense of *Sphagnum* species. This in turn increases the rate of decomposition and, thus, diminishes

C-storage in peatlands. Warming is predicted in all regions with peatlands. However, increasing precipitation in temperate regions may counteract the effects of warming if water level in peatlands can be maintained (Limpens et al. 2008). For some peatland regions in North America and Southeast Asia, significant decreases in precipitation are predicted (Li et al. 2007), and droughts and heat waves are likely to occur more often almost everywhere on the globe (Christensen et al. 2007). Exceptional droughts can increase decomposition, and also increase the risk of fire. Frequency and severity of fire is important for the NECB of several ecosystems. Often, fires affect vegetation and only a few centimeters of the topsoil. However, this may be different in the case of organic soils. Globally, around 2 Pg C $year^{-1}$ (1997–2009) were emitted by fires, 3% of this was ascribed to tropical peat fires and 15% to extra-tropical forest fires (van der Werf et al. 2010). The latter might also include forest with peaty soil. Even in tropical montane cloud forests, organic soils are present due to wet and relatively cool conditions (Schawe et al. 2007) and forests fires are prominent (Roman-Cuesta et al. 2011). While vegetation recovers soon after burning, loss of C from organic soil has been reported to be as large as 0.8–1.8 Tg C $year^{-1}$ in the Andes for the period 2000–2008 (Roman-Cuesta et al. 2011). Not only in the tropics but also in mid-latitude (30–50°N, Poulter et al. 2006) and in northern peatland ecosystems (Turquety et al. 2007; Wieder et al. 2009), fire is an important factor in affecting the NECB. Poulter et al. (2006) estimated that the C-emission by fire was as high as 110 Mg C ha^{-1} for a single fire event in North Carolina. With a fire frequency of less than 20 years, the peat ecosystem would turn into a net-source of CO_2. For a Canadian bog, a fire return interval of <60 years was estimated to turn that ecosystem into a net-source of CO_2 (Wieder et al. 2009). Neither of these studies (Poulter et al. 2006; Wieder et al. 2009) took CH_4 emissions into account.

The examples described above show that an increase in fire frequency may turn peatland ecosystems into sources of CO_2. However, single catastrophic events in small regions can have even more severe impact on emissions. During an abnormally long, dry period caused by El Nino in Indonesia in the year 1997, the peat swamp forest of those regions caught fire. Page et al. (2002) estimated that on average 50 cm of the peat layer was destroyed and between 0.8 and 2.8 Pg C was released into the atmosphere from peat and the burnt vegetation. This emission was equivalent to between 13% and 40% of mean global C-emission from fossil fuel combustion in 1997. Further, the drained peatland was much more affected than that under undisturbed locations. This example shows that peat fires are a severe threat to climate, and emphasize the need for appropriate and careful management of peatland ecosystems.

7.15 Remote Sensing Possibilities to Capture Peat- and Wetland More Precisely

What are the options to improve scientific knowledge on the spatial distribution of peat- and wetlands without having to auger the entire globe?

As already stated in the previous sections, the knowledge about the global spatial coverage and latitudinal distribution of wetlands and in particular of peatlands is of utmost importance for a reliable estimate of the global C resources in soils. Remote sensing is a valuable means to map and monitor the status and changes of land surface properties at scales ranging from regional to global. Coarse to medium resolution remote sensing data deliver a globally consistent and objective source of information for a spatially explicit mapping of the global distribution of potential C stocks in terms of land cover type maps. However, there is still considerable uncertainty in estimates of the area and distribution of the relevant land cover types (e.g. peatlands, wetlands, wet forests) and hence of the stored C globally (Herold et al. 2008; Krankina et al. 2008).

These uncertainties are attributed to a number of limitations that are either determined by the technical specification of the sensor (wavelength, spectral and spatial resolution) or the derived data products (e.g. land cover maps). The following is a brief overview about some of these challenges. Despite its distinct canopy structure (a mix of tree canopy, a shrub layer, and a continuous layer of herbaceous vegetation with presence of mosses), mapping peatlands as a distinct type of land cover from remote sensing data is difficult. Most classifications of land cover focus on a dominant life form and are not well suited to mixtures of several life forms (e.g., Land Cover Classification System (LCCS), Di Gregorio 2005).

This is the reason why peatlands do not occur in most global land cover maps. Instead, they are often associated with water saturated conditions. Therefore, they are usually included in wetlands when defining thematic classes for land cover type mapping. However, it is well known that significant portions of peatlands, especially in the boreal zone do not experience prolonged inundation periods and furthermore, large parts of peatlands in some regions are drained. Considering wetlands, those areas are mapped as a separate class in most global land-cover maps. Because of the global availability of such maps, wetlands often dealt as a proxi to define the occurrence of peatlands in the global C cycling models (Krankina et al. 2008). An overview of the current global land cover products and the thematic representation of peatlands or wetlands respectively is given in Table 7.5.

Besides the occurrence of water, another proxy for peatlands is the vegetation cover. As stated above, peatlands have a characteristic canopy structure but on the other hand, this is also the main limitation when using the remote sensing data to map the spatial extent of peatlands. Although, the spectral signatures of peatland herbaceous vegetation and mosses distinctively differ from other vegetation types (e.g. lower spectral reflectance in the NIR and SWIR region) and they can be mapped from high to very high resolution satellite data, the spatial resolution of globally operating sensor systems and existing land cover maps is not adequate to detect single vegetation types and sometimes even omits the existence of small sized peatlands (Pflugmacher et al. 2007).

An approach to overcome the limitations in pixel size and categorical mapping is the modeling of "continuous fields" of quantitative surface parameters, e.g. the percentage of vegetation or tree cover within one single pixel (sub-pixel analysis, spectral unmixing Hansen et al. 2005). Another option is the synergistic use of passive and active remote sensing systems (e.g., radar or lidar Rosenqvist et al. 2007; Li and

Table 7.5 Global remote sensing based land cover products and representation of wetlands within the data

Product	Sensor	Year	Spatial resolution	Relevant class for wetlands	Scientific reference
GLCC	AVHRR	1992	1 km	Herbaceous wetland/wooded wetland	Loveland et al. (2000)
GLC2000	SPOT-VGT	2000	1 km	Regularly flooded shrub and/or herbaceous cover	Bartholomé and Belward (2005)
MODIS land-cover	MODIS	2000	1 km	Permanent wetlands	Friedl et al. (2002)
GlobCover	MERIS	2005	300 m	Closed to open (>15%) grassland or woody vegetation on regularly flooded or waterlogged soil	Arino et al. (2008)

Chen 2005). The advantage of radar sensors is that they can penetrate clouds and that they are sensitive to variations in top soil properties (e.g., soil moisture, grain size). However, global mapping based on active sensor systems is cost-intensive and time consuming and at present, there are no global products available with regard to wetland and/or peatland mapping.

Summarizing, the representation of peatlands in global land cover mapping as well as the definition and mapping accuracy of other relevant land cover types for the spatially explicit estimation of global C resources is challenging and should be an important objective for future global mapping campaigns. An effective and low-cost solution could be to build on existing global estimates of wetlands and other classes and use the crowdsourcing approach (Fritz et al. 2009) to validate and adjust global land cover maps for specific use in biogeochemical modeling.

7.16 Conclusions

Will peatland and other soil C-rich ecosystems maintain their C sink function remains to be the major climate change related question to be answered by soil scientists and other biogeochemists. Practical advice how to manage land use on soil rich in C cannot be given with simple answers which are applicable around the world. Rather than altogether ignoring these questions, it is important to address them systematically and objectively. It is difficult to exactly know how much vulnerable soil C recourses there are. For realistic bottom-up approaches to estimate potential feedbacks between soils and the climate systems, it is important to develop more precise inventories of soil C resources because net GHG exchanges are not only process-driven but also determined by area of OM (C and N) hotspots. This article has provided not only some rough global estimates on soil C estimates based on simple assumptions but also differentiated these into different zones (latitudes) and soil types. This information may trigger more research on the spatial distribution of soil C and attempts to divide these estimates into vulnerable and less vulnerable soil C stocks. In this context, the scientific communities of peatland research should join forces with all groups focusing on other C- rich soils. Furthermore a focus mainly on the northern (50–70° N) and the tropical latitudes (20°S–10°N) may not be sufficient for a global perspective.

References

AG-Boden (2005) Bodenkundliche Kartieranleitung, 5th edn. Bodenkundliche Kartieranleitung, Hannover

Arino OD, Gross F, Ranera L et al (2008) GlobCover: ESA service for global land cover from MERIS. 2007 IEEE Int Geosci Remote Sens Symp IGARSS:2412–2415

Aselmann I, Crutzen PJ (1989) Global distribution of natural freshwater wetlands and rice paddies, their net primary productivity, seasonality and possible methane emissions. J Atmos Chem 8:307–358

Augustin J, Chojnicki B (2008) Austausch von klimarelevanten Spurengasen, Klimawirkung und Kohlenstoffdynamik in den ersten Jahren nach der Wiedervernässung von degradiertem Niedermoorgrünland. In: Gelbrecht J, Zak D, Augustin (eds) Phosphor- und Kohlenstoff-Dynamik in wiedervernässtem Mooren des Peenetals in Mecklenburg-Vorpommern. Berichte des IGB 26/2008. Berlin, Germany

Bartholomé E, Belward AS (2000) A new approach to global land cover mapping from Earth observation data. Int J Remote Sens 26:1959–1977

Batjes NH (1996) Total carbon and nitrogen in the soils of the world. Eur J Soil Sci 47:151–163

Blackford J, Chambers FM (1993) Determining the degree of peat decomposition in peat-based palaeoclimatic studies. Int Peat J 5:7–24

Bridgeham SD, Pastor J, Dewey B et al (2008) Rapid carbon response of peatlands to climate change. Ecology 89:3041–3048

Caseldine CJ, Baker A, Charman DJ et al (2000) A comparative study of optical properties of NaOH peat extracts: implications for humification studies. Holocene 10:649–658

Charman D (2002) Peatlands and environmental change. Wiley, Chichester

Christensen JH, Hewitson B, Busuioc A et al (2007) Regional climate projections. In: Solomon S, Qin D, Manning M et al (eds) Climate change 2007: the physical science basis. Contribution of working group I to the fourth assessment report of the intergovernmental panel on climate change. Cambridge University Press, Cambridge

Clymo RS (1984) The limits to peat bog growth. Philos Trans R Soc Lond B 303:605–654

Collins ME, Kuehl RJ (2001) Organic matter accumulation and organic soils. In: Richardson JL, Vepraskas MJ (eds) Wetland soils. Genesis, hydrology, landscape and classification. Lewis Publishers, Boca Raton/London/New York/Washington, pp 137–162

Couwenberg J, Thiele A, Tanneberger F et al (2011) Assessing greenhouse gas emissions from peatlands using vegetation as a proxy. Hydrobiologia 674:67–89

D'Angelo EM, Reddy KR (1999) Regulators of heterothrophic microbial potentials in wetland soils. Soil Biol Biochem 31:815–830

Davidson EA, Janssens IA (2006) Temperature sensitivity of soil carbon decomposition and feedbacks to climate change. Nature 440:165–173

Di Gregorio A (2005) Land cover classification system, classification concepts and user manual for software (version 2). Environment and natural resources service series no. 8. FAO, Rome

Elberling B, Christiansen HH, Hansen BU (2010) High nitrous oxide production from thawing permafrost. Nat Geosci 3:332–335

FAO/UNESCO (1995, 2003) The digitized soil map of the world and derived soil properties (version 3.5). FAO land and water digital media series 1. FAO, Rome

Fargione J, Hill J, Tilman D et al (2008) Land clearing and the biofuel carbon debt. Science 319:1235–1238

Fiedler S, Holl BS, Jungkunst HF (2005) Methane budget of a Black Forest spruce ecosystem considering soil pattern. Biogeochemistry 76:1–20

Friedl MA, McIver DK, Hodges JCF et al (2002) Global land cover mapping from MODIS, algorithms and early results. Remote Sens Environ 83:287–302

Friedlingstein P, Cox P, Betts R et al (2006) Climate-carbon cycle feedback analysis: results from the C(4)MIP model intercomparison. J Climate 19:3337–3353

Fritz S, McCallum I, Schill C et al (2009) The use of crowdsourcing to improve global land cover. Remote Sens 1(3):345–354

Frolking S, Li C, Braswell R et al (2004) Short- and long-term greenhouse gas and radiative forcing impacts of changing water management in Asian rice paddies. Glob Change Biol 10:1180–1196

Frolking S, Talbot J, Jones MC et al (2011) Peatlands in the Earth's 21st century climate system. Environ Rev 19:371–396

Glatzel S (2011) Biogeochemische Stoffkreisläufe: Kohlenstoff- und Stickstoffkreislauf. In: Gebhardt H et al (Hrsg) Geographie. 2. Aufl. Spektrum, Heidelberg, pp S. 624–628

Glatzel S, Koebsch F, Beetz S et al (2011) Maßnahmen zur Minderung der Treibhausgasfreisetzung aus Mooren im Mittleren Mecklenburg. Telma 4:85–106

Gorham E (1991) Northern peatlands: role in the carbon cycle and probable responses to climate warming. Ecol Appl 1(2):182–195

Hansen MC, Townshend JRG, DeFries RS et al (2005) Estimation of tree cover using MODIS data at global, continental and regional/local scales. Int J Remote Sens 26:4359–4380

Harris ESJ (2008) Ethnobryology: traditional uses and folk classification of bryophytes. The Bryologist 111(2):169–217

Heimann M (2011) Enigma of the recent methane budget. Nature 476:157–158

Hergoualc'h K, Verchot LV (2011) Stocks and fluxes of carbon associated with land use change in Southeast Asian tropical peatlands: a review. Glob Biogeochem Cycles 25. doi:10.1029/2009GB003718

Herold M, Mayaux P, Woodcock CE et al (2008) Some challenges in global land cover mapping, an assessment of agreement and accuracy in existing 1 km datasets. Earth observations for terrestrial biodiversity and ecosystems special issue. Remote Sens Environ 112(5): 2538–2556

IPS (2011) Global peat resources. www.peatsociety.org/peatlands-and-peat/global-peat-resources-country. Accessed on 29 Dec 2011

Joosten H, Clarke D (2002) Wise use of mires and peatlands. Intl Mire Conserv Group, Intl Peat Soc, Vapaudenkatu

Jungkunst HF (2010) Soil science: Arctic thaw. Nat Geosci 3:306–307

Jungkunst HF, Fiedler S (2007) Latitudinal differentiated water table control of carbon dioxide, methane and nitrous oxide fluxes from hydromorphic soils: feedbacks to climate change. Glob Change Biol 13:2668–2683

Jungkunst HF, Flessa H, Scherber C et al (2008) Groundwater level controls CO_2, N_2O and CH_4 fluxes of three different hydromorphic soil types of a temperate forest ecosystem. Soil Biol Biochem 40:2047–2054

Kai FM, Tyler SC, Randerson JT et al (2011) Reduced methane growth rate explained by decreased Northern Hemisphere microbial sources. Nature 476:194–197

Kögel-Knabner I (2002) The macromolecular organic composition of plant and microbial residues as inputs to soil organic matter. Soil Biol Biochem 34:139–162

Kögel-Knabner I, Amelung W, Cao ZH et al (2010) Biogeochemistry of paddy soils. Geoderma 157:1–14

Krankina ON, Pflugmacher D, Friedl M et al (2008) Meeting the challenge of mapping peatlands with remotely sensed data. Biogeosciences 5:1809–1820

Lal R (2002) Soil carbon dynamics in cropland and rangeland. Environ Pollut 116:353–362

Lal R (2004) Soil carbon sequestration impacts on global climate change and food security. Science 304:1623–1627

Lane J, Minkkinen K, Trettin C (2009) Direct human impacts on the peatland carbon sink. Carbon cycling in northern peatlands. Geophys Mono Ser 184:71–78

Li JH, Chen WJ (2005) A rule-based method for mapping Canada's wetlands using optical, radar and DEM data. Int J Remote Sens 26:5051–5069
Li WH, Dickinson RE, Fu R et al (2007) Future precipitation changes and their implications for tropical peatlands. Geophys Res Lett 34:L08701. doi:10.1029/2006GL029022
Limpens J, Berendse F (2003) How litter quality affects mass loss and N loss from decomposing Sphagnum. Oikos 103:537–547
Limpens J, Berendse F, Blodau C et al (2008) Peatlands and the carbon cycle: from local processes to global implications – a synthesis. Biogeosciences 5:1475–1491
Loveland TR, Reed BC, Brown JF et al (2000) Development of a global land cover characteristics database and IGBP DISCover from 1-km AVHRR data. Int J Remote Sens 21:1303–1330
MacDonald GM, Beilman DW, Kremenetski KV et al (2006) Rapid early development of circumarctic peatlands and atmospheric CH_4 and CO_2 variations. Science 314:285–288
Marushchack ME, Pitkämäki A, Koponen H et al (2011) Hot spots for nitrous oxide emissions found in different types of permafrost peatlands. Glob Change Biol 17:2601–2614
Mikhailova EA, Post CJ (2006) Organic carbon stocks in the Russian Chernozems. Eur J Soil Sci 57:330–336
Mitsch WJ, Gosselink JG (2007) Wetlands. Wiley, Hoboken
Mitsch WJ, Gosselink JG, Anderson CJ et al (2009) Wetland ecosystems. Wiley, Hoboken
National Wetland Working Group (1997) The Canadian wetland classification system, 2nd edn. University of Waterloo, Waterloo
Nilsson M, Mikkela C, Sundh I et al (2001) Methane emission from Swedish mires: national and regional budgets and dependence on mire vegetation. J Geophys Res Atmos 106:20847–20860
Page SE, Siegert F, Rieley JO et al (2002) The amount of carbon released from peat and forest fires in Indonesia during 1997. Nature 420:61–65
Page SE, Rieley JO, Banks CJ (2011) Global and regional importance of the tropical peatland carbon pool. Glob Change Biol 17:798–818
Paul EA, Clark FE (1996) Soil microbiology and biochemistry, 2nd edn. Academic, San Diego
Pflugmacher D, Krankina ON, Cohen WB (2007) Satellite based peatland mapping, potential of the MODIS sensor, global planet. Change 56:248–257
Poulter B, Christensen NL, Halpin PN (2006) Carbon emissions from a temperate peat fire and its relevance to interannual variability of trace atmospheric greenhouse gases. J Geophys Res Atmos 111. doi:10.1029/2005JD006455
Prechtel A, von Luetzow M, Schneider BU et al (2009) Organic carbon in soils of Germany: status quo and the need for new data to evaluate potentials and trends of soil organic carbon sequestration. J Plant Nutr Soil Sci 172:601–614
Primavesi A (1984). Manejo ecológico del suelo. La agricultura en regiones tropicales. 5ta Ediciôn. El Ateneo, Rio de Janeiro, 499 pp
Ramsar Information Bureau (1971). http://www.ramsar.org/cda/en/ramsar-documents-texts-convention-on-20708/main/ramsar/1-31-38%5E20708_4000_0__
Repo ME, Susiluoto S, Lind SE et al (2009) Large N_2O emissions from cryoturbated peat soils in tundra. Nat Geosci 2:189–192
Richardson JL, Arndt JL, Montgomery JA (2001) Hydrology of wetland and related soils. In: Richardson JL, Vepraskas MJ (eds) Wetland soils. Genesis, hydrology, landscape and classification. Lewis Publishers, Boca Raton/London/New York/Washington, DC
Roman-Cuesta RM, Salinas N, Asbjornsen H et al (2011) Implications of fires on carbon budgets in Andean cloud montane forest: the importance of peat soils and tree resprouting. Forest Ecol Manage 261:1987–1997
Rosenqvist A, Finlayson CM, Lowry J et al (2007) The potential of long wavelength satellite borne radar to support implementation of the Ramsar Wetlands Convention. Aquat Conserv Mar Freshw Ecosyst 17(3):229–244
Roulet N, Moore T, Bubier J et al (1992) Northern fens-methane flux and climatic change. Tellus B 44:100–105
Rydin H, Jeglum J (2006) The biology of peatlands. Oxford University Press, Oxford
Sahrawat KL (2004) Organic matter accumulation in submerged soils. Adv Agron 81:169–201

Sauer D, Sponagel H, Sommer M et al (2007) Podzol: soil of the year 2007. A review on its genesis, occurrence, and functions. J Plant Nutr Soil Sci 170:581–597
SCEP (1970) Man's impact on the global environment: assessment and recommendations for action. The MIT Press, Cambridge
Schawe M, Glatzel S, Gerold G (2007) Soil development along an altitudinal transect in a Bolivian tropical montane rainforest: podzolization vs. hydromorphy. Catena 69:83–90
Schlesinger WH (1997) Biogeochemistry: an analysis of global change, 2nd edn. Academic, San Diego
Schmidt MWI, Torn MS, Abiven S et al (2011) Persistence of soil organic matter as an ecosystem property. Nat Geosci 478:49–56
Schulze E-D, Luyssaert S, Ciais P et al (2009) Importance of methane and nitrous oxide for Europe's terrestrial greenhouse-gas balance. CarboEuropeTeam 2:842–850
Strack M (ed) (2008) Peatlands and climate change. Intl Peat Soc, Vapaudenkatu
Tarnocai C, Canadell JG, Schuur EAG et al (2009) Soil organic carbon pools in the northern circumpolar permafrost region. Glob Biogeochem Cycles 23. doi:10.1029/2008GB003327
Thompson JA, Bell JC (1996) Color index for identifying hydric conditions for seasonally saturated mollisols in Minnesota. SSSA J 60:1979–1988
Thompson JA, Bell JC (2001) Hydric soils indicators in mollisol landscapes. In: Richardson JL Vepraskas MJ (ed) Wetland soils. Genesis, hydrology, landscape and classification. Lewis Publishers, Boca Raton, pp 371–382
Tiner RW (1998) In search of swampland: a wetland source book and field guide. Rutgers University Press, New Brunswick
Tiner RW (1999) Wetland indicators: a guide to wetland identification, delineation, classification and mapping. Lewis Publishers, Boca Raton
Tooth S, McCarthy TS (2007) Wetlands in drylands: geomorphological and sedimentological characteristics, with emphasis on examples from southern Africa. Prog Phys Geogr 31:3–41
Turquety S, Logan JA, Jacob DJ et al (2007) Inventory of boreal fire emissions for North America in 2004: importance of peat burning and pyroconvective injection. J Geophys Res Atmos 112. doi:10.1029/2006JD007281
US Soil Taxonomy (2006). http://soils.usda.gov/technical/classification/tax_ke
van der Werf GR, Randerson JT, Giglio L et al (2010) Global fire emissions and the contribution of deforestation, savanna, forest, agricultural, and peat fires (1997–2009). Atmos Chem Phys 10:11707–11735
Vicca S, Janssens IA, Flessa H et al (2009) Temperature dependence of greenhouse gas emissions from three hydromorphic soils at different groundwater levels. Geobiology 7:465–476
Waksman SA (1936) Humus – origin, chemical composition, and importance in nature. The Williams and Wilkins Company, Baltimore
Weltzin JF, Pastor J, Harth C et al (2000) Response of bog and fen plant communities to warming and water-table manipulations. Ecology 81:3464–3478
Wheeler BD, Shaw SC (1995) Restoration of damaged peatlands with particular reference to lowland raised bogs affected by peat extraction. Department of the Environment, University of Sheffield, London
Wieder RK, Vitt DH (eds) (2006) Boreal peatland ecosystems. Springer, Berlin
Wieder RK, Scott KD, Kamminga K et al (2009) Postfire carbon balance in boreal bogs of Alberta, Canada. Glob Change Biol 15:63–81
World Energy Council (2007) Survey of energy resources 2007. London, UK. pp 586 (ISBN: 0946 121 265)
WRB (2007) World reference base for soil resources 2006, first update 2007. FAO, Rome, pp 128
Yu Z, Loisel J, Brosseau DP et al (2010) Global peatland dynamics since the Last Glacial Maximum. Geophys Res Lett 110. doi:10.1029/2010GL043584
Yu Z, Beilman DW, Frolking S et al (2011) Peatlands and their role in the global carbon cycle. EOS 92(12):97–98
Zoltai SC, Tarnocai C (1975) Perennial frozen peatlands in the western Arctic and sub-Arctic of Canada. Can J Earth Sci 12:28–34

Chapter 8
Permafrost – Physical Aspects, Carbon Cycling, Databases and Uncertainties

Julia Boike, Moritz Langer, Hugues Lantuit, Sina Muster, Kurt Roth, Torsten Sachs, Paul Overduin, Sebastian Westermann, and A. David McGuire

Abstract Permafrost is defined as ground that remains below 0°C for at least 2 consecutive years. About 24% of the northern hemisphere land area is underlain by permafrost. The thawing of permafrost has the potential to influence the climate system through the release of carbon (C) from northern high latitude terrestrial ecosystems, but there is substantial uncertainty about the sensitivity of the C cycle to thawing permafrost. Soil C can be mobilized from permafrost in response to changes in air temperature, directional changes in water balance, fire, thermokarst, and flooding. Observation networks need to be implemented to understand responses of permafrost and C at a range of temporal and spatial scales. The understanding

J. Boike (✉) • M. Langer • H. Lantuit • S. Muster • P. Overduin
Alfred Wegener Institute for Polar and Marine Research,
Telegrafenberg A6, 14473 Potsdam, Germany
e-mail: julia.boike@awi.de; moritz.langer@awi.de; hugues.lantuit@awi.de; sina.muster@awi.de; paul.overduin@awi.de

K. Roth
Institute of Environmental Physics, Heidelberg University,
Heidelberg, Germany
e-mail: kurt.roth@iup.uni-heidelberg.de

T. Sachs
Inorganic and Isotope Geochemistry, GFZ German Research Centre for Geosciences,
Potsdam, Germany
e-mail: torsten.sachs@gfz-potsdam.de

S. Westermann
Department of Geosciences, University of Oslo, Oslo, Norway
e-mail: sebastian.westermann@geo.uio.no

A.D. McGuire
U.S. Geological Survey, Alaska Cooperative Fish and Wildlife Research Unit,
University of Alaska Fairbanks, Fairbanks, AK, USA
e-mail: admcguire@alaska.edu

R. Lal et al. (eds.), *Recarbonization of the Biosphere: Ecosystems and the Global Carbon Cycle*, DOI 10.1007/978-94-007-4159-1_8,

gained from these observation networks needs to be integrated into modeling frameworks capable of representing how the responses of permafrost C will influence the trajectory of climate in the future.

Keywords Permafrost • Carbon • Land cover • Surface energy balance • Arctic biome • Arctic vegetation • Greenhouse gases • Positive feedback • IPCC • Climate models • Soil organic matter • Soil organic carbon • Methane • Active layer • Little Ice Age • Surface energy balance • Thermokast • Fish habitats • Siberia • Cryoturbation

Abbreviations

AOGCMs	Atmosphere-Ocean General Circulation Models
C	carbon
CO_2	carbon dioxide
CAVM	Circum Arctic Vegetation Map
CALM	Circumpolar Active Layer Monitoring network
C4MIP	Coupled Carbon Cycle Climate Model Intercomparison Project
DOC	dissolved organic carbon
ECHAM	European Centre/Hamburg Model
GTN-P	Global Terrestrial Network for Permafrost
GHG	greenhouse gas
IPCC	Intergovernmental Panel on Climate Change
MAAT	mean annual air temperatures
CH_4	methane
OM	organic matter
Pg C	1 Pg = 1 billion metric tons = 10^{15} g = 1 Gt
RCM	Regional Climate Model
SOC	soil organic carbon
SOM	soil organic matter

8.1 Permafrost: A Phenomenon of Global Significance

Ongoing discussions on the degradation of permafrost are now reaching a broad scientific, public, and political audience since the consequences of permafrost degradation are expected to be felt not only locally (infrastructure) and regionally (water supply), but also globally due to the resulting greenhouse gas (GHG) emissions. Trenberth (2010) highlighted the need to include feedback mechanisms, such as GHG emissions from shrinking permafrost areas in climate models for the next Intergovernmental Panel on Climate Change (IPCC) assessment. Changes in carbon (C) cycling (the mobilization of frozen, formerly protected soil organic carbon (SOC) pools) and changes in surface energy partitioning (Chapin et al. 2005) are expected to be nonlinear and, once certain thresholds in a system have been reached, subsequent incremental changes have the potential to produce strong effects. However, changes to permafrost have not yet been taken into account in these "tipping elements" for the earth's climate system, in which a small change in control parameters can have large consequences for some system variables (Lenton et al. 2008). Recent studies have emphasized the role of permafrost as a crucial factor affecting the global C budget. Thawing of permafrost has the potential to release large C reservoirs (Schuur et al. 2008), but there is a lot of uncertainty about the sensitivity of the C cycle to changes in the Arctic (McGuire et al. 2009). This uncertainty is due to a limited knowledge of the large Arctic area in general, as well as to small- scale variability and the complexity of processes.

Friedlingstein et al. (2006) have shown that the current generation of global climate models lead to very large uncertainties in the magnitude of feedbacks to global change from the high northern latitudes. There is a large inter-model spread, with estimates of cumulative carbon dioxide (CO_2) exchange from high-latitude land surfaces by the year 2100 ranging from −80 (−: sink) to +20 (+: source) Pg C (1 Pg = 1 billion metric tons = 10^{15} g = 1 Gt) under the IPCC's SRES A2 scenario. It is possible that the model mean could even have the wrong sign, simply because the dynamics and size of the permafrost C reservoir were not correctly represented in the models used for the last round of the Coupled Carbon Cycle Climate Model Intercomparison Project (C4MIP). Moreover, the partitioning between methane (CH_4) and CO_2 emissions, which can have a major effect on global-scale climate feedback, was not even analyzed because most models simply did not consider CH_4 emissions. Recent modeling by Schaefer et al. (2011) projects that the Arctic permafrost areas will change from C sinks to C sources by the mid 2020s, amounting to a cumulative C flux to the atmosphere of 190 ± 64 Pg C by 2200. Schneider von Deimling et al. (2011) predict that by the year 2300 more than half of the potentially vulnerable C in the upper 3 m of soil of the northern permafrost region (600–1,000 Pg C) could be released as CO_2. Permafrost, therefore, needs to be included in all projections of future climate.

This chapter provides an overview of the current understanding of the distribution of permafrost, the physical processes, and the interactions between physical and biogeochemical cycles, C pools and fluxes. The overall objective is to discuss major gaps and uncertainties in the understanding of both negative and positive feedbacks between permafrost and climate.

8.2 Permafrost: Definition, Distribution and History

Permafrost is defined as ground that remains below 0°C for at least 2 consecutive years (van Everdingen 1998). About 24% (22.8 million km²) of the northern hemisphere land area is underlain by permafrost, amounting to about one fifth of the world's total land area (Brown et al. 1997). The material forming permafrost can

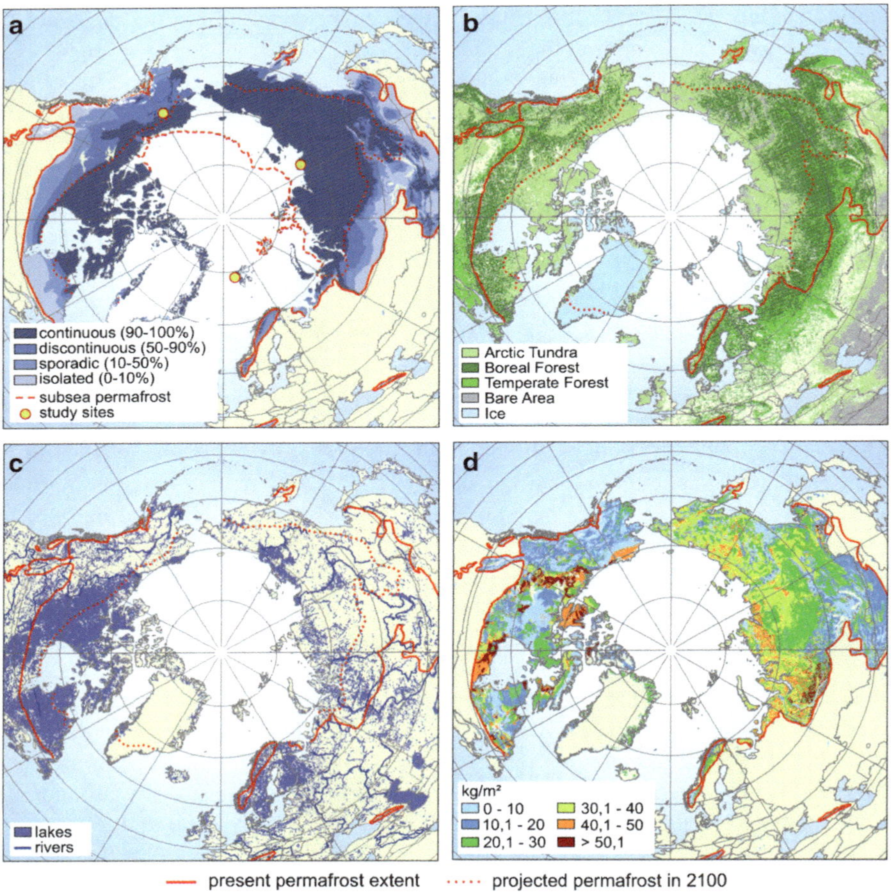

Fig. 8.1 (**a**) Permafrost map, including subsea permafrost (Brown et al. 1997) and predicted decrease by 2100 (ACIA 2004). (**b**) Major terrestrial ecozones (Modified after Global Land Cover 2000 database). (**c**) Lakes, wetlands and major rivers (Modified after Lehner and Döll 2004). (**d**) Soil organic carbon content map for the upper 100 cm of soils (Tarnocai et al. 2009). Sites introduced in Sect. 8.3 are shown as *yellow dots* in (**a**)

include bedrock, sediment of mineral and organic origin, and ice (or unfrozen water with a freezing point depression due to high salt content). Permafrost regions are classified as continuous (with permafrost underlying 90–100% of the landscape), discontinuous (50–90%), or sporadic (10–50%) and isolated (0–10%) permafrost, on the basis of areal distribution (Brown et al. 1997, Fig. 8.1a). The thickness of the permafrost layer varies from less than 1 m to up to 1,600 m (recorded in Siberia: Romanovskii et al. 2004). The extreme thickness is explained by large regions in Northeast Siberia and North Alaska having remained unglaciated during much of the Pleistocene, which allowed greater heat loss from the ground. In addition, loess deposition and syngenetic permafrost (permafrost that formed through a rise of the permafrost table during the deposition of additional sediments or other earth material on the ground surface (van Everdingen 1998)) resulted in the formation of permafrost to a depth of several hundreds of meters, with a high ice content sporadically (>70 vol% at depths of up to 60 m in Yedoma sediments: Romanovskii et al. 2004). These late Pleistocene, ice-rich sediments "survived" the Holocene temperature maximum (ca. 5–9 kyear before present), when temperatures were up to 3°C higher than today. A large portion of these ice-rich sediments was degraded by thermokarst, and riverine and coastal erosion. The exact amount of the fraction that "survived" is unknown, but first-order estimates assume that 1 million km^2 still exist, compared to an estimated more than two million during the last glacial maximum (Zimov et al. 2006; Walter et al. 2007). In contrast, the southern discontinuous permafrost zone is largely prone to thawing with warmer temperatures (Zhang et al. 2008).

8.3 Physical Factors Affecting the Permafrost Thermal Regime

8.3.1 Permafrost Temperatures

The ground thermal regime of permafrost areas has an important influence on their stability and resistance to environmental change. Permafrost temperature is a result of the glaciation history, the large scale surface energy balance (past and present), and the thermal properties (conductivity and capacity) of surface and subsurface material (soil, snow, vegetation) and water bodies, including subsurface waters (Yershov 2004). Below the permafrost base, temperatures are above the freezing point of water due to geothermal heat from the interior of the earth (Yershov 2004).

Significant variations are observed in permafrost temperatures within the Arctic. High Arctic areas have permafrost temperatures in the −5°C to −10°C range, with some sites below −15°C (Romanovsky et al. 2010). The permafrost thermal state and spatial variability for three sites are shown as examples in Fig. 8.2. Imnavait Creek (northern foothills of the Brooks Range, Alaska) was glaciated during the Pleistocene and continuous permafrost extends to depths of about 240 m (Lachenbruch et al. 1982), with present-day mean annual air temperatures (MAAT) of −7.4°C (Fig. 8.2a). Of the three sites, the coldest permafrost temperature is recorded from Eastern Siberia (Fig. 8.2b), resulting from the lack of glaciation during the Pleistocene together with extreme climatic conditions during the Late Pleistocene and still today, with a present-day MAAT of about −14.9°C. The warmest high Arctic permafrost exists to depths of only 100 m in Svalbard (Fig. 8.2c) as a result of repeated glaciations during the Quaternary and a relatively "warm" maritime climate with a present-day MAAT of −5.5°C.

Almost all temperature measurements in permafrost show an increasing trend during the last 20–25 years of the twentieth century, and into the first few years of the twenty-first century, caused by atmospheric warming and/or changes in snow

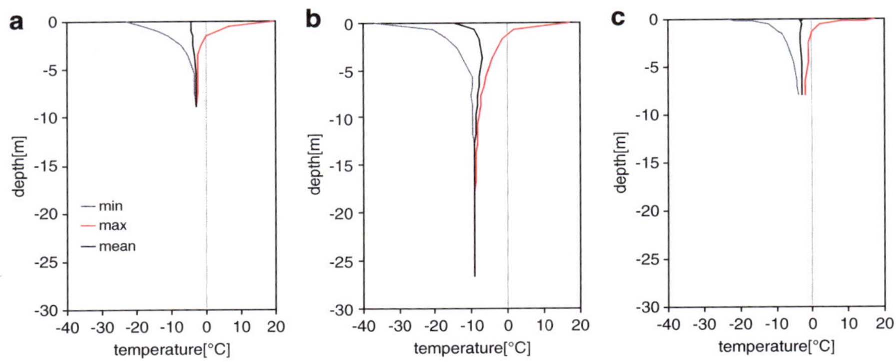

Fig. 8.2 Mean, maximum, and minimum permafrost temperatures recorded in selected boreholes in Alaska (**a**, Imnavait Creek, northern foothills of the Brooks Range), Siberia (**b**, Samoylov, Lena Delta), and Svalbard (**c**, Bayelva). For locations see Fig. 8.1a. The same depth and temperature scales are used to demonstrate the differences between warm (Svalbard) and cold (Siberia) permafrost. Analysis based on August 2009 – August 2010 (Svalbard), August 2007 – August 2008 (Siberia) and February 2006 – February 2007 (Alaska) (Data for the Alaska site courtesy of L. Hinzman (Hinzman et al. 2008))

depth (Romanovsky et al. 2010). For example, in Alaska the permafrost temperatures at 20 m depth increased by 1.5°C over a period of 15 years (Osterkamp 2007; Hinzman et al. 2005). In warmer, discontinuous permafrost such warming will eventually convert permafrost ice to water and potentially mobilize previously frozen material containing C. Through the Global Terrestrial Network for Permafrost (GTN-P, www.gtnp.org), a network of permafrost observatories has been initiated since 1990 for detecting changes in permafrost temperature and active layer thickness (Circumpolar Active Layer Monitoring network, or CALM). However, these observatories are not distributed evenly; especially the Russian Arctic is an area of large data paucity.

8.3.2 Active Layer Dynamics

The upper soil layer in permafrost areas, characterized by seasonal thawing and freezing is called the *active layer,* and can range from several cm to several meters in thickness (van Everdingen 1998). The annual freeze-thaw cycle of the active layer is illustrated using temperature and soil moisture data over 2 years (2004, 2005; Fig. 8.3) from the polygonal tundra study site in Siberia (Fig. 8.4e).

The annual temperature range is the highest at the surface, ranging between −30°C and +15°C, and it decreases with depth. During the short spring and summer period from June to September, the active layer progressively warms and thaws to a maximum depth of about 40 cm. Autumnal isothermal conditions (at 0°C) and freeze back start in September but can take more than a month for the active layer

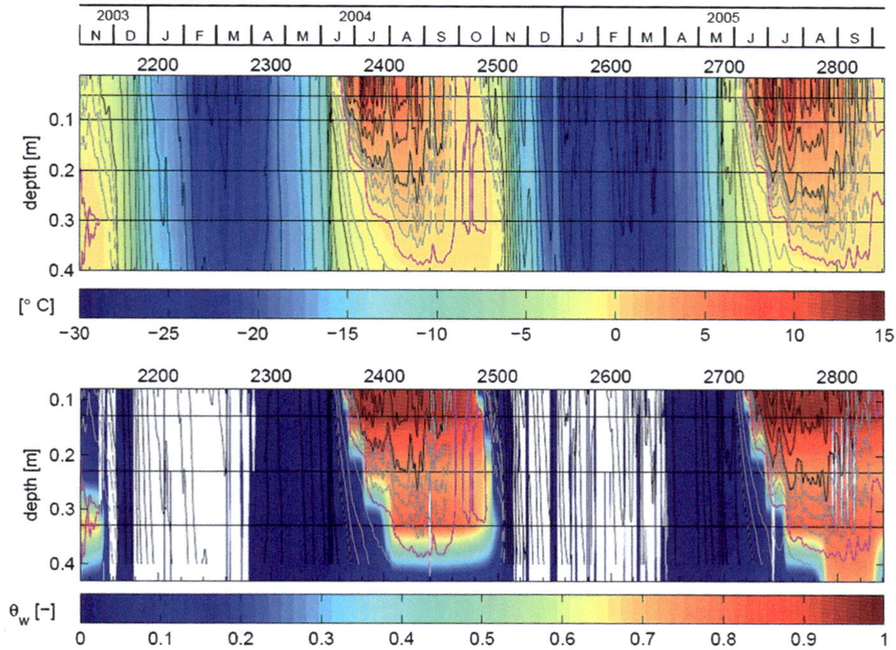

Fig. 8.3 Example of thermal and hydrologic dynamic of the active layer at the Siberian permafrost site (Lena Delta, Fig. 8.4e). Profile plots of soil temperature (*upper plate*) and volumetric water content using Time Domain Reflectometry (*lower plate*) over 2 years for the polygonal tundra site. The *straight, horizontal black lines* indicate the probe depths while the *curved black lines* are isothermal lines at intervals of 2°C. The 0°C isotherm is shown as a *magenta line*, with intermediate isotherms (at intervals of 0.5°C) shown as *gray lines*

to refreeze (Fig. 8.3). During winter, the ground temperature stays well below 0°C and almost all water in the soil is frozen.

Only minor differences in thaw depth, soil temperature and volumetric water content occurred between 2004 and 2005, as a result of similar climatic and hydrologic conditions. Once the peat soil (with a porosity > 80) is thawed, the soil's volumetric water content remains at maximum (between 0.9 and 1 vol%). An exception to this are only drier years when evapotranspiration exceeds summer rainfall and the water level drops below the ground surface (Boike et al. 2008).

The active layer is of special significance for the C cycle since many of the biogeochemical processes take place in this layer and the release of C in the gas, liquid or particulate form occurs via this layer. Processes operating in the active layer include freezing and thawing and associated volume changes and sediment sorting due to frost heave and thaw subsidence, migration of water as a result of gradients (topographic, matric potential, temperature, vapour pressure, solute concentrations), and mechanical mixing through cryoturbation and cracking. These processes are complex and patterned ground results from a combination of many individual

Fig. 8.4 Examples of permafrost patterned ground, photographed at various sites. The size of the patterns increases from meters (*top row*) to several tens and hundreds of meters (*bottom row*). (**a**) Non-sorted circles (Galbraith Lake, North Slope, Alaska); (**b**, **c**) circles and polygons (Brøgger Peninsula, Svalbard); (**d**) circles and polygons (Howe Island, Alaska); (**e**) water-filled polygons (Lena Delta, Siberia); (**f**) thermokarst lake; (**g**) pingo (Seward Peninsula, Alaska); (**h**) Yedoma sediments with high ground ice content, exposed by erosion (Lena Delta, Siberia)

processes (Fig. 8.4; Washburn 1979). For example, non-sorted circles (Fig. 8.4a) are characterized by a circular, bare, almost flat soil surface (~1–3 m diameter) and lack a border of stones. These features form by differential frost heave and ice lens formation during winter (Washburn 1979). Cryoturbation, i.e., the mechanical mixing of material as a result of seasonal freeze-thaw, mobilizes soil C both upwards and downwards (see examples in Fig. 8.5). Tarnocai et al. (2009) and Ping et al. (2008) have shown that cryoturbated soils generally have the highest mean SOC stocks. Koven et al. (2009) included soil organic matter (SOM) and a simplified vertical mixing in their C cycle model. The soil's thermal regime is changed by the change of the thermal properties (due to changes in SOC) which, in turn, affects the residence time of SOC. The effects of thermal insulation by organic matter (OM) and of cryoturbation lead to a significant increase in SOC stocks, in agreement with estimates of high-latitude soil C stocks (Tarnocai et al. 2009).

Predicting the vulnerability of permafrost C to climate change requires simulation of the active layer's annual dynamics coupled with the C cycle (Hollesen et al. 2010), as well as the soil water status which determines aerobic or anaerobic decomposition of OM. Predicting the hydrologic conditions of the surface is challenging due to difficulties associated with predicting regional changes of temperature,

Fig. 8.5 Examples of the effect of cryoturbation on the surface and subsurface. (**a**) Tundra site with ice wedge polygonal pattern (Samoylov, Lena Delta, Siberia). The *polygons* are about 10–20 m across. (**b**) Soil profile across a polygonal rim showing thawed soil and frozen ice wedge below. The effect of cryoturbation is visible by the organic material pulled downwards and the distortion of the horizontal soil layers. The *tube* is used to drain water from the otherwise waterlogged profile. (**c**) *Non-sorted circles* (Bayelva, Svalbard). The diameter of the circles is about 1 m. (**d**) Soil profile across a *non-sorted circle*. Cryoturbation results in downward migration of material at the *edges* of the circle and upward migration in the *center*

precipitation and drainage in global circulation models. To date, only a few small-scale local processes, such as cryoturbation, have been identified as globally significant for C cycling and incorporated into climate models.

8.3.3 Land Cover

Land cover affects the biogeophysical and biogeochemical properties of the permafrost surface (hydrology, albedo, biomass, and vegetation type) which in turn determine the exchange of energy, water and C between the surface and the atmosphere (Bonan et al. 1995; Chapin et al. 2005). With a very low population density (0.32 per km^2), the anthropogenic influence on the land cover of permafrost areas is usually small and concentrated around towns or industrial areas in the south (Arctic Human Development Report: Einarsson et al. 2004). There is a limited use of land for cultivating crops, pasture, or forests. For example, less than 0.5% of the total land area covered by the Lena river drainage basin in Siberia is used for forest harvest and agriculture (McGuire et al. 2010).

Figure 8.1b shows the present-day distribution of ecozones in permafrost areas, covering a wide spectrum from barren ground surface in high Arctic areas to tundra and boreal forests in the south. The Circum Arctic Vegetation Map (CAVM 2003) by Walker et al. (2005) is based on 1 km resolution satellite data and provides a more detailed map of Arctic vegetation types. The CAVM is, however, restricted to the Arctic tundra, covering an area of about 5.05×10^6 km^2 (Walker et al. 2005) while forests cover an area of 14.6×10^6 km^2 (ACIA 2004), with large regions in central Siberia extending into the continuous permafrost zone. Boreal forests have been predicted to decrease and shift northwards; forest–steppe and steppe ecosystems have been predicted to be the dominant vegetation type rather than forests over half of Siberia in warmer and drier climate scenarios by 2080 (Tchebakova et al. 2009).

Water surfaces show a distinctly different surface energy balance than the surrounding vegetated or barren land surfaces (Gutowski et al. 2007; Rouse et al. 2007). Lakes and wetlands are abundant in permafrost landscapes (Fig. 8.1c). They show the greatest number and surface area compared to temperate and tropical lakes (Lehner and Döll 2004), but little difference in lake number and size has been found between areas of continuous, discontinuous, and/or sporadic permafrost (Smith et al. 2007). Current databases (such as the Global Lakes and Wetlands Database by Lehner and Döll 2004) only capture larger lakes (>10 ha), with smaller water bodies being not visible by coarse-resolution mapping (Grosse et al. 2008; Muster et al. 2012). Moreover, land cover classifications on similar scales show considerable diversity in Arctic regions, especially with respect to the extent of water bodies and wetlands (Frey and Smith 2007).

Thermokarst is a process that drastically alters the surface structures in permafrost terrains. The process of thermokarst involves the thawing of ice-rich permafrost, and subsidence of the ground surface (see example in Fig. 8.6).

Fig. 8.6 Example of drastic landscape changes caused by thawing permafrost (thermokarst) on Alaska's North Slope during August 2004. (**a**) Helicopter aerial photo (Courtesy of L. Hinzman), (**b**–**e**) ground-based photos. It is not clear what caused the thaw and subsequent collapse of the surface, which was followed by retrogressive erosion of the stream. In the short term, sediment and solute transport was significantly enhanced into the nearby Toolik lake (~10 km to the north) affecting fish habitats (Bowden et al. 2008). Over the long term, the landscape surface and drainage characteristics were changed

Thermokarst lakes and ponds are formed through water accumulating in the resulting depressions. In discontinuous permafrost areas, further thawing, however, may lead to the draining of lakes when thaw bulbs (taliks) underneath a lake completely penetrate the permafrost (Yoshikawa and Hinzman 2003). The quantification of thermokarst lake dynamics of three circumpolar peatland sites shows that lake drainage and new thermokarst lake formation is most pronounced in the sporadic permafrost zone (northern Sweden) compared to the continuous (central Canada) and discontinuous (northeastern European Russian) permafrost zone (Sannel and Kuhry 2011).

Changes in land cover (due to thermokarst or fires), earlier snow melt, and later snow cover result in a change in surface albedo which alters the radiative feedback from the surface to the atmosphere. Vegetation and organic soil layers may act as insulators, protecting permafrost from warmer temperatures. For example, the expansion of shrub cover observed on tundra in Alaska (Sturm et al. 2001; Tape et al. 2006), reduced the mean annual permafrost temperature by several degrees (Blok et al. 2010). This effect may be offset by an increase in snow cover associated with shrub expansion (Sturm et al. 2005) which insulates permafrost from cold winter temperatures (Blok et al. 2010).

Documenting the current state of land cover establishes a baseline for monitoring land cover changes due to climate warming. None of the existing land cover classifications are able to resolve the heterogeneity of Arctic land surfaces which can alter the surface properties on a scale of meters. Regional studies show promising results as they can process higher resolution data such as Landsat data (Schneider et al. 2009), but the lack of uniformity in classification approaches means that it is difficult to make comparisons across the Arctic (Walker et al. 2005).

The accuracy of available databases is largely compromised by their low resolution and limited field validation or ground truthing. This is especially evident in the case of small water bodies or patterned ground.

8.3.4 Surface Energy Balance

The permafrost's surface temperature is determined by the surface energy balance composed of radiation (short and long wave), turbulent heat fluxes (sensible and latent), and heat flux into the ground, snow, or water bodies. Several descriptive seasonal studies are available for Alaskan, Canadian, and Scandinavian sites, as well as a few for Siberian sites (Ohmura 1982; Eugster et al. 2000; Lloyd et al. 2001), but quantitative long-term studies are almost non-existent. Annual energy balance studies and studies from Siberia are particularly scarce but are of great importance for the validation of climate and permafrost surface schemes within climate models. In this section, the differences in annual surface energy budget are demonstrated for the Svalbard (Westermann et al. 2009) and Siberian sites (Langer et al. 2011a, b) introduced in Sect. 8.3.1. At both locations, the surface energy balance is determined largely by radiation, i.e., net short-wave radiation during summer as an energy source and net long-wave radiation during the winter as an energy sink (Fig. 8.7).

The latent heat flux is a factor of two higher than the sensible heat flux at the wet tundra Siberian site while at the Svalbard site with drier surface conditions, the sensible and latent heat fluxes are nearly equal. The seasonal thawing of the active layer in July and August takes up to 20% of the net radiation at the Siberian site. Of importance is the fact that the winter ground heat flux forms a significant component of the surface energy balance at the Siberia site, with a relative contribution of up to 60%. The high contribution of this ground heat flux is due to the strong soil temperature gradient, the high ice content, and the large annual surface temperature range which is related to the extreme climatic conditions. During winter, the importance of water bodies, in this case small thermokarst ponds (~100 m^2: Fig. 8.4e) is twofold. First the energy storage close to the surface in the form of unfrozen water is greater than in the surrounding tundra so that the heat flux supplied to the atmosphere is higher by a factor of two. Second, ponds require substantially more time (several months) to fully refreeze so that biogeochemical processes are active for a much longer period of the year (Langer et al. 2011b). Predicting C emissions through permafrost and climate models requires the accurate representation of the energy

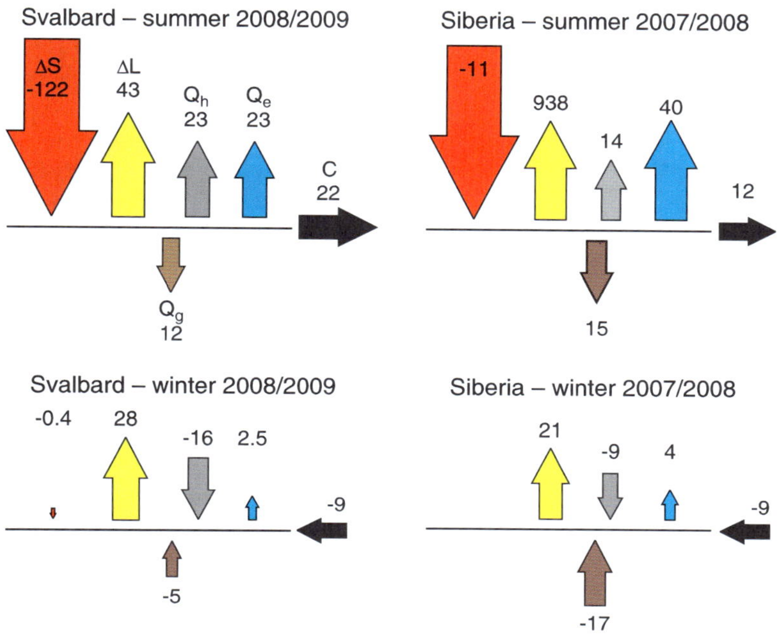

Fig. 8.7 Comparison of typical mean summer (*upper two figures*) and winter (*lower two figures*) energy fluxes (W m^{-2}) for the Svalbard and Siberian sites. Fluxes are scaled relative to each other. ΔS net shortwave radiation, ΔL net longwave radiation, Q_h sensible heat flux, Q_e latent heat flux, Q_g ground heat flux, C closure term (Databases: Westermann et al. 2009; Langer et al. 2011a, b)

balance, including freeze-thaw processes, as well as representation of sub-grid cell variability in the landcover, especially with regard to water bodies.

8.4 Carbon Stocks and Carbon Mobilization

8.4.1 Carbon Stocks of Soils and Deeper Permafrost

Tarnocai et al. (2009) estimated that 495 Pg C are stored in the upper meter of soil, and 1,024 Pg C in the upper 3 m of soil in the northern circumpolar permafrost region (Fig. 8.1d). By including an 407 Pg C for deeper amounts stored in Yedoma (ice-rich and C-rich loess deposits formed in the Pleistocene) and 241 Pg C stored in alluvial deposits (Tarnocai et al. 2009) a total of 1,672 Pg C was estimated. The organic soils (peatlands), cryoturbated soils, and Yedoma deposits being the soils with the highest mean organic C contents. The C content of Arctic permafrost areas is, therefore, significant when compared to global SOC stocks in the upper meter of soil of 1,462–2,344 Pg C (Batjes 1996; Denman et al. 2007).

Comparisons of SOC stocks between regions are difficult because of the different methods and upscaling techniques used. Using the large scale circumpolar data set for cryoturbated soils (Turbels) for the Thule peninsula, Greenland, 0.018 Pg C was estimated. But applying a combined high resolution data set of field and satellite observations, only 0.004 Pg C was calculated (using a correlation between SOC from field profiles and high resolution satellite picture determined normalized difference vegetation index; Howarth Burnham and Sletten 2010). For the central Canadian Arctic region, mean SOC storage estimates have been revised from previous estimates to show higher contents overall as well as large spatial variability. The peatlands (mainly bogs) comprised the highest SOC pool with 56% of the total SOC, but cryoturbated soil pockets in turbic cryosols contributed 17% to the total SOC stock (Hugelius et al. 2010).

The estimation of C stocks through upscaling attempts to compensate for limited data sets of SOC content, poor accuracy in geolocation of older datasets, low resolution of available soil maps, and high local variabilities in SOC. The estimation of permafrost C stocks is based on about 3,530 pedons and confidence levels are higher for the North American regions to 1-m depth level data, while for the Eurasian sector they are low to medium (33–66%) (Tarnocai et al. 2009). The lowest confidence (<33%) is observed for deeper soil layers. It is, therefore, not surprising that these estimates need to be revised and updated. For example, a recent publication by Schirrmeister et al. (2011) suggested that the SOC stocks in deeper Siberian deposits may be smaller than that previously estimated based on revised bulk density data. Still, large uncertainties related to spatial distribution of soils, deeper stocks in Yedoma and peatland deposits render the total C stock highly uncertain (Tarnocai et al. 2009).

8.4.2 Carbon Mobilization

Carbon in permafrost can be mobilized in response to pressure related disturbances (slow but persistent) and pulse disturbances (rapid but local) (Grosse et al. 2011). Pressure disturbances include changes in air temperature and directional changes in water balance, while pulse disturbances include fire, thermokarst, and flooding. Responses to these disturbances can result in paludification, aridification, top-down permafrost thaw, subsidence, changes in soil processes and vegetation cover, lake drainage, inundation, and erosion (Grosse et al. 2011; Kuhry et al. 2010). All of these disturbances can have effects on either vertical or horizontal C mobilization resulting in CO_2 or CH_4 exchange with the atmosphere resulting in a direct feedback on the climatic system or, in the case of lateral transport, for freshwater systems.

When permafrost thaws, large quantities of otherwise temperature-protected organic C becomes available for decomposition. The production and consumption of CO_2 and CH_4 within the pedosphere and biosphere have been documented in permafrost regions (Sachs et al. 2008; Friborg et al. 2000), but the role of storage and the timing between production, consumption and actual release of CO_2 and CH_4

is less clear. In addition to frequently reported spring emission peaks, Mastepanov et al. (2008) also reported high CH_4 emissions during autumn freeze-back that likely included significant amounts of CH_4 produced earlier during the growing season. Wagner et al. (2007), on the other hand, reported evidence of recent methanogenesis under *in situ* conditions in permafrost deposits at temperatures down to –6°C. This microbial CH_4 production at subzero temperatures and recent data on spring and autumn emissions suggest that these traditionally understudied periods of the year may be particularly important in the context of the annual GHG budget of tundra ecosystems. Although it can be assumed that, over long time scales, the GHGs produced eventually reach the atmosphere, many process-based models now run on time steps where intermittent storage and short-term atmospheric processes may play an important role. For example, using eddy covariance methods, Sachs et al. (2008) and Wille et al. (2008) reported that atmospheric conditions (wind speed, pressure) play a dominant role in tundra-atmosphere CH_4 exchange running on daily time steps.

However, quantification of the actual CO_2, and, in particular, CH_4 land-atmosphere exchange remains localized with very few and often clustered flux towers providing data for only a small part of the vast terrestrial ecosystems in the Arctic (Fan et al. 1992; Friborg et al. 2000; Kutzbach et al. 2007; Sachs et al. 2008, 2010; Wille et al. 2008). Most research is concentrated in the Alaskan Arctic with relatively few sites in the Canadian Arctic and even fewer in the remote northern regions of the Russian Federation. Year-round operation of flux observation sites remains an exception rather than the rule, despite the importance of the winter, spring and fall seasons in annual C budgets. Consequently, most large-scale assessments of Arctic C fluxes rely on process-based models, which include models that use drivers derived from remote sensing, or atmospheric inverse modeling. McGuire et al. (2010) analyzed the decadal C balance of the entire Arctic Basin from 1997 to 2006 using several model-based tools. The terrestrial and marine Arctic has been found to be a net CO_2 sink of 0.109 Pg C year^{-1}, but that the Arctic Basin as a whole is a source of GHG radiative forcing due to the terrestrial net source of 0.042 Pg CH_4 year^{1}. There is also a concern about a possible weakening of the high-latitude terrestrial CO_2 sink due to enhanced decomposition and increased fire frequency (Hayes et al. 2011).

Fire emissions may account for much of the variability in C sources and sinks between Arctic watersheds from 1997 to 2006 (McGuire et al. 2010). Although the proportion of burned area was small (e.g., ~4% for the Lena watershed), the actual area covered was large with a 25 year fire return interval. Increased fire frequency and intensity have been observed in the second half of the twentieth Century in Canada, Alaska and northern Eurasia (McGuire et al. 2004, 2007). Wildfires can destroy the insulating organic surface layer and warm the soil, increasing the rates of permafrost thaw and the active layer thickness (Yoshikawa et al. 2003; Johnstone et al. 2010). With subsequent regrowth of mosses the soils cool again, but with a warmer and drier climate, increased fire frequency and intensity may trigger a positive feedback loop between the loss of SOC and subsequent warming and thawing of permafrost soils (O'Donnell et al. 2011; McGuire et al. 2007).

Inland waters (lakes, wetlands, streams, and rivers) also process large quantities of organic C. Thermokarst lakes have been identified as a major source of CO_2 and CH_4 in permafrost terrain, from which the highest CH_4 fluxes are reported in Yedoma lakes, in which older C stocks are being mobilized (Zimov et al. 1997; Walter et al. 2006). However, a recent thermokarst lake modeling study by van Huissteden et al. (2011) reduced these flux estimates by an order of magnitude due to some hydrological effects, and lake drainage in particular, that limit lake expansion. Small polygonal ponds and lakes can release large amounts of C, accounting for 80% of landscape-scale net CO_2 emissions during September at a polygonal tundra landscape in Siberia, with roughly half of this CO_2 outgassing from ponds (Abnizova et al. 2012, in review).

The major Arctic rivers drain C-rich peatlands and soils and (Fig. 8.1b) transport large quantities of organic C to the Arctic Ocean (McGuire et al. 2010). The most important period for dissolved organic carbon (DOC) export is during spring snowmelt, when 60% of the annual DOC flux occurs, half of which is estimated to be only 1–5 years old (Raymond et al. 2007). Raymond and colleagues conservatively estimated the total input of DOC into the Arctic Ocean to be 0.025–0.026 Pg C, which is 2.5 times more when compared to estimates of river export of temperate watersheds with similar basin size and discharge.

8.4.3 Arctic Coasts, Subsea Permafrost, and Gas Hydrates

The Arctic coastal system is defined as extending landwards as far as the influence of the marine realm, and seawards as far as terrestrial influences to the edge of the continental shelf (Lantuit et al. 2011). Where the Arctic coastline is eroding, this coastal zone is expanding landward, with inundation of onshore permafrost transforming it into offshore or subsea permafrost (Fleming et al. 1998).

Terrestrial permafrost formed on Arctic continental shelves that were sub aerially exposed due to a lowered sea level during the Pleistocene glacial stages and not covered by continental ice masses. Subsequent inundation during the deglacial and Holocene sea level rise turned much of the terrestrial shelf permafrost into subsea permafrost. Today, most of the shelf area potentially affected by permafrost (>80%) lies on the broad and gently inclined shelves of the Laptev and East Siberian Seas (Fig. 8.1a). Various models have shown that permafrost on the Siberian coastal plains has probably existed without interruption for at least the last 400,000 years, with cyclic cooling and warming corresponding to glacial and interglacial periods (Nicolsky and Shakhova 2010; Romanovskii et al. 2004).

Through warming of permafrost by geothermal heat flux from below and atmospheric cooling at the surface, CH_4 and other volatiles may migrate into permafrost and be retained in pore space, either in gas form or as gas hydrates. The temperature-pressure fields within and below permafrost result in gas hydrate stability (Romanovskii et al. 2005). Significant deposits of GHGs, and especially CH_4, can thus be found on the Arctic shelf (Ginsburg and Soloviev 1995; McGuire et al.

2009). Warming of the shelf seabed may lead to thawing of the permafrost and the potential release of GHG into the water column and the atmosphere.

These sources of GHGs have not been well quantified, especially for the eastern Siberian shelf seas (the Laptev, East Siberian and Chukchi seas), and fluxes are not yet included in estimates of the Arctic C cycle (Semiletov and Pipko 2007). High concentrations of dissolved CH_4 in relatively shallow shelf sea waters, together with evidence of atmospheric venting, indicate that CH_4 flux from shelf sediments is high (Shakhova et al. 2010). Nonetheless, there remain uncertainties with regard to the amount of GHGs involved, what physical and chemical form it takes, and how changes to the shelf seas and underlying permafrost will affect its release.

Relatively stable sea levels in the Arctic slow the onshore to offshore permafrost transition, but coastal erosion continues through thermal retreat of the coastal cliffs (Overduin et al. 2007). The highest coastal erosion rates in the Arctic are observed where the modern coastline cuts through unconsolidated, ice-rich permafrost deposits. Since two-thirds of the Arctic coastline is composed of such material, a considerable length of coastline is affected (Lantuit et al. 2011). Historical data on coastal change in the Arctic is not as available as it is for more populated temperature regions, and the critically relevant question along the Arctic coast is to determine the current trajectory and rate of change in coastline position. Coastal erosion of Arctic permafrost may contribute as much as 0.0069 Pg C per year of particulate organic C (Rachold et al. 2003), which is of the same order of magnitude as the contribution from all Arctic rivers. This amount, however, needs to be re-evaluated because of the previous lack of data on the contribution of riverine and coastal DOC and as new data for SOC content become available (see Sect. 8.4.1).

8.5 Modeling Permafrost and Carbon Cycling Under a Changing Climate

8.5.1 Modeling Permafrost and Implementing Physical Permafrost Processes in Global Models

Since even the most advanced subsurface schemes employed in Atmosphere-Ocean General Circulation Models (AOGCMs) rarely include explicit treatment of the ground below 10 m depth (Roeckner et al. 2003; Cox et al. 1999), the thermal regime of permafrost cannot be obtained directly from model outputs. Specialized permafrost models which use AOGCM output variables, such as near-surface air temperature and snowfall, are therefore, used to obtain predictions on the future state of permafrost. Empirical or semi-empirical methods (degree days of freezing and thawing, factors to account for temperature offset due to snow cover) have generally been used for pan-Arctic projections of permafrost (Anisimov and Nelson 1996), revealing projected wide-spread degradation during this century. Such approaches, however, do not allow any detailed insight in the dynamics of permafrost processes. The state-of-the-art in permafrost modeling is represented by transient models which numerically solve the one-dimensional heat transfer equation for temperature in a soil domain between the surface and a specific depth. The heat transfer equation contains two effective parameterizations of the material, the volumetric heat capacity and the thermal conductivity. They both depend on the water content and also on temperature since they implicitly represent the energetics of freezing and thawing. Furthermore, both the initial state of the system and the forcing data for the upper and lower boundary conditions must be defined. Despite the simplicity of the governing physical laws, modeling the thermal regime of permafrost is an intricate task since:

In most permafrost simulations, a constant (geothermal) heat flux is used as a lower boundary condition (Riseborough et al. 2008; Zhang et al. 2003). Thus, the model domain must extend to depths at which a stable temperature gradient exists during the target period of the simulation. Therefore, the longer the time period to be covered by the model, the deeper the chosen soil domain must be. For example, for a modeling period of 100 years a soil domain of more than 100 m is chosen for many studies (Zhang et al. 2003).

Only in very few areas where deep boreholes are available can the initial condition be derived from direct measurements (Marchenko et al. 2008). Elsewhere, a sufficiently long time series of forcing data is required to calculate the recent temperature distribution in the ground, especially if a deep soil domain is chosen. The required time series of this "model spin-up" can exceed 100 years in permafrost modeling studies (Zhang et al. 2008).

The forcing data for permafrost models can be derived from different sources such as meteorological observations, satellite measurements, or the output of atmospheric models. However, using the output of AOGCMs is the only possibility for future projections. Deficiencies in the output of these models are therefore, directly reflected and possibly emphasized in the permafrost models (Chapman and Walsh 2007). Reproducing the seasonal and perennial insulating snow cover in an adequate way is a particularly challenging task but critical for the ground thermal regime.

The results of permafrost simulations strongly depend on the thermal properties assumed for the soil. Although diverse approaches exist for calculating these parameters for permafrost soils on the basis of soil composition (de Vries 1952; Farouki 1981), virtually no ground truthing is available. Another major uncertainty relates to the parameterization of vegetation cover which displays distinctively different thermal properties compared to the underlying soil (Wania et al. 2009a, b).

The thermal properties of both soil and the surface cover can exhibit a strong spatial variability over distances ranging from meters to few kilometers which is considerably less than the grid-scales of typical gridded data sets for soil and surface properties as well as those of atmospheric models (Wilson and Henderson-Sellers 1985). Programs have been initiated to measure subgrid variability in permafrost areas, but these are not yet reflected in current modeling approaches (Westermann et al. 2011; Langer et al. 2010).

Input data sets and the thermal properties of soil can generally be tuned on a local or point scale to yield a satisfactory to excellent agreement with measured data

(Romanovsky and Osterkamp 1997; Nicolsky et al. 2009). On a larger scale, where validation data sets are generally sparse to non-existent, permafrost modeling must rely on gridded data sets but coarse features such as the annual temperature amplitude which can usually be sufficiently reproduced (Oelke and Zhang 2004). Most permafrost modeling has been accomplished on local to regional scales. Zhang et al. (2008) demonstrated the use of a transient permafrost model for Canada that is based on the output of six AOGCM-generated climate scenarios. They predicted a reduction in the permafrost area of up to 20% and a significant thickening of the active layer in the remaining areas by 2100. Their permafrost model was initialized during the Little Ice Age at which time the permafrost was assumed to be in a steady state defined by the average atmospheric temperature and the geothermal gradient. The "model spin-up" for calculating the present temperature distribution in the ground was accomplished using time series interpolated from meteorological observations. A similar study has been performed for Alaska by Marchenko et al. (2008), who initialized their model using soil temperature measurements from boreholes. They modeled a widespread permafrost degradation until the end of the century, except in the most northerly areas. Both studies were operated on coarse spatial resolutions of $0.5° \times 0.5°$. To moderate the impact of resolution-related inaccuracies in the data sets used for forcing the model, Stendel et al. (2007) used dynamic downscaling with a Regional Climate Model (RCM). They demonstrated significant improvements in the precipitation (and thus snow depth) pattern for eastern Siberia, resulting in marked improvements in the modeling of the thermal regime of the permafrost.

While future permafrost conditions derived from modeling appear to be converging towards a permafrost reduction on the order of 25% and a significant thickening of the active layer during the remainder of this century, there has, to date, been no comprehensive pan-Arctic study based on a transient permafrost model. Furthermore, future permafrost models could be improved by including up- and downscaling algorithms to take into account subgrid variability of the key input and output variables.

8.5.2 Permafrost-Atmosphere Feedback Through a Modified Surface Energy Balance

Future predictions on permafrost are necessarily based on AOGCMs. The "European Centre/Hamburg Model" (ECHAM, Roeckner et al. 2003), for instance, contains a total of five soil layers to a depth of 10 m but does not include the freezing of soil water (Roeckner et al. 2003). Thus, a considerable proportion of the energy exchange between soil and atmosphere is not accounted for (Boike et al. 1998; Romanovsky and Osterkamp 2000). This raises the question whether it is valid to base permafrost predictions on the output of a model that does not take into account the physics of frozen ground. In other words, is there a feedback between permafrost processes and atmospheric near-surface variables, which could influence/impact the regional climate?

A number of studies have pointed out the influence that selected soil parameters have on the results of atmospheric modeling (Peters-Lidard et al. 1998). Viterbo et al. (1999) observed that the inclusion of soil freezing leads to a significant improvement in modeled near-surface air temperatures for periods when a freezing front is close to the surface. Rinke et al. (2008) reported that including a low-conductivity organic layer for Arctic land masses has implications for modeled near-surface air temperatures, and even for the modeled regional circulation patterns. Furthermore, long-term monitoring of the surface energy balance in Siberia has demonstrated that the ground heat flux resulting from refreezing of the active layer and subsequent soil cooling compensates for more than half of the radiative losses during winter (Langer et al. 2011b). Permafrost processes must therefore be considered to be a driving force behind the wintertime surface energy balance in this area, and, thus, an accurate description in a land-surface scheme is highly desirable.

Since such evidence suggests a more active role for permafrost processes in the climate system, increased effort has been put into incorporating permafrost processes into land-surface schemes (Nicolsky et al. 2007), so that coupled runs with AOGCMs seem to be only a matter of time. However, such new schemes can only be considered a step forward if they are validated for sites across the entire range of climatic and ecological conditions found in permafrost areas.

8.5.3 Modeling the Permafrost-Carbon Feedback

The release of GHGs from permafrost soils is influenced by a number of factors, in particular the soil temperature, water content (as the aerobic/anaerobic state of the soil influences losses the types of GHG loss), SOC amount and quality, the availability of nutrients, and the composition of microbial communities. Models for soil C cycling based on these parameters are readily available (Jenkinson and Coleman 2008), although there is controversy about the classification of soil C stocks into pools with different turnover times, i.e., SOC quality (Davidson and Janssens 2006). For these modeling approaches to be applied in Arctic permafrost areas, field studies are required to validate their performance and to improve the mathematical representations between these factors and SOC dynamics. However, GHG emissions can only be successfully simulated if accurate data on the SOC concentration and distribution are available and if soil temperatures and soil water content are inferred from land-surface models. Furthermore, C cycling models are *a priori* formulations for a point in three-dimensional space, and application over large grid cells in terms of an area-averaged formulation may be problematic.

For these reasons, first estimates of the magnitude of permafrost C feedback are associated with considerable uncertainties. Off-line simulations with simplified climate scenarios (Schaefer et al. 2011) or simplified permafrost representation (Schneider von Deimling et al. 2011) have predicted a sizable permafrost C feedback after 2100, even under moderate warming scenarios. This lag is a consequence of the large thermal inertia of the frozen ground which delays thawing and the microbial degradation of the organic material.

A promising approach to a physically-based permafrost-C model has been presented by Khvorostyanov et al. (2008), but to date has only been applied offline for selected regions.

There remains a need for a fully coupled simulation with an AOGCM that includes permafrost C feedback. Furthermore, all the above-mentioned studies have pointed out that the applied methodology cannot account for crucial processes occurring on subgrid scale, such as the formation of thermokarst or modification of the local hydrological regime (Wania et al. 2009a, b). The development of up- and downscaling techniques for the most important variables must, therefore, be considered a necessary prerequisite for reliable modeling of GHG release from permafrost areas.

8.6 Conclusions and Recommendations

According to model projections, permafrost degradation will affect almost half of the current permafrost area in the northern hemisphere by 2100. Prediction of the sensitivity of the C cycle to climate change and permafrost thaw is complicated by complex interactions between hydrology, soil thermal regimes, and vegetation. These factors can result in both positive and negative feedbacks to permafrost and C exchange. Changes in land cover, such as vegetation type and distribution, or the areal extent of water bodies and drainage systems, will affect the vertical and horizontal fluxes of water, energy, and C. However, land-surface modules of most state-of-the-art coupled general circulation models include only considerably simplified descriptions of the thermal and hydrological effects of soil freezing and the related processes and properties (snow cover, high-latitude vegetation). They generally neglect the effects of sub-grid variability in landforms, soil types, etc., which have a strong influence on the large-scale effects of these processes.

Given the size of organic C stocks in permafrost, vulnerabilities associated with thawing and, decomposition of OM, and the production of GHG have the potential for strong positive feedbacks to the climate system. However, there are large uncertainties associated with estimates derived from the potential for both positive and negative climate feed backs as well as missing spatial data. In marked contrast to their recognized importance, comprehensive observations of Arctic soil, snow, and atmospheric quantities (climate and surface energy balance components) are extremely sparse.

For monitoring and modeling the present and future state of the permafrost, most approaches are designed to use averaged quantities for large grid cells. Such averaged quantities may, however, be inappropriate for some monitoring tasks. A spatial average of the active layer thickness is adequate for modeling distributed quantities like the background emissions of GHG. However, the monitoring of erosion or natural hazards due to permafrost degradation (which would initially occur at few localized "weak points" by capturing localized emission hotspots such as thermokarst lakes) is only feasible if information is available at the sub-grid level.

Implementing a potential feed-back mechanism by GHG-emission from warming permafrost in regional and global models requires the correct parameterization of a sufficient and critical set of thermal and hydrologic processes and parameters, such as the thermal properties of snow and soil. However, this would require: (i) accurate estimates of state variables such as SOC or thermal and hydraulic conductivities, (ii) both vertically and horizontally upscaled data on these state variables (Ciais 2010), (iii) an understanding of the key physical and biochemical processes in permafrost, and (iv) an understanding of the interaction and feedback mechanisms between permafrost and climate.

The primary objective of understanding permafrost and its role in the earth's climate system, including feed back mechanisms, requires new model developments and upscaling strategies:

Field studies should be conducted at representative sites to systematically monitor key parameters and processes over the long term (e.g. >10 years), thereby improving the understanding of permafrost dynamics at a range of scales.

The development of conceptual and numerical permafrost landscape models is required, including suitable upscaling methods ranging from local to global scales, and remote sensing products must be used to test, validate and monitor (1) and (2) above.

All strategies require field-based knowledge of the surface characteristics, key processes and monitoring data for a few key parameters. Obtaining funds for long term process monitoring is often difficult due to political and technical limitations. Such funds are often cut from budgets since the benefits derived from field studies are often not apparent until after a longer period of time (Nisbet 2007). It is, therefore, urgent and timely to initiate efforts at various locations across the climatic and ecological gradients in permafrost areas, to eventually establish a pan-Arctic data base. Such a compilation would be of outstanding importance for improving the understanding of the sensitivities of permafrost and high-latitude ecosystems, and their susceptibility to climate change.

References

Abnizova A, Siemens J, Langer M et al (2012) Small ponds with major impact: the relevance of ponds and lakes in permafrost landscapes to carbon dioxide emissions. Glob Biogeochem Cycles (in review)

ACIA (2004) Impacts of a warming Arctic. Cambridge University Press, Cambridge

Anisimov O, Nelson F (1996) Permafrost distribution in the Northern Hemisphere under scenarios of climatic change. Global Planet Change 14(1–2):59–72

Batjes NH (1996) Total carbon and nitrogen in the soils of the world. Eur J Soil Sci 47(2):151–163

Blok D, Heijmans PD, Schaepman-Strub G et al (2010) Shrub expansion may reduce summer permafrost thaw in Siberian tundra. Global Change Biol 16:1296–1305. doi:10.1111/j.1365-2486.2009.02110.x

Boike J, Roth K, Overduin PP (1998) Thermal and hydrologic dynamics of the active layer at a continuous permafrost site (Taymyr Peninsula, Siberia). Water Resour Res 34(3):355–363

Boike J, Wille C, Abnizova A (2008) The meteorology, and energy and water balances of polygonal tundra in the Lena Delta, Siberia during wet and dry years. J Geophys Res 113:G03025. doi:10.1029/2007JG000540

Bonan GB, Chapin FS, Thompson SL (1995) Boreal forest and tundra ecosystems as components of the climate system. Clim Chang 29(2):145–167

Bowden WB, Gooseff MN, Balser A et al (2008) Sediment and nutrient delivery from thermokarst features in the foothills of the North Slope, Alaska: potential impacts on headwater stream ecosystems. J Geophys Res 113:G02026. doi:10.1029/2007JG000470

Brown J, Ferrians O Jr, Heginbottom J et al (1997) Circum-Arctic map of permafrost and ground-ice conditions. National Snow and Ice Data Center/World Data Center for Glaciology. Digital Media, Boulder

CAVM (2003) Circumpolar arctic vegetation map, scale 1:7,500,000, conservation of Arctic Flora and Fauna (CAFF), Map No. 1. U.S., Fish and Wildlife Service, Anchorage, AK

Chapin FS III, Sturm M, Serreze MC et al (2005) Role of land-surface changes in Arctic summer warming. Science 310(5748):657–660. doi:10.1126/science.1117368

Chapman W, Walsh J (2007) Simulations of Arctic temperature and pressure by global coupled models. J Climate 20(4):609–632

Ciais P (2010) Soil map digs under the tundra. Nature 467(7311):30–31. doi:10.1038/467030a

Cox P, Betts R, Bunton C et al (1999) The impact of new land surface physics on the GCM simulation of climate and climate sensitivity. Clim Dyn 15(3):183–203

Davidson EA, Janssens IA (2006) Temperature sensitivity of soil carbon decomposition and feedbacks to climate change. Nature 440(7081):165–173. doi:10.1038/nature04514

de Vries D (1952) The thermal conductivity of soil. Mededelingen van de Landbouwhogeschool te Wageningen 52(1):1–73

Denman KL, Brasseur G, Chidthaisong A et al (2007) Couplings between changes in the climate system and biogeochemistry. In: Solomon S et al (eds) Climate change 2007: the physical science basis. Contribution of Working Group I to the fourth assessment report of the Intergovernmental Panel on Climate Change. Cambridge University Press, Cambridge/New York

Einarsson N, Larsen E, Nilsson A et al (2004) Arctic human development report. Stefansson Arctic Institute, Akureyi

Eugster W, Rouse WR, Pielke RA et al (2000) Land-atmosphere energy exchange in arctic tundra and boreal forest: available data and feedbacks to climate. Global Change Biol 6(1):84–115

Fan SM, Wofsy SC, Bakwin PS et al (1992) Micrometeorological measurements of CH_4 and CO_2 exchange between the atmosphere and subarctic tundra. J Geophys Res 97(D15):16627–16643. doi:10.1029/91JD02531

Farouki O (1981) The thermal properties of soils in cold regions. Cold Reg Sci Technol 5(1):67–75

Fleming K, Johnston P, Zwartz D et al (1998) Refining the eustatic sea-level curve since the Last Glacial Maximum using far- and intermediate-field sites. Earth Planet Sci Lett 163(1–4):327–342

Frey KE, Smith LC (2007) How well do we know northern land cover? Comparison of four global vegetation and wetland products with a new ground-truth database for West Siberia. Global Biogeochem Cycles 21:GB1016. doi:10.1029/2006GB002706

Friborg T, Christensen T, Hansen B et al (2000) Trace gas exchange in a high-arctic valley 2. Landscape CH_4 fluxes measured and modeled using eddy correlation data. Global Biogeochem Cycles 14(3):715–723

Friedlingstein P, Cox P, Betts R et al (2006) Climate–carbon cycle feedback analysis: results from the C4MIP model intercomparison. J Climate 19:3337–3353

Ginsburg G, Soloviev V (1995) Submarine gas hydrate estimation: theoretical and empirical approaches. In: 27th annual OTC, Houston, TX, 1–4 May 1995

Global Land Cover (2000) Modified Global Land Cover 2000 database. European Commission Joint Research Centre. http://bioval.jrc.ec.europa.eu/products/glc2000/glc2000.php. Accessed 21 Sept 2011

Grosse G, Romanovsky V, Walter K et al (2008) Distribution of thermokarst lakes and ponds at three yedoma sites in Siberia. In: Proceedings of the 9th international conference on permafrost, University of Alaska Fairbanks, June 29–July 3 2008, pp 551–556

Grosse G, Harden J, Turetsky M et al (2011) Vulnerability of high-latitude soil organic carbon in North America to disturbance. J Geophys Res 116:G00K06. doi:10.1029/2010JG001507

Gutowski WJ, Wei H, Vörösmarty CJ et al (2007) Influence of arctic wetlands on arctic atmospheric circulation. J Climate 20(16):4243–4254. doi:10.1175/JCLI4243.1
Hayes D, McGuire A, Kicklighter D et al (2011) Is the northern high-latitude land-based CO_2 sink weakening? Global Biogeochem Cycles 25(3):GB3018. doi:10.1029/2010GB003813
Hinzman LD, Bettez ND, Bolton WR et al (2005) Evidence and implications of recent climate change in Northern Alaska and other Arctic regions. Clim Chang 72:251–298. doi:10.1007/s10584-005-5352-2
Hinzman LD, Gieck RE, Kane DL (2008) Spatial and temporal variation of soil temperatures and arctic hydrology in the Kuparuk River Basin, Alaska. In: Proceedings of the 9th international conference on permafrost, University of Alaska Fairbanks, June 29–July 3 2008, pp 713–715
Hollesen J, Elberling B, Jansson PE (2010) Future active layer dynamics and carbon dioxide production from thawing permafrost layers in Northeast Greenland. Global Change Biol 17(2):911–926
Howarth Burnham J, Sletten RS (2010) Spatial distribution of soil organic carbon in northwest Greenland and underestimates of high Arctic carbon stores. Global Biogeochem Cycles 24:GB3012. doi:10.1029/2009GB003660
Hugelius G, Kuhry P, Tarnocai C et al (2010) Soil organic carbon pools in a periglacial landscape: a case study from the central Canadian Arctic. Permafr Periglac Process 21(1):16–29. doi:DOI: 10.1002/ppp. 677
Jenkinson D, Coleman K (2008) The turnover of organic carbon in subsoils. Part 2. Modelling carbon turnover. Eur J Soil Sci 59(2):400–413
Johnstone JF, Chapin FS, Hollingsworth TN et al (2010) Fire, climate change, and forest resilience in interior Alaska. Can J Forest Res 40(7):1302–1312
Khvorostyanov DV, Ciais P, Krinner G et al (2008) Vulnerability of east Siberia's frozen carbon stores to future warming. Geophys Res Lett 35:L10703. doi:10.1029/2008GL033639
Koven C, Friedlingstein P, Ciais P et al (2009) On the formation of high-latitude soil carbon stocks: effects of cryoturbation and insulation by organic matter in a land surface model. Geophys Res Lett 36:1–5
Kuhry P, Hugelius G, Dorrepaal E et al (2010) Short communication: potential remobilization of belowground permafrost carbon under future global warming. Permafr Periglac Process 21(2):208–214
Kutzbach L, Wille C, Pfeiffer EM (2007) The exchange of carbon dioxide between wet arctic tundra and the atmosphere at the Lena River Delta, Northern Siberia. Biogeosci Discuss 4(3):1953–2005
Lachenbruch AH, Sass JH, Lawver LA et al (1982) Depth and temperature of permafrost on the Alaskan arctic slope: preliminary results. U.S. Geological Survey, Reston
Langer M, Westermann S, Boike J (2010) Spatial and temporal variations of summer surface temperatures of wet polygonal tundra in Siberia – implications for MODIS LST based permafrost monitoring. Remote Sens Environ 114(9):2059–2069. doi:10.1016/j.rse.2010.04.012
Langer M, Westermann S, Muster S et al (2011a) The surface energy balance of a polygonal tundra site in northern Siberia – part 1: Spring to fall. The Cryosphere 5:151–171
Langer M, Westermann S, Muster S et al (2011b) The surface energy balance of a polygonal tundra site in northern Siberia – part 2: Winter. The Cryosphere 5:509–524
Lantuit H, Overduin P, Couture N et al (2011) The Arctic Coastal Dynamics database. A new classification scheme and statistics on arctic permafrost coastlines. Estuaries Coasts. doi:10.1007/s12237-010-9362-6
Lehner B, Döll P (2004) Development and validation of a global database of lakes, reservoirs and wetlands. J Hydrol 296:1–22
Lenton TM, Held H, Kriegler E et al (2008) Tipping elements in the Earth's climate system. Proc Natl Acad Sci USA 105(6):1786–1793
Lloyd CR, Harding RJ, Friborg T et al (2001) Surface fluxes of heat and water vapour from sites in the European Arctic. Theor Appl Climatol 70:19–33
Marchenko S, Romanovsky V, Tipenko G (2008) Numerical modeling of spatial permafrost dynamics in Alaska. In: Proceedings of the ninth international conference on Permafrost, Fairbanks, AK, 29 June–3 July 2008, pp 1125–1130

Mastepanov M, Sigsgaard C, Dlugokencky EJ et al (2008) Large tundra methane burst during onset of freezing. Nature 456:628–631. doi:10.1038/nature07464
McGuire AD, Apps M, Chapin FS et al (2004) Land cover disturbances and feedbacks to the climate system in Canada and Alaska. In: Gutman G, Janetos AC, Justice CO et al (eds) Land change science. Springer, Dordrecht, pp 139–161
McGuire AD, Chapin FS, Wirth C et al (2007) Responses of high latitude ecosystems to global change: potential consequences for the climate system. Terrestrial ecosystems in a changing world, global change – the IGBP series. Springer, New York. doi:10.1007/978-3-540-32730-1_24
McGuire AD, Anderson LG, Christensen TR et al (2009) Sensitivity of the carbon cycle in the arctic to climate change. Ecol Monogr 79(4):523–555
McGuire AD, Hayes DJ, Kicklighter DW et al (2010) An analysis of the carbon balance of the Arctic Basin from 1997 to 2006. Tellus B 62:455–474
Muster S, Langer M, Heim B et al (2012) Scaling land cover of arctic polygonal tundra and its effects on evapotranspiration. Tellus B (in press)
Nicolsky DJ, Shakhova N (2010) Modeling sub-sea permafrost in the East Siberian Arctic Shelf: the Dmitry Laptev Strait. Environ Res Lett 5:015006. doi:10.1088/1748-9326/5/1/015006
Nicolsky DJ, Romanovsky VE, Alexeev VA et al (2007) Improved modeling of permafrost dynamics in a GCM land-surface scheme. Geophys Res Lett 34(8):L08501. doi:10.1029/2007GL029525
Nicolsky DJ, Romanovsky VE, Panteleev G (2009) Estimation of soil thermal properties using insitu temperature measurements in the active layer and permafrost. Cold Reg Sci Technol 55(1):120–129
Nisbet E (2007) Earth monitoring: Cinderella science. Nature 450(6):789–790. doi:10.1038/450789a
O'Donnell JA, Harden JW, McGuire AD et al (2011) Exploring the sensitivity of soil carbon dynamics to climate change, fire disturbance and permafrost thaw in a black spruce ecosystem. Biogeosciences 8:1367–1382. doi:10.5194/bg-8-1367-2011
Oelke C, Zhang T (2004) A model study of circum-Arctic soil temperatures. Permafr Periglac Process 15:103–121
Ohmura A (1982) Climate and energy balance on the Arctic tundra. J Climatol 2:65–84
Osterkamp TE (2007) Causes of warming and thawing permafrost in Alaska. Eos Trans AGU 88(48):522–523
Overduin PP, Hubberten H-W, Rachold V et al (2007) The evolution and degradation of coastal and offshore permafrost in the Laptev and East Siberian Seas during the last climatic cycle. In: Harff J, Hay WW, Tetzlaff DM (eds) Coastline changes: interrelation of climate and geological processes, Geological Society of America, special paper 426, pp 97–111. doi:10.1130/2007.2426(07)
Peters-Lidard C, Blackburn E, Liang X et al (1998) The effect of soil thermal conductivity parameterization on surface energy fluxes and temperatures. J Atmos Sci 55:1209–1224
Ping CL, Michaelson GJ, Jorgenson MT et al (2008) High stocks of soil organic carbon in the North American Arctic region. Nat Geosci 1:615–619. doi:10.1038/ngeo284
Rachold V, Eicken H, Gordeev VV et al (2003) Modern terrigenous organic carbon input to the Arctic Ocean. In: Stein R, Macdonald RW (eds) Organic carbon cycle in the Arctic Ocean: present and past. Springer, Berlin, pp 33–55
Raymond PA, McClelland JW, Holmes RM et al (2007) Flux and age of dissolved organic carbon exported to the Arctic Ocean: a carbon isotopic study of the five largest arctic rivers. Global Biogeochem Cycles 21(4):GB4011. doi:10.1029/2007GB002934
Rinke A, Kuhry P, Dethloff K (2008) Importance of a soil organic layer for Arctic climate: a sensitivity study with an Arctic RCM. Geophys Res Lett 35:L13709. doi:10.1029/2008GL034052
Riseborough D, Shiklomanov N, Etzelmuller B et al (2008) Recent advances in permafrost modeling. Permafr Periglac Process 19(2):137–156
Roeckner E, Bäuml G, Bonaventura L et al (2003) The atmospheric general circulation model ECHAM5. Part I: Model description. Max Planck Institute for Meteorology report 349, 127 pp
Romanovskii NN, Hubberten H, Gavrilov AV et al (2004) Permafrost of the east Siberian Arctic shelf and coastal lowlands. Quartenary Sci Rev 23:1359–1369
Romanovskii NN, Hubberten H, Gavrilov AV et al (2005) Offshore permafrost and gas hydrate stability zone on the shelf of East Siberian Seas. Geo Mar Lett 25:167–182. doi:10.1007/s00367-004-0198-6

Romanovsky V, Osterkamp T (1997) Thawing of the active layer on the coastal plain of the Alaskan Arctic. Permafr Periglac Process 8(1):1–22

Romanovsky VE, Osterkamp TE (2000) Effects of unfrozen water on heat and mass transport processes in the active layer and permafrost. Permafr Periglac Process 11:219–239

Romanovsky VE, Smith SL, Christiansen HH (2010) Permafrost thermal state in the polar Northern Hemisphere during the international polar year 2007–2009: a synthesis. Permafr Periglac Process 21(2):106–116. doi:10.1002/ppp. 689

Rouse WR, Blanken PD, Duguay CR et al (2007) Climate–lake interactions. In: Woo MK (ed) Cold region atmospheric and hydrologic studies: the Mackenzie GEWEX experience, vol. 2: hydrologic processes. Springer, New York

Sachs T, Wille C, Boike J et al (2008) Environmental controls on ecosystem-scale CH_4 emission from polygonal tundra in the Lena River Delta, Siberia. J Geophys Res 113:G00A03. doi:10.1029/2007JG000505

Sachs T, Giebels M, Boike J et al (2010) Environmental controls on CH_4 emission from polygonal tundra on the microsite scale in the Lena river delta, Siberia. Global Change Biol 16:1–16. doi:10.1111/j.1365-2486.2010.02232.x

Sannel ABK, Kuhry P (2011) Warming-induced destabilization of peat plateau/thermokarst lake complexes. J Geophys Res 116:G03035. doi:10.1029/2010JG001635

Schaefer K, Zhang T, Bruhwiler L et al (2011) Amount and timing of permafrost carbon release in response to climate warming. Tellus B 63:165–180. doi:10.1111/j.1600-0889.2011.00527.x

Schirrmeister L, Grosse G, Wetterich S et al (2011) Fossil organic matter characteristics in permafrost deposits of the northeast Siberian Arctic. J Geophys Res 116:G00M02. doi:10.1029/2011JG001647

Schneider von Deimling T, Meinshausen M, Levermann A et al (2011) Estimating the permafrost-carbon feedback on global warming. BG Discuss 8:4727–4761. doi:10.5194/bgd-8-4727-2011

Schneider J, Grosse G, Wagner D (2009) Land cover classification of tundra environments in the Arctic Lena Delta based on Landsat 7 ETM+ data and its application for upscaling of methane emissions. Remote Sens Environ 113:380–391. doi:10.1016/j.rse.2008.10.013

Schuur EAG, Bockheim JG, Canadell JG et al (2008) Vulnerability of permafrost carbon to climate change: implications for the global carbon cycle. Bioscience 58(8):701–714. doi:10.1641/B580807

Semiletov I, Pipko I (2007) Sinks and sources of carbon dioxide in the Arctic Ocean: results of direct instrumental measurements. Dokl Earth Sci 414(1):642–645

Shakhova N, Semiletov IP, Salyuk A et al (2010) Extensive methane venting to the atmosphere from sediments of the east Siberian Arctic Shelf. Science 327:1246–1250. doi:10.1126/science.1182221

Smith LC, Sheng Y, MacDonald GM (2007) A first pan-Arctic assessment of the influence of glaciation, permafrost, topography and peatlands on northern hemisphere lake distribution. Permafr Periglac Process 18:201–208. doi:10.1002/ppp. 581

Stendel M, Romanovsky VE, Christensen JH et al (2007) Using dynamical downscaling to close the gap between global change scenarios and local permafrost dynamics. Global Planet Change 56(1–2):203–214

Sturm M, Racine C, Tape K (2001) Increasing shrub abundance in the Arctic. Nature 411:546–547. doi:10.1038/35079180

Sturm M, Douglas T, Racine C et al (2005) Changing snow and shrub conditions affect albedo with global implications. J Geophys Res 110:G01004. doi:10.1029/2005JG000013

Tape K, Sturm M, Racine C (2006) The evidence for shrub expansion in Northern Alaska and the Pan-Arctic. Global Change Biol 12:686–702. doi:10.1111/j.1365-2486.2006.01128.x

Tarnocai C, Canadell JG, Schuur EAG et al (2009) Soil organic carbon pools in the northern circumpolar permafrost region. Global Biogeochem Cycles 23(2):GB2023. doi:10.1029/2008GB003327

Tchebakova NM, Parfenova E, Soja AJ (2009) The effects of climate, permafrost and fire on vegetation change in Siberia in a changing climate. Environ Res Lett 4:1–9

Trenberth K (2010) More knowledge, less certainty. Nature 4:20–21. doi:10.1038/climate.2010.06

van Everdingen R (1998) Multi-language glossary of permafrost and related ground-Ice terms, revised May 2005. National Snow and Ice Data Center/World Data Center for Glaciology, Boulder
van Huissteden J, Berrittella C, Parmentier FJW et al (2011) Methane emissions from permafrost thaw lakes limited by lake drainage. Nat Climate Change 1(2):119–123. doi:10.1038/nclimate1101
Viterbo P, Beljaars A, Mahfouf J et al (1999) The representation of soil moisture freezing and its impact on the stable boundary layer. Q J R Meteor Soc 125:2401–2426
Wagner D, Gattinger A, Embacher A et al (2007) Methanogenic activity and biomass in Holocene permafrost deposits of the Lena Delta, Siberian Arctic and its implication for the global methane budget. Global Change Biol 13:1089–1099. doi:10.1111/j.1365-2486.2007.01331.x
Walker A, Raynolds MK, Daniels DJ et al (2005) The cirumpolar Arctic vegetation map. J Veg Sci 16:267–282
Walter KM, Zimov SA, Chanton JP et al (2006) Methane bubbling from Siberian thaw lakes as a positive feedback to climate warming. Nature 443(7):71–75. doi:10.1038/nature05040
Walter KM, Edwards M, Grosse G et al (2007) Thermokarst lakes as a source of atmospheric CH_4 during the last deglaciation. Science 318:633–636. doi:10.1126/science.1142924
Wania R, Ross I, Prentice IC (2009a) Integrating peatlands and permafrost into a dynamic global vegetation model: 1. Evaluation and sensitivity of physical land surface processes. Global Biogeochem Cycles 23(3):1–19
Wania R, Ross I, Prentice IC (2009b) Integrating peatlands and permafrost into a dynamic global vegetation model: 2. Evaluation and sensitivity of vegetation and carbon cycle processes. Global Biogeochem Cycles 23:1–15
Washburn AL (1979) Geocryology – a survey of periglacial processes and environments. Edward Arnold, London
Westermann S, Lüers J, Langer M et al (2009) The annual surface energy budget of a high-arctic permafrost site on Svalbard, Norway. The Cryosphere 3:245–263
Westermann S, Langer M, Boike J (2011) Spatial and temporal variations of summer surface temperatures of high-arctic tundra on Svalbard – implications for MODIS LST based permafrost monitoring. Remote Sens Environ 115(3):908–922
Wille C, Kutzbach L, Sachs T et al (2008) Methane emission from Siberian arctic polygonal tundra: eddy covariance measurements and modeling. Global Change Biol 14:1395–1408. doi:10.1111/j.1365-2486.2008.01586.x
Wilson M, Henderson-Sellers A (1985) A global archive of land cover and soils data for use in general circulation climate models. Int J Climatol 5(2):119–143
Yershov ED (2004) General geocryology. Cambridge University Press, Cambridge
Yoshikawa K, Hinzman LD (2003) Shrinking thermokarst ponds and groundwater dynamics in discontinuous permafrost near Council, Alaska. Permafr Periglac Process 14:151–160
Yoshikawa K, Bolton WE, Rbomanovsky VR et al (2003) Impacts of wildfire on the permafrost in the boreal forests of Interior Alaska. J Geophys Res 108(D1):8141. doi:10.1029/2001JD000438
Zhang Y, Chen W, Cihlar J (2003) A process-based model for quantifying the impact of climate change on permafrost thermal regimes. J Geophys Res 108:4695. doi:10.1029/2002JD003354
Zhang Y, Chen W, Riseborough DW (2008) Disequilibrium response of permafrost thaw to climate warming in Canada over 1850–2100. Geophys Res Lett 35:L02502. doi:10.1029/2007GL032117
Zimov SA, Voropaev YV, Semiletov IP et al (1997) North Siberian lakes: a methane source fueled by Pleistocene carbon. Science 277(5327):800–802. doi:10.1126/science.277.5327.800
Zimov SA, Schuur EAG, Chapin FS III (2006) Permafrost and the global carbon budget. Science 312:1612–1613

Chapter 9
Carbon Sequestration in Temperate Forests

Rattan Lal and K. Lorenz

Abstract Temperate forests, located between 25 and 55° N and S of equator, are highly diverse in species, soils, and the ecosystems' carbon (C) pool. Their composition and characteristics change among regions. Principal forest types are broad-leaved deciduous, broad-leaved evergreen, coniferous, and mixed. Temperate forests are primarily located in North America, Central and Western Europe, north-eastern Asia, southern Chile, New Zealand and the Mediterranean. Principal soils of the temperate forests are Alfisols, Inceptisols, Mollisols, Spodosols, and Ultisols. These are generally fertile soils with high soil organic C (SOC) pool. Typical temperate forest soils contain about 100 Mg C ha^{-1} in the soil profile, and often more. Total ecosystem C pool in biomes and soils of temperate forest is equivalent to, and sometimes even more, than that of the tropical rainforest ecosystems. The projected change in climate may shift the temperate forest biome polewards, alter species composition, and change the ecosystem C pool. With favorable climate characterized by four distinct seasons and relatively fertile soils, the temperate forest biomes have a high C sink capacity. Thus, sustainable forest management, planting and rehabilitation can contribute to recarbonize the biome previously disturbed by deforestation, degradation and poor forest management, and create draw down in the atmospheric concentration of carbon dioxide (CO_2) Fast growing temperate trees can accumulate about 20 Mg of wood ha^{-1} $year^{-1}$. The strategy is also to preserve the old-growth forests, and establish new forests on degraded lands and agriculturally marginal soils through afforestation and reforestation. In addition to

R. Lal (✉)
Carbon Management and Sequestration Center, The Ohio State University,
2021 Coffey Road, Kottmann Hall 422b, Columbus, OH 43210, USA
e-mail: lal.1@osu.edu

K. Lorenz
Global Soil Forum, IASS Institute for Advanced Sustainability Studies e.V.,
Berliner Strasse 130, 14467 Potsdam, Germany
e-mail: klaus.lorenz@iass-potsdam.de

R. Lal et al. (eds.), *Recarbonization of the Biosphere: Ecosystems and the Global Carbon Cycle*, DOI 10.1007/978-94-007-4159-1_9,

biomass, C can also be sequestered in soils. The rate of soil C sequestration is lower than that in the biomass, and depends on soil type, antecedent pool, species and other natural and marginal factors. While trading C credits can promote adoption of an appropriate forest land use and management to enhance C sequestration, how to account for changes in forest C pools (soil and biota) remains a contentious issue. Additional research is needed in understanding processes and practices to sequester C in soils and vegetation of temperate forests, and to develop methods of measurement, monitoring and verification of C pool and changes over short periods of 2–5 years.

Keywords Deciduous forests • Evergreen forests • Coniferous forests • Mixed forests • Alfisols • Mollisols • Inceptisols • Spodosols • Ultisols • Climate change • Hardwood forests • Marginal soils • Afforestation • Reforestation • Soil carbon sequestration • Trading carbon credit • Soil carbon pools • Hydrologic cycle • Net primary production • Mean residence time • Gross primary production • Fire • Steppe • Greenhouse gases • Missing carbon sink • CO_2 • Fertilization effects • Forestry-based off-sets

Abbreviations

R_a	autotrophic respiration
C	carbon
CO_2	carbon dioxide
FACE	Free Air Atmospheric Carbon Dioxide Enrichment
GCC	global carbon cycle
GHGs	greenhouse gases
HAC	high activity clays
LAC	low activity clays
MRT	mean residence time
NBP	net biome productivity
NEP	net ecosystem production
NPP	net primary production
SOC	soil organic C
SOM	soil organic matter content

9.1 Introduction

World forests play an important role in climate and the environment (Streck and Scholz 2006). Temperate forests cover a global area of about 767 Mha (Pan et al. 2011), and have an important role in the global carbon cycle (GCC), are characterized by relatively cool temperatures, high precipitation, and high humidity

Table 9.1 Characteristics of the temperate forests

Characteristics	Descriptions
1. Geography	Temperate forests extend from and occur in North America, 25–55°N, middle Europe, southwest Russia, Japan, eastern China, southern Chile, middle east coast of Paraguay, New Zealand and southern Australia. Total land area is about 767 Mha.
2. Climate	Temperature varies from –30°C to 30°C. The mean annual temperature is 5–10°C, and the mean annual precipitation is 50–150 cm. There are four distinct seasons. Decidious trees drop their leaves during the winter. The growing season is 140–200 days with 4–6 frost-free months.
3. Canopy	It comprises of five zones in intact mostly primary forests: (i) the tree stratum, (ii) the small tree and sampling zone, (iii) the shrub zone, (iv) the herb zone, and (v) the ground zone.
4. Types of vegetation	Temperature forests comprise of several categories: (i) moist conifer and evergreen with wet winter and dry summer, (ii) dry conifer: high altitude, low precipitation, (iii) temperate conifer: high precipitation, mild winter, (iv) broad leaved forests: mild frost-free winters and high precipitation, (v) Mediterranean: winter rainfall of ~1,000 mm $year^{-1}$.

Table 9.2 Examples of climate of some temperate forests (Adapted from Luyssaert et al. 2007; Keith et al. 2009)

Parameter	Humid evergreen	Humid deciduous	Semiarid evergreen	Overall mean
Mean winter temperature (°C)	4±5	2±9	0±5	1.5
Mean summer	17±4	20±5	14±3	18.9
Precipitation winter (mm)	449±337	183±164	356±182	404 (minimum, annual)
Precipitation summer (mm)	194±234	356±259	81±99	5,000 (maximum, annual)
Net radiation winter (W m^{-2})	147±92	150±100	152±141	–
Net radiation summer (W m^{-2})	437±104	425±78	502±95	–
Mean humidity winter (%)	84±11	79±11	85±18	–
Mean humidity summer (%)	67±12	77±5	50±6	-

(Table 9.1). These biomes have four distinct seasons with wide variation in temperatures and precipitation among seasons. The annual precipitation ranges from 50 to 200 cm, and comprises of both rain and snow. Temperate forests are characterized by warm or mild summers and cool or cold winters. The mean continental temperature may range from 30°C to –30°C, and the maritime from –10°C to 20°C. The mean annual rainfall in continental climate may range from 15 to 100 cm and in the maritime regions from 50 to 200 cm. Examples of some climates of temperate forests are shown in Table 9.2.

Geographically, temperate forests are located in North America, Central and Western Europe, north-eastern Asia, southern Chile, New Zealand and the Mediterranean. These biomes comprise a wide range of vegetation with a large variety of species. Important temperate deciduous tree genera include *Acer*,

Ailanthus, *Albizzia*, *Betula*, *Carya*, *Castanopis*, *Fagus*, *Fraxinus*, *Juglans*, *Liriodendron*, *Magnolia*, *Nothofagus*, *Populus*, *Quercus*, *Tilia*, and *Zelkova*. Common temperate coniferous tree genera include *Abies*, *Picea*, *Pinus*, *Pseudotsuga*, *Thuja*, and *Tsuga*. The genera *Agathis*, *Dacrycarpus*, *Eucalyptus*, *Nothofagus*, *Podocarpus*, and *Quercus* are common to temperate broadleaf evergreen forests. Yet, the species diversity in temperate forests is much lower than that in tropical rainforests. Trees are good indicators of site-specific climate. For example, birch (*Betula spp.*) and juniper (*Juniperus spp.*) grow under cold climate, alder (*Alnus spp.*) and willow (*Salix spp.*) on poorly drained soils. Holly (*Ilex spp.*) grows in western Europe and north-eastern USA but along coastal regions because it does not survive under the extreme continental climate. Similarly, ivys (*Hedera spp.*) have specific climatic adaptation. Thus, the projected change in climate may alter the species composition, and also influence soil properties.

The canopy of intact temperate forest areas without signs of significant human transformation driven by intrinsic tree population processes is distinctly stratified into diverse zones. The top tree zone may consist of hardwood trees. The second tree zone includes smaller trees. The third canopy zone is made of shrubs, and the fourth zone comprises of herbs. The fifth zone may be comprised of lichens and mosses (Table 9.1).

The temperate forest biome has four distinct seasons. The summer season is characterized by long day light hour and warm temperatures (~30°C maximum). Most deciduous trees drop leaves during the fall. The winter temperature may be –30°C (the minimum). Thus, vegetation regrowth and the flowering season begins with spring. Trees, compared with grasses, exert a strong influence on the hydrological cycle, and to a large extent promote their own growth hydrologically and suppress fire. In contrast, grasses take hold and promote their own growth through fire (Mayer and Khalyani 2011). Increase in frequency and intensity of fire decreases the dominance of trees.

9.2 Soils of Temperate Forests

Principal soils of temperate forests are Alfisols, Inceptisols, Entisols, Mollisols, Spodosols, and Ultisols, (Table 9.3) (FAO-UNESCO 1974, 1988). These soils have a high inherent fertility (Martin et al. 2001; Gower et al. 2003), soil organic matter content (SOM), and favorable moisture and temperature regimes during the growing season. Temperate forest biomes have high net primary production (NPP). Thus, soils of temperate forests have the capacity to support a large amount of biomass production, and a large amount of C is stored in soils and the vegetation. The soil C density to 1-m depth (kg C m^{-2}) for temperate forests in the U.S. ranges from 8 to 9 in Alfisols, Ultisols and Entisols to >20 in Spodosols (Table 9.3), and correspond with the soil organic carbon (SOC) pool of ~80–90 Mg C ha^{-1} in Entisols/Alfisols, 150 Mg ha^{-1} in Inceptisols, 170 Mg ha^{-1} in Mollisols, and 250 Mg ha^{-1} in Spodosols (Table 9.3).

Table 9.3 Soils of temperate forests (Adapted from Brady and Weil 2002; Johnson and Kern 2003)

Order	Area in the temperate regions (10^6 ha)	Description	SOC density to 1-m in forest soils of the U.S. (kg m^{-2})
Alfisols	165	Soils with argillic, kandic or nitric horizon or a fragipan with clay skins, developed under a hardwood forest cover	9.2
Entisols	–	Soils of recent origin and less development	8.2
Inceptisols	–	Recent soils which exhibit mineral horizon development, and containing cambic, calcic mollic, umbric or histic epipedon	15.6
Mollisols	–	Soils with high organic matter content and containing mollic epipedon	17.2
Spodosols	252	Soils containing amorphous mixtures of organic matter and Al, with or without Fe, with and overlying eluvial horizon of grey to light color	24.1
Ultisols	330	Soils with an argillic or kandic horizon and fragipan commonly known as red clay soils	8.5
Total	767		

Soil description is from Brady and Weil (2002), SOC density from Johnson and Kern (2003), estimated of area under different soil orders are taken as the difference between the total area (Brady and Weil 2002) and subtracted for the area under tropics (van Wambeke 1992)

Table 9.4 Productivity of some temperate forest (Mg C ha^{-1} $year^{-1}$) and biomass (Mg C ha^{-1}) (Recalculated and adapted from Luyssaert et al. 2007)

Parameter	Humid evergreen	Humid deciduous	Semiarid evergreen
Gross primary production	17.6 ± 0.6	13.8 ± 0.6	12.3 ± 0.3
Net primary production	7.8 ± 0.6	7.4 ± 0.6	3.5 ± 0.3
Net ecosystem production	4.0 ± 0.4	3.1 ± 0.4	1.3 ± 0.5
Above ground biomass (AGB)	149.3 ± 135.6	108.8 ± 56.7	62.8 ± 55.5
Below ground biomass (BGB)	42.3 ± 46.7	25.7 ± 26.1	22.4 ± 17.3
Ratio BGB:AGB	0.28	0.24	0.36

Because of a high rate of biosequestration, temperate forests play an important role in the GCC. Ecosystem productivity has four distinct but related components (Table 9.4). The gross primary production (GPP) represents the total uptake of carbon dioxide (CO_2) by photosynthesis. However, some of the biome's photosynthesis is used in autotrophic respiration (R_a). The balance (GPP-R_a) is the NPP. Some of the NPP is lost in heterotrophic metabolism and microbial respiration. The remainder is called the net ecosystem production (NEP) and the C comprising of NEP has a longer mean residence time (MRT). A large fraction of NEP, however, is subject to losses through harvest, herbivory, fire and other perturbations. The final remaining fraction is called the net biome productivity (NBP). Some examples of these components of the temperate forest biomes given in Table 9.4 indicate that NPP and NEP of these biomes is equivalent to that of the tropical humid evergreen forest

Table 9.5 Biomass (Mg C ha^{-1}) carbon in some temperate forests (Adapted and recalculated from Keith et al. 2009)

Parameter	Cool moist	Cool dry	Cool montane
Above ground biomass (live)	377 ± 182	176 ± 102	147
Root & dead biomass	265 ± 162	102 ± 77	–
Total living & dead biomass	642 ± 294	278 ± 173	153

Mean annual temperature = 1.5°C <min, 18.9°C maximum
Mean annual precipitation = 404 mm min, 5,000 mm, maximum
Gross ecosystem production (Mg C ha^{-1} year^{-1}) = 2.51 min, 4.75 maximum

(Luyssaert et al. 2007; Huston and Wolverton 2009). Keith et al. (2009) observed that the world's highest known total biomass C density (living and detritus) of 1,867 Mg C ha^{-1} may occur in Australian temperate moist forest (*Eucalyptus regans* F. Muell.). Furthermore, temperate moist forests contain diverse forest types characterized by a wide range of mature C pools. Some temperate moist forests have higher biomass C density than those of boreal and tropical forests (Keith et al. 2009). However, the data on biomass C in the below-ground or root allocation is limited, and is generally obtained through modeling (Rasse et al. 2001). Temperate forests with particularly high biomass density (Table 9.5) are dominated by *Tsuga spp.*, *Picea spp.*, *Pseudostuga spp.* and *Abies spp.* especially in the Pacific Northwest of USA and Canada (Keith et al. 2009). These forests contain the above-ground biomass of 224–587 Mg C ha^{-1} and total biomass of 568–694 Mg C ha^{-1} (Keith et al. 2009). The highest total biomass density has been measured at 600–982 Mg C ha^{-1} for the *Agathis spp.* in New Zealand, and 326–571 Mg C ha^{-1} (for *Nothofagus, Fitzroye, Philgerodendron* and *Laureliopsis spp.*) in Chile. High C density in the biomass plays an important role in the GCC (Houghton et al. 2009). Thus, deforestation and conversion to agricultural land use can drastically reduce the temperate forest ecosystem C pool. Total ecosystems C pool, that in soils and vegetation combined, is high in temperate forests (Table 9.5). In general, in contrast with the tropical rainforests, temperate forests may have relatively more C in soils than in the biomass/vegetation. Somewhat higher C proportion in soils may be due to a slower decomposition rate in soils of temperate than those of the tropics.

In intact temperate forests where vegetation dynamic is driven by intrinsic tree population processes, the density of SOC is extremely heterogeneous (Lorenz and Lal 2010). The heterogeneity is related to climate, soil type, vegetation, and drainage. Soils of the temperate forests have often a high C density, and total C pool (Eswaran and van den Berg 1993; Batjes 2010, 2011). With due consideration to concerns about the true and false interpretation (Powlson et al. 2011) and over optimism (Schlesigner 2000), soils of the temperate forest ecosystems have a high SOC pool and C sink capacity. Even to 30 cm depth, SOC pool under natural vegetation cover can be as much as 35–140 Mg C ha^{-1} under warm moist climates and 50–128 Mg C ha^{-1} under cool moist regions. Naturally, the SOC pool varies widely among soils. The SOC pool is large in soils containing high activity clays (HAC) compared to that in rocky and sandy soils, and those containing low activity clays (LAC). Further, SOC pool and its vertical distribution indicate high magnitude in soils of cool compared to warm climates (Table 9.6, Jenny 1941; Jobbágy and

Table 9.6 Organic carbon pool in soils of temperate forests to 30-cm depth (Adapted from Batjes 2010, 2011)

	SOC pool in temperate forest (0–30 cm, Mg C ha^{-1})			
Soil	Warm moist	Warm dry	Cool moist	Cool dry
High activity clay (HAC)	64±33	24±16	81±40	43±24
Low activity clay (LAC)	55±29	19±10	76±48	–
Sandy soils	36±26	10±5	51±39	13±7
Spodic soils	143±65	–	128±61	–
Volcanic soils	138±56	84±88	126±52	–
Wetland soils	135±101	75±45	128±55	–

Jackson 2000). It is important to understand the mechanisms of stabilization of SOC pool, which differ among soils in relation to vegetation, climate, physiography and other factors (von Lützow et al. 2006; Schmidt et al. 2011).

9.3 Impact of Fire on Ecosystem Carbon Pool

Temperate forests are generally not fire-dependent ecosystems, and are less subject to fire than boreal and dry tropical forests, and the grass (steppe) ecosystems. Nonetheless, fire can affect both vegetation structure and C pools at spatial scales ranging from tree to stand to landscape scale. The magnitude of C lost from a temperate forest by fire depends on a range of factors such as the climate, vegetation, fire regimes and the post disturbance weather. Trees killed by fire may have a long turn overtime of 580±180 years because of the charred biomass created by fire (Keith et al. 2009).

Fire can also impact the soil C pool. Bormann et al. (2008) assessed the impact of natural fires on the soil C pool in southwestern Oregon, USA. Bormann and colleagues observed the fire-related changes in SOC pool were extremely high and estimated at 23 Mg C ha^{-1} for organic and mineral soil layers, of which 60% were lost from the mineral horizon. The severe damage, especially to the soil quality, indicate not only the emission of greenhouse gases (GHGs) by fire but also the reduction of future forest productivity and decline in C sequestration in forest biomass and the soil. Savanna and the forest ecosystems can co-exist depending on the intensity and frequency of fire. The specific pattern of fire-caused discontinuities in the biome type also depends on the tree species, climate, soils, and the fire (Staver et al. 2011).

9.4 Factors Affecting Carbon Sequestration in Forest Ecosystems

There is a strong evidence of the high NEP of temperate forests, as high as those of the boreal and moist tropical forests. Barford et al. (2001) reported the net ecosystem exchange (NEE) of the Harvard Forest (42.5°N, 72.2° W) by the eddy-covariance techniques at 2.0 ± 0.4 Mg C ha^{-1} $year^{-1}$, with inter-annual variation exceeding 50%. Factors affecting NBP and NEE are climate growing season length, cloudiness, soil type, vegetation, perturbation (fire), any legacy of prior disturbance, and stand management. Principal factors affecting C uptake in soils under temperate forests are primarily those related to litter decomposition (Prescott 2010). Over and above the strong effects of temperature and moisture regimes, liter type, and relative preservation of recalcitrant compounds are other imported variables. In addition to total solum depth, the depth distribution (stratification) is also an important factor. Total annual soil respiration is strongly related to mean annual soil temperature, respective proportion of the type of ecosystems and biome (Bahn et al. 2010). The temperature-dependence, increase in respiration by increase in mean annual temperature by 10°C (Q_{10}), is especially high for non-water limiting ecosystems. The Q_{10} value was 2.2 for *Fagus sylvatica* L. forest (MAP of 1,100 mm, 41°52′ N, 13°38′ E), 3.4 for *Picea abies* (L.) H. Karst and *Pinus cembra* L. forests (MAP of 1,010 mm, 46°35′ N, 22°26′ E), 1.9 for *Pinus sylvestris* L. and *Quercus robur* L. forests (MAP of 750 mm, 51°18′ N and 4°31′ E), 4.0 for *F. sylvatica* L. forest (MAP of 600 mm, 56°00′ N, 12°20′ E), and 3.8 for mixed hardwood Harvard Forest (MAP of 1,089 mm, 42°32′ N, 72°11′ W), and 3.4 for mixed evergreen forest (MAP of 1,005 mm, 45°12′ N, 68°44′ W) (Bahn et al. 2010).

The rate of forest respiration decreases substantially in response to N deposition (Janssens et al. 2010). There are numerous factors which affect the rate of litter decomposition (Prescott 2010), and the knowledge of these factors can be used to enhance the MRT and total C pool in forest ecosystems. For example, there is a strong effect of drying and wetting on respiration and decomposition (Borken et al. 2003; Schmitt and Glaser 2011), and of the compounds of low molecular weight (van Hees et al. 2005).

9.5 Temperate Forests and the Missing/Unidentified Carbon Sink

The term "missing C" proposed by Broecker et al. (1979), refers to "the amount of C released to the atmosphere from combustion of fossil fuel that is unaccounted for by the increase of C in the atmosphere and the ocean" (Houghton 1993). The awareness about the problem of balancing the C budget dates back to early 1970s (Woodwell and Pecan 1973; Study of the Critical Environmental Problems 1970; Studies of the Man's Impact on Climate 1971). The magnitude of missing C sink was initially reported to be as much as 1–2 Pg C $year^{-1}$ (Houghton 1993; Tans et al. 1990). However, there also exist temporal trends in the magnitude of land and ocean C sinks, and the magnitude of missing sink is determined by the difference or the default value. For example, Sarmiento et al. (2010) estimated the net land C sink of 0.27 Pg C $year^{-1}$ between 1960 and 1988, increase to 0.88–1.55 Pg C $year^{-1}$ between 1989, and between 2003 and 2007. The land-based C sink in the northern hemisphere was estimated at 1.7 Pg C $year^{-1}$ over the period 2000–2004 (Ciais et al. 2010).

However, the concept of "missing C" sink is questionable if not completely invalid in view of the fact that several sources and sinks are unaccounted for (Table 25.2 in Chap. 25).

Possible sites of the missing C sink have also been debated. Pacala et al. (2001) reported for 1980–1989 the magnitude of missing C sink in conterminous U.S. between 0.30 and 0.58 Pg C $year^{-1}$. Similar claims have been made for Siberia and the Amazon forest, and by the C transported into the oceans (Stallard 1998), and that contained in char in the fire-prone ecosystems. The uncertainties also depend on the methodological approach to reconstruct the GCC (House et al. 2003): ground-based inventories *vs.* model calculations. A major issue which needs to be addressed is that can the land-based sinks (presumably in the northern forest biome) be reconciled from the known data? (Ciais et al. 2010). Using forestry inventory data, Pan et al. (2011) estimated a total global forest sink of 2.4±0.4 Pg C $year^{-1}$ for 1990–2007. Of this, C sink (Pg C $year^{-1}$) for the temperate forest zone was estimated at 0.67±0.08 for 1990s, 0.78±0.09 for 2000–2007, and 0.72±0.08 for 1990–2007. Pan and colleagues estimated that temperate forests contributed 0.7±0.1 and 0.8±0.1 Pg C $year^{-1}$ (27% and 34%) to the global C sinks in established forests for 1990s and 2000–2007. The reasons for increase in C sink capacity of the temperate forests may be: (i) increasing forest density, (ii) reduced biomass removal with harvest, (iii) increase in the forest area, (iv) CO_2 fertilization effect, and (v) N deposition. For example, growing density was responsible for substantially increasing sequestered C in European and North American forests during 1990–2010 despite smaller changes in forest area (Rautiainen et al. 2011).

9.6 Climate Change and Carbon Storage in Temperate Forests

The projected climate change can strongly affect the ecosystem C pool of temperate forests. There are numerous factors that can determine the magnitude of the ecosystem C pool (Fig. 9.1). The growing season may be extended in the temperate zone (Menzel and Fabian 1999). In the Canadian prairies, for example, the growing season may increase by 10 days for every 1°C increase in temperature. In addition, the area under temperate forests may increase through encroachment into the boreal forests. Beniston and Tol (1998) observed that 0.8°C warming over Europe during the twentieth century has shifted the climatic isotherms by an average of 120 km. Such shifts have also been observed for alpine plants (Grabherr et al. 1994). However, if global warming happens at a faster rate (>0.1°C per decade) then trees can disperse to new ecoregions with more suitable climate, the composition of forest may change with poor survival of migrating species. In the mid latitude, pole ward shift of biomes is expected to be 200–300 km for every °C of warming (UNEP 1992).

Increase in atmospheric CO_2 concentration may also enhance NBP and NEP due to the CO_2 fertilization effect. The global increase in NPP attributed to the CO_2 fertilization effect was estimated at 0.5–2.0 Pg C $year^{-1}$ (Davidson and Hirsch 2001).

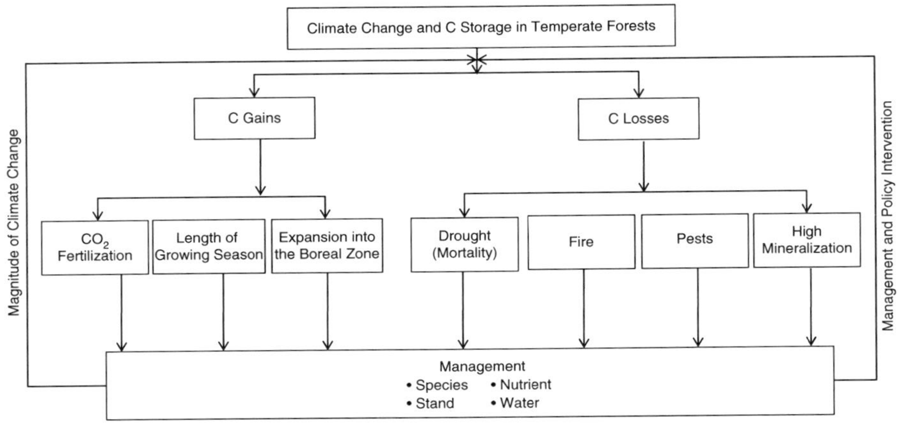

Fig. 9.1 Climate-induced changes in the C sink capacity of temperate forests

For example, the Free Air Atmospheric Carbon Dioxide Enrichment (FACE) experiments conducted in the southern U.S. showed that in comparison with the control photosynthesizing 0.3 kg $C\,m^{-2}$ $year^{-1}$, treatments receiving a higher CO_2 concentration sequestered additional 0.2 kg $C\,m^{-2}$ $year^{-1}$ (Davidson and Hirsch 2001). Furthermore, the MRT of the C sequestered ranged from 20 to 200 years in wood and soil, respectively. However, the CO_2 fertilization effect may be strongly limited by drought stress and/or nutrient deficiency. There is also a strong influence of the forest age on the CO_2 fertilization effect (Teneva and Gonzalez-Meler 2008). Specifically, the net response of photosynthesis, forest growth and C sequestration in the temperate forest biomes depends on the effects of elevated CO_2 during the entire cycle of forest stand development (Lorenz and Lal 2010). Further, some studies indicate that mature temperate trees may acclimate to the projected higher atmospheric CO_2 concentrations (Körner 2006). In summary, NPP in forest FACE experiments is increased by elevated CO_2, but the response can diminish over time (Norby and Zak 2011). Carbon accumulation is driven by the distribution of C among plant and soil components with differing turnover rates and by interactions between the C and N cycles. Plant community structure in forests may change, but elevated CO_2 has only minor effects on microbial community structure (Norby and Zak 2011).

The projected climate change can also exacerbate losses of C from forest ecosystems. The rate of mineralization/decomposition and erosion may increase with increase in temperature. There may be more frequent fires, and the incidence of pests and pathogens may increase (Fig. 9.1). The microbial activities in temperate forest soils are also affected by elevated CO_2 (Zheng et al. 2010).

The losses of C from the temperate forest biome affect the positive feedback between the climate change and the GCC. It is argued that global warming will reduce the efficiency of the C cycle to store anthropogenic CO_2 in the land and ocean sinks, and will exacerbate the positive feedback in the climate-GCC system. The magnitude of the feedback, however, may depend on: (i) climate-sensitivity of CO_2, (ii) the GCC-sensitivity of CO_2 and (iii) GCC sensitivity to climate change (Friedlingstein et al. 2003).

9.7 Potential of Temperate Forests to Recarbonization of the Biosphere

Most temperate forests are not at maximum C storage because of natural disturbances and harvesting (Dixon et al. 1994). Thus, temperate forests potentially store more C after forest management changes. The increase in the forest C pool by enhancing sequestration is a novel intention of forest management but the knowledge about forest production and forest growth is still incomplete (Andersson et al. 2000). A quantitative understanding of how forest management simultaneously enhances wood production and C sequestration is lacking for most forest types (Gough et al. 2008). In particular, the potential for conserving C through changes in management of temperate forests has not been estimated (Ciccarese et al. 2005). Thus, it is unclear how the forest C sink can be managed to mitigate atmospheric increases in CO_2 (Canadell and Raupach 2008). Also, management strategies to increase forest C storage in one area may have unintended negative or positive consequences on C storage elsewhere (Magnani et al. 2009). Further, biophysical effects of forest management on the amount and forms of energy transfer to the atmosphere may also occur (Anderson et al. 2011).

Heath et al. (2003) estimated the potential of C sequestration in soils under forest ecosystems in the U.S. However, the data on potential C sink capacity in soils and biomass of the temperate forests is not known at regional, national or global scale. The high C pool in soils and the biomass, as indicated by the data in Tables 9.4 and 9.7, can be increased even more through management and policy interventions. The importance of compensation to land managers through payments for ecosystem services cannot be over-emphasized. Despite the large potential, little research information exists on the processes and practices, for soil/site/climate specific situations, with regards to the rates of C sequestration in soils and biota through land use conversion and management. Scientific understanding is especially scanty with regard to the rates of SOC sequestration and decomposition in changing climates.

Whereas payments to land managers/foresters can provide the much needed incentives, forest C markets are not in place. How to account for changes in forest C (soils and vegetation) has been contentious (Carnell 2010). There are numerous concerns about forestry-based off-sets especially with regards to the credibility related to “additional”, “verifiable” and “permanent” off-sets (Gorte and Ramseur 2010). There is also an issue of “double-counting” the off-sets, implying that sellers may try to sell the same off-sets to multiple buyers (Gorte and Ramseur 2010).

Table 9.7 Carbon pool in temperature forest of the U.S. (Recalculated from Birdsey and Lewis 2003)

	Carbon density ($Mg\ ha^{-1}$)					
State	Trees	Soil	Forest floor	Understory	Total	Soil:tree ratio
I. Northeast	59.3	127.6	20.2	1.2	208.3	1.6
II. North central	47.3	123.9	20.5	1.0	196.1	1.7
III. Great plains	39.8	90.4	20.8	0.8	151.8	1.5
IV. Southwest	85.2	97.6	42.4	1.7	226.9	0.75

Substantial uncertainties also exist about the technical skills to accurately quantify, monitor and verify the amount of C sequestered by soil and biomass components of the temperate forest ecosystems (Gorte and Johnson 2010). Incorporation of forest C off-sets in any emission reduction program necessitates a firm/credible basis for measuring C pools and fluxes over a short period.

9.8 Conclusions

The atmospheric C pool may increase from about 390 ppm in 2010 to 700–1,000 ppm by 2100. The attendant future climate change may adversely affect human wellbeing. Thus, there is a need to identify natural sinks which can decarbonize the atmosphere. Temperate forests, covering a total land area of about 767 Mha, between mid latitude and ~55° N and S of equator, have favorable climate, fertile soils, and diverse species. A large proportion of temperate forests are managed to produce timber and other wood products. Others have been converted to agroecosystems. Thus, there is a potential to enhance the ecosystem C pool through conversion to a restorative land use for optimizing C capture and storage in temperate forests.

Realization of the potential C sink capacity, through land use and management, can be advanced by payments to land managers/foresters for ecosystem services such as C credits as off-sets towards anthropogenic emissions. However, there are numerous scientific uncertainties which must be addressed. Important among these are: (i) credible information on C pool in biomass and soil, and on the management-induced changes in these pools over short periods of 2–4 years, (ii) process and practices which affect C pools in soils and biota, such as CO_2 fertilization effect, interaction with water and nutrients, and the effects of global warming, (iii) the role of wild fires on emissions and changes in soil C and other properties which affect succession, (iv) leakage related to changes in land use and soil/stand management practices, and (v) the feedback to climate change.

Despite the uncertainties, C sequestrated in temperate forest has a larger sink capacity and longer MRT than that sequestered under croplands and other biomes. Thus, scientific programs and policy interventions must be in place to realize the potential C sink capacity of the temperate forest biome.

References

Anderson RG, Canadell JG, Randerson JT et al (2011) Biophysical considerations in forestry for climate protection. Front Ecol Environ 9:174–182

Andersson FO, Ågren GI, Führer E (2000) Sustainable tree biomass production. For Ecol Manage 132:51–62

Bahn M, Reichstein M, Davidson EA et al (2010) Soil respiration at mean annual temperature predicts annual total across vegetation types and biomes. Biogeosciences 7:1247–1257

Barford CC, Wofsy SC, Goulden ML et al (2001) Factors controlling long and short-term sequestration of atmospheric CO_2 in a mid-latitude forest. Science 294:1688–1690
Batjes NH (2010) A global framework of soil organic carbon stocks under native vegetation for use with the simple assessment option of the carbon benefits project system. Report 2010-10. ISRIC, Wageningen, Holland
Batjes NH (2011) Soil organic carbon stocks under native vegetation-revised estimates for use with the simple assessment option of the carbon benefits project system. Agric Ecosyst Environ 142:365–373
Beniston M, Tol RSJ (1998) Europe. In: Watson RT, Zinyowers MC, Moss RH (eds) Regional impacts of climate change. Cambridge University Press, Cambridge
Birdsey RA, Lewis GM (2003) Carbon in U.S. Forests and wood products, 1987–1997: state by state estimates. USDA-F.S., Newton Square
Borken W, Davidson E, Savage K et al (2003) Drying and wetting effects on carbon dioxide release from organic horizons. Soil Sci Am J 67:1888–1896
Bormann B, Homann P, Darbyshire R et al (2008) Intense forest wildfire sharply reduces mineral soil C and N: the first direct evidence. Can J For Res 38:2771–2783
Brady N, Weil R (2002) Nature and properties of soil. Prentice Hall, Upper Saddle River
Broecker WS, Takahashi T, Simpson HH et al (1979) Fate of fossil fuel carbon dioxide and the global measures. Science 206:409–418
Canadell JG, Raupach MR (2008) Managing forests for climate change mitigation. Science 320:1456–1457
Carnell R (ed) (2010) The role of forests in carbon capture and climate change. Nova, New York
Ciais P, Canadell J, Luyssaert S et al (2010) Can we reconcile atmospheric estimates of the Northern terrestrial carbon sink with land-based accounting? Curr Opin Environ Sustain 2:225–230
Ciccarese L, Brown S, Schlamadinger B (2005) Carbon sequestration through restoration of temperate and boreal forests. In: Stanturf JA, Madsen P (eds) Restoration of boreal and temperate forests. CRC Press, Boca Raton
Davidson E, Hirsch A (2001) Fertile forest experiments. Science 411:431–433
Dixon RK, Brown S, Houghton RA et al (1994) Carbon pools and flux of global forest ecosystems. Science 263:185–190
Eswaran H, van den Berg RP (1993) Organic carbon in soils of the world. SSSA J 57:192–194
FAO (1988) FAO-UNESCO soil map of the world, revised legend with correction and updates. Technical paper 20. FAO and ISRIC, Wageningen, Holland
FAO-UNESCO (1974) Soil map of the world. 1:5,000,000, vol 1 – legend. United Nations Educational, Scientifics and Cultural Organization, Paris, France
Friedlingstein P, Dufresne J, Cox P et al (2003) How positive is the feedback between climate changes and the carbon cycle? Tellus 55B:692–700
Gorte RW, Johnson R (2010) Measuring and monitoring carbon in the agricultural and forestry sectors. In: Carnell R (ed) The role of forest in carbon capture and climate change. Nova, New York
Gorte RW, Ramseur JL (2010) Forests carbon markets: potential and drawbacks. In: Carnell R (ed) The role of forest in carbon capture and climate change. Nova, New York
Gough CM, Vogel CS, Schmid HP, Curtis PS (2008) Controls on annual forest carbon storage: lessons from the past and predictions for the future. Bioscience 58:609–622
Gower ST, Landsberg JJ, Bisbee RL (2003) Forest biomes of the world. In: Young RA, Giese RL (eds) Forest ecosystem science and management. Wiley, Hoboken
Grabherr G, Gottfried M, Pauli H (1994) Climate effects on mountain plants. Nature 369:448
Heath S, Kimble JM, Binsey RA et al (2003) The potential of U.S. forest soils to sequester carbon. In: Kimble JM, Heath LS, Birdsey RA et al (eds) The potential of U.S. forest soils to sequester carbon and mitigate the greenhouse effect. Lewis Publishers, Boca Ratan
Houghton RA (1993) Is carbon accumulating in the northern temperate zone? Glob Biogeochem Cycles 7(3):611–617

Houghton RA, Hall F, Goetz SJ (2009) Importance of biomass in the global carbon cycle. J Geophys Res 114. http://dx.doi.org/10.1029/2009JG000935
House J, Prentice I, Ramankutty N et al (2003) Reconciling apparent inconsistencies in estimates of terrestrial CO_2 sources and sinks. Tellus 55B:345–363
Huston MA, Wolverton S (2009) The global distribution of net primary production: resolving the paradox. Ecol Appl 79:343–377
Janssens IA, Dieleman W, Luyssaert A et al (2010) Reduction of forest soil respiration in response to nitrogen deposition. Nat Geosci 3:315–322
Jenny H (1941) Factors of soil formation: a system of quantified pedology. McGraw-Hill, New York (reprinted in 1994)
Jobbagy E, Jackson R (2000) The vertical distribution of soil organic carbon and its relation to climate and vegetation. Ecol Appl 10:423–436
Johnson MG, Kern JS (2003) Quantifying the organic carbon held in forested soils of the U.S. and Puerto Rico. In: Kimble JM, Heath LS, Bindsey Ra et al (eds) The potential of U.S. forests soils to sequester carbon and mitigate the greenhouse effect. Lewis Publishers, Boca Raton
Keith H, Mackey B, Lindenmayer D (2009) Re-evaluation of forest biomass carbon stocks and lessons from the world's most carbon-dense forests. Proc Natl Acad Sci USA 106:11635–11640
Körner C (2006) Plant CO_2 responses: an issue of definition, time and resource supply. New Phytol 172:393–411
Lorenz K, Lal R (2010) Carbon sequestration in forest ecosystems. Springer, Dordrecht
Luyssaert S, Inglima I, Jungs M et al (2007) CO_2 balance of boreal, temperate, and tropical forests derived from a global database. Global Change Biol 13:2509–2537. doi:10.1111/j.1365-2486.2007.01439.x
Magnani F, Dewar RC, Borghetti M (2009) Leakage and spillover effects of forest management on carbon storage: theoretical insights from a simple model. Tellus 61B:385–393
Martin PH, Nabuurs GJ, Aubinet M et al (2001) Carbon sinks in temperate forests. Annu Rev Energy Environ 26:435–465
Mayer A, Khalyani A (2011) Grass trumps trees with fire. Science 334:188–189
Menzel A, Fabian P (1999) Growing season extended in Europe. Nature 397:659
Norby RJ, Zak DR (2011) Ecological lessons from free-air CO_2 enrichment (FACE) experiments. Annu Rev Ecol Evol Syst 42:181–203
Pacala S, Hurtt G, Baker D et al (2001) Consistent land- and atmosphere-based U.S. carbon sink estimates. Science 292:2316–2320
Pan Y, Birdsey R, Fang J et al (2011) A large and persistent carbon sink in the world's forests. Science 333:988–993
Powlson DS, Whitmore AP, Goulding KWT (2011) Soil carbon sequestration to mitigate climate change: a critical re-examination to identify the true and the false. Eur J Soil Sci 62:42–55
Prescott C (2010) Litter decomposition: what controls it and how can we alter it to sequester more carbon in forest soils? Biogeochemistry 101:133–149
Rasse D, Longdoz B, Ceulemans R (2001) TRAP: a modelling approach to below-ground carbon allocation in temperate forests. Plant Soil 229:281–293
Rautiainen A, Wernick I, Waggoner PE et al (2011) A national and international analysis of changing forest density. PLoS One 6(5):e19577. doi:10.1371/journal.pone.0019577
Sarmiento J, Gloor M, Gruber N et al (2010) Trends and regional distributions of land and ocean carbon sinks. Biogeosciences 7:2351–2367
Schlesigner WH (2000) Carbon sequestration in soils: some cautions amidst optimism. Agric Ecosyst Environ 82:121–127
Schmidt MWI, Torn MS, Abiven S et al (2011) Persistence of soil organic matter as an ecosystem property. Nature 478:49–56
Schmitt A, Glaser B (2011) Organic matter dynamics in a temperate forest soil following enhanced drying. Soil Biol Biochem 43:478–489
Stallard RF (1998) Terrestrial sedimentation and the carbon cycle: coupling weathering and erosion to carbon burial. Global Biogeochem Cycles 12:231–257. doi:10.1029/98GB00741
Staver AC, Archibald S, Levin SA (2011) The global extent and determinants of savanna and forest as alternative biome states. Science 334:230–232

Streck C, Scholz S (2006) The role of forests in global climate change: whence we come and where we go. Int Affairs 82:861–879

Study of Critical Environmental Problems (1970) Man's impact on the global environment: assessment and recommendations for action. In study of critical environmental problems. MIT Press, Cambridge, MA

Study of Critical Environmental Problems (1971) Inadvertent climate modification, report. MIT Press, Cambridge, MA

Tans PP, Fung IY, Takahashi T (1990) Observational constraints on the global atmospheric CO_2 budget. Science 247:1431–1438

Teneva L, Gonzalez-Meler M (2008) Decomposition kinetics of soil carbon of different age from a forest exposed to 8 years of elevated atmospheric CO_2 concentration. Soil Biol Biochem 40:2670–2677

UNEP (1992) The impacts of climate change on agriculture. Fact sheet 101, information unit on Climate change. Palais des Nations, Geneva, Switzerland

van Hees P, Jones D, Finlay R et al (2005) The carbon we do not see – the impact of low molecular weight compounds on carbon dynamics and respiration in forest soils: a review. Soil Biol Biochem 37:1–13

Van Wambeke A (1992) Soils of the tropics. McGraw Hill, New York

von Lützow M, Kögel-Knabner I, Ekschmitt K et al (2006) Stabilization of organic matter in temperate soils: mechanisms and their relevance under different soil conditions – a review. Eur J Soil Sci 57:426–445

Woodwell FM, Pecan EV (eds) (1973) Carbon and the biosphere. In: Proceedings of symposium, symposium series 30. Brookhaven National lab. Upton, NY, May 1972. US Atomic Energy Comm CONF-720510, 400 p

Zheng J, Han S, Zhou Y et al (2010) Microbial activity in a temperate forest soil as affected by elevated atmospheric $CO(_2)$. Pedosphere 20:427–435

Chapter 10
Decarbonization of the Atmosphere: Role of the Boreal Forest Under Changing Climate

Jagtar Bhatti, Rachhpal Jassal, and T. Andy Black

Abstract The boreal forest, with an area of about 11.4 million km^2, is the second largest terrestrial biome and plays a critical role in the global carbon (C) cycle. Its role in either accelerating or slowing climate change depends on whether the boreal forest is a net C source or a net C sink. The boreal forest stores 715.2 Pg C with 430.2 Pg present in peatlands and the remaining in forest ecosystems. In forest, about 60% of total C is present in the soil. The boreal forest sequesters C in surface vegetation, and has accumulated and conserved annual increments of C for millennia in soils, permafrost deposits, wetlands and peatlands. The net annual C sink of the boreal forest increased significantly over the last 20 years, from 0.54 Pg C year^{-1} to 1.07 Pg C year^{-1}. The future C balance of the boreal forest largely depends on the frequency and intensity of different disturbances, changes in species composition, forest management regimes and alterations to the nutrient and moisture regimes under changing climate conditions. The role of the boreal forest in the decarbonization of the atmosphere can be strengthened through techniques that reduce the time for stand establishment (such as site preparation, planting, and weed control) or increase the available nutrients for growth, or through the selection of species that are more productive. Fire- and insect-protection activities have a strong impact on the C sink strength of the boreal landscape. Therefore, reducing the area prone to fire and insect mortality, and extending the rotation age for holding C longer in older age classes will strongly increase the capacity of the boreal forest to decarbonize the atmosphere.

J. Bhatti (✉)
Northern Forestry Centre, Canadian Forest Service,
Edmonton, AB, Canada T6H 3S
e-mail: Jagtar.Bhatti@NRCan-RNCan.gc.ca

R. Jassal • T.A. Black
Biometeorology and Soil Physics Group, University of British Columbia,
Vancouver, BC, Canada V6T 1Z4
e-mail: rachhpal.jassal@ubc.ca; andrew.black@ubc.ca

R. Lal et al. (eds.), *Recarbonization of the Biosphere: Ecosystems and the Global Carbon Cycle*, DOI 10.1007/978-94-007-4159-1_10,

Keywords Decarbonization • Boreal forest • Global carbon cycle • Peatlands • Wetlands • Carbon sink • Forest management regimes • Fire • Rotation age • Canada • Russia • Scandinavia • Arctic air mass • Arctic front • Tropical forests • Bogs • Fens • Swamps • Global change • Chronosequence • Heterotrophic respiration • Land use change • Flooding • Carbon sequestration • Forest management • Drought • Mitigation policies • Herbivory

Abbreviations

BOREAS	Boreal Ecosystem-Atmosphere Study
C	carbon
CO_2	carbon dioxide
DOY	day of year
R	ecosystem respiration
EC	eddy-covariance
FLA	Flakaliden, Sweden
FYS	Fyedorovskoye, European Russia
GHGs	greenhouse gases
GDP	gross domestic product
GEP	gross ecosystem productivity or photosynthesis
HYT	Hyytiälä, Finland
P	mean annual precipitation
T	mean annual soil temperature
Ta	air temperature
CH_4	methane
NEE	net ecosystem exchange
NEP	net ecosystem productivity
NPP	net primary productivity
N	nitrogen
N_2O	nitrous oxide
NOPEX	Northern Hemisphere Climate Processes Land-surface Experiment
NOBS	Northern Old Black Spruce, Manitoba, Canada
NOR	Norunda, Sweden
OM	organic matter
PFT	plant functional type
R_h	heterotrophic respiration
SOA	Southern Old Aspen, Saskatchewan, Canada
SOBS	Southern Old Black Spruce, Saskatchewan, Canada
SOJP	Southern Old Jack Pine, Saskatchewan, Canada
ZOP	Zotino, Central Siberia

10.1 Introduction

The boreal forest occupies much of the circumpolar region between 50°N and 70°N (mostly from 52°N to 66°N) running through most of Canada, Russia and Scandinavia (Fig. 10.1). Its extent is associated with the location of the summertime arctic air mass, with the boreal forest's southern boundary corresponding to the southern limit of the arctic air mass in the winter. Thus, the boreal forest is located between the summer and winter positions of the arctic front. The boreal forest covers more than 10% of the world's land area and accounts for about one third of the total forest area. With an area of about 11.4 million km^2, the boreal forest is the second largest terrestrial biome, next to the tropical forest with an area of 19.5 million km^2 (Pan et al. 2011). Despite ongoing disturbance, e.g., fires, the boreal forest area has somewhat increased over the last two decades (Table 10.1). This contrasts to a significant and consistent decrease in the area of tropical forests.

The boreal forest has historically played a vital role in the development of societies and their economies. Settlements were only possible through the support of

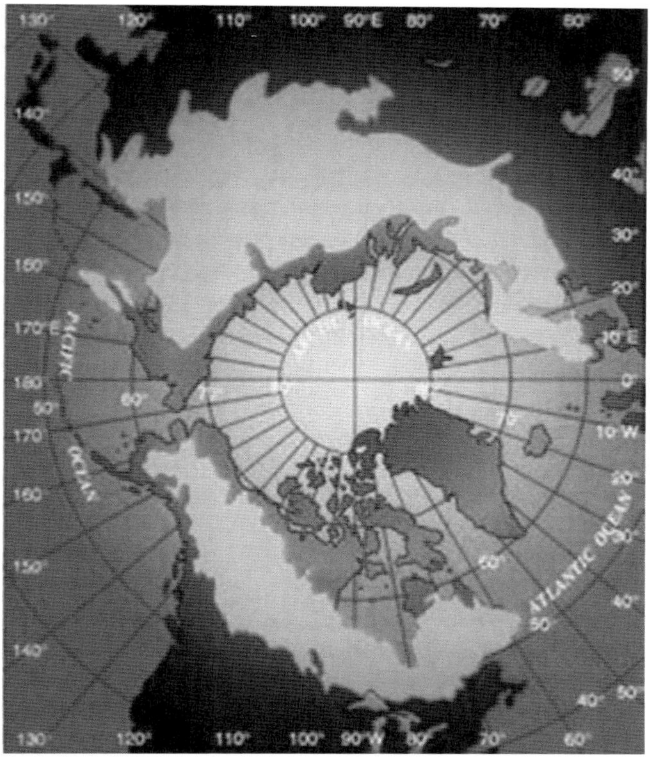

Fig. 10.1 The circumpolar range of the boreal forest (http://www.borealforest.org/; Hare and Ritchie 1972)

Table 10.1 Total area (Mha) of the world's forest biomes in 1990, 2000 and 2007 (Pan et al. 2011)

Biome	1990	2000	2007
Boreal	1102.7	1110.6	1135.2
Temperate	733.6	746.1	766.7
Tropical	2123.0	2009.2	1949.4
Total	3959.3	3865.9	3851.3

wood resources required for buildings, heating, cooking, tools and transportation. Mining, ship-building, charcoal production and tar extraction were major wood consumers until the end of the nineteenth century. The timber industry evolving during the twentieth century resulted in the rapid exploitation of the boreal forest, accounting for 3–7% of the gross domestic product (GDP) of boreal region countries, though this economic importance as a proportion of overall economic activity has declined since the post-World War II period (Nilsson 1996). However, future development of bioenergy may largely determine future utilization of boreal forest resources. There is a large potential as well as strong political demand to use wood for energy as bioenergy is being increasingly viewed as an alternative to fossil fuel use in the future. Forest sector managers in the boreal countries are aiming to intensify utilization of forest resources for energy production. Also, there are more than one million Aboriginal people living in the boreal zone whose cultures and economies are intricately adapted to their natural environment and depend upon it for survival. The "***lebensraum***" (quality of life) of Aboriginal people could probably be compromised with increased utilization of the boreal forests unless special actions are taken in the future. It is also important to take into account the other ecosystem services such as water protection, water supply, soil protection and climate change mitigation.

10.1.1 Climate

The boreal forest corresponds to regions of subarctic and cold continental climate. Long, severe winters (up to 6 months with mean temperatures below freezing) and short summers (50–100 frost-free days) are characteristic, as is a wide range of temperatures between the lows of winter and highs of summer. For example, Verkhoyansk, Russia, has recorded extremes of −32°C and 32°C. Mean annual precipitation is not high, generally ranging from 450 to 800 mm, but low evaporation rates make it a humid atmosphere. Winters are long and severe while summers are short though often warm. Because of the low temperatures, decomposition is slow, and soil organism activity limited. The highly lignified needles of the dominant conifer trees decompose slowly (Preston et al. 2006), creating a mat over the soil. Slow decomposition rates cause a strong nitrogen (N) deficiency as a limitation to plant productivity.

Climatically, the boreal zone can be divided into maritime, continental and high continental sub-zones, of which the continental area is the largest. Table 10.2

Table 10.2 Climate of the boreal forest (Black et al. 2005)

Region	Coldest monthly temperature (°C)	Mean annual precipitation (mm)
Maritime	2 to –5	400–1,000
Continental	–20 to –10	400–600
High continental	< –25	300–400

summarizes the mean monthly air temperature and annual precipitation in the three sub-zones. Soil temperature at the 5-cm depth stays within –3°C to 15°C in the maritime and continental zones while permafrost occurs in the high latitude continental zone.

In the maritime sub-zone, the range of climatic extremes is relatively small. Winters are comparatively mild and summers are often cool. The mean temperature of the warmest month is 10–15°C and of the coldest month –3°C to 2°C. The annual precipitation, much of it as snow, ranges from 400 mm to 800 mm but can be 1,000 mm or more in western Norway and Newfoundland.

Continental sub-zone winters are long and cold and there is abundant snow for 5–7 months. Considerable variation of the monthly mean temperature, especially in winter, is a marked feature of the zone, requiring adaptability of the trees. Desiccating winds and temperatures of –20°C to –40°C can be lethal for trees in the northern parts of the zone. Warming in the spring is rapid. However, there is a great variation in the time at which photosynthesis begins. Summer weather is comparatively warm but it can be very variable. The vegetation period, measured by the number of days when the mean daily temperature is more than 6°C, lasts from 100 to 150 days. The mean temperature of the warmest month ranges from 10°C to 20°C. The annual precipitation varies from 400 to 600 mm, with the major part of it falling during the summer months.

In the high continental sub-zone as in eastern Siberia, the winter is very long, extremely cold and dry. The mean annual temperature varies from –7°C to –10°C. The range of the mean monthly temperature can be more than 40°C and the lowest temperature can be –50°C to –60°C. The mean temperature of the coldest month can be below –25°C. Spring comes rapidly and summer is short and comparatively warm but frost is possible every night, even during the summer. At around 300–400 mm, annual precipitation is generally lower than in the other sub-zones. The greatest part of the precipitation falls during the growing season but the ratio of precipitation to evapotranspiration is <1 during the warm summer months.

10.1.2 *Landscape and Plant Species*

Overlying formerly glaciated areas and areas of patchy permafrost on both continents, the boreal landscape is a remarkable mosaic of forests of different ages reflecting the effects of fires. Also characteristic of the boreal forest are water bodies such as bogs, fens, marshes, shallow lakes, rivers and wetlands, mixed in among the forest and holding a vast amount of water (Fig. 10.2). Muskegs – low lying, water filled

Fig. 10.2 The boreal forest mosaic (http://atlas.nrcan.gc.ca/site/)

depressions or bogs – are common throughout the boreal forest (Fig. 10.3), occurring in poorly drained, glacial depressions. Sphagnum moss (*Sphagnum* sp.) forms a spongy mat on ponded water. There are relatively few tree species in the boreal forest. This is because of the relatively short time since the last glaciation and adaptation to harsh cold environment. Species include spruce (*Picea*), pine (*Pinus*), larch (*Larix*), fir (*Abies*), hemlock (*Tsuga*), poplar (*Populus*), birch (*Betula*), willow (*Salix*) and alder (*Alnus*) (Table 10.3). Understory species range from hazelnut (*Corylus cornuta*) in aspen (*Populus tremoloides*) stands, sphagnum moss and feather mosses (e.g., *Hylocomium splendens* and *Pleurozium schreberi*) in spruce stands to lichen (*Cladina* spp.) in jack pine (*Pinus banksiana*) stands. Soils include poorly drained organic soils supporting black spruce (*Picea mariana*), moderately well drained fine-textured soils supporting white spruce (*Picea glauca*) and aspen, and droughty sandy soils supporting jack pine. Ground-level vegetation mostly consists of feather mosses with sparse Labrador tea (*Ledum groenlandicum*) along with species of the tundra such as cotton grass (*Eriophorum Scheuchzer)* and shrubs of the heath family. Black spruce and larch ring the edge of fens and bogs? Sphagnum moss may enhance the water logging; once established it has the ability to hold up to 4,000% of its dry weight as water (Larsen 1980). Sphagnum often limits what species can establish once it gains a foothold. Some of the trees can reproduce by layering, since the probability of seeds germinating is low. The severe winters and short growing season favour evergreen tree species. These trees are also able to shed snow in the winter, which keeps them from breaking under the snow load, and to begin photosynthesis early in the spring, when the weather becomes more favourable.

Fig. 10.3 Bogs as part of the boreal landscape

Table 10.3 Dominant tree species and extent of their land cover in different regions of the boreal forest (Black et al. 2005)

Region (reference)	Species	% Land cover
North America	*Picea (mariana, glauca)*	39
	Pinus banksiana	14
	Abies balsamea	11
	Populus & Betula	21
Scandinavia	*Picea abies*	65
	Pinus sylvestris	26
	Populus & Betula	8
Siberia	*Larix gmelinii*	35
	Pinus sylvestris	17
	Picea abies	11
	Populus & Betula	17

Boreal peatlands, comprising bogs (67%), fens (32%), and swamps and marshes (1%) occupy about 72 Mha within the boreal forest (Bhatti and Tarnocai 2009). These ecosystems are created and develop in conditions with high water table levels, low temperature and consequently low rates of oxygen diffusion into the surface peat soil. These processes contribute to slow decomposition rates of organic matter (OM) and lead to the accumulation of organic carbon (C) as peat. However, some methane (CH_4) production in the anoxic soil layers and small amounts of nitrous oxide (N_2O) may also be produced. Thus, peatlands affect the chemical composition of the atmosphere through the exchange of these greenhouse gases (GHGs). Pristine peatlands are usually open habitats, with dominating field and bottom-layer vegetation, and often with sparse tree stands, and, thus, have much lower productivity than forests and other natural ecosystems (Vitt 2006). Carbon accumulation in peatlands represents the balance between above and belowground net primary productivity (NPP) and decomposition in both the upper, aerobic (acrotelm) and the underlying, anaerobic (catotelm) peat layers. Peat accumulation is enhanced by cool, wet conditions that limit OM decomposition (Moore et al. 1998).

10.2 Carbon Balance of the Boreal Forest

International negotiations to limit GHG emissions of CO_2 and CH_4 require understanding the current and potential future role of forest C emissions and sequestration in both managed and unmanaged forests. Forest biomass and annual production are of particular importance to the boreal forests in the global C budget (Kurz et al. 1993; Shvidenko et al. 1996). Total biomass, distribution of biomass, and productivity differ depending on site conditions, species composition and stand age. For example, on cold sites moss species may have levels of annual productivity equal to or greater than that of the over-story woody species (Wenping Yuan, personal Communication College of Global Change and Earth System Science, Beijing Normal University, Beijing). Warming in the boreal zone and extreme drought, co-incident with fires, represent the greatest risks to the C sink in the boreal forest. A better understanding of the role of forests in biosphere C fluxes and mechanisms responsible for forest C stock changes is critical for projecting future atmospheric carbon dioxide (CO_2) growth and guiding the design and implementation of mitigation policies.

10.2.1 Carbon Stocks

The primary C stocks in the boreal forest biome are forest biomass, forest soils including litter and coarse woody debris and peatlands. A summary of estimated pool sizes is provided in Table 10.4. In North America, peatlands account for 73% of the total boreal C stock as compared to Russia where peatlands account about 50% of this pool. Boreal forests account for about 32% (272 ± 23 Pg C) (1 Pg = 10^{15} g) of the total C stored in the world's forest biomes with a C stock density of 239 Mg C ha^{-1}, which is similar to that in tropical forests (242 Mg C ha^{-1}). However, compared to 56% of C stored in biomass and 32% stored in soil in tropical forests, boreal forests have most (60%) of the C stored in the soil (Table 10.4). Theede (2007) compared C stocks measurements made in 2004 at the Southern Old Aspen (SOA) and Southern Old Jack Pine (SOJP) sites in Saskatchewan with those made in 1994 during the Boreal Ecosystem-Atmosphere Study (BOREAS, see Sellers et al. 1997). At SOA, the change in C stocks was dominated by an increase in the forest floor

Table 10.4 Contemporary carbon stocks in the circumpolar boreal region

	Carbon stocks (Pg C)				
Boreal region	Forest biomass	Detritus	Forest soils	Peatlands	Total
Alaska	2[a]	1[a]	11[a]	15.9[b]	28.9
Canada	14.0[c]	15.7[c]	19.7[c]	152.3[b]	202.7
Russia	37.5[c]	23.9[c]	146.8[c]	220.2[d]	429.9
Europe	2.5[c]	1.5[c]	7.9[c]	41.8[e]	53.7

[a]Dixon et al. (1994)
[b]Bridgeham et al. (2008)
[c]Pan et al. (2011)
[d]Turner et al. (1998)
[e]Christensen and Friborg (2004)

Table 10.5 Total C stocks (Pg C) in the world's forest biomes in 1990, 2000 and 2007 (Pan et al. 2011)

Forest biome	1990	2000	2007
Boreal	258.6	264.0	271.5
Temperate	107.6	113.1	118.6
Tropical intact[a]	451.0	412.6	393.3
Tropical	501.9	477.3	471.0
Total	868.1	854.4	861.1

[a]Tropical forest not substantially affected by direct human activities

whereas at SOJP, the change in C stocks was dominated by an increase in tree biomass. The measured change in C stocks from 1994 to 2004 corresponded closely with estimates from eddy-covariance (EC) at both sites (Theede 2007). Norris et al. (2009) reported that in boreal jack pine forests, both fire and harvesting activities decreased soil C stocks and that a recovery to pre-disturbance levels took 30 years in both studied chronosequences.

10.2.2 *Carbon Fluxes*

The net annual C sink of the boreal forest increased significantly over the periods 1990–1999 and 2000–2007 from 0.54 to 1.07 Pg C year^{-1} (Pan et al. 2011) (Table 10.5). In comparison, it increased from 0.55 to 0.79 Pg C year^{-1} in the temperate forest. In contrast, the tropical forest was a net annual C source in both periods. In the case of the intact tropical forest, i.e., forest not substantially affected by direct human activities, the C source decreased from 3.84 Pg C year^{-1} in the first period to 2.76 Pg C year^{-1} in the second period, while for the entire tropical forest, it decreased from 2.46 to 0.90 Pg C year^{-1}. A model analysis (White et al. 2000) indicated that boreal forests had been a steady sink of 0.4 Pg C year^{-1} and that this sink could grow to 0.8–1.0 Pg C year^{-1} by the second half of this century and persist undiminished until 2100. However, the sink dynamics, associated with natural disturbances and forest management, vary among countries and regions. For example, the C sink has increased by 35% in European Russia during the last two decades, which has been attributed mainly to increased areas of forests after agricultural abandonment, reduced harvesting, and changes in forest age class distribution to more productive stages, particularly for the deciduous forests (Shvidenko et al. 2007). In contrast, the C sink in Canadian managed forests was reduced by half between the two periods, mainly due to biomass loss from intensive wildfires and insect outbreaks (Kurz et al. 2008), while a net loss of soil C in northern Europe has been attributed to the draining of water-logged soils (Kauppi et al. 2010).

During the last two decades, with the availability of fast response sonic anemometers and infrared gas analyzers, the EC technique has been widely used to measure CO_2 exchange between the atmosphere and terrestrial ecosystems. This method enables flux measurements to be made for periods of a half hour to years without disturbance to the ecosystem, mostly unattended and in climates as harsh as in the boreal forest (e.g., Baldocchi 2008). The first major initiatives to measure CO_2 exchange between boreal forests and the atmosphere using EC occurred during the early 1990s with BOREAS in North America and the Northern hemisphere climate Processes land-surface Experiment (NOPEX) in Sweden, and several campaigns in Siberia. The EC technique provides half-hourly or hourly values of net ecosystem exchange (NEE) of CO_2 between the ecosystem and the atmosphere, with positive values (upward fluxes) corresponding to losses and negative values corresponding to gains by the ecosystem. With knowledge of NEE-climate relationships, NEE is partitioned into ecosystem respiration (R) and gross ecosystem productivity or photosynthesis (GEP), and with a change in sign is a good approximation of net ecosystem productivity (NEP).

In the boreal forest, EC measurements have been made at 38 sites, i.e., 21 in North America, 5 in North Europe and 12 in Siberian Euro Russia (Black et al. 2005). Sites included evergreen conifers (black spruce, jack pine, Norway spruce (*Pices abies*)), Scots pine (*Pinus sylvestris*), a deciduous conifer (Dahurian larch (*Larix gmelinii*)), deciduous broadleaf trees (aspen, aspen-poplar, Norway birch (*Betula papyrifera*), white birch (*Betula alba*)) and fens and bogs. At nine sites, these flux measurements have been long-term year round thus making it possible to understand the effects of climate on interannual variability of NEP. Results in Table 10.6 show that mean NEP varies with stand age and plant functional type (PFT) as well as with climate or latitude with the YFY and NOR sites being C sources, the NOBS site being near neutral, and all other sites being C sinks. Temporal trends of annual NEP at NOBS indicated that NEP varied mainly with mean annual soil temperature (T_s) and that the effects of variation in mean annual precipitation (P) were not strong (Fig. 10.4). However, an increase in growing season rainfall decreased the water table depth, reduced respiratory losses from the soil and, thus, increased NEP (Black et al. 2005; Dunn et al. 2006) while the loss of C during 1995–1997 at NOBS has been attributed to the decomposition of old C deep in the soil profile (Goulden et al. 1998; Hirsch et al. 2002).

For the deciduous stand, SOA, annual NEP (Fig. 10.5) was characterized by a large interannual variability due to varied responses of GEP and R to climate variables. The 2001–2003 drought in the western plains of North America affected GEP

Table 10.6 Mean annual net ecosystem productivity (NEP) along with standard error derived from long-term eddy covariance measurements made at forest sites in North America, Northern Europe and Siberian boreal forests

Site	Location	Species	Avg. age (years)	Measurement period	NEP (g C m^{-2} $year^{-1}$)	Reference
Saskatchewan, Canada (SOA)	53°37′N 106°11′W	Trembling aspen	92	1994–2010	127 ± 93	Black et al. (2005), This study
Saskatchewan, Canada (SOBS)	53°59′N 105°70′W	Black spruce	138	2000–2010	45 ± 21	Black et al. (2005), This study
Saskatchewan, Canada (SOJP)	53°54′N 104°41′W	Jack pine	88	2000–2006	5 ± 50	Zha et al. (2009), This study
Manitoba, Canada (NOBS)	55°52′N 98°29′W	Black spruce	167	1995–2006	−2 ± 41	Dunn et al. (2006)
Norunda, Sweden (NOR)	60°05′N 17°28′E	Scots pine, Norway spruce	110	1995–2002	−52 ± 50	Lagergren et al. (2008)
Flakaliden, Sweden (FLA)	64°07′N 19°27′W	Norway spruce	50	1997–2002	132 ± 52	Janssens et al. (2001), Lindroth et al. (2008)
Hyytiälä, Finland (HYT)	61°51′N 24°17′E	Scots pine	48	1997–2007	219 ± 43	Lagergren et al. (2008), Kolari et al. (2009)
Zotino, Central Siberia (ZOP)	60°45′N 89°23′E	Scots pine	15–220	1999–2000	154 ± 9	Lloyd et al. (2002)
Fyedorovskoye, European Russia (FYS)	56°27′N 32°55′E	Norway spruce	10,160	1999–2000	−318 ± 6	Milyukova et al. (2002)

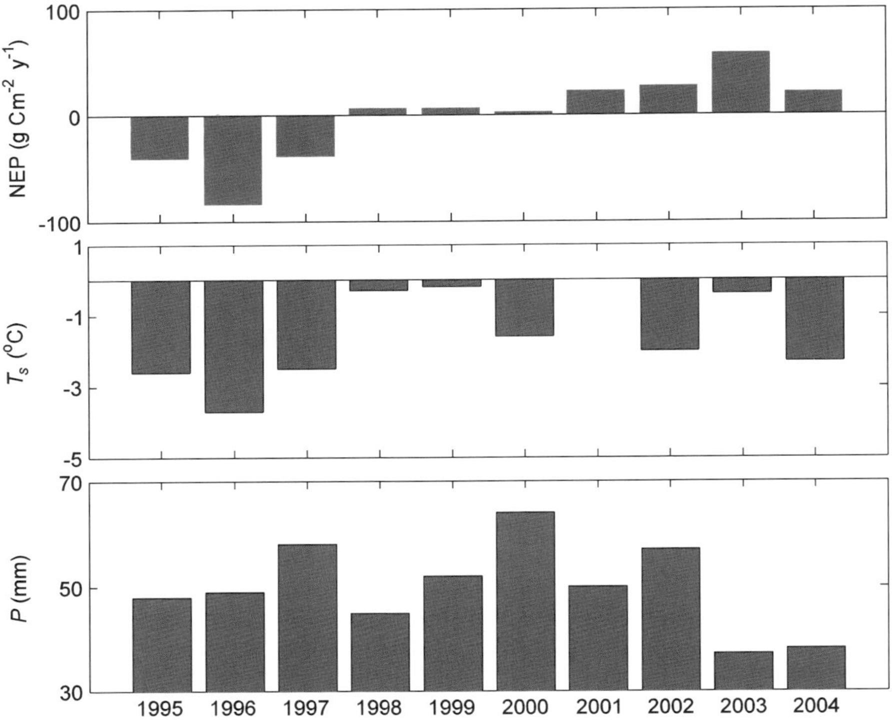

Fig. 10.4 Temporal changes in mean annual net ecosystem productivity(*NEP*), mean annual soil temperature (Ts) and mean annual monthly precipitation (*P*) at the Northern Old Black Spruce (*NOBS*) site in Manitoba, Canada (Redrawn from Dunn et al. 2006)

more than *R* and, thus, decreased NEP. Similarly, the marked decrease in GEP with a surprisingly small decrease in *R* in the cold La-Niña year 2007 resulted in the lowest NEP measured at SOA. In general, NEP was related to date of leaf emergence, which in turn was somewhat related to spring temperature (Figs. 10.6, 10.7) (Black et al. 2005). The earlier the start of the growing season, the higher was the NEP. Using regression analysis, Barr et al. (2009) identified spring temperature as the primary factor controlling annual NEP at SOA, SOBS and SOJP through its influence on the onset of the growing season. EC measurements in a post-burn chronosequence of jack pine stands in Saskatchewan (Fig. 10.8) show increasing annual NEP with age with stands becoming C neutral at about 10 years old. At 30 years old, these stands sequester about 90 g C m^{-2} $year^{-1}$ while at 90 years old, they have either a small uptake or loss of C depending upon climate. In comparison, for a chronosequence of post-burn black spruce stands in Manitoba, Bond-Lamberty et al. (2004) observed that younger (~36-year-old) stands were moderate annual C sink of about 100 g C m^{-2} $year^{-1}$, the middle-aged (~70-year-old) stands were relatively strong sinks at 100–300 g C m^{-2} $year^{-1}$, and the oldest (~130-year-old) stands were near C neutrality.

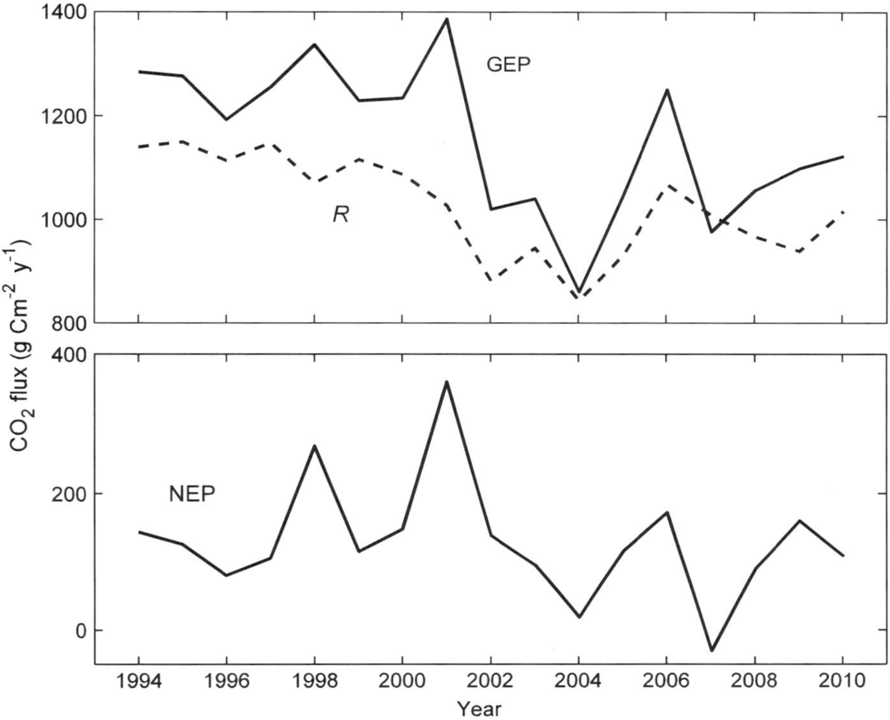

Fig. 10.5 Long-term variation in the carbon balance components at the Southern Old Aspen (*SOA*) site in Saskatchewan, Canada (Modified from Black et al. 2005)

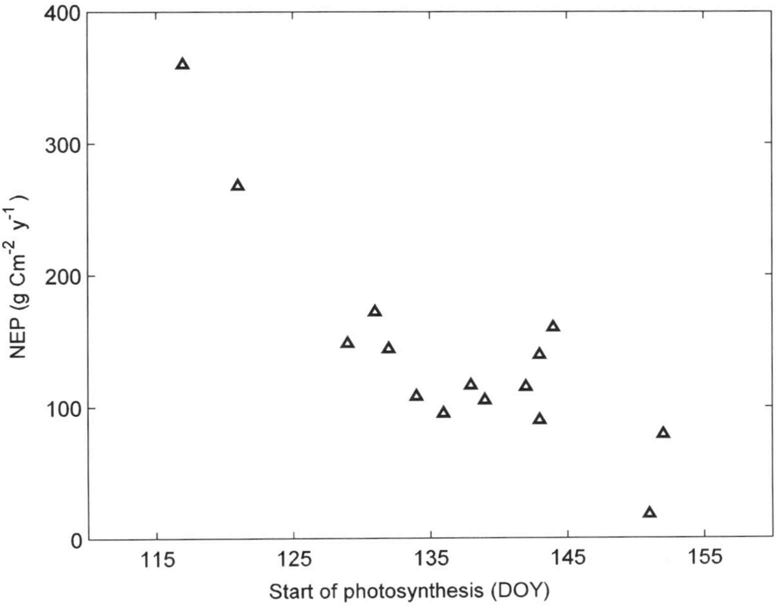

Fig. 10.6 Relationship between net ecosystem productivity (*NEP*) and the start of leaf photosynthesis (*DOY*, day of year) for 1994–2010 at the Southern Old Aspen (*SOA*) site in Saskatchewan, Canada (Modified from Black et al. 2005)

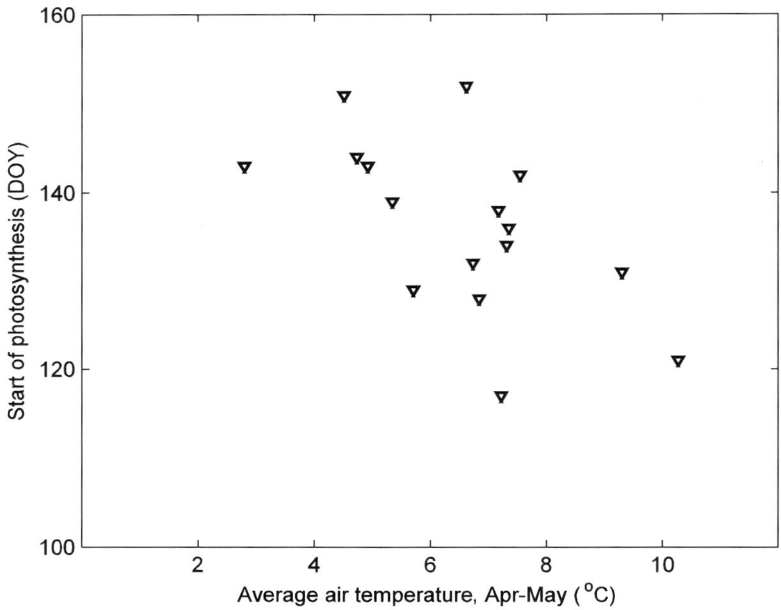

Fig. 10.7 Relationship between the start of leaf photosynthesis and average spring-season (April–May) air temperature for 1994–2010 at the Southern Old Aspen (*SOA*) site in Saskatchewan, Canada (Modified from Black et al. 2005)

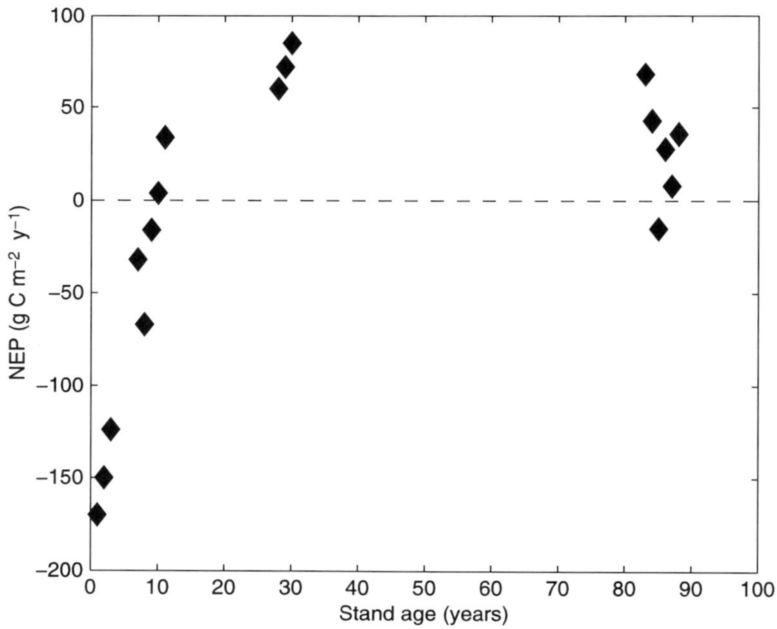

Fig. 10.8 Effect of stand age since burn on annual net ecosystem productivity (*NEP*) of a boreal jack pine forest chronosequence (Redrawn from Zha et al. 2009)

10.3 Carbon Balance of Boreal Peatlands

It is especially important to understand the C balance of boreal peatlands because they contain large amounts of organic C that is vulnerable to release to the atmosphere as CO_2 and CH_4 in response to climate warming. Peatland vegetation deposits various amounts of organic material (litter) annually on the peatland surface. Reader and Stewart (1972) observed that the amount of litter (dry biomass) deposited annually on the bog surface in boreal peatlands in Manitoba, Canada varied from 489 to 1,750 g C m^{-2}. Approximately 25% of the original litter fall decomposed during the following year. Reader and Stewart (1972) also reported that the average annual litter accumulation rate was 10% of the annual NPP. The gradual buildup of peat has been ongoing in peatlands during the last 5,000–8,000 years, resulting in peat deposits that are an average of 2–3 m thick, and in some cases, up to 10 m thick. At this stage, regardless of global change, peatlands can act as very effective C sinks for many thousands of years. Boreal peat lands are adding 2–11 cm per 100 years (National Wetlands Working Group 1988). Regional rates of C accumulation in Canadian peatlands range from 0.14 to 0.28 Mg C ha^{-1} $year^{-1}$ (Gorham 1991; Mäkilä 1997; Vitt et al. 2000) and vary with changes in soil moisture, soil temperature, reduction–oxidation conditions (Reader and Stewart 1972), acidity and alkalinity (Thormann et al. 1999), species composition (Johnson and Damman 1991), and litter quality (Updegraff et al. 1995; Yavitt et al. 1997). Joosten and Clarke (2002) reported that pristine fens presently take up 2.5 Mg C ha^{-1} $year^{-1}$ of CO_2 and release 2.97 Mg C ha^{-1} $year^{-1}$ as CH_4, while bogs currently take up 3.1 Mg C ha^{-1} $year^{-1}$ of CO_2 and release 0.53 Mg C ha^{-1} $year^{-1}$ as CH_4. While these figures indicate that pristine fens act as C sources and bogs as C sinks, total sequestration of 2.92 Mg C ha^{-1} $year^{-1}$ of CO_2 and release of 1.26 Mg C ha^{-1} $year^{-1}$ as CH_4 indicates that peatlands are a net C sink.

10.3.1 Forestation of Peatlands

Peatland drainage is a common management prescription to improve forest growth; as a result, over ten million ha of peatlands have been drained for forestry in Finland and Russia (Paavilainen and Päivänen 1995). A lowering of the water table level, induced by drainage through ditching, initiates a succession, which leads to a decrease in the abundance and biomass of species adapted to wet conditions, such as many sedges and herbs, and to an increase in the cover of shrubs and trees (Choi et al. 2007; Laine et al. 1995a, b; Laiho et al. 2003). The resultant increase in tree stand growth naturally results in an increased stand volume and biomass, and leads to canopy closure (e.g., Laine et al. 2006). Increases are highest at originally wet, nutrient-rich sites where predrainage tree stands are either sparse or absent (Keltikangas et al. 1986). Choi et al. (2007) observed that drainage improved radial growth of trees (C gain), while increasing soil N mineralization rates in minerotrophic peatland ecosystem. They further suggested that increased radial growth resulting from drainage was due to improved water use efficiency and N use efficiency. However, the change in vegetation cover from sedge and grass species to perennial shrubs and trees, especially conifers, results in decreased albedo, which increases radiative forcing, i.e., contributing to global warming (Betts 2000). Furthermore, drainage of peatlands is likely to result in rapid aerobic decomposition of organic matter and, consequently, to high CO_2 emissions and to a gradual depletion of the peat C pool (e.g., Turetsky and Louis 2006). At the same time, CH_4 emissions typically cease, and in some cases even leading to CH_4 uptake by the forest soil (Minkkinen et al. 2007) while N_2O emissions may increase in minerotrophic but not in ombrotrophic peatlands (Martikainen et al. 1993; Regina et al. 1996). In a recent study on C sequestration, GHG emissions and albedo effects of drainage in four peatland ecosystems in Finland, Lohila et al. (2010) reported that net radiative forcing was reduced at three of the sites with a greater reduction at nutrient-rich sites. EC measurements in a forestry-drained peatland ditched 35 years ago in Finland showed an annual C sink of 175 g C m^{-2} $year^{-1}$ (Lohila et al. 2011).

10.4 Global Change and the Boreal Forest

Global change is much more than merely climate change, and predominantly involves human population and human behavior. Most important for boreal forests in the future will be human behavior and how it affects the global environment – economic behavior (e.g. consumption patterns), social behavior (e.g., attitudes), and political behavior (e.g. cooperation or conflict). Forests are highly influenced by climate which determines their distribution, structure and composition, and much of their ecological function. Climate also affects forests indirectly through its impact on disturbance regimes, such as fire, insects, diseases, and windstorms, all of which modify the natural regeneration of conifers. In turn, forests have an impact on global climate through their influence on surface roughness, albedo, the hydrologic cycle, and the C cycle. The boreal biome plays a critical role in the global C cycle and has the capacity for either accelerating or slowing climate change to some degree depending on whether the forest acts as a net C source or a net C sink.

10.4.1 Interaction with Climate Change

If global mean temperature in the twenty-first century increases by 2.8°C, most of the boreal forest belt will experience 4–5°C warming and some northern parts of Canada even more (Price et al. 2011). The increase in winter temperatures will be far more than in summer temperatures. These figures originate from the rather moderate scenarios in the IPCC's Fourth Assessment Report (IPCC 2007). At present, actual emissions are increasing at a rate exceeding those in the IPCC's highest emission scenarios (Global Carbon Project 2008). Furthermore, IPCC's projections do not take into account the possible or probable positive feedback effects on global warming from large-scale, climate-induced changes in the Earth system, such as permafrost thawing or forest die-back. A recent study using HadCM3-QUMP, which includes climate change feedback loops, indicates that a global warming of 4°C would mean a 10–12°C rise in large parts of the boreal forest region. Regional differences are considerable with the largest temperature increases in eastern Canada and central Russia and the smallest (6–8°C) in the Scandinavian boreal region (Sanderson et al. 2011).

Changes in precipitation are more difficult to predict but several models suggest regional changes in summer and winter precipitation of ±20% for the boreal region (Kirschbaum and Fishlin 1996; Amiro 2001; Price et al. 2011). For example, precipitation is expected to decrease by 20% in northern Alberta, Saskatchewan, and Manitoba (Amiro 2001; Flannigan et al. 2001), while parts of eastern Canada (Flannigan et al. 2001; Amiro 2001) and Fennoscandia (IPCC 2001) are expected to experience an increase in precipitation. Short-term effects of climate change may be seen in increased rates of disturbance. Longer-term effects of climate change will be manifested in changes in tree growth and hence the volume of timber available for harvest. Higher temperatures increase the rate of both C uptake (photosynthesis) and loss (respiration), so the effect of higher temperatures depends on the net balance between these processes. Similarly, the net effect of changes in precipitation is determined by its seasonal distribution (e.g., spring versus summer) relative to the demand for water from the vegetation.

Higher temperatures result in higher evaporation rates, and hence soils in some boreal regions are expected to be drier in the summer (Flannigan et al. 1998; Amiro 2001). Increased drought stress would result in the boreal region becoming a long-term source of atmospheric CO_2 (Neilson 1993), and greater losses of boreal forest in the south (due to grassland and agricultural encroachment) than gains in the north (due to forest migration into tundra), which would lead to a net reduction in C stocks (Apps et al. 1993). However, some models (e.g., Smith et al. 1993) suggest that once the boreal forest achieves equilibrium with the new (and assumed stable) climate, the C stocks may be greater due to an increase in forest area from the migration of forest species into the presently un-forested tundra as well as increased forest productivity. Perez-Garcia et al. (1996) estimated that the boreal zone will have a substantially (15–20%) increased NPP.

Furthermore, warmer temperatures associated with climate change will accelerate both decomposition of soil C and permafrost melt. The latter will have profound effects on hydrology, vegetation, and C storage of the boreal forest. Climate change is also expected to affect both the distribution and the character of the boreal forest through changes in temperature, precipitation, and natural disturbance (fire, insects, disease, and wind throw) patterns (IPCC 2001). Natural disturbance could reduce forest C stocks by releasing large quantities of C to the atmosphere. Furthermore, accelerated decomposition as a result of warming will likely result in improved N nutrition of plants or a shift in species composition to those requiring higher N supplies and having higher N concentrations in their foliage, both expected to increase

Table 10.7 Area (Mha) of major disturbances in the boreal forest (Modified from Black et al. 2005)

Parameters	Eurasia	N. America	References
Boreal forest area	876	230	Pan et al. (2011)
Area burned annually	2–13[a]	1–8[b]	[a]Dixon and Krankina (1993) and Conard et al. (2002) [b]Stocks et al. (2002) and Murphy et al. (2000)
Area with moderate to severe insect infestation annually		10–25	Volney and Fleming (2000)
Area of forest harvested annually		1	CCFM (2002)

a and b are the respective source references

surface albedo, representing negative feedbacks to warming (Ollinger et al. 2008). Similar effects can be expected from the use of N fertilizers and anthropogenic N deposition, which also serve to increase foliage N content (Magill et al. 2004). Also, increasing levels of CO_2 in the future will likely increase tree growth with a higher rate of photosynthesis and greater light use efficiency (Long et al. 2004). The peatland sensitivity model of Tarnocai (2006) indicates that next to the subarctic region, the second largest effect of global climate change will occur in the boreal region.

10.4.2 Effects of Disturbance

Disturbances are part of the natural cycle, often releasing nutrients back to the forest to continue supporting the next generation of vegetation, and dictate stand renewal in most of the boreal forest. The major disturbance processes are fire, insect herbivory, disease, harvesting and wind throw. Estimates of the extent of these disturbances are given in Table 10.7. This chapter focuses on the severe events that have a well-defined impact on C exchange: mostly fire. Fire plays a major role in boreal forests with an average return period of about 120 years. This explains why the boreal forest is a mosaic of different-aged stands of various species. Fires burned an average of two million ha (Mha) of boreal forest annually in Canada from 1959 to 1999 releasing an average 27 Tg C year^{-1} (Amiro 2001). However, 115 Tg C were released in 1995 alone. Conard et al. (2002) estimated that 13 Mha burned in Siberia in 1998, and perhaps 12 Mha burned annually is a conservative estimate for Russia (Conard and Ivanova 1997). In Canada, the area burned has increased during the past four decades (Stocks et al. 2002) and will likely continue to increase with a warming climate (Flannigan et al. 1998). It is expected that climate change will alter the timing, intensity, frequency and extent of disturbances, including their interactions (Dale et al. 2001; Volney and Fleming 2000).

Several studies have focused on measuring C fluxes following fire and harvesting. Flux measurement data following insect infestation, major disease and wind throw events in the boreal forest are largely lacking. Post-fire EC CO_2 flux measurements along a black spruce fire chronosequence in northern Manitoba by Litvak et al. (2003) showed increasing growing season NEP with stand age, peaking at about 36 years with lower uptake at 70 years and release at 130 years following fire. Working on the same chronosequence, Bond-Lamberty et al. (2004) reported (see Sect. 10.2.2), using their NPP – R_h (heterotrophic respiration) method, that the youngest stands were moderate annual C sink at about 100 g C m^{-2} $year^{-1}$, the middle-aged stands relatively strong sinks at 100–300 g C m^{-2} $year^{-1}$, and the oldest stands about neutral.

Peatlands, and their ability to sequester C, are very sensitive to natural and anthropogenic disturbances, including wildfire, road construction, drainage, peat harvesting, and overburden removal from any form of open pit mining, including oil-sands (Turetsky and St. Louis 2006). Consequences of peatland disturbance may be direct, where the disturbance itself removes C from the peatland, or indirect, through reduced photosynthesis or increased decomposition resulting in an increased release either in gaseous or aqueous form. Decomposition, through the process of microbial respiration, converts previously stored C to CO_2 and CH_4, which are then released into the atmosphere or as dissolved organic C in streams flowing through peatlands (Freeman et al. 2004). In general, the peatland disturbances described above all cause C sequestration to be either reduced or eliminated (Turetsky and St. Louis 2006). Protection of C pools from intensifying and recurring disturbance events, solely as a mitigation strategy,v is likely neither efficient nor effective as a long-term measure (Vitt and Wieder 2006).

10.4.3 Land Use Change

Rapid expansion of agriculture along the southern border of the boreal forest has been a recognized risk for more than 50 years (Davidson 1998). The conversion of native upland and lowland into agriculture and urban lands has escalated over last 50 years, resulting in the contemporary patchwork of ecosystem types (Houghton 2000). In the prairie provinces of Canada alone, it has been estimated that there was a net deforestation of 12.5 Mha between 1869 and 1992 (Ramankutty and Foley 1999). Using the Canadian Land Inventory Database to estimate changes between 1966 and 1994, Hobbs and Theobald (2001) estimated that forests of the southern boreal plains of Saskatchewan declined from 1.8 to 1.35 Mha, an overall conversion of 24% of the boreal transition zone to agriculture over the 28-year period. Other studies have shown that forest land is being converted into agriculture, industrial, and urban development at the rate of 1,215 ha $year^{-1}$ along the southern boreal zone of Canada (Fitzsimmons 2002). This rate is approximately three times the world average. These estimates suggest that with the current rate of conversion, all the wetland and forested areas in the boreal transitional zone will be lost by 2050 unless purposeful action is taken to reverse the present trend. In addition to removal of natural vegetation cover, agricultural activities also deplete the soil C pool through reduction of biomass inputs and cause changes in soil temperature and moisture regimes, which further accelerate decomposition. Soil drainage, aimed at managing the water table, and soil cultivation intended to control weeds and prepare the seed bed, also accelerate soil erosion and mineralization of the organic C pool. Changes in land use also alter the uptake of CH_4 by soils, and different agricultural practices differ in their CH_4 emissions (Moss et al. 2005).

Flooding of forest and wetland areas in the boreal zone for hydroelectric reservoirs generates massive fluxes of dissolved organic C into the water, accelerates peat decomposition, and increases CH_4 and CO_2 fluxes to the atmosphere (Munn and Maarouf 1997). For example, experimental flooding of a boreal wetland in Ontario changed the site from a sink of 6.6 g C m^{-2} $year^{-1}$ to a source of 130 g C m^{-2} $year^{-1}$ (Kelly et al. 1997). Turetsky et al. (2002) estimated that 0.8 ± 0.2 Pg C $year^{-1}$ is released from peatlands occupying approximately 780 km^2 within hydroelectric reservoirs across the western boreal forest in Canada.

10.5 Increasing C Sequestration in the Boreal Forest

The vast area and large C stocks of the boreal forest with its C sink strength will continue to play a major role in the global C budget, though its ability to contribute to the decarbonization of the atmosphere is influenced by many factors and therefore difficult to predict. The future C balance of the boreal forest will largely depend on the frequency and intensity of different disturbances, changes in species composition, and alterations to the nutrient and moisture regimes under changing climate conditions. It will also depend on forest management practices that affect the disturbance regime as well as soil edaphic environment and nutrient availability which affects atmospheric C uptake in the long lasting woody biomass. However, relying on biospheric C sequestration is not without risk, since such sequestration is reversible due to climate change-induced and anthropogenic disturbances.

10.5.1 Management

With industrial forestry affecting inputs of nutrient and water quality of watersheds, and fluxes of GHGs into the atmosphere, forest management must reconcile demand for wood supply with non-timber values of forests. There is insufficient knowledge available to guide land managers in specific situations to change forest management practices to increase C sequestration, and there is some uncertainty about the longevity of effects of management on boreal forests (Caldeira et al. 2004). This reflects a gap in the availability of inexpensive techniques for measuring, monitoring, and predicting changes in ecosystem C stocks at the smaller scales appropriate for managers which requires conducting forest inventories more frequently than at present. Data gaps complicate analyses of the potential for policies to influence natural, social, and economic drivers that can change C stocks and fluxes. Forest management needs to pay greater attention to natural disturbance dynamics and must consider a variety of silvicultural systems. A combination of silvicultural approaches may be required to create stand compositions and age structures consistent with the effects of disturbances such as fire and harvesting. Use of stand-level ecosystem models as well as dynamic global vegetation models can help in assessing the effects of future climate and management activities on disturbance and forest productivity.

The forest management activities that have most influenced the boreal forests as a whole are forest harvesting and forest fire management. Both of these activities change forest dynamics at all scales of resolution, from altering the landscape structure to microsite characteristics and have introduced externally controlled dynamics versus natural dynamics. Protecting the forest from fire and insects can temporarily preserve forest C stocks. Protection of the entire boreal forest area, however, is not practical (Weber and Stocks 1998), nor may it be ecologically desirable since the boreal forest is a disturbance-dependent ecosystem. When disturbances occur, the seeding or planting of trees in the disturbed area can hasten forest regeneration and diminish the gap between the time of the disturbance and the time when the regenerating forest again recaptures C from the atmosphere (Kurz et al. 1995). Planting faster-growing tree species such as hybrid poplar (*Populus* spp.) appears to be one of the promising means to increase C sequestration rates (van Kooten 2000), though, however, planting species adapted to the local climate may be more effective for long-term C sequestration (Schroeder and Kort 2001).

Management practices such as reducing the time for stand establishment (e.g., site preparation, planting, and weed control), or increasing available nutrients for growth, or through the selection of species that are more productive for a particular area, may increase C uptake. Application of fertilizers to increase forest growth and, thus, C sequestration would be advisable where C sequestration benefits exceed the fossil-fuel footprint of fertilizer manufacturing and transportation (Jassal et al. 2008). Decreasing C losses can be accomplished through minimizing the time forests are a source to the atmosphere following disturbance, modification of harvesting practices such as engaging in low-impact harvesting (to reduce soil disturbance and damage to residual trees), increasing harvesting efficiency (and, thus, reducing logging residue), choosing the appropriate mix of species to plant, managing residues to leave C on site (Binkley et al. 1997), and protecting peatlands from drainage and land use change.

10.6 Conclusions

The boreal landscape, due to its low biological productivity compared to other ecosystems, may only be able to make a limited contribution to the decarbonization of the atmosphere. However, with its vast area and large C stocks, the boreal forest could play a major role in the global C budget. The future C balance in the boreal forest will largely depend on the type and frequency of disturbance, changes in species composition, and alterations to the nutrient and moisture regimes under changing climate conditions. It will also depend on forest management practices that affect both the disturbance regime and nutrient status. With projected climate change scenarios for the boreal region generally predicting warmer and somewhat drier conditions, disturbance patterns are also expected to change, which may release more C from the ecosystem than accumulate C in the growing vegetation. For example, the net GHG balance resulting from large-scale drainage of peatlands to improve forest growth will be one of the factors determining the role of boreal ecosystems in the decarbonization of the atmosphere in the coming years. The precise balance of C uptake and release depends on the controlling processes; and quantifying the life-cycle dynamics of C in boreal ecosystems is essential for projecting future changes. Whether the net effect of such changes results in positive feedback to climate change and thereby accelerate global warming is a critical issue that needs to be addressed.

Acknowledgements We would like to thank Rattan Lal and Klaus Lorenz for their critical review of an earlier version of this chapter.

References

Amiro BD (2001) Paired-tower measurements of carbon and energy fluxes following disturbance in the boreal forest. Glob Change Biol 7:253–268

Apps MJ, Kurz WA, Luxmoore RJ et al (1993) Boreal forests and tundra. Water Air Soil Pollut 70:39–53

Baldocchi D (2008) Breathing of the terrestrial biosphere: lessons learned from a global network of carbon dioxide flux measurement systems. Aust J Bot 56:1–26

Barr AG, Black TA, McCaughey H (2009) Climatic and phenological controls of the carbon and energy balances of three contrasting boreal forest ecosystems in Western Canada. In: Noormets A (ed) Phenology of ecosystem processes. Springer, New York

Betts RA (2000) Offset of the potential carbon sink from boreal forestation by decreases in surface albedo. Nature 408:187–190

Bhatti JS, Tarnocai C (2009) Influence of climate and land use change on carbon in agriculture, forest, and peatland ecosystems across Canada. In: Soil carbon sequestration and the greenhouse effect, 2nd edn, SSSA special publication 57. Soil Science Society of America, Madison, pp 47–70

Binkley CS, Apps MJ, Dixon RK et al (1997) Sequestering carbon in natural forests. Crit Rev Environ Sci Technol 27S1:23–45

Black TA, Gaumont-Guay D, Jassal RS et al (2005) Measurement of carbon dioxide exchange between the boreal forest and the atmosphere. In: Griffiths H, Jarvis P (eds) Carbon balance of forest biomes. BIOS Scientific Publishers, New York

Bond-Lamberty B, Wang C, Gower ST (2004) Net primary production and net ecosystem production of a boreal black spruce wildfire chronosequence. Glob Change Biol 10:473–487

Bridgeham SD, Megonigal JP, Keller JK et al (2008) Wetlands. In: King AW, Dilling DF et al (eds) The first state of the carbon cycle report (SOCCR): the North American carbon budget and implications for the global carbon cycle. A report by the U.S. Climate Change Science Program and the subcommittee on global change research. National Oceanic and Atmospheric Administration Climate Program Office, Silver Spring

Caldeira K, Morgan MG, Baldocchi D et al (2004) A portfolio of carbon management options. In: Field CB, Raupach MR (eds) Global carbon cycle: integrating humans, climate, and the natural world. CCFM (Canadian Council of Forest Ministers). SCOPE 62. Island Press, Washington, DC

CCFM (Canadian Council of Forest Ministers) (2002) Compendium of Canadian forestry statistics. Natural Resources Canada, Ottawa, Canada

Choi W-J, Chang SX, Bhatti JS (2007) Drainage affects tree growth and C and N dynamics in minerotrophic peatland as revealed by tree-ring and C and N isotopic analysis. Ecology 88:243–253

Christensen TR, Friborg T (eds) (2004) EU peatlands: current carbon stocks and trace gas fluxes. European Commission, CarboEurope-GHG report 7, specific study 4

Conard SG, Ivanova GA (1997) Wildfire in Russian boreal forests – potential impacts of fire regime characteristics on emissions and global carbon balance estimates. Environ Poll 98:305–313

Conard SG, Sukhinin AI, Stocks BJ et al (2002) Determining effects of area burned and fire severity on carbon cycling and emissions in Siberia. Clim Chang 55:197–211

Dale VH, Joyce LA, McNulty S et al (2001) Climate change and forest disturbances. Bioscience 51:723–734

Davidson C (1998) Issues in measuring landscape fragmentation. Wildl Soc Bull 26:32

Dixon RK, Krankina ON (1993) Forest fires in Russia: carbon dioxide emissions to the atmosphere. Can J For Res 23:700–705

Dixon RK, Brown S, Houghton RA et al (1994) Carbon pools and flux of global forest ecosystems. Science 263:185

Dunn AL, Barford CA, Wfsy SC et al (2006) A long-term record of carbon exchange in a boreal black spruce forest: means, responses to interannual variability, and decadal trends. Glob Change Biol 12:1–14

Fitzsimmons M (2002) Effects of deforestation and reforestation on landscape spatial structure in boreal Saskatchewan. Can J For Res 32:843

Flannigan MD, Bergeron Y, Engelmark O et al (1998) Future wildfire in circumboreal forests in relation to global warming. J Veg Sci 9:469–476

Flannigan MD, Campbell ID, Wotton M et al (2001) Future fire in Canada's boreal forest: paleoecology, GCM and RCM results. Can J For Res 31:854–864

Freeman C, Fenner N, Ostle NJ (2004) Export of dissolved organic carbon from peatlands under elevated carbon dioxide levels. Nature 430:195–198

Global Carbon Project (2008) Carbon budget and trends 2007. www.globalcarbonproject.org. Accessed 26 Sept 2008

Gorham E (1991) Northern peatlands: role in the carbon cycle and probable responses to climatic warming. Ecol Appl 1:182–195

Goulden ML, Wofsy SC, Harden JW et al (1998) Sensitivity of boreal forest carbon balance to soil thaw. Science 279:214–217

Hare FK, Ritchie JC (1972) The boreal microclimates. Geogr Rev 62:333–365

Hirsch AI, Trumbore SE, Goulden ML (2002) Direct measurement of the deep soil respiration accompanying seasonal thawing of a boreal forest soil. J Geophys Res 108:1–10

Hobbs NT, Theobald DM (2001) Effects of landscape change on wildlife habitat: applying ecological principles and guidelines in the western United States. In: Dale VH, Haeuber RA (eds) Applying ecological principles to land management. Springer, New York

Houghton RA (2000) Interannual variability in the global carbon cycle. J Geophys Res 105:20121–20126

IPCC (2001) Climate change 2001: impacts, adaptation and vulnerability. In: McCarthy JJ, Canziani OF, Leary NA et al (eds) Contribution of working group II to the third assessment report of the Intergovernmental Panel on Climate Change. Cambridge University Press, Cambridge

IPCC (2007) Chapter 4: ecosystems, their properties, goods and services. Fourth assessment report, climate change. Working group II report: impact, adaptation and vulnerability

Janssens IA, Lankreijer H, Matteucci G et al (2001) Productivity overshadows temperature in determining soil and ecosystem respiration across European forests. Glob Change Biol 7:269–278

Jassal RS, Black TA, Chen B et al (2008) N_2O emissions and carbon sequestration in a nitrogen-fertilized Douglas-fir stand. J Geophys Res Biogeosci. doi:10.1029/2008JG000764

Johnson LC, Damman AWH (1991) Species-controlled Sphagnum decay on a south Swedish raised bog. Oikos 61:234–242
Joosten H, Clarke D (2002) Wise use of mires and peatlands: background principles including a framework for decision-making. International Mire Conservation Group and International Peat Society, Saarijärvi
Kauppi PE, Rautiainen A, Korhonen KT et al (2010) Changing stock of biomass carbon in a boreal forest over 93 years. For Ecol Manage 259:1239–1244
Kelly CA, Rudd JWM, Bodaly RA (1997) Increases in fluxes of greenhouse gases and methyl mercury following flooding of an experimental reservoir. Environ Sci Technol 31:1334–1345
Keltikangas M, Laine J, Puttonen P et al (1986) Peatlands drained for forestry during 1930–1978: results from field surveys of drained areas. Acta For Fenn 193:1–94
Kirschbaum M, Fischlin A (1996) Climate change impacts on forests. Climate Change 1995: Impacts; Adaptations and mitigation of Climate Change. In: Watson R, Zinyowera MC, Moss RH (eds) Scientific-Technical Analysis. Contribution of Working Group II to the Second Assessment Report of the Intergovernmental Panel of Climate Change. Cambridge University Press, Cambridge pp 95–129
Kolari P, Kulmala L, Pumpanen J et al (2009) CO_2 exchange and component CO_2 fluxes of a boreal Scots pine forest. Boreal Environ Res 14:761–783
Kurz WA, Apps MJ, Webb TM et al (1993) The carbon budget of the Canadian forest sector: phase I. Information report NOR-X-326. Northwest Region, Northern Forestry Center, Forestry Canada, Edmonton, Alberta, Canada
Kurz WA, Apps MJ, Beukema SJ et al (1995) 20th century carbon budget of Canadian forests. Tellus 47B:170–177
Kurz WA, Stinson G, Rampley GJ et al (2008) Risk of natural disturbances makes future contribution of Canada's forests to the global carbon cycle highly uncertain. Proc Natl Acad Sci USA 105:1551
Lagergren F, Lindroth A, Dellwik ED et al (2008) Biophysical controls on CO_2 fluxes of three northern forests based on long-term eddy covariance data. Tellus 60B:143–152
Laiho R, Vasander H, Penttilä T et al (2003) Dynamics of plant-mediated organic matter and nutrient cycling following long-term water-level drawdown in boreal peatlands. Glob Biogeochem Cycles. doi:10.1029/2002GB002015
Laine J, Vasander H, Laiho R (1995a) Long-term effects of water level drawdown on the vegetation of drained pine mires in southern Finland. J Appl Ecol 32:785–802
Laine J, Vasander H, Sallantaus T (1995b) Ecological effects of peatland drainage for forestry. Environ Rev 3:286–303
Laine J, Laiho R, Minkkinen K et al (2006) Forestry and boreal peatlands. In: Kelman WR, Vitt DH (eds) Boreal Peatland Ecosystem. Ecological Studies, vol 188. Springer, Berlin
Larsen JA (1980) The boreal ecosystem. Academic, New York
Lindroth A, Klemedtsson L, Grelle A et al (2008) Measurement of net ecosystem exchange, productivity and respiration in three spruce forests in Sweden shows unexpectedly large soil carbon losses. Biogeochem 89:43–60
Litvak M, Miller S, Wofsy SC et al (2003) Effect of stand age on whole ecosystem CO_2 exchange in the Canadian boreal forest. J Geophys Res 108(D3):8225
Lloyd J, Shibistova O, Zolotoukhine D et al (2002) Seasonal and annual variations in the photosynthetic productivity and carbon balance of a central Siberian pine forest. Tellus 54B:590–610
Lohila A, Minkkinen K, Aurela M et al (2010) Forestation of boreal peatlands: impacts of changing albedo and greenhouse gas fluxes on radiative forcing. J Geophys Res. doi:10.1029/2010JG001327
Lohila A, Minkkinen K, Aurela M et al (2011) Greenhouse gas flux measurements in a forestry-drained peatland indicate a large carbon sink. Biogeosci 8:3203–3218
Long JN, Dean TJ, Roberts SD (2004) Linkages between silviculture and ecology: examination of several important conceptual models. For Ecol Manage 200(1–3):249–261. doi:10.1016/j.foreco.2004.07.005
Magill AH, Aber JD, Currie WS et al (2004) Ecosystem response to 15 years of chronic nitrogen additions at the Harvard Forest LTER, Massachusetts, USA. For Ecol Manage 196:7–28

Mäkilä M (1997) Holocene lateral expansion, peat growth, and carbon accumulation on Haukkasuo, a raised bog in southeastern Finland. Boreas 26:1–14
Martikainen PJ, Nykänen H, Crill P, Silvola J (1993) Effect of water on nitrous oxide fluxes from northern peatland. Nature 366:51–53
Milyukova IM, Kolle O, Varlagin AV et al (2002) Carbon balance of a southern taiga spruce stand in European Russia. Tellus 54B:429–442
Minkkinen K, Penttilä T, Laine J (2007) Tree stand volume as a scalar for methane fluxes in forestry-drained peatlands in Finland. Boreal Environ Res 12:127–132
Moore TR, Roulet NT, Waddingtot JM (1998) Uncertainty in predicting the effect of climatic change on the carbon cycling of Canadian peatlands. Clim Change 40:229–245
Moss AR, Jouany J-P, Newbold J (2005) Methane production by ruminants: its contribution to global warming. Ann Zootech 49:231–240
Munn RE, Maarouf AR (1997) Atmospheric issues in Canada. Sci Total Environ 203:1–14
Murphy PJ, Mudd JP, Stocks BJ (2000) Historical fire records in the North American boreal forest. Ecological studies, ISSU 138. Springer, New York
National Wetlands Working Group (1988) Wetlands of Canada. Ecological land classification series, no. 24. Sustainable Development Branch, Environment Canada/Polyscience Publications Inc, Ottawa/Montreal
Neilson RP (1993) Vegetation redistribution: a possible biosphere source of CO_2 during climatic change. Water Air Soil Pollut 70:659–673
Nilsson S (ed) (1996) Boreal forests – the role of research congress report. In: Caring for the forest: research in a changing world. Congress report, vol II. Proceedings of the 1995 IUFRO XX World Congress, 1995 Aug 6–12, Tampere, Finland, pp 399–409
Norris CE, Quideau SA, Bhatti JS et al (2009) Soil carbon stabilization in jack pine stands along the Boreal Forest Transect Case Study. Glob Change Biol 17:480–494
Ollinger SV, Richardson AD, Martin ME et al (2008) Canopy nitrogen, carbon assimilation, and albedo in temperate and boreal forests: functional relations and potential climate feedbacks. Proc Natl Acad Sci USA 55:19336–19341
Paavilainen E, Päivänen J (1995) Peatland forestry – ecology and principles. Springer, New York
Pan Y, Birdsey RA, Fang J et al (2011) A large and persistent carbon sink in the world's forests. Science. doi:10.1126/science.1201609
Perez-Garcia J, Lippke B, Commick J, Manriquez C (2006) An assessment of carbon pools, storage, and wood products market substitution using life-cycle analysis results. Wood and Fiber Sci 37:140–148
Preston CM, Bhatti JS, Flanagan LB et al (2006) Stocks, chemistry, and sensitivity to climate change of dead organic matter along the Canadian Boreal Forest Transect Case Study. Clim Change 74:223–251
Price DT, McKenney DW, Joyce LA et al (2011) High-resolution interpolation of climate scenarios for Canada derived from general circulation model simulations. Information report NOR-X-421. Natural Resources Canada, Canada Forest Service, North Forestry Centre, Edmonton, AB
Ramankutty N, Foley JA (1999) Estimating historical changes in land cover: North American croplands from 1850 to 1992. Global Ecol Biogeogr 8:381–394
Reader RJ, Stewart JM (1972) The relationship between net primary production and accumulation for a peatland in southeastern Manitoba. Ecology 53:1024–1037
Regina K, Nykänen H, Silvola J et al (1996) Fluxes of nitrous oxide from boreal peatlands as affected by peatland type, water table level and nitrification capacity. Biogeochemistry 35:401–418
Sanderson MG, Hemming DL, Betts RA (2011) Regional temperature and precipitation changes under high-end (≥4°C) global warming. Philos Trans R Soc A 369:85–98
Schroeder W, Kort J (2001) Temperate agroforestry: adaptive and mitigative roles in a changing physical and socio-economic climate. In: Proceedings of the 7th Biennial conference on agroforestry in North America, 12–15 Aug 2001, Regina SK

Sellers PJ, Hall FG, Kelly RD et al (1997) BOREAS in 1997: scientific results, experiment overview and future directions. J Geophys Res 102:28731–28770
Shvidenko AZ, Nilsson S, Roshkov V et al (1996) Carbon budget of the Russian boreal forests: a systems analysis approach to uncertainty. In: Apps M, Price D (eds) Forest ecosystems, forest management and the global carbon cycle. Springer, New York, pp 145–162
Shvidenko AZ, Schepashchenko DG, Vaganov EA et al (2007) Net primary production of forest ecosystems of Russia: a new estimate. Doklady Earth Sci 421A:1009–1012
Smith TM, Cramer W, Dixon RK et al (1993) The global terrestrial carbon cycle. Water Air Soil Pollut 70:19–37
Stocks BJ, Mason JA, Todd JB et al (2002) Large forest fires in Canada, 1959–1997. J Geophys Res 10:1029
Tarnocai C (2006) The effect of climate change on carbon in Canadian peatlands. Glob Planet Change 53:222–232
Theede AD (2007) Biometric and eddy-covariance estimates of ecosystem carbon storage at two boreal forest stands in Saskatchewan: 1994–2004. MSc thesis, University of Saskatchewan
Thormann MN, Szumigalski AR, Bayley SE (1999) Aboveground peat and carbon accumulation potentials along a bog-fen-marsh gradient in southern boreal Alberta, Canada. Wetlands 19:305–317
Turetsky MR, St. Louis V (2006) Disturbance in boreal peatlands. In: Wieder RK, Vitt DH (eds) Boreal peatland ecosystems. Ecological study series. Springer, Heidelberg
Turetsky MR, Weider RK, Halsey LA et al (2002) Current disturbance and the diminishing peatland carbon sink. Geophys Res Lett 29:1526
Turner DPJK, Winjum TP, Kolchugina TS et al (1998) Estimating the terrestrial C carbon pools of the former Soviet Union, the conterminous United States and Brazil. Climate Res 9:183–196
Updegraff K, Pastor J, Bridgham SD et al (1995) Environmental and substrate controls over carbon and nitrogen mineralization in northern wetlands. Ecol Appl 5:151–163
van Kooten GC (2000) Economic dynamics of tree planting for carbon uptake on marginal agricultural lands. Can J Agric Econ 48:51–65
Vitt DH (2006) Peatlands: Canada's Past and Future Carbon Legacy. In: Bhatti JS, Lal R, Apps MJ, Price MA (eds) Climate change and managed ecosystems. CRC Press, Taylor & Francis Group, Boca Raton, Florida, USA, pp 201–216
Vitt DH, Halsey LA, Zoltai SC (2000) The changing landscape of Canada's western boreal forest: the current dynamics of permafrost. Can J For Res 30:283–287
Vitt DH, Wieder RK (2006) Boreal peatland ecosystems: our carbon heritage. In: Wieder RK, Vitt DH (eds) Boreal peatland ecosystems, ecological study series. Springer, Berlin
Volney WJA, Fleming RA (2000) Climate change and impacts of boreal forest insects. Agric Ecosyst Environ 82:283–294
Weber MG, Stocks BJ (1998) Forest fires and sustainability in the boreal forests of Canada. Ambio 27:545–550
White A, Cannel MGR, Friend AD (2000) The high latitude terrestrial carbon sink: a model analysis. Glob Change Biol 6:227–245
Yavitt JB, Williams CJ, Wieder RK (1997) Production of methane and carbon dioxide in peatland ecosystems across North America: effects of temperature, aeration, and organic chemistry of peat. Geomicrobiol J 14:299–316
Zha T, Barr AG, Black TA et al (2009) Carbon sequestration in boreal jack pine stands following harvesting. Glob Change Biol 15:1475–1487

Chapter 11
Recarbonization of the Humid Tropics

Michelle Venter, Oscar Venter, Susan Laurance, and Michael Bird

Abstract To curb the effects of anthropogenic carbon (C) emissions, it is essential to explore the capacity of natural ecosystems to store and sequester atmospheric C. Tropical humid forests store most of their carbon in their above ground biomass which can be rapidly depleted by extractive industries or rapidly replenished through natural regeneration or reforestation. This rapid exchange of carbon dioxide between the forest and the atmosphere make humid tropical forests one of the most dynamic C pools on Earth. Tropical humid forests act as a store and sink for C, but are also the source of almost all anthropogenic C emissions from the land-use and land-use change. The influence of tropical humid forests also extends far beyond the edges of their canopies, affecting global C cycles, biodiversity, hydrological cycles, and the livelihoods of millions of rural people. These interlinked factors are what make tropical humid forests one of the most compelling actors in the recarbonization of the biosphere. This report presents the current state of knowledge on C stocks and fluxes in humid tropical forests, and investigate management interventions for the recarbonization the humid tropics. The C sequestration potential of seven recarbonization options is explored as well as co- benefits, risks and costs. Finally, relevant policies that could be harnessed to encourage recarbonization in humid tropics are reviewed.

Keywords Tropical humid forest • REDD • Carbon sequestration • Agroforestry • Restoration plantings • Humid tropics • Extractive industry • Carbon pool • Anthropogenic carbon emissions • Sink for atmospheric carbon • Management interventions • Carbon sequestration potential • Reduced impact logging • Plantation

M. Venter (✉) • S. Laurance • M. Bird
School of Earth and Environmental Sciences, James Cook University,
PO Box 6811, Cairns Queensland, 4870, Australia
e-mail: michelle.venter@my.jcu.edu.au; susan.laurance@jcu.edu.au; michael.bird@jcu.edu.au

O. Venter
School of Marine and Tropical Biology, James Cook University, Cairns, Australia
e-mail: oventer@jcu.edu.au

R. Lal et al. (eds.), *Recarbonization of the Biosphere: Ecosystems and the Global Carbon Cycle*, DOI 10.1007/978-94-007-4159-1_11,

• Mitigation • Fires • Deforestation • Degraded lands • Carbon pools • Global carbon cycle • Land use change • Pedologic pool • Missing terrestrial sink • Carbon fluxes • Carbon sink capacity • Biomass carbon • CO_2 fertilization • Logging

Abbreviations

ANR	accelerated natural regeneration
C	carbon
CO_2	carbon dioxide
EP	ecological planting
EP	Enrichment planting
GHG	greenhouse gas
MRV	Measurement Reporting and Verification
RIL	Reduced impact logging techniques
REDD+	Reducing emission from Deforestation forest Degradation
CDM	The Clean Development Mechanism
UNFCC	United Nations Framework Convention on Climate Change

11.1 Introduction

Tropical forests are intimately linked with the global climate system, and carbon (C) management within them holds enormous potential for mitigating anthropogenic climate change. Tropical forests account for a major fraction the total terrestrial C pool (Dixon et al. 1994; Phillips et al. 1998; Malhi et al. 1999; Malhi and Grace 2000; Malhi 2010; Pan et al. 2011). In addition, they act as a C sink by actively sequestering C from the atmosphere via transfer of photosynthically fixed C into biomass and soil (Malhi 2010). Moreover, tropical forests have a direct role in mitigating local warming through evaporative cooling and by reducing the absorption of radiant heat (Sagan et al. 1979; Bounoua et al. 2002; Defries et al. 2002; Bonan 2008).

In the twentieth century, there has been an unprecedented loss of tropical forest area (Malhi and Phillips 2005; Laurance 2007; van der Werf et al. 2009). An estimated 3.5 million km^2 tropical forests have been cleared and 5 million km^2 have been degraded, primarily through industrial logging, conversion to agriculture and fuel wood harvesting (ITTO 2002). The conversion and degradation of tropical forests make them the largest single source of greenhouse gas (GHG) emissions from the biosphere to the atmosphere (Achard et al. 2004; Clark 2004; Hansen et al. 2008; van der Werf et al. 2009; Malhi 2010). In addition roads created for extractive forest industries cause indirect C emission feedbacks by fragmenting the forest, increasing desiccation rates in and in turn the intensity and frequency of fires (Nepstad et al. 1991; Laurance and Williamson 2001; Kauffman et al. 2009).

Logging and agriculture, the main drivers of deforestation, play an important role in the economic development of most tropical countries. In some cases, forestry practices have been sustainable and productive agricultural lands have been established (Lamb 2011). However in many areas, tropical forests landscapes were altered through poor management practices and exist in a degraded sate, providing reduced economical value, productivity or ecosystem services (Lamb 2011; Sasaki et al. 2011). Degraded lands are widespread in the humid tropics covering about 40% humid tropical areas (ITTO 2002). Degraded lands and forests range from fragmented primary forests to secondary forests or to completely altered ecosystems such as non-native grasslands. The rates at which biomass recovers in degraded lands are variable and hence so are the rates of C sequestration. Even so, in many tropical areas biomass cannot recover effectively without human intervention (Chazdon 2008).

Part of the reason for the past unsustainable use of tropical forests is the lack of economic incentives to improve land management and avoid land degradation (Ebeling and Yasué 2008; Sandbrook et al. 2010). Because land-use change in the tropics typically reduces the C density of areas that were once humid forests (Fearnside 2000; Pongratz et al. 2009; Asner et al. 2010), placing economic value on forest C stocks can provide incentives to avoid reducing the C densities of forest areas as well as incentivizing efforts to increase the C density of other areas.

Recarbonization of the humid tropics is defined in this chapter as any land management strategy that aims to maintain or increase the C density of humid tropic regions. This chapter focuses on humid tropics because of their uniquely important role in the global C cycle and a potentially important role in ongoing efforts to mitigate anthropogenic climate change. The chapter begins by broadly describing humid tropical forests, followed an account of their C stocks and flux and then explores seven land-based options for the recarbonization in detail, and finally briefly reviews polices that are relevant to the recarbonization of the humid tropics under the United Nations Framework Convention on Climate Change (UNFCC).

11.1.1 Humid Tropical Forest

Humid tropical forests are the most productive and most ecologically complex of the terrestrial ecosystems (Myers 1984; Huston and Wolverton 2009). Tropical forests provide habitat for more than half of the world's species (Wilson 1988) and their rapid loss is an important driver of the global biodiversity crisis (Rockstrom et al. 2009; Wilcove and Koh 2010). In addition, 500 million people live in or near tropical forests or depend on them for food, non-timber forest products and traditional values (Chomitz 2007).

Humid tropical forests are largely concentrated within 10° latitude of the equator but may reach as far as 25°C on the east coast of continents and Pacific Islands. They are the characteristic vegetation type of the tropics and cover most of the land surface where climate is hot and rainfall is heavy and non-seasonal, except for swampy or volcanic areas (Richards 1952). In the lowlands, tropical forests are grand covered by trees up to 60 m in height, ornamented with epiphytes, thick-stemmed lianas and a sub canopy formed by a variety of palms, tree ferns and other plants (Schimper et al. 1903). They are found up to 3,500 m.a.s.l., though and tree stature and species diversity diminishes with elevation.

High levels of plant productivity, and hence C sequestration, of humid tropical forest is attributable to the year-round growing season. Tropical humid climate has monthly precipitation exceeding 60 mm, annual rainfall above 2,000 mm and average monthly temperatures above 18°C at sea level (Richards 1952; Huston and Wolverton 2009). The transition between humid forest and dryer landscapes are seldom static and are driven by environmental gradients or cyclicity in rainfall, temperature and anthropogenic disturbances (Murphy and Lugo 1986; Malhi and Wright 2004; Hirota et al. 2010).

11.2 Current State of Knowledge of C Stocks and Fluxes in the Humid Tropics

11.2.1 C Pools

The global C cycle is driven by the dynamics of five major C pools, in order of magnitude these pools are: (i) the oceanic pool, (ii) the geological pool (comprising fossil fuels), (iii) the pedologic, (iv) the atmospheric pool and (v) the biotic pool (Lal 2008) (Fig. 11.1). The bulk of anthropogenic C emissions are derived from the geological pool (from the use of fossil fuel) and the terrestrial biotic/pedologic pool (from land-use change).

The atmosphere, oceans and the terrestrial biosphere act as a sink for anthropogenic carbon dioxide (CO_2) emissions (Fig. 11.1). Of the C emitted by humans between the year 2000 and 2008, 45% (±4%) remained in the atmosphere, 26 (±5%) was absorbed by the oceans, and the remaining 29% (±6%) was absorbed by the terrestrial biosphere (Le Quere et al. 2009). The terrestrial sink was previously referred to as the "missing terrestrial sink" because of the lack of clarity of its location, size and trends (Dixon et al. 1994; Myneni et al. 2001). Recent research indicates that most of the terrestrial sink is located in the world's forests, and that tropical forests probably make up half of the terrestrial sink (Malhi 2010; Pan et al. 2011).

As seen in Fig. 11.1, the C enters the terrestrial biosphere via photosynthesis, and then makes its way to the more stable soil C pool by biomass decomposition and mineralization, after accounting for losses in soil erosion and soil respiration. About 56% of the C in humid tropical forests is stored in biomass, while 32% is stored in soil up to 1 m of depth and 12% in necromass (Pan et al. 2011). Storing C in biomass rather than soil has the effect that humid tropical forests accumulate or release significant quantities of C over shorter time periods compared to higher latitude ecosystems (Phillips et al. 2004; Houghton 2005; Le Quere et al. 2009; Gumpenberger et al. 2010; Pan et al. 2011).

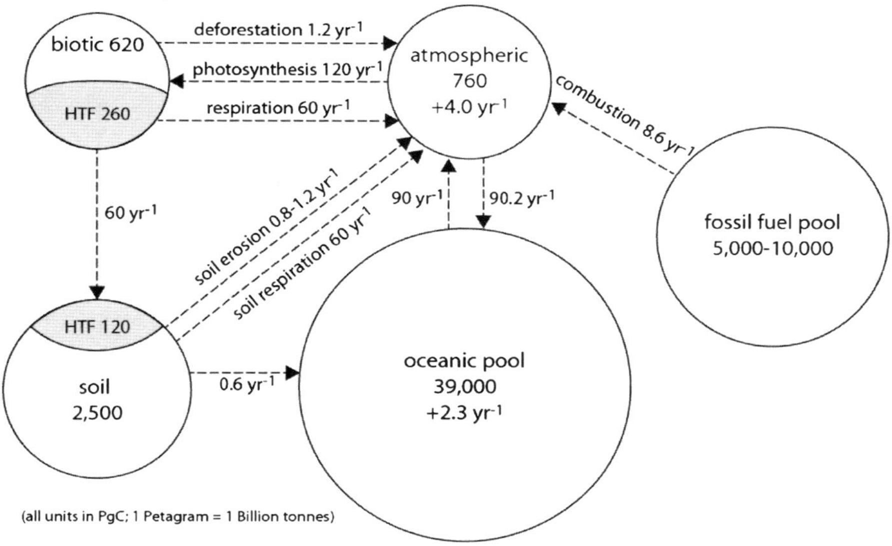

Fig. 11.1 Principal C pools and fluxes between them, adapted from a figure in Lal (2008) outlining the humid tropical forests (HTF) C pool. Updated C pools come from Lal (2010), deforestation rates from Pan et al. (2011), fossil fuel emissions from JRCNEA (2011), humid tropical forest stock from Scharlemann et al. (2010) in soil to 1 m, and based 32:64 soil to living C ratio in Pan et al. (2011)

11.2.2 C Fluxes

Sound knowledge of C stocks and fluxes in tropical humid forest are the cornerstone of effective recarbonization strategies. There are three main factors that determine whether tropical forests act as a net carbon sink or source. These are (i) carbon emissions from biomass burning, decomposition and soil erosion (ii) net carbon flux within intact forests and (iii) carbon uptake from forest regrowth.

The area of humid tropical forest is about 11 million km^2 (1,100 million ha), each hectare storing between 120 and 470 Mg C in biomass (Brown 1997; Fittkau and Klinge 1973; Brown and Lugo 1992; Malhi et al. 1999; Fearnside 2000; Houghton et al. 2000, 2001; Hughes et al. 2000; Nascimento and Laurance 2002; Saatchi et al. 2007; Saatchi et al. 2011; Keith et al. 2009) and between 123 and 186 Mg C in the first meter of soil (Malhi et al. 1999; Jobbágy and Jackson 2000). Total C pools in humid tropical forests are about 381 Pg C (Scharlemann et al. 2010).

When forests are degraded through logging or converted to non-forest vegetation types, the C stored in the biomass is rapidly released into the atmosphere. In comparison, soil C loss happens more slowly through erosion after land-use change (Achard et al. 2004). Between 2000 and 2005, the degradation and conversion of humid tropical forests released 1.03 Pg C $year^{-1}$, excluding peatland emissions (van der Werf et al. 2009). To put this figure into context, the total anthropogenic C emissions during the same period were 8.6 Pg C $year^{-1}$ (JRCNEA 2011).

Although, unlike carbon emissions from fossil fuels, emissions from tropical humid forests are reversible. Humid tropical forests continuously uptake CO_2 from the atmosphere through photosynthesis. A number of recent studies have argued that the C sink capacity within intact humid tropical forest might be increasing (Lewis et al. 2009; Luyssaert et al. 2008). The evidence supporting this argument comes from long term observation of increased biomass within forest plots. Globally, tropical humid forests seem to be reaching a higher biomass state. Lewis et al. (2009) compiled biomass collected in forest inventory plots for intact humid tropical forest and showed a mean increase of 0.49 Mg C ha^{-1} $year^{-1}$ between 1987 and 1997. There are two main schools of thought to explain this increase in biomass. The first is a response to CO_2 fertilization from increased anthropogenic carbon emissions (Lewis et al. 2009; Lewis 2006; Malhi 2010; Phillips et al. 1998) and the second is a recovery response from a widespread disturbance, such as drought, that happened in the past (Wright 2005). The extent to which this increase in biomass will continue and how it will influence soil C stocks is still unknown (Lal 2010).

Another carbon sink in tropical humid forest is located in forests that are growing after being disturbed or cleared. Humid tropical forests can sequester 6.2 Mg C ha^{-1} $year^{-1}$ over their first 20 years of growth, and 2.9 Mg C ha^{-1} $year^{-1}$ over their first 80 years (Silver et al. 2000). Sinks from regrowth are usually included within emission scenarios, but when independently estimated they sequester at total of 1.6 Pg C $year^{-1}$ (Pan et al. 2011).

When considering the combined effects of deforestation and degradation and sinks from intact forest and forest regrowth, humid and dry tropical forest are a net source of C to the atmosphere, emitting 1.3 (± 0.3) Pg C and sequestering 1.1 (±0.2) Pg C (Malhi 2010). However it is important to note that the net balance reported in this study is not statistically different from zero (Malhi 2010). Encouraging regrowth could have the power to turn humid tropical forests from a CO_2 source to a CO_2 sink.

11.3 Options for Recarbonizing the Humid Tropics

Recarbonization activities in the humid tropics include strategies that protect forests C pools, reduce the impacts of degrading activities, rehabilitate degraded forests and convert non-forest lands into forests (Brown and Lugo 1994; Lamb 1998; Silver et al. 2000; Niles et al. 2002; Lamb et al. 2005; Laurance 2008; Zomer et al. 2008; Torres et al. 2010). In Table 11.1, seven recarbonization options are presented. The option best suited for each recarbonization strategy depends on the land type. Additionally, the implementation of an option is usually governed by benefits or the desired outcomes of the stakeholders relative to the risks and costs. Each option is discussed within a following section.

11.3.1 Protecting Existing Forest by Reducing Deforestation

The most obvious way to reduce deforestation is to protect the forest from the activities that threaten them. The types of humid tropical forest available for protection have a wide range of biomass carbon, often dependent on their level of degradation.

Table 11.1 Recarbonization options for different land types in humid tropical regions and associated risks, costs, C sequestration and other benefits for each of option

Land type	Impediments to C sequestration potential in land type	Recarbonization option (optimize C sequestration)	Risks associated to recarbonization option	Costs (labor and capital)	C sequestration	Co-benefits
Primary or degraded forest	Timber and fuelwood harvest fires, drought, storms, climate change	1. Forest protection (Sect. 11.3.1)	Leakage Weak governance Reliance on C economy	High start-up Low running	High	Full range of ecosystem services Cultural conservation
	Competition with pioneer trees, weeds and vines Soil loss, compaction and reduced fertility	2. Reduced impact logging (RIL) (Sect. 11.3.2)	Transaction costs of collaborating with industry Other threats (later logging, fire) Option with lowest risk	Low	Low	Employment Biodiversity benefits
		3. Accelerated natural regeneration (ANR) (Sect. 11.3.3)	Tree mortality Other activities more profitable	Medium	Medium	Employment
Degraded non-forest lands	Altered disturbance regime Lack of seed stock Soil loss, lack of micro-fauna Altered hydrology, dryer habitat, new stable state	4. Agroforestry (Sect. 11.3.4.1)	Very few risks because practice is well established	Low	Low	Food production Cash economy Local biodiversity improvement
		5. Monoculture plantation (Sect. 11.3.4.2)	Species extinctions Market dependent	Medium	Low	Income generation
		6. Polyculture plantation (Sect. 11.3.4.3)	Access to markets Long return on investments	Medium	Medium	Income generation Propagation of native species and improved habitat for wildlife
		7. Restoration plantings (Sect. 11.3.4.4)	Fire, poor seedling establishment, lack of knowledge on ecosystem assembly, few goods Option with highest risk	High	High	Full range of ecosystem services

Degraded forests typically store less C than primary forests, and through industrial-scale logging, these forests are quickly becoming the dominant forest type in many tropical areas (Brown and Lugo 1992). If the clearing of tropical forests were to continue unabated, 87–130 Pg C will be released into the atmosphere by 2100 (Houghton 2005). Conversely, if business as usual deforestation rates were reduced by half by 2050 and then held constant until 2100, 50 Pg C of emissions would be avoided (Gullison et al. 2007). The Eliasch Review (Eliasch 2008) estimated that including reductions to tropical deforestation in global climate initiatives would half the cost of reducing global anthropogenic C emissions (Eliasch 2008), making this a likely political and economically attractive strategy.

Keeping forests standing can have many additional, non-carbon benefits as intact natural forests provide ecosystem services that contribute to human well-being. These services include the provision of non-timber forest products, the filtration of water and air (Sheil and Murdiyarso 2009), mitigation of floods (Bradshaw et al. 2007) and the provision of pollinators for adjacent farm crops (Ricketts et al. 2004). Natural forests also provide habitat for many species. C payments to reduce tropical deforestation could yield substantial benefits for biodiversity conservation, especially if the payments favor areas of high biodiversity (Venter et al. 2009).

Though reducing emissions from deforestation is now backed by substantial political support (UNFCCC 2009), there are challenges that may yet prevent broad implementation. A number of these issues are technical, such as establishing accurate reference emissions levels against which emissions reductions can be measured (Olander et al. 2008; Griscom et al. 2009). Accurately measuring forest cover change across large areas and in a timely manner remains complex and expensive (Gibbs et al. 2007; Grainger 2009) and techniques for minimizing or quantifying the 'leakage' or displacement of deforestation into other areas remains elusive (Ewers and Rodrigues 2008). Other issues are political, such as dealing with corruption (HRW 2009), and socio-economic, such as minimizing the opportunity costs to stakeholders, especially forest dependent communities (Pfaff et al. 2007).

11.3.2 Reducing Forest Degradation Through Reduced Impact Logging

Forest degradation is defined here as the loss or removal of forest canopy cover not resulting in deforestation, which is typically considered to occur at canopy cover levels below 10–30% (Sasaki and Putz 2009). Forest degradation can occur directly and indirectly due to human activities. Direct degradation occurs primarily through selective timber and fuel-wood harvesting, and is a major cause of forest cover change in the tropics (Wright 2010). A recent study using a variety of remote sensing data and literature sources estimated that roughly 20% of the humid tropics underwent some form of timber harvesting in the period 2000–2005 (Asner 2009) and that in the humid tropics, the footprint of selective logging is 20 times larger than that of deforestation (Asner et al. 2009).

Despite the intensity of the impacts from logging, forest degradation and associated emissions are difficult to measure over broad areas. While its impacts vary, in Southeast Asia, where selective timber harvest rates are highest, selective logging removes 33–56% of the biomass stored in forests (Pinard and Putz 1996). Early estimates concluded that forest degradation across the tropics was responsible for 4.4% of emissions from land-use change (Achard et al. 2004). However a more recent estimate attributed roughly 20% of emissions from the Amazon to selective logging (Asner et al. 2005). At the global scale, selective logging is responsible for the release of 0.5 Pg C $year^{-1}$ into the atmosphere (Putz et al. 2008). No data are available to quantify the emissions from other degrading activities, such as fuel wood harvesting (Griscom 2009).

Two options exist for reducing emissions from forest degradation. The first is to stop the extractive activity by protecting the forests. This is similar to protecting forests from deforestation, as discussed in Sect. 11.3.1, but with lower emissions benefits because degradation is being averted, not full deforestation. The second option is to work with the degrading activity to reduce its impacts on the forest. In the case of industrial logging, much of the emissions come from the building of roads, mistakenly harvesting unsuitable trees and collateral damage to non-target trees during felling and skidding (Bertault and Sist 1997). Reduced impact logging techniques (RIL) can reduce overall emissions from logging by 30% (Pinard and Cropper 2000; Keller et al. 2004). RIL involves careful planning and implementation of logging operations in order to minimize impact on the residual stand. Implementing RIL techniques across the production in the humid tropics has the potential to reduce GHG emissions by 0.16 Pg C (Putz et al. 2008).

Among the options for the recarbonization of the tropics, RIL to reduce emissions from forest degradation may have both the lowest costs and risks. The reason is that though its impacts are reduced, the extractive activity is allowed to continue. This has several important implications. First, it minimizes or even avoids opportunity costs (Bertault and Sist 1997). Second, it increases the permanence of emissions reductions. Compared to protecting forests where the threat of forest clearance will persist, the threat is allowed to occur and disappear. Finally, leakage cannot occur, as no activity has been stopped (Ewers and Rodrigues 2008). The main drawback, however, is that compared to other options, RIL has a far lower mitigation potential per hectare than other options, reducing C emissions by approximately 30% less than conventional logging (Putz et al. 2008).

The widespread extent of rainforest logging makes it a crucial priority for managing C emissions, however there are significant biomass losses occurring in forests that are degrading naturally due to habitat fragmentation. Large canopy trees suffer higher rates of mortality and damage in forest fragments and in forest edges adjacent to large-scale clearings (Laurance et al. 2000). These forests can lose up to 36% of their biomass and are considered an important source of atmospheric C emissions (Laurance et al. 1997). Accurately estimating the area of these naturally degrading forests is complicated by pooling them with logged areas, but earlier estimates suggested that naturally degraded forest areas are often equivalent in size to annually cleared areas (Broadbent et al. 2008).

11.3.3 Forest Rehabilitation Through Accelerated Natural Regeneration

Degraded forests make up almost half the 11 million km^2 of humid tropical forest (ITTO 2002). The capacity and rate for humid forest to regenerate depends on the intensity of the disturbance and the degree of degradation (Lamb et al. 2005; Ramankutty et al. 2007). Because of the current extent of degraded forest, management that accelerates forest regeneration, thereafter called accelerated natural regeneration (ANR) poses a great opportunity to increase C density across the tropics (Parrotta et al. 1997). ANR techniques can be implemented as part of certification for sustainable logging practices or through carbon payment mechanism such as REDD+ (Reducing emission from Deforestation, forest Degradation, plus the conservation, sustainable management and enhancement of forest carbon stocks).

In highly degraded forests competition with weeds and pioneer species barriers prevent biomass recovery can cause long-lasting ecosystem changes (Pariona et al. 2003; Wright 2005). Recovery is particularly slow following a disturbance that has severely altered soils and vegetation communities (Chazdon 2003). There are a variety of silvicutural techniques that can overcome these barriers by assisting natural regeneration and therefore increasing C density. Enrichment planting (EP), a technique that involves planting existing forest with trees is commonly used in forestry to increase the density of target species under canopy gaps (Lamb et al. 2005). This practice it is often supplemented by other treatments that accelerate C sequestration by reducing competition from vines (liberation cuttings) and certain pioneer trees (thinning) (Kuusipalo et al. 1997; de Graaf et al. 1999; Pariona et al. 2003; Peña-Claros et al. 2008). Forestry practices that have included EP and liberation cutting during second and third timber harvest rotations have reported increased productivity (Keefe et al. 2009).

The additional C sequestration that might result from EP and liberation cutting could potentially gain credits from C payment mechanisms (Edwards et al. 2010), and provide steady employment to rural communities (Paquette et al. 2009). ANR practices could also target tree species of conservation value, and assist in the propagation exploited species such as mahogany (*Swietenia macrophylla*) and those of large fruiting species that cannot self-propagate (Lamb et al. 2005; Keefe et al. 2009). Other benefits could include the improvement wildlife and riparian corridors and decreasing the risk of fires (Parrotta et al. 1997; Brown et al. 2004; Chazdon et al. 2009). Aside from mortality of planted trees or failure to compete with more profitable activities, there is very little risk associated with ANR relative to the small costs of labor, equipment, and training and site preparation (Keefe et al. 2009). Possibly accelerate natural regeneration is the most economically feasible recarbonization option over large areas (du Toit et al. 2004).

Fig. 11.2 (**a**) Photo of a mosaic of primary lowland forest fragments and *Imperata* grasslands in Morobe province of Papua New Guinea. (**b**) Photo taken within of one of the lowland primary forest fragments (Photos by Michelle Venter)

11.3.4 Converting Degraded Non-forest Lands to Forests

There are a few land management practices can be applied to convert degraded non-forest lands to forested lands and are amongst the most ambitious of the recarbonization strategies available in the humid tropics. In most tropical regions, secondary forest can regenerate naturally if there are residual seedbanks or nearby native forests. Although, grasslands can be slow to revert to forest if frequent fires and grazing pressures are present. Shade-intolerant grasses and ferns become well established in areas where forest once existed if disturbance regimes are not altered (Cramer et al. 2008) Therefore, despite being surrounded by forest some anthropogenic grassland persists (MacDonald 2004).

Anthropogenic grasslands are common throughout the tropical humid regions (Garrity et al. 1996). Carbon density in the tropics would increase if these grasslands were to be prioritized for conversion to agriculture or for forestry instead of forests. Figure 11.2 shows a photo of fragmented forest surrounded by anthropogenic grasslands and a photo of a tree found within one of the forest fragments. This figure highlights the potential to increase forest C pools by in Papua New Guinea by promoting the advancement of forest into grasslands.

The term 'reforestation' is used when trees are planted on sites where forest has been present during the last 50 years, and 'afforestation' refers to planting trees on sites where forest was absent for longer than 50 years (FAO 2000b). There are many re and afforestation techniques but they all require seeds or seedlings to be planted which makes them costly recarbonization options (Lamb 2011). The range of options for re and afforestation includes; agroforestry for crop production (Sect. 11.3.4.1), monoculture plantations on short rotations for fiber production (Sec. 11.3.4.2) longer rotation of mixed hardwoods for timber production (Sect. 11.3.4.3) and finally ecological planting (EP) for enhanced ecosystem services (Sect. 11.3.4.4).

11.3.4.1 Agroforestry

Agroforestry with C sequestration goals involves planting trees as part of a system that will provide food or timber for subsistence or cash income (Lamb 2011). They can be established by planting seedlings in the early fallow stage of a shifting cultivation cycle or on degraded land (de Jong 2002). Agroforests are multi-strata systems with a complex mixture of native and exotic trees and crop species which distinguishes then from monocultures crops for food production (such as oil palm plantations). They are a commonly cultivated in many parts of rural tropics, usually as home or community gardens (de Jong 2002; Albrecht and Kandji 2003).

The reported range for biomass and soil C storage in agroforests across the tropics range from 12 to 228 Mg C ha^{-1}, with a median value of 95 Mg C ha^{-1} (Albrecht and Kandji 2003) and an average of 50 Mg C ha^{-1} (Schroeder 1994). The C sequestration rates varies with types of species planted, the density to which they are plated and rotation length (Albrecht and Kandji 2003). While C stocks are relatively low in agroforests compared to other tree plantations, they usually have higher C sequestration rates than pastures, field crops and degraded lands (Ramachandran Nair et al. 2009).

Permaculture techniques used in agroforestry originate from a mix of traditional knowledge and agricultural sciences. For example, in Papua New Guinea, planting N fixing tree (*Casuarina oligodon*) to increase soil fertility is a common gardening practice (Bourke 1997). Permaculture techniques are also used to reducing weeds and dependence on pesticide e and composting is a common practice to increase soil fertility and in turn soil C stocks (Michon et al. 2007; Vieira et al. 2009).

Even if C sequestration benefits are low compared to other re and afforestation option, agroforestry is a viable recarbonizing activity, especially in areas where food security is a priority (Vieira et al. 2009). Very little risk is associated with this option because agroforestry systems are proven successful. The major drawbacks are caused by low nutrient in top soils in the humid tropics (Schroth et al. 2002), and the unpredictable impact of a changing climate on food production performance (Cacho et al. 2005). Agroforests have a rich understory of plants which can provide habitat for birds and invertebrates as well as contributing to landscape biodiversity (Bhagwat et al. 2008).

11.3.4.2 Monocultures in Short Rotations

Tropical forest plantations represent about 25% of global forest plantations (FAO 2006), and are expanding rapidly. From 1980 to 2000 they almost quadrupled in area, increasing from 18 million ha to more than 70 million ha (FAO 2001). Most of the plantations established after 1980 have been for wood fiber production on short rotation cycles (5–10 years) dominated by *Acacia, Eucalyptus, Picea* and *Pinus* (Brown et al. 1986). They usually are privately owned in partnership with large corporations (Lamb 2011).

Pulpwood species have rapid growth rates and can be a good temporal option for mitigating CO_2 emissions if established on degraded lands (Silver et al. 2000). On average, biomass C stocks in 13 year old plantation s are 62 Mg C ha^{-1} (Kanowski and Catterall 2010) and accumulates at a rate of 1.9 Mg C ha $year^{-1}$ (FAO 2000a; Elias and Potvin 2003; Liao et al. 2010). However the effect of plantation establishment on soil C stocks is debatable (Laganière et al. 2010). Some studies have shown that they cause a decrease in soil C pools (Guo and Gifford 2002). Evidence from *Pinus* plantations in particular, has shown a loss in soil C during initial stage of plantation growth (Paul et al. 2003), and soil C can then take many decades to recover (Richards et al. 2007). However, the overall soil C change is generally positive (Guo and Gifford 2002; Paul et al. 2010).

Monoculture are easy to establish and to maintain to over large areas, and provide income for small holders (Putz and Redford 2009; Bremer and Farley 2010). If natural regeneration is permitted in the understorey, plantations may contribute additional habitat for wildlife and act as corridors for wildlife migration. For example, in a 4.5-year-old exotic monoculture plantation, 19 secondary forest species established naturally in the plantation understory. Species included large fruited trees indicating the importance of trees for frugivorous bats and birds as facilitators of secondary forest species colonization (Parrotta 1995). These benefits, however are temporary, lasting only until the next harvest.

However, monocultures are the most controversial among re and afforestation strategies. Mostly, because they are characterized by limited ecosystem services, low biodiversity values and lower C benefits (Bremer and Farley 2010). These topics are especially relevant when they replace native forests. Even if replacing degraded grasslands with monocultures is a viable recarbonization strategy, grasslands can sometime perform better than plantations with respect to watershed protection (McElwee 2009) and they hold great value to the local people who harvest products from them. In Vietnam, reforestation has been strongly promoted by the government in a project know as the '5M hectare project'. This re and afforestation program has helped to reverse the trend of forest loss but it has had a multitude of unforeseen negative consequences. The worst of these was the increase in poverty amongst the poorest people in the region (McElwee 2009). Careful consideration of social and cultural welfare should be assessed before ambitious re and afforestion projects are undertaken.

11.3.4.3 Polycultures in Long Rotations

Polyculture plantations typically have rotations of 20 years or more and can include a mix of native and exotic tree species. Because polycultures are more difficult to manage than monocultures and have higher start up costs they are generally less favored by industry and cover very limited areas in the humid tropics (Lamb 2011). C sequestration rate vary with plantation types, though C stocks in polycultures range from 110 to 173 Mg C ha^{-1} (Lamb 2011). Longer rotations in polycultures plantations compared to monoculture plantations lead to higher average C sequestration over time (Paul et al. 2002). In general, plantations are most likely to contribute to biodiversity if a mixture of native trees is used (Bremer and Farley 2010).

Aside from improving local biodiversity, native tree plantation are as profitable to some highly valuable timbers such as teak (*Tectona grandis*) (Griess and Knoke 2011). Using mixed-species plantations is probably one of the best options to optimize C sequestration, livelihood benefits and biodiversity in heavily degraded sites. Because mixed species plantation have less stringent harvest times and provide a wider range of goods they are more flexible to markets responses (Lamb 2011)

11.3.4.4 Restoration Plantings

There are many factors that impede natural regeneration of rainforests on degraded lands but most fit within two categories, namely dispersal and recruitment limitation. Dispersal limitation accounts for all the factors that limit the arrival of a seed at a site, including seed source and production and seed dispersal. Recruitment limitation encompasses the next life history stage where the seed has arrived at the site but cannot germinate or successfully establish because of factors such as limited soil nutrients and obligate fungal or bacterial root symbionts, competition, and seed and seedling predation (Young et al. 1987; Francis and Read 1994; Parrotta et al. 1997; Chazdon 2003, 2008; Shono et al. 2007).

The primary goal of restoration plantings is to overcome these impediments and restore the natural processes which will, in time, produce a forest of similar structure and community to native forests. Techniques that restore forests on degraded tropical ecosystems include planting an assortment of native species in high densities or planting a small number of nurse trees to mimic natural regeneration (Lamb 2011). Restoration plantings compared with other recarbonization options areas have the highest overall potential to sequester C in both above- and belowground pools in the humid tropics. Because the duration of project involving restoration plantings are usually unlimited C sequestration can continue to be positive over 80 years during forest establishment, and possibly longer (Silver et al. 2000).

The practical barriers to RP are the time it takes to collect and germinate seeds and then to plant them. Also, land and resources must be committed over long periods. In the initial stages, risks from wild fires are important and fire prevention can be very costly (Brown et al. 2004). Other barriers to implementation come from a lack of understanding about assembling ecosystems. Creating any viable ecosystem requires deliberation. But trying to replicate a specific ecosystem, such as the one present before clearing, can be almost impossible. Moreover, with changing climates, the persistence of certain tree species will be compromised, and planning for future climates and unknown outcomes is even more challenging. This method is extremely intensive in terms of labor and funding, and currently is mostly used for rehabilitating strategic locations in the landscape such as wildlife corridors and creek banks (Goosem and Tucker 1995). The potential benefits to arise from EP, especially in the conservation of biodiversity and natural heritage, outweigh many

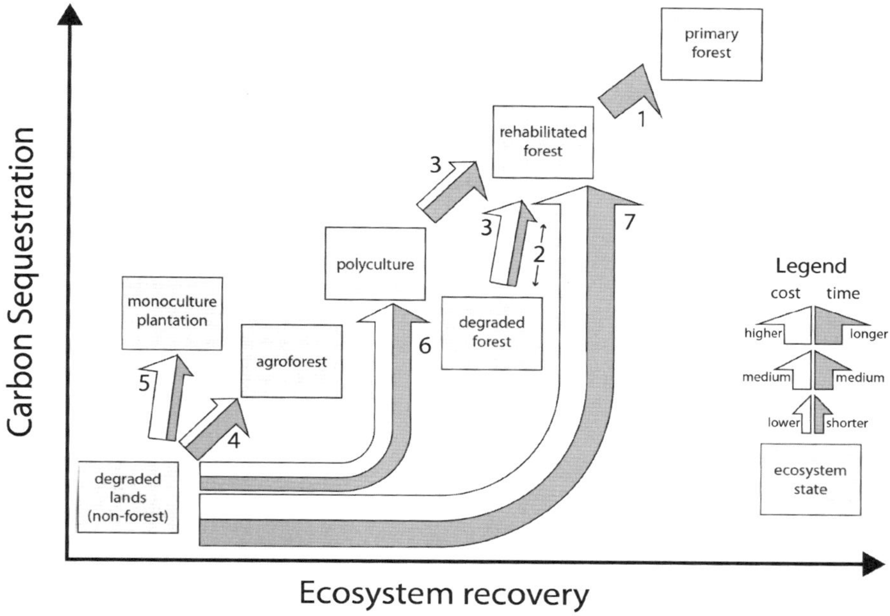

Fig. 11.3 Carbon sequestration benefits of recarbonization strategies relative to their contribution to ecosystem recovery. The arrows indicate the relative time and costs for the implementation of recarbonization options. The numbers refer to following recarbonization option; *1* forest protection, *2* reduced impact logging, *3* accelerated natural regeneration, *4* agroforestry, *5* monoculture plantation, *6* multi-species plantation, *7* restoration plantings

of the other recarbonization option. The success of RP will probably depend in the motivation and engagement of multiple parties seeking multiple benefits in the long term (Chazdon 2008).

11.3.5 Recarbonization Options Discussed

Figure 11.3 shows the tradeoffs between C benefits, ecosystem integrity and costs and time for each recarbonization option. With infinite resources and time, most natural systems could be restored to their full C potential. In reality, the recarbonization options depend on the state of land degradation and the resources available to restore it. The most costly and logistically complex of scenarios would be to start with marginal grassland and use restoration plantings to convert it to a secondary forest with structure and composition similar to primary a forest. Even thought costly, this would yield the highest C and biodiversity benefits (Table 11.1).

On the other hand, one of the fastest and cheapest options would be to start with degraded or secondary forest and, restore it through ANR techniques. In this case,

C sequestration and other benefits would be less favorable. Plantations are good options for the recarbonization of non-forest lands, especially over large areas. Plantations don't provide the same level of ecosystem services as natural ecosystems, but they are economically profitable and less risky.

The best options are often guided by tradeoffs in land suitability, access to markets and resources and stakeholder objectives. Nonetheless, careful planning must be an integral part of any recarbonization project to optimize benefits as well as to avoid unforeseen environmental or social consequences.

11.4 Recarbonizing Policies Under United Nations Framework Convention on Climate Change (UNFCCC)

Land management including forest conservation, forest restoration, reforestation, afforestation and agriculture that aims to reduce C emissions or increase C density, is central to international efforts to mitigate climate change (Parrotta et al. 1997; Laurance 2007; Ebeling and Yasué 2008; Miles and Kapos 2008; Burney et al. 2010). However, land-use change to maintain or augment C density in tropical biomes often competes with increasing demands for land to produce food, timber, and bio-fuels (Lambin et al. 2001; Gibbs et al. 2010). In this section we review policies that may encourage a shift towards recarbonizing the tropics.

The Kyoto protocol, adopted in Japan in 1997, is the first global climate agreement (UNFCCC 1997). This agreement binds industrialized countries to reduce their emissions domestically or to offset their emissions through projects in developing countries; this marked-based mechanism is termed The Clean Development Mechanism (CDM). This mechanism is designed to give some flexibility to industrialized countries in how they meet their emissions reduction targets as well as facilitating sustainable development and poverty alleviation in developing countries. Although the CDM has the potential to stimulate reforestation and afforestation in the tropics, of the 1,600 projects registered under the mechanism in 2010, only four were forest projects (Thomas et al. 2010). About 75% of the CDM projects to date are related to renewable energy and energy efficiency (UNEP 2009). The two main reasons for the lack of forest-related projects in the CDM is the high transaction costs associated with these relatively small land-use change projects and the difficulty in accounting for leakage, which is the displacement of emissions from one place to another (Chomitz 2002; Schwarze et al. 2002).

Since 2005, there has been increasing pressure to include REDD+ in any successor agreement to the Kyoto Protocol. At present, REDD+ has broad support from international climate negotiators and seems set to play a major role in future efforts to mitigate climate change (UNFCCC 2009). Because REDD+ will set baselines and award credits at the national scale, it has the potential to allow tropical forests to play a much larger role in mitigating climate change, essentially sidestepping the effects of within-country leakage and reducing the transaction costs of C credits by creating a single Measurement, Reporting and Verification (MRV) scheme to cover an entire country. Individual site-based projects can still participate in national-scale REDD+ through sub-national baselines and MRV schemes, but presumably these would be less stringent and costly than the national scale scheme and those required under Kyoto's CDM.

The scope of REDD+ has been a major issue throughout its policy development. Originally it was intended only to reduce emissions from deforestation, but it quickly expanded to consider emissions from forest degradation. In 2009, REDD further expanded to include the role of forest conservation, the sustainable management of forests and the enhancement of forest C pools in developing countries, and in so doing became REDD+. While its exact scope remains to be determined, REDD+ has the potential to go beyond simply maintaining forest C pools to also include incentives to improve industrial logging operations, and promote forest C sequestration through re\afforestation. The expanded scope of REDD+ and the scale of the incentives breathes new life into ongoing efforts to re-carbonize the tropics. For instance, Norway has already committed US$5 billion to REDD+.

11.5 Concluding Remarks

The influence of tropical rainforests reaches far beyond the edges of their canopies, interacting with global C cycles, global biodiversity, hydrological cycles, and the livelihoods of some of the poorest people on Earth. These interactions are what make humid tropical forest a compelling actor in the recarbonization of the biosphere.

The potential role of humid tropical forests in recarbonizing the biosphere is determined by their rates of C emissions and capacity to sequester and store carbon. In theory, the recarbonization capacity of humid tropical forest equates to the historic depletion of C from the biome (Rhemtulla et al. 2009). Positive or negative deviations from this potential could arise due to increased CO_2 fertilization from anthropogenic emissions (Lewis et al. 2009), or changes in global or regional climates (Briant et al. 2010).

Humid tropical forests have some the most dynamic C pools on Earth. They are responsible for almost all of the C emission from the biosphere and at the same time they form the majority of the terrestrial carbon sink (van der Werf et al. 2009; Malhi 2010; Pan et al. 2011). Although there are still many uncertainties regarding the exact figure of C fluxes in the humid tropics. These uncertainties are caused by from sheer extent of humid tropical forests the lack of on-the-ground studies that measure forest carbon stocks and the technical difficulties associated with measuring forest loss and gains (Ramankutty et al. 2007). However, the number of studies investigating the C fluxes in humid tropical forest as has increased during the last decade, with the latest research showing that humid tropical forests are likely a net source of C to the atmosphere. In some areas, humid tropical forests are being rapidly cleared, releasing huge quantities of carbon, while in other areas they are recovering from human disturbances and absorbing an almost equal amount of carbon.

By maintaining existing C stocks, reducing emissions from deforestation and forest degradation currently has the greatest potential among recarbonizing options to contribute to climate change mitigation. Implementing recarbonization options is only feasible when they align with economic and policy reality. Because of strong support from climate policy makers, a timely opportunity exists to harness REDD+ C payments to incentive the protection of humid tropical forests. Gaining better insight on the drivers for forest loss and expansion is probably the first step to achieving a long-term plan for REDD+ (Lambin et al. 2001).

In addition, given the extent of degraded humid tropical forests and non-forest lands, it is necessary to look beyond halting forest lost to consider options to accelerate the recovery of C stocks through increasing biomass. The sink capacity of humid tropical forests can be enhanced beyond the natural phenomenon of forest recovery. Many options exist and the best option to do this varies from region to region. The more degraded the land, the longer it will take or more costly it will be to restore C stocks. The scales at which recarbonization projects can potentially occur rely on the availability of resources and on management objectives.

While humid tropical forests are a net source of C emissions, a more in-depth assessment of the figures gives reason to be hopeful. The C sequestration in intact and recovering forests is almost equal to the net emissions from forest loss, and therefore, small reductions in forest loss or increases in forest sequestration could tip the balance, making humid topical forests part of the solution for climate change instead of part of the climate change problem.

References

Achard F, Eva HD, Mayaux P et al (2004) Improved estimates of net C emissions from land cover change in the tropics for the 1990s. Glob Biogeochem Cycle 18(2):GB2008

Albrecht A, Kandji ST (2003) C sequestration in tropical agroforestry systems. Agric Ecosyst Environ 99:15–27

Asner GP (2009) Tropical forest C assessment: integrating satellite and airborne mapping approaches. Environ Res Lett 4:034009

Asner GP, Knapp DE, Broadbent EN et al (2005) Selective logging in the Brazilian Amazon. Science 310:480–482

Asner GP, Rudel TK, Aide TM et al (2009) A contemporary assessment of change in humid tropical forests. Conserv Biol 23:1386–1395

Asner GP, Loarie SR, Heyder U (2010) Combined effects of climate and land-use change on the future of humid tropical forests. Conserv Lett 3:395–403

Bertault J-G, Sist P (1997) An experimental comparison of different harvesting intensities with reduced-impact and conventional logging in East Kalimantan. Indonesia For Ecol Manage 94:209–218

Bhagwat SA, Willis KJ, Birks HJB et al (2008) Agroforestry: a refuge for tropical biodiversity? Trends Ecol Evol 23:261–267

Bonan GB (2008) Forests and climate change: forcings, feedbacks, and the climate benefits of forests. Science 320:1444–1449

Bounoua L, DeFries R, Collatz GJ et al (2002) Effects of land cover conversion on surface climate. Clim Chang 52:29–64

Bourke RM (1997) Management of fallow species composition with tree planting in Papua New Guinea. Paper presented at the Research Management in Asia Pacific, Canberra

Bradshaw CJA, Sodhi NS, Peh KSH (2007) Global evidence that deforestation amplifies flood risk and severity in the developing world. Glob Chang Biol 13:2379–2395

Bremer LL, Farley KA (2010) Does plantation forestry restore biodiversity or create green deserts? A synthesis of the effects of land-use transitions on plant species richness. Biodiv Conserv 19:3893–3915

Brown S (1997) Estimating biomass and biomass change of tropical forests, vol Paper 134. Rome, Italy

Brown S, Lugo AE (1992) Above ground biomass estimates for tropical moist forests of the Brazilian Amazon. Interciencia 17:8–18

Brown S, Lugo AE (1994) Rehabilitation of tropical lands: a key to sustaining development. Restor Ecol 2:97–111

Brown S, Lugo AE, Chapman J (1986) Biomass of tropical tree plantations and its implications for the global C budget. Can J For Res 16:390–394

Brown RT, Agee JK, Franklin JF (2004) Forest restoration and fire: principles in the context of place Restauración de Bosques y Fuego—principios en el contexto del sitio. Conserv Biol 1:903–912

Burney JA, Davis SJ, Lobell DB (2010) Greenhouse gas mitigation by agricultural intensification. Proc Natl Acad Sci USA 107:12052–12057

Cacho OJ, Marshall GR, Milne M (2005) Transaction and abatement costs of carbon-sink projects in developing countries. Environ Dev Econ 10:597–614

Chazdon RL (2003) Tropical forest recovery: legacies of human impact and natural disturbances. Perspect Plant Ecol Evol Syst 6:51–71

Chazdon RL (2008) Beyond deforestation: restoring forests and ecosystem services on degraded lands. Science 320:1458–1460

Chazdon RL, Harvey CA, Komar O et al (2009) Beyond reserves: a research agenda for conserving biodiversity in human-modified tropical landscapes. Biotropica 41:142–153

Chomitz KM (2002) Baseline, leakage and measurement issues: how do forestry and energy projects compare? Clim Policy 2:35–49

Chomitz KM (2007) At loggerheads? agricultural expansion, poverty reduction, and environment in the tropical forests. In: Kenneth M. Chomitz et al (eds) World Bank policy research report, vol. World Bank, Washington, DC. Accessed from http://nla.gov.au/nla.cat-vn3912783

Clark DA (2004) Sources or sinks? The responses of tropical forests to current and future climate and atmospheric composition. Philios Trans R Lond Ser B Biol Sci 359:477–491

Cramer VA, Hobbs RJ, Standish RJ (2008) What's new about old fields? Land abandonment and ecosystem assembly. Trends Ecol Evol 23:104–112

de Graaf NR, Poels RLH, Van Rompaey RSAR (1999) Effect of silvicultural treatment on growth and mortality of rainforest in Surinam over long periods. For Ecol Manage 124:123–135

de Jong BHJ (2002) Forest products and local forest management in West Kalimantan, Indonesia implications for conservation and development, Tropenbos-Kalimantan Series 6. Tropenbos International, Wageningen

Defries RS, Bounoua L, Collatz GJ (2002) Human modification of the landscape and surface climate in the next fifty years. Glob Chang Biol 8:438–458

Dixon RK, Brown S, Houghton RA et al (1994) C Pool and flux of global forest ecosystem. Science 263:185–190

du Toit JT, Walker BH, Campbell BM (2004) Conserving tropical nature: current challenges for ecologists. Trends Ecol Evol 19:12–17

Ebeling J, Yasué M (2008) Generating C finance through avoided deforestation and its potential to create climatic, conservation and human development benefits. Philios Trans R Lond Ser B Biol Sci 363:1917–1924

Edwards DP, Fisher B, Boyd E (2010) Protecting degraded rainforests: enhancement of forest C stocks under REDD+. Conserv Lett 3:313–316

Elias M, Potvin C (2003) Assessing inter- and intra-specific variation in trunk C concentration for 32 neotropical tree species. Can J For Res 33:1039–1045
Eliasch J (2008) Climate change: financing global forests: the Eliasch review. Report to the UK government. Office of Climate Change, London
Ewers RM, Rodrigues ASL (2008) Estimates of reserve effectiveness are confounded by leakage. Trends Ecol Evol 23:113–116
FAO (2000a) The global outlook for future wood supplies from forest plantations. FAO Working Paper. Food and Agriculture Organization, Rome
FAO (2000b) On definitions of forest and forest change. Working paper. FRA, Rome
FAO (2006) Global forests resource assessment 2005: progress towards sustainable forest management. Report No. 147. Food and Agriculutre Oganization, Rome
Fearnside PM (2000) Global warming and tropical land-use change: greenhouse gas emissions from biomass burning, decomposition and soils in forest conversion, shifting cultivation and secondary vegetation. Clim Chang 46:115–158
Fittkau EJ, Klinge H (1973) On biomass and trophic structure of central Amazonian rain forest ecosystems. Biotropica 5:2–14
Francis R, Read D (1994) The contributions of mycorrhizal fungi to the determination of plant community structure. Plant Soil 159:11–25
Garrity D, Soekardi M, van Noordwijk M et al (1996) The imperatta grasslands of tropical Asia: area, distribution, and typology. Agrofor Syst 3:3–29
Gibbs HK, Brown S, Niles JO (2007) Monitoring and estimating tropical forest C stocks: making REDD a reality. Environ Res Lett 2:1748–9326
Gibbs HK, Ruesch AS, Achard F (2010) Tropical forests were the primary sources of new agricultural land in the 1980s and 1990s. Proc Natl Acad Sci USA 107:16732–16737
Goosem SP, Tucker NIJ (1995) Repairing the rainforest: theory and practice of rainforest re-establishment in north Queensland's wet tropics. Wet Tropics Management Authority, Cairns
Grainger A (2009) Towards a new global forest science. Int For Rev 11:126–133
Griess VC, Knoke T (2011) Can native tree species plantations in Panama compete with Teak plantations? An economic estimation. New For 41:13–39
Griscom B (2009) The hidden frontier of forest degradation: a review of the science, policy, and practice of the second "D" in REDD. The Nature Conservancy, Arlington
Griscom B, Shoch D, Stanley B et al (2009) Sensitivity of amounts and distribution of tropical forest C credits depending on baseline rules. Environ Sci Policy 12:897–911
Gullison RE, Frumhoff PC, Canadell JG et al (2007) Tropical forests and climate policy. Science 316:985–986
Gumpenberger M et al (2010) Predicting pan-tropical climate change induced forest stock gains and losses—implications for REDD. Environ Res Let 5:014013
Guo LB, Gifford RM (2002) Soil C stocks and land use change: a meta analysis. Glob Chang Biol 8:345–360
Hansen MC, Stehman SV, Potapov PV et al (2008) Humid tropical forest clearing from 2000 to 2005 quantified by using multitemporal and multiresolution remotely sensed data. Proc Natl Acad Sci USA 105:9439–9444
Hirota M, Nobre C, Oyama MD et al (2010) The climatic sensitivity of the forest, savanna and forest–savanna transition in tropical South America. New Phytol 187:707–719
Houghton RA (2005) Aboveground forest biomass and the global C balance. Glob Chang Biol 11:945–958
Houghton RA, Skole DL, Nobre CA et al (2000) Annual fluxes of C from deforestation and regrowth in the Brazilian Amazon. Nature 403:301–304
HRW (2009) Wild money: the human rights consequences of illegal logging and corruption in Indonesia's forestry sector. Human Rights Watch, New York
Hughes RF, Kauffman JB, Jaramillo VJ (2000) Ecosystem-scale impacts of deforestation and land use in a humid tropical region of Mexico. Ecol Appl 10:515–527
Huston MA, Wolverton S (2009) The global distribution of net primary production: resolving the paradox. Ecol Monogr 79:343–377

ITTO (2002) Guidelines for the restoration, management and rehabilitation of degraded and secondary tropical forests. International Tropical Timber Organization, Yokohama

Jobbágy EG, Jackson RB (2000) The vertical distribution of soil organic C and its relation to climate and vegetation. Ecol Appl 10:423–436

JRCNEA (2011) The emissions database for global atmospheric research (EDGAR). Available from: http://edgar.jrc.ec.europa.eu/index.php, Inspra, Italy

Kanowski J, Catterall C (2010) Carbon stocks in above-ground biomass of monoculture plantations, mixed species plantations and environmental restoration plantings in north-east Australia. Ecol Manage Restor 11:119

Kauffman JB, Hughes RF, Heider C (2009) C pool and biomass dynamics associated with deforestation, land use, and agricultural abandonment in the neotropics. Ecol Appl 1:1211–1222

Keefe K, Schulze MD, Pinheiro C et al (2009) Enrichment planting as a silvicultural option in the eastern Amazon: case study of Fazenda Cauaxi. For Ecol Manage 258:1950–1959

Keith H, Mackey BG, Lindenmayer DB (2009) Re-evaluation of forest biomass C stocks and lessons from the world's most carbon-dense forests. Proc Natl Acad Sci USA 106:11635–11640

Keller MK, Asner GP, Silva N et al (eds) (2004) Sustainability of selective logging of upland forests in the Brazilian Amazon: C budgets and remote sensing as tools for evaluating logging effects. Working forests in the Neotropics, Columbia/New York

Kuusipalo J, Hadengganan S, Ådjers G et al (1997) Effect of gap liberation on the performance and growth of dipterocarp trees in a logged-over rainforest. For Ecol Manage 92:209–219

Laganière J, Angers DA, Paré D (2010) Carbon accumulation in agricultural soils after afforestation: a meta-analysis. Glob Chang Biol 16:439–453

Lal R (2008) C sequestration. Philios Trans R Lond Ser B Biol Sci 363:815–830

Lal R (2010) Managing soils and ecosystems for mitigating anthropogenic carbon emissions and advancing global food security. Bioscience 60:708–721

Lamb D (1998) Large-scale ecological restoration of degraded tropical forest lands: the potential role of timber plantations. Restor Ecol 6:271–279

Lamb D (2011) Regreening the bare hills. In: Tropical forest restoration in the Asia-Pacific region, vol 8, World forest. Springer, Dordrecht

Lamb D, Erskine PD, Parrotta JA (2005) Restoration of degraded tropical forest landscapes. Science 310:1628–1632

Lambin EF, Turner BL, Geist HJ et al (2001) The causes of land-use and land-cover change: moving beyond the myths. Glob Environ Chang 11:261–269

Laurance WF (2007) A new initiative to use C trading for tropical forest conservation. Biotropica 39:20–24

Laurance WF (2008) Can C trading save vanishing forests? Bioscience 58:286–287

Laurance WF, Williamson GB (2001) Positive feedbacks among forest fragmentation, drought, and climate change in the Amazon. Retroalimentaciones positivas entre fragmentación de bosques, sequía y cambio climático en el Amazonas. Conserv Biol 15:1529–1535

Laurance WF, Laurance SG, Ferreira LV et al (1997) Biomass collapse in Amazonian forest fragments. Science 278:1117–1118

Laurance WF, Delamonica P, Laurance SGW et al (2000) Rainforest fragmentation kills big trees. Nature 404:836

Le Quere C, Raupach MR, Canadell JG et al (2009) Trends in the sources and sinks of C dioxide. Nat Geosci 2:831–836

Lewis SL (2006) Tropical forests and the changing earth system. Philios Trans R Lond Ser B Biol Sci 361:195–210

Lewis SL, Lopez-Gonzalez G, Sonke B et al (2009) Increasing C storage in intact African tropical forests. Nature 457:1003–1006

Liao CZ, Luo YQ, Fang CM et al (2010) Ecosystem C stock influenced by plantation practice: implications for planting forests as a measure of climate change mitigation. PLoS One 5(5). doi:e10867 10.1371/journal.pone.0010867

Luyssaert S, Schulze ED, Borner A et al (2008) Old-growth forests as global C sinks. Nature 455:213–215

MacDonald GE (2004) Cogongrass (Imperata cylindrica)—biology, ecology, and management. Crit Rev Plant Sci 23:367–380
Malhi Y (2010) The C balance of tropical forest regions, 1990–2005. Curr Opin Environ Sustain 2:237–244
Malhi Y, Grace J (2000) Tropical forests and atmospheric C dioxide. Trends Ecol Evol 15:332–337
Malhi Y, Phillips O (2005) Tropical forests and global atmospheric change. Oxford University Press, Oxford
Malhi Y, Wright J (2004) Spatial patterns and recent trends in the climate of tropical rainforest regions. Philos Trans R Lond Ser B Biol Sci 359:311–329
Malhi Y, Baldocchi DD, Jarvis PG (1999) The C balance of tropical, temperate and boreal forests. Plant Cell Environ 22:715–740
McElwee P (2009) Reforesting "Bare Hills" in Vietnam: social and environmental consequences of the 5 million hectare reforestation program. AMBIO J Hum Environ 38:325–333
Michon G, Foresta H, Levang P (2007) Domestic forests: a new paradigm for integrating local communities' forestry into tropical forest science. Ecol Soc 12(2)
Miles L, Kapos V (2008) Reducing greenhouse gas emissions from deforestation and forest degradation: global land-use implications. Science 320:1454–1455
Murphy PG, Lugo AE (1986) Ecology of tropical dry forest. Ann Rev Ecol Syst 17:67–88
Myers N (1984) The primary source: tropical forests and our future. W.W. Norton, New York
Myneni RB, Dong J, Tucker CJ et al (2001) A large C sink in the woody biomass of northern forests. Proc Natl Acad Sci USA 98:14784–14789
Nascimento HEM, Laurance WF (2002) Total aboveground biomass in central Amazonian rainforests: a landscape-scale study. For Ecol Manage 168:311–321
Nepstad DC, Uhl C, Serrao EAS (1991) Recuperation of a degraded Amazonian landscape – forest recovery and agricultural restoration. Ambio 20:248–255
Niles JO, Brown S, Pretty J et al (2002) Potential C mitigation and income in developing countries from changes in use and management of agricultural and forest lands. Philios Trans R Lond Ser A: Math Phys Eng Sci 360:1621–1639
Olander LP, Gibbs HK, Steininger M et al (2008) Reference scenarios for deforestation and forest degradation in support of REDD: a review of data and methods. Environ Res Lett 3:025011
Pan Y, Birdsey RA, Fang J et al (2011) A large and persistent C sink in the world's forests. Science. doi:10.1126/science.1201609
Paquette A, Hawryshyn J, Senikas AV et al (2009) Enrichment planting in secondary forests: a promising clean development mechanism to increase terrestrial C sinks. Ecol Soc 14(1)
Pariona W, Fredericksen TS, Licona JC (2003) Natural regeneration and liberation of timber species in logging gaps in two Bolivian tropical forests. For Ecol Manage 181:313–322
Parrotta JA (1995) Influence of overstory composition on understory colonization by native species in plantations on a degraded tropical site. J Veg Sci 6:627–636
Parrotta JA, Turnbull JW, Jones N (1997) Catalyzing native forest regeneration on degraded tropical lands. For Ecol Manage 99:1–7
Paul KI, Polglase PJ, Richards GP (2003) Predicted change in soil C following afforestation or reforestation, and analysis of controlling factors by linking a C accounting model (CAMFor) to models of forest growth (3PG), litter decomposition (GENDEC) and soil C turnover (RothC). For Ecol Manage 177:485–501
Paul M, Catterall CP, Pollard PC et al (2010) Recovery of soil properties and functions in different rainforest restoration pathways. For Ecol Manage 259:2083–2092
Peña-Claros M, Fredericksen TS, Alarcón A et al (2008) Beyond reduced-impact logging: silvicultural treatments to increase growth rates of tropical trees. For Ecol Manage 256:1458–1467
Pfaff A, Kerr S, Lipper L, Cavatassi R et al (2007) Will buying tropical forest C benefit the poor? Evidence from Costa Rica. Land Use Policy 24:600–610
Phillips OL, Malhi Y, Higuchi N et al (1998) Changes in the C balance of tropical forests: evidence from long-term plots. Science 282:439–442

Phillips OL, Baker TR, Arroyo L et al (2004) Pattern and process in Amazon tree turnover, 1976–2001. Philios Trans R Lond Ser B-Biol Sci 359:381–407
Pinard MA, Cropper WP (2000) Simulated effects of logging on C storage in dipterocarp forest. J Appl Ecol 37:267–283
Pinard MA, Putz FE (1996) Retaining forest biomass by reducing logging damage. Biotropica 28:278–295
Pongratz J, Reick CH, Raddatz T, Claussen M (2009) Effects of anthropogenic land cover change on the C cycle of the last millennium. Glob Biogeol Cycles 23. doi:Gb4001 10.1029/2009gb003488
Putz FE, Redford KH (2009) Dangers of carbon-based conservation. Glob Environ Chang 19:400–401
Putz FE, Zuidema PA, Pinard MA et al (2008) Improved tropical forest management for C retention. PLoS Biol 6:1368–1369
Ramachandran Nair PK, Mohan Kumar B, Nair VD (2009) Agroforestry as a strategy for C sequestration. J Plant Nutr Soil Sci 172:10–23
Ramankutty N, Gibbs HK, Achard F et al (2007) Challenges to estimating C emissions from tropical deforestation. Glob Chang Biol 13:51–66
Rhemtulla JM, Mladenoff DJ, Clayton MK (2009) Historical forest baselines reveal potential for continued C sequestration. Proc Natl Acad Sci USA 106:6082–6087
Richards PW (1952) The tropical rain forest: an ecological study. Cambridge University Press, Cambridge
Ricketts TH, Daily GC, Ehrlich PR et al (2004) Economic value of tropical forest to coffee production. Proc Natl Acad Sci USA 101:12579–12582
Rockstrom J, Steffen W, Noone K et al (2009) A safe operating space for humanity. Nature 461:472–475
Saatchi SS, Houghton RA, Dos Santos AlvalÁ RC et al (2007) Distribution of aboveground live biomass in the Amazon basin. Glob Chang Biol 13:816–837
Saatchi SS, Harris NL, Brown S et al (2011) Benchmark map of forest C stocks in tropical regions across three continents. Proc Nat Acad Sci USA. doi:10.1073/pnas.1019576108
Sagan C, Toon OB, Pollack JB (1979) Anthropogenic albedo changes and the earth's climate. Science 206:1363–1368
Sandbrook C, Nelson F, Adams WM et al (2010) Carbon, forests and the REDD paradox. Oryx 44:330–334
Sasaki N, Putz FE (2009) Critical need for new definitions of "forest" and "forest degradation" in global climate change agreements. Conserv Lett 2:226–232
Sasaki N, Asner G, Knorr W et al (2011) Approaches to classifying and restoring degraded tropical forests for the anticipated REDD+ climate change mitigation mechanism (Approaches to classifying and restoring degraded tropical forests for the anticipated REDD+ climate change mitigation mechanism). Biogeosci For 4:1–6
Scharlemann JPW, Kapos V, Campbell A et al (2010) Securing tropical forest carbon: the contribution of protected areas to REDD. Oryx 44:352–357
Schimper AFW, Groom P, Balfour IB (1903) Plant-geography upon a physiological basis. Clarendon, Oxford
Schroeder P (1994) C storage benefits of agroforestry systems. Agrofor Syst 27:89–97
Schroth G, D'Angelo SA, Teixeira WG et al (2002) Conversion of secondary forest into agroforestry and monoculture plantations in Amazonia: consequences for biomass, litter and soil C stocks after 7 years. For Ecol Manage 163:131–150
Schwarze R, Niles JO, Olander J (2002) Understanding and managing leakage in forest–based greenhouse–gas–mitigation projects. Philos Trans R Lond Ser A Math Phys Eng Sci 360:1685–1703
Sheil D, Murdiyarso D (2009) How forests attract rain: an examination of a new hypothesis. Bioscience 59:341–347
Shono K, Cadaweng EA, Durst PB (2007) Application of assisted natural regeneration to restore degraded tropical forestlands. Restor Ecol 15:620–626

Silver WL, Ostertag R, Lugo AE (2000) The potential for C sequestration through reforestation of abandoned tropical agricultural and pasture lands. Restor Ecol 8:394–407

Thomas S, Dargusch P, Harrison S et al (2010) Why are there so few afforestation and reforestation clean development mechanism projects? Land Use Policy 27:880–887

Torres AB, Marchant R, Lovett JC et al (2010) Analysis of the C sequestration costs of afforestation and reforestation agroforestry practices and the use of cost curves to evaluate their potential for implementation of climate change mitigation. Ecol Econ 69:469–477

UNEP (2009) The natural fix? The role of ecosystems in climate mitigation. A UNEP rapid response assessment. United Nations Environment Programme, UNEP, Cambridge

UNFCCC (1997) Decision 1/CP.3 Adoption of the Kyoto Protocol to the United Nations Framework Convention on Climate Change. UNFCCC document FCCC/CP/1997/7/Add.1

UNFCCC (2009) Copenhagen accord. FCCC/CP/2009/L.7. United Nations Framework Convention on Climate Change (UNFCCC), Bonn. Available from: http://unfccc.int/resource/docs/2009/cop15/eng/l07.pdf

van der Werf GR, Morton DC, DeFries RS et al (2009) CO_2 emissions from forest loss. Nat Geosci 2:737–738

Venter O, Laurance WF, Iwamura T et al (2009) Harnessing C payments to protect biodiversity. Science 326:1368–1368

Vieira DLM, Holl KD, Peneireiro FM (2009) Agro-successional restoration as a strategy to facilitate tropical forest recovery. Restor Ecol 17:451–459

Wilcove D, Koh L (2010) Addressing the threats to biodiversity from oil-palm agriculture. Biodiv Conserv 19:999–1007

Wilson EO (ed) (1988) Biodiversity. National Academy Press, Washington, DC

Wright SJ (2005) Tropical forests in a changing environment. Trends Ecol Evol 20:553–560

Wright SJ (2010) The future of tropical forests. Ann NY Acad Sci 1195:1–27

Young KR, Ewel JJ, Brown BJ (1987) Seed dynamics during forest succession in Costa Rica. Plant Ecol 71:157–173

Zomer RJ, Trabucco A, Bossio DA et al (2008) Climate change mitigation: a spatial analysis of global land suitability for clean development mechanism afforestation and reforestation. Agric Ecosyst Environ 126:67–80

Chapter 12
Carbon Cycling in the Amazon

Carlos Clemente Cerri, Martial Bernoux, Brigitte Josefine Feigl, and Carlos Eduardo Pellegrino Cerri

Abstract Brazil is the fifth largest country in the world by area (8,550,000 km^2), and represents one of the ten largest economies. The country has a diversified middle-income economy with wide variations in development levels and mature manufacturing, mining and agriculture sectors. Almost 60% of the Brazil's territory is occupied by the Legal Amazon or also known as Brazilian Amazon. In recent decades, human settlements in the Brazilian Amazon have grown at an exponential rate. Its population grew 130% since the 1970s, reaching about 19.5 in 2000 (last census in Brazil). The exponential settlement rate and associated land cover changes have caused impacts on environmental services such as biodiversity, carbon (C) storage and regulation of the water cycle. Therefore, the Amazon nations, particularly Brazil, face one of the grand challenges for society in the twenty-first century – how to foster economic growth while protecting and improving the environment. This is a complex challenge that requires a variety of inputs on socio-economic and scientific issues. This chapter addresses part of this rather complex issue, focusing more on the scenarios of land-use-changes in Brazil that have direct implications on

C.C. Cerri (✉) • B.J. Feigl
Laboratório de Biogeoquímica Ambiental, CENA Centro de Energia
Nuclear na Agricultura, USP Universidade de São Paulo, Av. Centenario, 303,
Caixa Postal 96, 13400-970, Piracicaba, SP, Brazil
e-mail: cerri@cena.usp.br; beduardo@cena.usp.br

M. Bernoux
IRD, UMR Eco&Sols (Montpellier-SupAgro, Cirad, Inra, IRD),
Bat. 12-2 Place Viala, 34060 Montpellier Cedex 2, France
e-mail: martial.bernoux@ird.fr

C.E.P. Cerri
Dept. Ciência do Solo, ESALQ Escola Superior de Agricultura Luiz de Queiroz,
USP Universidade de São Paulo, Av. Padua Dias 11, 13418-900
Piracicaba, SP, Brazil
e-mail: cepcerri@esalq.usp.br

R. Lal et al. (eds.), *Recarbonization of the Biosphere: Ecosystems and the Global Carbon Cycle*, DOI 10.1007/978-94-007-4159-1_12,

environmental and socio-economic aspects for the Brazilian Amazon. Finally, this chapter aims to estimate, using data from the literature, the effects of some land use change scenarios on soil C sequestration for the Brazilian Amazon and the entire area of the Amazon Tropical Rain Forest.

Keywords Brazilian Amazon • Land-use change • Soil C sequestration • Tropical rain forest • Pasture • Mining land cover change • Environmental services • Tropical rainforest • Primary forest • Secondary forests • Animal units (AU) • Global cycle • Terrestrial primary productivity • Cation exchange capacity • Degraded pastures • CO_2 fertilization • Nutrient cycling • Cattle grazing • Monoculture • Rubber tree • Brazil nut • Soil compaction • Erosion • Coffee • Rice • Bean • Soybean • Sugarcane • Castor • Oil palm • Sunflower • Beef production • Biofuels • Respiration • Drought • Ecotourism • Photosynthesis

Abbreviations

AU	animal units = the standard measure of an animal unit is a 1,000-pound beef cow
C	carbon
CO_2	carbon dioxide
CEC	cation exchange capacity
GHGs	greenhouse gases
Mha	million hectare
NEE	net ecosystem exchange
SOC	soil organic carbon
TRF	Tropical Rain Forest

12.1 Introduction

Brazil is the fifth largest country in the world by area (8,550,000 km^2) and represents one of the ten largest economies. The country has a diversified middle-income economy with wide variations in development levels and mature manufacturing, mining and agriculture sectors (IBGE 2011).

Almost 60% of the Brazil's territory is occupied by the Legal Amazon or also known as Brazilian Amazon. In recent decades, human settlements in the Brazilian Amazon have grown at an exponential rate. Its population grew 130% since the 1970s, reaching about 17 million in 2005 (IBGE 2005). The exponential settlement rate and associated land cover changes have caused impacts on environmental services such as biodiversity, carbon (C) storage and regulation of the water cycle.

These environmental impacts have caused concerns about the sustainability of land use in the Amazon region. Sustainable development of the Brazilian Amazon has become a topic of national and international debate that has involved policy makers, scientists and environmental advocates. Sustainable development is an important issue in the Amazon region, not only because it contains about 40% of the world's remaining Tropical Rain Forest (TRF), but also because it represents one of the richest biodiversity areas in the world (Laurance et al. 2001a).

On the other hand, further expansion of agriculture in the region is expected as Brazil and other Amazon nations are still building their economies to meet the needs of their society. The growing world demand for beef, soybeans (*Glycine max L.*) and biofuels may reshape the landscape of Amazon region in the near term, with the intensification of extant agriculture and the expansion of new pastures and croplands. Agriculture in the Brazilian Amazon has proved highly productive, and has contributed to Brazil's growth into one of the world's major producers and exporters of grains and beef.

Intensive agriculture in the Brazilian Amazon has increased in the last three decades. Since this land use is recent, the region is susceptible to an array of economic and social pressures. Pasture is the main land use in the Amazon region, because it represents one of the most inexpensive agricultural alternatives after deforestation and requires a relatively small and largely unskilled labor force (IBGE 2007).

Development activities, including new roads, electrical power distribution, financial incentives, and improvement of river transportation and ports have made cleared land in the Amazon more valuable and promoted the trend of converting pastures to croplands. Nowadays, soybean cultivation, originally concentrated in the southeast and central parts of Brazil, has been extended through the Amazon region especially in the states of Rondônia and Mato Grosso. The new road connecting Cuiabá to Santarém (BR-163) and improvements in the port infrastructures in Santarém and Porto Velho will reduce soybean exportation costs and will accelerate the cultivation of soybean on former pasture lands and even promote the clearing of native vegetation, mainly from Cerrado vegetation for pasture cultivation (Brasil 2009).

Therefore, the Amazon nations, particularly Brazil, face one of the grand challenges for society in the twenty-first century – how to foster economic growth while protecting and improving the environment. This is a complex challenge that requires a variety of inputs on socio-economic and scientific issues. This chapter addresses part of this rather complex issue, focusing more on the scenarios of land-use-changes in Brazil that have direct implications on environmental and socio-economic aspects for the Brazilian Amazon.

Land use change usually alters land cover and the terrestrial change in C stocks (Bolin and Sukumar 2000). The change from one ecosystem to another could occur naturally or through human activity. Anthropogenic changes are prominent in TRF which is a diverse and complex ecosystem and occupies approximately 17% of the world's area and is a habitat for about 40–50% of the earth's species (Meyers 1981).

The TRF represent significant sources/sinks of greenhouse gases (GHGs), and the exchange of carbon dioxide (CO_2) between forest and the atmosphere is an important component of the global C cycle. The economic damage through global warming from tropical deforestation alone is estimated at $1.4–10.3 billion per year (Pearce and Brown 1994). Fearnside (1996) estimated the economic damage by global warming in the Amazon region at approximately US$1,200–8,600 per ha.

Despite the small fraction (about 14%) of cleared forest in the Amazon, the total deforested area is larger than the area of France or of Texas, USA. In addition to the large area, this ecoregion is also characterized by favorable climatic conditions for plant growth, expressed by high temperatures throughout the year and well-distributed precipitation. Therefore, the Amazonian TRF has, in principle, great potential for soil C sequestration, despite its poor to moderate soil fertility.

The objective of this chapter is to estimate, using data from the literature, the effects of some land use change scenarios on soil C sequestration for the Brazilian Amazon and the entire area of the Amazon TRF.

12.2 The Brazilian Amazon General Characterization

The Amazonian TRF (Fig. 12.1) is the world's largest region of continuous intact tropical forest. The total area of lowland, humid Amazon forest in 1990 was 710.5 ha (Mha) (Phillips et al. 1998), of which 500.5 ha of which was closed canopy forest (Houghton 1997).

The Amazon Basin covers an area of about 700 Mha (Pires and Prance 1986) and occupies large portions of the national territories of Venezuela, Colombia, Peru, Guyanas, Bolivia, Ecuador and Brazil (Fig. 12.1). The basin is confined in the west by the Andean Mountains, in the north by the crystalline mass of Guyana and the savannas of Colombia and Venezuela, in the south by the plateau of Mato Grosso, and in the east by the Atlantic Ocean. The length of the Basin is more than 3,000 km from west to east, and its width ranges from 300 km in the west, to 800 km in the east. Its central part is almost entirely located within the Brazilian territory, and forms the Brazilian Amazon area (Fig. 12.1). The latter is mainly developed on sediments from the Pleistocene. On the whole, the Amazon has a hot and humid climate, characterized by only small variations in the diurnal and monthly temperatures. However, because of its vast size and geomorphological heterogeneity, this region represents a wide range of local climates, with different annual rainfall distribution and often different temperature extremes (Marengo and Nobre 2001).

This region had the highest rates of deforestation in the world (Skole and Tucker 1993). For instance, deforestation has been higher in Latin America than in Asia or Africa not only in area (4.3 Mha year^{-1}) but also in percentage of forest area cleared (0.64% year^{-1}) (Anderson 1990). Of the Latin American forest area of 1850, 370 Mha, or 28% had been cleared by 1985. Of cleared area, 44% was converted to pasture, 25% to cropland, 20% had been degraded, and 10% had been changed to shifting cultivation (Houghton 1991).

In the Brazilian Amazon, estimated deforestation rates range from 1.1 to 2.9 Mha year^{-1} and the total area cleared is about 55 Mha, which is about 14% of its total area (INPE 2004). Laurance et al. (2001a, b) suggested several reasons for this

Fig. 12.1 The Amazonian tropical rain forest and the Brazilian Amazon

rapid rate of deforestation. First, non-indigenous populations in the Brazilian Amazon have increased tenfold since the 1960s, from about 2–20 million people, as a result of immigration from other areas of Brazil and high rates of intrinsic growth. Second, industrial logging and mining are growing dramatically in importance, and road networks are expanding that sharply increase access to forests for ranchers and colonists. Third, the spatial patterns of forest loss are changing; past deforestation has been concentrated along the densely populated eastern and southern margins of the basin, but new highways, roads, logging projects, and colonization are now penetrating deep into the heart of the area. Finally, human-induced wildfires are becoming an increasingly important cause of forest loss, especially in logged or fragmented areas (Laurance et al. 2001a). More recently, this region has received much more attention from national political authorities at all levels, non-governmental organizations, and various other national and international stakeholders. The deforestation rate for 2009 was estimated at about 7 Mha $year^{-1}$, and, therefore, the average rate for the 2006–2009 period was 12.2 Mha $year^{-1}$.

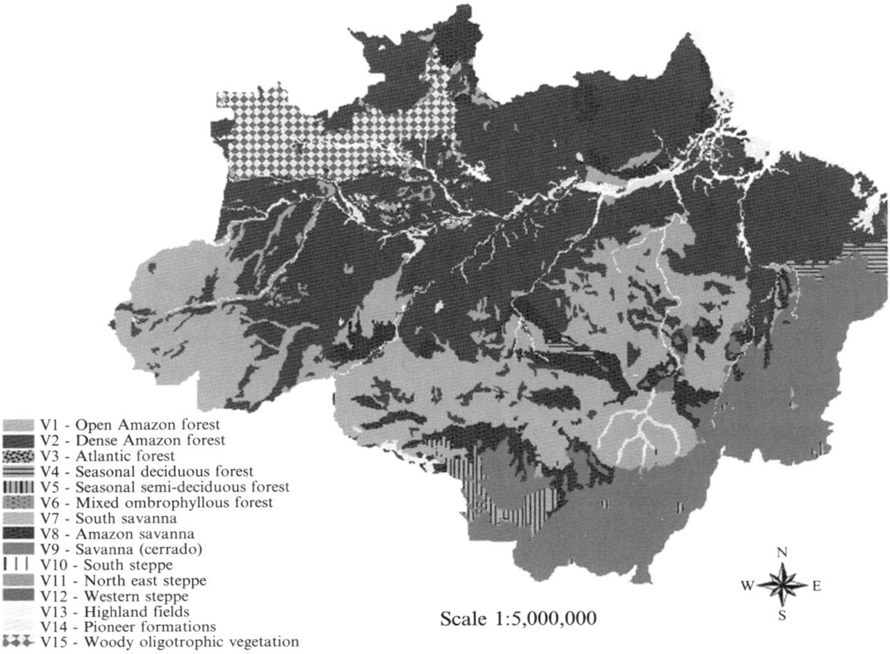

Fig. 12.2 Native vegetation types in the Brazilian Amazon (Bernoux et al. 2002)

Despite the magnitude of deforested area, the Brazilian Amazon still contains about 40% of the world's remaining tropical rainforest and plays vital roles in maintaining biodiversity, regional hydrology and climate, and the terrestrial C storage (Laurance et al. 2001a). Native vegetation types are represented in Fig. 12.2. The forest accounts for about 10% of the world's terrestrial primary productivity and for a similar fraction of the C stored in land ecosystems (Keller et al. 1997). Cattle pastures dominate this once-forested land in most of the basin (Pires and Prance 1986; Skole and Tucker 1993; Fearnside and Barbosa 1998; Dias-Filho et al. 2001).

Jacomine and Camargo (1996) reported that the two main soil divisions of the Brazilian soil classification, Latossolos (Oxisols) and Podzólicos (mainly Ultisols), cover nearly 75% of the area in the Amazon Basin (Fig. 12.3). The remaining area comprises 13 soil divisions, only two of which cover more than 5% of the Amazon area: Plintossolos (Inceptisols, Oxisols, and Alfisols) and Gleissolos (Entisols and Inceptisols) representing 7.4% and 5.3%, respectively (Fig. 12.4).

The Latossolos (Oxisols according to the US Soil Taxonomy), are old, deep, permeable and well-drained soils. The clay mineral component is predominantly kaolinite, a low-activity clay, with varying amounts of iron and aluminum oxides (Cerri et al. 2000). Generally, the cation exchange capacity (CEC) is only partially saturated with bases and the exchangeable Al^{3+} content is relatively high. The Podzólicos are mainly Ultisols according to the US Soil Taxonomy. They contain also low-activity clays and unsaturated CEC. Eutrophic soils (Alfisols) occur only

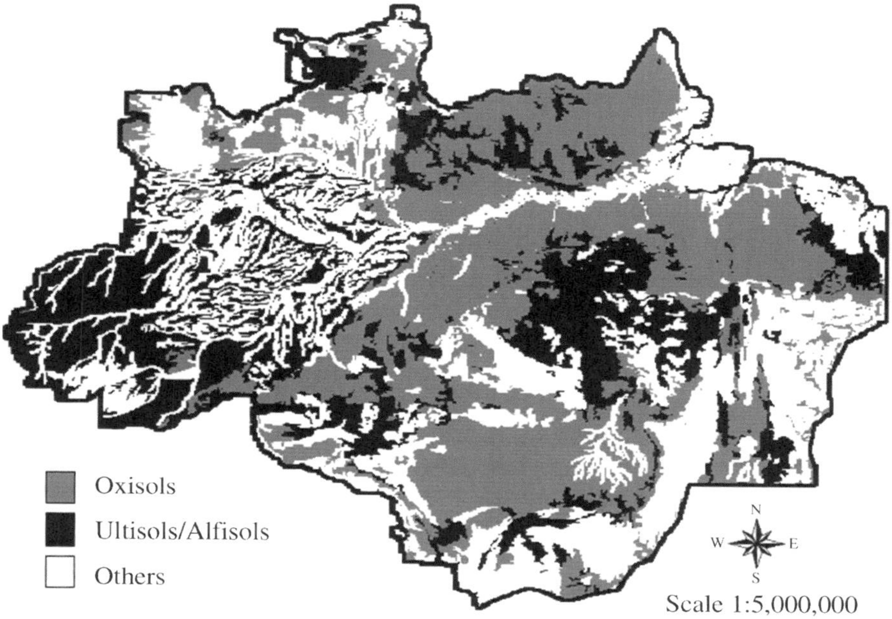

Fig. 12.3 Simplified soil map of the Brazilian Amazon

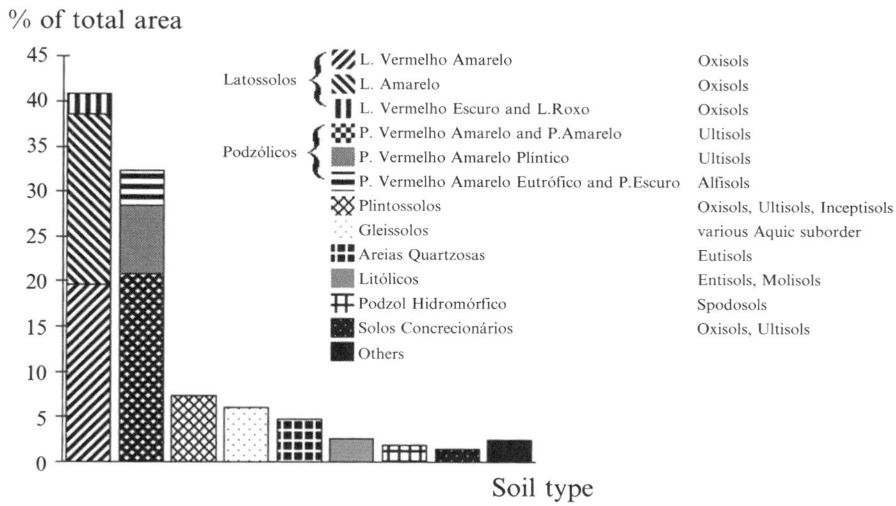

Fig. 12.4 Relative distribution of main soil types in the Amazon (Cerri et al. 2000)

sparingly in the region. Ultisols usually occupy younger geomorphic surfaces than Oxisols with which they are often associated in the landscape. These are deep mineral soils with profiles often deeper than 2 m (Moraes et al. 1996). Moraes et al. (1995) estimated that approximately 47 Pg (1 Pg = 10^{15} g) C are contained in the soils of the Brazilian Amazon to 2 m depth. Of that, 21 Pg C is contained in the top 20 cm where changes in soil C stocks that follow land use conversions are most rapid. Cerri et al. (2000) estimated a C stock of 41 Pg C in the 0–100 cm, and 23.4 Pg C (i.e. 57%) in the 0–30 cm depth.

Ecological characteristics of the Brazilian Amazon described by many researchers (see e.g., McClain et al. 2001) report high plant biomass and concentration of nutrients within the plant biomass (Bernoux et al. 2001), rapid rates of nutrient recycling (Cuevas 2001), high annual rainfall with little seasonal variation in temperature and humidity (Marengo and Nobre 2001), and a relatively closed system for nutrient and water cycling (Melack and Forsberg 2001).

12.3 Scenarios of Soil Carbon Sequestration in the Amazon

The objective of this section is to estimate, using data from the literature, the effects of some land use change scenarios on soil C sequestration for the Brazilian Amazon and then broadly extrapolate it to the entire area of the Amazon TRF. This objective is achieved by collating the available data from the literature on soil C accumulation rates for the following categories of land use and land conversion: (i) primary forest (avoided deforestation), (ii) conversion of forest to well managed pasture, (iii) conversion of degraded pasture to well managed pasture, (iv) conversion of degraded pasture to secondary forest (abandonment) and existing secondary forests, and (v) conversion of degraded pasture to agroforestry.

The primary forest is defined as the native forest before it is cleared for other land uses. Well-managed pasture implies the land used for productive cattle grazing mainly under African grasses (often from the genus *Brachiaria* and *Panicum*) where weeds are controlled and stocking rate is low. Degraded pastures are dominated by weeds and are prone to soil erosion. Secondary forest is the vegetation that develops naturally on abandoned land formerly used for other purposes, mainly pasture. Agroforestry is a dynamic management system that diversifies and sustains production for increased social, economic and environmental benefits for land users at all levels, through the integration of trees in farmland and rangeland.

12.3.1 Primary Forest (Avoided Deforestation)

Undisturbed, mature forests were once thought to be in a steady state in terms of CO_2 flux, with CO_2 uptake by photosynthesis being balanced by CO_2 releases through respiration and biomass decay. However, there is a strong evidence suggesting that intact forests are actually responding to the increasing levels of CO_2 in the atmosphere caused by anthropogenic emissions of CO_2, the so-called "CO_2 fertilization effect" (ECCM 2002). The extent to which CO_2 fertilization affects productivity of natural forests depends on a number of factors including the availability of water and nutrients. However, even a small increase in productivity (Norby et al. 1999 reported a 0.3% increase in productivity per year) could result in a substantial increase in C uptake by forests. For instance, Malhi and Grace (2000) estimated that a 0.3% increase would translate to an annual uptake of 1 Mg (1 Mg = 10^6 g) C ha^{-1} $year^{-1}$ by the Amazon forests.

Eddy-covariance analysis (monitoring of gas micro-turbulences in air flows inside and immediately above the forest) allows direct measurement of net changes in ecosystem C stocks [net ecosystem exchange (NEE)]. However, eddy covariance studies are limited in temporal and spatial scales and there are claims that C uptake may be overestimated due to various limitations in the method. Several studies have been carried out in Amazon forests using the eddy-covariance technique, forest inventory, and C model approaches (ECCM 2002). Grace et al. (1995), Fan et al. (1990) and Malhi et al. (1998) concluded that Amazon forests were accumulating 1.0, 2.2 and 5.9 Mg C ha^{-1} $year^{-1}$, respectively. Higuchi et al. (1997) estimated an uptake of 1.2 Mg C ha^{-1} $year^{-1}$ in 3 ha of forest growth measurements over the 1986–1996 period near Manaus in the Central Amazon. Phillips et al. (1998) concluded that Amazon forests were accumulating 0.62 ± 0.37 Mg C ha^{-1} $year^{-1}$. In Central Amazon, Chambers et al. (2001) reported an accumulation rate of 0.5 Mg C ha^{-1} $year^{-1}$. Tian et al. (1998) and Prentice and Lloyd (1998) estimated that average uptake by Amazon forests was 0.3 Mg C ha^{-1} $year^{-1}$. Lal and Kimble (1999) reported the Amazon region has a C sequestration rate of 2–3 Mg ha^{-1} $year^{-1}$. This sequestration rate may be attributed to CO_2 fertilization effect, tree mortality due to catastrophic events and effects of El Niño in exacerbating tree mortality. The authors mentioned that, apparently, the rate of new growth exceeds the decomposition and mineralization of dead biomass and soil organic carbon (SOC).

The literature reviewed herein suggests that C uptake in the Amazon ranges from 0.3 to 5.9 Mg C ha^{-1} $year^{-1}$ (mean value around 1 Mg C ha^{-1} $year^{-1}$). This is the range used in the present study to assess the potential soil C sequestration. These rates must to be used with caution, because some studies show considerable year-to-year variation. This is particularly important in the modeling studies by Tian et al. (1998) and Prentice and Lloyd (1998) who conclude that in some years Amazon forests acted as a net source of C at the rate of 0.3 and 0.6 Mg C ha^{-1} $year^{-1}$, respectively. These variations are primarily due to changes in rainfall and temperature associated with the El Niño weather phenomenon which causes drought in some area of the Amazon (Phillips et al. 2009).

The future rate of C uptake by Amazon forests may, therefore, depend on the extent to which forests continue to respond to CO_2 fertilization. Some studies (e.g., Cox et al. 2000) suggest that climate change may cause significant drying in the Amazon, creating a C source as forests decline. In contrast, there is also evidence to suggest that CO_2 fertilization may continue to result in C uptake for many years to come (Chambers et al. 2001).

Given the vast area of standing forest in the Brazilian Amazon, estimated to be about 344 Mha (Laurance et al. 2001b), and also using the uptake range of 0.3–5.9 Mg C ha^{-1} $year^{-1}$ leads to a potential of C sequestration of 103–2,030 Tg C $year^{-1}$ (1 Tg = 10^{12} g).

It is important to stress that large spatial coverage is needed in order to draw better conclusions, as uptake at one site may be balanced by emissions at other sites. Fearnside (1999) argued that time scale is also important. Over the long term, "mature" forest cannot continue to grow at the same rates in biomass, but imbalances over periods of years or decades are still important to understanding the global C dynamics, including clarifications of the terrestrial sink. An uptake would increase the impact of deforestation by eliminating part of the sink. While the amount of sink loss due to deforestation in a single year may appear modest compared to the emissions from forest biomass caused by the clearing, the fact that the sink represents an annual flux rather than a one-time emission means that it would have significant consequences over the long term if the sink can exist over several decades.

12.3.2 Conversion of Forest to Well Managed Pasture

Pastures represent the largest single use of cleared forest land in most of the Brazilian Amazon. As much as 70% of the deforested land has been converted to pastures at one stage or another (Serrão and Toledo 1990; Dias-Filho et al. 2001). Fearnside and Barbosa (1998) observed that about 45% of the Brazilian Amazon is occupied by actively grazed cattle pasture that accounts for approximately 24.7 Mha. Similar numbers were reported by Homma (1994), Kitamura (1994) and Camarão and Souza Filho (1999). Farmers were motivated to convert lands cleared from forest into pasture because of the resultant real or perceived increases in land value. Farmers not only maintain cattle as standing "bank accounts" and obtain cash from sales of animals and milk, but build savings by investing time and resources in pasture, fencing, corrals, and ponds (Fujisaka et al. 1996).

Despite the enormous scale of pasture expansion in the Amazon, there is yet no clear understanding of the direction of the resulting changes in soil C stocks. Fearnside and Barbosa (1998) reported that depending on management conversion of Amazon forest to pasture can produce a net soil C sink (well managed pasture) or a net C source (overgrazed pasture). Neill and Davidson (1999) observed that conversion of forest to pasture in the Amazon occurs on a range of soils and in regions that differ in the amount and timing of precipitation. The sequence leading to pasture development also differs. Some pastures are created by planting grasses directly into forest slash, while others are created after 1 or 2 years of annual cropping or after a cropping and fallow sequence. The choice of grass species and practices of interplanting with legumes also differ. These factors can influence whether a pasture soil will become source or sink of C. Once established, pasture management by stocking rate, burning frequency, effectiveness of weed control, fertilizing or disking may also affect soil C balance (Neill and Davidson 1999).

Therefore, in some locations, soil C stocks in pastures are lower compared with the original forest (Luizão et al. 1992; Desjardins et al. 1994). In other locations, pastures grass productivity declines in older pastures, but soil C concentrations remain relatively constant (Falesi 1976; Serrão et al. 1979; Buschbacher et al. 1988). Yet in other locations, inputs of C from roots of pasture grasses increase soil C stocks (Cerri et al. 1991, 2003; Bonde et al. 1992; Trumbore et al. 1995; Moraes et al. 1996; Neill et al. 1997; Bernoux et al. 1998).

Neill and Davidson (1999) synthesized the available literature on soil C stocks in pasture following deforestation in the Amazon. They reported that 19 out of 29 pastures accumulated C in surface soils and 10 showed C losses. They also observed a strong relationship between pasture grass species with the change in surface soil C stocks. Pasture planted to *Brachiaria humidicola* lost C and those planted to *Panicum maximum* and *Brachiaria brizantha* gained C.

Moraes et al. (1996) observed that total soil C stocks to 30 cm depth in 20-year old well managed pastures were 17–20% higher than in the original forest sites in the western Amazon. A comparison of C budgets for forest and pastures in the eastern Amazon was made by Trumbore et al. (1995). In a rehabilitated and fertilized pasture of *Brachiaria brizantha*, they estimated gains, of over 20 Mg C ha^{-1} in the top 1-m of soil relative to forest soil C stocks and a loss of about 0.5 Mg C ha^{-1} in the 1–8 m soil depth during the first 5 years following pasture rehabilitation. More than 50% of the forest-derived C in surface soils of pastures on converted Amazon forest turns over within 10–30 years (Choné et al. 1991; Trumbore et al. 1995). Cerri et al. (1999) reported C sequestration of 0.27 Mg C ha^{-1} year^{-1} for the 0–30 cm depth. Neill et al. (1997) reported annual soil C accumulation rate, in the range of 0.2–0.3 Mg ha^{-1} in the top 50 cm. Those results are within the range (0.2–3.9 Mg C ha^{-1} year^{-1}) of those reported by Watson et al. (2000) for pastures in wet tropical areas of the world.

For the present study, a range of 0.2–3.9 Mg C ha^{-1} year^{-1} (mean accumulation rate of 0.27 Mg C ha^{-1} year^{-1}) in the top 0–30 cm soil layer was assumed. Further assuming that 20% of the pastures are under well managed system (about 5 Mha), the potential of soil C sequestration is between 1 and 19.5 Tg C year^{-1}.

In contrast to annual crops, pasture grasses maintain a continuous vegetative cover on the soil, reduce soil temperatures, and often have high productivity and turnover rates which add organic matter, particularly belowground, to the soil (Brown and Lugo 1990). However, any lack of difference between soil C and N contents in forest and pasture ecosystems does not account for the fact that large aboveground C stocks (100–300 Mg C ha^{-1}) are lost when forests are converted to agricultural uses (Fearnside 2000). Dias-Filho et al. (2001) reported that forest-to-pasture conversion release 100–200 Mg C ha^{-1} from above-ground forest biomass to the atmosphere. There are also additional benefits of intact forests to ameliorating floods, conserving soils, maintaining stable regional climates, preserving biodiversity, and supporting indigenous communities and ecotourism industries (Laurance et al. 2001a). Therefore, any policy changes that reduce the rate of deforestation would have the greatest potential of reducing the net emission of GHGs. Moreover, it is also greatly desired that these policies enhance the rate of C sequestration in soil.

12.3.3 Conversion from Degraded to Well Managed Pasture

More than half of the cattle pasture areas in the Brazilian Amazon are degraded (Serrão and Toledo 1990; Dias-Filho et al. 2001), which represents about 13 Mha. Productivity of Amazonian pastures is often good during the first 3–5 years after establishment. A rapid decline in productivity of the planted grasses occurs after the period due to encroachment by herbaceous and woody invaders (Uhl et al. 1988; Serrão and Toledo 1990). If uncontrolled, invaders species gradually dominate and severely degrade pastures, a condition characterized by a complete dominance of the weedy community (Dias-Filho et al. 2001).

If the entire area under degraded pasture could be well managed, and assuming the accumulation range of soil C sequestration for well managed pastures of 0.2–3.9 Mg C ha^{-1} $year^{-1}$ in the top 0–30 cm soil layer, the potential of soil C sequestration of converting degraded to well managed pasture in the Brazilian Amazon varies from 2.6 to 51 Tg C $year^{-1}$.

There are numerous factors and processes that must be considered in estimating the direction and rate of change in SOC content by change in soil management. Post and Kwon (2000) reported that important factors for increasing C sequestration include: (i) increasing the input rates of organic matter, (ii) changing the decomposability of organic inputs increasing the light fraction organic C, (iii) placing organic matter deeper in the soil either directly by increasing belowground inputs or indirectly by enhancing surface mixing by soil organisms, and (iv) enhancing physical protection through formation of aggregates or organomineral complexes.

12.3.4 Conversion from Degraded Pasture to Secondary Forest (Abandonment) and Existing Secondary Forest

Secondary forest associated with agriculture in Amazon follows a clear pattern of development. During pasture use, burning and weeding delay succession, but the forest begins to regenerate once the field is abandoned. Secondary vegetation establishes itself through four main processes: regeneration of remnant individuals, germination from the soil seed bank, sprouting from cut or crushed roots and stems, and dispersal and migration of seed from other areas (Tucker et al. 1998). Variance in the speed of forest regrowth is evident across regions and along a soil fertility gradient in the Brazilian Amazon. The rate of forest succession is determined by several factors. Original floristic composition, neighboring vegetation, and soil fertility and texture may affect regrowth. In addition, farmers' land use decisions, such as clearing size, clearing procedures, crops planted, frequency and duration of use, influence tree establishment and path of secondary succession. At the regional scale, soil fertility and land use history are the critical factors influencing the rate of forest regrowth (Tucker et al. 1998).

Secondary forests in the Amazon have high rates of regeneration, both following slash-and-burn agriculture and after abandonment of degraded pasture. Brown and

Lugo (1990) reported that abandoned agricultural lands reverted to forests accumulated C at rates proportional to the initial forest biomass. Rates ranged from about 1.5 Mg C ha^{-1} $year^{-1}$ in forests with initial biomass of <100 Mg C ha^{-1} to about 5.5 Mg C ha^{-1} $year^{-1}$ for forests with biomass of >190 Mg C ha^{-1}. Woomer et al. (1999) observed a rate of 6.2 ± 1.3 Mg C ha^{-1} $year^{-1}$ of C sequestration in secondary forest regrowth in the Brazilian Amazon. Watson et al. (2000) suggested a range of C accumulation of 3.1–4.6 Mg C ha^{-1} $year^{-1}$ for tropical regions over 40 years. Schroth et al. (2002) reported that secondary forest on an infertile upland soil in central Amazon accumulated C in above- and belowground biomass and litter at a rate of about 4 Mg C ha^{-1} $year^{-1}$. The rate of accumulation in aboveground biomass reported by Nepstad et al. (2001) ranged from 2.5 to 5 Mg C ha^{-1} $year^{-1}$ for 20-year-old Amazon secondary forest. A study in the Central Amazon reported C accumulation of 128 Mg C ha^{-1} for a 12 year old secondary forest dominated by *Vismia ssp.* (Feldpausch et al. 2004). The secondary vegetation was regenerated on an abandoned severely degraded pasture near Manaus.

Estimates of the potential of C sequestration when degraded pastures in the Brazilian Amazon are abandoned for secondary forest regrowth, were made by multiplying 13 Mha (degraded pasture area) by the accumulation range of 1.5–5.5 Mg C ha^{-1} $year^{-1}$ or 19.5–71.5 Tg C $year^{-1}$. This potential of C sequestration accounts for C in soil + aboveground biomass. Further, secondary forests cover about one third of the area of the Brazilian Amazon that had been cleared for agriculture (Houghton et al. 2000) and if abandoned it can also sequester C in soil and plant biomass. Thus, multiplying the same accumulation range values (1.5–5.5 Mg C ha^{-1} $year^{-1}$) by the 18 Mha already under secondary forest regrowth, leads to an additional sequestration of 27–99 Tg C $year^{-1}$. Therefore, potential soil + biomass C sequestration in the Brazilian Amazon due to secondary forest regrowth is between 46.5 and 170.5 Tg C $year^{-1}$.

Forests that develop on abandoned land also counteract many of the deleterious impacts of forest conversion to agriculture and cattle pasture. They play an important role in the regional C budget, as they re-assimilate part of the C that was released upon cutting and burning of the original forest vegetation. They restore hydrological functions performed by mature forests, and reduce the flammability of landscapes. Secondary forests transfer nutrients from the soil to living biomass, thereby reducing the potential losses of nutrients from the land through leaching and erosion. They also allow the expansion of native plant and animal populations from mature forest remnants back into agricultural landscapes (Nepstad et al. 2001).

12.3.5 Conversion from Degraded Pasture to Agroforestry

Agroforestry is a possible option not only for C sequestration but also for the valorization of previously cleared forest land in the humid tropics (Fujisaka and White 1998; Nair et al. 2010). Where ever agroforestry succeeds in maintaining soil fertility at a satisfactory level and increases farmers' income, additional clearing of primary forest and accompanying C emissions are drastically reduced.

When established on degraded soils, timber and tree crops in these systems sequester C in the biomass and soil, and also provide firewood and charcoal as offsets for fossil fuel. On the other hand, when agroforestry systems or tree crop plantations are established on previously cleared fallow or secondary forest land, C is released from the fallow vegetation and the successional processes which would have accumulated C in biomass and litter. Instead, establishment of trees, timber trees, annual crops and eventually cover crops affect ecosystem C budget through soil management, fertilizer application and suppression of spontaneous vegetation through weeding (Schroth et al. 2002).

Few data are available on rate of C sequestration through agroforestry in the Amazon. A mean rate of 2.7 Mg C ha^{-1} $year^{-1}$ over 25–30 years is supported by the literature elsewhere, in which the values vary from 0.5 to 3.8 Mg C ha^{-1} $year^{-1}$. Woomer et al. (1999), measuring total system C in chronosequences in Brazilian Amazon, Cameroon, Indonesia and Peru, reported that agroforestry systems sequestered about 3.3 Mg C ha^{-1} $year^{-1}$ in soils and vegetation. Watson et al. (2000) mentioned a range of 0.5–1.8 Mg C ha^{-1} $year^{-1}$ of C accumulation for agroforest management in the tropics. McCaffery et al. (2000) found that agroforestry systems based on native fruits and palms, planted on a severely degraded pasture in the Central Amazon, accumulated up to 33 Mg C ha^{-1} on the aboveground biomass after 12 years of management. Biomass from degraded pasture was 9 Mg C ha^{-1}, indicating a net C uptake of 2 Mg C ha^{-1} $year^{-1}$. For the same systems, Rondon et al. (2000) reported soil C stocks (to 1 m depth) under agroforestry systems of 120 Mg C ha^{-1} as compared with the degraded pasture soil which stored 110 Mg C ha^{-1}. This resulted in soil C accumulation rates of 0.83 Mg C ha^{-1} $year^{-1}$. Schroth et al. (2002) reported that monocultures accumulated C at lower rates, i.e. 1.0 Mg ha^{-1} $year^{-1}$ for *Citrus*, 1.3 Mg ha^{-1} $year^{-1}$ for cupuaçu (*Theobroma grandiflorum*) and 2.5 Mg ha^{-1} $year^{-1}$ for rubber tree (*Hevea brasiliensis*). Multistrata agroforestry favors C sequestration more than monocultures. A fast-growing system in Central Amazon accumulated 3.8 and 3.0 Mg ha^{-1} $year^{-1}$ of C in full (N, P, K and dolomitic lime) and low-fertilization (only 30% of the fertilizer and lime of the former treatment) treatment, respectively. These high rates compared to most monocultures were due to a relatively high tree density and to the association of the smaller tree crops, such as cupuaçu and peach palm for palmito (*Bactris gasipaes*), with larger and faster-growing trees, i.e. rubber and Brazil nut trees (*Bertholleta excelsa*). Schroth and colleagues also observed that in all of the investigated plantation systems, there was more than twice as much C in the soil than in the biomass and litter combined. Changes in the SOM stocks could, therefore, being of crucial importance to the net C budget by land use transformations. However, Schroth et al. (2002) observed no effects of vegetation types and plant species on the SOC stocks to 2-m depth, although C content of the topsoil was affected. Authors proposed two possible explanations for this trend. First, the conversion of primary forest into different tree crop plantations may have affected the distribution of C in the soil, but not its total quantity. Such changes may occur through an altered distribution of root mass in the soil profile, or through differences in the abundance and activity of burrowing soil fauna between vegetation types and plant species.

Secondly, the total C stock in the soil to 2-m depth may be less sensitive than the C content of the topsoil as a measure for SOC loss over a relatively short time period (Schroth et al. 2002).

Schroth et al. (2002) also observed that the tree crops with low litter quality (e.g., Cupuaçu and Brazil nut) restored and maintained SOC levels in the topsoil comparable to those in the primary forest, even when they were grown in association with tree and cover crops that produced easily decomposable litter. When integrated into multistrata agroforestry systems, such tree crops reduce SOMC loss. Thus, these systems must be established on sites with low standing biomass (e.g., degraded pastures or other degraded soils) while preserving vigorously growing secondary forests. However, Amazonian pastures are often prone to topsoil compaction and erosion (Fearnside 1985), with adverse effects on growth and yields of tree crops.

Considering the complexities involved, it is apparent that the estimate of 6.5–49.4 Tg C year^{-1} upon conversion of degraded pasture to agroforestry is highly uncertain. This estimate is obtained by multiplying 13 Mha of degraded pasture area by the soil + biomass C accumulation range of 0.5–3.8 Mg C ha^{-1} year^{-1}. This estimate is based on the assumption that the entire area under degraded pasture is converted to agroforestry. Furthermore, SOC sequestration in agroforestry is affected by species. Though the total area in the Amazon dedicated to agroforestry seems relatively small compared to other management systems, agroforests are a viable alternative land use for the region, and the area under these systems has been steadily increasing since the mid 1980s.

12.4 Potential of Soil and Biomass Carbon Sequestration in the Brazilian Amazon

The potential of soil + biomass C sequestration in the Brazilian Amazon ranges from a minimum of 134–150 Tg C year^{-1} to a maximum of 2,197–2,220 Tg C year^{-1}, with mean values ranging from 421 to 470 Tg C year^{-1} (Table 12.1). Of this, soil C sequestration accounts for a minimum of 40 to a maximum of 667 Tg C year^{-1}, and the rest is related to C sequestration in plant biomass. These estimates involve the following categories of land use and land conversion: primary forest (avoided deforestation), forest to well managed pasture, degraded pasture to well managed pasture, degraded pasture to secondary forest (abandonment) and existing secondary forest, and degraded pasture to agroforestry. The largest contributor of C sequestration in the Brazilian Amazon is attributed to avoided deforestation, which represents about 75–80% of the total potential. Deforestation in Amazon releases large quantities of GHGs that are significant both in terms of their present impact and long-term contribution to global warming through continued clearing of the remaining forest. Therefore, it is important to avoid deforestation by intensification of agriculture on existing land, rehabilitation of degraded pastures, and adoption of recommended management practices.

Table 12.1 Potential of carbon sequestration in the Brazilian Amazon

Land use	Area (Mha)	C sequestration rate (Mg C ha^{-1} year^{-1})	Total Potential (Tg C year^{-1})
Primary forest (avoided deforestation)	344	0.3–5.9 (1.00)[a]	103–2030 (344)
Primary forest to well managed pasture	5	0.2–3.9 (0.27)[b]	1–19.5 (1.35)
Degraded to well managed pasture	13	0.2–3.9 (0.27)[b]	2.6–51 (3.50)
Degraded pasture to secondary forest	13	1.5–5.5 (4.00)[a]	19.5–71.5 (52)
Existing secondary forest	18	1.5–5.5 (4.00)[a]	27–99 (72)
Degraded pasture to agroforestry	13	0.5–3.8 (2.70)[a]	6.5–49.4 (35)
Total			134–2220 (508)

Figures in brackets are mean values
[a]Soil + biomass carbon sequestration
[b]Soil carbon sequestration

Extrapolating the data on C sequestration for the Brazilian Amazon to the entire area of the Amazonian TRF, the potential of soil C sequestration ranges from of 56 to 934 Tg C year^{-1}. Further, assuming that the current situation described in this study is maintained for 30 years, the total potential of soil C sequestration in the Amazonian TRF ranges from a minimum of 1.7 Pg C to a maximum of 28 Pg C, with best estimates ranging from 5.3 to 5.9 Pg C.

These estimates must be used with great caution because of numerous uncertainties. For example, the largest contributor of C sequestration (avoided deforestation) was estimated by multiplying a mean value taken from several data on annual C accumulation rates by estimated of the area under primary forest. Both assumptions can accentuate the error. There are also uncertainties in the primary data cited from other studies. For instance, current estimates of total C storage in Brazilian Amazon vary by more than a factor of 2, from 39 to 93 Pg C, largely as a result of uncertainty in the quantity and spatial distribution of forest biomass. Houghton (2003) reported that 60% of the uncertainty in their estimates of annual C flux from Brazilian Amazon resulted from variable estimates of the forest biomass.

There is a strong need for additional measurements across large expanses of the Amazonian TRF. Moreover, studies of specific management regimes maintained over the long term are needed to improve our understanding of the consequences of land use and management practices on soil C stocks, and related soil C and nutrient cycling processes.

12.5 Conclusions

Official Brazilian government data estimate that in 2007 the eight main crops occupied 54.6 Mha. This number is projected to increase to 14% in 2017–2018, rising to 62.2 Mha. Four crops (rice-*Oryza sativa*, bean-*Phaseolus vulgaris*, wheat-*Triticum spp*. and coffee-*Coffea canephora*) will decrease 1.1 Mha, but the other four crops (soybean-*Glycine max*, maize-*Zea mays*, sugarcane-*Saccharum officinarum* and cotton-*Gossypium hirsutum*) will increase 8.7 Mha. The demand for biodiesel needs an extra area of 2.6 Mha for oil crops (soybean-*Glycine max*, castor bean-*Ricinus communis*, palm-*Elaeis guineensis* and sunflower-*Helianthus annuus*). Reforestation crops for industrial uses need another 3.4 Mha. The total land-use-change to satisfy all these needs is estimated to be 14.7 Mha. It is highly desirable that this production growth is not linked to any new deforestation, but is associated with a better use of already installed pastures, which occupy currently 172 Mha. The result would be a total of 157.3 Mha of pastures by 2017. However, beef production should also increase 28% along the next 10 years. Based on the actual mean holding capacity of the pastures of 0.9 AU ha^{-1}, (AU = animal units = the standard measure of an animal unit is a 1,000-pound beef cow) 48 Mha of pasture would be necessary to accomplish this projection. To maintain the pasture extension in 157.3 Mha Brazil should invest in science and technology to increase the present 0.9 to at least 1.4 AU ha^{-1}. Our proposition to reach a national mean of 1.4 AU ha^{-1} is to: (i) Increase pasture productivity; (ii) Rehabilitate degraded pastures; (iii) Introduce integrated crop-livestock system; (iv) Partial cattle confine and others. Land-use-changes to attend the demand for food, fiber and biofuels will occur in all Brazilian territory. National and regional public policies to incentive these actions in already installed pasture lands in the Amazon region would bring important social and economic benefits. Nevertheless the pasture conversion to agricultural land has to be done carefully, using as much as possible best management practices to avoid environmental impacts on water streams, soil biodiversity, and air quality, among others. Part of the economic incentives could be provided by taxes derived from remuneration of avoided deforestation. Any further deforestation in the Brazilian Amazon should not be justified by the expansion of land use dedicated to produce grain, fiber, timber and beef productions in order to meet national goals.

Acknowledgements Researches that led to this study were supported partly by grants and fellowships from Conselho Nacional de Desenvolvimento Científico e Tecnológico (CNPq), Coordenação de Aperfeiçoamento de Pessoal de Nível Superior (CAPES), Comité Français d'Évaluation de la Coopération Universitaire et Scientifique avec le Brésil (COFECUB), Fundação de Amparo a Pesquisa do Estado de São Paulo (Fapesp) and support from University of São Paulo (USP-Brazil) and the Institut de Recherche pour le Développement (IRD-France).

References

Anderson A (1990) Deforestation in Amazônia: dynamics, causes, and alternatives. In: Anderson AB (ed) Alternatives to deforestation: steps towards sustainable use of the Amazon rain forest. Colombia University Press, New York

Bernoux M, Arrouays D, Cerri CC et al (1998) Modeling vertical distribution of carbon in Oxisols of the western Brazilian Amazon. Soil Sci 163:941–951

Bernoux M, Graça PMA, Cerri CC et al (2001) Carbon storage in biomass and soils. In: McClain ME, Victoria RL, Richey JE (eds) The biogeochemistry of the Amazon basin. Oxford University Press, New York, pp 84–105

Bernoux M, Carvalho MCS, Volkoff B et al (2002) Brazil's soil carbon stocks. Soil Sci Soc Am J 66:888–896

Bolin B, Sukumar R (2000) Global perspective. In: Watson RT, Noble IR, Bolin B et al (eds) Land use, land use change, and forestry. Cambridge University Press, Cambridge, pp 23–51

Bonde TA, Christensen BT, Cerri CC (1992) Dynamics of soil organic matter as reflected by natural ^{13}C abundance in particle size fractions of forested and cultivated Oxisols. Soil Biol Biochem 24:275–277

Buschbacher R, Uhl C, Serrão AS (1988) Abandoned pastures in eastern Amazonia. II. Nutrient stocks in the soil and vegetation. J Ecol 76:682–699

Brasil (2009) Ministério da Agricultura, Abastecimento e Agropecuária. http://www.agricultura.gov.br/. Visited May 2009

Brown S, Lugo AE (1990) Tropical secondary forests. J Trop Ecol 6:1–32

Camarão AP, Souza Filho AP (1999) Pastagens nativas da Amazonia. Embrapa Amazonia Ocidental, Belem, p 150

Cerri CC, Volkoff B, Andreux F (1991) Nature and behavior of organic matter in soils under natural forest, and after deforestation, burning and cultivation, near Manaus. For Ecol Manag 38:247–257

Cerri CC, Bernoux M, Feigl BJ et al (1999) Carbon dynamics in forest and pasture soils of the Brazilian Amazon. Workshop on tropical soils. Academia Brasileira de Ciências. Rio de Janeiro, pp 65–72

Cerri CC, Bernoux M, Arrouays D (2000) Carbon stocks in soils of the Brazilian Amazon. In: Lal R, Kimble J, Follet R, Stewart BA (eds) Global climate change and tropical ecosystems, Advances in soil science. CRC Press, Boca Raton, pp 33–50

Cerri CEP, Coleman K, Jenkinson DS et al (2003) Soil carbon dynamics at Nova Vida Ranch, Amazon, Brazil. Soil Sci Soc Am J 67:1879–1887

Chambers JQ, Higuchi N, Tribuzy ES et al (2001) Carbon sink for a century. Nature 410:429

Choné T, Andreux F, Correa JC et al (1991) Changes in organic matter in an oxisol from the central Amazonian forest during eight years as pasture, determined by ^{13}C composition. In: Berthelin J (ed) Diversity of environmental biogeochemistry. Elsevier, New York, pp 307–405

Cox PM, Betts RA, Jones CD et al (2000) Acceleration of global warming due to carbon-cycle feedbacks in a coupled climate model. Nature 408:184–187

Cuevas E (2001) Soil versus biological controls on nutrient cycling in terra firme forests. In: McClain ME, Victoria RL, Richey JE (eds) The biogeochemistry of the Amazon basin. Oxford University Press, New York, pp 53–67

Desjardins T, Andreux F, Volkoff B et al (1994) Organic carbon and ^{13}C contents in soils and soil size-fractions, and their changes due to deforestation and pasture installation in eastern Amazonia. Geoderma 61:103–118

Dias-Filho MB, Davidson EA, Carvalho CJR (2001) Linking biogeochemical cycles to cattle pasture management and sustainability in the Amazon basin. In: McClain ME, Victoria RL, Richey JE (eds) The biogeochemistry of the Amazon basin. Oxford University Press, New York, pp 84–105

ECCM (2002) Carbon sinks in the Amazon – the evidence. Produced for the World Land Trust by The Edinburgh Centre for Carbon Management. October 2002. www.worldlandtrust.org/carbon/carbonfigures.pdf

Falesi IC (1976) Ecosistema de pastagem cultivada na Amazônia Brasileira. Centro de Pesquisa Agropecuária do Trópico Úmido. Empresa Brasileira de Pesquisa Agropecuária. Boletim Técnico 1. p 193

Fan SM, Wofsy SC, Bakwin PS et al (1990) Atmosphere-biosphere exchange of CO_2 and O_3 in the central Amazon forest. J Geophys Res Atmos 95:16851–16864

Fearnside PM (1985) Agriculture in Amazonia. In: Prance GT, Lovejoy TE (eds) Key environments: Amazonia. Pergamon, New York, pp 393–418

Fearnside PM (1996) Human carrying capacity estimation in the Brazilian Amazon: research requirements to provide a basis for sustainable development. In: Lieberei R, Retsdorff G, Machado AD (eds) Interdisciplinary research on the conservation and sustainable use of the Amazonian Rain Forest and its information requirements. Report on a workshop held in Brasilia, 20–22 Nov 1995, pp 274–291

Fearnside PM (1999) Greenhouse gas emissions from land-use change in Brazil's Amazon region. In: Lal R, Kimble JM, Stewart BA (eds) Global climate change and tropical ecosystems. CRC Press, Boca Raton, pp 231–249

Fearnside PM (2000) Global warming and tropical land-use change: greenhouse gas emissions from biomass burning, decompositions and soils in forest conversion, shifting cultivation and secondary vegetation. Clim Chang 46:115–158

Fearnside PM, Barbosa RI (1998) Soil carbon changes from conversion of forest to pasture in Brazilian Amazon. For Ecol Manag 108:147–166

Feldpausch T, Rondon M, Fernandes E et al (2004) Carbon and nutrient accumulation in secondary forest regenerating from degraded pastures in Central Amazonia. Ecol Appl 14:164–176

Fujisaka S, White D (1998) Pasture or permanent crops after slash-and-burn cultivation? Land-use choice in three Amazon colonies. Agrofor Syst 42:45–59

Fujisaka S, Bell W, Thomas N et al (1996) Slash-and-burn agriculture, conversion to pasture, and deforestation in two Brazilian Amazon colonies. Agric Ecosyst Environ 59:115–130

Grace J, Lloyd J, McIntyre J et al (1995) Carbon dioxide uptake by an undisturbed tropical rain forest in southwest Amazonia, 1992 to 1993. Science 270:778–780

Higuchi N, der Santos J, Ribeiro RJ et al (1997) Crescimento e incremento de uma floresta amazônica de terra-firme manejada experimentalmente. In: Higuchi N, Ferraz JBS, Antony L et al (eds) Bionte: biomassa e nutrientes florestais. Final report. Instituto Nacional de Pesquisa da Amazônia (INPA), Manaus, pp 87–132

Homma AKO (1994) Amazônia: desenvolvimento econômico e questão ambiental. In: Vilhena EF, Santos LC (eds) Agricultura e meio ambiente. UFV-NEPEMA, Viçosa, pp 25–37

Houghton RA (1991) Tropical deforestation and atmospheric carbon-dioxide. Clim Chang 19:99–118

Houghton RA (1997) Terrestrial carbon storage: global lessons for Amazonian research. Ciência e Cultura. J Braz Assoc Adv Sci 49:58–72

Houghton RA (2003) Why are estimates of the terrestrial carbon balance so different? Glob Chang Biol 9:500–509

Houghton RA, Skole DL, Nobre CA et al (2000) Annual fluxes of carbon from deforestation and regrowth in the Brazilian Amazon. Nature 403:301–304

IBGE (2005) Instituto Brasileiro de Geografia e Estatística. Available at http://www.ibge.com.br

IBGE (2007) Instituto Brasileiro de Geografia e Estatística. Censo Agropecuário 2006: Resultados Preliminares. IBGE, Rio de Janeiro, pp 1–146

IBGE (2011) Instituto Brasileiro de Geografia e Estatística. Available at http://www.ibge.gov.br/paisesat/

INPE (2004) Instituto Nacional de Pesquisas Espaciais. Amazônia: desflorestamento 2002–2003. São José dos Campos, SP

Jacomine PKT, Camargo MN (1996) Classificação pedológica nacional em vigor. In: Alvarez VH, Fontes LEF, Fontes MPF (eds) Solos nos grandes domínios morfoclimáticos do Brasil e o desenvolvimento sustentado. SBCS-UFV, Viçosa, pp 675–689

Keller M, Melillo JM, deMello WZ (1997) Trace gas emissions from ecosystems of the Amazon basin. Cienc Cult J Braz Assoc Adv Sci 49:87–97

Kitamura PC (1994) A Amazônia e o desenvolvimento sustentável. Embrapa, Brasília, p 182

Lal R, Kimble JM (1999) What do we know and what needs to be known and implemented for C sequestration in tropical ecosystems. In: Lal R, Kimble JM, Stewart BA (eds) Global climate change and tropical ecosystems. CRC Press, Boca Raton, pp 417–431

Laurance WF, Cochrane MA, Bergen S et al (2001a) The future of the Brazilian Amazon. Science 291:438–439

Laurance WF, Cochrane MA, Bergen S et al (2001b). Re: the future of the Brazilian Amazon. Science online, www.sciencemag.org/cgi/eletters. Visited Feb 2001

Luizão RC, Bonde TA, Rosswall T (1992) Seasonal variation of soil microbial biomass: the effect of clearfelling a tropical rainforest and establishment of pasture in the central Amazon. Soil Biol Biochem 24:805–813

Malhi Y, Grace J (2000) Tropical forests and atmospheric carbon dioxide. Trends Ecol Evol 15:332–337

Malhi Y, Nobre A, Grace J et al (1998) Carbon dioxide transfer over a Central Amazonian rain forest. J Geophys Res 103:593–612

Marengo JÁ, Nobre CA (2001) General characteristics and variability of climate in the Amazon Basin and its links to the global climate system. In: McClain ME, Victoria RL, Richey JE (eds) The biogeochemistry of the Amazon basin. Oxford University Press, New York, pp 17–41

McCaffery K, Rondon M, Gallardo J et al (2000) Carbon and nutrient stocks in agroforestry systems and secondary forest in the Central Amazon. In: Proceedings of the second scientific conference of the LBA project, Atlanta

McClain ME, Victoria RL, Richey JE (2001) The biogeochemistry of the Amazon basin. Oxford University Press, New York, p236

Melack JM, Forsberg BR (2001) Biogeochemistry of Amazon Floodplain lakes and associated wetlands. In: McClain ME, Victoria RL, Richey JE (eds) The biogeochemistry of the Amazon basin. Oxford University Press, New York, pp 235–274

Meyers N (1981) Conversion rates in tropical moist forests. In: Mergen F (ed) Tropical forests utilization and conservation. Yale University, New Haven, pp 48–66

Moraes JFL, Cerri CC, Melillo JM et al (1995) Soil carbon stocks of the Brazilian Amazon basin. Soil Sci Soc Am J 59:244–247

Moraes JFL, Volkoff B, Cerri CC et al (1996) Soil properties under Amazon forest change due to pasture installation in Rondônia, Brazil. Geoderma 70:63–81

Nair PKR, Nair VD, Kumar BM et al (2010) Chapter Five – Carbon sequestration in agroforestry systems. Adv Agron 108:237–307

Neill C, Davidson EA (1999) Soil carbon accumulation for loss following deforestation for pasture in the Brazilian Amazon. In: Lal R, Kimble JM, Stewart BA (eds) Global climate change and tropical ecosystems. CRC Press, Boca Raton, pp 197–211

Neill C, Cerri CC, Melillo J et al (1997) Stocks and dynamics of soil carbon following deforestation for pasture in Rondonia. In: Lal R, Kimble JM, Follett RF et al (eds) Soil processes and the carbon cycle. CRC Press, Boca Raton, pp 9–28

Nepstad D, Moutinho PRS, Markewitz D (2001) The recovery of biomass, nutrient stocks, and deep-soil functions in secondary forests. In: McClain ME, Victoria RL, Richey JE (eds) The biogeochemistry of the Amazon basin. Oxford University Press, New York, pp 139–155

Norby RJ, Wullschleger SD, Gunderson CA et al (1999) Tree responses to rising CO_2 in field experiments: implications for the future forest. Plant Cell Environ 22:683–714

Pearce DW, Brown K (1994) Saving the world's tropical forests. In: Brown K, Pearce DW (eds) The causes of tropical deforestation: the economic and statistical analysis of factors giving rise to the loss of the tropical forests. University College London Press Limited, London, pp 1–26

Phillips OL, Malhi Y, Higuchi N et al (1998) Changes in the carbon balance of tropical forests: evidence from long-term plots. Science 282:439–442

Phillips OL, Aragão LEOC, Lewis SL et al (2009) Drought sensitivity of the Amazon rainforest. Science 323:1344–1347

Pires JM, Prance GT (1986) The vegetation types of the Brazilian Amazon. In: Prance GT, Lovejoy TM (eds) Amazonia. Pergamon Press, Oxford, pp 109–115

Post WM, Kwon KC (2000) Soil carbon sequestration and land-use change: processes and potential. Glob Chang Biol 6:317–327

Prentice IC, Lloyd J (1998) C-quest in the Amazon Basin. Nature 396:619–620

Rondon M, Fernandes E, Lima R et al (2000) Carbon storage in soils from degraded pastures and agroforestry systems in Central Amazonia: The role of charcoal. In: Proceedings of the second scientific meeting of the LBA project, Atlanta

Schroth G, D'Angelo SA, Teixeira WG et al (2002) Conversion of secondary forest into agroforestry and monoculture plantations in Amazonia: consequences for biomass, litter and soil carbon stocks after 7 years. For Ecol Manag 163:131–150

Serrão EAS, Toledo JM (1990) The search for sustainability in Amazonian pastures. In: Anderson AB (ed) Alternatives to deforestation: steps towards sustainable utilization of Amazon forests. Columbia University Press, New York, pp 195–214

Serrão EAS, Falesi IC, da Veiga JB et al (1979) Productivity of cultivated pastures on low fertility soils of the Amazon Basin. In: Sanchez PA, Tergas LE (eds) Pasture production in acid soils of the tropics. Centro Internacional de Agricultura Tropical, Cali, pp 195–225

Skole D, Tucker C (1993) Tropical deforestation and habitat fragmentation in the Amazon satellite data from 1978 to 1988. Science 260:1905–1910

Tian H, Mellilo JM, Kicklighter DW et al (1998) Effect of interannual climate variability on carbon storage in Amazonian ecosystems. Nature 396:664–667

Trumbore SE, Davidson EA, Camargo PB et al (1995) Below-ground cycling of carbon in forest and pastures of eastern Amazonia. Glob Biogeochem Cycle 9:515–528

Tucker JM, Brondizio ES, Moran EF (1998) Rates of forest regrowth in Eastern Amazonia: a comparison of Altamira and Bragantina regions, Para State, Brazil. Interciencia 23:64–73

Uhl C, Buschbacher R, Serrão EAS (1988) Abandoned pastures in Eastern Amazonia. I. Patterns of plant sucession. J Ecol 76:663–681

Watson RT, Noble IR, Bolin B et al (2000) Land use, land use change, and forestry. A special report of the IPCC. Cambridge University Press, Cambridge, p 377

Woomer PL, Palm CA, Alegre J et al (1999) Slash-and-burn effects on carbon stocks in the humid tropics. In: Lal R, Kimble JM, Stewart BA (eds) Global climate change and tropical ecosystems. CRC Press, Boca Raton, pp 99–115

Chapter 13
Grassland Soil Organic Carbon Stocks: Status, Opportunities, Vulnerability

Richard Theodore Conant

Abstract Grasslands, including rangelands, shrublands, pastureland, and cropland sown with pasture and fodder crops, covered approximately 3.5 billion ha in 2000. This represented 26% of the global ice-free land area and 70% of the agricultural area, and contained about 20% of the world's soil organic carbon (C) stocks. People rely heavily upon grasslands for food and forage production. Around 20% of the world's native grasslands have been converted to cultivated crops, and significant portions of milk and beef production occurs on grasslands managed solely for those purposes. The livestock industry – largely based on grasslands – provides livelihoods for about one billion of the world's poorest people and provides one-third of global protein intake. Much of the world's grasslands are under pressure to produce more livestock by grazing more intensively, particularly in Africa's rangelands, which are vulnerable to climate change and expected to satisfy most of the beef and milk demand in Africa. As a result of past practices, between 5% and 10% of the world's grasslands have been degraded by overgrazing. Portions of the grasslands on every continent have been degraded due to human activities with about 7.5% of grassland having been degraded because of overgrazing. More recently, the Land Degradation Assessment in Drylands (LADA) concluded that about 16% of rangelands are currently undergoing degradation and that rangelands comprise 20–25% of total land area currently being degraded affecting the livelihood of over 1.5 billion people.

Disturbance is an integral part of traditional grassland management systems to sustain yields of forage. However, disturbance through overgrazing, fire, and invasive species can also deplete grassland systems of C stocks. Harvesting a large

R.T. Conant (✉)
Natural Resource Ecology Laboratory, Colorado State University, 1499 Campus Delivery, Fort Collins, CO 80523-1499, USA

Institute for Sustainable Resources, Queensland University of Technology, Brisbane, Australia
e-mail: conant@nrel.colostate.edu

R. Lal et al. (eds.), *Recarbonization of the Biosphere: Ecosystems and the Global Carbon Cycle*, DOI 10.1007/978-94-007-4159-1_13,

proportion of plant biomass enhances yields of useful material (e.g., for forage or fuel), but decreases C inputs to the soil. Many management techniques intended to increase livestock forage production have the potential to increase soil C stocks by sequestering atmospheric carbon dioxide(CO_2) – as much as 0.2–0.8 Pg CO_2 $year^{-1}$. Methods of improved management include fertilization, irrigation, intensive grazing management, and sowing of more favorable forage grasses and legumes. Grassland management to enhance production (i.e., through sowing improved species, irrigation or fertilization), to minimize negative impacts of grazing, or to rehabilitate degraded lands can each lead to C sequestration. Improved grazing management (i.e., management that increases production) can also lead to an increase of soil C stocks as can rehabilitation of degraded lands.

Keywords Grassland • Soil carbon stocks • Grazing • Rangelands • Beef production • Pasture lands • Shrublands • Fodder crops • Milk production • Overgrazing • Land degradation • Forages • Invasive species • Carbon sequestration • Rehabilitation of degraded lands • Carbon emissions • Adaptation to climate change • Workable policies • Incentives • Land tenure • Governance • Policy implementations • Modeling • Developing countries • Net primary production • Greenhouse gases • Soil organic matter

Abbreviations

C	carbon
CO_2	carbon dioxide
CERs	Certified Emission Reductions
CDM	clean development mechanism
GHGs	greenhouse gases
LADA	Land Degradation Assessment in Drylands
CH_4	methane
NPP	net primary production
N_2O	nitrous oxide
COP7	Seventh Conference of the Parties
SOM	soil organic matter
UNFCCC	United Nations Framework Convention on Climate Change.

13.1 Introduction

Grassland C sequestration efforts face comparable challenges as those facing forestry and agricultural sequestration. However, grassland sequestration rates can be slower, ability to measure changes could be more difficult, benefits may be distributed across more landowners/land managers with less certain tenure, practices may be more varied, costs of implementation are more poorly quantified, and the scientific information to inform policy analysis is more incomplete. Similarly, the opportunities to benefit from grassland practices that sequester C can be greater too. The large population who depend directly on grasslands tend to be poor and vulnerable to climate variability and climate change. Implementing practices to build – or rebuild – soil C stocks in grasslands could lead to considerable mitigation, adaptation, and development benefits. Assessment of grassland C sequestration has lagged behind that of agriculture and forestry. Specifically, forestry is an important existing component of the Clean Development Mechanisms (CDM) of the Kyoto Protocol. This chapter discusses the challenges that grassland sequestration faces, and the substantial and diverse opportunities that arise with management practices that lead to C sequestration in grasslands. This chapter concludes by identifying key knowledge barriers and deriving a set of recommended activities and observations that can overcome them.

13.2 Background

13.2.1 Grasslands Cover Broad Areas, Contribute Substantially to Livelihoods, and Are Vulnerable

Grasslands, including rangelands, shrublands, pastureland, and cropland sown with pasture and fodder crops, covered approximately 3.5 billion ha in 2000. This represented 26% of the global ice-free land area and 70% of the agricultural area, and contained about 20% of the world's soil C stocks (FAO 2009b; Ramankutty et al. 2008; Schlesinger 1977) People rely heavily upon grasslands for food and forage production. About 20–25% of the world's native grasslands have been converted to cultivated crops (Fig. 13.1) (Ramankutty et al. 2008), and significant portions of world milk and beef production occurs on grasslands managed solely for those purposes. The livestock industry – largely based on grasslands – provides livelihoods for about one billion of the world's poorest people and one-third of global protein intake (Steinfeld et al. 2006).

The development challenges faced by the population of the world's dry grassland systems vividly illustrate the tightening linkage between ecosystem services and enhanced human well-being. Specifically, two billion people inhabit dryland regions but dryland regions have only 8% of the world's renewable water supply meaning that people have access to water that meets only two-thirds of the minimum per capita requirements. Further, population growth rates are faster in drylands than anywhere else but production potential is lower than anywhere else. Traditional socio-ecological systems have evolved to cope with climatic and economic uncertainty in dry grassland systems but population and economic pressures are increasingly taxing traditional systems (Verstraete et al. 2007).

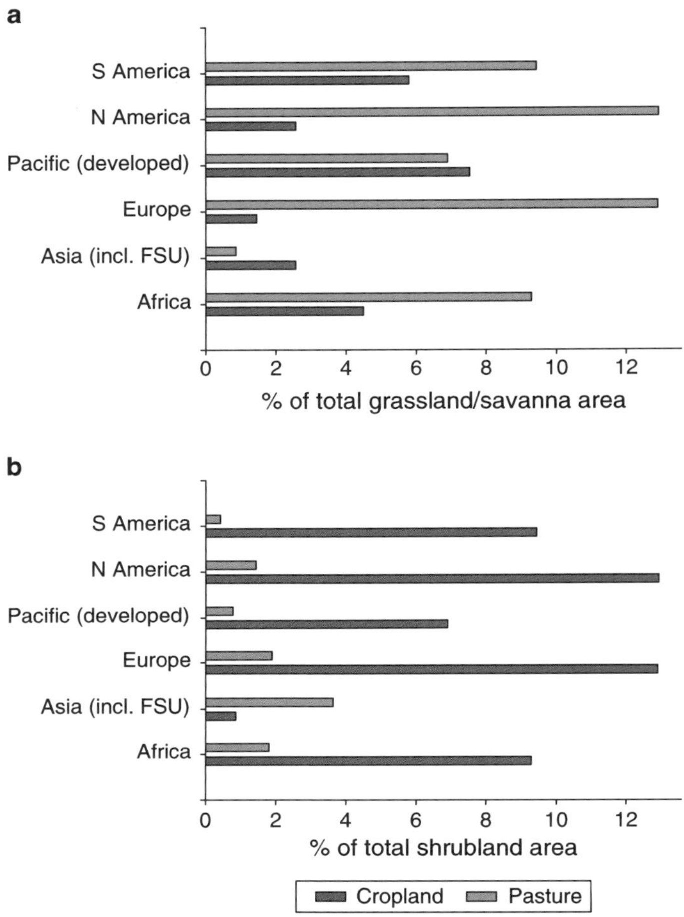

Fig. 13.1 Percentage of native grassland/savanna (**a**) and shrubland (**b**) that has been converted to cropland and pasture (All data are from Ramankutty et al. 2008)

Primary production in rangelands is relatively low, varies substantially from place to place, and is strongly limited by precipitation (Le Houerou 1984). Even where rainfall is high (i.e., some grassland areas receive as much as 900 mm of precipitation per year), almost all of the precipitation occurs during distinct rainy seasons and evapotranspiration demands exceed precipitation during most of the year. Moreover precipitation, and thus production, varies considerably from year to year, with coefficients of variation averaging 33% and as high as 60% in some of the drier areas (Ellis and Galvin 1994). Thus, grasslands are highly vulnerable to climate change induced alterations in precipitation (Thornton et al. 2007, 2009).

13.2.2 Grasslands Are Intensively Used and Degradation Is Widespread

Much of the world's grassland area is under pressure to produce more livestock by grazing more intensively, particularly in Africa's rangelands, which are vulnerable to climate change and expected to still satisfy most of the beef and milk demand in Africa (Reid et al. 2004). As a result of past practices, 5–10% of the world's grasslands have been degraded by overgrazing (Oldeman 1994). Previous research has documented that improved grazing management could lead to greater forage production, more efficient use of land resources, and enhanced profitability and rehabilitation of degraded lands (Oldeman 1994). The tightening linkage between ecosystem services and human well-being in the world's dryland systems acutely demonstrates the need for a new, integrated approach to diagnosing and addressing sustainable development priorities, including maintenance of the supply of critical ecosystem services.

One of the reasons for the intensive use of grasslands is the high natural soil fertility. Typically, grasslands have high inherent soil organic matter (SOM) content averaging 331 Mg SOM ha^{-1} (Schlesinger 1977). SOM is an important source of plant nutrients and can enhance production, increase soil aggregation, limit soil erosion, and increase cation exchange and water holding capacities (Kononova 1966; Allison 1973; Tate 1987; Miller and Donahue 1990), and, thus, a key regulator of grassland ecosystem processes. The key underlying goal of sustainable management of grassland ecosystems is therefore to maintain high levels of SOM.

Portions of the grasslands on every continent have been degraded due to human activities, with about 7.5% of grassland having been degraded because of overgrazing (Oldeman 1994). More recently, the Land Degradation Assessment in Drylands (LADA) concluded that about 16% of rangelands are currently undergoing degradation, and that rangelands comprise 20–25% of total land area currently being degraded affecting livelihoods of over 1.5 billion people globally (Bai et al. 2008). This more recent degradation is likely occurring in addition to historic degradation (Bai et al. 2008). Historically, intensive use of grasslands has resulted in the transfer of 993 Tg of soil C to the atmosphere in the form of CO_2 just from land use change in the United States (Kern 1994). SOM losses due to conversion of native grasslands to cultivation are both extensive and well documented (Donigian et al. 1994; Kern 1994; Follett et al. 2001a). Removal of a large amounts of aboveground biomass, continuous heavy stocking rates, and other poor grazing management practices are an important human-controlled factor that influences grassland production and has led to depletion of soil C stocks (Conant and Paustian 2002a; Ojima et al. 1993). However, best grassland management practices can potentially reverse historical soil C losses and sequester substantial amounts of C in soils.

13.3 Opportunities for Greenhouse Gas Mitigation in Grasslands

13.3.1 Carbon Sequestration in Grasslands

Disturbance – defined as removing biomass, changing the vegetation, or altering soil function – is an integral part of traditional grassland management systems that fosters dependable forage yields. However, disturbance through overgrazing, fire and invasive species can also deplete grassland systems of C stocks (Smith et al. 2008). Removing a large proportion of plant biomass enhances yield of useful material (e.g., for forage or fuel) but decreases C inputs to the soil (Fig. 13.2, Box 13.1) (Wilts et al. 2004). Primary production in overgrazed grasslands can decrease if herbivory decreases plant growth capacity, vegetation density, community biomass, or if community composition changes (Chapman and Lemaire 1993). If C inputs to the soil in these systems decrease because of decreased net primary production (NPP) or direct C removal by livestock, soil C stocks would decline. Like C sequestration in forests or agricultural land, sequestration in grassland systems – primarily, but not entirely in the soils – is brought about by increasing carbon inputs to soil. It is widely accepted that continuous excessive grazing is detrimental to plant

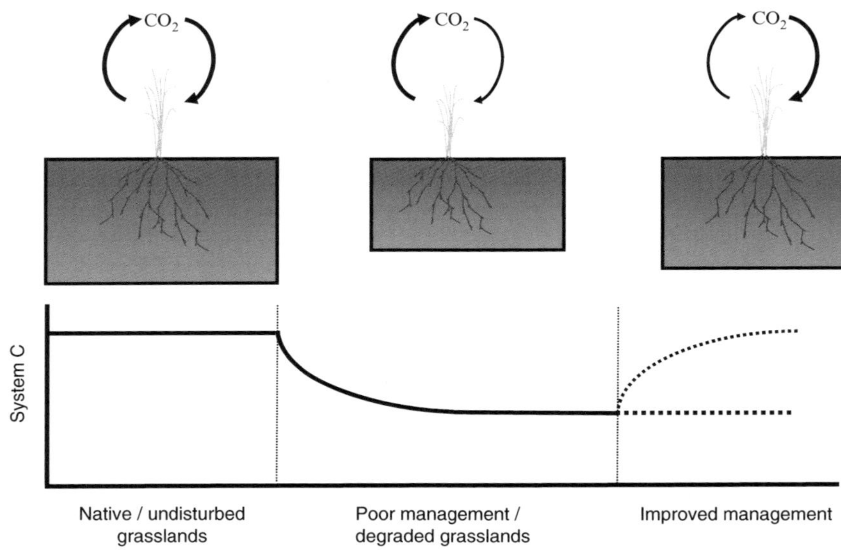

Fig. 13.2 Conceptual diagram illustrating how past land management can lead to depletion of grassland soil C stocks due to practices that decrease carbon uptake relative to carbon losses (indicated by with of *arrows* in the *upper* part of the diagram). Implementation of improved management practices can lead to enhanced C uptake, restoring ecosystem C stocks (*dotted line*) in comparison with continuation of the same management practices (*lower dotted line*), and sequester atmospheric CO_2 in grassland soils

communities (Milchunas and Lauenroth 1993) and soil C stocks (Conant and Paustian 2002a). When management practices that deplete soil C stocks are reversed, grassland ecosystem C stocks can be rebuilt by sequestering atmospheric CO_2 (Follett et al. 2001a).

Box 13.1 Carbon Stocks Are a Function of Carbon Inputs and Outputs

All terrestrial ecosystems – e.g., forested ecosystems, agroecosystems, grassland ecosystems – take up atmospheric carbon dioxide (CO_2) and mineral nutrients and transform them into organic products. In grasslands, carbon (C) assimilation is directed towards production of fiber and forage by manipulating species composition and growing conditions. Ecosystems are a major source and sink for the three main biogenic greenhouse gases (GHGs) – CO_2, nitrous oxide (N_2O), and methane (CH_4). In undisturbed ecosystems, the C balance tends to be positive as C uptake through photosynthesis exceeds losses from respiration, even in mature, old-growth forest ecosystems (Luyssaert et al. 2008; Gough et al. 2008; Stephens et al. 2007). Respiration is not the only flux of C from ecosystems, though it is the primary one; others include dissolved organic and inorganic C and C removed through erosion. Similarly, plant transfers of C to the soil are not the only C input, though others (like manure inputs, C from depositional erosion, etc.) tend to be smaller. Disturbance, such as fire, drought, disease, or excessive forage consumption by grazing, can lead to substantial losses of C from both soils and vegetation (Page et al. 2002; Ciais et al. 2005; Adams et al. 2009). Disturbance is a defining element of all ecosystems that continues to influence the C uptake and losses that determine the long-term ecosystem C balance (Randerson et al. 2002).

Human land use activities function much like natural activities in their influence on the ecosystem C balance. Carbon dioxide is produced when forest biomass is burned, and soil C stocks begin to decline soon after soil disturbances (Smith et al. 1999). Like natural disturbances such as fire and drought, land use change affects vegetation and soil dynamics, often prompting further increased C releases and decreased C uptake. Deforestation, degradation of native grasslands, and conversion of grassland to cropland has prompted losses of biomass and soil C of 100–200 Pg C – equivalent to 30–40% of cumulative fossil fuel emissions (Houghton 2010; Houghton et al. 1983; Olofsson and Hickler 2008; DeFries et al. 1999; Marland et al. 2000). Emissions from conversion of forests to cropland or other land use have dominated C losses from terrestrial ecosystems (DeFries et al. 1999), but substantial amounts of C have been lost from biomass and soils of grassland systems as well (Shevliakova et al. 2009).

(continued)

Box 13.1 (continued)

The basic processes governing the C balance of grasslands are the same as for other terrestrial ecosystems, i.e., the photosynthetic uptake and assimilation of CO_2 into organic compounds and the release of gaseous carbon through respiration (primarily CO_2 but also CH_4).

Biomass in grassland systems, being predominantly herbaceous is a small, transient C pool (compared to trees) and, thus, soils constitute the major C stock. Grassland systems can be productive ecosystems, but restricted growing season length, drought periods, and grazing-induced shifts in species composition or production can reduce C uptake relative to that in other ecosystems. Soil organic carbon stocks in grasslands have been depleted to a lesser degree than for cropland (Ogle et al. 2004) and in some regions biomass has increased due to suppression of disturbance and subsequent woody encroachment. Much of the C lost from agricultural soil and biomass pools can be recovered with changes in management practices that increase C inputs, stabilize C within the system, or reduce C losses, while still maintaining harvestable yields of fiber and forage.

Many management techniques intended to increase livestock forage production have the potential to increase soil C stocks (Table 13.1). Methods of improved management include fertilization, irrigation, intensive grazing management, and sowing of more productive forage grasses and legumes. Grassland management to enhance production (through sowing improved species, irrigation or fertilization), to minimize negative impacts of grazing, or to rehabilitate degraded lands can each lead to C sequestration (Follett et al. 2001b; Conant et al. 2001; Conant and Paustian 2002a). Improved grazing management (i.e., management that increases NPP), leads to an increase in soil C stocks by an average of 0.35 Mg C ha^{-1} $year^{-1}$ (Conant et al. 2001).

Using seeded grasses for cover cropping, catch crops, and more complex crop rotations all increase C inputs to the soil by extending the time over which plants are fixing atmospheric CO_2 in cropland systems. Rotations with grass, hay, or pasture tend to have the largest impact on soil C stocks (West and Post 2002). Adding manure to soil builds SOM in grasslands (Conant et al. 2001). The synthesis by Smith et al. (2008) suggests that adding manure or biosolids to soil could sequester between 0.42 and 0.76 Mg C ha^{-1} $year^{-1}$ depending on region (e.g., sequestration rates tend to be greater in moist regions than in dry). Rapid incorporation of manure into fields would reduce the time that manure decomposes in anaerobic piles and lagoons, reducing emissions of CH_4 and N_2O. Smith et al. (2007) estimate the technical potential for reduction of CH_4 emissions from manure to be 12.3 Tg C $year^{-1}$ by 2030. N_2O emissions could also be reduced. Adding manure in one place to build soil C stocks is offset by removal or what

Table 13.1 Mitigative effects of various aspects of grazingland improvement (Adapted from Smith et al. 2007)

	CO_2	CH_4	N_2O	Agreement	Evidence
Grazing intensity	+/–	+/–	+/–	*	*
Increased productivity (e.g., through fertilization)	+	?	+/–	**	*
Nutrient management	+	?	+/–	**	**
Fire management	+	+	+/–	*	*
Species introductions (incl. legumes)	+	?	+/–	*	**

A question mark indicates that the effects are unknown or poorly understood. A "+/–" denotes uncertain or variable response. Agreement refers to the relative degree of consensus in the literature (the more asterisks, the higher the agreement); evidence refers to the relative amount of data in support of the proposed effect (the more asterisks, the more evidence)

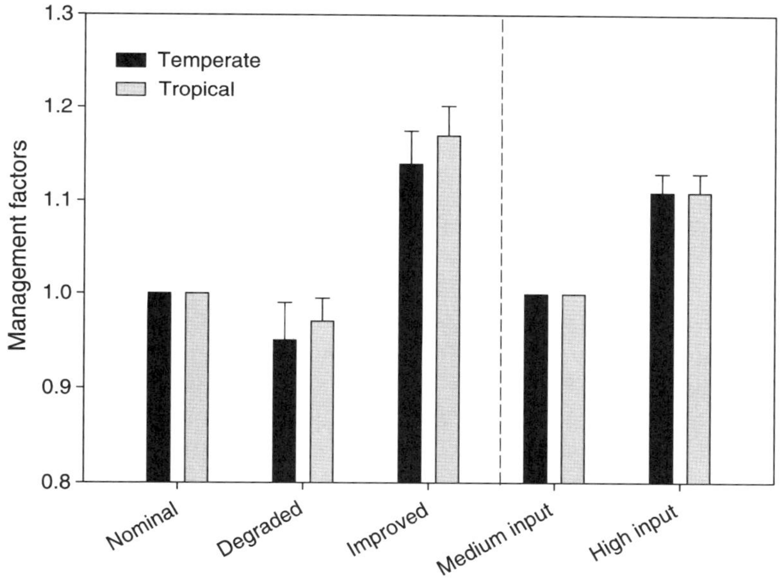

Fig. 13.3 Grassland management factors (with standard errors) for temperate and tropical regions. These factors estimate proportional C sequestration or loss (i.e., through degradation) given departure from nominal management practices. *Medium inputs* require one external inputs (e.g., fertilizer improved species, etc.) whereas *high inputs* require more than one external input. These management factors are presented as proportional increases in C stocks rather than C sequestration rates so that the factors can be applied to all soils (Figure reproduced from Ogle et al. 2004)

would be C inputs in another place (by forage or feed harvest). However, the balance between these has not been well characterized. Summary data synthesized by climate region are presented in Fig. 13.3.

Globally, an estimated 0.2–0.8 Pg CO_2 year^{-1} could be sequestered in grassland soils by 2030 given prices for CO_2 of 20–50 $US per metric ton (Smith et al. 2007). Though fertilization and fire management both could contribute to C sequestration,

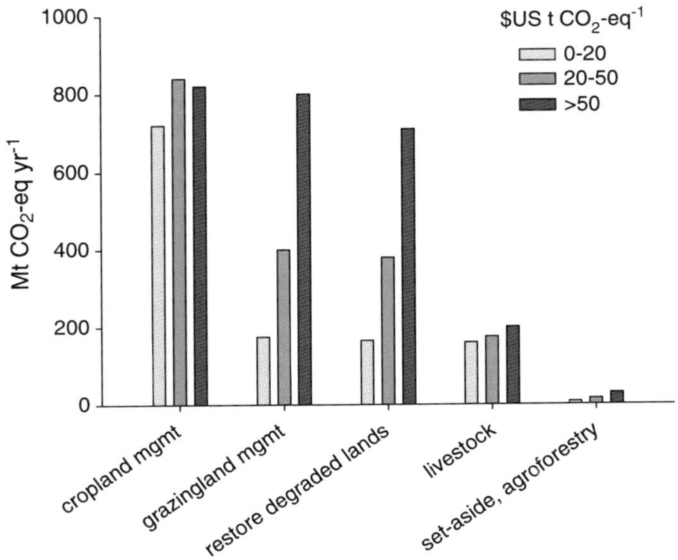

Fig. 13.4 Estimates of C sequestration potential for several mitigation measures at varying C prices (Data reproduced from Smith et al. 2007)

most of the potential sequestration in non-degraded grasslands is through changes in grazing management practices. Estimated rates of C sequestration per unit are lower than those for sequestration on agricultural land, but sequestration potential is comparable to that of croplands because grasslands cover a large portion of Earth's surface (Fig. 13.4). Nearly 270 million ha of grassland worldwide have been degraded to some degree by mismanagement (Oldeman 1994; Bridges and Oldeman 1999). Much of this land can be rehabilitated by enhancing plant productivity, capturing water resources and using them more efficiently, or improving soil fertility. Doing so could sequester about as much C as could be sequestered in grasslands (0.15–0.7 Pg CO_2 year^{-1} depending on C prices) (Smith et al. 2007).

13.3.2 Reduced Carbon Emissions Through Reduced Grassland Degradation

Grasslands contain a substantial amount of the world's soil organic carbon. Integrating data on grassland area (FAO 2009b) and grassland soil C stocks (Sombroek et al. 1993) results in a global estimate of about 343 Pg C – nearly 50% more than is stored in forest biomass worldwide (FAO 2007). Just as is the case for forest biomass C stocks, grassland soil C stocks are susceptible to loss upon conversion to other land uses (Paustian et al. 1997) or following activities that lead to grassland degradation (e.g., overgrazing). Current rates of C loss from grassland systems are not well-quantified. Over the last decade grassland area has been shrinking while arable land area has been growing, suggesting continued conversion of grassland to croplands (FAO 2009b). When grasslands are converted to agricultural land, soil C stocks tend to decline by an average of about 60% (Guo and Gifford 2002; Paustian et al. 1997). Grassland degradation has also expanded (Bai et al. 2008), likely prompting the loss of grassland soil and biomass C stocks. Halting grassland conversion and degradation would preserve grassland soil C stocks. The magnitude of the impact on atmospheric CO_2 is much smaller than that due to deforestation, but preserving grassland soil C stocks serves to maintain the productive capacity of these ecosystems that make a substantial contribution to livelihoods of many.

13.3.3 Practices That Sequester Carbon in Grasslands Often Enhance Productivity

An important argument in favor of grassland soil C sequestration is that implementation of practices to sequester C often lead to increased NPP and greater economic returns. Forage removal practices that disturb the system and prompt C losses usually reflect attempts to enhance forage utilization, but the complement is not necessarily true. Specifically, practices that sequester C do not necessarily result in reduced forage utilization. Reducing the amount of C inputs removed, or increasing production, C inputs, or belowground allocation could all lead to increasing soil C stocks (Conant et al. 2001). Grazing management can lead to decreased C removal if grazing intensities are reduced or if grazing is deferred while forage species are most actively growing (Kemp and Michalk 2007). Sustainable grazing management can, thus, increase C inputs and stocks without necessarily reducing forage production. Grazing management can also be used to restore productive forage species, also increasing C inputs and soil C stocks. Other practices that enhance production by sowing more productive species or supplying adequate moisture and nutrients, can enhance C uptake, ecosystem C stocks, and forage production (Conant et al. 2001).

Improved management techniques can increase forage production and reduce feed costs, financially benefitting producers. As forage production increases, an ancillary benefit may be increased sequestration of atmospheric CO_2. Indeed, Gifford et al. (1992) noted that improved pasture management is an important consideration when computing a national C budget. A variety of grassland management practices lead to near-term increases in both production and sequestration of C, and practices that sequester more C often enhance producer income. Practices that reduce C removal – through grazing or harvest – tend to enhance C inputs, building soil C stocks. Thus, grazing management practices that increase C inputs by increasing NPP can sequester C. Also, practices that increase production inputs by enhancing soil fertility or sowing more productive species can build soil C stocks.

Directly introducing more C to the system by organic matter (e.g., manure) additions will also lead to increased C stocks, though maintain those come at the expense of C inputs where feed crops are grown (Conant et al. 2001).

In addition to enhancing forage production and food security, many land management practices that sequester C result in changes in environmental processes that are beneficial to other reasons. For example, practices that sequester C in grassland soils tend to maximize vegetative cover, reducing wind and water-induced erosion (Follett et al. 2001b). Reducing sediment load increases water quality while reducing airborne particulate matter enhances air quality. Carbon sequestering practices can also enhance ecosystem water balance; building SOM stocks tends to enhance water infiltration and soil moisture status in arid/semiarid environments (Unger et al. 1991). In many cases practices that sequester C can lead to enhanced biodiversity (Bekessy and Wintle 2008).

Most grassland management practices with the potential to sequester C were developed to address issues other than C sequestration. For example, expanding grasslands through agricultural set asides and rehabilitating degraded rangelands are often intended to mitigate wind and water erosion (Lal 2009a). Practices that preserve habitat, like grassland preservation and rehabilitation, can preserve species and biodiversity. A variety of practices that integrate grass species into arable crop rotation (e.g., catch crops used to retain nutrients, cover crops to reduce erosion, grass crops in rotation) sequester C and also retain nutrients in agricultural systems, reducing downstream pollution (Stevens and Quinton 2009).

13.3.4 Practices That Sequester Carbon in Grasslands Can Enhance Adaptation to Climate Change

Mitigation investments are crucially important for reducing the impacts of climate change, but greenhouse gases (GHGs) concentration will continue to increase for decades despite implementation of even the most aggressive climate policies (Barker et al. 2007). Therefore, adaptation is an important response to climate change that should begin now (Barker et al. 2007). Because yield reductions under drought, heat stress, floods, and other extreme events will be the most consequential negative impacts of climate change, efforts to adapt to a changing climate should focus on increasing resilience of management systems (FAO 2008; Henry et al. 2007). Increasing the frequency of droughts in the drylands and longer-duration droughts (Thornton et al. 2008) are expected to have a substantial negative effect on the sustainability and viability of livestock production systems in semiarid regions. Grassland management practices maximize the infiltration, capture, and utilization of precipitation for production (Woodfine 2009). In cases where sustainable grazing management increases soil C stocks, soil water holding capacity increases. Both facets of enhancing water balance will increase drought resilience.

Grassland management practices that sequester C tend to make systems more resilient to climate variation and climate change. Specifically, increased SOM increases yields (Vallis et al. 1996; Pan et al. 2006), SOM also enhances soil fertility, reducing reliance on external nitrogen inputs (Lal 2009b), surface cover and mulch, and SOM decrease interannual variation in yields (Lal et al. 2007), and practices that diversify cropping systems, including grass and forage crops in rotation, sequester C and enhance yield consistency.

Agricultural practices intended to mitigate GHG emissions could increase vulnerability to climate variation and climate change if they increase energy supply from food production systems (e.g., to supply biomass energy) or prevent arable land from being cultivated (e.g., afforestation). Similarly, actions intended to foster adaptation could lead to increased emissions, e.g., increased nitrogen fertilization (and N_2O release) to enhance yields or harvest of stover for producing biofuels (Barker et al. 2007). However, practices which minimize soil disturbance, maintain ground cover, rebuild soil C stocks and related soil biological activity, diversify crops, and integrate crop/livestock production will tend to increase soil C stocks, and enhance resilience to climate change (Woodfine 2009).

13.4 Challenges to Greenhouse Gas Mitigation Through Grassland Management

13.4.1 Challenges to Developing Workable Policies and Incentives

The principle of "common but differentiated responsibilities" in the Kyoto Protocol regulates emissions on Annex I countries, but encourages developing country participation through the clean development mechanism (CDM). The current rules for the land use, land use change, and forestry projects under the CDM, adopted at the Seventh Conference of the Parties (COP7) in 2001 resulted in an agreement that permits afforestation and reforestation C offset projects in developing countries, but with complex monitoring and reporting requirements and exclusion of emissions from deforestation or credits for agricultural or grassland sequestration (Schlamadinger et al. 2007). Emissions from afforestation, reforestation, and deforestation since 1990 are reported as part of United Nations Framework Convention on Climate Change (UNFCCC) official National Communications that will determine compliance with Kyoto Protocol emission reduction targets. The CDM is designed to lower costs for achieving that goal while encouraging participation of non-Annex I countries and helping foster sustainable development (Paulsson 2009). Many developing countries strongly supported inclusion of sinks in anticipation that emission caps would substantially increase the flow of aid – in the form of emission offset projects – from developed countries (Boyd et al. 2008). Inclusion of sinks through the CDM allows participation of a wide range of actors in emission reduction efforts, but puts strict limits on only a subset of those participants. Balancing emission reductions for large emitters with mechanisms that engage small emitters remains a key component of international negotiations.

13.4.2 Demonstrating Additionality Is a Formidable Challenge

Under the Marrakesh Accords, projects that reduce GHG emissions "below those that would have occurred in the absence of the registered CDM project activity" are eligible for credit under the CDM (UNFCCC 2001). Key challenges for projects from uncapped countries – for all types of offset projects, not just sequestration projects (Reilly and Asadoorian 2007) – is proving the counter-factual, i.e., convincingly demonstrating what would have been done in the absence of C sequestration incentives (additionality – is the practice *additional* to what would have been done?) and how implementing a new practices has affected the behavior of other actors (leakage – are afforestation projects driving deforestation?). Methods of assessment have been developed (Chomitz 2002) and various rules have been proposed (Wiley and Chameides 2007) and applied (see Paulsson 2009; Palm et al. 2008) to address additionality and leakage. To-date, results of C emission offsets under the Kyoto Protocol have been mixed (Paulsson 2009). Several projects of dubious emission reduction value have been approved (Wara 2007) and few sequestration projects have been accepted. Research addressing the feasibility of the CDM continues to address this issue (Paulsson 2009).

Demonstrating additionality requires information other than sampling of biomass or soil C stocks (Conant 2011). Policies that incentivize adoption of behavioral (i.e., land management) changes are confronted by additionality and the potential for perverse incentives – which in the case of forestry and agricultural sequestration could encourage land owners to get rid of ecosystem C through tillage, fire, or harvest so that they could then be paid to re-sequester it. All policies, grants, or investments that fund or incentivize some action implicitly assume that the action would not have taken place in the absence of policy implementation. The difficulty is compounded in terrestrial C sequestration projects because the direct, human-induced changes in C stocks must be distinguished from changes in C stocks driven by natural processes (e.g., biomass C stock recovery after a fire) and indirectly by human actions (e.g., enhanced biomass C stocks driven by CO_2 fertilization or nitrogen deposition; increased soil C stocks driven by shifts in species composition) (Lovbrand 2004). In theory, such changes could be documented by sampling, but disentangling drivers of C stock changes remains challenging (Alexandrov and Yamagata 2004; Canadell et al. 2007; Smith 2005).

The anticipated low costs of grassland C sequestration are intimately intertwined with the additionality issue – if barriers (i.e., costs) are low for adopting practices that sequester C, they are more likely to be adopted in the absence of policies to promote them. Documenting changes in biomass or soil C stocks will require some kind of measurement coupled with extrapolation or interpolation (Conant et al. 2011a). These measurements differ from those required for other types of offset projects, they contribute more significantly to project costs, and economies of scale may not be as effective at reducing costs. Enacting a project in which several land owners enhance C sequestering practices would require documenting the effect of those practices (collectively or individually) on each parcel. These difficulties lie not in measuring C stocks, but in devising measurement/monitoring/verification systems that are accurate yet cost-effective (Conant et al. 2011a).

13.4.3 Carbon Sequestered in Grassland Systems Is Subject to Reversals

Disturbance can cause rapid reversals of previously sequestered C (Galik and Jackson 2009). Such disturbances can be large or small, intentional or unintentional (e.g. Page et al. 2002). The CDM has dealt with this issue by developing temporary Certified Emission Reductions (CERs) for 5 or 20 year periods (Dessai et al. 2005) while other standards reduce emission reduction credits to buffer against losses.[1] Impermanence decreases the value of sequestration projects compared with emission reduction projects and increase uncertainty and transaction costs (van Kooten 2009). Resolution of additionality, leakage, and permanence issues are critical for acceptance of reducing emission from deforestation, forest degradation (REDD) and terrestrial sequestration in a post-2012 climate agreement, as is identification of a pre-agreement baseline against which deforestation/degradation reductions can be evaluated (Karsenty 2008). There are benefits that are unique to C sequestration activities despite the fact that they are not permanent. To achieve these benefits, policies must assure accurate value of temporary C sequestration and minimizing costs associated with transactions (Marland and Marland 2009).

13.4.4 Well-Intentioned Policies Do Not Necessarily Lead to Good Practices

Scientific information is lagging behind the desire to craft robust terrestrial C sequestration policies. Some argue that there are too many uncertainties to proceed. For example, conservation tillage is one of the largest potential sources of

[1] e.g., The Voluntary Carbon Standard (http://www.v-c-s.org/)

Box 13.2 Which Grassland Management Practices Increase Carbon Stocks?

1. Grazing management can be improved to reverse practices that continually removed a very large proportion of aboveground biomass. Implementing a grazing management system that maximizes NPP rather than C removal can increase C inputs and sequester C.
2. Sowing improved species can lead to increased production through species that are better adapted to local climate, more resilient to grazing, more resistant to drought, or enhance soil fertility (i.e., nitrogen fixing crops). Enhancing production leads to greater C inputs and sequestration.
3. Direct inputs of water, fertilizer, or organic matter can enhance water and nitrogen balances, increase NPP and C inputs, potentially sequestering C. Inputs of water, nitrogen, and organic matter all tend to require energy and can each enhance fluxes of N_2O, likely offsetting C sequestration gains.
4. Restoring degraded lands enhances production in areas with low productivity, increasing C inputs, and sequestering C.
5. Including grass in rotation in arable lands can increase production, return organic matter (when grazed as a forage crop), and reduce disturbance to the soil through tillage. Thus, integrated grasses into crop rotations can enhance C inputs and reduce decomposition losses of C, each of which leads to C sequestration.

GHG mitigation within the agricultural sector (Smith et al. 2008) and, coupled with associated declines in fuel use, could make an immediate, substantial contribution to offsetting and reducing GHG emissions (Caldeira et al. 2004; CAST 2004). However, implementation of reduced- or no-tillage does not always lead to significant increases in soil C stocks (Ussiri and Lal 2009; Blanco-Canqui and Lal 2008). In some cases depletion of soil C stocks at depth offset gains in surface soils. The mechanism driving this process is not well-understood (Angers and Eriksen-Hamel 2008; Baker et al. 2007). There is also uncertainty about how practices that sequester C impact local climate through albedo and water balance (Denman et al. 2007). Practices that lead to reduced GHG concentrations could promote local warming (Chapin et al. 2008). Further, practices that sequester soil C could also lead to increased N_2O (e.g., fertilization to enhance C inputs) or CH_4 (e.g., flooding to preserve organic soils; see Box 13.2) (Schlesinger 2000). The contribution of erosion to depletion of soil C stocks, and the fate of eroded C are additional important uncertainties (Berhe et al. 2007). Finally, disturbances are stochastic and often unpreventable processes that can lead to C losses (Smith 2005) sand ecosystem and socio-economic feedbacks (i.e., leakage, unintended consequences) are capable of undermining the intended benefits of sequestration projects (Jack et al. 2008).

13.4.5 Land Tenure and Governance Issues Complicate Policy Implementation

Small-holder households represent a serious challenge for documenting C sequestration (Coomes et al. 2008) because aggregation across a variety of landowners increases monitoring transaction costs, implying that the cost-effectiveness of C sequestration projects conflicts with poverty alleviation goals (Lipper and Cavatassi 2004; Jack et al. 2008). Pastoralists occupy substantial portions of the land area in many parts of the world with potential to sequester C in grasslands, but pastoralists are often socially marginalized and with tenuous land tenure, making it very difficult for participation in C markets (Neely et al. 2009). In many of the places identified as having low-cost sequestration options, a large percentage of people make their living from the land. Compensation for foregoing land development could be financially beneficial, but may be of limited long-term development value. Uncertainty about land-tenure amongst small-holders and weak institutions are key issues that discourage potential participants from adopting C sequestering practices (Grieg-Gran et al. 2005). Also, practices that sequester C are not inherently coupled with other environmental benefits. For example, Nelson et al. (2008) found that in the Northwestern US sequestration policies did not necessarily achieve forest conservation goals and none of the conservation policies studied sequestered C. Similarly, the CDM has not yet led to forestry mitigation that successfully foster adaptation to climate change (Reyer et al. 2009).

13.4.6 Systems for Documenting Carbon Stocks Changes Have Not Been Agreed Upon

Methods for analyzing soil C concentration of a given sample are well-established and easily carried out with high precision and small analytical error (Nelson and Sommers 1996). However, soil C stocks vary as a function of soil texture, landscape position, drainage, plant productivity, and bulk density, all of which vary spatially and create heterogeneity that makes it difficult to quantify changes in soil C stocks over time (VandenBygaart 2006; Robertson et al. 1997; Cambardella et al. 1994). Sampling error can be large and "the cumulative effects of managing small net sinks to mitigate fossil-fuel emissions will have to be understood, analyzed, monitored, and evaluated in the context of larger, highly variable, and uncertain sources and sinks in the natural cycle," (Birdsey et al. 2007). Thus, the main challenge in documenting plot-level changes in soil C stocks is not with measuring C, but rather in designing an efficient, cost-effective sampling and C stock estimation system. Given higher rates of soil C sequestration, relatively low initial amounts of soil C, and modest spatial variability, the standard approach for a project – sampling and then future re-sampling of soil cores – would still require collection and analysis of dozens of soil samples to detect changes within a given field over a 5–10 year period that might be used for verification in an agricultural offset project (Conant and Paustian 2002b; Yang et al. 2008). Quantifying soil C changes at national or regional scales requires much more modest sampling densities (Makipaa et al. 2008; Saby et al. 2008), but such sampling precludes attribution of C credits to a particular practice or plot of land.

13.4.6.1 Practice-Based Estimates of Soil Carbon Sequestration

One common approach to assessing changes in soil C stocks is to use information synthesized from previously published studies on how changes in management practices impact soil C stocks. Offsets can be verified by monitoring agronomic practices (e.g., monitoring no-tillage by surveying residue coverage on the soil surface). Such verification is already established for other conservation programs and can be relatively inexpensive. Syntheses of existing field experiments (Ogle et al. 2005) provide empirical estimates of the average soil C change for a particular practice within a broad region (e.g., Fig. 13.3). However, studies of management impacts on soil C stocks are sparse enough that relying upon them for sequestration rates for a specific farm or group of farms in a given region, which are unlikely to be well-represented by published studies, will lead to substantial uncertainty. This uncertainty is difficult to quantify using statistical methods with limited data. Moreover, the rates are typically based on relative changes in soil C stock changes, which could differ from the actual rates if there are other environmental drivers, such as climate change, that are also contributing to significant changes in soil C stocks. If uncertainty is high, permitted soil C offsets may be substantially discounted relative to estimated C sequestered in order to limit the risk that the offsets do not represent net reductions in CO_2 emissions relative to the atmosphere (VCS 2008) Another limitation of a broad practice-based approach is that it is economically inefficient (Antle et al. 2003). Because of heterogeneity in the response of soils to a specific management practices (due to differences in soils, climate conditions, land use history), broadly-based payments by practice will over-compensate poorer performance and under compensate better performance (hence dis-incentivizing their participation). Thus, even if the practice-based credit was an accurate estimate for the average performance within the region, the actual benefits achieved would be overestimated, and this inefficiency would increase as a function of the degree of spatial heterogeneity in soil response (Antle et al. 2003). Hence, an estimation system that can account for more of the local variability in soil responses to a particular management practice will increase the economic efficiency of the mitigation policy and provide a better estimate of the actual mitigation benefits achieved.

13.4.6.2 Combining Measurement with Mechanistic Modeling

Terrestrial soil C offsets can be quantified using a mechanistic ecosystem model. A dynamic system comprised of a measurement database that is updated as new measured soil C offset data became available, could integrate measurements with state-of-the-art knowledge about ecosystem function and enable up-to-date calculation of model uncertainty estimates using established statistical methods (Ogle et al. 2007). A system that combines measurement of soil C with models would have a number of unique benefits not possible with modeling or measurement alone. Systems that discount or withhold reserve credits to account for uncertainty, such as the Voluntary Carbon Standard[2] could use uncertainty derived from the model analysis associated with a particular offset activity to determine reserve requirement. These systems would have the flexibility of a model-based approach, being able to account for all types of terrestrial offsets unlike the measurement approach that is likely to have gaps, but would be reliable because the associated uncertainty is determined from on-the-ground observations. As a system, such a modeling-measurement approach would be robust because it would be continually updated as new sample data are made available, and it could be used to direct sampling towards those areas where uncertainty is largest relative to offset activity. Such systems could also potentially encourage more innovation by agricultural producers because new measurements would be incorporated from the latest management options, while using the model to allow all producers to receive credit from the latest innovations without necessarily requiring new measurements on each farm. Finally, a combined system could make use of published information on how other factors (like global change, widespread land use changes, changes in land use prompted by terrestrial soil C offset programs, etc.) affect soil C stocks both on- and off-site, to account for shifting baselines, additionality, and leakage.

13.4.6.3 Data on Management Impacts on Carbon Stocks Is Limited in Developing Countries

Systems that integrate measurement and mechanistic modeling require robust sources of data that reflect the range of potential management practices. Across the developed world, a variety of efforts are underway to develop, test, and implement such systems. But all syntheses document that in the developing world, observations of management-induced changes in soil C stocks are relatively rare (Smith et al. 2006; Conant et al. 2001). Lack of accurate information can lead to greater uncertainty in estimates of soil C stock changes and could result in climate-driven bias because developed-country studies are more common in temperate regions. More importantly, practices that could be most beneficial could be left outside of schemes to encourage C sequestration because the practices are not widely familiar to the developed-world scientists and policy makers developing quantification tools. This paucity of data from developed countries presents a challenge to developing robust accounting systems that offer the same utility for quantifying soil C sequestration in developed and developing countries.

---

[2]http://www.v-c-s.org/

13.5 The Way Forward

13.5.1 Foundations for Sound Policies

Current yields and economic returns can often be maximized by practices that maximize forage harvest, deplete soil nutrients, and reduce the long-term productive capacity of grassland systems. Indeed, economic pressures to "adopt unsustainable practices as yields drop" in response to a changing climate, "may increase land degradation and resource use," (Easterling et al. 2007) further motivating support for policies and programs that encourage implementation of sustainable grassland management practices. Identifying and understanding situations in which short-term interests in harvest trump long-term interests in maintaining productive capacity, and developing technical solutions that involve research, education, and technical assistance in implementing sustainable practices should be a high priority. A key challenge is the large number of small-holders and pastoralists who may be among the hardest hit by climate change (FAO 2009a). Their challenge is often enlarged because uncertain land tenure discourages investments that pay dividends in the longer-term. Thus, efforts to spread knowledge on sustainable grassland management practices essential for ensuring successful implementation and must address tenure-related motivations to implement sustainable practices.

Not all categories of producers have the same potential for implementing sustainable land management practices, and some producers will benefit more and sooner than others. Development-mitigation-adaptation strategies must be evaluated within the framework of local environmental conditions, institutions and capacities. Priority should be given to investments in sustainable land management practices that:

- have strong evidence of enhancing near- and longer-term productivity and profitability for farmers and pastoralists
- offer opportunities to enhance production, mitigate GHG emissions, and enable adaptation to climate change
- develop incentives that foster sustainability of existing resources – soil, water, air, labor
- rehabilitate lands that can be improved with modest cost and low-tech changes in management practices
- support research and education on best practices for maintaining fertility and production
- align with existing investment programs.

Despite win-win situations in which practices that sequester C also lead to enhanced productivity and substantial biological potential to sequester c in grasslands, policies to encourage adoption of practices that sequester C in grasslands lag behind policies for forest and agricultural lands. Like forestry and agricultural sequestration, policies that promote carbon sequestration in rangelands could form an important part of a "no regrets" climate strategy. This is particularly true for practices that promote increased NPP or livestock production, and practices that halt rangeland degradation. In addition to sequestering C, implementing practices that sequester C can help achieve the strategic objectives of the UN Convention to Combat Desertification, i.e., improving livelihoods, improving productivity, and generating global benefits. Reducing emissions from grassland degradation is likely to pay dividends in maintaining soil C stocks but also in sustaining the livelihoods people making a living on grasslands.

13.5.2 Grassland Carbon Sequestration in Context

Much of the world's grassland, a disproportionately large share of the degraded grassland, and a majority of grassland sequestration potential is found in the developing world. More importantly, the fate of large portions of the populations in these areas is intimately tied to livestock production systems directly dependent upon grasslands. Sustaining productivity and rehabilitating degraded grassland systems are crucially important to people right now. It is also clear that there are synergistic effects with other development agendas. For example, Kandji and Verchot (2007) point out several ways that developing countries in semiarid East Africa will be adversely impacted by climate change and the relationship of those impacts to the Millennium Development Goals. These include reduced vulnerability to extreme events (Goal 1), enhanced environmental sustainability (Goal 7), and enhanced ability for governments to invest in key socio-economic sectors (Goals 17). Synergies between environmental, development, and agricultural activities indicate opportunities for engagement from multiple sectors.

13.5.3 Research Priorities

A key barrier to identifying priority investments is lack of knowledge on the impacts of grassland management in most of the developing world. Despite a large estimated potential in the developing world, lack of direct observations makes these estimates highly uncertain (Conant and Paustian 2002a; Ogle et al. 2004). Moreover, best management practices are typically based on those identified in other regions, limiting the breadth of management alternatives and possibly overlooking those that could do more to build or rebuild soil C stocks and enhance NPP. Efforts to build capacity while enhancing environmental benefits, such as the participatory practice capture used by the World Overview of Conservation Approaches and Technologies, can simultaneously facilitate identification and implementation of best practices. Building soil C stocks through implementation of improved/more sustainable management practices is just one component of developing more productive and efficient livestock production systems. Increasing livestock production could lead to increased CH_4 emissions, but improving feed quality by improving pasture management to produce forage with more balanced quality (Leng 1993) could concurrently sequester C and increase milk or meat production. If implemented in coordination with grazing practices that encourage consumption of a quality mixed diet, CH_4 emissions per unit product could decline. To the extent that improved technologies can facilitate better breeding, reducing the number of replacement heifers, reaching slaughter weight at an earlier age, increased milk production, higher pregnancy rates, etc. greenhouse gas emissions per unit product, despite the fact that none of those practices directly reduce emissions (Boadi et al. 2004). A systems perspective is crucial. Research to assess C sequestration alone could miss important interactions with factors that control ruminant CH_4 emissions – one of the largest sources of radiative forcing from GHG emissions in developing countries.

Successful pilot projects carried out in collaboration with national scientists, grassland managers, and development actors will be a key component of demonstrating the feasibility of new practices. At the same time, pilot projects are necessary for expansion of information on the efficacy of grassland management practices as a mitigation strategy. Understanding the institutional requirements and testing C accounting procedures are crucial next steps for legitimizing mitigation through grassland management. Investing in pilot projects will engage community leaders, farmers and other resource uses in program development and build technical, organizational, and human capacities (Pender et al. 2009). An important component of a pilot program will be conducting desk reviews and collecting additional information on current and projected GHG emissions from other grassland projects and pilot studies. Outputs from this work built around a series of pilot study programs could include:

- a comprehensive database of estimates of GHG factors by region and complete grassland emission inventory
- a focus on documenting C sequestration responses for areas or practices which are understudied, in particular in developing countries
- an analysis of different global and regional scenarios for grasslands under different C constraints (different policy measures and prices for C), financing and crediting arrangements and the development of supporting models and tools
- an analysis of the marginal costs of C sequestration in grasslands driven by changes in management practices plus their implications for food security and livelihoods
- policy and technical guidance for Nationally-Appropriate Mitigation Actions that may affect grassland production and food security
- scientific underpinning in support of international (post-Kyoto) agreements on climate change.

References

Adams HD, Guardiola-Claramonte M, Barron-Gafford GA et al (2009) Temperature sensitivity of drought-induced tree mortality portends increased regional die-off under global-change-type drought. Proc Natl Acad Sci USA 106:7063–7066

Alexandrov G, Yamagata Y (2004) Verification of carbon sink assessment: can we exclude natural sinks? Clim Chang 67(2–3):437–447

Allison FE (1973) Soil organic matter and its role in crop production, vol 3, Developments in soil science. Amsterdam, Elsevier

Angers DA, Eriksen-Hamel NS (2008) Full-inversion tillage and organic carbon distribution in soil profiles: a meta-analysis. Soil Sci Soc Am J 72:1370–1374

Antle JM, Capalbo SM, Mooney S et al (2003) Spatial heterogeneity, contract design, and the efficiency of carbon sequestration policies for agriculture. J Environ Econ Manag 46(2):231–250

Bai ZG, Dent DL, Olsson L et al (2008) Global assessment of land degradation and improvement. 1. Identification by remote sensing. ISRIC – World Soil Information, Waginengen

Baker JM, Oschsner TE, Venterea RT et al (2007) Tillage and soil carbon sequestration—what do we really know? Agric Ecosyst Environ 118(1–4):1–5

Barker T, Bashmakov I, Bernstein L et al (2007) IPCC working group III: technical summary. In: Metz B, Davidson OA, Bosch PR et al (eds) Climate change 2007: mitigation. Contribution of working group III to the fourth assessment report of the Intergovernmental Panel on Climate Change. Cambridge University Press, Cambridge

Bekessy SA, Wintle BA (2008) Using carbon investment to grow the biodiversity bank. Conserv Biol 22(3):510–513. doi:10.1111/j.1523-1739.2008.00943.x

Berhe AA, Harte J, Harden JW (2007) The significance of the erosion-induced terrestrial Carbon sink. Bioscience 57(4):337–346

Birdsey RA, Jenkins JC, Johnston M et al (2007) North American forests. In: King AW, Dilling L, Zimmerman GP et al (eds) State of the carbon cycle report. Department of Energy, Washington, DC

Blanco-Canqui H, Lal R (2008) No-tillage and soil profile carbon sequestration: an on-farm assessment. Soil Sci Soc Am J 72(3):693–701

Boadi D, Benchaar C, Chiquette (2004) Mitigation strategies to reduce enteric methane emissions from dairy cows: update review. Can J Anim Sci 84(3):319–335

Boyd E, Corbera E, Estrada M (2008) UNFCCC negotiations (pre-Kyoto to COP-9): what the process says about the politics of CDM-sinks. Intl Environ Agreem Polit Law Econ 8(2):95–112. doi:10.1007/s10784-008-9070-x

Bridges EM, Oldeman LR (1999) Global assessment of human-induced soil degradation. Arid Soil Res Rehab 13(4):319–325

Caldeira K, Granger Morgan M, Baldocchi D et al (2004) A portfolio of carbon management options. In: Field CB, Raupach MR (eds) The global carbon cycle: integrating humans, climate, and the natural world. Island Press, Washington, DC

Cambardella CA, Moorman TB, Novak JM et al (1994) Field-scale variability of soil properties in central Iowa soils. Soil Sci Soc Am J 58(5):1501–1511

Canadell JG, Kirschbaum MUF, Kurz WA et al (2007) Factoring out natural and indirect human effects on terrestrial carbon sources and sinks. Environ Sci Policy 10(4):370–384

CAST (2004) In: Paustian K, Babcock BA, Hatvield J et al (eds) Climate change and greenhouse gas mitigation: challenges and opportunities for agriculture. Council for Agricultural Science and Technology (CAST), Ames

Chapin FS III, Randerson JT, McGuire AD (2008) Changing feedbacks in the climate-biosphere system. Front Ecol Environ 6(6):313–320

Chapman DF, Lemaire G (1993) Morphogenic and structural determinants of plant regrowth after defoliation. In: Baker MJ (ed) Grasslands for our world. SIR, Wellington

Chomitz KM (2002) Baseline, leakage and measurement issues: how do forestry and energy projects compare? Clim Policy 2(1):35–49

Ciais P, Reichstein M, Viovy N et al (2005) Europe-wide reduction in primary productivity caused by the heat and drought in 2003. Nature 437:529–533

Conant RT (2011) Sequestration through forestry and agriculture. Wiley Interdiscip Rev Clim Change 2:238–254

Conant RT, Ogle SM, Paul EA et al (2011a) Measuring and monitoring soil organic carbon stocks in agricultural lands for climate mitigation. Front Ecol Environ 3(9):169–173

Conant RT, Paustian K (2002a) Potential soil carbon sequestration in overgrazed grassland ecosystems. Glob Biogeochem Cycle 16:1143
Conant RT, Paustian K (2002b) Spatial variability of soil organic carbon: Implications for detecting change at different scales. Environ Pollut 116(Supplement):127–135
Conant RT, Paustian K, Elliott ET (2001) Grassland management and conversion into grassland: Effects on soil carbon. Ecol Appl 11(2):343–355
Coomes OT, Grimard F, Potvin C et al (2008) The fate of the tropical forest: carbon or cattle? Ecol Econ 65(2):207–212. doi:10.1016/j.ecolecon.2007.12.028
DeFries RS, Field CB, Fung I et al (1999) Combining satellite data and biogeochemical models to estimate global effects of human-induced land cover change on carbon emissions and primary productivity. Glob Biogeochem Cycle 13(3):803–815
Denman KL, Brasseur G, Chidthaisong A et al (2007) Couplings between changes in the climate system and biogeochemistry. In: Solomon S, Qin D, Manning M et al (eds) Climate change 2007: the physical science basis. Fourth assessment report of the Intergovernmental Panel on Climate Change. Cambridge University Press, Cambridge/New York
Dessai S, Schipper ELF, Corbera E et al (2005) Challenges and outcomes at the ninth session of the conference of the parties to the United Nations framework convention on climate change. Int Environ Agreem Polit Law Econ 5(2):105–124
Donigian AS Jr, Barnwell TO, Jackson RB et al (1994) Assessment of alternative management practices and policies affecting soil carbon in agroecosystems of the central United States. US EPA, Athens
Easterling WE, Aggarwal PK, Batima P et al (2007) Food, fibre and forest products. In: Parry ML, Canziani OF, Palutikof JP et al (eds) Climate change 2007: impacts, adaptation and vulnerability. Contribution of working group II to the fourth assessment report of the Intergovernmental Panel on Climate Change. Cambridge University Press, Cambridge
Ellis JE, Galvin KA (1994) Climate patterns and land-use practices in the dry zones of Africa. Bioscience 44(5):340–349
FAO (2007) State of the World's forests 2007. Food and Agricultural Organization of the United Nations, Rome
FAO (2008) Climate change, water and food security. High-level conference on food security: the challenges of climate change and bioenergy. Food and Agricultural Organization of the United Nations, Rome
FAO (2009a) Agriculture and environmental challenges of the twenty-first century: a strategic approach for FAO. Food and Agricultural Organization of the United Nations, Rome
FAO (2009b) FAO STAT statistical database 2007. FAO, Rome
Follett RF, Kimble JM, Lal R (2001a) The potential of US grazing lands to sequester carbon and mitigate the greenhouse effect. CRC Press, Chelsea
Follett RF, Kimble JM, Lal R (2001b) The potential of US grazing lands to sequester soil carbon. In: Follett RF, Kimble JM, Lal R (eds) The potential of US grazing lands to sequester soil carbon. CRC Press, Chelsea
Galik CS, Jackson RB (2009) Risks to forest carbon offset projects in a changing climate. For Ecol Manage 257(11):2209–2216. doi:10.1016/j.foreco.2009.03.017
Gifford RM, Cheney NP, Noble JCR et al (1992) Australian land use, primary production of vegetation and carbon pools in relation to atmospheric carbon dioxide concentration. In: Gifford RM, Barson MM (eds) Australia's renewable resources: sustainability and global change. Australian Government Publishing Service, Canberra
Gough CM, Vogel CS, Schmid HP et al (2008) Controls on annual forest carbon storage: Lessons from the past and predictions for the future. Bioscience 58(7):609–622. doi:10.1641/b580708
Grieg-Gran M, Porras I, Wunder S (2005) How can market mechanisms for forest environmental services help the poor? Preliminary lessons from Latin America. World Dev 33(9):1511–1527. doi:10.1016/j.worlddev.2005.05.002
Guo LB, Gifford RM (2002) Soil carbon stocks and land use change: a meta analysis. Glob Change Biol 8(4):345–360

Henry B, McKeon G, Syktus J et al (2007) Climate variability, climate change, and land degradation. In: Sivakumar MVK, Ndiang'ui N (eds) Climate and land degradation. Springer, Berlin
Houghton RA (2010) How well do we know the flux of CO_2 from land-use change? Tellus B 62(5):337–351. doi:10.1111/j.1600-0889.2010.00473.x
Houghton RA, Hobbie JE, Melillo JM et al (1983) Changes in the carbon content of terrestrial biota and soils between 1860 and 1980: A net release of CO_2 to the atmosphere. Ecol Monogr 53:235–262
Jack BK, Kousky C, Sims KRE (2008) Designing payments for ecosystem services: lessons from previous experience with incentive-based mechanisms. Proc Natl Acad Sci USA 105(28):9465–9470. doi:10.1073/pnas.0705503104
Kandji ST, Verchot LV (2007) Impact of adaptation to climate variability and climate change in the East African community: A focus on the agricultural sector. World Agroforestry Centre, Nairobi
Karsenty A (2008) The architecture of proposed REDD schemes after Bali: facing critical choices. Intl For Rev 10(3):443–457
Kemp DR, Michalk DL (2007) Towards sustainable grassland and livestock management. J Agric Sci 145:543–564. doi:10.1017/s0021859607007253
Kern JS (1994) Spatial patterns of soil organic matter in the contiguous United States. Soil Sci Soc Am J 58:439–455
Kononova MM (1966) Soil organic matter. Its nature, its role in soil formation and in soil fertility, 2nd edn. Pergamon Press, Oxford
Lal R (2009a) The plow and agricultural sustainability. J Sustain Agric 33(1):66–84
Lal R (2009b) Soils and food sufficiency: a review. Agron Sustain Dev 29:113–133
Lal R, Follett F, Stewart BA (2007) Soil carbon sequestration to mitigate climate change and advance food security. Soil Sci 172(12):943–956
Le Houerou HN (1984) Rain use efficiency: a unifying concept in arid-land ecology. J Arid Environ 7:213–247
Leng RA (1993) Quantitative ruminant nutrition – a green science. Aust J Agric Res 44:363–380
Lipper L, Cavatassi R (2004) Land-use change, carbon sequestration and poverty alleviation. Environ Manage 33:S374–S387. doi:10.1007/s00267-003-9146-x
Lovbrand E (2004) Bridging political expectations and scientific limitations in climate risk management – on the uncertain effects of international carbon sink policies. Clim Chang 67(2–3):449–460
Luyssaert S, Schulze ED, Borner A et al (2008) Old-growth forests as global carbon sinks. Nature 455(7210):213–215
Makipaa R, Hakkinen M, Muukkonen P et al (2008) The costs of monitoring changes in forest soil carbon stocks. Boreal Environ Res 13:120–130
Marland G, Marland E (2009) Trading permanent and temporary carbon emissions credits. Clim Chang 95(3):465–468
Marland G, Boden TA, Andres RJ (2000) Global, regional, and national CO_2 emissions. Trends: a compendium of data on global change. Carbon Dioxide Information Analysis Center, Oak Ridge National Laboratory, Oak Ridge
Milchunas DG, Lauenroth WK (1993) Quantitative effects of grazing on vegetation and soils over a global range of environments. Ecol Monogr 63(4):327–366
Miller RW, Donahue RL (1990) Soils, 6th edn. Prentice Hall, Englewood Cliffs
Neely C, Bunning S, Wilkes A (2009) Review of evidence on drylands pastoral systems and climate change: implications and opportunities for mitigation and adaptation. Food and Agricultural Organization of the United Nations, Rome
Nelson DW, Sommers LE (1996) Total carbon, organic carbon, and organic matter. In: Sparks DL (ed) Methods of soil analysis. Part 3: chemical methods. Soil Science Society of America, Madison
Nelson E, Polasky S, Lewis DJ et al (2008) Efficiency of incentives to jointly increase carbon sequestration and species conservation on a landscape. Proc Natl Acad Sci USA 105(28):9471–9476. doi:10.1073/pnas.0706178105

Ogle SM, Conant RT, Paustian K (2004) Deriving grassland management factors for a carbon accounting method developed by the intergovernmental panel on climate change. Environ Manage 33(4):474–484

Ogle SM, Breidt FJ, Paustian K (2005) Agricultural management impacts on soil organic carbon storage under moist and dry climatic conditions of temperate and tropical regions. Biogeochemistry 72(1):87–121

Ogle SM, Breidt FJ, Easter M et al (2007) An empirically based approach for estimating uncertainty associated with modeling carbon sequestration in soils. Ecol Model 205:453–463

Ojima DS, Parton WJ, Schimel DS et al (1993) Modeling the effects of climatic and CO_2 changes on grassland storage of soil C. Water Air Soil Pollut 70(1–4):643–657

Oldeman LR (1994) The global extent of soil degradation. In: Greenland DJ, Szabolcs I (eds) Soil resilience and sustainable land use. CAB International, Wallingford

Olofsson J, Hickler T (2008) Effects of human land-use on the global carbon cycle during the last 6,000 years. Veg Hist Archaeobot 17:605–615. doi:10.1007/s00334-007-0126-6

Page SE, Siegert F, Rieley JO et al (2002) The amount of carbon released from peat and forest fires in Indonesia during 1997. Nature 420:61–65

Palm M, Ostwald M, Reilly J (2008) Land use and forestry based CDM in scientific peer-reviewed literature pre-and post-COP 9 in Milan. Intl Environ Agreem Polit Law Econ 8(3):249–274

Pan Y, Birdsey RA, Hom J et al (2006) Improved satellite estimates of net primary productivity from MODIS satellite data at regional and local scales. Ecol Appl 16(1):125–132

Paulsson E (2009) A review of the CDM literature: from fine-tuning to critical scrutiny? Intl Environ Agreem Polit Law Econ 9(1):63–80. doi:10.1007/s10784-008-9088-0

Paustian K, Collins HP, Paul EA (1997) Management controls on soil carbon. In: Paul EA, Paustian K, Elliot ET et al (eds) Soil organic matter in temperate agroecosystems. CRC Press, Boca Raton

Pender J, Ringler C, Magalhaes M, Place F (2009) The role of sustainable land management for climate change adaptation and mitigation in sub-Saharan Africa. The World Bank, Washington, DC

Ramankutty N, Evan AT, Monfreda C et al (2008) Farming the planet: 1. Geographic distribution of global agricultural lands in the year 2000. Glob Biogeochem Cycle 22:GB1003. doi:1010.1029/2007GB002952

Randerson JT, Chapin FS III, Harden JW et al (2002) Net ecosystem production: a comprehensive measure of net carbon accumulation by ecosystems. Ecol Appl 12(4):937–947

Reid R, Thornton P, McCrabb G et al (2004) Is it possible to mitigate greenhouse gas emissions in pastoral ecosystems of the tropics? Environ Dev Sustain 6(1):91–109

Reilly JM, Asadoorian MO (2007) Mitigation of greenhouse gas emissions from land use: creating incentives within greenhouse gas emissions trading systems. Clim Chang 80(1–2):173–197. doi:10.1007/s10584-006-9151-1

Reyer C, Guericke M, Ibisch PL (2009) Climate change mitigation via afforestation, reforestation and deforestation avoidance: and what about adaptation to environmental change? New For 38(1):15–34. doi:10.1007/s11056-008-9129-0

Robertson GP, Klingensmith KM, Klug MJ et al (1997) Soil resources, microbial activity, and primary production across an agricultural ecosystem. Ecol Appl 7(1):158–170

Saby NPA, Bellamy PH, Morvan X et al (2008) Will European soil-monitoring networks be able to detect changes in topsoil organic carbon content? Glob Chang Biol 14(10):2432–2442

Schlamadinger B, Bird N, Johns T et al (2007) A synopsis of land use, land-use change and forestry (LULUCF) under the Kyoto Protocol and Marrakech Accords. Environ Sci Policy 10(4):271–282. doi:10.1016/j.envsci.2006.11.002

Schlesinger WH (1977) Carbon balance in terrestrial detritus. Ann Rev Ecol Systemat 8:51–81

Schlesinger WH (2000) Carbon sequestration in soils: some cautions amidst optimism. Agric Ecosyst Environ 82(1–3):121–127

Shevliakova E, Pacala SW, Malyshev S et al (2009) Carbon cycling under 300 years of land use change: Importance of the secondary vegetation sink. Glob Biogeochem Cycle 23:16. doi:Gb202210.1029/2007gb003176

Smith P (2005) An overview of the permanence of soil organic carbon stocks: influence of direct human-induced, indirect and natural effects. Eur J Soil Sci 56(5):673–680. doi:10.1111/j.1365-2389.2005.00708.x
Smith P, Falloon P, Coleman K et al (1999) Modeling soil carbon dynamics in tropical ecosystems. In: Lal R, Kimble J, Levine E et al (eds) Global climate change and tropical ecosystems, Advances in soil science. CRC Press, Boca Raton
Smith P, Martino D, Cai Z et al (2006) Policy and technological constraints to implementation of greenhouse gas mitigation options in agriculture. Agric Ecosyst Environ 118(1–4):6–28
Smith P, Martino D, Cai Z et al (2007) Agriculture. In: Metz B, Davidson OR, Bosch PR et al (eds) Climate change 2007: mitigation. Contribution of working group III to the fourth assessment report of the Intergovernmental Panel on Climate Change. Cambridge University Press, Cambridge
Smith P, Martino D, Cai Z et al (2008) Greenhouse gas mitigation in agriculture. Philos Trans R Soc B Biol Sci 363(1492):789–813
Sombroek WG, Nachtergaele FO, Hebel A (1993) Amounts, dynamics and sequesteting of carbon in tropical and subtropical soils. Ambio 22(7):417–426
Steinfeld H, Gerber P, Wassenaar T et al (2006) Livestock's long shadow: environmental issues and options. Food and Agricultural Organization of the United Nations, Rome
Stephens BB, Gurney KR, Tans PP et al (2007) Weak northern and strong tropical land carbon uptake from vertical profiles of atmospheric CO_2. Science 316(5832):1732–1735
Stevens CJ, Quinton JN (2009) Diffuse pollution swapping in arable agricultural systems. Crit Rev Environ Sci Technol 39(6):478–520. doi:10.1080/10643380801910017
Tate RL (1987) Soil organic matter. Biological and ecological effects. Wiley, New York
Thornton PK, Boone RB, Galvin KA (2007) Coping strategies in livestock-dependent households in east and southern Africa: A synthesis of four case studies. Hum Ecol 35:461–476
Thornton PK, van de Steeg J, Notenbaert A et al (2008) The livestock-climate-poverty nexus. International Livestock Research Institute, Nairobi, p 76
Thornton PK, van de Steeg J, Notenbaert A et al (2009) The impacts of climate change on livestock and livestock systems in developing countries: A review of what we know and what we need to know. Agricult Syst 101:113–127
UNFCCC (2001) Report to the conference of the parties on its seventh session. Part Two: action taken by the conference of the parties. United Nations Framework Convention on Climate Change, Marrakesh
Unger PW, Stewart BA, Parr JF et al (1991) Crop residue management and tillage methods for conserving soil and water in semiarid regions. Soil Tillage Res 20(2–4):219–240
Ussiri DAN, Lal R (2009) Long-term tillage effects on soil carbon storage and carbon dioxide emissions in continuous corn cropping system from an alfisol in Ohio. Soil Tillage Res 104(1):39–47
Vallis I, Parton WJ, Keating BA et al (1996) Simulation of the effects of trash and N fertilizer management on soil organic matter levels and yields of sugarcane. Soil Tillage Res 38(1–2):115–132
van Kooten GC (2009) Biological carbon sequestration and carbon trading re-visited. Clim Chang 95:449–463. doi:10.1007/s10584-009-9572-8
VandenBygaart AJ (2006) Monitoring soil organic carbon stock changes in agricultural landscapes: issues and a proposed approach. Can J Soil Sci 86(3):451–463
VCS (2008) Voluntary Carbon Standard. www.v-c-s.org/, www.carbonfootprint.com/vsc.html
Verstraete MM, Scholes RJ, Stafford Smith M (2007) Climate and desertification: looking at an old problem through new lenses. Front Ecol Environ 8(7):421–428
Wara M (2007) Is the global carbon market working? Nature 445:595–596
West TO, Post WM (2002) Soil organic carbon sequestration rates by tillage and crop rotation: a global data analysis. Soil Sci Soc Am J 66(6):1930–1946
Wiley Z, Chameides W (2007) Harnessing farms and forests in the low-carbon economy. Duke University Press, Durham

Wilts AR, Reicosky DC, Allmaras RR et al (2004) Long-term corn residue effects: harvest alternatives, soil carbon turnover, and root-derived carbon. Soil Sci Soc Am J 68(4):1342–1351

Woodfine A (2009) The potential of sustainable land management practices for climate change mitigation and adaptation in sub-Saharan Africa. Food and Agricultural Organization of the United Nations, Rome

Yang XM, Drury CF, Wander MM et al (2008) Evaluating the effect of tillage on carbon sequestration using the minimum detectable difference concept. Pedosphere 18(4):421–430

Chapter 14
Cropland Soil Carbon Dynamics

Klaus Lorenz and Rattan Lal

Abstract Humans began thousands of years ago to cultivate land for growing crops after clearing the previous vegetation cover and plowing the soil. The soil disturbance altered soil carbon (C) dynamics which has been recently exacerbated by the increase in crop intensification (i.e., fertilization, irrigation, mechanization). For example, conversion to croplands may release up to 36% of soil organic carbon (SOC) to 27-cm depth in temperate regions, and up to 30% of SOC to 48-cm depth in tropical regions. In 2000, about 12% of Earth's ice-free land surface or 15 million km^2 were covered by croplands. Climate, geology and land and crop management practices control the size of the cropland soil C pool. A major fraction (25–70%) of the carbon dioxide (CO_2) fixed during plant photosynthesis in croplands by gross primary production (GPP) is respired autotrophically (R_a) back to the atmosphere. Globally, cropland GPP is about 14.8 Pg C year^{-1} (1 Pg = 10^{15} g). The remaining net primary production (NPP = GPP–R_a) is the main natural C input into cropland soils aside addition of manure and organic residues. Cropland NPP includes the production of biomass in foliage, shoots and roots, weed and seed production, root exudation, the C transfer to microorganisms that are symbiotically associated with roots, and the volatile organic carbon (VOC) emissions that are lost from leaves to the atmosphere. NPP enters soil by rhizodeposition and decomposition of plant litter but the major fraction is heterotrophically converted back to CO_2 by soil respiration and some lost as methane (CH_4). Aside decomposition, C losses from croplands

K. Lorenz (✉)
Global Soil Forum, IASS Institute for Advanced Sustainability Studies,
Berliner Strasse 130, 14467 Potsdam, Germany
e-mail: klaus.lorenz@iass-potsdam.de

R. Lal
Carbon Management and Sequestration Center, The Ohio State University,
2021 Coffey Road, Kottmann Hall 422b, Columbus, OH 43210, USA
e-mail: lal.1@osu.edu

R. Lal et al. (eds.), *Recarbonization of the Biosphere: Ecosystems and the Global Carbon Cycle*, DOI 10.1007/978-94-007-4159-1_14,

occur also by fire, erosion, leaching, and most importantly harvest removing about 2.2 Pg C $year^{-1}$ in the 1990s. Thus, a small amount of fixed C remains in cropland soils and accumulates in the SOC pool due to a combination of short- and long-term stabilization processes. Stabilization processes include physical protection of organic matter (OM) against decomposers and their enzymes, stabilization by organomineral complexes and organo-metal interactions, and some as biochemically recalcitrant black carbon (BC). Soil aggregation, in particular, may be the most important stabilization process in cropland topsoils. Site-specific factors including climate, physicochemical characteristics, soil and vegetation management determine the balance between C input and losses. Cropland soils can be recarbonized to some extent through adoption of recommended management practices (RMPs) such as conservation tillage, residue mulching and use of cover crops, practices which all contribute to soil C accumulation and sequestration by an additional transfer of C from the atmosphere to the soil. Whether cultivation of SOC-accreting crops can also contribute to the recarbonization of cropland soils needs additional research.

Keywords Soil organic carbon • Gross primary production • Net primary production • Recommended management practices

Abbreviations

AM	Arbuscular mycorrhiza
AUR	Acid-unhydrolyzable residue
BC	Black carbon
BIO	Microbial biomass
CQT	Carbon quality-temperature
DIC	Dissolved inorganic carbon
DOC	Dissolved organic carbon
DPM	Decomposable plant material
ECM	Ecto-mycorrhiza
ERM	Ericoid mycorrhiza
EU	European Union
FAO	Food and Agriculture Organization of the United Nations
GPP	Gross primary production
HI	Harvest index
HUM	Humified organic matter
NBP	Net biome production
NPP	Net primary production
NT	No-tillage
OM	Organic matter
PT	Plow tillage
PTF	Plant functional type
R_a	Autotrophic respiration

RMP	Recommended management practices
RothC	Rothamsted carbon model
RPM	Resistant plant material
SIC	Soil inorganic carbon
SOC	Soil organic carbon
SOM	Soil organic matter
UK	United Kingdom
USA	United States of America
VOC	Volatile organic carbon

14.1 Introduction

The global cropland area has increased dramatically as crops have been cultivated by humans for thousands of years on land reclaimed by converting forests, savannas and grasslands causing the largest emissions of carbon (C) from land-use change (Houghton 2010; Pielke et al. 2011). Predominantly in Asia, forests have been cleared for croplands, and a substantial amount of savanna and grasslands have been converted to croplands in North America, Africa, and the Former Soviet Union (Ramankutty et al. 2008). About 80% of new croplands are currently replacing forests in the tropics (Gibbs et al. 2010). Between 1985 and 2005, global cropland area increased by only about 2.4% but harvested area by about 7% (Foley et al. 2011). This cropland intensification is associated with large increases in energy, water and fertilizer consumption, and considerable losses in biodiversity (Foley et al. 2005). In 2000, about 12% of Earth's ice-free land surface or 15 million km^2 were covered by croplands (Ramankutty et al. 2008). The greatest cropland areas were located in South Asia, the Former Soviet Union and Tropical Africa (2.22, 2.07 and 1.94 million km^2, respectively). Cropland area can also be distinguished by biomes such as residential irrigated (i.e., irrigated cropland with substantial human populations), residential rainfed mosaic (i.e., mix of trees and rainfed cropland with substantial human populations), populated irrigated, populated rainfed and remote croplands (Ellis and Ramankutty 2008). In 2008, residential, rainfed mosaic was by far the most extensive among the cropland biomes with 4 million km^2 covered by crops, most abundant in Africa and Asia. Areas covered by crops in the other cropland biomes were 2.2, 0.97, 0.51, and 0.25 million km^2 in populated rainfed, residential irrigated, remote and populated irrigated cropland biomes, respectively (Ellis and Ramankutty 2008).

In 2000, the globally harvested areas for cereals, oil crops and forage were about 6.6, 1.8, and 1.4 million km^2, respectively (Monfreda et al. 2008) Annual grains were planted on 75% of the global cropland area. The largest cereal-harvested areas were located in Asia (2.7 million km^2), and Europe and the Former Soviet Union (in total 1.3 million km^2). On about 0.7 million km^2 oil crops were harvested in Asia, and on similar-sized area forage in combined Europe and the Former Soviet Union. Perennial crops can be distinguished from annual crops based on longevity.

Specifically, crops that live for more than one year under typical cultivation practices can be classified as perennials. Distinguishing between annual and perennial crop area is important with regard to soil organic carbon (SOC) sequestration. Specifically, perennial crops allocate a higher proportion of photosynthetically-fixed C belowground compared to annual crops by maintaining a permanent vegetation cover and a high root turnover with a high SOC input (Don et al. 2011). Further, compared with annual counterparts, perennial crops tend to have longer growing seasons and deeper rooting depths, and intercept, retain, and utilize more precipitation (Glover et al. 2010). About 1.8 million km^2 of harvested crop area in 2000 was under perennial orchards, grasses and pastures (Monfreda et al. 2008). High proportions of perennial crops were located in the tropics, particularly in Indonesia, Thailand, Malaysia, and Papua New Guinea. Often perennial crops produce high-value products like coffee (*Coffea arabica* L.), cocoa (*Theobroma cacao* L.), fruit, oil palm (*Elaeis guineensis* or *E. oleifera*), and nuts for export.

For the assessment of SOC sequestration potential it is useful to classify the cropland also based on other plant functional types (PFTs) aside annual/perennial PFTs (Smith et al. 1993). For example, the crop life form or physiognomy classification herbaceous (forbs and graminoids) dominated 91% of all harvested crop area in 2000 (12.3 million km^2; Monfreda et al. 2008). Annual crops, in particular, are exclusively herbaceous. The remaining 9% nonherbaceous crop area were under the life forms wood perennial shrubs and trees (0.4 and 0.8 million km^2, respectively). Shrubs were predominant in the coffee and cocoa growing regions of western and eastern Africa, the Central American countries south of Mexico, the eastern coast of Brazil, northern South America in Ecuador, Colombia, and Venezuela, and pockets of southeast Asia. Grapes (*Vitis* spp.) and berry bearing shrubs covered small areas in temperate countries. Further, herbaceous crops occupied much more area than either shrub or tree crops throughout the temperate zones. Although fruit bearing orchards were present in the humid tropics, coconuts (*Cocos nucifera* L.) and oil palm dominated. Specifically, Indonesia cultivated one quarter of the global coconuts area, and Indonesia and Malaysia together grew one half of all oil palm trees in 2000 (Monfreda et al. 2008).

Classification of croplands with respect to SOC sequestration is also possible by distinguishing crop functional types C_3 and C_4 based on photosynthesis pathways. Specifically, the proportion of C_3 and C_4 vegetation in a region affects the flux of carbon dioxide (CO_2) between plants and the atmosphere (Still et al. 2003). In 2000, the C_4 crops corn (*Zea mays* L.), sorghum [*Sorghum bicolor* (L.) Moench], millet, sugar cane (*Saccharum* spp.) and some grasses comprised 3.2 million km^2 or a disproportionate 24% of all harvested area (Monfreda et al. 2008). The greatest proportion of C_4 crops was located in central Mexico through the Yucatan Peninsula, the eastern portion of southern Africa extending from South Africa to Mozambique and Zambia, and a huge swath of the Sahel running from Mauritania in the west to Ethiopia in the east (Monfreda et al. 2008). Further, at least 50% of all crops planted in the U.S. Corn Belt and of the crops grown in northern Argentina and southern Brazil were C_4 crops. In these regions, corn is grown in rotation with soybeans [*Glycine max* (L.) Merr.] and fodder grass to produce animal feedstock.

The C_4 crops are, in particular, critical in arid, agriculturally marginal regions (Monfreda et al. 2008).

Legumes are the second most important source for nitrogen (N) in the terrestrial biosphere contributing about 40 Tg N year^{-1} (1 Tg = 10^{12} g), or about half of the amount released through synthetic fertilizer (Monfreda et al. 2008). In 2000, leguminous crops occupied 2.4 million km^2 or 18% of the global harvested crop area which included all pulses plus groundnuts, soybeans, alfalfa (*Medicago sativa* L.), and other leguminous forage crops, and half of the area of mixed grasses and legumes. Specifically, soybeans and alfalfa covered much of the central U.S., the region near the border of Argentina and Brazil, and southern Australia. Soybeans and alfalfa were primary protein source for livestock in these regions and were grown in rotation with corn, which supply livestock with the bulk of their energy requirements (Monfreda et al. 2008).

14.2 Cropland Soil Organic Carbon Pool

The fate of C recently fixed during plant photosynthesis (i.e., gross primary production or GPP) is the major determinant of the cropland SOC pool (Sanderman et al. 2010). A major fraction of GPP (25–70%) is expended during plant autotrophic respiration (R_a; Lambers et al. 2005). Only the fraction remaining after accounting for C losses by R_a (i.e., the net primary production or NPP = GPP – R_a) is stored in new plant biomass. Before reaching the soil, crop NPP is lost by biomass removal during harvest (i.e., grains, pulses, and other harvestable products). For example, 30–50% of the aboveground dry mass is harvested from cereal croplands (Johnson et al. 2006). In the 1990s, global crop harvest was about 2.2 Pg C year^{-1} (1 Pg = 10^{15} g) (Bondeau et al. 2007). Herbivory by insects and mammals, and emissions of volatile organic compounds (VOCs) also contribute to biomass loss from croplands (Ciais et al. 2010). And some cropland C is lost by fire, erosion and leaching (Torn et al. 2009). For example, about 1.6 kg CO_2 is emitted per kg of crop residues burned (Andreae 2004). Globally, about 1,200 Tg of crop residues was burnt in the late 1990s. Especially in the tropics, the ignition of fires is a common human activity to clear land for shifting cultivation by converting forests into croplands (Crutzen and Andreae 1990). Also, dry vegetation is removed by human-induced fires to promote crop productivity. However, burning of crop residues in developed countries has been greatly reduced during the industrial era due to the large-scale expansion of intensive cropping and active fire management (Marlon et al. 2008). Still being burned are crop residues in developing countries but comprehensive contemporary global data on crop burning are lacking (Crutzen and Andreae 1990; Pechony and Shindell 2010).

Some of the fixed C in non-erosional croplands remaining after accounting for the losses by fire, harvest, herbivory and VOCs is deposited as above- and belowground residues. Aboveground but more importantly belowground residues are SOC sources (Rasse et al. 2005). Other belowground input aside residues is GPP transferred into the soil as root and mycorrhizal exudates including turnover of fine roots (i.e., <2 mm in diameter) and fungal hyphae, and as other dead microbial cells. Transfer of crop C into soils occurs also by the activity of the mesofauna (e.g., earthworms), hydrological flows of soluble C, and root senescence and mortality. About 50–70% of C fixed in croplands remains aboveground, and 20–30% is transferred belowground (Johnson et al. 2006). Aside by natural processes, organic C is also entering cropland soils by plant and soil management (i.e., addition of manure, and of non-harvested and non-burned residues).

The soil C inputs are substrate for decomposition and subject to losses by heterotrophic, primarily microbial respiration. The remaining SOC pool consists of organic C compounds in plant, microbial and faunal residues at various stages of decomposition with residence times ranging from days to millennia (Sanderman et al. 2010). Natural processes, environmental factors (i.e., climate, soil parent material), and vegetation and soil management (i.e., crop type, fertilization, irrigation, harvest, residue management, tillage) determine amount and composition of the SOC pool. The persistence of SOC is largely due to complex interactions between organic matter (OM) and its environment, such as the interdependence of compound chemistry, reactive mineral surfaces, climate, water availability, soil acidity, soil redox state and the presence of potential degraders in the immediate microenvironment (Schmidt et al. 2011). Among natural processes affecting the SOC pool in croplands are (i) C allocation and partitioning among plant organs (i.e., leaves, stems, roots), (ii) C fluxes within plant organs (i.e., to respiration, storage compounds, defensive compounds, structural components) and (iii) C fluxes among soil pools (U.S. Department of Energy 2008). The fraction of organic C accumulating in croplands (i.e., net biome production or NBP) which enters the soil (NBP_{soil}) primarily determines the amount of SOC sequestered (Schulze et al. 2010).

14.2.1 Cropland Soil Carbon Input

The SOC pool is primarily derived from photosynthetically fixed C (Sanderman et al. 2010). Thus, cropland GPP or the annual photosynthetic C uptake of all plant tissues (i.e., crops, weeds) over a specified cropland area determines the potential C input into cropland soils (U.S. Department of Energy 2008). Globally, cropland GPP is about 14.8 Pg C $year^{-1}$ (Beer et al. 2010). From a crop production perspective, photosynthesis includes all the events from light interception to the export of photosynthate for biomass accumulation and grain production (Murchie et al. 2009). Thus, GPP is the basis for food, fiber and biofuel production, and one of the major processes providing the capacity of croplands to partly offset anthropogenic CO_2 emissions (Beer et al. 2010). The net C stored in croplands as new plant material before harvest and other losses is the NPP. The portion of crop NPP at non-erosional croplands remaining after accounting for losses by harvest, herbivory and VOCs is deposited as above- and belowground residues. Decomposition of aboveground residues but more important of belowground residues together with root exudates, and biotic and hydrologic belowground C transfer are the major natural cropland soil C input processes (Lorenz and Lal 2005). Addition of manure and residues during land use and soil management may result in additional direct C input into cropland soils.

14.2.1.1 Natural Input Processes

The crop residue input depends on the biomass that grows from a unit input of solar radiation (Amthor 2010). The crop functional types C_3 and C_4 differ in their efficiency in converting solar energy into biomass. Specifically, crops using the C_4 photosynthetic pathway are generally more efficient than C_3 crops but this difference varies with solar radiation, temperature and water supply (Jansson et al. 2010; Sanderman et al. 2010). The reduced photosynthetic efficiency of C_3 crops arises, in particular, from reduced performance (Murchie et al. 2009). However, in spite of the greater capacity and water use efficiency of the C_4 photosynthetic pathway C_4 crops may be equally or even more sensitive to water stress than C_3 crops (Ghannoum 2009). Many economically important agricultural crops such as wheat (*Triticum* L.), rice (*Oryza sativa*) and soybean are C_3 crops (Schulze et al. 2005). The C_4 photosynthetic pathway occurs in the economically important crops such as corn, sugar cane, millet and sorghum.

Gross Primary Production

The GPP is controlled by leaf area, nitrogen (N) supply, season, temperature, light and atmospheric CO_2 concentration (Chapin et al. 2002). Differences in annual GPP among croplands depend primarily on the quantity of leaf area and the length of time this leaf area is photosynthetically active which are both ultimately determined by the interacting effects of soil resources, climate, vegetation, and disturbance regime (Falge et al. 2002).

Previously, plant- and stand-level GPP was calculated as two times biomass production with considerable variation among biomes and sites (Beer et al. 2010). However, the assumption that the NPP/GPP ratio is consistent regardless of ecosystem type has been challenged (Zhang et al. 2009). For example, under the assumption that NPP/GPP equals 2, global cropland GPP is estimated to be 8.2 Pg C $year^{-1}$, but to be 14.8 Pg C $year^{-1}$ based on eddy covariance flux data and diagnostic models (Saugier et al. 2001; Beer et al. 2010). GPP can be measured but is generally estimated from simulation models as it is impossible to measure the net C exchange of all leaves of a cropland in isolation from other components (e.g., soil respiration; Chapin et al. 2002). Soil respiration is the flux of CO_2 from the soil surface to the atmosphere (Bond-Lamberty and Thomson 2010). Thus, estimations of GPP are based on (i) observations, (ii) diagnostic modeling approaches, (iii) process-oriented models and (iv) satellite-data based models (Yang et al. 2007; Beer et al. 2010). Field studies, for example, use tower eddy covariance systems to calculate seasonal and inter-annual dynamics of cropland GPP (Peng and Gitelson 2011). Some examples of estimated cropland GPP for different time periods, regions and crop species are given in the following section.

Global mean annual cropland GPP for the period 2000–2005 was estimated to be 765 gC m^{-2} $year^{-1}$ based on the first continuous satellite-derived dataset monitoring global vegetation productivity (Zhao et al. 2010). For the period 1982–2004, global crop GPP simulated by a terrestrial biosphere model ranged from 340 to 788 gC m^{-2} $year^{-1}$ (Bonan et al. 2011). For EU-25 (i.e., Austria, Belgium, Cyprus, Czech Republic, Denmark, Estonia, Finland, France, Germany, Greece, Hungary, Ireland, Italy, Latvia, Lithuania, Luxembourg, Malta, Netherlands, Poland, Portugal, Slovakia, Slovenia, Spain, Sweden and UK), mean cropland GPP was estimated to be 1,120 gC m^{-2} $year^{-1}$ for the period 2000–2005 (Schulze et al. 2009). Schulze et al. (2009) used a combination of top-down estimates based on atmospheric observations and bottom-up estimates derived from ground-based measurements. Based on a process-oriented and a remote sensing model, cropland GPP for EU-25 was 1,360 and 879 gC m^{-2} $year^{-1}$ over the period 1990–1999, respectively (Ciais et al. 2010). By applying a NPP/GPP ratio of 0.55, cropland GPP of Europe excluding Russia, Belarus and Ukraine was estimated to be 1,591 gC m^{-2} $year^{-1}$ (calculated based on Beer et al. 2007).

Seasonal changes in crop GPP can be characterized by CO_2 flux measurements. For mixed croplands, CO_2 fluxes within a season were as low as 0.2 μmol CO_2 m^{-2} s^{-1} and as high as 33.5 μmol CO_2 m^{-2} s^{-1} (Falge et al. 2002). The maximum flux was higher for C_4 than that for C_3 crops (60.5 *vs.* 27.3 μmol CO_2 m^{-2} s^{-1}). Estimated annual GPP based on these measurements were 599 gC m^{-2} $year^{-1}$ for C_3 crops, 1,101–1,396 gC m^{-2} $year^{-1}$ for mixed croplands, and 1,471 gC m^{-2} $year^{-1}$ for C_4 crops. Often, a second maximum of CO_2 emissions developed after harvest due to photosynthetic activity of inter-crops or weeds (Falge et al. 2002).

Over 6 years, GPP of irrigated and rainfed corn-soybean cropping systems were estimated based on continuous CO_2 measurements in eastern Nebraska, USA (Suyker and Verma 2010). Peak daily GPP values for irrigated and rainfed corn, and for irrigated and rainfed soybean were 24.9 and 22.9 gC m^{-2} day^{-1}, and 15.4 and 14.4 gC m^{-2} day^{-1}, respectively. Mean growing season totals of GPP for irrigated corn and soybean were 1,738 and 996 gC m^{-2}, respectively. Corresponding values for rainfed corn and soybean were 1,553 and 895 gC m^{-2}, respectively (Suyker and Verma 2010). Carbon fluxes were also measured in the North China Plain in a typical irrigated wheat/corn rotation cropland during four seasons (Lei and Yang 2010). The seasonal GPP for wheat ranged between 782 and 1,114 gC m^{-2}, and for corn between 872 and 880 gC m^{-2}, respectively. The annual GPP values for this rotational cropland were 2,008 gC m^{-2} $year^{-1}$ for the period 2006–2007, and 1,668 gC m^{-2} $year^{-1}$ for the period 2007–2008 (Lei and Yang 2010). In a corn cropland in Northeast China, GPP was simulated for a 3 years period using (i) a photosynthesis model and (ii) estimated based on tower CO_2 flux data (Wang et al. 2010). Annual simulated GPP in 2004, 2005 and 2006 were 310, 464 and 360 gC m^{-2} $year^{-1}$, respectively. In contrast, annual estimated GPP were 392, 504 and 437 gC m^{-2} $year^{-1}$, respectively, for the same time period (Wang et al. 2010).

For a winter wheat/summer maize cropping system in China, GPP varied between 1,370 and 1,531 g C m^{-2} $year^{-1}$ during 2 years (Yu et al. 2006). Depending on crop development stage GPP for winter wheat at a site in Belgium varied between 0.92 and 18.72 g C m^{-2} day^{-1} during three growing seasons (Dufranne et al. 2011). Growing season GPP at this site was as low as 1,568 g C m^{-2} and as high as 1,716 g C m^{-2}. Satellite-based remote sensing technologies can also be used to characterize GPP of croplands during the growing season. For example, Peng et al. (2011) used chlorophyll-related vegetation indices to estimate corn GPP over a period of 8 years at eight rainfed and irrigated sites in Nebraska, USA. GPP followed seasonal changes in chlorophyll content. Daytime growing season GPP varied strongly between 0 and 35 g C m^{-2} day^{-1} depending on weather conditions and growth stage with mean values between 10.4 and 15.1 g C m^{-2} day^{-1} (Peng et al. 2011).

Net Primary Production

The bulk of NPP in croplands is allocated to the production of biomass in foliage, shoots and roots. Cropland NPP includes also weed and seed production, root exudation (i.e., the soluble organic compounds that diffuse or are secreted by roots into the soil), the C transfer to microorganisms that are symbiotically associated with roots (e.g., mycorrhizae and N-fixing bacteria), and the VOC emissions that are lost from leaves to the atmosphere (Chapin et al. 2002; Ciais et al. 2010). Further, NPP includes also the biomass removed by herbivory often accounting for 5–10% of NPP. The processes of NPP loss are among the reasons why direct measurements of cropland NPP are not possible as not all of the biomass produced remains in croplands (Ciais et al. 2010). In addition, other components are rarely measured such as weed production, seed production, emission of VOCs to the atmosphere, exudation from roots and C transfer to root symbionts. Specifically, all components of cropland NPP have not yet been measured in a single study. However, satellites provide a tool for estimating cropland NPP globally (Monfreda et al. 2008). Other important estimation methods are statistical and process-based models. However, cropland NPP estimates mainly rely on census and survey data but these data measure agricultural production and not NPP (Ciais et al. 2010). The estimates of cropland NPP from yield inventory statistics are biased due to uncertainties in NPP definition, allometry, cropland area and input yield data themselves (Ciais et al. 2010). Some examples of cropland NPP for different regions and species are given in the following section.

Globally, cropland NPP for 1991 was 344 g C m^{-2} $year^{-1}$ derived from FAO agricultural production through application of crop-specific coefficients such as dry matter content, harvest index (HI), root production, and C and N content (Goudriaan et al. 2001). Productivity ranged from as low as 181 g C m^{-2} $year^{-1}$ for pulses to as much as 801 g C m^{-2} $year^{-1}$ for sugar crops. In 2000, average global cropland NPP was 397 g C m^{-2} $year^{-1}$ extrapolated by using harvest indices and harvest statistics (Haberl et al. 2007). The average EU-25 cropland NPP for the decade 1990–1999 ranged from 586 to 646 gC m^{-2} $year^{-1}$ based on yield data, from 482 to 585 gC m^{-2} $year^{-1}$ based on process-oriented and from 419 to 510 gC m^{-2} $year^{-1}$ based on remote sensing models, respectively (Ciais et al. 2010). Agricultural statistics (yields and cropped area) were also used to estimate cropland NPP for the conterminous U.S. (Hicke et al. 2004). The cropland NPP in the U.S. increased from about 350 gC m^{-2} $year^{-1}$ in 1972 to about 490 gC m^{-2} $year^{-1}$ in 2001. This gain was explained by more effective fertilization and pest management, higher yielding cultivars, more favorable climate, shifts to productive crop types (e.g., wheat to corn), and economic factors. Further, a substantial interannual variation in cropland NPP in the conterminous U.S. was likely driven by changes in climate, but economic and management decisions (e.g., increases in irrigation) may have also contributed (Hicke et al. 2004). A 41% increase in cropland NPP in the southern U.S. was reported by Tian et al. (2010) for the period 1895–2007. Specifically, the annual mean cropland NPP based on a process-based model increased from 369 gC m^{-2} $year^{-1}$ for the period 1895–1950 to 520 gC m^{-2} $year^{-1}$ for the period 1951–2007. Drought events were identified as major threats to cropland NPP in this study (Tian et al. 2010). Using plant C allocation coefficients obtained from studies published after 1970, Bolinder et al. (2007) estimated a NPP of 537 gC m^{-2} $year^{-1}$ for small-grain cereals in Canada. However, estimating belowground NPP in this study was associated with considerable uncertainty.

Among species, paddy rice had a higher NPP per unit land globally in 1991 than wheat (376 *vs.* 271 gC m^{-2} $year^{-1}$) as rice is produced under more intensified growing conditions than wheat (Goudriaan et al. 2001). Further, small and coarse grains together accounted for 60% of global crop NPP whereas oil crop and sugar crops each accounted for 9%. Globally, the highest cultivated crop NPP was estimated for western Europe, eastern Asia including Japan, South Korea, and China, and the central United States (Monfreda et al. 2008). High crop NPP was also estimated for southern Brazil and northern Argentina. These are the only large cultivated areas with NPP greater than 1,000 gC m^{-2}. The highest rates of all exceed 2,000 gC m^{-2} in the countries of northern Europe bordering the North Sea. Especially intensive agriculture, usually associated with irrigation, attains comparable rates in a smaller areas scattered throughout the world, including New Zealand, Israel, Egypt, Saudi Arabia, California, Oregon, and Washington in the western United States, Java, and pockets of Columbia, and India, and Mexico (Monfreda et al. 2008).

Using a method based on production and crop-specific factors, Prince et al. (2001) estimated that in 1992 county-level NPP in the U.S. ranged from less than 200 gC m^{-2} $year^{-1}$ in North Dakota, Wisconsin and Minnesota to over 850 gC m^{-2} $year^{-1}$ in central Iowa, Illinois and Ohio (assuming 50% of dry matter is C). Further, corn grain NPP reached very high values (>850 gC m^{-2} $year^{-1}$) over extensive areas whereas NPP for corn silage was lower, and the lowest for wheat and barley (*Hordeum vulgare* L.). Large interannual variability in NPP (<350 to >700 gC m^{-2} $year^{-1}$) was observed in Iowa over a 15-year period from 1982 to 1996, with the lowest NPP in 1983 (which had an unusually wet spring), in 1988 (which was a drought year), and in 1993 (which experienced floods). In 1992, NPP in Iowa ranged from 232 gC m^{-2} $year^{-1}$ for soybean to 547 gC m^{-2} $year^{-1}$ for barley (Prince et al. 2001). The increases in total NPP for croplands in the Great Plains from 1991 to 1996 were largely related to high NPP for corn (Bradford et al. 2005).

Belowground Carbon Input

Surface plant residues are incorporated into the mineral soil by physical mixing and solubilisation, transport and subsequent adsorption (Lorenz and Lal 2005). The relative importance of these processes *vs.* root litter and rhizodeposition for profile SOC distribution and dynamics depend on climate, soil and vegetation types (Rumpel and Kögel-Knabner 2011). However, most studies on SOC dynamics in sub-soil and, in particular, on dissolved organic carbon (DOC), have been done in temperate forest soils, and there are only a few studies on DOC in temperate arable soils (Chantigny 2003). DOC represents only a small proportion of SOC but the most mobile and actively cycling SOC fraction (Bolan et al. 2011).

Tillage mixes surface crop residues into the mineral soil which may result in subsurface SOC formation (Wright et al. 2007). For example, stubble incorporated into the soil makes a significant direct contribution to SOC sequestration (Sanderman et al. 2010). Over and above the effects of crop management and translocation of DOC, crop rooting and belowground biomass also impact SOC formation in the subsoil. Further, in extensively managed croplands and, in particular, those with low or no tillage activity, there is a myriad of soil biota that affect directly and indirectly SOC inputs into subsoil (Wilkinson et al. 2009). For example, earthworms, termites, ants, arthropods but also roots efficiently bury SOC while forming voids in the form of burrows, nests, chambers, galleries and root channels. Direct inputs include litter sequestration into nests, termitaria, borrows, and bioturbator disposal of dead tissues. The earthworms and litter arthropods are effective in fragmenting and mixing surface residues into the soil. Indirect inputs of SOC into subsoils may occur by infilling of biogenic voids with litter, redistribution of SOC and subsurface mixing and burial (Rumpel and Kögel-Knabner 2011).

Plant Roots

Annual crops transfer about 20% of the C from photosynthates to the roots during the growing season (Whalen and Sampedro 2009). However, not all fluxes and the entire root biomass can be directly measured. In particular, inventory data on root biomass are uncertain due to spatial and temporal heterogeneity, uneven sampling and methodological differences among studies (Robinson 2007). For example, only 64% of wheat roots were recovered by direct sampling after ^{13}C-labelling (Subedi et al. 2006). Yet, direct measurements of complete root profiles are rarely reported in the primary literature (Schenk and Jackson 2005).

About 50% of the C translocated belowground in croplands is used for root growth (Johnson et al. 2006). The root biomass ranges between 6% of the shoot biomass for sunflower (*Helianthus annuus* L.) and 50% for barley (Whalen and Sampedro 2009). About 30% of the C translocated belowground moves into the rhizosphere or is released during root respiration (Johnson et al. 2006). Further, as much as 30% of NPP may be transferred to mycorrhizal fungi. In soybean, about 20–30% of photosynthates are allocated to support rhizobia in nodules (Whalen and Sampedro 2009).

The root C is preferentially stabilized in SOC relative to surface residue C (Rasse et al. 2005). For example, root-derived C from hairy vetch (*Vicia dasycarpa* Ten.) is stabilized three times more readily in soils under corn cropping systems than is surface residue-derived C (Kong and Six 2010). Similarly, Bolinder et al. (1999) and Rasse et al. (2006) estimated that the humification ratio (i.e., SOC/C input ratio) of corn residues is about 0.12 for shoots and 0.19 for roots. Further, the humification ratio for barley (*Hordeum vulgare* L.) residues is 0.12 for shoots and 0.16 for roots (Broadbent and Nakashima 1974). For different crops, the contribution factor of roots *vs.* shoots to total SOC (i.e., [root-derived C/total root C input]/[shoot derived C/total shoot C input]) is 2.4 (Rasse et al. 2005). Thus, roots are a more important SOC source, in particular, in subsoils than shoots. However, the root C flux into soil is poorly understood because accurate measurements of root exudation and root cell sloughing are difficult (Rumpel and Kögel-Knabner 2011). Further, it is not known what root properties influence SOC stability or rhizodeposition (Mendez-Millan et al. 2010). While soil C inputs in annual crops can be modeled based on root-to-shoot ratios the above- and belowground biomass in perennial species may not be very representative of C inputs to the soil (Gill et al. 2002).

To overcome difficulties in measurements of root-derived C, belowground C inputs are often estimated from root-to-shoot ratios (Sanderman et al. 2010). However, much current knowledge about root-shoot biomass relations is based on unreliable data. Specifically, by omitting unreliable data, higher estimates for median root-to-shoot ratios are calculated than previously obtained. Accordingly, vegetation-specific root-to-shoot ratios must be used to predict root biomass (Mokany et al. 2006). Also, with regard to belowground C inputs, the often used static measurement of live root biomass may result in inaccurate estimates because a significant fraction of total root biomass may be short-lived fine roots. For example, turnover times of tree fine roots is 1 year or less based mainly on minirhizotron experiments and between 1.3 and 32 years based on changes in C isotopic ratios with time (Pritchard and Strand 2008). Minirhizotron measurement generally underestimate the turnover times whereas estimated fine root turnover times based on isotopic studies are always too high (Guo et al. 2008). The non-woody roots of many plants in croplands which lack secondary growth and cell wall thickening are short-lived (Hodge et al. 2009).

Not all of the about 40% net fixed C allocated belowground is used for root growth. Specifically, about 50% of the C allocated belowground (19% of net fixed C) is retained in root biomass and roughly 27% of C allocated to roots (11% of the net fixed C) is rhizodeposition (Jones et al. 2009). For example, when the total belowground C allocation is considered, the estimated root-to-shoot ratios of 0.50, 0.30 and 0.33 for wheat, corn and soybean increase to 0.82, 0.55 and 0.62, respectively

(Johnson et al. 2006). However, most isotopic labelling studies used to quantify the amount of photosynthate partitioned belowground have focused on young plants at a vegetative stage but partitioning is strongly affected by plant age. For example, the partitioning of C to the rhizosphere decrease by 43%, 28%, and 20% for roots, rhizosphere respiration and soil residues, respectively, for annual plants between plant ages ranging from 28 to 600 days (Nguyen 2003). The root-to-shoot ratios for wheat, corn, soybean and lupin (*Lupinus* L.) decrease from about 0.4–0.6 during vegetative growth phases to as low as 0.1 at flowering (Gregory et al. 1997). Further, almost half of the published data on rhizodeposition are for wheat and ryegrass, and 76% of the studies are related to only five crop/grassland species. Thus, the knowledge of C rhizodeposition, in particular, in mixed plant communities is scanty (Jones et al. 2009).

Rhizodeposition

The C flow at the soil-root interface is bidirectional with C being lost from roots and taken up from the soil simultaneously (Jones et al. 2009). Rhizodeposition describes the release of organic C compounds by roots which results in dramatic changes in the physical, biological and chemical nature of the soil. Specifically, rhizodeposition fuels the soil microflora leading to an abundant microbial population in the rhizosphere which is involved in the great majority of soil biological activity (Bais et al. 2006; Nguyen 2003). Rhizodeposition processes include (i) loss of root cap and border cells, (ii) death and lysis of root cells, (iii) flow of C to root-associated symbionts living in the soils (e.g., mycorrhiza), (iv) gaseous losses, (v) leakage of solutes from living cells (root exudates), and (vi) insoluble polymer secretion from living cells (mucilage). Most C lost during root growth is in the form of complex polymers. For example, root exudates contain a diverse array of C-containing primary and secondary metabolites aside ions, free oxygen and water, enzymes and mucilage (Bais et al. 2006). However, the relative importance of each rhizodeposition process is unknown as it is extremely difficult to discriminate between them in both space and time. The C flow in the rhizosphere is extremely complex, being highly plant and environment dependent and also varying both spatially and temporally along the root. Thus, amount and type of rhizodeposits is highly context specific (Jones et al. 2009).

Rhizdeposition has direct and indirect effects on C sequestration in croplands as root exudation, in particular, may represent a significant C loss for crop plants. The magnitude of photosynthates secreted varies with soil type, age, and physiological state of the plant, and nutrient availability (Bais et al. 2006). The root exudates may affect plant-plant, plant-microbe and plant-nematode interactions. In particular, plant-microbe interactions in the rhizosphere may contribute to C sequestration in plant and soil as micronutrient acquisition is enhanced by the stimulation of microbial communities through exudates. Rhizodeposition may range between less than 10% of the net C assimilation by a plant to 44% of a nutrient-stressed plant's total C (Bais et al. 2006). However, many of the published data on C flow in both soil and roots are severely biased (Rees et al. 2005; Kuzyakov 2006). Nevertheless, from the about 1.5–2.2 Mg C ha^{-1} allocated belowground by cereals during the vegetation period roughly 0.4–0.6 Mg C ha^{-1} enters the soil in the form of rhizodeposition (Jones et al. 2009; Kuzyakov and Domanski 2000). However, these estimates are highly uncertain as the partitioning of rhizosphere respiration from mycorrhizal respiration is almost impossible. Also, whether root exudates directly contribute to the SOC pool is under discussion (Sanderman et al. 2010). Due to the rhizosphere priming effect the decomposition rate of soil organic matter (SOM) in the rhizosphere may increase three- to five-fold in response to root exudation but the response of the subsoil SOC pool is less well known (Kuzyakov 2002, 2010). For example, laboratory studies by Fontaine et al. (2007) indicated that adding an additional energy source similar those in rhizodeposits to the subsoil prime microbial decomposition of subsoil SOC. However, the stimulation of stable subsoil C decomposition in the field by addition of labile material may be small (Sanaullah et al. 2011). Further, the subsoil priming effect is not always observed but an increase in the subsoil SOC pool from inputs of root exudates is likely (Salomé et al. 2010). Also, complex compounds derived from root turnover may contribute indirectly to the SOC pool by enhancing aggregation (Rees et al. 2005).

Mycorrhizal Fungi

Mycorrhizal fungi live in symbiosis with the roots of plants. Almost all agricultural crops are mycorrhizal plants with the exception of the Brassicaceae family (cauliflower [*Brassica oleracea* L. Botrytis group], cabbage [*Brassica oleracea* L. Capitata group], canola [Brassica campestris L.], and others; Whalen and Sampedro 2009). Soil yeasts may affect the colonization of crop plant roots by arbuscular mycorrhizal (AM) fungi (Botha 2011).

AM fungi are important for sustainable crop production (Sen 2003). For example, AM fungi may facilitate uptake of up to 80% of the P, 25% of the N, 10% of the K, 25% of the Zn and 60% of the Cu requirements of the host plant (Marschner and Dell 1994). Other functional benefits provided by AM fungi are pathogen protection and facilitation of water uptake by the host plant (Verbruggen and Kiers 2010). AM fungi receive about 3–20% of photosynthate from their host plant, and the hyphae outside of the plant root play a major role in C translocation into the soil (Treseder and Cross 2006). This may enhance SOC sequestration as C is translocated away from the high respiratory activity around the root. The large network of hyphae outside of the root may represent up to 15% of the SOC pool (Leake et al. 2004).

AM fungi produce the glycoprotein glomalin within their hyphal walls (Wright and Upadhyaya 1996). Glomalin is deposited in the soil as the hyphae senesce, and may comprise 0.7–2.4% of the SOC pool in agroecosystems and enhance soil aggregation (Purin and Rillig 2007; Treseder and Turner 2007). Thus, glomalin may represent a reasonably large C influx to SOC in the order of tens to hundreds g C m^{-2} $year^{-1}$ with relatively slow turnover rates (Rillig et al. 2003; Treseder and Turner 2007). However, glomalin cannot be directly isolated and methods for measuring soil glomalin have limitations (Schindler et al. 2007). Thus, glomalin is present in measurable but modest quantities in most cropland soils. Important may be indirect effects of mycorrhizal fungi on SOC storage as hyphal growth improves soil structure and aggregate stability (Sanderman et al. 2010). Thus, AM fungal hyphal abundance, soil structure and C storage in croplands may be closely correlated (Wilson et al. 2009). In addition to contributing to SOC gain, mycorrhizal fungi also contribute to direct SOC loss through decomposition (Talbot et al. 2008). However, compared to AM fungi ericoid mycorrhizal (ERM) and ecto-mycorrhizal (ECM) fungi have more extensive enzymatic capabilities and are, thus, more efficient decomposers (Smith and Read 2008). In summary, mycorrhizal fungi may moderate SOC dynamics in croplands by enhancing SOC sequestration and by the formation and maintenance of soil aggregates (Talbot et al. 2008; Wilson et al. 2009). Further, AM fungi facilitate shifts in the structure and function of host plant communities by symbiotic interactions (Cameron 2010).

14.2.2 Cropland Soil Carbon Sequestration

The net balance between soil C inputs and losses determines the capacity of a crop soil to sequester C. The activity of soil microorganisms and fauna leads to decomposition of OM and mineralization to CO_2 (Whalen and Sampedro 2009). Abiotic processes such as photodegradation of surface OM may also contribute to decomposition. Photodegradation is the decomposition of OM compounds directly by solar irradiance which increases CO_2 fluxes through either microbial facilitation or more important photochemical mineralization (Rutledge et al. 2010). Photodegradation may be an important contributor to CO_2 loss from croplands where SOC, litter and/or standing dead material are exposed to solar irradiance. Thus, photodegradation occurs in croplands of arid and semi-arid areas, bare burnt areas, sparsely vegetated croplands like shrublands, croplands after cultivation or harvest especially when crop residues are left on the surface, during prolonged drought and croplands with large amount of exposed standing dead material like croplands under no-tillage (Rutledge et al. 2010). Photodegradation of SOM may also be an important process after crop residues have been removed or where erosion exposes SOM to solar radiation (Feng et al. 2011). For example, photo-oxidation increased the solubility of SOM but did not substantially affect the organo-chemical composition of corn litter. Thus, photo-oxidation may contribute to soil C loss through leaching and oxidation (Feng et al. 2011).

Net C losses from croplands may also occur by soil erosion. This involves (i) detachment and breakdown of soil aggregate, (ii) transport of soil particles by runoff water or wind, (iii) redistribution of eroded material over the landscape, and (iv) deposition of eroded material in depressional sites and protected areas (Lal 2004). Historically, crop soils have lost significant amounts of SOC by erosion (Lal 2003). However, eroded SOC can also be a net C sink relative to the atmospheric CO_2 pool (Van Oost et al. 2007). Specifically, erosion can induce changes in soil-atmosphere C exchange via altered patterns of oxidation of SOC remaining at the eroding site and of the SOC originally present at the depositional site as well as via SOC production at both locations (Billings et al. 2010). Three key mechanisms can alter the flux of C between the soil and the atmosphere. First, dynamic replacement describes the partial replacement of the depleted SOC pool at the eroding site by newly assimilated C. The continued C input and decrease in SOC available to decomposition can lead to a net C gain (Van Hemelryck et al. 2009). Secondly, topsoil SOC is buried and decomposition reduced. Specifically, the decomposition rate of SOC in depositional settings can be reduced due to a combination of physical and chemical processes, such as increased soil wetness, limited aeration, compaction and physical protection of the deposited soil material within newly formed aggregates, leading to a long-term preservation of buried C (Quinton et al. 2010). The third mechanism affecting the C flux is transport and increased decomposition. Specifically, aggregates may breakdown by the disruptive energy of forces applied to the soil by water erosion (raindrop impact, the shearing force of flowing water and collision with other aggregates). This process of disaggregation exposes previously protected SOC to microbial decomposition and combined with a relatively greater proportion of labile SOC within larger soil aggregates may lead to rapid mineralization of this easily decomposable C following water erosion (Van Hemelryck et al. 2009). Thus, erosion can induce a net C sink or source (maximum net source and sink of 1.1/3.1 Pg C $year^{-1}$, respectively), depending on management practices, the extent to which SOC oxidation and production characteristics change with erosion, and the fate of eroded SOC (Billings et al. 2010).

Some cropland C may be lost by emission of CH_4 (Denman et al. 2007). Among the biogenic CH_4 sources are the anaerobic decomposition of OM in wetland soils emitting 100–231 Tg CH_4 $year^{-1}$, flooded soils under rice cultivation emitting 31–112 Tg CH_4 $year^{-1}$, and crop residues under very wet field conditions (Greenhouse Gas Working Group 2010). Microbial methanogenesis is the dominant soil process generating CH_4 carried out by a group of anaerobic Archaea converting substrates produced by fermentation of organic macromolecules (Singh et al. 2010). However, a large proportion of CH_4 produced in drier soil is often consumed by methanotrophic bacteria before escaping to the atmosphere (Reay 2003).

Cropland soils may also lose small amounts of C by emitting VOCs, compounds which may also interact with C cycling in soils (Gray et al. 2010; Insam and Seewald 2010). Plants, microorganisms and animals are the main producers of VOCs in soils (Laothawornkitkul et al. 2009; Leff and Fierer 2008; Loreto et al. 2008). Decomposing litter including the microbial metabolism has been identified as a major VOC source in laboratory incubation studies (Gray et al. 2010). However, the relative contributions of abiotic and biotic sources to soil VOC emissions in croplands are not entirely known (Leff and Fierer 2008).

Carbon losses through leaching are important for the C balance of croplands (Kindler et al. 2011). For example, leaching losses of biogenic C (DOC plus biogenic dissolved inorganic carbon or DIC) increase the net C loss from European cropland soils by 25%. Leaching of DOC from European croplands can be up to 4 g C m^{-2} $year^{-1}$. However, DOC concentrations are much smaller than DIC concentrations, particularly in subsoils (Kindler et al. 2011). Potentially leachable C compounds are derived mainly from plant litter and SOM but also from root exudates and the microbial biomass (Kalbitz et al. 2000). Plant residues are the major DOC source in cultivated soils (Bolan et al. 2011). The addition of manure and sewage sludge increases the amount of DOC in soils. DOC is produced largely by the activity of the soil biota as litter is physically and chemically altered during decomposition, and substances are leached from litter and soluble compounds are formed. Abiotic processes (i.e., desorption and dissolution from the potentially leachable C) control the DOC concentration of the soil solution. DOC compounds can be grouped into labile DOC compounds such as simple carbohydrates, low molecular weight organic acids and proteins, and amino sugars (Marschner and Kalbitz 2003). Recalcitrant DOC compounds consist of polysaccharides, other plant compounds and/or microbially derived degradation products. Only the mobile DOC fraction in macro- and mesopores is subjected to convective transport by seepage. DOC in micropores is immobile and interacts with the mobile fraction by diffusion. DOC in the soil solution may be decomposed or removed from solution by various processes (Kalbitz et al. 2000). Thus, DOC concentrations strongly decrease with increase in soil depth and small amounts of leached C is advectively transported to aquatic ecosystems (Sanderman and Amundson 2008). Leaching of DOC from subsoils is controlled, in particular, by sorption to poorly crystalline Fe and Al (hydr) oxides with a high specific surface area (Kindler et al. 2011). Further, DOC leached from soils may partly be retained in the vadose zone before reaching aquifers. In general, recently deposited crop residues and application of organic amendments such as biosolids and manures are the most important sources of DOC in arable soils (Bolan et al. 2011). Minor sources are root decay, exudates and microbial metabolites. However, very few studies have been published on boreal and tropical agroecosystems (Chantigny 2003).

14.2.2.1 Stable Soil Organic Matter and Humic Substances

Historically, SOM has been thought to consist of parts with a molecular structure that decomposes easily and other dark, amorphous parts that decompose slowly if at all (Kleber and Johnson 2010). The molecular properties of these dark 'humus compounds' have been thought to render them refractory (Alexander 1965). During the humification process, humus compounds supposedly evolve into polymeric macromolecular humic substances, i.e., the part of OM in soils most resistant to microbial attack because of its complexity. However, it is questionable whether operationally defined humic substances artificially prepared by an alkaline extraction procedure can serve as models of SOM fractions (Baldock and Nelson 2000). Materials extracted from soil with alkali do not occur as such in natural soils and there is no molecular-level evidence for the existence of distinct humic molecules in soils (Kelleher and Simpson 2006; Lehmann et al. 2008). Thus, SOM can be rather defined conceptually as a mixture of organic compounds in various stages of decomposition (Kleber and Johnson 2010).

The humification process does not create stable organic macromolecules out of labile plant litter precursor materials. SOM is not a stable (recalcitrant, refractory) product of secondary syntheses in soil as previously suggested (Stevenson 1994), and not the true end product of decomposition which is CO_2 (Kleber and Johnson 2010). All plant residues decay rather rapidly in aerobic soils with adequate moisture and temperature, and even lignin is more or less transformed with only parts of lignin accumulating and potentially being stabilized in soils (Hofmann et al. 2009; Thevenot et al. 2010; Vancampenhout et al. 2009). Some fresh plant C is reworked by the biotic community, and the remnants of soil biota attach to mineral surfaces and may be protected against decomposition for centennial to millennial timescales (Kleber and Johnson 2010). Seemingly labile OM persists or is stable in soil because of sorptive protection, aggregation, occlusion and facultative nonutilization (Ekschmitt et al. 2008; Salomé et al. 2010). Thus, whether recalcitrant OM exists in soil aside black carbon (BC) due to inherent molecular characteristics is a matter of debate (Kleber 2010a,b; von Lützow and Kögel-Knabner 2010). However, important for SOC sequestration is the fact that simple and relatively fresh OM inputs can directly contribute to the stable SOM pool as a slow aging process to produce stable SOM is apparently not required (Sanderman et al. 2010).

14.2.2.2 Decomposition

Decomposition is second only to photosynthesis in driving the quantity of C cycled through ecosystems (McGuire and Treseder 2010). The unharvested remnants of crops such as leaf, stem and root tissue not removed from the field and the die-back of legumes such as leaves and stems are subject to decomposition (Whalen and Sampedro 2009). The residues may be decomposed rapidly by a bacterial food chain or slowly by a fungal food chain depending on tillage practices as litter placement can strongly influence the decomposer community composition and decomposition rates (Beare et al. 1992). In particular in no-till croplands, macrofauna such as earthworms fragment litter and redistribute it in the soil profile.

Decomposition is the physical and chemical breakdown of dead OM of animal, microbial and plant origin that emit C into the atmosphere and release nutrients in forms that can be used for microbial, plant and animal production (Chapin et al. 2002). Litter decomposition is the result of the three interlinked processes leaching, fragmentation and catabolism (Cotrufo et al. 2009; Sanderman et al. 2010). Decomposition is driven primarily by bacteria and fungi accounting for up to 90% of the total decomposer biomass, and about 85–90% of OM decomposition is mediated microbially. Fungi mainly decompose fresh plant material whereas gram-positive bacteria consume both fresh and older SOM (Amelung et al. 2008). About 10–15% of decomposition is performed by soil animals (Wolters 2000). Decomposition is a key ecological process for maintaining supply of most plant-essential nutrients. For example, nutrient recycling via decomposition may account for >90% of plant-available N and P, and for >70% of plant-available K and Ca in natural ecosystems (Chapin et al. 2002).

Leaching is most important during tissue senescence and when plant litter is deposited at the soil surface. Mainly labile compounds such as low-molecular weight soluble sugars, polyphenols and amino acids are leached from plant litter (Berg and McClaugherty 2008). Some of the leached and solubilised material may be absorbed by soil organisms. However, depending on adequate water supply a major portion of leached OM may be transported deeper into the mineral soil and adsorbed to SOM and soil minerals, or transported into adjacent aquatic ecosystems. Soil fauna also partially solubilises fresh plant and fragments residues, and facilitates the establishment of microbial decomposers, for example, by mixing residues into the soil for more intimate contact with decomposer microorganisms (Wolters 2000). However, soil fauna has only a limited ability to enzymatically decompose organic residues compared to soil microorganisms (Whalen and Sampedro 2009).

The main initial decomposers fungi are mostly concentrated closer to the soil surface (Chapin et al. 2002; Fierer et al. 2003). Fungal C is about 75% of total microbial residue C in arable soils, and fungal respiration is about 61% of microbial respiration in cropland soils (Joergensen and Wichern 2008). Fungal enzymes can break down virtually all classes of plant compounds, and hyphal fungi can adjust mycelium growth across considerable distances towards OM substrates (Ritz 1995). Fungi have higher C assimilation efficiencies (i.e., store more C than they metabolize) than bacteria (Singh et al. 2010). Thus, SOM in croplands favoring a fungal-dominated community is improved quantitatively (Six et al. 2006). In contrast to fungi, bacteria primarily decompose more labile substrates (Moorhead and Sinsabaugh 2006). However, bacterial biomass extends to deeper depths. Bacteria rely on solute transport and soil perturbation for their translocation towards organic residues (Jiang et al. 2005). Thus, the composition of the initial microbial community decomposing a residue may gradually shift through time concurrently with changes in residue composition (Hättenschwiler et al. 2005).

In addition to the primary resource plant OM, biomass of microorganisms and microfauna (Ø<0.2 mm) are themselves secondary resources for decomposition. The C polymers in fungal cell walls (chitin, melanin) are more resistant to decomposition than those in bacterial cell membranes and walls (phospholipids, peptidoglycan) (Singh et al. 2010). Thus, respiration rates in soils dominated by fungi are typically low which increases the potential for SOC sequestration (Six et al. 2006). Ultimately, however, most natural compounds are fully mineralized to inorganic forms under favorable environmental conditions (Kleber 2010a; Marschner et al. 2008). Specifically, about one-half to two-thirds of plant residues entering the soil may be decomposed in one year (Whalen and Sampedro 2009).

Less well known is the long-term fate of leaf and root residues in cropland soil as decomposition is often studied only for 1–2 years (Silver and Miya 2001). In contrast, some litter decomposition experiments in forests run for 10 years and longer (Harmon et al. 2009; Moore et al. 2011). Thus, studies monitoring crop litter decomposition for several years are required as some litter may decompose very slowly depending on environmental conditions and biochemical characteristics. For example, up to 65% of the initial *Triticum aestivum* L. leaf mass and up to

55% of the initial *Andropogon gerardii* Vitman root mass may remain after 10 years decomposing in a soil (Harmon et al. 2009). However, very slow mass loss in late-stage decomposition may occur only in cold northern latitudes or other suboptimal conditions as it does, for example, not appear to occur in warmer climates (Prescott 2010).

Most studies on above-ground litter decomposition use the litter bag method to follow the progressive loss of mass from fresh litter through time (Bocock and Gilbert 1957). This method is still widely accepted for quantifying and comparing litter decomposition rates (Torn et al. 2009). However, a compromise on mesh size based on the experimental question is needed and the method has its limitations as soil faunal activities may be excluded depending on mesh size (Prescott 2005). In particular, the process of litter fragmentation may be prevented by studying decomposition in mesh bags (Cotrufo et al. 2009). Another limitation is that decomposition is often studied using single-species litterbags but mixed-species litterbag decomposition studies indicate that ecosystem litter decomposition is not predictable from decay rates of component species (Gartner and Cardon 2004). Thus, available litter decomposition data for croplands should be used with caution and referred to the specific decomposition component process (i.e., leaching, fragmentation, catabolism) measured (Cotrufo et al. 2009). For example, the use of litter bags is suitable for testing the effects of plant protection products on OM decomposition in agroecosystems (Knacker et al. 2003).

14.2.2.3 Organic Matter Stabilization

Inputs of OM to soils can increase the SOC pool when they are stabilized and not completely mineralized to CO_2 (Kleber and Johnson 2010). Beyond the decadal timeframe selective preservation of relatively unaltered plant-derived compounds due to biochemical recalcitrance (i.e., aliphatic compounds such as lipids and waxes) may not be as an important long-term OM stabilization mechanism as previously thought (Krull et al. 2003; Lorenz et al. 2007). However, charring of OM may result in the formation of biochemically recalcitrant BC. The molecular structure of plant inputs and OM plays only a secondary role in determining C residence times over decades to millennia. The soil C stability is instead mainly an ecosystem property as it depends on the biotic and abiotic environment (Schmidt et al. 2011). The ^{14}C age of SOM fractions is not necessarily related to molecular structure or thermodynamic stability (Kleber et al. 2011). Thus, old C cannot be viewed as being composed of complex or recalcitrant compounds. Further, old and stable SOM is not necessarily biochemically recalcitrant (Kleber et al. 2011). Similarly, microbial derived OM may not be recalcitrant or composed of complex compounds. Thus, mean residence times of microbial biomarkers in soils do not exceed several hundred years (Amelung et al. 2008). Biomarkers are structurally unique biomolecules that retain their C skeleton information and can be used to determine their biological origins and/or environmental settings. However, microbial derived materials may play a special role in SOM stabilization as the SOM turnover appears to be a function of microbial ecology and the resource availability within a given physical soil environment (Kleber et al. 2011). Thus, processes which slow down mineralization are major centennial-scale stabilization mechanisms for SOM (Sanderman et al. 2010).

The two important groups of processes for long-term stabilization of OM are (i) processes which lead to physical protection, rendering OM spatially inaccessible to decomposers or their water-soluble degradative enzymes and (ii) organomineral complexes and organo-metal interactions, i.e., interactions of OM with minerals, metal ions, and other organic substances (Von Lützow et al. 2006). Physical protection may retard decomposition for decades to centuries whereas organomineral complexes or organo-metal interactions may be responsible for most of the highly stable (centuries to millennia) non-charred SOM. However, biochemical recalcitrance and physical protection may allow OM to remain in the soil longer, giving time for organomineral complexes to form (Six et al. 2000). Thus, SOM stabilization is a combination of short- and long-term processes (Nair et al. 2010). Any disruption of the stabilization process may result in decomposition of SOM even if it's thousands of years old (Ewing et al. 2006). In summary, the persistence of SOM is largely due to complex interactions between OM and its environment, such as the interdependence of compound chemistry, reactive mineral surfaces, climate, water availability, soil acidity, soil redox state and the presence of potential degraders in the immediate microenvironment (Schmidt et al. 2011).

Several processes may be the reason why OM is physically inaccessible to decomposers and extracellular enzymes. First, decomposers and OM are sparsely and heterogeneously distributed in soils but the mobility of most decomposer organisms is limited (Ekschmitt et al. 2008; Young et al. 2008). This process of spatial separation may control SOC dynamics, in particular, in the subsoil (Salomé et al. 2010). Thus, a proportion of SOC of any chemical composition located in biologically non-preferred soil spaces is not subject to decomposition (Ekschmitt et al. 2008). Secondly, pores <0.2 μm in diameter are too small for bacteria (typical dimensions 0.5–1 μm), fungi (3–10 μm), microfauna (diameter <0.2 mm), mesofauna (0.2–2 mm) and macrofauna (>2 mm) to enter (Wolters 2000; Young et al. 2008). As pore size decreases to <50 nm, enzymes may also be inhibited from entering (Zimmerman et al. 2004). Third, the hydrophobicity of OM, in particular, those of partially-oxidized OM may greatly reduce accessibility as microbial decomposition is governed by distribution of and accessibility to water (Bachmann et al. 2008). Fourth, soil aggregation at multiple spatial scales may cause inaccessibility of OM by creating strong diffusional limitations to enzymes, and also oxygen which is another important controlling factor of decomposition (von Lützow et al. 2006).

In cropland topsoils, soil aggregation may be most important for spatial separation between decomposers or their extracellular enzymes and OM (Sanderman et al. 2010). Thus, OM losses from crop soils increase when soil aggregates are disrupted (Rovira and Greacen 1957). Aggregates are formed when mineral particles combine with organic and inorganic substances (Bronick and Lal 2005). Microaggregates (20–250 μm in diameter) are formed from small microaggregates (<20 μm) primarily within macroaggregates (>250 μm; Oades 1984). The smallest microaggregate fraction <53 μm is usually referred to as silt+clay fraction. The size classes occur in an aggregate hierarchy. Macroaggregates are less stable with a life span of years, break up most easily by slaking, have the highest concentration of C and the youngest C on average. In contrast, the smallest microaggregates are most stable with a life span of decades, have the lowest concentration of C and the oldest C. The OM in microaggregates is relatively stable as their organomineral complexes are hydrophobic and clogging of micropores creates physical barriers for bacteria and enzymes which retards decomposition (Bachmann et al. 2008).

In cropland soils, fungal extracellular polysaccharides and hyphae are primarily responsible for the formation of macroaggregates which protect plant- and microbial-derived OM from decomposition (Six et al. 2006). Further, fine roots contribute to the physical enmeshment of soil particles in aggregates (Miller and Jastrow 1990). Creating and maintaining the stability of aggregates critically depends on soil fauna, especially earthworms (Edwards and Bohlen 1996). Stabilization of OM within macroaggregates may be restricted to surface horizons in croplands but stabilization within silt-size aggregates is also important in subsurface horizons (Moni et al. 2010). Protection of OM by aggregation against microbial decomposition is effective primarily for soils with a large pool of labile OM (Goebel et al. 2009).

Other processes contributing to OM stabilization in soils involve sorptive reactions with mineral surfaces (Fe-, Al-, Mn-oxides, phyllosilicates), metal ions, and other organic substances (von Lützow et al. 2006). The binding of OM on minerals differs in stability depending on the suite of soil minerals and solution chemistry (Sanderman et al. 2010). For example, the quantity and characteristics of clay minerals, and the presence of polyvalent cations such as Ca^{2+} or Mg^{2+} facilitate sorptive OM stabilization (Whalen and Sampedro 2009). Multiple layers of OM may range outward from the mineral surface with decreasing strength of association (Kleber et al. 2007). Thus, the outer layers may be most actively cycling mineral-stabilized OM (Torn et al. 2009). The destabilization of SOM at the mineral surface may occur by mineral dissolution or soil evolution rather than by direct degradation. For example, the evolution of metastable non-crystalline minerals into crystalline minerals with lower surface area and charge density during soil development reduces the ability to stabilize SOM (Torn et al. 1997). Most of the SOM in soils of croplands with poor structure after cultivation for many years is often associated with mineral surfaces and stabilized in organomineral complexes (Basile-Doelsch et al. 2009; Flessa et al. 2008). However, sandy soils do contain less organomineral complexes than clayey soils and their aggregates are typically weaker (Sarkhot et al. 2007).

14.2.2.4 Controls on Decomposition and Stabilization

Organic substrates for decomposition in cropland soils are plant, microbial and animal residues, rhizodeposits, animal manure and SOM (Whalen and Sampedro 2009). Litter decomposition is influenced by the physical-chemical environment, decomposer organisms and substrate quality (Swift et al. 1979).

Globally, rates of plant litter decomposition are mainly controlled by variations in litter quality (i.e., N, C:N, lignin (%), lignin:N, and P, K, Ca, Mg), microclimate, soil properties and microbial community composition (Zhang et al. 2008). Early rates of litter mass loss can be predicted from climate and litter chemistry (Prescott 2010). Litter chemistry has the most direct influence on decay rates with the acid-unhydrolyzable residue (AUR):N ratio as the most consistent predictor. AUR was formerly known as "lignin". Further, leaf dry matter content and specific leaf area are useful predictors of mass loss rates. Apparently, thresholds exist at which specific factors have an overriding influence on decomposition (Prescott 2010). In contrast to leaf litter decomposition, root chemistry is the primary controller of root decomposition whereas climate and environmental factors play only secondary roles (Silver and Miya 2001). Most of the plant litter compounds decompose rapidly whereas the remaining recalcitrant compounds decompose slowly. However, plant tissues are not decay-resistant as, for example, lignin, cutin and suberin have residence times in years to decades (Prescott 2010). The C dynamics in aboveground litter *vs.* SOM in mineral soil bear little resemblance as both litter quality and physical-chemical mechanisms control SOM protection and degradability (Giardina and Ryan 2000). Thus, plant litter decomposition does not produce SOM with slow turnover time as the link between litter quality and SOM is not well established (Torn et al. 2009).

Under the same environmental conditions, litter quality controls the decomposition rate. Litter of high quality decomposes faster than that of lower quality. For example, alfalfa (*Medicago sativa* L.) residues decompose faster than residues of grain sorghum (*Sorghum bicolor* [L.] Moench) and winter wheat (*Triticum aestivum* L. emend. Thell.) under the same environmental conditions (Schomberg et al. 1994). Thus, alfalfa litter supposedly has a higher quality than sorghum and wheat litter. However, there is neither a common definition nor a quantitative index of 'quality' (Cotrufo et al. 2009). As high energy and nutrient supply for microbial use is required for fast decomposition, a high quality litter has high nutrient (i.e., N and P) concentrations, a high proportion of easily degradable C-compounds (e.g., sugars), and low concentrations of substances inhibiting microbial activity (Swift et al. 1979). Thus, indices of litter quality under discussion are N, C:N, AUR:N, holocellulose:lignocellulose, leaf width and specific leaf area (Cotrufo et al. 2009; Prescott 2010). In later decomposition stages, factors controlling microbial degradation of lignin such as concentrations of N and Mn become key quality parameters (Berg and McClaugherty 2008). However, a general and quantitative index of litter quality/decomposability is lacking (Cotrufo et al. 2009).

Litter decomposition is a biological process (Prescott 2010). Thus, primary controls of decomposition are also microbial activity and ultimately enzymatic activity, except for the abiotic process of photodegradation, and initial litter communication and mixing by soil fauna (Sanderman et al. 2010). Decomposition occurs as a result of secretion of extracelluar enzymes by microorganisms (Prescott 2010). Thus, decomposition depends on the successful diffusion of enzymes to appropriate OM and successful diffusion of decay products back into a microorganism.

Adequate temperature, a water film, aeration, suitable pH, enzymes, diffusion conditions, and an accessible and susceptible substrate in the same place at the same time are requirements that decomposition occurs (Prescott 2010). The size, diversity and activity of the microbial community as well as interactions with large soil biota affect the decomposition rate (Whalen and Sampedro 2009). Soil fauna comprises of 10–20% of total soil biomass and control the abundance and diversity of the microbial community by predation (Cotrufo et al. 2009). However, the effects of fauna such as microarthropods on litter decomposition are poorly understood (Kampichler and Bruckner 2009).

Fungi use C more efficiently than bacteria, and fungi are composed of more recalcitrant structural C. Thus, fungal-dominated food webs in croplands may be associated with higher amounts of SOC than those dominated by bacteria (van der Heijden et al. 2008). The optimal soil pH for bacteria-mediated decomposition is 6.5–8.0 whereas optimal soil pH ranges between 5.5 and 6.5 for fungi-mediated decomposition. Due to a variety of interacting factors, decomposition is more rapid in neutral than in acidic soils (Chapin et al. 2002). Soil bacteria are more tolerant to anaerobic conditions than fungi and are, thus, probably responsible for most decomposition in anaerobic soils. Soil texture determines the habitat available for microorganisms and the amount of predation by larger soil organisms. Further, the activities of microbial decomposer may be limited by insufficient quantities of essential nutrients such as N, P, K and others required to sustain microbial growth (Whalen and Sampedro 2009).

Soil fauna mediate decomposition and typically increase rates of mass loss from litter and SOM (Ayres et al. 2009). Faunal species in soil include collembolans, mites, enchytraeids, isopods and earthworms. Soil fauna enhances decomposition by (i) partial digestion of OM and returning it to the soil, (ii) fragmentation of plant litter thereby increasing the surface area available for microbial colonization, (iii) bringing microbes and OM into direct contact (during gut passage and by dragging litter from the surface into the soil), and (iv) grazing on soil microbes (Bardgett 2005). After gut passage, OM may be either readily available or less available for decomposition (Fox et al. 2006; Osler and Sommerkorn 2007). However, only the effects of grazing on decomposition are better characterized (Ayres et al. 2009). OM is partially degraded in the gut of saprotrophic soil fauna and some OM is returned to the soil as faecal pellets. The OM returned with collembolan, isopod and termite faeces, and with earthworm casts is typically easier to decompose than the original OM which leads to accelerated decomposition. However, decomposition rates of earthworm casts decline rapidly and casts may enhance below-ground C sequestration over longer time scales (Martin 1991). Whether other faunal groups also contribute to fragmentation resulting in increased surface area available for microbial colonization is less well known.

Earthworms may increase the amount of OM in soil aggregates and, thus, stabilize soil C (Bossuyt et al. 2005). Earthworms play a key role in decomposition by their partial digestion and fragmentation of OM, by dragging litter into the soil and bringing microbial decomposers into direct contact with OM (Wolters 2000). Further, grazing by microarthropods and nematodes may alter decomposition rates by influencing activity and growth of soil bacteria and fungi, and through selective feeding (Ayres et al. 2009). For example, the oribatid mite *Scheloribates moestus* (Acari: Oribatida) which is abundant in many U.S. ecosystems, stimulate extracellular enzyme activity, enhance microbial respiration rates and increase water-extractable organic C during decomposition of corn litter (Wickings and Grandy 2011). The mites decrease the relative abundance of polysaccharides in decomposing litter. Further, the feces have a higher relative abundance of polysaccharides and phenols and a lower relative abundance of lignin compared to unprocessed corn litter. Thus, *S. moestus* may play a key role in soil C cycling dynamics (Wickings and Grandy 2011).

Soil properties are secondary controls but climate has an overriding control on OM decomposition and stabilization (Sanderman et al. 2010). Specifically, temperature and rainfall (soil moisture) exert important direct controls on soil microbial activity (Whalen and Sampedro 2009). Water addition through irrigation, for example, may increase decomposition rates of crop residues (Schomberg et al. 1994). In general, faster litter decomposition rates are measured under warmer and wetter conditions when soil temperature and moisture conditions are ideal for microbial activity. However, which climatic index is the best predictor of decomposition rate is under discussion (Cotrufo et al. 2009). In soils with high moisture content, C accumulation is the greatest as decomposition is more restricted than is NPP (Chapin et al. 2002). Further, soil C respiration rate indicative of SOM decomposition roughly doubles for a 10°C warming when microbial activity is not limited by substrate availability or soil moisture (Davidson and Janssens 2006). The carbon quality-temperature (CQT) theory of the temperature sensitivity of OM decomposition links the temperature sensitivity of old SOM to an increase in molecular complexity. However, old SOM fragments are not large, complex, polymeric humic macromolecules. Thus, old SOM may not be particularly sensitive to temperature increases as predicted by the CQT theory (Kleber et al. 2011). Indirectly, temperature may affect decomposition by altering soil moisture, and quantity and quality of OM inputs (Chapin et al. 2002). In summary, over large geographical areas climate (i.e., temperature and soil moisture) exerts the strongest controls on C losses from soil (Sanderman et al. 2010).

Any control or processes that constrain microbial and enzymatic activity such as aggregation and association with mineral surfaces affect the decomposition rate (Sanderman et al. 2010). The soil-aggregate formation and SOC stabilization are affected, in particular, by soil type, climate, landscape position, ecology, and anthropogenic factors (Christensen 1996). Soil texture, for example, plays a large role in the number and kind of primary organomineral complexes formed. Specifically, soils high in clay content exhibit strong aggregate formation and stability. Thus, clay is positively correlated with SOC and the importance of clay for SOC stabilization increases with increase in soil depth (Jobbágy and Jackson 2000). Texture exerts strong controls on soil C losses by reducing apparent decomposition rates through a range of stabilization mechanisms (von Lützow et al. 2006). Clay minerals reduce the decomposition rate of SOM as the higher the mineral surface area the higher may be the amount of OM that can be adsorbed and protected from microbial enzymes (Chapin et al. 2002). However, clay content is not the most important determinant of SOC storage in contrast to the common assumption (Kutsch et al. 2009). Rather than texture *per se* the reactivity of the mineral surface determines the protection of OM (Kleber et al. 2007). For example, SOC in tropical soils such as Oxisols is poorly correlated with soil aggregates as oxides are the key binding factors (Six et al. 2002). Thus, SOM storage and turnover are often closely related to the mineral properties, in particular, the poorly crystalline phases (Torn et al. 2009). Subsoils have a larger potential for long-term stabilization due to abundance of reactive surfaces, smaller and less diverse biotic communities, and limitations on oxygen diffusion compared to topsoils (Rumpel and Kögel-Knabner 2011). In contrast, environmental conditions in topsoils are more often favorable for decomposition because of large and diverse biotic community, adequate moisture, available nutrients, good aeration (Sanderman et al. 2010). However, the topsoil environment is less favorable for C stabilization due to scarcity of reactive mineral surfaces.

Temperature and soil moisture affect OM breakdown in aggregates through their effect on microbial activity. For example, aggregate turnover is slower in cold or dry climates but faster in moist and/or warm climates. However, very wet climates may reduce OM breakdown under anaerobic soil conditions. In clayey soils, wetting and drying cycles can increase aggregation (Horn and Smucker 2005). In temperate climates, freeze thaw cycles may also promote aggregate formation and SOC stabilization (Chen et al. 1997). Plant species in croplands can also affect aggregation. For example, N fixing species and mycorrhizal associations may boost microbial populations and result in higher levels of aggregate stability. Further, molecular-level properties of SOM under different plant species may have an impact on resistance to degradation in microaggregates (Bachmann et al. 2008). Higher levels of SOC and aggregation are observed under plants with great root density (Nair et al. 2010).

In summary, decomposition of a substrate and destabilization of SOM are accelerated when: (i) conditions are suitable for microbial activity, enzyme production and diffusion, (ii) OM is vulnerable to enzymatic degradation, (iii) microbial transformations that lead to the creation of recalcitrant products are constrained, (iv) chemical reactions that lead to formation of stabilized SOM are constrained, (v) physical protection of OM in soil aggregates is weak and, (vi) chemical protection of OM in soil is also weak (Prescott 2010).

14.2.3 The Net Balance of Cropland Soil Organic Carbon

At steady state and under similar soil and vegetation management, the C inputs to cropland soil and losses from it are approximately balanced depending on site-specific factors (Sanderman et al. 2010). Quantitative descriptions of SOM dynamics were previously based on the assumption that SOM consists of a single homogenous pool decomposing at varying relative rates (Shibu et al. 2006). Others consider SOM as comprising of heterogeneous components and decomposition of the components occurs at different relative rates. The components are represented by arbitrary SOM pools. For example, the 'active pool' decomposes on time scale of hours to months to years, and may consist of root exudates, microbial cell contents and some fresh litter compounds (Torn et al. 2009). The 'intermediate' or 'slow' SOM pool has turnover times in the range of decades to centuries, and may consist of structural components of plants more resistant to decay or OM stabilized by association with soil minerals or aggregate structures. The 'passive' or 'millennial cycling' SOM pool persists in soils for thousands of years and consists of highly stabilized OM, typically associated with soil minerals or stable aggregates (Torn et al. 2009). However, the molecular model for stable SOM is under discussion but the specific structure of OM may be responsible for long-term preservation of OM in soil mineral fractions (Clemente et al. 2011; Kleber et al. 2011).

In comprehensive SOM models essential elements are thoroughly understood based on the scientific knowledge (Shibu et al. 2006). In contrast, essential aspects are formulated in less detail in summary SOM models and these models are more suitable for applicative and predictive purposes. For example, the Rothamsted carbon model (RothC) and the CENTURY model are two of the most widely used summary SOM models (Coleman and Jenkinson 1995; Parton et al. 1987). RothC is purely concerned with soil processes and not linked to a plant production model (Falloon and Smith 2009). In contrast, CENTURY is part of a larger ecosystem model than simulates crop, grass and tree growth, and the effects of management practices on both plant production and SOM.

In RothC, quality of residue/litter entering the soil is defined using the ratio of decomposable plant material (DPM) to resistant plant material (RPM) (Falloon and Smith 2009). In contrast, CENTURY uses the lignin:N ratio to define litter quality. In RothC, the SOC is split into four active compartments that decompose by a first-order process and have their own characteristic rate constants. In the order of decreasing maximum decomposition rates, the active compartments are DPM, RPM, microbial biomass (BIO) and humified organic matter (HUM). RothC assumes that a small amount of SOC is resistant to decomposition (inert organic matter [IOM]). In contrast, the CENTURY SOM sub-model includes three SOM pools, two surface and two sub-surface litter pools, and a surface microbial pool. All pools decompose by first-order kinetics and have characteristic decomposition rate constants. The pools can be ordered in decreasing maximum decomposition rates as the soil metabolic litter pool, the surface metabolic litter pool, the active SOM pool, the surface microbial pool, the soil structural litter pool, the surface structural litter pool, the slow SOM pool and the passive SOM pool. CENTURY also simulates leaching of OM whereas RothC is not capable of modeling DOC loss from soils (Falloon and Smith 2009).

Both RothC and CENTURY are applicable to cropland ecosystems (Falloon and Smith 2009). Both simulate the effects of farmyard manure application on SOM but CENTURY can also simulate the impact of tillage, harvesting, organic amendments, irrigation, erosion and fire on SOM. Both models have rarely been applied to assess short-term aspects of the C cycle and soil C cycling on a small scale. The majority of applications have focused on evaluating RothC and CENTURY against datasets of changes in SOC over decadal to century time scales. Both models can simulate the effects of management on SOC in the long term. Large scale applications of both models have also been performed (Falloon and Smith 2009).

Both RothC and CENTURY models were originally developed to predict turnover of C in topsoils. However, RothC has been modified to a multi-layer model to describe also the turnover of C in the top meter of soil (Jenkinson and Coleman 2008). It is recognized that the soil profile cannot be treated as a homogenous unit to predict the acceleration of SOC decomposition by global warming. Both RothC and CENTURY fail to account for pH effects on SOM turnover (Falloon and Smith 2009). Most SOM models are unable to simulate SOC changes in permanently waterlogged, very dry, highly organic and recent volcanic soils. The SOM models may be further limited in their applicability to tropical croplands. For evaluating SOM models, there are relatively few long-term experiments related to land use change rather than land-management changes (Falloon and Smith 2009).

The confidence in SOM model predictions is particularly limited as SOC measurements from available long-term experiments are rarely replicated (Falloon and Smith 2003). Challenges in measuring SOC include obtaining representative undisturbed soil cores for different layer depths, accurately measure bulk density for conversion to SOC pool, and addressing the high spatial variation in SOC (Falloon and Smith 2009). Accurately measuring litter lignin concentration which is among the input variables for CENTURY is also challenging (Preston et al. 1997). Most importantly, the SOM pools in models are theoretical without measurable counterparts. Whether a totally inert IOM pool (RothC) receiving no C inputs and being resistant to decomposition exists is unclear as was discussed previously regarding the lability of BC and SOC stabilization (Hammes et al. 2008; Von Lützow et al. 2006). Further, uncertainties are associated with the use of radiocarbon (^{14}C) age as input data in CENTURY and output data produced by RothC (Falloon and Smith 2009). Specifically, old (^{14}C age) and stable SOM is not necessarily chemically recalcitrant (Kleber et al. 2011). Also, the simplifying assumption in many models that SOC pools are near equilibrium has been challenged (Wutzler and Reichstein 2007). For example, agricultural soils may never reach a theoretical equilibrium SOC level because of changing conditions (i.e., climate change, land-management change) and partial resets by disturbances such as erosion (Polyakov and Lal 2004; Bell et al. 2011).

It has been hypothesized that the SOC pool in a particular soil eventually saturates at a maximum C-saturation level depending on inherent physicochemical characteristics (Stewart et al. 2007). In soils close to their saturation value, any additional increases in C inputs may then remain as unprotected POM and rapidly cycled back to the atmosphere (Stewart et al. 2008). However, while data from some long-term agricultural field experiments indicate that SOC saturation does occur others did not observe saturation behavior even after many years of additions of OM by manure application (Stewart et al. 2007; Blair et al. 2006a, b). Even though the whole soil may not be saturated, the chemically and biochemically protected pools may be influenced by C-saturation behavior (Stewart et al. 2009). Once the chemically protected SOC pool is filled, added C may accumulate in the physically and in the non-protected fractions. Mineral-associated SOC pools, in particular, eventually saturate. The reactive mineral surface area is a finite resource in topsoils and C-saturation may occur (Séquaris et al. 2010). Thus, arbitrarily defined soil fractions may have different C-saturation dynamics (Stewart et al. 2009).

In summary, recent advances in mechanistic understanding of soils have not yet been incorporated into widely used models of SOM cycling (Schmidt et al. 2011). Current models assume a pool of organic material that will have an intrinsic decay rate. These models rely on simple proxies such as soil texture as a surrogate for sorption and other organo-mineral interactions. Further, litter quality (e.g., lignin:N ratios or structural C groupings) is used a proxy of partitioning plant inputs into pools of different turnover times. However, these parameters are not consistent with the observations that emerge. Also, global models largely ignore deep mineral soils (Schmidt et al. 2011).

14.2.4 Anthropogenic Drivers of Cropland Soil Organic Carbon

The SOC balance of croplands is a function of past and present agricultural technology and farming practice (Ciais et al. 2011). Examples are manure and tillage management. Technology may impact the soil C input directly by affecting the management of harvest residues and indirectly via effects on yield and NPP. Anything that increases crop biomass production such as rotations, nutrients, improved cultivars and irrigation (i.e., agricultural intensification) impacts the soil C input (Robbins 2011). As croplands are often intensively managed they offer opportunities to deliberately alter the SOC dynamics (Smith et al. 2008). However, C budgeting studies over croplands are scarce in comparison to studies on crop yields and on processes controlling plant and soil fertility (Ciais et al. 2011).

Improving cropland management, restoring degraded crop lands and cultivated organic soils greatly affect SOC dynamics (Smith et al. 2008). For example, improved agronomic practices that increase yields and generate higher inputs of residue C can lead to increased SOC storage. Practices include (i) using improved crop varieties, (ii) extending crop rotations, notably those with perennial crops which allocate more C below-ground, and (iii) avoiding or reducing use of bare fallow (Lal 2004; Smith and Conen 2004). Adding more nutrients, when deficient, can also promote soil C gains (Alvarez 2005). By providing temporary vegetative cover between agricultural crops, ‘catch’ or ‘cover’ crops also add C to cropland soils (Freibauer et al. 2004). Since soil disturbance tends to stimulate soil C losses through enhanced decomposition and erosion, reduced- or no-till agriculture often results in soil C gain, though not always (Govaerts et al. 2009). Cropland systems that retain crop residues (e.g., avoiding the burning of residues) also tend to increase soil C because these residues are the precursors for SOM, the main store of C in the soil (Smith et al. 2008).

The management of croplands is an important factor in altering SOC dynamics (Franzluebbers 2010). Conservation practices, in particular, have a great potential to increase SOC. Guiding conservation principles that can be globally applied are to (i) minimize soil disturbance, (ii) maximize surface cover by managing crops and crop residues and, (iii) stimulate biological activity through crop rotations, cover crops and integrated nutrient and pest management (Franzluebbers 2010). However, the mechanisms that govern changes in SOC after reducing tillage operations are less clear (Govaerts et al. 2009). Among factors that may play a role are root development and rhizodeposits, baseline soil C content, bulk density and porosity, climate, landscape position, and erosion/deposition history. Further, altering crop rotation may influence SOC by changing quantity and quality of OM input (Govaerts et al. 2009).

Cropland irrigation may affect SOC dynamics by altering crop yields and residue returns (Lal 2004). About 25% of the global harvested crop area (i.e., area of all major food crops and cotton (*Gossypium* L.) as well as those of perennial, annual and fodder grasses) was irrigated in 2000 (Portmann et al. 2010). Further, the irrigated harvested area for rice was 1 million km^2 while it was 0.7 and 0.3 million km^2 for irrigated wheat and corn, respectively. The average crop yield of irrigated cereals was 442 Mg km^2 while average yield of rainfed cereals was 266 Mg km^2 (Siebert and Döll 2010). Crop yields and residue returns and, thus, soil C input may decrease when crop irrigation is discontinued. For example, the global production of dates (*Phoenix dactylifera* L.), rice, cotton, citrus (*Citrus*) and sugar cane would decrease by 60%, 39%, 38%, 32% and 31%, respectively, if currently irrigated crops were not irrigated. Further, cereal production on irrigated land would decrease by 47% without irrigation (Siebert and Döll 2010). In contrast, drainage of agricultural lands in humid regions can promote productivity and hence SOC accrual (Smith et al. 2008).

Agroforestry refers to the practice of purposeful growing of trees and crops, and/or animals, in interacting combinations, for a variety of benefits and services such as increasing crop yields, reducing food insecurity, enhancing environmental services and resilience of agroecosystems (Ajayi et al. 2011; Nair et al. 2008). In the Tropics, agroforestry systems includes alley cropping, homegardens, improved fallows, multipurpose trees on farms and rangelands, silvopastoral grazing systems, shaded perennial-crop systems, shelterbelts, windbreaks, and taungya (i.e., growing agricultural crops during early stages of establishment of forestry plantations). In temperate regions, agroforestry practices include alley cropping, forest farming, riparian buffer strips, silvopasture and windbreaks (Nair et al. 2009). In particular, planting trees may affect SOC dynamics and the SOC pool may be higher compared to that of croplands, pastures or natural grasslands replaced by the agroforestry system (Nair et al. 2010).

Organic soils contain high C densities because of delayed decomposition under flooded conditions. Thus, when they are used as croplands organic soils must be drained (Freibauer et al. 2004). SOC is lost, specifically, from organic soils by deep drainage and intensive mechanical disturbance such as deep plowing. Potential alternative uses for organic soils are maintaining a more shallow water table and avoiding deep plowing as well as avoiding cultivation with potatoes (*Solanum tuberosum* L.) and sugar beets (*Beta vulgaris* L.) and instead cropping to permanent cultures (Freibauer et al. 2004).

14.3 Recarbonization of Cropland Soils

The depletion of the SOC pool by cropland cultivation can be partially reversed by recarbonization. Soil C losses may occur when land under other uses is converted for the cultivation of crops. For examples, about 25–30% of the SOC stored in the top meter of soil is released by cultivation of native soils, whether under forest or prairie vegetation (Houghton 2010). Specifically, in temperate regions conversion from forest to cropland and from grassland to cropland may cause the loss of 31 ± 20% SOC to 28.5 ± 13.5 cm depth and of 36 ± 5% SOC to 27.1 ± 11.1 cm depth, respectively (Poeplau et al. 2011). New cropland SOC equilibrium may be reached if at all 23 years after conversion from forest and 17 years following conversion from grassland (Bell et al. 2011). In tropical regions, SOC losses of 25% to 36 ± 4 cm depth and of 30% to 48 ± 8 cm depth occurred by conversion of primary forest to cropland or perennial crops, respectively (Don et al. 2011). Losses of 21% SOC to 39 ± 5 cm depth occurred when secondary tropical forest was converted to cropland but no changes were observed to 51 ± 9 cm depth when converted to perennial crops. When tropical grassland was converted to cropland, 10.4 ± 6.1% of SOC were lost to 38 ± 11 cm depth (Don et al. 2011). However, cropping practices, irrigation, use of fertilizers and different types of tillage affect changes in SOC density. Thus, both conversions of native soils to crop soils and cultivation of croplands may decrease SOC pools. Croplands can partially recarbonized through adoption of recommended management practices (RMPs) such as conservation tillage, residue mulching and use of cover crops, practices which all contribute to soil C accumulation and sequestration by an additional transfer of C from the atmosphere to the soil (Lal 2007; Powlson et al. 2011).

Agricultural systems can be specifically managed to enhance C sequestration (Power 2010). For example, a net gain of the SOC pool can be achieved by conversion of plow tillage (PT) to no till (NT) and other conservation tillage practices, along with crop residue mulch or cover crops (Lal 2009). The transfer of C from the atmosphere into both the soil inorganic carbon (SIC) and SOC pool for enhancement of soil C sinks can be accelerated (Macías and Arbestain 2010). This may be achieved by (i) favouring growth of crop biomass which is the major source for SOC, (ii) promoting and facilitating carbonation processes to increase the SIC pool, (iii) reducing erosional C loss from croplands and favouring pedogenesis for build-up of the soil profile C pool, (iv) developing OM-rich horizons, and/or (v) recovering degraded or contaminated crop soils to restore the soil C sink. Some of the C lost in the past from cropland soils by changes in land use and cultivation can be recovered through improved management, thereby withdrawing atmospheric CO_2 (Smith et al. 2008). Most agricultural soils contain 30–75% less C than their potential capacity as determined by soil, climate, terrain, drainage, land use, and soil and crop management practices (Lal and Follett 2009b). The term 'soil C sequestration' implies that the total C pool in the soil profile is increasing through managerial interventions aimed at transferring atmospheric CO_2 to the soil C pool by moderating either organic and/or inorganic transformations (Lal and Follett 2009b). Thus, practices that retains or returns more of the C captured by growing plants increases the soil C pool (Sanderman and Baldock 2010). Carbon sequestration in crop soil implies an additional transfer of C from the atmosphere to the soil and, thus, a genuine contribution to climate change mitigation (Powlson et al. 2011).

The rate of C sequestration in croplands with adoption of RMPs depends on soil texture and structure, rainfall, temperature, farming system, and soil management (Lal 2004). The SOC pool in croplands can be enhanced by increasing use efficiency of input, decreasing losses by erosion and leaching, and improving soil structure (Lal and Follett 2009b). Further, the SIC pool can be enhanced by application of biosolids, liming/application of cations, and conserving water in the root zone. However, the potential of SIC sequestration by pedogenic carbonate formation is less well known. Soils of irrigated croplands may sequester both SIC and SOC. The rate of soil C sequestration ranges from about 100–1,000 kg ha^{-1} $year^{-1}$ for SOC and 5–15 kg ha^{-1} $year^{-1}$ for SIC (Lal and Follett 2009a). However, some agricultural field trials indicate that the relative increase in soil C pool with adoption of RMPs is not an actual increase but rather due to a reduction or cessation of soil C losses (Sanderman and Baldock 2010). The global C sink capacity of agricultural pools is estimated to be up to 78 Pg C and can be filled at the potential maximum rate of about 1 Pg C $year^{-1}$. However, the attainable and actual cumulative global rate of soil C sequestration may be lower because of managerial, economic, and policy constraints (Lal and Follett 2009a). The duration of soil C sequestration may be 25–50 years (Lal 2004). With increasing saturation of the soil C pool the sink activity diminishes but sequestration may continue due to climate change and land-management change (Bell et al. 2011).

Common RMPs for SOC sequestration are mulch farming, conservation tillage, agroforestry and diverse cropping systems, cover crops, and integrated nutrient management, including the use of manure, compost, and biosolids (Lal 2004). Irrigation can be used to buffer against soil moisture deficits and to sustain crop productivity. Where irrigation is constrained, the selection of drought-resistant crops promotes SOC sequestration. Further, crop production can also be adapted to temperature. For example, wheat in North America is now cultivated in environments once considered too arid, too variable, and too harsh to cultivate (Olmstead and Rhode 2011). Rising atmospheric CO_2 concentrations may directly alter crop yield and the SOC pool (Ainsworth and McGrath 2010). Free-air CO_2 enrichment experiments indicate that soybean and rice grain yield may increase by 13% at 550 ppm CO_2. In contrast, grain yield of sorghum and corn are not expected to increase at elevated CO_2 when water supply is adequate. However, in the long term responses of row crop agroecosystems to management practices may be more important than responses to elevated CO_2 (Moran and Jastrow 2010).

14.4 Conclusions

By converting land under other uses to cropland, by cropland management and its intensification (i.e., fertilization, irrigation, mechanization) humans have altered SOC dynamics on 12% of Earth's ice-free land area. Up to 36% of SOC may have been lost from cropland topsoils. Thus, most cropland soils contain less C than their potential capacity as determined by soil, climate, terrain, drainage, land use, and soil and crop management practices. Photosynthesis is the major natural C input into cropland soils while direct input occur by addition of manure and organic residues. Harvest removes a major proportion of cropland NPP before it enters the soil. Natural crop soil C losses occur by decomposition, erosion and leaching. Cropland SOC can be increased by mulch farming, conservation tillage, agroforestry and diverse cropping systems, cover crops, and integrated nutrient management, including the use of manure, compost, and biosolids. Recent advances have improved our understanding of SOC dynamics and SOC persistence. However, the deep mineral cropland soils have been largely ignored. Further, it remains to be studied whether SOC-accreting crops (e.g., crop plants with a bushy and deep root system, perennial crops) can be cultivated to recarbonize cropland soils.

References

Ainsworth EA, McGrath JM (2010) Direct effects of rising atmospheric carbon dioxide and ozone on crop yields. In: Lobell D, Burke M (eds) Climate change and food security. Springer, Dordrecht, pp 109–130

Ajayi OC, Place F, Akinnifesi FK, Sileshi GW (2011) Agricultural success from Africa: the case of fertilizer tree systems in southern Africa (Malawi, Tanzania, Mozambique, Zambia and Zimbabwe). Int J Agric Sustain 9:129–136

Alexander M (1965) Biodegradation: problems of molecular recalcitrance and microbial fallibility. Adv Appl Microbiol 7:35–80

Alvarez R (2005) A review of nitrogen fertilizer and conservative tillage effects on soil organic storage. Soil Use Manage 21:38–52

Amelung W, Brodowski S, Sandhage-Hofmann A, Bol R (2008) Combining biomarker with stable isotope analyses for assessing the transformation and turnover of soil organic matter. Adv Agron 100:155–250

Amthor JS (2010) From sunlight to phytomass: on the potential efficiency of converting solar radiation to phyto-energy. New Phytol 188:939–959

Andreae MO (2004) Assessment of global emissions from vegetation fires. Int For Fire News 31:112–121

Ayres E, Wall DH, Bardgett RD (2009) Trophic interactions and their implications for soil carbon fluxes. In: Kutsch WL, Bahn M, Heinemeyer A (eds) Soil carbon dynamics: an integrated methodology. Cambridge University Press, Cambridge, UK, pp 187–206

Bachmann J, Guggenberger G, Baumgartl Th, Ellerbrock R, Urbanek E, Goebel M-O, Kaiser K, Horn R, Fischer WR (2008) Physical carbon-sequestration mechanisms under special consideration of soil wettability. J Plant Nutr Soil Sci 171:14–26

Bais HP, Weir TL, Perry LG, Gilroy S, Vivanco JM (2006) The role of root exudates in rhizosphere interactions with plants and other organisms. Annu Rev Plant Biol 57:233–266

Baldock JA, Nelson PN (2000) Soil organic matter. In: Sumner ME (ed) Handbook of soil science. CRC Press, Boca Raton, pp B25–B84
Bardgett RD (2005) The biology of soil: a community and ecosystem approach. Oxford University Press, Oxford
Basile-Doelsch I, Brun T, Borschneck D, Masion A, Marol C, Balesdent J (2009) Effect of landuse on organic matter stabilized in organomineral complexes: a study combining density fractionation, mineralogy and $\delta^{13}C$. Geoderma 151:77–86
Beare MH, Parmelee RW, Hendrix PF, Cheng W, Coleman DC, Crossley DA (1992) Microbial and faunal interactions and effects on litter nitrogen and decomposition in agroecosystems. Ecol Monogr 62:569–591
Beer C, Reichstein M, Ciais P, Farquhar GD, Papale D (2007) Mean annual GPP of Europe derived from its water balance. Geophys Res Lett 34:L05401. doi:10.1029/2006GL029006
Beer C, Reichstein M, Tomelleri E, Ciais P, Jung M, Carvalhais N, Rödenbeck C, Altaf Arain M, Baldocchi D, Bonan GB, Bondeau A, Cescatti A, Lasslop G, Lindroth A, Lomas M, Luyssaert S, Margolis H, Oleson KW, Roupsard O, Veenendal E, Viovy N, Williams C, Ian Woodward F, Papale D (2010) Terrestrial gross carbon dioxide uptake: global distribution and covariation with climate. Science 329:834–838
Bell MJ, Worrall F, Smith P, Bhogal A, Bhogal A, Black H, Lilly A, Barraclough D, Barraclough D, Merrington G (2011) UK land-use change and its impact on SOC: 1925–2007. Glob Biogeochem Cycle 25:GB4015. doi:10.1029/2010GB0038
Berg B, McClaugherty C (2008) Plant litter: decomposition, humus formation, carbon sequestration. Springer, Berlin
Billings SA, Buddemeier RW, deB Richter D, Van Oost K, Bohling G (2010) A simple method for estimating the influence of eroding soil profiles on atmospheric CO_2. Glob Biogeochem Cycle 24:GB2001. doi:10.1029/2009GB003560
Blair N, Faulkner RD, Till AR, Körschens M, Schulz E (2006a) Long-term management impacts on soil C, N and physical fertility. Part II: Bad Lauchstädt static and extreme FYM experiments. Soil Till Res 91:39–47
Blair N, Faulkner RD, Till AR, Poulton PR (2006b) Long-term management impacts on soil C, N and physical fertility Part I: Broadbalk experiment. Soil Till Res 91:30–38
Bocock KL, Gilbert O (1957) The disappearance of leaf litter under different woodland conditions. Plant Soil 9:179–185
Bolan NS, Adriano DC, Kunhikrishnan A, James T, McDowell R, Senesi N (2011) Dissolved organic matter: biogeochemistry, dynamics, and environmental significance in soils. Adv Agron 110:1–75
Bolinder MA, Angers DA, Giroux M, Laverdiere MR (1999) Estimating C inputs retained as soil organic matter from corn (*Zea mays* L.). Plant Soil 215:85–91
Bolinder MA, Janzen HH, Gregorich EG, Angers DA, VandenBygaart AJ (2007) An approach for estimating net primary productivity and annual carbon inputs to soil for common agricultural crops in Canada. Agric Ecosyst Environ 118:29–42
Bonan GB, Lawrence PJ, Oleson KW, Levis S, Jung M, Reichstein M, Lawrence DM, Swenson SC (2011) Improving canopy processes in the Community Land Model version 4 (CLM4) using global flux fields empirically inferred from FLUXNET data. J Geophys Res 116:G02014. doi:10.1029/2010JG001593
Bondeau A, Smith P, Zaehle S, Schaphoff S, Lucht W, Cramer W, Gerten D, Lotze-Campen H, Müller C, Reichstein M, Smith B (2007) Modelling the role of agriculture for the 20th century global terrestrial carbon balance. Glob Change Biol 13:679–706
Bond-Lamberty B, Thomson A (2010) A global database of soil respiration data. Biogeosciences 7:1915–1926
Bossuyt H, Six J, Hendrix PF (2005) Protection of soil carbon by microaggregates within earthworm casts. Soil Biol Biochem 37:251–258
Botha A (2011) The importance and ecology of yeasts in soil. Soil Biol Biochem 43:1–8
Bradford JB, Lauenroth WK, Burke IC (2005) The impact of cropping on primary production in the U.S. Great Plains. Ecology 86:1863–1872

Broadbent FE, Nakashima T (1974) Mineralization of carbon and nitrogen in soil amended with carbon-13 and nitrogen-15 labeled plant material. Soil Sci Soc Am Proc 38:313–315
Bronick CJ, Lal R (2005) Soil structure and management: a review. Geoderma 124:3–22
Cameron DD (2010) Arbuscular mycorrhizal fungi as (agro)ecosystem engineers. Plant Soil 333:1–5
Chantigny MH (2003) Dissolved and water-extractable organic matter in soils: a review on the influence of land use and management practices. Geoderma 113:357–380
Chapin FS III, Matson PA, Mooney HA (2002) Principles of terrestrial ecosystem ecology. Springer, New York
Chen Z, Pawluk S, Ng J (1997) Impact of variations in granular structures on carbon sequestration in two Alberta mollisols. In: Lal R, Kimble JM, Follett RF, Stewart BA (eds) Soil processes and the carbon cycle. CRC Press, Boca Raton, pp 225–243
Christensen BT (1996) Carbon in primary and secondary organomineral complexes. In: Carter MR, Stewart BA (eds) Structure and organic matter storage in agricultural soils. CRC Press, Boca Raton, FL, pp 97–165
Ciais P, Wattenbach M, Vuichard N, Smith P, Piao SL, Don A, Luyssaert S, Janssens IA, Bondeau A, Dechow R, Leip A, Smith PC, Beer C, Van DerWerf GR, Gervois S, Van Oost K, Tomelleri E, Freibauer A, Schulze E-D, CARBOEUROPE Synthesis Team (2010) The European carbon balance. Part 2: Croplands. Glob Change Biol 16:1409–1428
Ciais P, Gervois S, Vuichard N, Piao SL, Viovy N (2011) Effects of land use change and management on the European cropland carbon balance. Glob Change Biol 17:320–338
Clemente JS, Simpson AJ, Simpson MJ (2011) Association of specific organic matter compounds in size fractions of soils under different environmental controls. Org Geochem 42:1169–1180
Coleman K, Jenkinson DS (1995) ROTHC-26.3. A model for the turnover of carbon in soil. Model description and user's guide. Lawes Agricultural Trust, Harpenden
Cotrufo MF, Del Galdo I, Piermatteo D (2009) Litter decomposition: concepts, methods and future perspectives. In: Kutsch WL, Bahn M, Heinemeyer A (eds) Soil carbon dynamics: an integrated methodology. Cambridge University Press, Cambridge, pp 76–90
Crutzen PJ, Andreae MO (1990) Biomass burning in the tropics: impact on atmospheric chemistry and biogeochemical cycles. Science 250:1669–1678
Davidson EA, Janssens IA (2006) Temperature sensitivity of soil carbon decomposition and feedbacks to climate change. Nature 440:165–173
Denman KL, Brasseur G, Chidthaisong A, Ciais P, Cox PM, Dickinson RE, Hauglustaine D, Heinze C, Holland E, Jacob D, Lohmann U, Ramachandran S, da Silva Dias PL, Wofsy SC, Zhang X (2007) Couplings between changes in the climate system and biogeochemistry; Chapter 7. In: Intergovernmental Panel on Climate Change (ed) Climate change 2007: the physical science basis, Cambridge University Press, Cambridge
Don A, Schumacher J, Freibauer A (2011) Impact of tropical land-use change on soil organic carbon stocks – a meta-analysis. Glob Change Biol 17:1658–1670
Dufranne D, Moureaux C, Vancutsem F, Bodson B, Aubinet M (2011) Comparison of carbon fluxes, growth and productivity of a winter wheat crop in three contrasting growing seasons. Agric Ecosyst Environ 141:133–142
Edwards CA, Bohlen PJ (1996) Biology and ecology of earthworms. Chapman and Hall, London
Ekschmitt K, Kandeler E, Poll C, Brune A, Buscot F, Friedrich M, Gleixner G, Hartmann A, Kästner M, Marhan S, Miltner A, Scheu S, Wolters V (2008) Soil-carbon preservation through habitat constraints and biological limitations on decomposer activity. J Plant Nutr Soil Sci 171:27–35
Ellis EC, Ramankutty N (2008) Putting people in the map: anthropogenic biomes of the world. Front Ecol Environ 6:439–447
Ewing SA, Sandermann J, Baisden WT, Wang Y, Amundson R (2006) Role of large-scale soil structure in organic carbon turnover: evidence from California grassland soils. J Geophys Res 111:G03012. doi:10.1029/2006JG000174
Falge E, Baldocchi D, Tenhunen J, Aubinet M, Bakwin P, Berbigier P, Bernhofer C, Burba G, Clement R, Davis KJ, Elbers JA, Goldstein AH, Grelle A, Granier A, Guomundsson J, Hollinger D, Kowalski AS, Katul G, Law BE, Malhi Y, Meyers T, Monson RK, Munger JW, Oechel W,

Paw UKT, Pilegaard K, Rannik U, Rebmann C, Suyker AE, Valentini R, Wilson K, Wofsy S (2002) Seasonality of ecosystem respiration and gross primary production as derived from FLUXNET measurements. Agric For Meteorol 113:53–74

Falloon P, Smith P (2003) Accounting for changes in soil carbon under the Kyoto Protocol: need for improved long-term data sets to reduce uncertainty in model projections. Soil Use Manage 19:265–269

Falloon P, Smith P (2009) Modelling soil carbon dynamics. In: Kutsch WL, Bahn M, Heinemeyer A (eds) Soil carbon dynamics: an integrated methodology. Cambridge University Press, Cambridge, pp 221–244

Feng X, Hills KM, Simpson AJ, Whalen JK, Simpson MJ (2011) The role of biodegradation and photo-oxidation in the transformation of terrigenous organic matter. Org Geochem 42:262–274

Fierer N, Schimel JP, Holden PA (2003) Variations in microbial community composition through two soil depth profiles. Soil Biol Biochem 35:167–176

Flessa H, Amelung W, Helfrich M, Wiesenberg GLB, Gleixner G, Brodowski S, Rethemeyer J, Kramer C, Grootes PM (2008) Storage and stability of organic matter and fossil carbon in a Luvisol and Phaeozem with continuous maize cropping: a synthesis. J Plant Nutr Soil Sci 131:36–51

Foley JA, DeFries R, Asner GP, Barford C, Bonan G, Carpenter SR, Chapin FS, Coe MT, Daily GC, Gibbs HK, Helkowski JH, Holloway T, Howard EA, Kucharik CJ, Monfreda C, Patz JA, Prentice IC, Ramankutty N, Snyder PK (2005) Global consequences of land use. Science 309:570–574

Foley JA, Ramankutty N, Brauman KA, Cassidy ES, Gerber JS, Johnston M, Mueller ND, O'Connell C, Ray DK, West PC, Balzer C, Bennett EM, Carpenter SR, Hill J, Monfreda C, Polasky S, Rockström J, Sheehan J, Siebert S, Tilman D, Zaks DPM (2011) Solutions for a cultivated planet. Nature 478:337–342

Fontaine S, Barot S, Barré P, Bdioui N, Mary B, Rumpel C (2007) Stability of organic carbon in deep soil layers controlled by fresh carbon supply. Nature 450:277–281

Fox O, Vetter S, Ekschmitt K, Wolters V (2006) Soil fauna modifies the recalcitrance-persistence relationship of soil carbon pools. Soil Biol Biochem 38:1353–1363

Franzluebbers AJ (2010) Will we allow soil carbon to feed our needs? Carbon Manage 1:237–251

Freibauer A, Rounsevell M, Smith P, Verhagen A (2004) Carbon sequestration in the agricultural soils of Europe. Geoderma 122:1–23

Gartner TB, Cardon ZG (2004) Decomposition dynamics in mixed-species leaf litter. Oikos 104:230–246

Ghannoum O (2009) C_4 photosynthesis and water stress. Ann Bot 103:635–644

Giardina CP, Ryan MG (2000) Evidence that decomposition rates of organic carbon in mineral soil do not vary with temperature. Nature 404:858–861

Gibbs KH, Ruesch AS, Achard F, Clayton MK, Holmgren P, Ramankutty N, Foley JA (2010) Tropical forests were the primary sources of new agricultural land in the 1980s and 1990s. Proc Natl Acad Sci USA 107:16732–16737

Gill RA, Kelly RH, Parton WJ, Day KA, Jackson RB, Morgan JA, Scurlock JMO, Tieszen LL, Castle JV, Ojima DS, Zhang XS (2002) Using simple environmental variables to estimate belowground productivity in grasslands. Glob Ecol Biogeogr 11:79–86

Glover JD, Reganold JP, Bell LW, Borevitz J, Brummer EC, Buckler ES, Cox CM, Cox TS, Crews TE, Culman SW, DeHaan LR, Eriksson D, Gill BS, Holland J, Hu F, Hulke BS, Ibrahim AMH, Jackson W, Jones SS, Murray SC, Paterson AH, Ploschuk E, Sacks EJ, Snapp S, Tao D, Van Tassel DL, Wade LJ, Wyse DL, Xu Y (2010) Increased food and ecosystem security via perennial grains. Science 328:1638–1639

Goebel M-O, Woche SK, Bachmann J (2009) Do soil aggregates really protect encapsulated organic matter against microbial decomposition? Biologia 64:443–448

Goudriaan J, Groot JJR, Uithol PWJ (2001) Productivity of agro-ecosystems. In: Roy J, Saugier B, Mooney HA (eds) Terrestrial global productivity. Academic, San Diego, pp 301–313

Govaerts B, Verhulst N, Castellanos-Navarrete A, Sayre KD, Dixon J, Dendooven L (2009) Conservation agriculture and soil carbon sequestration: between myth and farmer reality. Crit Rev Plant Sci 28:97–122
Gray CM, Monson RK, Fierer N (2010) Emissions of volatile organic compounds during the decomposition of plant litter. J Geophys Res 115:G03015. doi:10.1029/2010JG001291
Greenhouse Gas Working Group (2010) Agriculture's role in greenhouse gas emissions & capture. Greenhouse Gas Working Group Rep. ASA, CSSA, and SSSA, Madison
Gregory PJ, Palta JA, Batts GR (1997) Root systems and root: mass ratio – carbon allocation under current and projected atmospheric conditions in arable crops. Plant Soil 187:221–228
Guo D, Li H, Mitchell RJ, Han W, Hendricks JJ, Fahey TJ, Hendrick RL (2008) Fine root heterogeneity by branch order: exploring the discrepancy in root turnover estimates between minirhizotron and carbon isotopic methods. New Phytol 177:443–456
Haberl H, Erb K-H, Krausmann F, Gaube V, Bondeau A, Plutzar C, Gingrich S, Lucht W, Fischer-Kowalski M (2007) Quantifying and mapping the human appropriation of net primary production in earth's terrestrial ecosystems. Proc Natl Acad Sci USA 104:12942–12947
Hammes K, Torn MS, Lapenas AG, Schmidt MWI (2008) Centennial black carbon turnover in a Russian steppe soil. Biogeosciences 5:1339–1350
Harmon ME, Silver WL, Fasth B, Chen H, Burke IC, Parton WJ, Hart SC, Currie WS, LIDET (2009) Long-term patterns of mass loss during the decomposition of leaf and fine root litter: an intersite comparison. Glob Change Biol 15:1320–1338
Hättenschwiler S, Tiunov AV, Scheu S (2005) Biodiversity and litter decomposition in terrestrial ecosystems. Annu Rev Ecol Evol Syst 36:191–218
Hicke JA, Lobell DB, Asner GP (2004) Cropland area and net primary production computed from 30 years of USDA agricultural harvest data. Earth Interact 8:1–21
Hodge A, Berta G, Doussan C, Merchan F, Crespi M (2009) Plant root growth, architecture and function. Plant Soil 321:153–187
Hofmann A, Heim A, Christensen BT, Miltner A, Gehre M, Schmidt MWI (2009) Lignin dynamics in two ^{13}C-labelled arable soils during 18 years. Eur J Soil Sci 60:205–257
Horn R, Smucker A (2005) Structure formation and its consequences for gas and water transport in unsaturated arable and forest soils. Soil Till Res 82:5–14
Houghton RA (2010) How well do we know the flux of CO_2 from land-use change? Tellus 62B:337–351
Insam H, Seewald MSA (2010) Volatile organic compounds (VOCs) in soils. Biol Fertil Soils 46:199–213
Jansson C, Wullschleger SD, Kalluri UC, Tuskan GA (2010) Phytosequestration: carbon biosequestration by plants and the prospects of genetic engineering. Bioscience 60:685–696
Jenkinson DS, Coleman K (2008) The turnover of organic carbon in subsoils. Part 2: Modelling carbon turnover. Eur J Soil Sci 59:400–413
Jiang GM, Noonan MJ, Buchan GD, Smith N (2005) Transport and deposition of *Bacillus subtilis* through an intact soil column. Aust J Soil Res 43:695–703
Jobbágy EG, Jackson RB (2000) The vertical distribution of soil organic carbon and its relation to climate and vegetation. Ecol Appl 10:423–436
Joergensen RG, Wichern F (2008) Quantiative assessment of the fungal contribution to microbial tissue in soil. Soil Biol Biochem 40:2977–2991
Johnson JMF, Allmaras RR, Reicosky DC (2006) Estimating source carbon from crop residues, roots and rhizodeposits using the national grain-yield database. Agron J 98:622–636
Jones DL, Nguyen C, Finlay RD (2009) Carbon flow in the rhizosphere: carbon trading at the soil-root interface. Plant Soil 321:5–33
Kalbitz K, Solinger S, Park J-H, Michalzik B, Matzner E (2000) Controls on the dynamics of dissolved organic matter in soils: a review. Soil Sci 165:277–304
Kampichler C, Bruckner A (2009) The role of microarthropods in terrestrial decomposition: a meta-analysis of 40 years of litterbag studies. Biol Rev 84:375–389
Kelleher BP, Simpson AJ (2006) Humic substances in soils: are they really chemically distinct? Environ Sci Technol 40:4605–4611

Kindler R, Siemens J, Kaiser K, Walmsley DC, Bernhofer C, Buchmann N, Cellier P, Eugster W, Gleixner G, Grûnwald T, Heim A, Ibrom A, Jones SK, Jones M, Klumpp K, Kutsch W, Larsen KS, Lehuger S, Loubet B, McKenzie R, Moor E, Osborne B, Pilegaard K, Rebmann C, Saunders M, Schmidt MWI, Schrumpf M, Seyfferth J, Skiba U, Soussana J-F, Sutton MA, Tefs C, Vowinckel B, Zeeman MJ, Kaupenjohann M (2011) Dissolved carbon leaching from soil is a crucial component of the net ecosystem carbon balance. Glob Change Biol 17:1167–1185

Kleber M (2010a) What is recalcitrant soil organic matter? Environ Chem 7:320–332

Kleber M (2010b) Response to the Opinion paper by Margit von Lützow and Ingrid Kögel-Knabner on 'What is recalcitrant soil organic matter?' by Markus Kleber. Environ Chem 7:336–337

Kleber M, Johnson MG (2010) Advances in understanding the molecular structure of soil organic matter: implications for interactions in the environment. Adv Agron 106:77–142

Kleber M, Sollins P, Sutton R (2007) A conceptual model of organo-mineral associations in soils: self-assembly of organic molecular fragments into zonal structures on mineral surfaces. Biogeochemistry 85:9–24

Kleber M, Nico PS, Plante A, Filley T, Kramer M, Swanston C, Sollins P (2011) Old and stable soil organic matter is not necessarily chemically recalcitrant: implications for modeling concepts and temperature sensitivity. Glob Change Biol 17:1097–1107

Knacker T, Förster B, Römbke J, Frampton GK (2003) Assessing the effects of plant protection products on organic matter breakdown in arable fields-litter decomposition test systems. Soil Biol Biochem 35:1269–1287

Kong AYY, Six J (2010) Tracing root vs. residue carbon into soils from conventional and alternative cropping systems. Soil Sci Soc Am J 74:1201–1210

Krull ES, Baldock JA, Skjemstad JO (2003) Importance of mechanisms and processes of the stabilisation of soil organic matter for modelling carbon turnover. Funct Plant Biol 30:207–222

Kutsch WL, Bahn M, Heinemeyer A (2009) Soil carbon relations: an overview. In: Kutsch WL, Bahn M, Heinemeyer A (eds) Soil carbon dynamics: an integrated methodology. Cambridge University Press, Cambridge, pp 1–15

Kuzyakov Y (2002) Review: factors affecting rhizosphere priming effects. J Plant Nutr Soil Sci 165:382–396

Kuzyakov Y (2006) Sources of CO_2 efflux from soil and review of partitioning methods. Soil Biol Biochem 38:425–448

Kuzyakov Y (2010) Priming effects: interactions between living and dead organic matter. Soil Biol Biochem 42:1363–1371

Kuzyakov Y, Domanski G (2000) Carbon input by plants into the soil. Rev J Plant Nutr Soil Sci 163:421–431

Lal R (2003) Soil erosion and the global carbon budget. Environ Int 29:437–450

Lal R (2004) Soil carbon sequestration to mitigate climate change. Geoderma 123:1–22

Lal R (2007) Farming carbon. Soil Till Res 96:1–5

Lal R (2009) Soils and food sufficiency. A review. Agron Sustain Dev 29:113–133

Lal R, Follett RF (2009a) Preface. In: Lal R, Follett RF (eds) Soil carbon sequestration and the greenhouse effect, vol 57, 2nd edn, SSSA special publication. Soil Science Society of America, Madison, pp xi–xii

Lal R, Follett RF (2009b) Soils and climate change. In: Lal R, Follett RF (eds) Soil carbon sequestration and the greenhouse effect, vol 57, 2nd edn, SSSA special publication. Soil Science Society of America, Madison, pp xxi–xxviii

Lambers H, Robinson SA, Ribas-Carbo M (2005) Regulation of respiration in vivo. In: Lambers H, Ribas-Carbo M (eds) Plant respiration: from cell to ecosystem. Springer, Dordrecht, pp 43–61

Laothawornkitkul J, Taylor JE, Paul ND, Hewitt CN (2009) Biogenic volatile organic compounds in the Earth system. New Phytol 183:27–51

Leake J, Johnson D, Donnelly D, Muckle G, Boddy L, Read D (2004) Networks of power and influence: the role of mycorrhizal mycelium in controlling plant communities and agroecosystem functioning. Can J Bot 82:1016–1045

Leff JW, Fierer N (2008) Volatile organic compound (VOC) emissions from soil and litter samples. Soil Biol Biochem 40:1629–1636

Lehmann J, Solomon D, Kinyangi J, Dathe L, Wirick S, Jacobsen C (2008) Spatial complexity of soil organic matter forms at nanometre scales. Nat Geosci 1:238–242

Lei H-M, Yang D-W (2010) Seasonal and interannual variations in carbon dioxide exchange over a cropland in the North China Plain. Global Change Biol 16:2944–2957

Lorenz K, Lal R (2005) The depth distribution of soil organic carbon in relation to land use and management and the potential of carbon sequestration in subsoil horizons. Adv Agron 88:35–66

Lorenz K, Lal R, Preston CM, Nierop KGJ (2007) Strengthening the soil organic carbon pool by increasing contributions from recalcitrant aliphatic bio(macro)molecules. Geoderma 142:1–10

Loreto F, Kesselmeier J, Schnitzler JP (2008) Volatile organic compounds in the biosphere–atmosphere system: a preface. Plant Biol 10:2–7

Macías F, Arbestain MC (2010) Soil carbon sequestration in a changing global environment. Mitig Adapt Strateg Glob Change 15:511–529

Marlon JR, Bartlein PJ, Carcaillet C, Gavin DG, Harrison SP, Higuera PE, Joos F, Power MJ, Prentice IC (2008) Climate and human influences on global biomass burning over the past two millennia. Nat Geosci 1:697–702

Marschner H, Dell B (1994) Nutrient uptake in mycorrhizal symbiosis. Plant Soil 159:89–102

Marschner B, Kalbitz K (2003) Controls of bioavailability and biodegradability of dissolved organic matter in soils. Geoderma 113:211–235

Marschner B, Brodowski S, Dreves A, Gleixner G, Gude A, Grootes PM, Hamer U, Heim A, Jandl G, Ji R, Kaiser K, Kalbitz K, Kramer C, Leinweber P, Rethemeyer J, Schäffer A, Schmidt MWI, Schwark L, Wiesenberg GLB (2008) How relevant is recalcitrance for the stabilization of organic matter in soils? J Plant Nutr Soil Sci 171:91–110

Martin A (1991) Short- and long-term effects of the endogenic earthworm *Millsonia anomala* (Omodeo) (Megascolecidae: Oligochaeta) of tropical savanna, on soil organic matter. Biol Fertil Soils 11:234–238

McGuire KL, Treseder KK (2010) Microbial communities and their relevance for ecosystem models: decomposition as a case study. Soil Biol Biochem 42:529–535

Mendez-Millan M, Dignac MF, Rumpel C, Rasse DP, Derenne S (2010) Molecular dynamics of shoot vs. root biomarkers in an agricultural soil estimated by natural abundance ^{13}C labelling. Soil Biol Biochem 42:169–177

Miller RM, Jastrow JD (1990) Hierarchy of root and mycorrhizal fungal interactions with soil aggregation. Soil Biol Biochem 22:579–584

Mokany K, Raison RJ, Prokushkin AS (2006) Critical analysis of root:shoot ratios in terrestrial biomes. Glob Change Biol 12:84–96

Monfreda C, Ramankutty N, Foley JA (2008) Farming the planet: 2. Geographic distribution of crop areas, yields, physiological types, and net primary production in the year 2000. Glob Biogeochem Cycle 22:GB1022. doi:10.1029/2007GB002947

Moni C, Rumpel C, Virto I, Chabbi A, Chenu C (2010) Relative importance of sorption versus aggregation for organic matter storage in subsoil horizons of two contrasting soils. Eur J Soil Sci 61:958–969

Moore TR, Trofymow JA, Prescott CE, Titus BD, CIDET Working Group (2011) Nature and nurture in the dynamics of C, N and P during litter decomposition in Canadian forests. Plant Soil 339:163–175

Moorhead DL, Sinsabaugh RL (2006) A theoretical model of litter decay and microbial interaction. Ecol Monogr 76:151–174

Moran KK, Jastrow JD (2010) Elevated carbon dioxide does not offset loss of soil carbon from a corn–soybean agroecosystem. Environ Pollut 158:1088–1094

Murchie EH, Pinto M, Horton P (2009) Agriculture and the new challenges for photosynthesis research. New Phytol 181:532–552

Nair PKR, Gordon AM, Mosquera-Losada M-R (2008) Agroforestry. In: Jorgensen SE, Fath BD (eds) Ecological engineering, encyclopedia of ecology, vol 1. Elsevier, Oxford, pp 101–110

Nair PKR, Kumar BM, Nair VD (2009) Agroforestry as a strategy for carbon sequestration. J Plant Nutr Soil Sci 172:10–23

Nair PKR, Nair VD, Kumar BM, Showalter JM (2010) Carbon sequestration in agroforestry systems. Adv Agron 108:237–307

Nguyen C (2003) Rhizodeposition of organic C by plants: mechanisms and controls. Agronomie 23:375–396

Oades JM (1984) Soil organic matter and structural stability, mechanisms and implications for management. Plant Soil 76:319–337

Olmstead AL, Rhode PW (2011) Adapting North American wheat production to climatic challenges, 1839–2009. Proc Natl Acad Sci USA 108:480–485

Osler GHR, Sommerkorn M (2007) Toward a complete soil C and N cycle: incorporating the soil fauna. Ecology 88:1611–1621

Parton WJ, Schimel DS, Cole CV, Ojima DS (1987) Analysis of factors controlling soil organic matter levels in Great Plains grasslands. Soil Sci Soc Am J 51:1173–1179

Pechony O, Shindell DT (2010) Driving forces of global wildfires over the past millennium and the forthcoming century. Proc Natl Acad Sci USA 107:19167–19170

Peng Y, Gitelson AA (2011) Application of chlorophyll-related vegetation indices for remote estimation of maize productivity. Agric For Meteorol 151:1267–1276

Peng Y, Gitelson AA, Keydan G, Rundquist DC, Moses W (2011) Remote estimation of gross primary production in maize and support for a new paradigm based on total crop chlorophyll content. Remote Sens Environ 115:978–989

Pielke RA, Pitman A, Niyogi D, Mahmood R, McAlpine C, Hossain F, Goldewijk KK, Nair U, Betts R, Fall S, Reichstein M, Kabat P, de Noblet N (2011) Land use/land cover changes and climate: modeling analysis and observational evidence. WIREs Clim Change 2:828–850

Poeplau C, Don A, Vesterdal L, Leifeld J, Van Wesemael B, Schumacher J, Gensior A (2011) Temporal dynamics of soil organic carbon after land-use change in the temperate zone – carbon response functions as a model approach. Glob Change Biol 17:2415–2427

Polyakov V, Lal R (2004) Modeling soil organic matter dynamics as affected by soil water erosion. Environ Int 30:547–556

Portmann FT, Siebert S, Döll P (2010) MIRCA2000—Global monthly irrigated and rainfed crop areas around the year 2000: a new high resolution data set for agricultural and hydrological modeling. Glob Biogeochem Cycles 24:GB1011. doi:10.1029/2008GB003435

Power AG (2010) Ecosystem services and agriculture: tradeoffs and synergies. Philos Trans R Soc B 365:2959–2971

Powlson DS, Whitmore AP, Goulding KWT (2011) Soil carbon sequestration to mitigate climate change: a critical re-examination to identify the true and the false. Eur J Soil Sci 62:42–55

Prescott CE (2005) Do rates of litter decomposition tell us anything we really need to know? For Ecol Manage 220:66–74

Prescott CE (2010) Litter decomposition: what controls it and how can we alter it to sequester more carbon in forest soils? Biogeochemistry 101:133–149

Preston CM, Trofymow JA, Sayer BG, Niu J (1997) ^{13}C nuclear magnetic resonance spectroscopy with cross-polarization and magic-angle spinning investigation of the proximate-analysis fractions used to assess litter quality in decomposition studies. Can J Bot 75:1601–1613

Prince SD, Haskett J, Steininger M, Strand H, Wright R (2001) Net primary production of U.S. Midwest croplands from agricultural harvest yield data. Ecol Appl 11:1194–1205

Pritchard SG, Strand AE (2008) Can you believe what you see? Reconciling minirhizotron and isotopically derived estimates of fine root longevity. Plant Phytol 177:287–291

Purin S, Rillig MC (2007) The arbuscular mycorrhizal fungal protein glomalin: limitations, progress, and a new hypothesis for its function. Pedobiologia 51:123–130

Quinton JN, Govers G, Van Oost K, Bardgett RD (2010) The impact of agricultural soil erosion on biogeochemical cycling. Nat Geosci 3:311–314

Ramankutty N, Evan AT, Monfreda C, Foley JA (2008) Farming the planet: 1. Geographic distribution of global agricultural lands in the year 2000. Glob Biogeochem Cycle 22:GB1003. doi:10.1029/2007GB002952

Rasse DP, Rumpel C, Dignac M-F (2005) Is soil carbon mostly root carbon? Mechanisms for a specific stabilization. Plant Soil 269:341–356
Rasse DP, Mulder J, Moni C, Chenu C (2006) Carbon turnover kinetics with depth in a French loamy soil. Soil Sci Soc Am J 70:2097–2105
Reay DS (2003) Sinking methane. Biologist 50:15–19
Rees RM, Bingham IJ, Baddeley JA, Watson CA (2005) The role of plants and land management in sequestering soil carbon in temperate arable and grassland ecosystems. Geoderma 128:130–154
Rillig MC, Ramsey PW, Morris S, Paul EA (2003) Glomalin, an arbuscular-mycorrhizal fungal soil protein, responds to land-use change. Plant Soil 253:293–299
Ritz K (1995) Growth responses of some fungi to spatially heterogeneous nutrients. FEMS Microbiol Ecol 16:269–280
Robbins M (2011) Crops and carbon. Earthscan, Oxon
Robinson D (2007) Implications of a large global root biomass for carbon sink estimates and for soil carbon dynamics. Proc R Soc B 274:2753–2759
Rovira AD, Greacen EL (1957) The effect of aggregate disruption on the activity of microorganisms in soil. Aust J Agric Res 8:659–673
Rumpel C, Kögel-Knabner I (2011) Deep soil organic matter—a key but poorly understood component of terrestrial C cycle. Plant Soil 338:143–158
Rutledge S, Campbell DI, Baldocchi D, Schipper LA (2010) Photodegradation leads to increased carbon dioxide losses from terrestrial organic matter. Glob Change Biol 16:3065–3074
Salomé C, Nunan N, Pouteau V, Lerch TZ, Chenu C (2010) Carbon dynamics in topsoil and in subsoil may be controlled by different regulatory mechanisms. Glob Change Biol 16:416–426
Sanaullah M, Chabbi A, Leifeld J, Bardoux G, Billou D, Rumpel C (2011) Decomposition and stabilization of root litter in top- and subsoil horizons: what is the difference? Plant Soil 338:127–141
Sanderman J, Amundson R (2008) A comparative study of dissolved organic carbon transport and stabilization in California forest and grassland soils. Biogeochemistry 89:309–327
Sanderman J, Baldock JA (2010) Accounting for soil carbon sequestration in national inventories: a soil scientist's perspective. Environ Res Lett 5:034003
Sanderman J, Farquharson R, Baldock J (2010) Soil carbon sequestration potential: a review for Australian agriculture – a report prepared for department of climate change and energy efficiency, CSIRO land and water. http://www.csiro.au/resources/Soil-Carbon-Sequestration-Potential-Report.html
Sarkhot DV, Comerford NB, Jokela EJ, Reeves JB, Harris WG (2007) Aggregation and aggregate carbon in a forested southeastern coastal plain Spodosol. Soil Sci Soc Am J 71:1779–1787
Saugier B, Roy J, Mooney HA (2001) Estimations of global terrestrial productivity: converging toward a single number? In: Roy J, Saugier B, Mooney HA (eds) Terrestrial global productivity. Academic, San Diego, pp 543–557
Schenk HJ, Jackson RB (2005) Mapping the global distribution of deep roots in relation to climate and soil characteristics. Geoderma 126:129–140
Schindler FV, Mercer EJ, Rice JA (2007) Chemical characteristics of glomalin-related soil protein (GRSP) extracted from soils of varying organic matter content. Soil Biol Biochem 39:320–329
Schmidt MWI, Torn MS, Abiven S, Dittmar T, Guggenberger G, Janssens IA, Kleber M, Kögel-Knabner I, Lehmann J, Manning DAC, Nannipieri P, Rasse DP, Weiner S, Trumbore SE (2011) Persistence of soil organic matter as an ecosystem property. Nature 478:49–56
Schomberg HH, Steiner JL, Unger PW (1994) Decomposition and nitrogen dynamics of crop residues: residue quality and water effects. Soil Sci Soc Am J 58:372–381
Schulze E-D, Beck E, Müller-Hohenstein K (2005) Plant ecology. Springer, Berlin
Schulze E-D, Luyssaert S, Ciais P, Freibauer A, Janssens IA et al (2009) Importance of methane and nitrous oxide for Europe's terrestrial greenhouse-gas balance. Nat Geosci 2:842–850
Schulze E-D, Ciais P, Luyssaert S, Schrumpf M, Janssens IA, Thiruchittampalam B, Theloke J, Saurat M, Bringezu S, Lelieveld J, Lohila A, Rebmann C, Jung M, Bastviken D, Abril G,

Grassi G, Leip A, Freibauer A, Kutsch W, Don A, Nieschulze J, Börner A, Gash JH, Dolman AJ (2010) The European carbon balance. Part 4: Integration of carbon and other trace-gas fluxes. Glob Change Biol 16:1451–1469
Sen R (2003) The root–microbe–soil interface: new tools for sustainable plant production. New Phytol 157:391–394
Séquaris J-M, Guisado G, Magarinos M, Moreno C, Burauel P, Narres H-D, Vereecken H (2010) Organic-carbon fractions in an agricultural topsoil assessed by the determination of the soil mineral surface area. J Plant Nutr Soil Sci 173:699–705
Shibu ME, Leffelaar PA, Van Keulen H, Aggarwal PK (2006) Quantitative description of soil organic matter dynamics – a review of approaches with reference to rice-based cropping systems. Geoderma 137:1–18
Siebert S, Döll P (2010) Quantifying blue and green virtual water contents in global crop production as well as potential production losses without irrigation. J Hydrol 384:198–217
Silver WL, Miya RK (2001) Global patterns in root decomposition: comparisons of climate and litter quality effects. Oecologia 129:407–419
Singh BK, Bardgett RD, Smith P, Reay DS (2010) Microorganisms and climate change: terrestrial feedbacks and mitigation options. Nat Rev Microbiol 8:779–790
Six J, Elliott ET, Paustian K (2000) Soil macroaggregate turnover and microaggregate formation: a mechanism for C sequestration under no-tillage agriculture. Soil Biol Biochem 32:2099–2103
Six J, Feller C, Denef K, Ogle SM, de Moraes Sa JC, Albrecht A (2002) Soil organic matter, biota and aggregation in temperate and tropical soils – effects of no-tillage. Agronomie 22:755–775
Six J, Frey SD, Thiet RK, Batten KM (2006) Bacterial and fungal contributions to carbon sequestration in agroecosystems. Soil Sci Soc Am J 70:555–569
Smith KA, Conen F (2004) Impacts of land management on fluxes of trace greenhouse gases. Soil Use Manage 20:255–263
Smith SE, Read DJ (2008) Mycorrhizal symbiosis. Academic, London
Smith TM, Shugart HH, Woodward FI, Burton PJ (1993) Plant functional types. In: Soloon AM, Shugart HH (eds) Vegetation dynamics and global change. Chapman and Hall, New York, pp 272–292
Smith P, Martino D, Cai Z, Gwary D, Janzen H, Kumar P, McCarl B, Ogle S, O'Mara F, Rice C, Scholes B, Sirotenko O, Howden M, McAllister T, Pan G, Romanenkov V, Schneider U, Towprayoon S, Wattenbach M, Smith J (2008) Greenhouse gas mitigation in agriculture. Philios Trans R Soc B 363:789–813
Stevenson FJ (1994) Humus chemistry: genesis, composition, reactions. Wiley, New York
Stewart CE, Paustian K, Conant RT, Plante AF, Six J (2007) Soil carbon saturation: concept, evidence and evaluation. Biogeochem 86:19–31
Stewart CE, Plante AF, Paustian K, Conant RT, Six J (2008) Soil carbon saturation: linking concept and measurable carbon pools. Soil Sci Soc Am J 72:379–392
Stewart CE, Paustian K, Conant RT, Plante AF, Six J (2009) Soil carbon saturation: implications for measurable carbon pool dynamics in long-term incubations. Soil Biol Biochem 41:357–366
Still CJ, Berry JA, Collatz GJ, DeFries RS (2003) Global distribution of C_3 and C_4 vegetation: carbon cycle implications. Glob Biogeochem Cycle 17:1006. doi:10.1029/2001GB001807
Subedi KD, Ma BL, Liang BC (2006) New method to estimate root biomass in soil through root-derived carbon. Soil Biol Biochem 38:2212–2218
Suyker AE, Verma SB (2010) Coupling of carbon dioxide and water vapor exchanges of irrigated and rainfed maize–soybean cropping systems and water productivity. Agric For Meteorol 150:553–563
Swift MJ, Heal OW, Anderson JM (1979) Decomposition in terrestrial ecosystems. University of California Press, Berkeley
Talbot JM, Allison SD, Treseder KK (2008) Decomposers in disguise: mycorrhizal fungi as regulators of soil C dynamics in ecosystems under global change. Funct Ecol 22:955–963

Thevenot M, Dignac M-F, Rumpel C (2010) Fate of lignins in soils: a review. Soil Biol Biochem 42:1200–1211

Tian H, Chen G, Liu M, Zhang C, Sun G, Lu C, Xu X, Ren W, Pan S, Chappelka A (2010) Model estimates of net primary productivity, evapotranspiration, and water use efficiency in the terrestrial ecosystems of the southern United States during 1895–2007. For Ecol Manage 259:1311–1327

Torn MS, Trumbore SE, Chadwick OA, Vitousek PM, Hendricks DM (1997) Mineral control of soil organic carbon storage and turnover. Nature 389:170–173

Torn MS, Swanston CW, Castanha C, Trumbore SE (2009) Storage and turnover of natural organic matter in soil. In: Senesi N, Xing B, Huang PM (eds) Biophysico-chemical processes involving natural nonliving organic matter in environmental systems. Wiley, Hoboken, pp 219–272

Treseder KK, Cross A (2006) Global distributions of arbuscular mycorrhizal fungi. Ecosystems 9:305–316

Treseder KK, Turner KM (2007) Glomalin in ecosystems. Soil Sci Soc Am J 71:1257–1266

U.S. Department of Energy (2008) Carbon cycling and biosequestration: integrating biology and climate through systems science. Report from the March 2008 workshop, DOE/SC-108, U.S. Department of Energy Office of Science. http://genomicsgtl.energy.gov/carboncycle/

van der Heijden MGA, Bardgett RD, van Straalen NM (2008) The unseen majority: soil microbes as drivers of plant diversity and productivity in terrestrial ecosystems. Ecol Lett 11:296–310

Van Hemelryck H, Fiener P, Van Oost K, Govers G (2009) The effect of soil redistribution on soil organic carbon: an experimental study. Biogeosci Discuss 6:5031–5071

Van Oost K, Quine TA, Govers G, De Gryze S, Six J, Harden JW, Ritchie JC, McCarty GW, Heckrath G, Kosmas C, Giraldez JV, Marques Da Silva JR, Merckx R (2007) The impact of agricultural soil erosion on the global carbon cycle. Science 318:626–629

Vancampenhout K, Wouters K, De Vos B, Buurman P, Swennen R, Deckers J (2009) Differences in chemical composition of soil organic matter in natural ecosystems from different climatic regions – a pyrolysis-GC/MS study. Soil Biol Biochem 41:568–579

Verbruggen E, Kiers ET (2010) Evolutionary ecology of mycorrhizal functional diversity in agricultural systems. Evol Appl 3:547–560

Von Lützow M, Kögel-Knabner I (2010) Response to the concept paper: 'What is recalcitrant soil organic matter?' by Markus Kleber. Environ Chem 7:333–335

Von Lützow M, Kögel-Knabner I, Ekschmitt K, Matzner E, Guggenberger G, Marschner B, Flessa H (2006) Stabilization of organic matter in temperate soils: mechanisms and their relevance under different soil conditions – a review. Eur J Soil Sci 57:426–445

Wang Z, Xiao X, Yan X (2010) Modeling gross primary production of maize cropland and degraded grassland in northeastern China. Agric For Meteorol 150:1160–1167

Whalen JK, Sampedro L (2009) Primary production. In: Whalen JK, Sampedro L (eds) Soil ecology and management. CAB International, Wallingford, pp 109–133

Wickings K, Grandy AS (2011) The oribatid mite *Scheloribates moestus* (Acari: Oribatida) alters litter chemistry and nutrient cycling during decomposition. Soil Biol Biochem 43:351–358

Wilkinson MT, Richards PJ, Humphreys GS (2009) Breaking ground: pedological, geological, and ecological implications of soil bioturbation. Earth Sci Rev 97:254–269

Wilson GWT, Rice CW, Rillig MC, Springer A, Hartnett DC (2009) Soil aggregation and carbon sequestration are tightly correlated with the abundance of arbuscular mycorrhizal fungi: results from long-term field experiments. Ecol Lett 12:452–461

Wolters V (2000) Invertebrate control of soil organic matter stability. Biol Fert Soils 31:1–19

Wright SF, Upadhyaya A (1996) Extraction of an abundant and unusual protein from soil and comparison with hyphal protein of arbuscular mycorrhizal fungi. Soil Sci 161:575–586

Wright AL, Dou F, Hons FM (2007) Crop species and tillage effects on carbon sequestration in subsurface soil. Soil Sci 172:124–131

Wutzler T, Reichstein M (2007) Soils apart from equilibrium - consequences for soil carbon balance modelling. Biogeosciences 4:125–136

Yang F, Ichii K, White MA, Hashimoto H, Michaelis AR, Votava P, Zhu A, Huete A, Running SW, Nemani RR (2007) Developing a continental-scale measure of gross primary production by combining MODIS and AmeriFlux data through support vector machine approach. Remote Sens Environ 110:109–122

Young IM, Crawford JW, Nunan N, Otten W, Spiers A (2008) Microbial distribution in soils: physics and scaling. Adv Agron 100:81–121

Yu GR, Fu Y, Sun X, Wen X, Zhang L (2006) Recent progress and future directions of ChinaFLUX. Sci China Ser D 49(Suppl 1):1–23

Zhang D, Hui D, Luo Y, Zhou G (2008) Rates of litter decomposition in terrestrial ecosystems: global patterns and controlling factors. J Plant Ecol 1:85–93

Zhang Y, Xu M, Chen H, Adams J (2009) Global pattern of NPP to GPP ratio derived from MODIS data: effects of ecosystem type, geographical location and climate. Glob Ecol Biogeogr 18:280–290

Zhao M, Running S, Heinsch FA, Nemani R (2010) MODIS-derived terrestrial primary production. In: Ramachandran B, Justice CO, Abrams MJ (eds) Land remote sensing and global environmental change. Springer, Dordrecht, pp 635–659

Zimmerman AR, Chorover J, Goyne KW, Brantley SL (2004) Protection of mesopore-adsorbed organic matter from enzymatic degradation. Environ Sci Technol 38:4542–4548

Chapter 15
The Carbon Cycle in Drylands

Penélope Serrano-Ortiz, Enrique P. Sánchez-Cañete, and Cecilio Oyonarte

Abstract Drylands are characterized by an aridity index (ratio of annual rainfall to potential evapotranspiration) lower than 0.65, and occupy nearly a third of the total land surface. Globally, the organic and inorganic carbon (C) storage in such water-limited systems is about 20–30% of the terrestrial global total. The total soil organic C (SOC) stored in drylands is approximately 230 Pg. The C content in dryland biomass is about four times lower than that stored as SOC (65 Pg). The soil inorganic C (SIC) pools are estimated to be more than twice the SOC pools for drylands and may exceed SOC by a factor of 10 in some arid lands. These statistics can be modified significantly taking into account anthropogenic practices. Ideally, NT management may potentially increase the SOC by 20%, while non-grazing in grassland could increase SOC storage by about 45%. These ecosystems are highly vulnerable to climatic changes and susceptible to desertification, leading to reduction in the C pool. In addition, due to arid conditions and the large percentage of bare soil, some other processes besides photosynthesis and respiration contribute to C sequestration or gaseous emissions to the atmosphere. These include geochemical

P. Serrano-Ortiz (✉)
Estación Experimental de Zonas Àridas (EEZA, CSIC),
04120 Almería, Spain

Department of Applied Physics, Universidad de Granada,
Avda. del Hospicio, s/n C.P., Granada 18071, Spain
e-mail: penelope@ugr.es

E.P. Sánchez-Cañete
Estación Experimental de Zonas Àridas (EEZA, CSIC),
04120 Almería, Spain
e-mail: enripsc@urg.es

C. Oyonarte
Departamento de Edafología & Química Agrícola, Universidad de Almería,
04120 Almería, Spain
e-mail: coyonart@ual.es

R. Lal et al. (eds.), *Recarbonization of the Biosphere: Ecosystems and the Global Carbon Cycle*, DOI 10.1007/978-94-007-4159-1_15,

processes, formation of secondary carbonates, bio-sequestration, subsoil ventilation, erosion and photodegradation and can even dominate the ecosystem C exchange during the dry season. Such contributions limit the use of biological models to provide estimates of C pool in drylands. And the differences in the measured net C exchange with the atmosphere, ranging from −106 to 145 gC m^{-2} for deserts and from −190 to 140 gC m^{-2} for grasslands are not easily explained. A better understanding of C cycle in drylands is highly relevant to the Kyoto Protocol in order to prevent degradation and the C emissions to the atmosphere. This survey reviews measured C pools and annual C sink capacities in drylands, together with an analysis of principal processes involved and, finally, summarizes suggestions for management practices with the capability to reduce C losses from drylands.

Keywords Drylands • Pools • Desertification • Geochemical processes • Ventilation • Soil inorganic C pool • Weathering • Erosion • Biosequestration • Photodegradation • Net erosion exchange • Global C cycle • Aridity index • Evapotranspiration • Ventilation • Anthropogenic emissions • Missing sink • Terrestrial ecosystems • Mitigation • Photosynthesis process • Rangeland • Desertification • Arid • Hyper-arid • Desert biomes • Caliche • No-till • Minimum tillage • Lithogenic carbonates • Primary carbonates • Secondary carbonates • Pedogenic carbonates • Silicate weathering • Calcite • FLUXNET • Flux measurements • Carboniferous rocks • Biomineralization • Phytoliths • Eddy covariance

Abbreviations

P	annual rainfall
P/E_P	aridity index
C	carbon
CO_2	carbon dioxide
GCC	global carbon cycle
NEE	net ecosystem CO_2 exchange
E_P	potential evapotranspiration
SIC	soil inorganic C
SOC	soil organic C
UV	Ultraviolet

15.1 The Global Carbon Cycle

The global carbon (C) cycle (GCC) depends on feedbacks among a number of source and sink processes occurring among different systems: ocean, atmosphere, soil and biosphere. These processes operate at different time scales modifying the C composition of components (Boucot and Gray 2001). In the last decades, the increase of atmospheric C via anthropogenic carbon dioxide (CO_2) emissions and changes in land use has produced a climatic perturbation inducing changes in temperature and rainfall regimes (IPCC 2007; Keeling 1960). Moreover, the effect of such perturbations may also be altering other systems beside the atmosphere such as the soil and biosphere. The annual increase of the atmospheric CO_2 concentration is only half that expected from anthropogenic activities, implying a terrestrial or oceanic sink absorbing CO_2. Isotopic studies reveal that air–sea CO_2 exchange is too small to explain the "missing sink" which must, therefore, be accounted for by terrestrial ecosystems (Schimel et al. 2001; Tans et al. 1990). Thus, a better understanding of the role of the biosphere in the current global C budget as well as the potential of soil as a C storage medium is needed in order to enable the mitigation of human impacts.

The GCC is strongly related to the C balance of terrestrial ecosystems due to the capacity of the biomass and soil to store C. The biosphere, via the photosynthetic process, captures CO_2 from the atmosphere and stores it in the living biomass. Then, soil microorganisms degrade the non recalcitrant compounds of dead biomass emitting CO_2 to the atmosphere. The organic matter can be stabilized and stored in the soil at long time scales via spatial inaccessibility to decomposer organisms or interactions with minerals and metal ions (von Lützow et al. 2008). Thus, the soil is the largest pool of organic C in terrestrial ecosystems, representing a reserve of more than 1,500 Pg C (1 Petagram = 10^{15} g) (Safriel et al. 2005). In addition, soil contains more than 900 Pg C in inorganic forms such as calcite or dolomite (Safriel et al. 2005; Vande Walle et al. 2001), while living biomass represents a C reserve of about 600 Pg. Therefore, terrestrial ecosystems with deeper soils and greater biomass, such as forests, present higher potential to sequester CO_2 from the atmosphere and contribute to amelioration of anthropogenic CO_2 emissions and, thus, climate change.

15.2 Main Characteristics of Drylands

Drylands are characterised by patches of vegetation and bare soil exposed to erratic rainfall events producing water-stressed vegetation during the drought period (Domingo et al. 1999). These water-limited ecosystems exist on every continent and comprise nearly a third of the total land surface corresponding to 60 million km^2 (Okin 2001; Schlesinger 1990). Dryland rangelands support about 50% of the world's livepool and provide forages for both domestic animals and wildlife (Puigdefábregas 1998). Although drylands withstand extreme climatic conditions, they are very sensitive to perturbations such as drought, fires or climate change, leading to desertification (Mouat and Lancaster 2006). This process can be defined as land degradation in arid, semi-arid, and dry sub-humid areas resulting from climatic variations and human activities (UNEP 1997). Desertification reduces the potential for plant C assimilation, degrades soil and, thus, decreases dryland C pools. One of the major trends in the degradation of dryland ecosystems is the replacement of grass by shrubs (Puigdefábregas 1998). This trend alters soil properties and can modify the amount of C stored in biomass and soil.

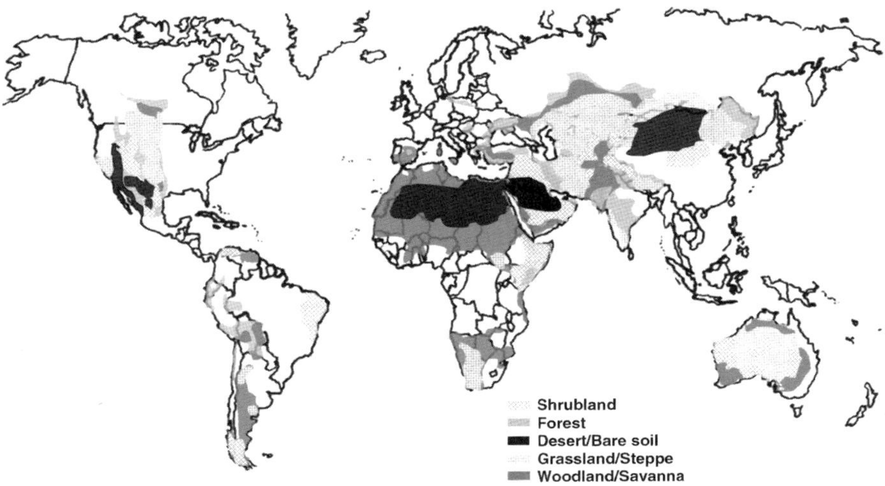

Fig. 15.1 Global distribution of main ecosystem types in drylands

According to the World Atlas of desertification (UNEP 1997), a terrestrial ecosystem is considered a dryland when the ratio of annual rainfall (P) to potential evapotranspiration (E_p), i.e., the aridity index (P/E_p), is lower than 0.65. Drylands are divided into four subtypes, in order of increasing aridity or moisture deficit: dry sub-humid (11% of total dryland surface), semi-arid (31%), arid (34%) and hyper-arid (24%). Desert biomes or bare soil predominate in hyper-arid and arid subtypes, while grasslands typify semi-arid, and woodland or forest survive in dry sub-humid drylands; about 35% and 40% of semi-arid and dry sub-humid subtypes, respectively, are cultivated (mainly croplands) (Safriel et al. 2005, Fig. 15.1).

Generally, drylands represent a limited potential to store C due to their poor soils and sporadic and dispersed vegetation. However, their large extent around the terrestrial land surface explains the estimated soil organic C (SOC) pool content of about 400 Pg C, and a contribution of about 20–30% to the terrestrial organic and inorganic C (Eswaran et al. 2000; Rasmussen 2006; Safriel et al. 2005). In this context, the main characteristic of drylands is the capacity to store about 95% of the soil inorganic C (SIC) globally via "caliche formation" (Marion et al. 2008). However, the role of the SIC pool in relation to climate change is less well understood (Lal and Kimble 2000b). Most carbonate rocks are found in drylands (mainly in the Mediterranean, China and Africa) where other processes besides photosynthesis and respiration (biological processes) also contribute to C sequestration or C emission to the atmosphere (Serrano-Ortiz et al. 2010).

This chapter of the C cycle in drylands reviews measurements of C pools and annual C sink capacities together with analysis of the main drivers controlling the potential C sequestration capacity and principal processes involved. Finally, some suggestions for management practices to improve the role of drylands in the storage of SOC are summarized.

15.3 Carbon Pools

15.3.1 *Soil Organic Carbon*

Several studies have been published related to SOC pools in drylands (Table 15.1).

According to published information regarding SOC pool in some studied deserts and bare soil, it ranges from 1 to 3 kg C m^{-2} in Africa and North and Central of America (Rasmussen 2006; Woomer et al. 2004), while it ranges from 2 to 7 kg C m^{-2} in Asia (Wiesmeier et al. 2011). For shrublands, SOC ranges from 2 to 4 kg C m^{-2} except on the Asian continent where SOC higher than 6 kg C m^{-2} has been reported (Chen et al. 2007; Wiesmeier et al. 2011). The average value of SOC in grassland is 5 ± 2 kg C m^{-2}, but can be more than double for non-grazed systems (He et al. 2008). However, there is little information for woodland or savannas. Some studies have reported a range of about 3–5 kg C m^{-2} (Chen et al. 2007; Noellemeyer et al. 2006; Shukla et al. 2006). Forests are not a common vegetation in drylands. For SE Spain, Martínez-Mena et al. (2008) reported a value of 14 kg C m^{-2} while a study in Kenyan dryland forest indicated an average value of 2.3 kg C m^{-2} mainly due to differences in climate conditions and species (Glenday 2008). About 30% of drylands are cultivated. Dry croplands have an average SOC pool of 3.5 kg m^{-2}. These are mostly located in dry-subhumid and semiarid ecosystems, and store globally about 40 Pg SOC. Many studies suggest an improvement in C sequestration if non-tillage (NT) or minimum-tillage practices are used instead of conventional ploughing (López-Fando and Pardo 2009, 2011; Sombrero and de Benito 2010).

Globally, about 230 Pg of total SOC is stored in drylands according to the coverage of drylands per continent occupied by the different ecosystem types (Fig. 15.1), with average SOC values summarized in Table 15.1. The average SOC is in the same range of values published by Lal (2004) (241 Pg C) but higher than those published by IPCC (1990) and Bolin et al. (2001) (191 and 159 Pg C, respectively). However, these estimates are highly uncertain mainly because of low sample numbers used for global upscaling and assumptions on mean soil depths (Rodeghiero et al. 2009). In addition, global values can be easily modified considering anthropogenic practices such as taking into account the percentage of cultivated lands (SOC reduced by 15%), an ideal situation of NT which would increase the total SOC in drylands by 20% (258 Pg C), and by prohibiting grazing in grassland which could increase the value by about 45% (313 Pg C).

15.3.2 *Soil Inorganic Carbon*

The SIC pool consists of primary inorganic carbonates or lithogenic inorganic carbonates, and secondary inorganic carbonates or pedogenic inorganic carbonates (Saharawat 2003). Extra inputs of Ca^{2+} due to atmospheric deposition and/or silicate weathering combined with the negative water balance may result in calcite precipitation

Table 15.1 Soil organic carbon (kg C m^{-2}) from different dryland ecosystems together with mean annual temperature and precipitation

Continent	Location	Reference	Mean annual temperature (°C)	Mean annual precipitation (mm)	Soil depth (cm)	Vegetation	Management	SOC (kg C m^{-2})
Europe	SE Spain	Oyonarte (personal communication)	18	200	0–50	Cropland	–	3
						Shrubland	–	3
						Grassland	–	4.5
		Martínez-Mena et al. (2008)	16.6	300	–	Forest	–	14
						Abandoned	–	8.2
						Olive	Non irrigated	7.2
	NE Spain	Plaza-Bonilla et al. (2010)	–	430	0–40	Cropland	No Tillage	3–3.5
	Center Spain	López-Fando and Pardo (2009, 2011)	–	400	0–30	Cropland	No Tillage	5
							Conventional Tillage	4
		Sombrero and de Benito (2010)	–	448	0–30	Cropland	Conventional Tillage	4
							Minimum Tillage	5
							No tillage	10
Asia	Center China	Chen et al. (2007)	34 – (–27)	427	0–40	Cropland	–	3
						Grassland	–	5
						Shrubland	–	6
						Woodland	–	4
	North China	He et al. (2008)	1.1	345	0–100	Grassland	Grazing	6–14
		Wang et al. (2009)	(–19) – (–23)	350	0–30	Cropland	28 year cropland	1
							42 year cropland	2
		Wiesmeier et al. (2011)	0.7	350	0–100	Arable	–	11 ± 4
						Bare	–	5 ± 2
						Steppe	–	14 ± 5
						Sand Dunes	–	5 ± 3
	India	Singh et al. (2007)		100–400	0–100	Bare	–	2.4

America	Mexico	Shukla et al. (2006)	15	400	0–20	Oak	–	8
						Juniper	–	4
	USA (Arizona)	Rasmussen (2006)	–	–	–	Arid	–	1.4–2.8
		Emmerich (2003)	17	356	0–30	Grassland	–	2.6 ± 0.5 (Spring) 2.3 ± 0.1 (Fall)
						Shrubland	–	4.0 ± 0.5 (Spring) 2.9 ± (0.3) (Fall)
	USA (Nebraska, Colorado)	Denef et al. (2008)	9.5	500	0–75	Native grassland	–	5–9
						Dryland cultivation	–	6–7
						Pivot Irrigated	–	7–8
	Argentina	Noellemeyer et al. (2006)	16	480	0–18	Grassland	–	1
						Grass + shub + trees	–	2.7
	Chile	Perez-Quezada et al. (2011)	26–5	153	0–50	Shrubland	–	2–4
		Muñoz et al. (2007)	–	695	–	Afforested Shrubland	–	4.2
Africa	Sahel desert	Woomer et al. (2004)	–	–	0–40	Desert	–	2
	Kenya	Glenday (2008)	–	–	–	Forest	–	2.3
	Tanzania	Birch-Thomsen et al. (2007)	20	542	0–50	Maize cultivation	–	2–4

forming secondary carbonates (caliche) and contribute to SIC sequestration (Marion 1989; Schlesinger 1985). The contribution of inorganic C formations from non-carbonate material (caliche) may range from 0.12 to 0.42 gC m^{-2} $year^{-1}$ (Marion et al. 2008; Schlesinger 1985). Since SIC is relatively stable, with turnover periods >1,000 years (Amundson et al. 1994), the C stocks in soil are generally similar following land use and management changes and it is usually not considered in soil C dynamics (Allen et al. 2010). What is more, inorganic C formed from re-precipitation of calcareous material may not be involved in C sequestration in the soil.

The SIC pools are estimated to be more than twice the SOC pools for drylands (Eswaran et al. 2000; Lal and Kimble 2000a). Further, SIC pools may exceed SOC by a factor of 10 in some arid lands (Schlesinger 1985, 2006). What is more, a study in a site located in Southeasth Spain reveal an average of SIC pool 17 times that of SOC content (134 kg m^{-2}) (Díaz-Hernández et al. 2003). Since SIC Although, there is no clear evidence to confirm an effect of SIC on SOC, soils with caliche formations are almost twice as rich in SOC as in those of a similar depth but without such horizons (Díaz-Hernández and Barahona Fernádez 2008). Soils with high SIC are mainly located in hyper-arid and arid regions with a pool of about 732 Pg C (Safriel et al. 2005). Pools of SIC in semi-arid and dry sub-humid systems are almost four times lower.

15.4 Biomass Organic Carbon

The C content in dryland biomass is about four times lower than that stored as SOC (Eswaran et al. 2000). The vegetation is mainly comprised of grass, steppe and woody species with a large proportion of bare soil and, thus, a low capacity to store C. Hyper-arid lands (deserts) in Asia and Africa have the capacity to store 0.04–0.40 kg Cm^{-2} in biomass (Fan et al. 2008; Woomer et al. 2004). Biomass C storage in shrublands is in the range of 0.08–0.40 kg m^{-2} depending on the percentage of bare soil and the degree of degradation (Perez-Quezada et al. 2011). The C content in woody species, mostly located in Africa, ranges from 0.9 to 2.6 kg Cm^{-2} depending on species and climatic conditions (Shackleton and Scholes 2011; Williams et al. 2008). Although forest occupies less than 15% of drylands, its capacity to store C can be about 4–5 kg Cm^{-2} (Glenday 2008). Finally, grasslands located mainly in Asia store around 1 kg Cm^{-2} with decreased potential depending on the grazing intensity (He et al. 2008).

Globally, average pool of 65 Pg of total C in the biomass is estimated based on the percent of land cover (Figs. 15.2 and 15.3) and average value of published C content of biomass during the last 5 years. This value can be modified significantly considering some anthropogenic practices such as grazing, which may reduce C pools by more than 10% of the given value. Degradation of the vegetation comprised of woody and shrubland types due to desertification may reduce C pool in the biomass by more than 20% (50 Pg C). The estimated potential biomass C based on the maximum estimated values is about 81 Pg C, and this is in accord with that reported by Safriel et al. (2005).

Fig. 15.2 Drylands in Arizona (USA): (**a**) Near Tombstone (by E. P. Sánchez-Cañete) and (**b**) Santa Rita Mesquite Savanna (by Russel L. Scott)

15.5 Main Natural Processes Involved in Carbon Sequestration and Loss

Estimates of C pools mentioned above are the result of several processes which contribute to the net ecosystem CO_2 exchange (NEE) with the atmosphere in drylands.

Fig. 15.3 Tabernas desert located in Almería, Southeast Spain (by C. Oyonarte)

15.5.1 Biological Processes

The SOC stored in soils and that in the biomass are mainly due to the balance between photosynthesis (net CO_2 uptake) and respiration (net CO_2 release via decomposition, degradation and diffusion processes). Such biological processes are mainly responsible for annual NEEs in most ecosystems (forest, wetlands, cropland, etc.). Thus, the FLUXNET community (Baldocchi et al. 2001) interprets CO_2 fluxes measured using micrometeorological techniques (Dabberdt et al. 1993) as a biological flux neglecting non-biological processes (Falge et al. 2002; Reichstein et al. 2005; Stoy et al. 2006; Valentini et al. 2000). However, many CO_2 flux measurements over drylands indicate contributions of abiotic processes to the NEE (Emmerich 2003; Ferlan et al. 2011; Hastings et al. 2005; Inglima et al. 2009; Mielnick and Dugas 2000; Were et al. 2010; Wohlfahrt et al. 2008; Xie et al. 2008). These processes can dominate the flux during the dry season (Kowalski et al. 2008) with annual contributions >50% depending mainly on meteorological conditions (Serrano-Ortiz et al. 2009).

15.5.2 Weathering Processes

Some soils are derived from carboniferous (calcareous) rocks and/or include additional carbonates (secondary carbonates or caliche) as a result of weathering and precipitation processes (Eq. 15.1).

$$H_2O_{(aq)} + CO_{2(g)} + CaCO_{3(s)} \Leftrightarrow 2HCO_3^- + Ca^{2+}_{(aq)} \quad (15.1)$$

In terms of NEE, Eq. 15.1 specifies that for each molecule of $CaCO_3$ dissolved a molecule of atmospheric CO_2 is consumed. During precipitation, one molecule of CO_2 is released for every molecule of $CaCO_3$ deposited to the surface. Thus, either water lost by evapotranspiration or additional sources of Ca^{2+} would enrich aqueous concentrations and enhance the deposition of $CaCO_3$ from the aqueous solution, and release CO_2 to the atmosphere (Eq. 15.1 to the left).

Globally, and over long time scales, weathering processes are balanced with respect to CO_2 (Berner 2003; Lasaga et al. 1994). However, at annual and seasonal scales the predominance of dissolution or precipitation processes may be relevant to local NEEs (Serrano-Ortiz et al. 2010) and contribute to the observed CO_2 fluxes (Emmerich 2003; Mielnick et al. 2005). In addition, although precipitation processes imply CO_2 release at short time scales, caliche formation can be considered a net atmospheric CO_2 sink over long timescales: every two molecules of bicarbonate, previously formed during the growing season by dissolution of two molecules of CO_2, react with one molecule of additional Ca^{2+} to form one molecule of calcite and release one molecule of CO_2. The formation of this secondary C form depends on land use and soil/crop management systems (Lal 2004). Addition of biomass, whose decomposition increases the partial pressure of CO_2 in the soil, together with irrigation, increases SIC in agricultural soils (Entry et al. 2004; Lal 2004). While its contribution to the total annual atmospheric CO_2 sink may be less than 10% (Eswaran et al. 2000; Gombert 2002; Liu and Zhao 2000; Mermut et al. 2000), it is unclear how SIC responds to rainfall and temperature changes predicted under the climate change scenarios (Rasmussen 2006).

15.5.3 Bio-sequestration

Fungi, lichens, and cyanobacteria play a prominent part in calcite dissolution and precipitation by biomineralization processes (Verrecchia et al. 1999). In addition, one inert form of organic C can be bio-sequestered within plants and accumulates in soil after the decomposition of that vegetation (phytolith-occluded C) (Parr and Sullivan 2005). Although phytoliths are highly resistant to oxidation and very stable in soil, they can be involved in SOC dynamics in response to land use and management change (Allen et al. 2010). There are many studies regarding the ability of different microorganisms to precipitate carbonate in drylands (Delgado et al. 2008; Li et al. 2011; Parraga et al. 2004; Rivadeneyra et al. 1997). However, little is known about its contribution in the soil C sequestration.

15.5.4 Ventilation

Drylands over carbonate rocks with cracks, pores and cavities, together with soils with deep vadose-zones, show a high capacity to store CO_2 belowground. Since such CO_2 storage may represent as much as 60% of the annual atmospheric sink (Serrano-Ortiz et al. 2011), the subsurface can be considered a temporal depot for CO_2 coming from different processes (mainly weathering and respiration) (Serrano-Ortiz et al. 2010). In addition to diffusion processes, such soils have the potential to emit the stored CO_2 via ventilation (Sanchez-Cañete et al. 2011) and contribute to ecosystem CO_2 exchange observed. Ventilation is a transport process due to net movements of air in and out of an enclosed space. The behaviour of ventilation processes, in caves for example, is controlled by the degree of connection between the cavities and the aboveground system (Cuezva et al. 2011) and, thus, such processes have only been detected when the soil is dry (Cuezva et al. 2011; Sanchez-Cañete et al. 2011; Serrano-Ortiz et al. 2009; Were et al. 2010). The main meteorological drivers controlling soil CO_2 ventilation due to pressure pumping are wind speed and turbulence (Jassal et al. 2005; Lewicki et al. 2010; Subke et al. 2005; Takle et al. 2004). Therefore, the non-negligible role of subsurface as a temporal depot of CO_2, along with seasonal ventilation can contribute to the annual net ecosystem C balance (Serrano-Ortiz et al. 2010).

15.5.5 Erosion

Erosion is a natural process that occurs when a liquid (air or water) moves into and/or across a soil surface with subsequent transport of the detached particles to another location (Flanagan 2006) reducing the amount of SOC in the eroded soil. This process is more important in drylands with lower percentages of vegetation cover, and human activities associated with agricultural practices usually enhancing erosion. There is a lack of agreement whether water erosion induces net release of C to the atmosphere (Jacinthe and Lal 2001; Lal et al. 2004) or net C sequestration (Boix-Fayos et al. 2009; Harden et al. 1999; Van Oost et al. 2007), with estimates ranging from a source of 1 Pg C $year^{-1}$ to a sink of the same magnitude. The SOC displaced in terrestrial ecosystems and mineralization during water transport can lead to CO_2 emissions, with global estimates of 56–168 gC m^{-2} $year^{-1}$ and 6–52 gC m^{-2} $year^{-1}$, respectively (Jacinthe and Lal 2006). On the other hand, SOC exported from the eroded areas is replaced by additional C derived from the atmosphere providing a sink of atmospheric CO_2 (Van Oost et al. 2007). A global C sink of 0.12 Pg C $year^{-1}$ is estimated to result from erosion in the world's agricultural lands (Van Oost et al. 2007).

15.5.6 Photodegradation

The direct breakdown of organic matter by ultraviolet (UV) light (photodegradation) contributes to litter mass loss (emission of CO_2) in water-limited ecosystems receiving intense sunlight (Austin and Vivanco 2006; Rutledge et al. 2010).

Therefore, future climate changes in radiation due to decreased cloudiness or increased stratospheric ozone depletion may have an effect on the C balance in such ecosystems (Austin and Vivanco 2006). Although photodegradation contribute to organic matter decomposition via microbial facilitation, the direct breakdown of organic matter to CO_2 can occur in the absence of microbial activity (Brandt et al. 2009). Thus, organic matter decomposition is not restricted to periods of high moisture availability as is plant production (Gallo et al. 2009). During midday in summer, the CO_2 efflux due to photodegradation contributes around 90% of the total half-hourly CO_2 flux from an arid grassland (Rutledge et al. 2010). However, the relevance of photodegradation and its contribution to the total CO_2 losses at ecosystems scales is still unknown. While Rutledge et al. (2010) estimates a C loss of 16 $g\,m^{-2}$ for the dry season in an arid grassland located in the lower foothills of the Sierra Nevada (USA) using chambers and *eddy covariance* measurements, an extrapolated laboratory study to field conditions reveal an annual C emission of 4 $g\,m^{-2}$ due to photodegradation for a desert grassland located in New Mexico (Brandt et al. 2009). Further studies are needed to increase the understanding, importance and drivers of photodegradation.

15.6 Carbon Sink Capacity at Ecosystem Level

The processes mentioned above (weathering, ventilation and/or erosion processes) act together in drylands and contribute to the measured annual net C exchange (Table 15.2). For deserts located in southwest of the U.S.A. and Baja California, published studies have determined that the most important driver controlling CO_2 flux is the not the amount of rainfall but mostly its timing (Hastings et al. 2005; Mielnick et al. 2005; Wohlfahrt et al. 2008). However, the published data on annual net C exchanges do not support this hypothesis. While two desert shrubland located in the Mojave Desert and Baja California with similar annual precipitation act as annual net C sinks of 106 ± 70 and 52 $g\,C\,m^{-2}\,year^{-1}$ respectively, the Chihuahuan Desert site emits ~145 $g\,C\,m^{-2}$ annually. Grasslands located in North America (New Mexico, Arizona and California) and Europe (Southeast Portugal and Southwest of Spain) are C sources ranging from 141 (source) to −190 (sink) $g\,C\,m^{-2}\,year^{-1}$ depending mostly on the total amount of rainfall (Aires et al. 2008; Anderson-Teixeira et al. 2011; Emmerich 2003; Ma et al. 2007; Scott et al. 2006) and also wind speed for the particular site located in Southwest of Spain (Rey et al. 2012). While ecosystem C sink capacity in grasslands located in Northern Asia also depend on optimal temperature in summer (10–20°C) (Kato et al. 2006; Wang et al. 2008). For savannas, Scott et al. (2009) measured an annual net C releases ranging from 14 to 95 $g\,C\,m^{-2}\,year^{-1}$ in a semiarid savanna in southern Arizona, while Ma et al. (2007) measured an annual net C uptake ranging from 56 to 155 $g\,C\,m^{-2}\,year^{-1}$ in a savanna site located in California with higher annual precipitation and lower temperature. Finally, shrublands in drylands act mostly as small sinks for atmospheric CO_2 (uptake from 2 to 75 $g\,C\,m^{-2}\,year^{-1}$) (Anderson-Teixeira et al. 2011; Luo et al. 2007;

Table 15.2 Annual net ecosystem C exchange (g C m^{-2}) measured mostly using the eddy covariance technique, together with annual temperature and rainfall

Vegetation	Location	Experimental Site	Reference	Mean annual temperature (°C)	Mean annual precipitation (mm)	Net ecosystem C exchange (g C m^{-2} year^{-1})
Desert and bare soil	Mojave Desert (USA)	Desert on the Nevada Test Site, 120 km northwest of Las Vegas	Wohlfahrt et al. (2008)	20	210	−106 ± 70
	Chihuahuan Desert (USA)	About 40 km northeast of Las Cruces, New Mexico	Mielnick et al. (2005)	–	272	145[a]
	Baja California (Mexico)	15 km west of the city of La Paz and 1.5 km from the Bay of La Paz (CIBNOR)	Hastings et al. (2005)	24[b]	147	−39
					197	−52
Grassland	Southwest Spain	*"Cabo de Gata Natural Park" Almería (Andalucía)*	Rey et al. (2012)	17	210	66
					251	144
					294	92
	Southeast Portugal	*Monte do Tojal*, Évora in Southern Portugal	Aires et al. (2008)	14.7	364	49
					751	−190
	New Mexico (USA)	Sevilleta LTER in Central New Mexico	Anderson-Teixeira et al. (2011)	13	244	30
	Arizona (USA)	The Kendall grassland Agricultural Research Service Walnut Gulch Experimental Watershed	Scott et al. (2010)	14.5	313	−69
					312	−98
					274	−55
					246	−47
					162	21
			Emmerich (2003)	17	356	126[a]
		Foodplain terraces along the San Pedro River	Scott et al. (2006)	17	234	−63
	California (USA)	Foothills of the *Sierra Nevada*	Ma et al. (2007)	17 ± 1	562 ± 193	(−88)–141
	Qinghai-Tibetan Plateau (China)	Alpine meadow	Kato et al. (2006)	−0.65	561[c]	−79
				−0.91		−92
				−1.53		−173
	Northern China	Inner Mongolia Autonomous region	Wang et al. (2008)	2.5	297	10
				1.3	174	30
				1.7	215	−15
	Central Mongolia	Hentiy province of Mongolia	Li et al. (2005)	1.2	196	−41

Savanna	Arizona (USA)	The Santa Rita mesquite savanna site located on the Santa Rita Experimental Range (SRER)	Scott et al. (2009)	19	285	60
				20	335	14
				20	289	95
				19	330	30
	California (USA)	Foothills of the *Sierra Nevada*	Ma et al. (2007)	17±1	562±193	(−155) – (−56)
Shrubland	Southwest Spain	Mediterranean plateau, 25 km from the coast	Serrano-Ortiz et al. (2009)	12	475	−2±23
	New Mexico (USA)	Sevilleta LTER in Central New Mexico	Anderson-Teixeira et al. (2011)	14	244	−30
	San Diego (USA)	Southern California, 75 km east of Pacific Ocean	Luo et al. (2007)	15	349	−52
	Arizona(USA)	Foodplain terraces along the San Pedro River	Scott et al. (2006)	17	234	−212
		Lucky Hills, Agricultural Research Service Walnut Gulch Experimental Watershed	Emmerich (2003)	17	256	144[a]
	Qinghai-Tibetan Plateau (China)	Alpine *Potentilla fruticosa* at the Haibei Research Station	Zhao et al. (2006)	2.3	542	−59
				2.2	493	−75

Negative sign (−) for net ecosystem exchange indicates a C sink

[a]Data not measured by the eddy covariance technique

[b]Estimated data using information of Fig. 15.1 of the cited reference

[c]Annual average precipitation for 1981–2000

Serrano-Ortiz et al. 2009; Zhao et al. 2006), with some exceptions. In riparian areas where shrubs and woody plants have the capacity to exploit water resources by growing deep roots (Domingo et al. 1999), annual NEE can be higher than 200 g C m^{-2} $year^{-1}$ (Scott et al. 2006). On the contrary, the annual C loss estimation for the Lucky Hills site located in Arizona is 144 g C m^{-2} $year^{-1}$. The source of this C appears to be from the large SIC pool in these soils (Emmerich 2003). In summary, according to published studies, the C sink capacity at ecosystem level in drylands is highly variable depending on the ecosystem type, SIC pool and mostly on rainfall timing and temperature during the growing season.

15.7 Management Practices

Human activities are directly or indirectly responsible for dryland degradation, but have also the capacity to alter natural processes involved in C sequestration with the potential to ameliorate poverty-provoking desertification highly linked to poverty (Glantz 1994; Mouat and Lancaster 2006). Cultivated lands are about 30% of drylands contributing 20% of the total SOC pool. Thus, better management practices in cultivated lands (mainly croplands) could improve the role of drylands in the storage of SOC. Sequestration of C in croplands can be improved if NT or minimum-tillage practices are applied instead of the conventional systems (Table 15.1) (López-Fando and Pardo 2009, 2011; Sombrero and de Benito 2010). In addition, crop residues left on the soil surface instead of being removed or incorporated into the soil may increase the SOC storage by more than a 30% (Álvaro-Fuentes and Paustian 2011). Also, continuous cultivation instead of leaving land fallow could increase SOC pools by more than twice (Álvaro-Fuentes and Paustian 2011). To prevent soil erosion, croplands with slopes greater than 15% should be converted to grasslands. In this context, land use conversion from cropland to shrubland or wild grassland would be better for SOC sequestration than tree plantation in semi-arid lands (Chen et al. 2007). For grasslands, SOC storage decreases substantially by grassland degradation due to long-term heavy grazing. At least two decades of grazing prohibition would be appropriate for restoring grasslands from degraded to undisturbed natural SOC conditions (He et al. 2008). However, such direct interventions and control policies should be based on reliable ecological and economic arguments (Puigdefábregas 1998).

15.8 Conclusions

Analyses of the published literature concerning the carbon cycle over drylands support the following conclusions:

1. Although numerous studies have been published related to SOC pools and C in biomass, a conclusive global analysis using models is needed to provide credible estimates of C pool in drylands.

2. However, an approximate value of 230 Pg of SOC in drylands is widely accepted. This value is reduced by 15% taking into account the percentage of cultivated lands.
3. Also, an approximate value of 65 Pg of total C content in biomass is widely reported. This value can also be modified significantly taking into account the anthropogenic practices.
4. Implications for the drylands C sink capacity of geochemical processes (relevant for SIC) or ventilation and erosion are poorly understood.
5. Thus, the differences in the measured net C exchange with the atmosphere, ranging from −106 to 145 g C m^{-2} for deserts and from −190 to 140 g C m^{-2} for grasslands are not easily explained.
6. Less drastic anthropogenic land use such as NT or minimum-tillage agricultural practices, leaving residues on the soil surface, or reducing fallow croplands and temporary grazing exclusions in grassland may improve SOC sequestration in drylands. Ideally, NT management may potentially increase the SOC by 20%, while non-grazing in grassland could increase SOC storage by about 45%.

Acknowledgements PSO is a Juan de la Cierva fellow, funded by the Spanish Ministry of Science and Innovation. English revision was provided by Professor Kowalski. Comments from A. S. Kowalski, Ana Were, Francisco Domingo, Gabriel Delgado and Russel L. Scott improved this paper.

References

Aires LMI, Pio CA, Pereira JS (2008) Carbon dioxide exchange above a Mediterranean C3/C4 grassland during two climatologically contrasting years. Glob Chang Biol 14:539–555

Allen DE, Pringle MJ, Page KL et al (2010) A review of sampling designs for the measurement of soil organic carbon in Australian grazing lands. Rangel J 32:227–246

Álvaro-Fuentes J, Paustian K (2011) Potential soil carbon sequestration in a semiarid Mediterranean agroecosystem under climate change: quantifying management and climate effects. Plant Soil 338:261–272

Amundson R, Wang Y, Chadwick O et al (1994) Factors and processes governing the 14C content of carbonate in desert soils. Earth Planet Sci Lett 125:385–405

Anderson-Teixeira KJ, Delong JP, Fox AM et al (2011) Differential responses of production and respiration to temperature and moisture drive the carbon balance across a climatic gradient in New Mexico. Glob Chang Biol 17:410–424

Austin AT, Vivanco L (2006) Plant litter decomposition in a semi-arid ecosystem controlled by photodegradation. Nature 442:555–558

Baldocchi DD, Falge E, Gu L et al (2001) FLUXNET: a new tool to study the temporal and spatial variability of ecosystem-scale carbon dioxide, water vapor, and energy flux densities. Bull Am Meteorol Soc 82:2415–2434

Berner RA (2003) The long-term carbon cycle, fossil fuels and atmospheric composition. Nature 426:323–326

Birch-Thomsen T, Elberling B, Fog B et al (2007) Temporal and spatial trends in soil organic carbon stocks following maize cultivation in semi-arid Tanzania, East Africa. Nut Cycl Agroecosyst 79:291–302

Boix-Fayos C, de Vente J, Albaladejo J et al (2009) Soil carbon erosion and stock as affected by land use changes at the catchment scale in Mediterranean ecosystems. Agric Ecosyst Environ 133:75–85

Bolin B, Sukumar R, Ciais P et al (2001) The global perspective. In: Noble B, Bolin B et al (eds) IPCC special report on land use, land-use change and forestry. Cambridge University Press, Cambridge
Boucot AJ, Gray J (2001) A critique of Phanerozoic climatic models involving changes in the CO_2 content of the atmosphere. Earth Sci Rev 56:1–159
Brandt LA, Bohnet C, King JY (2009) Photochemically induced carbon dioxide production as a mechanism for carbon loss from plant litter in arid ecosystems. J Geophys Res 114. doi:10.1029/2008JG0772
Chen L, Gong J, Fu B et al (2007) Effect of land use conversion on soil organic carbon sequestration in the loess hilly area, loess plateau of China. Ecol Res 22:641–648
Cuezva S, Fernandez-Cortes A, Benavente D et al (2011) Short-term CO (g) exchange between a shallow karstic cavity and the external atmosphere during summer: role of the surface soil layer. Atmos Environ 45:1418–1427
Dabberdt WF, Lenschow DH, Horst TW et al (1993) Atmosphere-surface exchange measurements. Science 260:1472–1481
Delgado G, Delgado R, Párraga J et al (2008) Precipitation of carbonates and phosphates by bacteria in extract solutions from a semi-arid saline soil. Influence of Ca^{2+} and Mg^{2+} concentrations and Mg^{2+}/Ca^{2+} molar ratio in biomineralization. Geomicrobiol J 23:1–13
Denef K, Stewart CE, Brenner J et al (2008) Does long-term center-pivot irrigation increase soil carbon stocks in semi-arid agro-ecosystems? Geoderma 145:121–129
Díaz-Hernández JL, Barahona Fernádez E (2008) The effect of petrocalcic horizons on the content and distribution of organic carbon in a Mediterranean semiarid landscape. Catena 74:80–85
Díaz-Hernández JL, Barahona Fernádez E, Linares González J (2003) Organic and inorganic carbon in soils of semiarid regions: a case study from the Guadix–Baza basin (Southeast Spain). Geoderma 114:65–80
Domingo F, Villagarcía L, Brenner AJ et al (1999) Evapotranspiration model for semi-arid shrublands tested against data from SE Spain. Agric For Meteorol 95:67–84
Emmerich EW (2003) Carbon dioxide fluxes in a semiarid environment with high carbonate soils. Agric For Meteorol 116:91–102
Entry JA, Sojka RE, Shewmaker GE (2004) Irrigation increases inorganic carbon in agricultural soils. Environ Manage 33:S309–S317
Eswaran H, Reich PF, Kimble JM et al (2000) Global carbon stocks. In: Lal R et al (eds) Global climate change and pedogenic carbonates. Lewis Publishers, Boca Raton
Falge E, Baldocchi DD, Tenhunen J et al (2002) Seasonality of ecosystem respiration and gross primary production as derived from FLUXNET measurements. Agric For Meteorol 113:53–74
Fan J, Zhong H, Harris W et al (2008) Carbon storage in the grasslands of China based on field measurements of above- and below-ground biomass. Clim Chang 86:375–396
Ferlan M, Alberti G, Eler K et al (2011) Comparing carbon fluxes between different stages of secondary succession of a karst grassland. Agric Ecosyst Environ 140:199–207
Flanagan DC (2006) Erosion. Encycl Soil Sci. doi:10.1081/E-ESS-120042669
Gallo ME, Porras-Alfaro A, Odenbach KJ et al (2009) Photoacceleration of plant litter decomposition in an arid environment. Soil Biol Biochem 41:1433–1441
Glantz MH (1994) Drought follows the plow: cultivating marginal areas. Cambridge University Press, Cambridge
Glenday J (2008) Carbon storage and emissions offset potential in an African dry forest, the Arabuko-Sokoke Forest, Kenya. Environ Monit Asses 142:85–95
Gombert P (2002) Role of karstic dissolution in global carbon cycle. Glob Planet Chang 33:177–184
Harden JW, Sharpe JM, Parton WJ et al (1999) Dynamic replacement and loss of soil carbon on eroding cropland. Glob Biogeochem Cycle 13:885–901
Hastings SJ, Oechel WC, Muhlia-Melo A (2005) Diurnal, seasonal and annual variation in the net ecosystem CO_2 exchange of a desert shrub community (Sarcocaulescent) in Baja California, Mexico. Glob Chang Biol 11:1–13
He N, Yu G, Wu L et al (2008) Carbon and nitrogen store and storage potential as affected by land-use in a Leymus chinensis grassland of northern China. Soil Biol Biochem 40:2952–2959

Inglima I, Alberti G, Bertolini T et al (2009) Precipitation pulses enhance respiration of Mediterranean ecosystems: the balance between organic and inorganic components of increased soil CO_2 efflux. Glob Chang Biol 15:1289–1301
IPCC (1990) In: Houghton JT, Jenkins GJ, Ephraums JJ (eds) Climate change: the IPCC scientific assessment. Cambridge University Press, Cambridge
IPCC (2007) In: Solomon S, Qin D, Manning M, Solomon S, Qin D, Manning M et al (eds) Climate change 2007: the physical science basis. Contribution of working group I to the fourth assessment report of the Intergovernmental Panel on Climate Change. Cambridge University Press, Cambridge
Jacinthe PA, Lal R (2001) A mass balance approach to assess carbon dioxide evolution during erosional events. Land Degrad Dev 12:329–339
Jacinthe PA, Lal R (2006) Erosion and carbon dioxide. Encycl Soil Sci. doi:10.1081/E-ESS-120001611
Jassal R, Black A, Novak M et al (2005) Relationship between soil CO_2 concentrations and forest-floor CO_2 effluxes. Agric For Meteorol 130:176–192
Kato T, Tang Y, Gu S et al (2006) Temperature and biomass influences on interannual changes in CO_2 exchange in an alpine meadow on the Qinghai-Tibetan Plateau. Glob Chang Biol 12:1285–1298
Keeling CD (1960) The concentration and isotopic abundance of carbon dioxide in the atmosphere. Tellus 12:200–203
Kowalski AS, Serrano-Ortiz P, Janssens IA et al (2008) Can flux tower research neglect geochemical CO_2 exchange? Agric For Meteorol 148:1045–1054
Lal R (2004) Carbon sequestration in dryland ecosystems, environmental management. Springer, New York, pp 528–544
Lal R, Griffin M, Apt J et al (2004) Ecology: managing soil carbon. Science 304:393
Lal R, Kimble JM (2000a) Inorganic carbon and the global C cycle: research and development priorities. In: Lal R, Kimble JM, Eswaran H et al (eds) Global climate change and pedogenic carbonates. Lewis Publishers, Boca Ratón
Lal R, Kimble JM (2000b) Pedogenic carbonates and the global carbon cycle. In: Lal R, Kimble JM, Eswaran H et al (eds) Global climate change and pedogenic carbonates. Lewis Publishers, Boca Ratón
Lasaga AC, Soler JM, Ganor J et al (1994) Chemical weathering rate laws and global geochemical cycles. Geochim Cosmochim Acta 58:2361–2386
Lewicki JL, Hilley GE, Dobeck L et al (2010) Dynamics of CO_2 fluxes and concentrations during a shallow subsurface CO_2 release. Environ Earth Sci 60:285–297
Li S-G, Asanuma J, Eugster W et al (2005) Net ecosystem carbon dioxide exchange over grazed steppe in central Mongolia. Glob Chang Biol 11:1941–1955
Li W, Liu L-P, Zhou P-P et al (2011) Calcite precipitation induced by bacteria and bacterially produced carbonic anhydrase. Curr Sci 100:502–508
Liu Z, Zhao J (2000) Contribution of carbonate rock weathering to the atmospheric CO_2 sink. Environ Geol 39:1053–1058
López-Fando C, Pardo MT (2009) Changes in soil chemical characteristics with different tillage practices in a semi-arid environment. Soil Tillage Res 104:278–284
López-Fando C, Pardo MT (2011) Soil carbon storage and stratification under different tillage systems in a semi-arid region. Soil Tillage Res 11:224–230
Luo H, Oechel WC, Hansting SJ et al (2007) Mature semiarid chaparral ecosystems can be a significant sink for atmospheric carbon dioxide. Glob Chang Biol 13:386–396
Ma S, Baldocchi DD, Xu L et al (2007) Inter-annual variability in carbon dioxide exchange of an oak/grass savanna and open grassland in California. Agric For Meteorol 147:157–171
Marion GM (1989) Correlation between long-term pedogenic $CaCO_3$ formation rate and Modern precipitation in deserts of the America southwest. Quat Res 32:291–295
Marion GM, Verburg PS, McDonald EV et al (2008) Modeling salt movement through a Mojave Desert soil. J Arid Environ 72:1012–1033. doi:1010.1016/j.jaridenv.2007.1012.1005
Martínez-Mena M, Lopez J, Almagro M et al (2008) Effect of water erosion and cultivation on the soil carbon stock in a semiarid area of South-East Spain. Soil Tillage Res 99:119–129

Mermut AR, Amundson R, Cerling TE (2000) The use of stable isotopes in studying carbonate dynamics in soils. In: Lal R (ed) Global climate change and pedogenic carbonates. Lewis Publishers, Boca Raton

Mielnick P, Dugas WA (2000) Soil CO_2 flux in a tallgrass prairie. Soil Biol Biochem 32:221–228

Mielnick P, Dugas WA, Mitchell K et al (2005) Long-term measurements of CO_2 flux and evapotranspiration in a Chihuahuan desert grassland. J Arid Environ 60:423–436

Mouat DA, Lancaster J (2006) Desertification: impact. Encycl Soil Sci. doi:10.1081/E-ESS-120001732

Muñoz C, Ovalle C, Zagal E (2007) Distribution of soil organic carbon stock in an alfisol profile in Mediterranean Chilean ecosystems. J Soil Sci Plant Nutr 7:15–27

Noellemeyer E, Quiroga AR, Estelrich D (2006) Soil quality in three range soils of the semi-arid Pampa of Argentina. J Arid Environ 65:142–155

Okin GS (2001) Chapter 1: Wind-driven desertification: process modeling, remote monitoring, and forecasting. California Institute of Technology, Pasadena, p 12

Parr JF, Sullivan LA (2005) Soil carbon sequestration in phytoliths. Soil Biol Biochem 37:117–124

Parraga J, Rivadeneyra MA, Martin-Garcia JM et al (2004) Precipitation of carbonates by bacteria from a saline soil, in natural and artificial soil extracts. Geomicrobiol J 21:55–66

Perez-Quezada JF, Delpiano CA, Snyder KA et al (2011) Carbon pools in an arid shrubland in Chile under natural and afforested conditions. J Arid Environ 75:29–37

Plaza-Bonilla D, Cantero-Martínez C, Álvaro-Fuentes J (2010) Tillage effects on soil aggregation and soil organic carbon profile distribution under Mediterranean semi-arid conditions. Soil Use Manage 26:465–474

Puigdefábregas J (1998) Ecological impact of global change on drylands and their implications for desertification. Land Degrad Dev 9:393–406

Rasmussen C (2006) Distribution of soil organic and inorganic carbon pools by biome and soil taxa in Arizona. Soil Sci Soc Am J 70:256–265

Reichstein M, Falge E, Baldocchi DD et al (2005) On the separation of net ecosystem exchange into assimilation and ecosystem respiration: review and improved algorithm. Glob Chang Biol 11:1–16

Rey A, Belelli-Marchesini L, Were A et al (2012) Wind as a main driver of the net ecosystem carbon balance of a semiarid Mediterranean steppe in the South East of Spain. Global Change Biol 18:539–554

Rivadeneyra MA, Delgado G, Ramos-Cormenzana A et al (1997) Precipitation of carbonates by Deleya halophila in liquid media: pedological implications in saline soils. Arid Soil Res Rehabil 11:35–47

Rodeghiero M, Heinemeyer A, Schrumpf M et al (2009) Determination of soil carbon stocks and changes. In: Kutsch WL, Bahn M, Heinemeyer A (eds) Soil carbon dynamics: an integrated methodology. Cambridge University Press, Cambridge

Rutledge S, Campbell DI, Baldocchi DD et al (2010) Photodegradation leads to increased carbon dioxide losses from terrestrial organic matter. Glob Chang Biol 16:3064–3074

Safriel U, Adeel Z, Niemeijer D (2005) Dryland systems. In: Safriel U, Adeel Z, Niemeijer D, El-Kassas M, Ezcurra E (eds) Ecosystems and human well-being: current state and trends. Cambridge University Press, Cambridge

Saharawat KL (2003) Importance of inorganic carbon in sequestering carbon in soils of the dry regions. Curr Sci 84:864–865

Sanchez-Cañete EP, Serrano-Ortiz P, Kowalski AS et al (2011) Subterranean CO_2 ventilation and its role in the net ecosystem carbon balance of a karstic shrubland. Geophys Res Lett 38:L09802. doi:09810.01029/02011GL047077

Schimel DS, House JI, Hibbard KA et al (2001) Recent patterns and mechanisms of carbon exchange by terrestrial ecosystems. Nature 414:169–172

Schlesinger WH (1985) The formation of caliche in soils of the Mojave Desert, California. Geochim Cosmochim Acta 49:57–66

Schlesinger WH (1990) Evidence from chronosequence studies for a low carbon-storage potential of soil. Nature 348:232–234

Schlesinger WH (2006) Inorganic carbon and the global C cycle. Encycl Soil Sci. doi:10.1081/E-ESS-120042705

Scott RL, Hamerlynck EP, Jenerette GD et al (2010) Carbon dioxide exchange in a semidesert grassland through drought induced vegetation change. J Geophys Res 115:G03026. doi:03010.01029/02010JG001348

Scott RL, Huxman TE, Williams DG et al (2006) Ecohydrological impacts of woody-plant encroachment: seasonal patterns of water and carbon dioxide exchange within a semiarid riparian environment. Glob Chang Biol 12:311–324

Scott RL, Jenerette D, Potts DL et al (2009) Effects of seasonal drought on net carbon dioxide exchange from a woody-plant-encroached semiarid grassland. J Geophys Res 114:G04004. doi:04010.01029/02008JG000900

Serrano-Ortiz P, Domingo F, Cazorla A et al (2009) Interannual CO_2 exchange of a sparse Mediterranean shrubland on a carbonaceous substrate. J Geophys Res 114:G04015. doi:04010.01029/02009JG000983

Serrano-Ortiz P, Roland M, Sánchez-Moral S et al (2010) Hidden, abiotic CO_2 flows and gaseous reservoirs in the terrestrial carbon cycle: review and perspectives. Agric For Meteorol 150:321–329

Serrano-Ortiz P, Roland M, Sanchez-Moral S et al (2011) Corrigendum to hidden, abiotic CO_2 flows and gaseous reservoirs in the terrestrial carbon cycle: review and perspectives. Agric For Meteorol 151:529

Shackleton CM, Scholes RJ (2011) Above ground woody community attributes, biomass and carbon stocks along a rainfall gradient in the savannas of the central Lowveld, South Africa. S Afr J Bot 77:184–192

Shukla MK, Lal R, Ebinger M et al (2006) Physical and chemical properties of soils under some piñon-juniper-oak canopies in a semi-arid ecosystem in New Mexico. J Arid Environ 66:673–685

Singh SK, Singh AK, Sharma BK et al (2007) Carbon stock and organic carbon dynamics in soils of Rajasthan, India. J Arid Environ 68:408–421

Sombrero A, de Benito A (2010) Carbon accumulation in soil. Ten-year study of conservation tillage and crop rotation in a semi-arid area of Castile-Leon, Spain. Soil Tillage Res 107:64–70

Stoy PC, Katul G, Juang J-Y et al (2006) An evaluation of models for partitioning eddy covariance-measured net ecosystem exchange into photosynthesis and respiration. Agric For Meteorol 141:2–18

Subke JA, Reichstein M, Tenhunen JD (2005) Explaining temporal variation in soil CO_2 efflux in a mature spruce forest in Southern Germany. Soil Biol Biochem 35:1467–1483

Takle ES, Massman WJ, Brandle JR et al (2004) Influence of high-frequency ambient pressure pumping on carbon dioxide efflux from soil. Agric For Meteorol 124:193–206

Tans PP, Fung IY, Takahashi T (1990) Observational constraints on the global atmospheric CO_2 budget. Science 247:1431–1438

UNEP (1997) World atlas of desertification, 2nd edn. United Nations Environment Programme, Nairobi

Valentini R, Matteucci G, Dolman AJ et al (2000) Respiration as the main determinant of carbon balance in European forests. Nature 404:861–865

Van Oost K, Quine TA, Govers G et al (2007) The impact of agricultural soil erosion on the global carbon cycle. Science 318:626–629

Vande Walle I, Mussche S, Samson R et al (2001) The above- and belowground carbon pools of two mixed deciduous forest stands located in East-Flanders (Belgium). Ann For Sci 58:507–517

Verrecchia EP, Dumont J-L, Rolko KE (1999) Do fungi building limestones exist in semi-arid regions? Naturwissenschaften 77:584–586 von Lützow M, Kögel-Knabner I, Ludwig B et al (2008) Stabilization mechanisms of organic matter in four temperate soils: Development and application of a conceptual model. J Plant Nutr Soil Sci 171:111–124

Wang Y, Zhou G, Wang Y (2008) Environmental effects on net ecosystem CO_2 exchange at half-hour and month scales over Stipa krylovii steppe in northern China. Agric For Meteorol 148:714–722

Wang Q, Zhang L, Li L et al (2009) Changes in carbon and nitrogen of Chernozem soil along a cultivation chronosequence in a semi-arid grassland. Eur J Soil Sci 60:916–923

Were A, Serrano-Ortiz P, Moreno de Jong C, Villagarcía L et al (2010) Ventilation of subterranean CO_2 and Eddy covariance incongruities over carbonate ecosystems. Biogeosciences 7:859–867

Wiesmeier M, Barthold F, Blank B et al (2011) Digital mapping of soil organic matter stocks using Random Forest modeling in a semi-arid steppe ecosystem. Plant Soil 340:7–24

Williams M, Ryan CM, Rees RM et al (2008) Carbon sequestration and biodiversity of re-growing miombo woodlands in Mozambique. For Ecol Manage 254:145–155

Wohlfahrt G, Fenstermaker LF et al (2008) Large annual net ecosystem CO_2 uptake of a Mojave Desert ecosystem. Glob Chang Biol 14:1475–1487

Woomer PL, Touré M, Sall M (2004) Carbon stocks in Senegal's Sahel Transition Zone. J Arid Environ 59:499–510

Xie J, Li Y, Zhai C et al (2008) CO_2 absorption by alkaline soils and its implication to the global carbon cycle. Environ Geol 56:953–961

Zhao L, Li Y, Xu S et al (2006) Diurnal, seasonal and annual variation in net ecosystem CO_2 exchange of an alpine shrubland on Qinghai-Tibetan plateau. Glob Chang Biol 12:1940–1953

Chapter 16
Carbonization of Urban Areas

Galina Churkina

Abstract Re-carbonization of forests and agricultural land along with carbon dioxide (CO_2) capture technologies have been considered as options for reducing atmospheric CO_2 concentrations. Can also land ecosystem created by humans like urban areas be carbonized? The urbanization trends show that the area and importance of urban areas continue to increase. Share of urban population will increase from present 50–70% by 2050 globally. The share of urban land will be also progressively increasing to accommodate the growing number of urbanites. This chapter discusses strategies to carbonize urban areas, so that more carbon (C) per capita can be stored there. It starts with a brief review of the urban C cycle. Two features distinguish urban C cycle from other ecosystems: (1) C cycling of a city and its footprint are intimately linked and (2) natural and anthropogenic components of urban C cycle are equally important and interdependent. Then major pools and mechanisms for C storage in both anthropogenic and natural components of the urban C cycle are reviewed. Soil is the largest potential C pool followed by vegetation, landfills, and buildings. Although in settlements with low build-up density soil may store more than 60% of total C, in densely build-up cities soils and buildings may store equal amounts of C (~40% each). Potential of C storage in landfills is controversial, because of accompanying methane emissions and groundwater pollution. Human-driven mechanisms for C accumulation in cities such as import of C containing materials supersede the natural ones. At the current level of technology any option for carbonization of cities is associated with CO_2 emissions. Cities' carbonization cannot be considered as a pure increase in C storage per capita, but as an increase in C storage per capita per unit of emitted CO_2.

G. Churkina (✉)
Geography Department, Humboldt Universität zu Berlin,
Rudower Chaussee 16, 12489 Berlin, Germany
e-mail: galina@churkina.org

R. Lal et al. (eds.), *Recarbonization of the Biosphere: Ecosystems and the Global Carbon Cycle*, DOI 10.1007/978-94-007-4159-1_16,

Keywords Carbon sequestration • Urban ecosystems • Carbonization • Megacities • Urbanization • Carbon footprint • Urban carbon cycle • Soil organic matter • Urban infrastructure • Greenhouse gases • Carbon uptake • Above ground carbon storage • Below ground carbon storage • Carbon costs • Green roofs • CO_2 emissions • Anthropogenic carbon pools • Natural carbon pools • Carbon density • Exurban areas • Landfills • Backfills

Abbreviations

NH_3	ammonia
$CaCO_3$	calcite
C	carbon
CO_2	carbon dioxide
CO	carbon monoxide
GHGs	greenhouse gases
CH_4	methane
NO_x	nitrogen oxides
N_2O	nitrous oxide
SOM	soil organic matter
VOC	volatile organic compounds

16.1 Introduction

A reduction of atmospheric carbon dioxide (CO_2) concentration is needed to minimize risks of further climate change. This reduction can be achieved by carbon (C) sequestration in the ocean and on land. CO_2 could be drawn down using technologies for CO_2 capture (Matter and Kelemen 2009) or natural processes. On land, C sequestration can be naturally attained by increasing plant C uptake and C storage in plants and soil. A potential drawdown of 50 ppm of atmospheric CO_2 by 2100 can be achieved by re-carbonization of agricultural and forest ecosystems as well as of peat lands (Hansen et al. 2008). For instance, reforestation could mostly compensate for deforestation over the past few hundred years, which net emissions amount to 60 ± 30 ppm (Hansen et al. 2008). Can substantial amount of C also be stored in ecosystems which have been artificially created such as in urban areas?

Given the urbanization trends, the prominence of urban areas as a land cover type has been increasing and these trends are most likely to continue in the future. The world population is becoming progressively more urban. The urban share of the world's population increased from one-third in 1950 to one-half in 2009. Because most of the future population growth is predicted in urban areas, the share of urban population is likely to increase to 70% by 2050 (UN 2008).

Table 16.1 Urbanization trends in ten European cities from 1950s to 1990s based on Moland database (Lavalle et al. 2002)

City	Total area (km^2)	Total urbanized area (km^2) 1950s	1990s	Change in urbanized area (%)
Algarve	781.5	32.2	119.1	270
Brussels	1,308.8	318.6	560.3	76
Copenhagen	665.0	242.7	386.1	59
Dublin	676.8	163.1	319.3	96
Dresden	1,256.7	231.1	314.1	36
Helsinki	1,041.5	135.0	326.0	141
Munich	797.8	246.7	357.0	45
Padua-Venice	515.5	69.7	188.9	171
Prague	797.6	186.9	288.4	54
Tallinn	1,070.1	88.3	182.1	106
Vienna	841.8	249.7	341.1	37

The area covered by urban land is also likely to increase. It is to happen not only because more people need more space, but also because people prefer to live in less densely build-up areas once they can afford it. The recent urbanization trends in Europe and the USA support this statement. Since mid-1950s, European cities have expanded on average by 78%, whereas the population has only grown by 33% (EEA 2006). Changes in urbanized areas of individual cities have large differences: sometimes several orders of magnitude (Table 16.1). Over the same time period urban and exurban lands increased by 400%, while urban population increased only by 100% in the USA (Brown et al. 2005). Increases in personal wealth and mobility combined with land availability are behind these trends. What implications do these trends have for C cycle?

An undisputable effect of cities on C cycle is their CO_2 emissions. Cities are currently responsible for ~70% of the global CO_2 emissions from energy production (IEA 2008). This implies that increasing share of urban population will most likely result in greater share of global CO_2 emissions coming from cities. The question arises if urbanites can be responsible not only for CO_2 emissions, but also for carbonization of urban areas.

This chapter explores the potential of urban areas to store more C, or carbonization. It also discusses where and how more C can be stored in the cities. Given the rising number of urbanites the opportunities to store more C per capita in urban areas should not be overlooked.

16.2 Carbon Storage in Urban Areas

To explore the carbonization potential of urban areas a review of the distinct features of urban C cycle is followed by a discussion of the urban C pools as well as mechanisms of C accumulation in these pools.

16.2.1 What Distinguishes C Cycle of Urban Areas from Other Ecosystems?

Two features discern C cycle of urban areas from other ecosystems. First, C cycle of an urban area and its footprint are intimately linked. Second, the C cycle of an urban area has natural and anthropogenic components, which are equally important and interdependent.

16.2.1.1 Urban Area and Its Footprint

Urban area is never self-supporting. It is dependent on the resources drawn from its footprint. Urban footprint is the area required to meet demands of urban population in terms of consumption and waste accumulation and the area affected by urban pollution and changes in climate (Churkina 2008). The material flows through urban areas are estimated to be so huge that the area of ecosystems, which are required to assimilate and process these flows, is 400–1,000-times larger than the size of the cities themselves (Decker et al. 2000). A review of eight metropolitan regions across five continents shows that the metabolism of cities is increasing in volume (Kennedy et al. 2007). Most regions exhibit escalating per capita metabolism with respect to water, wastewater, energy, and materials flows. For instance water and waste water flows were larger for studies in 1990s than those in the early 1970s and several cities have become more material intensive.

The carbon cycle of a city is not limited to the city itself, it is closely linked with the C cycle of the city's footprint (Fig. 16.1). These connections are established through the flows of food and fiber from the footprint into the city, the flow of trash from city into its footprint, and the effect of urban pollution on the C uptake and release by ecosystems in the footprint.

16.2.1.2 Natural and Anthropogenic Components of the Urban C Cycle

Both natural and anthropogenic components contribute to the C cycling of an urban area. Cycling of C through urban vegetation and soil constitute the natural component. The uptake, storage, and release of C associated with transportation and the artifacts such as buildings comprise the anthropogenic component.

16.2.2 Carbon Pools

Total amount of C stored in urban areas is determined by the size and number of pools as well as by the C density of these pools. Cities have more diverse C pools than natural ecosystems. In addition to soil and vegetation, cities store C also in

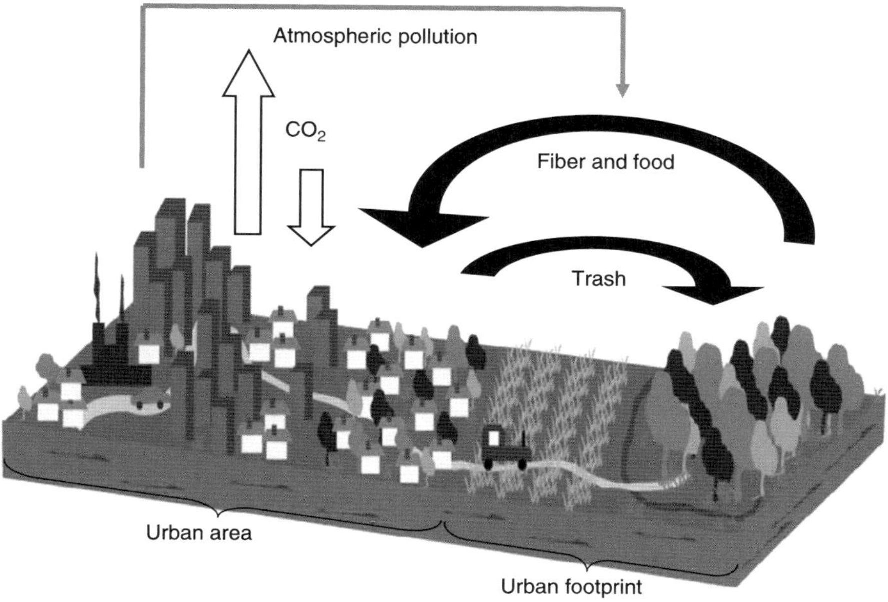

Fig. 16.1 C fluxes of an urban areas and its connections with urban footprint (Churkina 2008). C fluxes driven by anthropogenic activities are shown in *black*. C fluxes driven natural processes are shown in *white*

buildings, furniture, books, clothes, footwear, landfills, people, and pets (Fig. 16.2). Previous studies (Churkina et al. 2010) show that the largest of these are soils, vegetation, buildings, and landfills. People and pets store drastically smaller amounts of C than the four pools mentioned above. In the conterminous U.S. the estimated C stored in humans was at least two orders of magnitude lower (0.0024 PgC) than that stored in the buildings (0.4–1.3 PgC) or in landfills (1.8–2.5 PgC) (Churkina et al. 2010). The C storage in pets is several orders of magnitudes smaller than in humans (Bramryd 1980).

Similar to boreal forests, urban areas store approximately 70% of the C below and 30% above-ground (Fig. 16.3). Also above- and below-ground C densities of urban areas are similar to those of boreal forests (Lorenz and Lal 2010). In contrast, grasslands and croplands store most of their C belowground. All different land cover types (Fig. 16.3) have below-ground C density of the same magnitude. In the estimate for urban areas in the Fig. 16.3 the sum of C storage in vegetation and buildings is counted towards above-ground storage. Below-ground storage is calculated based on C storage in soils and landfills (30% and 70%, respectively).

The relative importance of one C pool or another depends on the build-up density of urban areas. In cities and towns with low build-up density such as suburban or exurban areas, soil C pool may be the largest followed by C pools in vegetation and landfill. However, in densely build-up urban areas buildings can store as much C as

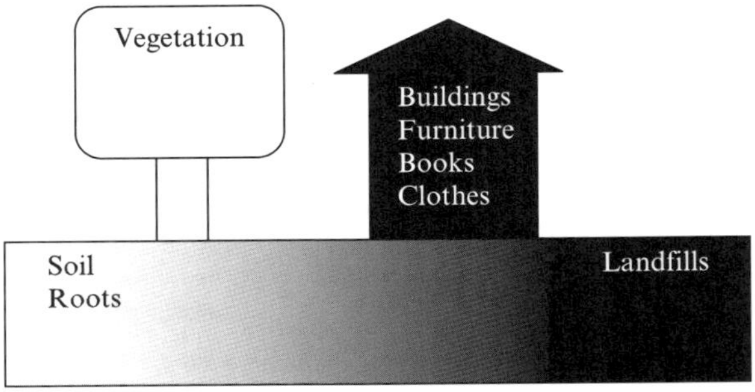

Fig. 16.2 Natural (*white fill*) and anthropogenic (*black fill*) carbon pools of urban areas. Urban soils can be of natural and anthropogenic origins (*white-to-black gradient fill*). Urban soils vary with regard to degree of their disturbance from man-influenced and man-changed to man-made soils (e.g., landfill) (Lehmann and Stahr 2007)

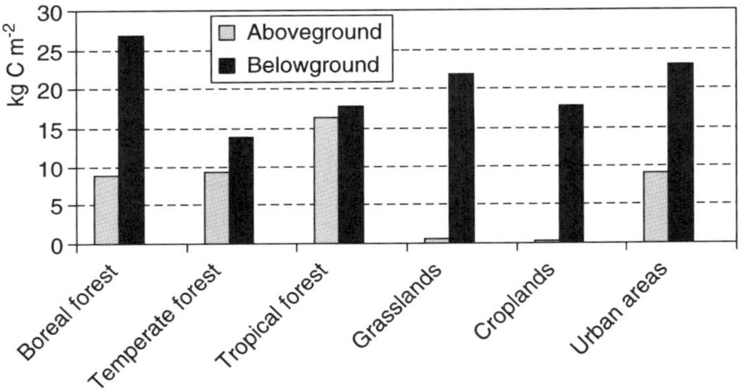

Fig. 16.3 Comparison of mean carbon densities of urban areas and other land cover types. Urban areas (Churkina et al. 2010) like forests (Lorenz and Lal 2010) store C *above-* and *belowground.* Grasslands and croplands store 95% of C *belowground* (Jobbágy and Jackson 2000; Jackson et al. 1996)

soils. For instance, in the conterminous U.S. (Fig. 16.4), soils and buildings in urban areas store equal amount of C (~42%), but soils are the largest C pool in exurban areas (67%).

16.2.3 Mechanisms of Carbon Accumulation

16.2.3.1 Vegetation and Soil

Several factors determine the total accumulation of C in urban vegetation and soil. The most essential one is the difference between C uptake by plants and C release by plant and soil respiration. This difference is modified by the C displacement by

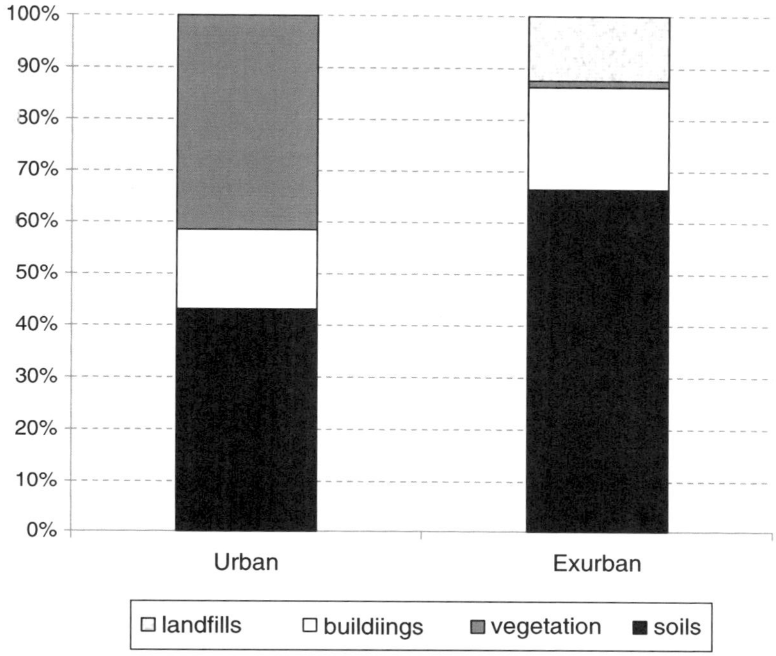

Fig. 16.4 Relative carbon pool sizes (%) in urban and exurban areas of the conterminous U.S. in 2000 (Churkina et al. 2010)

humans, such as removal of fallen leaves in autumn or occasional burial of trash. Deposition of elemental C also contributes to total C accumulated in urban soil (Lorenz et al. 2006), but its contribution is poorly understood. Elemental C is a product of incomplete combustion and/or pyrolysis of either organic matter or fossil fuels. Because landfills are entirely created by humans, they are discussed in the next section as part of urban artifacts.

Atmospheric C uptake occurs as a result of the photosynthesis. It is controlled by light intensity, air temperature, atmospheric CO_2 concentration, as well as water and nutrient availabilities (Larcher 2002; Chapin et al. 2002). Air and soil pollution are important additional factors which may enhance or inhibit photosynthesis of urban plants. Typical urban pollution includes CO_2, nitrogen oxides (NO_x), carbon monoxide (CO), methane (CH_4), and volatile organic compounds (VOC). Less typical pollutions are nitrous oxide (N_2O) and ammonia (NH_3) Emissions of N_2O from cities are small in comparison to the ones from agriculture. These chemical compounds are emitted mostly as a result of fossil fuel burning by vehicles and power generation. CH_4 and CO_2 are also emitted from landfills. NH_3 is emitted from fertilizer application in urban gardens and parks. Dry or wet deposition of the emitted N compounds increases N availability and therefore increase plant C uptake. At daylight, airborne CO, CH_4, and VOC can react with other chemical compounds and lead to enhanced ozone concentrations in the troposphere, which at high levels inhibit photosynthesis.

The gaseous release of C in urban ecosystems is mostly a result of soil organic matter (SOM) decomposition as well as plant's maintenance and growth respiration. Air and soil temperatures as well as plant tissue chemistry and growth stage mainly control plant respiration (Reich et al. 1998). Decomposition of SOM is driven by soil temperature and moisture, soil physical characteristics (e.g., texture, aggregation), chemical composition of SOM (e.g., carbon to nitrogen ratio, fraction of elemental C), as well as the characteristics of soil microbial community and fauna (Chapin et al. 2002). In the cities humans influence all these drivers directly, for instance, by lawn irrigation or by leaves removal and indirectly by enhanced air temperatures. Irrigation increases soil moisture and therefore plant growth, litter input, and SOM decomposition rates. Removal of leaves in fall reduces leaf litter input into soil. Litter inputs are dominated by dead roots. Consequently the chemical composition of SOM is altered. Several studies report that C content of urban soils is higher than that of natural soils (Pouyat et al. 2006; Lorenz et al. 2006; Townsend-Small and Czimczik 2010).

Although solar energy remains the main driver behind C accumulation in vegetation and soils, in some cities it is complemented by fossil fuel energy needed for irrigation, fertilizer production, its application, and other vegetation/soil management activities.

16.2.3.2 Urban Artifacts

Two mechanisms are behind the C accumulation in artifacts in urban areas. The first mechanism is the direct uptake of CO_2 by concrete urban infrastructures. This process is called carbonation. Atmospheric CO_2 reacts with CaO in concrete to form calcite ($CaCO_3$). This is the reverse reaction of the calcination process used in cement making. The main controls behind CO_2 uptake in concrete are atmospheric CO_2 concentrations, air temperature, air humidity as well as water content, chemical composition, and porosity of materials (Gajda and Miller 2000). The carbonation process is relatively slow as atmospheric CO_2 has to diffuse into the solid material and to dissolve in its pore fluid. For instance, the total amount of C which can be captured in the American concrete infrastructures in one year (0.0004 Pg C $year^{-1}$, Gajda 2001) is two orders of magnitudes smaller than that captured by urban forests in the U.S. (0.02 Pg C $year^{-1}$, Nowak and Crane 2002).

The second mechanism behind the C accumulation in urban artifacts is import of C containing materials necessary for construction of urban infrastructures and for maintaining people's lifestyles. These include construction materials, furniture, books, clothes, and food. Some of this C ends up as being transported outside of urban areas as municipal soil waste, construction and demolition derbies, or sludge. The C accumulation rate in artifacts can be estimated as a difference between C brought in with food and fiber flows minus C removed with trash. The drivers behind C accumulation in artifacts are not well understood. Most likely the amount of C accumulated is driven by wealth, household size, as well as by life span of

Table 16.2 Annual carbon uptake, storage above and belowground of urban grasses trees, and green roofs

Storage	Annual C uptake ($g\,C\,m^{-2}$)	C storage aboveground ($g\,C\,m^{-2}$)	C storage belowground ($g\,C\,m^{-2}$)
Grass	–80 to 140[a]	n/a	1,200–14,400[b]
Trees	15–94[c]	2,100–5,900	7,700–11,600
Green roof	73–276[d]	64–239	37–185

[a]Hiller et al. (2011), Townsend-Small and Czimczik (2010)
[b]Pouyat et al. (2006)
[c]Nowak and Crane (2002)
[d]Getter et al. (2009)

products and buildings. Given the slow rates of C uptake by concrete buildings, most C accumulated in artifacts is brought into the cities by people.

In contrast to accumulation of C in vegetation and soil, actual solar energy is insufficient for accumulation of C in artifacts. Energy from burning of fossil fuel, which is stored solar energy from millions years ago, is used for transportation of materials in and out the cities and their maintenance.

Mechanisms of anthropogenic C releases in urban areas are fossil fuel burning, waste incineration, and decomposition of trash at landfills. These C releases currently dominate the ongoing changes in the global C cycle.

16.2.4 How to Store More Carbon in Urban Areas?

16.2.4.1 Vegetation and Soils

Urban vegetation and soils have a large potential to store more C especially in cities and towns with low build-up density. This potential is limited by climate conditions and the level of urban pollution.

Cities can be designed to store more C in soils and vegetation. Studies of C sequestration in urban vegetation show that urban vegetation is a C sink (Table 16.2, Fig. 16.2), but its C sequestration is offset by C release with landscape maintenance such as irrigation, fertilization, lawn mowing, as well as tree and shrub pruning (Townsend-Small and Czimczik 2010; Jo and McPherson 1995). Woody plants are more beneficial for long-term C sequestration than herbaceous plants, because of higher gaseous C emissions associated with grass maintenance, even though a mowed lawn, often provides greater storage of organic C than the natural cover it replaces (Townsend-Small and Czimczik 2010). Mowing of grasslands in Chicago, USA, released 1.5 times more C than the grasslands sequestered annually (Jo and McPherson 1995).

Less intensive grass management is recommended to reduce C release. Recycling of plant litter and especially grass clippings in cities should be encouraged.

Fig. 16.5 Two types of green roofs: (**a**) shrubs and trees on the top of Rockefeller Center, New York, the U.S. (Photo: D. Shankbone) and (**b**) herbaceous vegetation on the roofs of houses near Oslo, Norway (Photo: G. Churkina)

This recycling would return nutrients from litter to ecosystem and reduce fertilizer use and associated emissions of greenhouse gases (GHGs). For instance, less fertilizer is needed if grass clippings are left to decompose on the turf surface rather than composted or bagged and sent to landfill.

Green roofs may also contribute to C storage in urban areas, but are unlikely to sequester large amounts of C. There are two types of green roofs. The first one includes shrubs and trees (Fig. 16.5a). The second one has herbaceous perennial or annual plants (Fig. 16.5b). The first one is similar to garden landscaping found on the ground. It is maintenance intensive and requires relatively deep 'soil' or substrate to grow. The second one requires minimal maintenance and grows on shallow substrate. Because the second type of green roof is more common, a recent study (Getter et al. 2009) evaluated its C sequestration potential for roofs of the second type in Michigan and Maryland in the U.S., both located in temperate climate zone. Michigan has a continental climate with hot summer and cold winter, while Maryland's climate is influenced by the ocean and therefore milder. The green roofs sequestered 375 g C m^{-2} over 2 years of study in Michigan. It was calculated that after the roof maintenance costs were calculated in, 9 years would be needed to offset the roof establishment and maintenance costs.

Green fancy roofs of the modern buildings are not a recent invention. In some countries like Norway there is a tradition of green roofs (Fig. 16.5b). For hundreds of years houses in Norway have been covered with turf or sod roofs. Turf roofs provide good insulation and stability to the houses.

Along with emissions of CO_2 from energy production needed for urban vegetation management, emissions of other greenhouse gases such as N_2O are often associated with enhanced C storage in urban soils (Townsend-Small and Czimczik 2010). These emissions are associated with frequent fertilization of urban lawns. "Carbon Costs" associated with emissions of CO_2 as well as other GHGs have to be calculated while estimating benefits of C storage in urban vegetation and soils.

16.2.4.2 Buildings

C storage in buildings could also contribute to carbonization of urban areas. Using more wood or other plant derived materials like straw in house construction materials and in furniture could be an important means of increasing organic C storage. The three most abundant construction materials containing wood include lumber, oriented strand board, and plywood. Storage of organic C in buildings can be increased if wooden furniture is used instead of plastic and natural materials like wool, cotton, and silk are preferred for interior design. Use of wood in buildings, instead of brick, aluminum, steel, and concrete, can increase C storage in human settlements and reduce emissions of GHGs related to construction (Buchanan and Levine 1999) and life cycle of buildings (Upton et al. 2008). The emissions of GHGs are reduced because, in contrast, to fabrication of wooden construction materials, production of bricks and concrete is much more energy intensive and accompanied by high CO_2 emissions from burning of fossil fuel.

Refined models are necessary to account for material residence time in buildings. The residence time is influenced by renovation activities, building service life, and the fate of waste materials in landfill reservoirs which also requires modeling of GHG emissions from landfill decomposition processes (EPA 2006).

16.2.4.3 Landfills

Landfills can accumulate appreciable amounts of C over time. Solid waste deposited at landfills composed mostly of municipal solid waste with possible additions of combustion ash, sludge, and industrial process waste. In landfills C is stored long term in biologically recalcitrant materials (e.g., resistant to biological degradation) such as plastic rubber, leather, as well as in slowly decomposing parts of paper, wood, and food waste. In these materials C is stored in form of cellulose (30–50%), hemicelluloses (7–12%), and lignin (14–30%) (Warith et al. 1995). Cellulose and hemicellulose represent the major degradable components of waste. Lignin is essentially recalcitrant under anaerobic conditions, which are typical for most landfills, and may prevent other carbohydrate materials from degradation (Warith et al. 1995; Barlaz 1998). The amount of C stored in a landfill depends on the waste composition, age of landfill, and landfill treatment. It is controversial if it is beneficial to store C in landfills because of accompanying emissions of CH_4 and groundwater pollutions.

Barlaz (1998) estimated the amount of C that remains in long-term storage after anaerobic decomposition of municipal solid waste in landfills. The total C sequestration in global landfills was 119 Tg (1 Tg = 10^{12} g) C year^{-1}. It was calculated based on waste generation rates for 1994 for the United States and the early 1990s for other countries as well as with the assumptions that waste at the landfills achieves maximum decomposition and that C sequestration factor does not change from country to country.

16.3 Conclusions

Any option for increasing C storage in cities should be assessed together with the supplementary benefits or issues it may bring to urban dwellers.

In addition to offering C storage and uptake urban vegetation provides a range of further rewards to urbanites, which include heat island mitigation and energy saving. City trees cool surfaces by up to 10°C through shading and transpiration of water. This reduces energy use in summer and CO_2 emissions from energy production for air conditioning and fans. Green roofs lower the demand of buildings for heating and air conditioning use and therefore decrease release of CO_2 from energy production (Sailor 2008).

Higher C storage in buildings can be achieved by higher wood use in building construction. Any increase in wood use in building construction however would have implications for its production. Rising demand for wood must be accompanied by an increase in the area of forest or higher productivity of existing forest being managed for long-term sustainable timber production. Wooden houses are more prone to fire than those made of brick or concrete, which means that potential CO_2 emissions from burning are an issue in settlements where wooden houses dominate.

One of the undisputable advantages of storing C in urban areas is that one can control and monitor C accumulation in cities easier than in forests or in the ocean. One of the reasons is that it is easier to monitor processes which are closer to one's home. The other reason is related to the industrial processes that are partially behind C accumulation in the cities. These processes are more predictable and controllable than C uptake and release of natural ecosystems.

At the current level of technology any option for carbonization of cities is associated with CO_2 emissions. Therefore carbonization of the cities cannot be considered as a pure increase in C storage per capita. Instead it is an increase in C storage per capita per unit of emitted CO_2.

References

Barlaz MA (1998) Carbon storage during biodegradation of municipal solid waste components in laboratory-scale landfills. Glob Biogeochem Cycle 12(2):373–380

Bramryd T (1980) Fluxes and accumulation of organic carbon in urban ecosystems on a global scale. In: Bornkamm R, Lee JA, Seaward MRD (eds) Urban ecology. Blackwell Scientific Publications, Oxford, pp 3–12

Brown DG, Johnson KM, Loveland TR, Theobald DM (2005) Rural land-use trends in the conterminous United States. Ecol Appl 15(6):1851–1863

Buchanan AH, Levine SB (1999) Wood-based building materials and atmospheric carbon emissions. Environ Sci Policy 2(6):427–437

Chapin SF III, Mooney HA, Chapin MC et al (2002) Principles of terrestrial ecosystem ecology. Springer, New York

Churkina G (2008) Modeling the carbon cycle of urban systems. Ecol Model 216(2):107–113

Churkina G, Brown D, Keoleian GA (2010) Carbon stored in human settlements: the conterminous US. Glob Chang Biol 16:135–143. doi:10.1111/j.1365-2486.2009.02002.x

Decker EH, Elliot S, Smith FA et al (2000) Energy and material flow through the urban ecosystem. Ann Rev Energy Environ 25:685–740

EEA (2006) Urban sprawl in Europe – the ignored challenge. European Environment Agency (EEA), Copenhagen

EPA (2006) Solid waste management and greenhouse gases. A life cycle assessment of emissions and sinks, 3rd edn. The U. S. Environmental Protection Agency, Washington, DC

Gajda J (2001) Absorption of atmospheric carbon dioxide. Portland Cement Association, Skokie

Gajda J, Miller FM (2000) Concrete as a sink for atmospheric carbon dioxide: a literature review and estimation of CO_2 absorption. Portland Cement Association, Skokie

Getter KL, Rowe DB, Robertson GP et al (2009) Carbon sequestration potential of extensive green roofs. Environ Sci Technol 43(19):7564–7570. doi:10.1021/es901539x

Hansen J, Sato M, Kharecha P et al (2008) Target atmospheric CO_2: where should humanity aim? Open Atmos Sci 2:217–231

Hiller R, McFadden J, Kljun N (2011) Interpreting CO_2 fluxes over a suburban lawn: the influence of traffic emissions. Bound-Lay Meteorol 138(2):215–230. doi:10.1007/s10546-010-9558-0

IEA (2008) World energy outlook 2008: executive summary. International Energy Agency (IEA), Paris

Jackson RB, Canadell J, Ehleringer JR et al (1996) A global analysis of root distributions for terrestrial biomes. Oecologia 108:389–411

Jo HK, McPherson GE (1995) Carbon storage and flux in urban residential greenspace. J Environ Manage 45:109–133

Jobbágy EG, Jackson RB (2000) The vertical distribution of soil organic carbon and its relation to climate and vegetation. Ecol Appl 10(2):423–436. doi:10.1890/1051-0761(2000) 0102.0.CO;2
Kennedy C, Cuddihy J, Engel-Yan J (2007) The changing metabolism of cities. J Indust Ecol 11(2):43–59
Larcher W (2002) Physiological plant ecology, 4th edn. Springer, Berlin
Lavalle C, Demicheli L, Kasanko M et al (2002) Towards an urban atlas. Assessment of spatial data on 25 European cities and urban areas. European Environmental Agency, Copenhagen
Lehmann A, Stahr K (2007) Nature and significance of anthropogenic urban soils. J Soils Sed 7(4):247–260. doi:10.1065/jss2007.06.235
Lorenz K, Lal R (2010) Carbon dynamics and pools in major forest biomes of the world. In: Lorenz K, Lal R (eds) Carbon sequestration in forest ecosystems. Springer, Dordrecht
Lorenz K, Preston CM, Kandeler E (2006) Soil organic matter in urban soils: estimation of elemental carbon by thermal oxidation and characterization of organic matter by solid-state 13C nuclear magnetic resonance (NMR) spectroscopy. Geoderma 130(3–4):312–323
Matter JM, Kelemen PB (2009) Permanent storage of carbon dioxide in geological reservoirs by mineral carbonation. Nat Geosci 2(12):837–841
Nowak DJ, Crane DE (2002) Carbon storage and sequestration by urban trees in the USA. Environ Pollut 116:381–389
Pouyat RV, Yesilonis ID, Nowak DJ (2006) Carbon storage by urban soils in the United States. J Environ Qual 35(4):1566–1575
Reich PB, Walters MB, Ellsworth DS, Vose JM, Volin JC, Gresham C, Bowman WD (1998) Relationships of leaf dark respiration to leaf nitrogen, specific leaf area and leaf life-span: A test across biomes and functional groups. Oecologia 114:471–482
Sailor DJ (2008) A green roof model for building energy simulation programs. Energy Build 40(8):1466–1478
Townsend-Small A, Czimczik CI (2010) Carbon sequestration and greenhouse gas emissions in urban turf. Geophys Res Lett 37(2):L02707. doi:10.1029/2009gl041675
UN (2008) World urbanization prospects: the 2007 revision (trans: United Nations Department of Economic and Social Affairs PD). United Nations, New York
Upton B, Miner R, Spinney M et al (2008) The greenhouse gas and energy impacts of using wood instead of alternatives in residential construction in the United States. Biomass Bioenergy 32(1):1–10. doi:10.1016/j.biombioe.2007.07.001
Warith M, Li X, Jin H (1995) Bioreactor landfills: state-of-the-art review. Emir J Eng Res 10(1):1–14

Chapter 17
Potential Carbon Emission Trajectories of Shanghai, China from 2007 to 2050

Ru Guo, Xiaojing Cao, Jing Zhang, Fengting Li, and Hongtao Wang

Abstract A system analysis approach on energy-related carbon (C) emission projection at urban level was developed by use of international local government greenhouse gas (GHG) emissions analysis protocol (IEAP) and long-range energy alternative planning (LEAP) model, taking Shanghai as a case. Based on the historical change of the energy-related C emission of Shanghai, four potential C trajectories from 2007 to 2050 were analyzed. The results show that the energy-related C emission of Shanghai may peak around 249–324 Tg CO_2 in year 2030 under three scenarios. Meanwhile, Shanghai can achieve absolute decline in total C emission compared to the level of 2007 when the energy structure is changed to a low-C system, and the C capture and storage (CCS) technology is widely used. Based on these results, several strategic suggestions for developing low-C economy in Shanghai have been proposed, including total emission control, low-C manufacturing industry, low-C transport, low C building, industrial restructure, sustainable consumption and climate change adaptation.

Keywords Climate change • GHG • Carbon emission trajectory • LEAP • Low carbon • Urban emissions • Energy structure • Low-carbon economy • Low-carbon transport • Low-carbon building • Sustainable consumption • Climate change adaptation

---

R. Guo • X. Cao • J. Zhang • F. Li (✉) • H. Wang
UNEP-Tongji Institute of Environment for Sustainable Development, Tongji University, 1239, Siping Road, 200092 Shanghai, China

College of Environmental Science and Engineering, Tongji University, 1239, Siping Road, 200092 Shanghai, China
e-mail: ruguo@tongji.edu.cn; jeanne.cao1985@gmail.com; jingzhang@tongji.edu.cn; fengting@tongji.edu.cn; wanght010@gmail.com

R. Lal et al. (eds.), *Recarbonization of the Biosphere: Ecosystems and the Global Carbon Cycle*, DOI 10.1007/978-94-007-4159-1_17,

Abbreviations

C	carbon
CO_2	carbon dioxide
CCS	carbon capture and storage
EE	energy efficiency
ER	energy restructure
GHG	greenhouse gas
IEAP	international local government GHG emissions analysis protocol
LC	low carbon
LEAP	long-range energy alternative planning
PCCE	per capita carbon emission

17.1 Introduction

Urban population in China reached 607 million in 2008 and the number of Chinese cities grew quickly from 193 in 1978 to 655 in 2008 (Dai 2009). Many developed cities in the eastern China are vulnerable to sea level rise and climate change. It is crucial for cities to actively participate in mitigating climate change since urban population use two-thirds of the global energy and emit around 70% of the global greenhouse gases (GHGs). Therefore, understanding the GHG emissions of cities is important (Nel et al. 2003; Dhakal and Peter 2008; Ramaswami et al. 2008).

China has played an important role in responding to the risks of climate change (China National Development and Reform Commission 2007; Richerzhagen and Scholz 2008; Yan and Crookes 2009). In 2009, China set the target of reducing per unit GDP CO_2 by 40–45% from 2005 to 2020. It is essential for cities to consider a low carbon (C) development path alongside the sustainable development perspective. Shanghai, the most developed city in China, was already affected by global warming with the fact that its average temperature increase is 1.93 times that of global level between 1873 and 2007. Meanwhile, Shanghai is facing the challenges of energy security and emission control (Gnansounou et al. 2004; Chen et al. 2007; Guo et al. 2010; Li et al. 2010).The methodology used in this study can offer an efficient tool to help Shanghai and other similar cities to know their future trajectory of C emission and make better decisions to manage GHG emissions.

17.2 Methodology

Energy-related C emissions are the majority of total emissions in many regions of China. A system analysis approach on energy-related C emission projection at urban level was developed by use of *international local government GHG emissions analysis protocol* (IEAP). It was developed by ICLEI-Local Governments for Sustainability (2009) and long-range energy alternative planning (LEAP) model developed by Stockholm Environment Institute. IEAP approach provides an easily implemented set of guidelines to assist local governments in quantifying the GHG emissions from both their internal operations and from the whole communities within their geopolitical boundaries. In this paper, the energy related C emission is divided into three scopes referring to IEAP (Table 17.1).

Due to limited data availability, only emissions from scope 1 and scope 2 are estimated in this paper. The C emission is calculated by use of fuel consumed and C emission coefficient.

$$\mathrm{C} = A \times EF \tag{17.1}$$

where, C is the total energy-related C emission, Mg CO_2-eq; A is the quantity of fuel used(expressed in mass), Mg; EF is C emission coefficient, Mg CO_2-eq/Mg.

LEAP is an integrated energy-environment modeling system that is based on transparent accounting and simulation approach. The model has a broad scope of application, including analysis on demand, transformation, resource extraction, GHG and local air pollutant emissions. But LEAP is only a framework model and need to be localized according to local features and objectives. The structure of LEAP model for Shanghai was constructed based on historical change of energy-related C emission and local characteristics, as shown in Fig. 17.1. The types of energy demand are divided into industrial sector and household sector according to current statistical system of China. Industrial sector is further divided into five sub-sectors, including agriculture, manufacturing industry, construction, public transportation, commercial and public (including wholesale and retail trade, hotels and catering services, education and culture industry etc.). Household sector is divided into two sub-sectors including private transportation and residential building. The types of energy supply are divided into electricity supply and heating supply.

Scenario analysis is a useful method that describes future's energy trajectory (Ghanadan and Koomey 2005). In this paper, four scenarios were designed and calculated to forecast the future C emission trajectories of Shanghai by use of system analysis method and LEAP model. The description of each scenario is given in Table 17.2.

17.3 Current and Future Energy-Related Carbon Emission of Shanghai

The energy related CO_2 emission of year 2007 was estimated according to IEAP for Shanghai. The results show that Shanghai emitted 184.89 Tg CO_2 in total with the per capita emission of 9.95 Mg CO_2 in 2007 without including shipping and aviation, while emission of 214.67 Tg CO_2 with the per capita emission of 11.55 Mg CO_2 if including shipping and aviation. The sectoral analysis showed that industry is the major emitter among all sectors (Fig. 17.2), covering 65.6% of total emission

Table 17.1 Community emission sources guidance (ICLEI-Local Governments for Sustainability 2009)

UNFCCC sector		Scope 1 emissions	Scope 2 emissions	Scope 3 emissions
Energy	Stationary energy	Utility-delivered fuel consumption	Utility-delivered electricity/heat/steam cooling consumption	Upstream/downstream emissions(e.g., mining/transport of coal)
		Decentralized fuel consumption	Decentralized electricity/heat/steam consumption	
		Utility-consumed fuel for electricity/heat generation		
	Transport	Tailpipe emissions from on-road vehicles	Electricity consumption associated with vehicle movement within the community (e.g., light rail)	Tailpipe emissions from vehicles used by community residents
		Tailpipe emissions from rail, sea, airborne and non-road vehicles operating within the community		Upstream/downstream emissions(e.g., mining/transport of oil)
				Tailpipe emissions from rail, sea, and airborne vehicles departing from or arriving into the community
	Fugitive emissions	Fugitive emissions not already accounted for	Not applicable	Upstream/downstream emissions

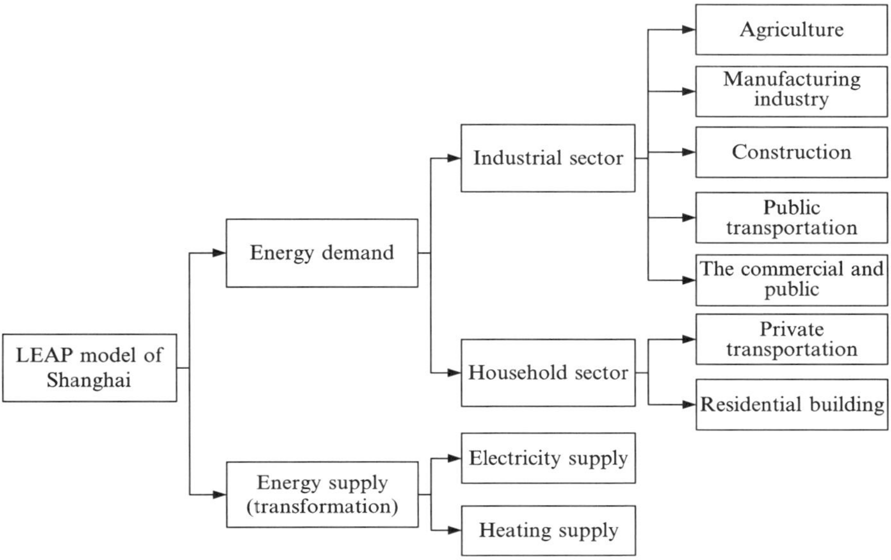

Fig. 17.1 The structure of Shanghai's LEAP model

Table 17.2 Four scenarios of Shanghai

Scenario Name	H-EE	L-EE	L-ER	L-LC
Activity level	High	Low	Low	Low
Emission coefficient	Industrial structure is adjusted toward low-carbon industry, energy efficiency is improved gradually.	Industrial structure is adjusted toward low-carbon emission, energy efficiency is improved gradually.	Based on the EE scenario, energy structure is adjusted with low-carbon or zero-carbon energy; CCS is primarily used	Based on the ER scenario, energy structure is adjusted further with low-carbon or zero-carbon energy; CCS is widely used

Note: *H* denotes high-speed growth of population and GDP, *L* denotes low-speed growth of population and GDP, *EE* denotes energy efficiency, *ER* denotes energy restructure, *LC* denotes low carbon. The detailed parameters are given in Tables 17.3, 17.4, 17.5 and 17.6

(excluding shipping and aviation sector) or 56.5% of total emission (including shipping and aviation sector). Building sector is the second largest emitter among all sectors when excluding shipping and aviation sector, covering 21.9% of total emission. However, transportation sector become the second largest emitter when including shipping and aviation sector, covering 24.1% of total emission.

The simulation results of four scenarios are given in Fig. 17.3. It's clear that population and GDP growth rate have an obvious effect on total C emission. For instance, the emission of scenario H-EE in 2050 is 1.40 times as that of scenario L-EE. Furthermore, Shanghai may reach C emission peak around 2030 under

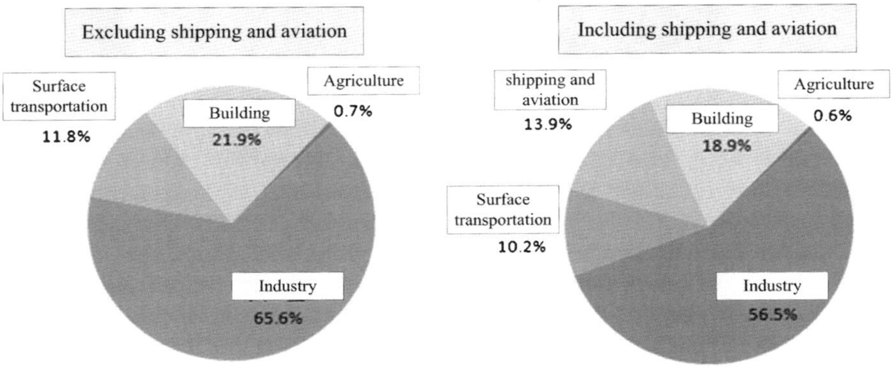

Fig. 17.2 Sectoral emission in 2007

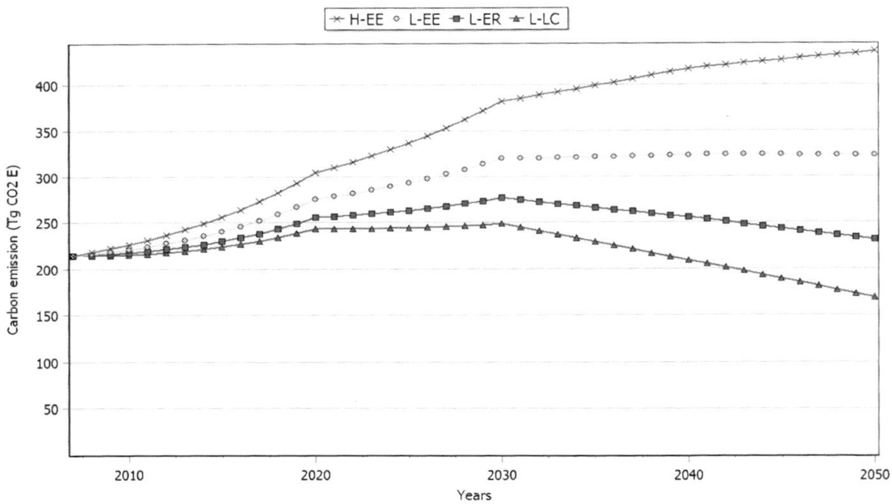

Fig. 17.3 Simulation results of four emission scenarios for Shanghai from year 2007 to year 2050

scenario L-ER and L-LC. The carbon emission of Shanghai will gradually increase between year 2020 and year 2030 in scenario L-LC, which implies that Shanghai can possibly reach emission peak earlier in case there are more inputs in C emission reduction. Furthermore, Shanghai can achieve absolute decline in total C emission compared with year 2007 if the energy structure is changed to a low-C system, and if the C capture and storage (CCS) technology is widely used (scenario L-LC).

Regarding sectoral emission, taking L-LC as an example, it's clear that emission from manufacturing industry may decrease gradually (Fig. 17.4). Meanwhile, emission from the commercial and public sector and residential building are the principal source in 2050. Furthermore, emission from private transportation may have the highest growth rate from year 2007 to year 2050.

Table 17.3 Population and GDP growth rate

	Population annual growth rate (%)				GDP annual growth rate (%)			
Scenario	2007–2020	2020–2030	2030–2040	2040–2050	2007–2020	2020–2030	2030–2040	2040–2050
H-EE	2	1.8	1.5	1	8	7.5	6	4.5
L-EE	1.22	1.02	0.72	0.57	7.2	6.7	5.2	4.1
L-ER	1.22	1.02	0.72	0.57	7.2	6.7	5.2	4.1
L-LC	1.22	1.02	0.72	0.57	7.2	6.7	5.2	4.1

Table 17.4 Value-added share for each sector from 2007 to 2050 (%)

Sector	2007	2020	2030	2040	2050
Primary industry	0.8	0.6	0.5	0.4	0.3
Secondary industry	46.6	31.4	24.5	19.6	14.7
# manufacture	93.3	86.0	82.0	78.0	73.0
# construction	6.7	14.0	18.0	22.0	27.0
Tertiary industry	52.6	68.0	75.0	80.0	85.0
# public transportation	11.3	9.7	8.1	6.6	5.0
# commercial and public	88.7	90.3	91.9	93.4	95.0

Note: # means sub sector

Table 17.5 Energy intensity for each sector

	Energy consumption per unit of value added (ton sce[a])/10,000 Chinese Yuan)	Annual decline rate of energy consumption per unit of value added[b] (%)			
Sector	2007	2007–2020	2020–2030	2030–2040	2040–2050
Agriculture	0.65	2.5	2.0	1.0	1.0
Manufacturing industry	0.60	6.5	7.0	5.0	3.0
Construction	0.37	8.0	4.0	4.0	3.0
Public transportation	2.48	5.0	6.0	5.0	3.0
Commercial and public	0.14	3.0	5.0	5.0	3.0

[a]"sce" is the abbreviation of "standard coal-equivalent"

[b]The annual decline rate of energy consumption per unit of value added for each sector is assumed according to the historical trajectory of Shanghai from 2005 to 2010

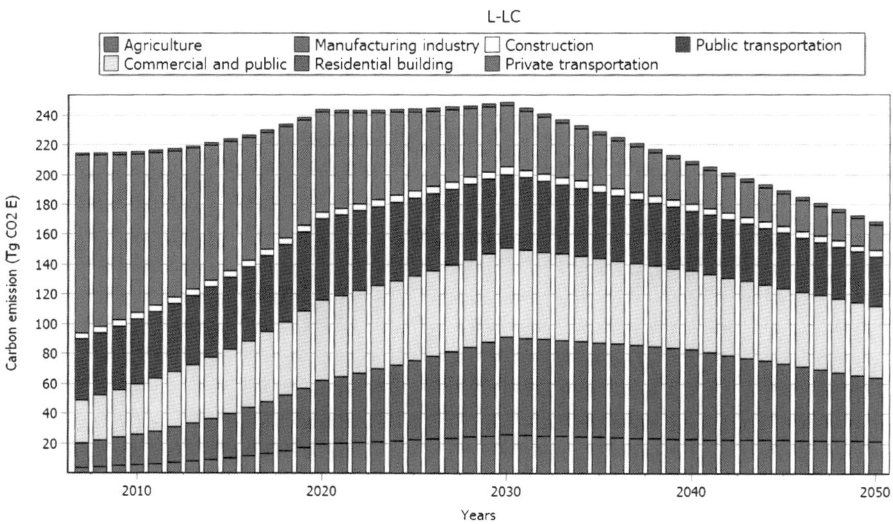

Fig. 17.4 Sectoral emission in scenario L-LC from 2007 to 2050

Table 17.6 Energy restructure parameters for end-users

Scenario		2020	2030	2040	2050
ER	Replacement of coal with natural gas	20% in RB; 10% in MI, CAP	40% in RB; 20% in MI, CAP	50% in RB; 30% in MI, CAP	60% in RB; 40% in MI, CAP
	Replacement of coal gas with natural gas	25% in RB; 15% in MI, CAP	50% in RB; 30% in MI, CAP	60% in RB; 50% in MI, CAP	70% in RB; 60% in MI, CAP
	Replacement of oil with renewable energy	5% in public and private transportation	10% public and private transportation	15% public and private transportation	20% public and private transportation
LC	Replacement of coal with natural gas	40% in RB; 20% in MI, CAP	80% in RB; 40% in MI, CAP	100% in RB; 60% in MI, CAP	100% in RB; 80% in MI, CAP
	Replacement of coal gas with natural gas	50% in RB; 30% in MI, CAP	100% in RB; 60% in MI, CAP	100% in RB; 100% in MI, CAP	100% in RB; 100% in MI, CAP
	Replacement of oil with renewable energy	10% public and private transportation	20% public and private transportation	30% public and private transportation	40% public and private transportation

Note: [a] Energy structure for end-users is pegged at 2007 level in the EE scenario

[b] *RB* is the abbreviation of residential buildings, *MI* is the abbreviation of manufacture industry, *CAP* is the abbreviation of commercial and public

[c] The oil includes gasoline, kerosene, and diesel

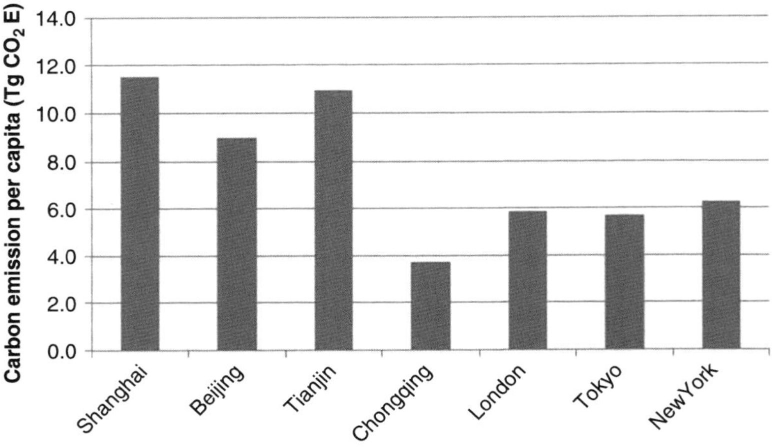

Fig. 17.5 Comparison of per capita C emission in different cities. *Note*: The data of Beijing, Chongqing and Tianjin are from Geng et al. (2011), the data of Shanghai, London, Tokyo, New York are from Guo et al. (2011)

Further analysis from Fig. 17.5 shows that the per capita C emission (PCCE) of Shanghai is relatively higher than that of many domestic and international cities, including Beijing, Tianjin, Chongqing, London, Tokyo and New York. Therefore, a major challenge facing Shanghai is how to gradually reduce its PCCE which is a signal of low carbon city.

17.4 Strategies for Low Carbon Development of Shanghai

Based on the current status and future trend of Shanghai's energy-related C emission, several strategies have been proposed to facilitate the development of low-C economy in Shanghai, including total emission control, low-C manufacturing industry, low-C transport, low-C building, industrial restructure, sustainable consumption and climate change adaptation.

17.4.1 Total Emission Control

In fact, total emission control is a comprehensive strategy for Shanghai. Since Shanghai is likely to increase its C emissions gradually in the coming decades, it's necessary for Shanghai to set the target of total C emission control. Shanghai can possibly achieve the peak C emission around 2030 under certain scenarios. And this can help local policy makers to consider the time schedule of total C emission control.

17.4.2 Low Carbon Manufacturing Industry

It's crucial to develop low C manufacturing in the coming future since manufacture industry is currently the major C emitter. The transformation of traditional manufacturing and development of advanced manufacturing should be considered simultaneously.

17.4.3 Low Carbon Transport

Transport policies and actions should encourage the integration of land use and transport planning to avoid travel demand, the construction of public transport, improvement of transportation management, and development of low-C vehicles through technological innovations.

17.4.4 Low Carbon Building

Building policies and actions should encourage building energy conservation, the construction of low-C building, and the development of low-C building technologies.

17.4.5 Industrial Restructure

Industrial restructure is the fundamental way to develop a low-C city. Shanghai should focus more on the development of high value-added and emerging low-C industry, such as C finance, energy management, alternative fuel, etc.

17.4.6 Sustainable Consumption

Everyone wants to have the best in the way of food, clothing, equipment, etc. Therefore, sustainable consumption should be encouraged and guided since the carrying capacity of the earth is limited, especially in cities like Shanghai. A detailed analysis and comparison of emission profiles for different social groups should be conducted to develop tailored emission control polices for high-emitting groups.

17.4.7 Climate Change Adaptation

From a long-term perspective, it's necessary to consider adaptation policies and actions to respond to climate change since Shanghai is a coastal city. Shanghai should consider climate change mitigation and adaptation simultaneously. Specifically, more attention should be paid to the C management of Chongming Island since Chongming is adjacent to East China sea and also the main carbon sinks of Shanghai.

Acknowledgements The authors would like to express our sincere thanks for the financial support of HSBC Environmental Funds, National scientific and technological support project (2009BAC62B02), Shanghai Municipal Science and Technology Commission Funds (10230712200), Fund of the Tongji University for Promoting Academic Exchange with Germany.

References

Chen CH, Chen BH, Wang BH et al (2007) Low-carbon energy policy and ambient air pollution in Shanghai, China: a health-based economic assessment. Sci Total Environ 373(1):13–21

China National Development and Reform Commission (2007) China's National Climate Change Programme. China National Development and Reform Commission, Beijing

Dai F (2009) Report on urban development of China 2008. China City Press, Beijing

Dhakal S, Peter D (2008) Climate change and cities: the making of a climate friendly future. Urban energy transition. Elsevier, Amsterdam

Geng YH, Peng CH, Tian MZ (2011) Energy use and CO_2 emission inventories in the four municipalities of China. Energy Procedia 5:370–376

Ghanadan R, Koomey JG (2005) Using energy scenarios to explore alternative energy pathways in California. Energy Policy 33:1117–1142

Gnansounou E, Dong J, Bedniaguine D (2004) The strategic technology options for mitigating CO_2 emissions in power sector: assessment of Shanghai electricity-generating system. Ecol Econ 50(1–2):117–133

Guo R, Cao XJ, Yang XY et al (2010) The strategy of energy-related carbon emission reduction in Shanghai. Energy Policy 38(1):633–638

Guo R, Cao XJ, Li FT (2011) 2050 energy related carbon emission in Shanghai. Tongji University Press, Shanghai (In Chinese)

ICLEI-Local Governments for Sustainability (2009) International Local Government GHG Emissions Analysis Protocol (Version 1.0)

Li L, Chen C, Xie S et al (2010) Energy demand and carbon emissions under different development scenarios for Shanghai, China. Energy Policy 38(9):4797–4807

Nel J, Nel G, Snyman E (2003) Towards sustainability: reducing energy consumption and carbon emission footprints – examples and lessons learned from Potchefstroom, a medium sized African city. In: Proceedings of the 1st international conference on sustainable energy, planning and technology in relationship to the environment, Halkidiki, Greece

Ramaswami A, Hillman T, Janson B et al (2008) A demand-centered, hybrid life-cycle methodology for city-scale greenhouse gas inventories environmental. Sci Technol 42(17):6455–6461

Richerzhagen C, Scholz I (2008) China's capacities for mitigating climate change. World Dev 36(2):308–324

Yan X, Crookes RJ (2009) Reduction potentials of energy demand and GHG emissions in China's road transport sector. Energy Policy 37(2):658–668

Chapter 18
Processes of Soil Carbon Dynamics and Ecosystem Carbon Cycling in a Changing World

Felix Heitkamp, Anna Jacobs, Hermann F. Jungkunst, Stefanie Heinze, Matthias Wendland, and Yakov Kuzyakov

Abstract Climate change is evident and increases of carbon dioxide concentration (CO_2), temperature and extreme weather events are predicted. To predict the effects of such changes on carbon (C) cycling, the processes and mechanisms determining the magnitude of C storage and fluxes must be well understood. The biggest challenge is nowadays to quantify belowground components of the C-cycle. Soil respiration accounts for ~70% of total annual ecosystem respiration. However, the CO_2 flux from soil originates from several sources, such as root respiration, rhizomicrobial respiration, mineralization of litter and mineralization of soil organic matter (SOM). Increasing atmospheric CO_2 concentrations will generally increase plant growth, thus C-input to soil. This higher C-input will be accompanied by higher SOM mineralization due to warming. However, mineralization of more stable pools

F. Heitkamp (✉)
Landscape Ecology, Faculty of Geoscience and Geography, Georg August-University, Goldschmidtstr. 5, 37077 Göttingen, Germany

Carbon Sequestration and Management Center, School of Environment and Natural Resources, The Ohio State University, 2021 Coffey Road, Cloumbus, OH 43210, USA
e-mail: fheitka@uni-goettingen.de

A. Jacobs
Department of Agronomy, Institute for Sugar Beet Research, Holtenser Landstr. 77, 37079 Göttingen, Germany
e-mail: jacobs@ifz-goettingen.de

H.F. Jungkunst
Landscape Ecology, Faculty of Geoscience and Geography, Georg August-University, Goldschmidtstr. 5, 37077 Göttingen, Germany
e-mail: hjungku@uni-goettingen.de

S. Heinze
Department of Soil Science & Soil Ecology, Geographical Institute, Ruhr-University Bochum, Universitätsstrasse 150, 44780 Bochum, Germany
e-mail: stefanie.heinze@rub.de

R. Lal et al. (eds.), *Recarbonization of the Biosphere: Ecosystems and the Global Carbon Cycle*, DOI 10.1007/978-94-007-4159-1_18,

may be affected more by warming compared to mineralization of labile pools. The importance of cropland management is demonstrated in a model scenario. Crop residue incorporation increased C-storage in the soil markedly. However, under the assumption of a higher temperature sensitivity of mineralization of stable C-pools the net-sink of C under recommended management practice is severely reduced. Precise predictions are hampered due to the lack of quantitative, mechanistic knowledge. It is discussed that a more interdisciplinary scientific approach will increase the speed in generating urgently needed understanding of belowground processes of C-cycling.

Keywords Climate change • Respiration • Temperature sensitivity • CO_2 fertilization • Soil organic matter • Ecosystem C cycling • Autotrophic organisms • Soil respiration • Litter decomposition • Priming effect • Rhizosphere respiration • Rhizodeposition • Mean residence time • Soil fauna • Root litter • Stabilization of soil organic carbon • Biochemical recalcitrance • Spatial inaccessibility • Organo-mineral associations • Soil organic matter fractions • Spectroscopic methods • Thermal stability • Depolymerization • FACE experiments • Temperature sensitivity (Q_{10}) • Van't Hoff equation • Rate constant • Arrhenius equation • Extreme weather events • Substrate • Roth C model • SOC dynamics • Residues incorporation • CO_2 fertilization effect

Abbreviations

AGBDM	Aboveground biomass dry matter
AUR	Acid unsoluble residue
BIO	Microbial biomass, model pool in RothC
C	Carbon
CH_4	Methane
CI	Confidence interval
CO_2	Carbon dioxide
CO_2-fert	Max-CC and CO_2 fertilization of crops, climate scenario for the modelling example
CON	Control treatment in the Puch experiment

M. Wendland
Institut für Agrarökologie, ökologischen Landbau und Bodenschutz,
Bayrische Landesanstalt für Landwirtschaft, Lange Point 12,
85354 Freising, Germany
e-mail: matthias.wendland@lfl.bayern.de

Y. Kuzyakov
Department of Soil Science of Temperate Ecosystems, Büsgen Institute,
Georg August-University, Büsgenweg 2, 37077 Göttingen, Germany
e-mail: kuzyakov@uni-goettingen.de

DJF	December, January, February
DPM	Decomposable plant material, model pool in RothC
ETP	Evapotranspiration
FACE	Free air carbon dioxide enrichment
GHG	Greenhouse gas
GPP	Gross primary production
HUM	Humified organic matter, model pool in RothC
IOM	Inert organic matter, model pool in RothC
IOSDV	"Internationale organische Stickstoff-Dauerdüngungsversuche" (German) International organic long-term nitrogen fertilization experiment
JJA	June, July, August
MAM	March, April, May
MAP	Mean annual precipitation
MAT	mean annual temperature
Max-CC	Maximal climate change, climate scenario for the modelling example
MRT	Mean residence time
N	Nitrogen
NECB	Net ecosystem carbon balance
NEP	Net ecosystem production
No-CC	No climate change climate, scenario for the modelling example
NPP	Net primary production
OM	Organic matter
ppm	Parts per million
R_A	Respiration by autotrophic organisms
R_E	Ecosystem respiration
RES	Residue incorporation treatment in the Puch experiment
R_H	Respiration by heterotrophic organisms
RMSE	Root mean square error
RPM	Resistant plant material model pool in RothC
SOC	Soil organic carbon
SOM	soil organic matter
SON	September October, November

18.1 Introduction

It is evident that atmospheric carbon dioxide (CO_2) concentrations rose drastically from 280 ppm during the preindustrial era to about 390 ppm in 2010 (Conway and Tans 2011). Similar drastic increases of other greenhouse gases (GHGs) are also evident. The resulting increase in temperature due to radiative forcing was in the range of 0.10–0.16°C per decade (1956–2005), which is likely the strongest warming since the last 1,300 years (Solomon et al. 2007). Future projections of CO_2 concentration increase and warming until 2100 depends on underlying emission scenarios. The atmospheric concentration of CO_2 is projected to increase to up to

1,000 ppm, and global surface warming is estimated to increase by 1.1–4.0°C, with higher values over land compared to oceans. Moreover, extreme weather events such as heat waves, droughts and heavy precipitation are likely to increase in most regions (Solomon et al. 2007).

Uncertainties in climate change modeling for a given emission scenario result mainly from unknown feedback effects between warming and the carbon (C) cycle (Friedlingstein et al. 2006). On one hand, CO_2 fertilization of plants results in higher uptake of CO_2 (negative feedback), thereby more biomass and increased storage of soil organic carbon (SOC) (Heimann and Reichstein 2008). On the other hand, increasing temperature induced by rising GHG concentrations accelerates mineralization of SOC, which in turn results in higher atmospheric CO_2 concentrations (positive feedback). Moreover, drying and rewetting as well as freezing and thawing cycles of the soil may increase or decrease in frequency or severity, with uncertain effects on the global C cycle.

The ecosystem C-cycle begins with C-assimilation by autotrophic organisms, which are the higher plants in most terrestrial ecosystems. The rate of CO_2 uptake depends mainly on light energy (photosynthetically active radiation) and ambient CO_2 concentration, but also on water availability, temperature, nutritional status and plant species. The sum of assimilated C in an ecosystem, typically expressed on annual basis per square meter, is the gross primary production (GPP). A fraction of the assimilated C is used for growth or reserve, in other words for buildup of the biomass, which is the net primary production (NPP). Another fraction is respired by plants to meet energy demands for growth and maintenance. This CO_2 flux is known as respiration by autotrophs (R_A). Cannel and Thornley (2000) reported that the portion of NPP as GPP normally ranges between 0.4 and 0.6, especially when observed over time scales of several weeks or longer.

In natural ecosystems, most of annual NPP enters the decomposition cycle as leaf litter, root litter, rhizodeposition (exudates, exfoliates) or, woody debris. Most of the decomposition process takes place on or in the soil. The C cycle is closed by mineralization of organic C to CO_2 by microorganisms. This part of the CO_2 flux is termed respiration by heterotrophic organisms (R_H). Thus, the CO_2 flux from the ecosystem back into the atmosphere is the sum of R_A and R_H, and is termed ecosystem respiration (R_E).

A first step to determine if a particular ecosystem gains ("CO_2-sink") or looses ("CO_2-source") C over time is the balance, or imbalance, between NPP and R_H (equals the balance between GPP and R_E). Chapin et al. (2006) defined this balance as the net ecosystem production (NEP). Valentini et al. (2000) showed for 15 forest ecosystems (latitudes ranging from 41°N to 64°N) that GPP is similar across all locations. Thus, NEP is primarily determined by respiration and we will focus on this topic below. Clearly, the C-balance is also determined by gains and losses not induced by photosynthesis or respiration. This includes e.g. leaching, fire, harvested products, methane (CH_4) flux, erosion, herbivory and organic fertilization. For net changes of C in ecosystem the term net ecosystem C balance (NECB) has been proposed.

Measuring *net* C-fluxes in ecosystems is relatively straightforward (although expensive) by using eddy-flux towers and automated chambers for measuring soil respiration, and supplementing the data with independent biomass and SOC measurements (Baldocchi 2003). However, predicting the effects of environmental changes on fluxes and pools of C, necessitate understanding of how the components of the *gross* fluxes affect the pools. The biggest challenge in understanding components of C-fluxes is to quantify the belowground processes (Schulze et al. 2009). Therefore, the focus of this chapter is to provide an overview on the sources of soil respiration, mechanisms of litter decomposition and processes of C stabilization by the soil matrix. Finally, a model is also used as an example to illustrate how changes in temperature and CO_2 concentration may influence the SOC dynamics.

18.2 Mechanisms and Processes of Belowground Carbon Cycling

Soil respiration accounts for the second largest CO_2 flux after GPP, and amounts for ~70% of total annual R_E (Yuste et al. 2005). Although soil respiration contributes considerably to annual CO_2 emissions there is a lack of knowledge with regards to the abiotic and biotic impacts on respiratory activity of soils and the true sources of soil derived CO_2 (Kuzyakov 2006; Trumbore 2006).

Soil respiration is highly variable temporarily, but can be measured on very fine time scales by using automated chambers. However, measured fluxes represent a mixture of R_H and R_A with the portions of the sources varying among seasons, depending on plant state, substrate supply to heterotrophs as well as temperature and moisture regimes (Ryan and Law 2005). Thus, the biggest challenge in understanding components and fluxes affecting the NECB is quantification of the different sources of soil respiration.

Flux of CO_2 from the soil into the atmosphere originates from different sources. On a basic functional level, respiration is divided into respiration by autotrophs and by heterotrophs. Dominant autotrophic organisms in terrestrial ecosystems are plants. Heterotrophic organisms include various animals and microorganisms. However, contribution of animals is in general of minor importance, only representing a few percent of total respiration by heterotrophs. In general, mean annual R_H accounts for 54% of soil respiration (Hanson et al. 2000).

Quantification of different sources of soil respiration is important, but remains to be a work in progress (see Box 18.1 for methods). Kuzyakov (2006) identified basically three main compartments as a source of soil respiration: (i) the rhizosphere, (ii) plant residue or litter and (iii) soil organic matter (SOM). While the respiration from litter and SOM is mainly driven by heterotrophic organisms, that from the rhizosphere is driven by C-allocation of plants to roots (Kuzyakov and Gavrichkova 2010).

Box 18.1 Overview on Methods for Partitioning of Soil Respiration

For detailed descriptions readers are referred to reviews by e.g. Hanson et al. (2000) and Kuzyakov (2006). For *component integration* all compartments of interest have to be separated by e.g. sieving and handpicking. Commonly, roots, litter and soil is divided by this method. The components are then measured for their specific flux rate, and their contribution to total soil respiration is calculated by the mass balance. Clearly, this method is accompanied by high disturbance of the system, which may lead to a shift in the proportion of contribution of components to total CO_2 flux. This method provides only relative values.

Root exclusion techniques include basically root removal, root trenching and gap analysis. All techniques have to deal with the problem to alter microclimatic conditions and nutrient budgets within the soil and as well as with decaying roots, which contribute to respiration

Isotope tracers are used for partitioning of CO_2fluxes from soil without strong disturbance of the system. The principle of all isotopic approaches for CO_2 partitioning is based on differences in C isotopic signature of various SOM pools and living or dead roots. Both, the radioactive ^{14}C and stable ^{13}C isotopes as well as their combination are used successfully for partitioning CO_2 fluxes. The differences in isotopic signature of SOM pools may originate from natural processes (radioactive decay of ^{14}C; natural changes of vegetation) or can be artificially induced. The natural processes usually change the isotopic signature too slowly and therefore, were seldom used (Kuzyakov 2011). The artificially induced changes of SOM pools and root-derived CO_2 were used in the most CO_2 partitioning studies up to now and can be grouped into the following approaches: Continuous or pulse labeling of plants in $^{13}CO_2$ or $^{14}CO_2$ enriched or depleted atmosphere (Werth and Kuzyakov 2008), ^{13}C natural abundance (Heitkamp et al. 2012a; Rochette et al. 1999), and bomb ^{14}C approach (Wang and Hsieh 2002). The isotopic methods are precise and less invasive, but are expensive and provide usually results for small areas, only.

18.2.1 Rhizosphere Respiration

The rhizosphere is the soil directly influenced by the root and often comprises of only a few millimeter distance to the root. The rhizosphere is different from surrounding soil by the presence of rhizosphere organisms (e.g. mycorrhiza), and the strong influence of rhizodeposition (Jones et al. 2004; Kuzyakov 2006). The rhizosphere respiration consists of heterotrophic (rhizomicrobial respiration) and autotrophic (root respiration) components. However, with current methodology, these components are hardly distinguishable. Mycorrhizal fungi, for example, are located inside and outside the roots, and directly utilize C from plant metabolites. Therefore, even isotopic labeling fails to identify the source of respiration (Kuzyakov and Larionova 2005). Rhizomicrobial and root respiration are often lumped together as rhizosphere respiration due to methodological problems in separating the CO_2 fluxes (Chapin et al. 2006).

Rhizodeposition is the release of organic compounds from living roots into the surrounding soil, the rhizosphere. Rhizodeposition occurs in the intercellular space of roots (endorhizosphere), on the root surface (rhizoplane) and outside the root (ectorhizosphere). Released compounds, such as starch, glucose, carboxylic acids and amino acids, are often low in molecular weight and are easily degradable by microorganisms (Fischer et al. 2007; Jones et al. 2004; Schenck zu Schweinsberg-Mickan et al. 2010). Microorganisms in the rhizosphere take up C and N from exudates and exfoliates of roots quickly within a few millimeter distance to roots (Schenck zu Schweinsberg-Mickan et al. 2010) and turnover times are within hours up to weeks (Kuzyakov 2006). The root exudates are mainly produced during daylight through stimulation of photosynthetic plant activity. Dilkes et al. (2004) showed by ^{14}C labeling that rhizodeposition of wheat (*Triticum aestivum* L.) was the highest 3 h after C-uptake. On average, few hours are necessary for grasses and herbs and about 4 days for mature trees for the release of rhizodeposits from roots after CO_2 assimilation in leafs (Kuzyakov and Gavrichkova 2010).

Due to their low mean residence time (MRT), rhizodeposition does not contribute significantly to C storage in soil. However, contribution to respiration during daylight hours might be substantial (Kuzyakov 2006). Furthermore, the labile nature of rhizodeposits can influence activity and enzyme production of microorganisms and, therefore, accelerate or retard mineralization from SOM or litter (i.e. priming effect, Kuzyakov et al. 2000). Priming can significantly alter mineralization kinetics. For example, Seiffert et al. (2011) showed under laboratory conditions that after addition of glucose microorganisms incorporated and mineralized black slate, a low grade metamorphic rock formed from shale. Therefore, increased rhizodeposition can induce mineralization of the stabilized SOC pool (Fontaine et al. 2007).

18.2.2 Decomposition of Litter

In a broad sense, litter includes all solid debris such as leaves, roots, stems, stalks and wood (Zhang et al. 2008). However, most research has been conducted on leaf litter decomposition in forest ecosystems (Prescott 2010), and crop residues (Abiven et al. 2005; Jensen et al. 2005).

Litter decomposition includes chemical alteration of litter, assimilation by decomposers and mineralization to CO_2. Mass loss from the so termed litter bags (Box 18.2) without quantification of CO_2 flux is the common approach to measure litter decomposition under field conditions. Exposure of litter bags in the field includes losses by leaching and export by fauna. Therefore, rates of mass loss are higher than decomposition or mineralization rates but not *vice versa*. Besides these methodological shortcomings, the mass loss is closely related to C-mineralization, justifying to use "mass loss" as a proxy for "C-loss" and "C-mineralization" (Cotrufo et al. 2010).

In most modeling approaches it is accepted that for one litter type under constant conditions, mass loss or C-mineralization follows decay by the first order kinetics. Therefore mass loss can be described by Eq. 18.1:

$$Y(t) = \sum_{i=1}^{n} Y_i \times e^{-k_i t} \qquad (18.1)$$

Where, $Y(t)$ is the mass remaining at time t, Y_i is the initial mass of compartment i, k_i is the decay constant of compartment i. A model with one compartment (n = 1) is often successfully used in litter decomposition studies (Zhang et al. 2008), but two compartment models are also used to account for different decomposition stages (Gholz et al. 2000). The reciprocal value of the decay constant is termed MRT. After the time span of the MRT, approximately 2/3 of the initial mass is lost. In a global meta-analysis including 70 studies at 110 sites, MRT for litter of different biomes ranged from 0.2 to 10 years with a median of 3.3 years (Zhang et al. 2008). The wide range of MRTs is a result of climate, litter quality and decomposer community (Swift et al. 1979).

Climate influences the decomposition rate through the effects of soil temperature and moisture regimes (Swift et al. 1979). This influence is not always straightforward, but thresholds exist. For example if mean annual temperature (MAT) is lower than 10°C, the rate of decomposition is slow regardless of litter type. The same is true for the moisture contents below 30% and above 80% (Prescott 2010). Therefore, in studies at sites in Canada, MAT was the principal control of decomposition dynamics, whereas in tropical studies, moisture is relatively more important (Powers et al. 2009; Trofymow et al. 2002). One limiting factor can hence determine the decomposition kinetics: in the tropics temperature is high throughout and decomposition is governed by moisture conditions. Zhang et al. (2008) reported in their meta-analysis that MAT and mean annual precipitation (MAP) only explained 30% of variation in k-values. However, climate and litter quality are closely linked by the common vegetation in bioclimatic zones differing in the chemical decomposition of litter.

The chemical composition is often referred to as *litter quality*. Litter decomposition rates are often correlated with litter fractions obtained by stepwise chemical digestion, operationally defined as cellulose, hemicellulose or acid unsoluble residue (AUR, often referred to as "lignin"), and N-content or other nutrients (Berg and McClaugherty 2003; Swift et al. 1979). Prescott (2010) pointed out that a good correlation between litter quality and decomposition is likely over a range of intermediate values. If, e.g. the ratio of AUR-to-N is below 10 or above 40, other factors are likely to control the rate of decomposition and no significant correlation can be find between AUR-to-N ratio and MRT.

Using a global dataset, Zhang et al. (2008) observed that AUR-to-N, N-content and C-to-N ratio explained 73% of variation in decomposition rate constants, making litter quality the most influential factor in decomposition. However, global analyses

Table 18.1 Average mean residence time (MRT, in years) of above (Zhang et al. 2008) and belowground litter (Silver and Miya 2001) of broad life form categories

	Broadleaf	Conifer	Graminoid
Leaves or needles	1.3 (115)	2.9 (55)	0.9 (15)
Fine roots (< 2 mm)	2.2 (43)	5.9 (10)	0.7 (35)

Mean values across a broad range of biomes
Figures in brackets are the numbers of values (n)

Box 18.2 Methods for Measuring Litter Decomposition (Cotrufo et al. 2010)

Litter bag approach: Litter is placed in synthetic bags with varying mesh sizes either to include or exclude fauna of various sizes. The litter bags are then exposed in the field, either on the ground or buried in soil. Mass or nutrient loss is regularly determined by weighing harvested bags.

Litter input and standing litter can be measured and annual decomposition can be calculated by dividing input by standing mass. This approach provides estimates of MRT on annual basis. Furthermore, the calculated MRT integrates input and standing litter from all species present. This approach is only possible in ecosystems where MRT of litter is longer than 1 year.

Laboratory incubation studies with analysis of CO_2 dynamics are especially useful to compare one particular property or process under controlled conditions. This approach renders interpretation more straightforward than field studies. However, extrapolation to field conditions is difficult.

Isotope tracers can be used instead of, or in addition to, measuring the isotopic signal in respired CO_2 (Box 18.1). The exposed material is sampled and directly analyzed.

are subject to intercorrelation: the vegetation type is clearly influenced by climate and soil conditions. For example, the lowest decomposition rates have been reported for Tundra ecosystems were decomposition is slow due to low temperatures, frozen soil and often waterlogged conditions. Furthermore, common Tundra vegetation is inherently resistant to decomposition. It is important to recognize that a change in vegetation (i.e. litter quality) due to climate change may affect decomposition rates of litter (Table 18.1) perhaps as strong as increasing temperature (see also Sect. 18.3.2).

A still unresolved issue is the influence of *fauna* on litter decomposition. Whereas past studies have reported mostly the positive influence on decomposition rate, recent studies report mostly neutral or even slowing effects (Prescott 2010). This trend might be due to methodological issues. Moreover, fauna can alter litter composition and increase contact with soil particles by bioturbation. This in turn can lead to chemical or physical stabilization of litter, but can also increase initial decomposition due to favorable moisture conditions and higher nutrient availability

(Jacobs et al. 2011; Potthoff et al. 2005). Knowledge on litter decomposition, mostly in forest ecosystems, was greatly enhanced by cross-site studies. Wall et al. (2008) concluded that invertebrate fauna increased decomposition under temperate and tropical climates. In case of temperature or moisture limited decomposition rates, faunal effects became neutral. However, the standardized methodology across 30 sites did not allow mixing with soil particles. Therefore, the magnitude and rate of measured decomposition rates may differ from the true rates. Nevertheless, inclusion of fauna in decomposition models is an important and a challenging task.

Root litter has often higher MRTs as compared to leaf or needle litter (Table 18.1). It has been hypothesized that most of SOC is root derived (Rasse et al. 2005). A part of the higher MRT of roots can be explained by the quality. Roots often contain higher amounts of recalcitrant compounds, such as lignin, suberin, lignin and tannin. However, it is likely that physical or chemical protection by the soil matrix contributes to the higher MRT of roots. These mechanisms are discussed in Sect. 18.2.3.

The short overview presented above indicates that climate change may affect litter decomposition by increasing temperature, especially when rising above the threshold and by altering the duration of very wet or dry phases. Large effects may also result from changes in the vegetation pattern (i.e. in litter chemistry) and accompanying changes in faunal and microbial communities. Decomposition dynamics can even change without completely changing the vegetation. Jacob et al. (2010) showed that beech (*Fagus sylvatica* L.) leaf litter decomposed slower in the presence of litter from other tree species. Therefore, occurrence or absence of a few species can influence significantly the C-cycle. Nevertheless, even highly decomposed litter is not intrinsically stable, as has been shown by Harmon et al. (2009): even after 10 years of decomposition in litter bags the decay rate was an order of magnitude higher compared to that of SOC in mineral soil. This trend shows that studies of soil respiration, litter decomposition and stabilization of C in mineral soil should be linked more closely for better insight in the belowground C cycle (Fierer et al. 2009; Kuzyakov 2011; Ryan and Law 2005).

18.2.3 Stabilization of Soil Organic Carbon

The stabilization of SOC is defined as all mechanisms that protect it against decomposition and, thus, slow down mineralization (Baldock and Skjemstad 2000; von Lützow et al. 2006). Destabilization is defined as the reverse of stabilization, increases the susceptibility of SOC to decomposition (Sollins et al. 1996), and is, thus, one of the mechanisms regulating CO_2 emission from soils. Therefore, stabilization and destabilization are closely related to each other and a detailed knowledge of the mechanisms regulating them is required to better predict CO_2 efflux from soil (Schmidt et al. 2011; Trumbore 2006).

Over decades, several studies have been conducted to describe and to distinguish different mechanisms of SOC stabilization. The traditional theory of SOC stabilization is based on the understanding that dead organic matter once entered the soil is either mineralized or humified by the soil microorganisms. These biologically produced "humic substances" were assumed to be resistant to mineralization due to their biochemical structure (MacCarthy 2001). However, evidence against this hypothesis emerged (Burdon 2001; MacCarthy 2001; von Lützow et al. 2006). More recent studies take into account that C can be stabilized in the soil being biochemically relatively unaltered: SOC is in general a mixture of plant and microbial derived compounds (Burdon 2001). The current conceptual model of SOC stabilization is mainly based on Sollins et al. (1996), and has been synthesized by von Lützow et al. (2006) who provided an excellent and detailed description. For temperate zones, basically three main stabilization mechanisms are identified: (i) biochemical recalcitrance, (ii) spatial inaccessibility and (iii) organo-mineral association. Baldock et al. (2004) proposed that the capacity of the decomposer community must also be considered as a fourth mechanism which leads, when capacity is limited, to slow mineralization. It should be noted however, that not any of these stabilization mechanisms explains the origin and production of the humic substances which are ubiquitous in soil.

All stabilization mechanisms can occur simultaneously (spatially and temporally), they may affect each other, and co-limitation is possible (Heimann and Reichstein 2008; Wutzler and Reichstein 2008). The relevance of the respective stabilization mechanism differs among environmental, geographical, and land-use characteristics (von Lützow et al. 2006). Thus, a general classification and evaluation of the stabilization mechanisms is difficult. The scientific community is challenged to bridge between conceptual models and ecosystem-specific processes (Heimann and Reichstein 2008; Schmidt et al. 2011). Moreover, there is currently not method which is capable to isolate equivalents of the conceptual model pools (Box 18.3). However, the conceptual model of stabilization (Sollins et al. 1996; von Lützow et al. 2006) is the main basis of the recent understanding of SOC dynamics and, thus, is presented herein.

Box 18.3 Methods for Isolating Soil Organic Matter Fractions

Physical fractionation procedures separate the SOC due to physical properties, according to particle size, aggregate size, or density (light, heavy, free, and occluded organic particles). Thereafter, the mass of the fraction and the respective C concentrations are measured (Christensen 2001). Density fractionation is particularly useful because fractions influenced by the main stabilization mechanisms can be separated (Golchin et al. 1994).

Chemical fractionation means to extract (e.g. hot water, 6 M HCl, H_2O_2, NaOCl, $Na_2S_2O_8$) more labile fractions of total SOC (Balesdent 1996; Helfrich et al. 2006). The SOC is quantified before and after the procedure. Chemical treatments are not completely standardized and may differ in terms of concentration, duration, and external energy added making comparisons difficult.

Biological fractionation is applied to determine the CO_2 evolved during incubation of soil samples. The CO_2 emitted in a certain time is assumed to represent a SOC fraction with a respective turnover time. Pool sizes and cor-

(continued)

Box 18.3 (continued)

responding turnover times can be analyzed by curve fitting (e.g., Eq. 18.1, Paul et al. 2006). Because only labile SOC is mineralized in incubation studies, this approach is best supplemented by another fractionation approach (Haile-Mariam et al. 2008; Heitkamp et al. 2009).

Thermal stability of OM can be used as a measure for resistance to mineralization. By thermogravimetry mass loss of a sample over a range of temperature (typically 20–550°C for release of organic matter) is measured. The higher the energy needed to induce a reaction (i.e. oxidation of organic C to CO_2), the more stable (recalcitrant) the fraction (Rovira et al. 2008). At least two peaks, representing SOC of different stability, can be identified by thermogravemetry (Siewert 2004). However, by measuring directly the CO_2 release during heating it is possible to identify up to four clear peaks (H. F. Jungkunst, unpublished data).

Spectroscopic methods, e.g. infrared-spectroscopy or nuclear magnetic resonance spectroscopy deliver information of the chemical composition of SOC in a sample (Ellerbrock et al. 1999; Golchin et al. 1995). However, information is only useful when the turnover time of a respective substance or functional group is known.

Analytical techniques using C *isotopes* are a very helpful tool for determination of turnover times or the age of SOC fractions (Balesdent 1996; Wang and Hsieh 2002). Isotopic measurements are the only way to assign respiration to certain SOC fractions (Kuzyakov 2011).

Biochemically recalcitrant substances have a molecular structure which leads to a selective discrimination by the soil microorganisms. Such substances are: (i) not "attractive" to microorganisms since the net gain in energy by depolymerization is low (Fontaine et al. 2004; Wutzler and Reichstein 2008) and/or (ii) cannot be hydrolyzed by common enzymes (von Lützow et al. 2006). Biochemical recalcitrance is mainly caused by a complex macromolecular structure as aromatic and aliphatic compound, e.g., waxes, lipids, chitin (Derenne and Largeau 2001), while compounds of a low-molecular weight, e.g., sugars, amino acids, are more easily degradable (Sollins et al. 1996). However, in various experiments e.g., sugars with longer MRT than the SOC has on average were observed (Schmidt et al. 2011; Thevenot et al. 2010). Overall, no matter if plant or microbial derived, biochemical recalcitrance of certain SOC compounds leads to a selective preservation compared to easily degradable material (von Lützow et al. 2006). Recalcitrance is relevant to stabilization in the time frame of up to a few decades. An exception is so called "black carbon", which might be stable over millennial time frames (Hammes et al. 2007; Kuzyakov et al. 2009).

Spatial inaccessibility may be the result of occlusion of organic particles (particulate organic matter) within aggregates (Balesdent et al. 2000; Oades 1984).

Aggregate formation is induced by biotic activity: Organic and mineral particles are glued together either in the intestinal tract of soil fauna or by excreted metabolites of microorganisms as well as by root exudates (Elliott 1986; Oades 1984, 1993). Moreover, fungal hyphae are a major agent in formation of macroaggregates (>250 μm, Tisdall and Oades 1982). Since microbial activity drives aggregate formation, the latter also depends on litter quality (Martens 2000). Aggregates play a substantial role in the stabilization of SOC, for the formation and stability of the soil structure, and, thus, for fertility of cropland soil (Abiven et al. 2009; Trumbore and Czimczik 2008). The formation of water-repellent surfaces (hydrophobicity) is also a type of spatial inaccessibility of SOC to microorganisms (Lamparter et al. 2009; Piccolo et al. 1999).

Chemical reactions of SOM with the mineral surface have been considered to be strong and durable forms of stabilization since the oldest SOC is found often in *organo-mineral associations* (Eusterhues et al. 2003; von Lützow et al. 2006). Due to their variable or permanent negative charge of the surface, mainly clay particles, silicates, and oxides act as mineral sorbents (Sollins et al. 1996; Wiseman and Puttmann 2005). Positively and negatively charged organic groups can bond to the sorbent by ligand exchange and/or polyvalent cation bridges. The complexation and/or precipitation of SOM with metal ions, mainly Ca^{2+}, Al^{3+}, and Fe^{3+}, is a further process of inaccessibility (Kiem and Kögel-Knabner 2002; von Lützow et al. 2006). Further, bonding mechanisms are water bridging and van der Waals forces which are relatively weak (von Lützow et al. 2006). There is evidence that, depending on texture and environmental conditions, the capacity for organo-mineral associations in soils is limited (Baldock and Skjemstad 2000; Wiseman and Puttmann 2005).

Destabilization is the process of reversing physical or chemical protection of SOC, rendering SOC to microbial attack, i.e. mineralization. External factors, as ecosystem properties and soil management, are the major agents controlling timing and kinetics of SOC destabilization (Schmidt et al. 2011; von Lützow et al. 2006). Thus, the determination and evaluation of destabilization mechanisms is complicated and needs more detailed research (Trumbore and Czimczik 2008). Kuzyakov (2011) observed that it is crucial to directly link SOC fractions to CO_2 fluxes in future experiments. Thus, physico-chemical factors that control destabilization are briefly discussed herein.

Depolymerization, dissolution, and desorption are the reactions which reduce biochemical recalcitrance and organo-mineral associations. Organo-mineral associations may disintegrate due to changes in the pH, the redox potential and cation concentration (Sollins et al. 1996). Macroaggregates may disrupt when they are exposed to physical stress, as dry-wetting and thaw-freeze cycles and cropland soil management (Denef et al. 2002; Navarro-García et al. 2012). The input of easily degradable organic matter (OM) can initiate the decomposition of recalcitrant compounds (priming effect) (Kuzyakov et al. 2000; Trumbore 2006). Further, all factors that are able to enhance microbial activity (climate, availability of easily degradable compounds, availability of N) increase the susceptibility of SOM to mineralization.

18.3 Potential Alterations of the Carbon Cycle in a Changing World

In the next sections, an overview of the possible effects of climate change on the C-cycle is presented, with emphasis on those processes in soil which are the least understood. Firstly, the effect of elevated CO_2 concentrations on NPP and below-ground processes is reviewed. Next, uncertainties of how warming can effect C-mineralization are discussed, and finally the effects of extreme weather events, (i.e. rewetting and thawing) on C cycling are presented.

18.3.1 Elevated Atmospheric Carbon Dioxide Concentration

A reduction of the increase of atmospheric CO_2 concentrations is expected due to the CO_2-fertilization of plants (i.e. negative feedback, Friedlingstein et al. 2006; Heimann and Reichstein 2008). It has been shown that the light saturated uptake of CO_2 increases (in C3 plants) with increasing CO_2 concentration (Leakey et al. 2009).

Much effort is going on to test the effects of elevated CO_2 concentrations on the C-cycle. Globally, 36 free air CO_2 enrichment (FACE) experiments have been conducted, and some are still running. A list is available online (http://public.ornl.gov/face/global_face.shtml). FACE plots are surrounded by pipes injecting a CO_2 stream into the air. Concentrations of CO_2 of up to 600 ppm are tested by this method in forest, grassland, cropland and dessert ecosystems (Ainsworth and Long 2005). Albeit restricted plot size (up to 30 m diameter), this method provides the possibility to test the effect of elevated CO_2 concentration under field conditions. However, no forest experiments were conducted in boreal and tropical regions, and no FACE experiment fumigates mature forests (Hickler et al. 2008). Furthermore, many forest FACE are only running until 2011 (Ledford 2008).

Effects of changing climate and increasing CO_2 concentration on NPP are of high importance. Ainsworth and Long (2005) showed, by summarizing data from FACE experiments, that biomass and yield of plant species with C4 photosynthetic pathway are largely unaffected by CO_2 concentration. However, most C3 crops and juvenile trees showed increased aboveground biomass and crop yields (Table 18.2). FACE experiments have been extremely valuable, but they are implemented only at a very limited number of sites and for only a few plant species. De Graaff et al. (2006) summarized data from FACE in a meta-analysis and concluded that below-ground biomass may even increase by 34%. Thus, increased biomass production may increase C-input into soil, enhancing the SOC storage and mitigate the increased mineralization caused by warming. However, this may only be true if other nutrients will not limit plant growth (de Graaff et al. 2006).

Considering only the increases in C-input in the ecosystem is only one issue with respect to global change. For the C balance studies, the amount of C retained in the ecosystem is crucial. Specifically, if mineralization also increases with C-input the C balance may be unaffected. It has been documented that microbial growth rates in

Table 18.2 Aboveground dry matter and crop yield changes as affected by increased CO_2 concentration (ca. 500–600 ppm) at FACE experiments

			Change (%)		
	No. of species	No. of sites	Mean	Lower CI	Upper CI
ABGDM	34	6	17.0	14.5	19.6
Tree	7	2	28.0	6.4	54.1
C4 crop	1	1	6.7	−2.2	16.6
C3 grass	8	3	10.5	6.5	14.8
Legume	6	3	20.3	13.7	27.3
Crop yield	6	3	17.3	10.2	24.9
Cotton	–	1	42.2	23.7	63.6
Wheat	–	1	14.4	−1.6	33.1
Rice	–	1	10.4	−4.4	30.2
Beet	–	1	12.5	n.d.	n.d.

Data compiled from Ainsworth and Long (2005) and Manderscheid et al. (2010)
ABGDM Aboveground dry matter, *CI* 95% confidence interval, *n.d.* not determined

soil increase linearly with increasing atmospheric CO_2 concentration (Blagodatskaya et al. 2010). Bacterial respiration, but not that of saprotrophic fungi is enhanced under elevated CO_2 (Anderson and Heinemeyer 2011). This indicates the preferential increase of fast growing microorganisms (r-strategists), probably due to more rhizodeposition. Increased microbial growth on labile substrates can induce priming effects on SOC mineralization (Kuzyakov et al. 2000) and encounter the increased C-input of plants caused by CO_2-fertilization. On the other hand, Rillig and Allen (1999) showed an increase of glomalin produced by arbuscular mycorrhizal fungi after 3 years of elevated CO_2. Glomalin is a recalcitrant organic glycoprotein which is preferentially fixed in macroaggregates and, therefore, protected against microbial breakdown (Rillig and Allen 1999). However, direct quantification of glomalin is not possible up to now, all available methods having specific drawbacks (Rosier et al. 2006). Nevertheless, glomalin seems to have a relatively long MRT (few decades) and can, therefore, account for a significant C-pool in ecosystems (Treseder and Allen 2000; Treseder and Turner 2007). Whether priming or production of glomalin affects SOC storage in the long-term is unknown, and is likely to be ecosystem specific.

In fertilized agroecosystems the effect of priming may be less compared to N-limited systems such as forests. Anderson et al. (2011) showed that SOC stocks under cropland use increased by 10% in 6 years under elevated CO_2 (550 ppm) relative to ambient CO_2 concentration. In a warm temperate forest (Duke FACE) Drake et al. (2011) showed that C-fluxes increased under elevated CO_2 (ambient plus 200 ppm). This trend also increased the N-turnover, presumably increasing N mineralization from SOM. As a consequence, tree biomass increased (2003–2007), but SOC stocks remained unaffected. Thus, N or other nutrients may become a limiting factor for biomass increase in non-fertilized ecosystems (de Graaff et al. 2006). Thus, CO_2 fertilization may enhance the ecosystem C-sink, but only to a minor extent, which could also be offset by warming.

18.3.2 Increase in Temperature

Microbial decomposition is, as are all chemical or biochemical reactions, temperature dependent. Therefore, it implies that rising temperature can induce a positive feedback: C-mineralization will increase with temperature and the higher release of CO_2 will cause additional warming. Therefore, it is important to quantify the effect of temperature on respiration for improved predictions of effects on SOC storage.

Temperature sensitivity is often expressed as Q_{10} values by the van't Hoff equation (Eq. 18.2):

$$Q_{10} = (k_2 / k_1)^{(10/(T_2 - T_1))} \tag{18.2}$$

Where, k_2 and k_1 are rate constants of a certain process and T_2 and T_1 the corresponding temperatures. An often assumed Q_{10} of 2 means that respiration would increase twofold by raising the temperature from 10°C to 20°C. While this empirical relationship has been often used (Davidson and Janssens 2006; Vicca et al. 2009; von Lützow and Kögel-Knabner 2009), the theoretical basis is determined by thermodynamical laws. Arrhenius formulated an equation which relates the reaction rate constant (k) of a certain compound to its bio-chemical stability, i.e. activation energy (E_a). This relationship is presented in Eq. 18.3:

$$k = \alpha \times e^{(-E_a/RT)} \tag{18.3}$$

where, α is a pre-exponential factor, R the gas constant (8.324 J K^{-1} mol^{-1}) and T temperature (K). There are two important implications of the Arrhenius equation for the temperature sensitivity of C-mineralization. Firstly, the relation shows that Q_{10} values decrease with increase in temperature for a certain compound. That is, Q_{10} values are not constant even for pure substances over a range of different temperatures. Secondly, substances with higher activation energies (i.e., less reactive and more recalcitrant) exhibit higher sensitivity to changing temperatures (Fig. 18.1). This theoretical basis gives rise to the assumption that more stable (i.e. presumably more decomposed) SOC fractions are affected relatively more by increasing temperatures than labile fractions (Knorr et al. 2005). Figure 18.1 shows an example adapted from Davidson and Janssens (2006) for reaction of glucose ($E_a \approx 30$ kJ mol^{-1}) and tannin ($E_a \approx 70$ kJ mol^{-1}) relative to 10°C. The relative effect of temperature on k of tannin is much higher. However, the absolute change in reaction speed of glucose in relation to tannin (k_{GLU}/k_{TAN}) is negligible, given the order of magnitude presented in Fig. 18.1. On the other hand, labile pools or fractions normally form a minor part of the SOC stocks. Therefore, despite slow turnover, mineralization of stable pools can contribute significantly to CO_2 efflux from SOC to the atmosphere (Flessa et al. 2000). Concerns that stable SOC pools/fractions are more sensitive to warming (Knorr et al. 2005) may only be true if stability is induced by recalcitrance (see Sect. 18.2.3, Schmidt et al. 2011). Moreover, the Arrhenius equation is only

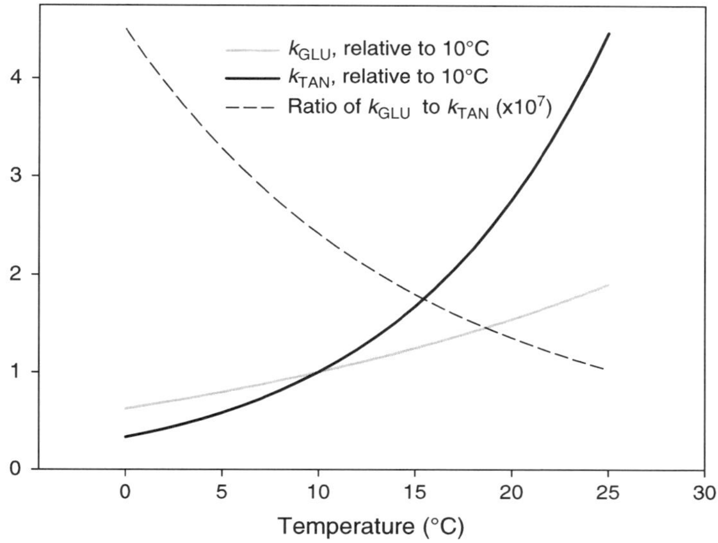

Fig. 18.1 Relative effect of temperature on reaction rate constant (*k*) of glucose and tannin and the ratio of the absolute rate constants. The reaction rate at 10°C is accepted as 1

valid under the assumption of unlimited substrate availability. Theoretically, the Arrhenius equation can be combined with Michaelis-Menten kinetics (i.e. describing reaction rates affected by substrate limitation) but such attempts are difficult in a complex media such as soil. Therefore, empirical observation and phenomenological description seem to be the only ways to determine temperature sensitivity of SOC, as well as of its fractions (Kirschbaum 2006).

The determination of Q_{10} values seem to be straightforward. Measuring respiration rates during laboratory or field studies and application of Eq. 18.2 should be easy. However, a wide range of Q_{10} values (1.4–6.9) have been reported (von Lützow and Kögel-Knabner 2009). Several methodological problems arise in determining Q_{10} values under field conditions. First of all, soil respiration consists of several components, which are not easy to distinguish (see Sect. 18.2.1). Moreover, increasing temperature is often accompanied with decreasing soil moisture. Therefore, respiration may not, or less, increase with temperature because moisture is limiting. This bias the determination of Q_{10} towards underestimation. Water logged conditions, freezing-thawing or drying-rewetting cycles and different substrate supply also hamper determination of temperature sensitivity under field conditions. For this reason, Kirschbaum (1995, 2006) recommended laboratory incubations under controlled conditions as the best method to determine temperature sensitivity of SOC mineralization (Box 18.4).

Box 18.4 Methodological Considerations for Determination of Q_{10} Values During Laboratory Incubations

A common way for determination of Q_{10} values is incubation of a soil sample at different temperatures under otherwise equal conditions (i.e. optimal moisture content) and measuring respiration rate at different times. Using such "parallel incubations", result in "apparent" Q_{10} values which are strongly biased by the substrate supply. Figure 18.2a illustrates a simplified example with pool sizes and turnover times taken from Heitkamp et al. (2009). Bulk respiration is modeled as contributions from pool 1 (0.4 Mg $C ha^{-1}$; k=0.059), pool 2 (3.2 Mg $C ha^{-1}$ k=0.002) and pool 3 (16.8 Mg $C ha^{-1}$; k=0.0001). The Q_{10} values of 2, 3 and 4 are assigned to pools 1, 2 and 3, respectively. Figure 18.2a shows that apparent Q_{10} values of all pools decrease with time and even fall below 1. That is the case when pool 1 is exhausted at higher temperature, but still contributes to respiration at lower temperature. A false conclusion from this pattern is that stable pools (i.e., contributing to respiration at later incubation time) are less sensitive to temperature than labile pools (i.e., contributing to respiration at early stages). Reichstein et al. (2000) tried to overcome this problem by determining pool sizes and decay constants at each temperature. Then, the pool sizes were hold constant, and the Q_{10} values were determined for the decay constant. This approach indeed overcomes the problems with substrate depletion, but the non-linear curve fitting approach for determining pool size and decay constant is itself not straightforward and results depend on incubation conditions (Böttcher 2004; Heitkamp et al. 2009; Sierra 1990). By applying the approach to specific compounds sampled during incubation, Feng and Simpson (2008) showed, in accordance with the Arrhenius equation, that lignin monomers exhibited higher temperature sensitivity than n-alkanoic compounds. Nevertheless, a Q_{10} could not be calculated for almost 50% of the dataset due to poor model fits.

Another solution is to incubate all samples at the same temperature and exposing the sample only for short period to different temperatures (Leifeld and Fuhrer 2005). This approach avoids *different* substrate supply at different temperature for the *same* pool. Nevertheless, pool sizes change with time (Fig. 18.2b) simultaneously affecting bulk apparent Q_{10} values (i.e. apparent Q_{10} of respired C). In the present example, the bulk intrinsic Q_{10} (i.e., Q_{10} value inherent to a compound due to its chemical properties) value is largely determined by pool 3, due to its large size. Therefore, bulk apparent Q_{10} values increase towards the bulk intrinsic Q_{10} value with time, but do not coincide during the short incubation time because respiration is largely determined by pool 2. The bulk intrinsic Q_{10} increases with incubation time due to depletion of more labile substrates with lower Q_{10} values. If there would be any possibility to measure contribution of pools directly, the "rotating incubation" method would be straightforward and yield intrinsic Q_{10} values for each pool at any time. Attempts using ^{13}C natural abundance after C3/C4 vegetation changes indeed indicate that "old" SOC is more sensitive to temperature compared to "young" SOC (Vanhala et al. 2007; Waldrop and Firestone 2004).

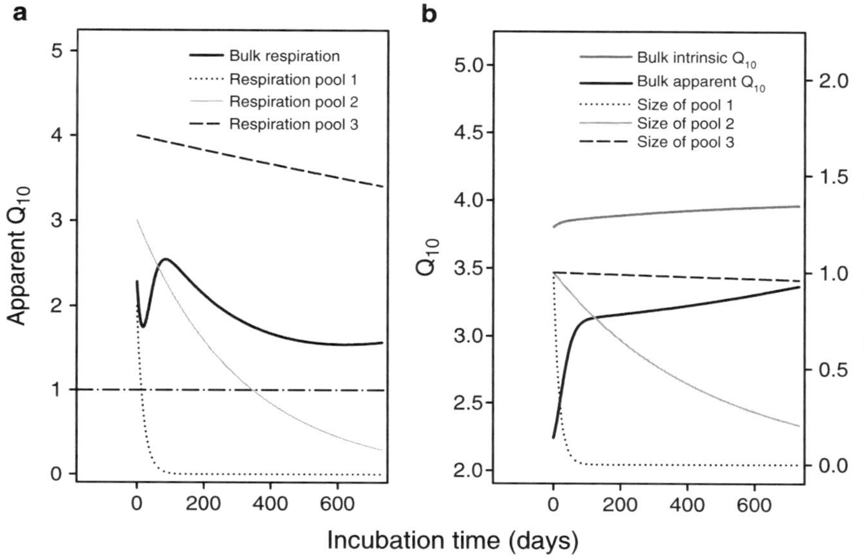

Fig. 18.2 Theoretical effect of incubation time on Q_{10} values calculated by respiration from different pools by parallel incubation (**a**) and relation between remaining pool size and calculated intrinsic and apparent Q_{10} value of bulk respiration with rotating incubation (**b**). Parallel incubation means incubation of samples at different temperatures throughout, whereas rotating incubation is incubation at one temperature and exposure to different temperature only for short time frames

Current knowledge indicates that recalcitrance does not lead to stabilization of SOC on millennial time scales (Kögel-Knabner et al. 2008). If stabilization of C in soil is a consequence of chemical protection against decomposition, the Arrhenius equation might not be relevant for temperature sensitivity of stabilized SOC. However, Craine et al. (2010) showed that physical or chemical stabilization may happen without altering temperature sensitivity. For example, mineralization data of soil and litter samples differed in their respiration rate by an order of magnitude (30 and 420 μg C $(g\,C\,h)^{-1}$, respectively), but not in their activation energies (59 kJ mol^{-1}). Thus, physical and chemical stabilization mechanisms seem to be less sensitive to temperature compared to biochemical stabilization (i.e. recalcitrance). In contrast, Gillabel et al. (2010) compared temperature sensitivity of topsoil and subsoil samples. In subsoil, the amount of chemically stabilized SOC is assumed to be relatively higher compared to topsoil (Rumpel and Koegel-Knabner 2011). Gillabel et al. (2010) observed that respiration from topsoil samples was in accordance with the Arrhenius equation, whereas subsoil respiration was not sensitive at all to temperature. It was hypothesized that chemical protection induced these trends. The Arrhenius equation only applies to conditions of unlimited substrate availability. Due to low substrate availability in subsoil, the effect of temperature might be cancelled out by processes described by Michaelis-Menten kinetics in the subsoil (von Lützow and Kögel-Knabner 2009). Therefore, the apparent temperature sensitivity is determined by substrate availability (i.e., abundance and availability of substrate and stabilization mechanism), whereas the intrinsic sensitivity (see Box 18.4) is determined by the chemistry of the compound. Future research is needed to distinguish between the effects of these different processes (Conant et al. 2011).

18.3.3 Frequency of Extreme Weather Events

Besides increasing temperature and changes in precipitation increases in extreme weather events are also predicted in a future climate (Christensen et al. 2007). Thus, increasing numbers of drying and wetting and/or freezing and thawing events are likely in most ecosystems.

A flush of CO_2 efflux occurs upon rewetting of a soil. This is termed the "Birch effect". Birch (1958) speculated that the CO_2 flush is derived from "solid organic material" and regulated by microbial state before and during rewetting. Death of microbial cells due to drying and subsequent re-utilization as substrate after rewetting is another explanation (Kieft et al. 1987). Whereas microbial death and re-utilization of cell debris after rewetting remain a common explanation, Fierer and Schimel (2003) reported that the CO_2 release can additionally be explained by accumulation of labile substrate and possibly also of enzymes. Disruption of aggregates, thus exposure of physical protected SOC to mineralization, may be in part responsible for increased respiration (Navarro-García et al. 2012). However, drying and rewetting can also increase aggregate stability in the long-term (Denef et al. 2002). Aggregate size and stability (i.e., soil structure) also determine gas diffusion. Therefore, oxygen supply and thus microbial activity can be influenced by changes in soil structure (Jäger et al. 2011).

After several cycles, the CO_2 flush after rewetting is reduced, indicating depletion of substrate affected by rewetting (Fierer and Schimel 2002). Moreover, the short flush may contribute only a small portion to annual emissions. Muhr et al. (2008) reported even decreased cumulative CO_2 emissions from soil samples with drying and rewetting cycles compared to continuously moist samples. If cumulative respiration is reduced depends on the duration of the dry phase, where microbial activity is limited by soil moisture. Furthermore, microbial respiration can be reduced after the rewetting flush (Fierer and Schimel 2002), probably because of substrate depletion and acclimation of the microbial community (Lundquist et al. 1999). Whereas fungal growth was not affected by drying and rewetting cycles, bacterial growth decreased after exposure to several cycles (Bapiri et al. 2010). An increase of fungal population may shift the specific respiration (i.e. respiration per unit microbial biomass) to lower values, since saprotrophic fungi are more effective in substrate utilization than bacteria (Joergensen and Wichern 2008). Physical, chemical and biological interactions apparently govern the net-effects of drying and rewetting on SOC mineralization (Kim et al. 2011). In the long term, the duration of dry phases (Bottner et al. 2000) and the number of cycles may determine the net effect on annual C-mineralization.

Similar to drying and rewetting cycles, a flush of CO_2 is also observed after thawing of a frozen soil (Kim et al. 2011; Matzner and Borken 2008). The flush is ascribed to microbial death and subsequent utilization of cell debris. Also, diffusion barriers might be involved. Specifically, microbial activity continues at unfrozen microsites and/or in subsoil. After thawing, gas can diffuse out of the soil (Teepe et al. 2001). Aggregate stability is often reduced after thawing, depending on the water content before the freezing event (Dagesse 2011). Reduced physical protection can, therefore, contribute to the CO_2 flush after thawing. Further, biology and chemistry of soil also changes after freezing and thawing. Schmitt et al. (2008) reported a decrease in fungal biomass, whereas bacteria were largely unaffected by freezing and thawing cycles. Besides the observed flush after thawing, the net-effect on soil respiration is not entirely clear. Matzner and Borken (2008) reported in their review that cropland soils seem to lose slightly more (<5%) C by respiration after freeze-and-thaw cycles compared to unfrozen soil. The opposite has been reported for natural vegetation. Comparison of studies is further complicated by methodological issues, such as freezing or thawing temperature, sampling time and experimental duration (Hugh 2007).

For both events, the net effect likely depends on the frequency of cycles, but seem to be small on annual basis (Fierer and Schimel 2002; Matzner and Borken 2008). However, the lack of understanding the processes involved hampers a general conclusion. Most knowledge is based on laboratory studies, which is also a consequence of methodological issues in measuring soil respiration in the field. Studies involving subsequent events of dry-and-rewet and freeze-and-thaw (e.g., effects of subsequent freezing and thawing during winter and subsequent drying and rewetting in spring) seem to be entirely missing, but are most important to elucidate effects of climate change on SOC mineralization.

18.4 Cropland Management, Elevated Carbon Dioxide, and Temperature Increase: A Model Scenario

The following section of this chapter exemplifies effects of CO_2-fertilization and warming on SOC stocks for two cropland management options with a modeling scenario. However, a model can hardly take into account all factors affecting predictions of SOC dynamics over the next 100 years. There will be changes in management, fertilization techniques, and plant cultivars. Further, climate change is not simply increasing temperature and precipitation but also cause increases in extreme weather conditions, which may lead to hardly predictable socio-economic and agro-ecological changes. Therefore, the model was applied as a tool to separate the possible quantitative influence of selected variables within a certain scenario on SOC dynamics. This approach is useful to identify the magnitude of some factors related to climate change.

The Rothamsted Carbon Model 26.3 (RothC) is chosen as a tool because it is useful in simulating SOC dynamics (Smith et al. 1997). Furthermore, RothC works well at the chosen site (Heitkamp et al. 2012b) and the model can easily be re-parameterized (Gu et al. 2004; Heitkamp et al. 2012b).

18.4.1 Site Conditions and Model Setup

By using regional projections of changing temperature and precipitation, the effects of climate change and residue management on SOC dynamics at a cropland site at Puch, Germany were modeled. The long-term (1960–1990) MAT of the site is 7.9°C and MAP is 922 mm. The soil is a Luvisol which developed on loess deposits, thus silt is the dominating particle size class (9% sand, 73% silt and 18% clay). The fertilization experiment was set up within the network of the "Internationale Organische Stickstoff Dauerduengungsversuche" (IOSDV) in 1983. Crop rotation consisted of sugar beet (*Beta vulgaris* L.), winter wheat and winter barley (*Hordeum vulgare* L.). This analysis was based on two out of several treatments: (i) removal of straw and beet leaves (CON) and (ii) incorporation of straw and beet leaves (RES). All treatments were under conventional tillage. The SOC content was measured episodically, in 1983 in samples composited among plots with different N-rates; in 1986, 1989 and 2003 bulked among field replicates, and in 1994 and 2004 for individual plots (n=3). Ploughing depth was 25 cm and bulk density was assumed to be 1.5 $g\,cm^{-3}$ for conversion of SOC concentration into stocks. Nitrogen was applied at a rate (kg $N\,ha^{-1}$) of 100 to beet, 80 to wheat and 60 to barley until 1998 and was raised by 20 for cereals thereafter. Straw for incorporation was weighed until 1998, thereafter a harvest index of 0.5 was applied. Beet leaves were weighed throughout until 2004. Residue input by stubbles, roots and rhizodeposition was estimated by linear regression of yield (grain and beet) and C-input as shown by Eq. 18.4:

$$I = (Y \times F + K) \times R \tag{18.4}$$

Where, I is the C-input (Mg ha^{-1}), Y is crop yield (fresh mass for beet, incl. 13% water in grain; Mg ha^{-1}), F (Mg C (Mg Y)$^{-1}$) and K (Mg $C\,ha^{-1}$) are crop-specific constants and R is a multiplier to account for rhizodeposition (Franko 1997; Ludwig et al. 2007). Crop yields were published (Hege and Offenberger 2006), F was set to 0.008 (winter cereals) or 0.0008 (sugar beet), K was set to 0.4 (winter cereals) or 0.16 (sugar beet), and R was set to 1.5 or 1.2 (winter cereals and sugar beet), respectively. The constants are published in Franko (1997) and for rhizodeposition see Domanski et al. (2001) and Ludwig et al. (2007).

The RothC model was used for modeling SOC dynamics (Coleman and Jenkinson 1999). The model consists of five pools with different turnover, is easy to calibrate and was proven useful for simulating SOC dynamics (Ludwig et al. 2007; Smith et al. 1997). Every pool has a specific decay constant which is modified by temperature, moisture, and plant cover. Partitioning between mineralization and humification is influenced by clay content (Coleman and Jenkinson 1999). The original temperature function of RothC was replaced by Eq. 18.2 to evaluate the effect of different temperature sensitivity of pools (Gu et al. 2004). The replaced function with the commonly assumed Q_{10} of 2 (Davidson and Janssens 2006) was tested against the original model and only minor differences occurred. In Model A, Eq. 18.2 was used with $Q_{10}=2$ for all pools. As stated above, more stable pools might have a higher sensitivity to temperature changes. Therefore, for Model B, a Q_{10} of 2 was

Table 18.3 Maximal projected increases in temperature and precipitation (Christensen et al. 2007) and potential evapotranspiration (Baguis et al. 2010) until the period of 2080–2099 for Northern Europe

	Temperature (°C)	Precipitation (%)	ETP (%)
DJF	8.2	25	40
MAM	5.3	21	20
JJA	5.4	16	33
SON	5.4	13	30

DJF December, January, February, *MAM* March, April, May, *JJA* June, July, August; *SON* September, October, November, *ETP* evapotranspiration

used for DPM and BIO (Davidson and Janssens 2006), a Q_{10} of 3 for RPM (Wetterstedt et al. 2010) and a Q_{10} of 4 for the HUM pool (Leifeld and Fuhrer 2005). The model was initialized with an equilibrium run to 1983. For this purpose, C-input and size of inert organic matter (IOM) pool were adjusted. From 1984 to 2004, the model was run with available weather data and C-input was measured or calculated independently (Eq. 18.4). From 2004 to 2099 different climate change scenarios were assumed (Table 18.3).

As a reference scenario (No-CC), monthly temperature, precipitation, and actual evapotranspiration were assumed to be constant at the mean value of the period 1983–2004. This period is close to that (1980–1999) used for modeling of regional climate change projections for 2080–2099 by the IPCC (Christensen et al. 2007). The input of C was calculated as mean value between 1984 and 2004. Assuming a scenario without changing temperature and precipitation was necessary because SOC stocks were not in equilibrium in 2004 and management effects must be separated from climate change effects.

The second scenario (Max-CC) represents a regional climate scenario with maximum assumed temperature increase and precipitation changes for Northern Europe (Christensen et al. 2007). All increases are supposed to be linear and are originally projected from the period of 1980–1999 to the period of 2080–2099 (Table 18.3).

The third scenario is the same as the second (Max-CC), but with consideration of CO_2 fertilization effects on crop growth. A linear increase in growth of up to 16% was assumed until 2050 (Leakey et al. 2009). From 2050 to 2100, an additional increase of 8% was assumed. Therefore the total increase was in the range reported from FACE (de Graaff et al. 2006).

Modeling of all scenarios was done in monthly resolution. For presentation of data the modeled SOC stocks were averaged over a crop rotation period of 3 years.

18.4.2 Effect of Residue Incorporation

During the observed period from 1983 to 2004, SOC stocks showed marked changes. For instance, SOC stocks (Mg ha^{-1} ± standard error) of CON were 35.3 ± 3.3 and stocks of RES were 39.9 ± 1.6. Model fits were satisfactory with root mean square

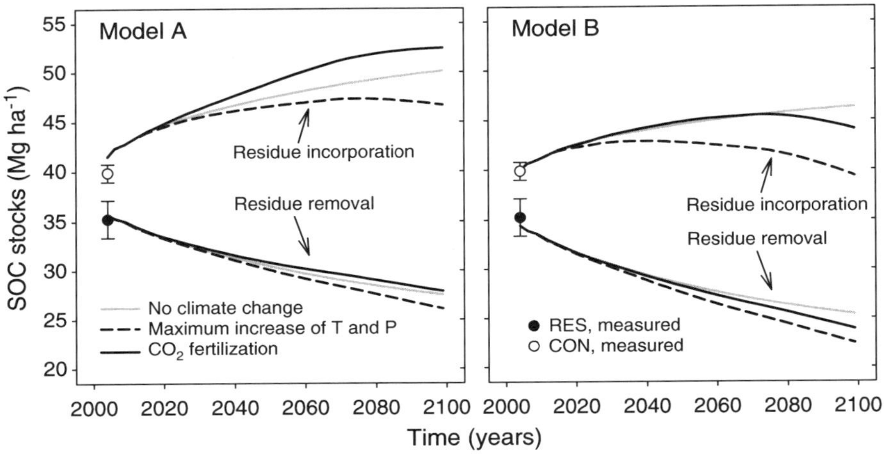

Fig. 18.3 Modeled SOC dynamics for residue removal (*CON*) and incorporation (*RES*) for the different climate change scenarios in Puch, Germany. In *model A* the same temperature sensitivity for mineralization is assumed for all model pools. In *model B*, temperature sensitivity increases with the MRT of pools (see Table 18.4). Measured stocks in 2004 are means with standard errors

errors (RMSE) between 5.0 and 7.8, and with the mean differences between observed and predicted values (–1.6 to 1.8) well in the range of the standard errors (Smith et al. 1997).

With both model parameterizations, the increasing gap in predicted SOC storage between cropland management with residue incorporation and with residue export (Fig. 18.3) became obvious. The SOC stocks in the RES treatment were about twice as large as those in CON treatments (Table 18.4) at the end of the modeled period. In the CON treatment modeled SOC stocks ranged from 26.1 to 27.9 Mg ha^{-1}, loosing between 7.8 and 9.6 Mg C ha^{-1}. The estimated SOC stocks were low, but considering the low C-input (Table 18.4) this is in a realistic range. For example, Rühlmann (1999) summarized SOC stocks for long-term bare fallow experiments. According to the empirical equation used by Rühlmann (1999), the SOC stock under bare fallow (i.e., no C-input) at the Puch site is estimated to be 17 Mg ha^{-1}. The modeled data indicate that SOC stocks of both treatments will not attain equilibrium until 2100, which contradicts observations of West and Post (2002), who estimated that a new equilibrium due to enhanced crop rotation will be reached after 40–60 years. One explanation may be the large difference in C-input between treatments at the Puch site (Table 18.4).

18.4.3 Effect of Climate Change Scenarios

Evaluation of climate change effects was done in comparison to a scenario where no climate change will be present. This was done by creating a scenario which used the

Table 18.4 Model results for the period 2005–2100 for different climate change scenarios and model parameterization

		Model A			Model B		
Scenario	C-input	SOC	Δ SOC	CC-effect	SOC	Δ SOC	CC-effect
CON							
No CC	106	27.5	−8.2	–	25.4	−9.0	–
Max CC	106	26.1	−9.6	−1.4	22.4	−12.0	−3.0
CO_2 fert	121	27.9	−7.8	+0.4	23.9	−10.5	−1.5
RES							
No CC	340	50.1	8.6	–	46.3	6.3	–
Max CC	340	46.7	5.2	−3.4	39.3	−0.6	−7.0
CO_2 fert	387	52.5	10.9	+2.3	44.1	4.1	−2.2

Given SOC stocks are modeled for the year 2100, whereas Δ SOC is the difference of the measured stocks in 2004 and the modeled stocks in 2100. Stocks of SOC at the start of the experiment in 1983 were 40.5 Mg ha^{-1}. All figures in Mg ha^{-1}

CON crop residue removed, *RES* crop residue incorporated, *No CC* scenario with average temperature and precipitation, no climate change, *Max CC* maximal climate change, scenario described in Table 18.3, *CO_2 fert* Max CC but with CO_2 fertilization of plants, *Model A* $Q_{10}=2$ for all pools, *Model B* $Q_{10}=2$ for DPM and BIO, $Q_{10}=3$ for RPM and $Q_{10}=4$ for HUM

monthly mean values of the Puch site during the experimental period 1984–2004. However, during that period an increase of MAP by 0.07°C $year^{-1}$ was reported by Hege and Offenberger (2006). Therefore, choosing that period as baseline is somewhat arbitrary because temperature already increased. However, this approach coincides with that used in the IPCC for regional climate scenarios (Christensen et al. 2007).

The data predicted by model A ($Q_{10}=2$ for all pools) showed only small effects of climate change scenarios on SOC stocks in the CON treatments. Loss of SOC in case of the Max-CC scenario was predicted to be 1.4 Mg C ha^{-1}. That difference was cancelled out under the assumption of CO_2-fertilization (Fig. 18.3, Table 18.4). The predicted differences were in a range hardly detectable under field conditions, given the effects of soil heterogeneity (Ellert et al. 2008; Heinze et al. 2010; Heitkamp et al. 2011). Predicted outcomes were different for the RES treatment. Modeled SOC loss induced by climate change (Max-CC) until the year 2100 was 3.4 Mg C ha^{-1}. Remarkably, there is a tipping point around the year 2075 (at 47.3 Mg C ha^{-1}) after which SOC stocks in the RES treatment are predicted to decrease. Inclusion of CO_2 fertilization in the scenario even increased the C-sink in the soil by 2.3 Mg C ha^{-1}, as compared to No-CC.

A different sensitivity of pools to temperature was incorporated in model B, changing the outcome of predictions markedly. Effect of warming (Max-CC) was predicted to be strong on SOC stocks of the RES treatment (Table 18.4). The tipping point from sink to source was predicted for the year 2039, turning the soil of the RES treatment for that scenario over the almost 100 years into a source of CO_2. Assuming CO_2 fertilization occurs led to almost identical predictions in SOC stocks of No-CC and CO_2-Fert scenarios until 2075. Therefore, the predicted source-sink tipping point was procrastinated by 36 years, as compared to the Max-CC scenario.

It has to be pointed out that model B predicted in general higher mineralization as compared to model A.

For the chosen scenarios changes of +2.3 to −7.0 Mg C ha^{-1}, were predicted to be induced by climate change from 2005 until 2100 for the Puch site. Therefore, the feedback between climate change and SOC balance is likely positive, as is also assumed in several global feedback simulations (Friedlingstein et al. 2006). However, when comparing the predicted (2005–2100) effects of climate change (+2.3 to −7.0 Mg C ha^{-1}) and residue incorporation or export (+12.6 and −12.8 Mg C ha^{-1}, respectively) it becomes obvious that appropriate management of cropland soils is of outstanding importance for reducing CO_2 emissions from these agroecosystems. Nevertheless, SOC accumulation by recommended management practices may be severely reduced by warming. This effect will be stronger if sensitivity of mineralization differs between pools of different stability.

18.5 Conclusions and Outlook

Soils are of major importance for C storage and vice versa. CO_2 can accumulate in soil as OM which is beneficial in terms of reduction of atmospheric CO_2 concentrations and improving soil fertility. Despite their outstanding importance for the C cycle, soils are still treated as a "black box" in most models. To predict feedbacks between soil, biosphere and atmosphere, progress is needed to shed light into this black box, i.e., to apply strategies for reducing SOC loss or re-accumulating SOC, quantitative predictions on the outcome of diverse strategies must be possible. There is no doubt that soils have the potential to reduce atmospheric CO_2 concentrations. However, many uncertainties in our understanding of C-cycling in soils hamper quantitative predictions of the sink potential and feedbacks between climate change and SOC dynamics. It must be certain that soil management does not turn soils into a net-source of CO_2. Therefore, research on this topic must continue, and increase, but it is clear that the mitigation potential of soil is not well enough understood to rely on it: reducing anthropogenic GHG emissions is essential to mitigate climate change.

The mechanistic understanding of processes and mechanisms on C-cycling in soils has vastly improved, but many uncertainties still remain. The analysis presented show that the temperature sensitivity of different SOC pools significantly affects the outcome of model predictions of SOC dynamics. Still, there is no consensus on a general applicable sensitivity of mineralization of SOC. In fact, today there is a lack of methodological tools to determine temperature effects, and there exists a strong need for new and well designed experiments (Conant et al. 2011). With no reliable quantitative model to assess the temperature response of SOC mineralization, it is difficult to predict the outcome of complex processes such as drying-and-rewetting or freezing-and-thawing. Such events are likely to increase in frequency in the future, and it is essential to improve the mechanistic knowledge of these processes. Even increased productivity of plants by CO_2 fertilization can potentially destabilize SOC by priming effects (Fontaine et al. 2007). Knowledge is generated in many different disciplines which are often only weakly linked in terms of exchanging results. This limits the progress at a time where fast action is required. However, the speed of progress may be advanced by the following considerations.

18.5.1 Connect Research Communities

Knowledge on C cycling was and is generated by many different disciplines. Atmospheric science, soil science, plant ecology, forestry, geography and many more disciplines are working on the topic. When reading the different subchapters, it becomes obvious that there is different terminology, even in related topics: studies in "litter decomposition" and "SOC stabilization" evolved largely independent from each other. Disconnect between "general ecology" and "soil ecology" was demonstrated recently by tracking citation between specialized journals (Barot et al. 2007). Knowledge transfer between the disciplines is slow, but is beginning to emerge (Prescott 2010; Schmidt et al. 2011). By exchanging results, concepts and ideas, scientific knowledge should increase faster and more effectively.

18.5.2 Connect Empirical Result with Models

Advances in mechanistic understanding of SOC stabilization are only poorly incorporated into quantitative models. To date, only one model exists which can simulate C-dynamics by inclusion of aggregate turnover (Yoo et al. 2011). Soil scientists use models only sparingly (Barot et al. 2007). Yet, the concepts must be quantified and validated. By using models as a tool to quantitatively synthesize various processes and mechanism, it is possible to better predict how changing conditions may influence the C cycle in general (Schmidt et al. 2011).

18.5.3 Connect Specificity and Generality

Under "specificity" empirical case studies are understood, whereas "generality" refers to the broad application of result and synthesis in theories. Both "specificity" and "generality" are essential to the scientific progress, but more merit should be given for synthesizing existing data. Entering e.g. "litter decomposition" in web of knowledge (7th November 2011) yielded 12,600 results, the first listed study being from 1930 (Melin 1930). Entering all these data into a global database and making it available for the scientific community will likely increase the precision of models describing litter decomposition across biomes. It is not said that that no experimental research will be needed after a global synthesis. Rather by developing a more general and quantitative model upon the vast data already existing it will be possible to identify gaps in knowledge and concentrate research efforts.

18.5.4 Using Long-Term Cross-Site Experiments

One reason why specificity seems to dominate research in soil science is the heterogeneous nature of soils. A lot of knowledge is generated under controlled laboratory conditions. However, it is also important to identify or design field studies, where the same treatment is implemented across several sites. Selecting or laying out such cross-site studies for long-term research will add to the knowledge generated in laboratory or single-site field studies, increase scientific output in many disciplines, and create excellent opportunities to test hypothesis and theories (Leuschner et al. 2009).

References

Abiven S, Recous S, Reves V et al (2005) Mineralisation of C and N from root, stem and leaf residues in soil and role of their biochemical quality. Biol Fert Soils 42:119–128

Abiven S, Menasseri S, Chenu C (2009) The effects of organic inputs over time on soil aggregate stability – A literature analysis. Soil Biol Biochem 41:1–12

Ainsworth EA, Long SP (2005) What have we learned from 15 years of free-air CO_2 enrichment (FACE)? A meta-analytic review of the responses of photosynthesis, canopy properties and plant production to rising CO_2. New Phytol 165:351–371

Anderson T-H, Heinemeyer WH-J (2011) Changes in the fungal-to-bacterial respiratory ratio and microbial biomass in agriculturally managed soils under free-air CO(2) enrichment (FACE): a six-year survey of a field study. Soil Biol Biochem 43:895–904

Baguis P, Roulin E, Willems P et al (2010) Climate change scenarios for precipitation and potential evapotranspiration over central Belgium. Theor Appl Climatol 99:273–286

Baldocchi DD (2003) Assessing the eddy covariance technique for evaluating carbon dioxide exchange rates of ecosystems: past, present and future. Glob Chang Biol 9:479–492

Baldock JA, Skjemstad JO (2000) Role of the soil matrix and minerals in protecting natural organic materials against biological attack. Org Geochem 31:697–710

Baldock JA, Masiello CA, Gelinas Y et al (2004) Cycling and composition of organic matter in terrestrial and marine ecosystems. Mar Chem 92:39–64

Balesdent J (1996) The significance of organic separates to carbon dynamics and its modelling in some cultivated soils. Eur J Soil Sci 47:485–493

Balesdent J, Chenu C, Balabane M (2000) Relationship of soil organic matter dynamics to physical protection and tillage. Soil Till Res 53:215–230

Bapiri A, Bååth E, Rousk J (2010) Drying-rewetting cycles affect fungal and bacterial growth differently in an arable soil. Microb Ecol 60:419–428

Barot S, Blouin M, Fontaine S et al (2007) A tale of four stories: soil ecology, theory evolution and the publication system. PloS One 2(11):e124810

Berg B, McClaugherty C (2003) Plant litter – decomposition, humus formation carbon sequestration. Springer, Berlin

Birch HF (1958) The effect of soil drying on humus decomposition and nitrogen availability. Plant Soil 10:9–31

Blagodatskaya E, Blagodatsky S, Dorodnikov M et al (2010) Elevated atmospheric CO_2 increases microbial growth rates in soil: results of three CO_2 enrichment experiments. Glob Chang Biol 16:836–848

Böttcher J (2004) Uncertainties of nonlinearly estimated parameters from incubations of soil organic matter. J Plant Nut Soil Sci 167:293–302

Bottner P, Couteaux MM, Anderson JM et al (2000) Decomposition of C-13-labelled plant material in a European 65–40 degrees latitudinal transect of coniferous forest soils: simulation of climate change by translocation of soils. Soil Biol Biochem 32:527–543
Burdon J (2001) Are the traditional concepts of the structures of humic substances realistic? Soil Sci 166:752–769
Cannel MGR, Thornley JHM (2000) Modelling the components of plant respiration: some guiding principles. Ann Bot 85:45–54
Chapin F, Woodwell G, Randerson J et al (2006) Reconciling carbon-cycle concepts, terminology, and methods. Ecosystem 9:1041–1050
Christensen BT (2001) Physical fractionation of soil and structural and functional complexity in organic matter turnover. Eur J Soil Sci 52:345–353
Christensen JH, Hewitson B, Busuioc A et al (2007) Regional climate projections. In: Solomon S, Qin D, Manning M et al (eds) Climate change 2007: the physical science basis. Contribution of Working Group I to the fourth assessment report of the Intergovernmental Panel on Climate Change. Cambridge University Press, Cambridge
Coleman K, Jenkinson DS (1999) ROTHC-26.3 a model for the turnover of carbon in soil. IACR, Rothamsted
Conant RT, Ryan MG, Agren GI et al (2011) Temperature and soil organic matter decomposition rates – synthesis of current knowledge and a way forward. Glob Chang Biol 17:3392–3404
Conway T, Tans P (2011) Trends in atmospheric carbon dioxide. NOAA/ESRL. http://www.esrl.noaa.gov/gmd/ccgg/trends. Accessed 27 Nov 2011
Cotrufo MF, Ngao J, Marzaioli F et al (2010) Inter-comparison of methods for quantifying above-ground leaf litter decomposition rates. Plant Soil 334:365–376
Craine JM, Fierer N, McLauchlan KK (2010) Widespread coupling between the rate and temperature sensitivity of organic matter decay. Nat Geosci 3:854–857
Dagesse D (2011) Effect of freeze-drying on soil aggregate stability. SSSA J 75:2111–2121
Davidson EA, Janssens IA (2006) Temperature sensitivity of soil carbon decomposition and feedbacks to climate change. Nature 440:165–173
de Graaff M-A, van Groenigen K-J, Six J et al (2006) Interactions between plant growth and soil nutrient cycling under elevated CO_2: a meta-analysis. Glob Chang Biol 12:2077–2091
Denef K, Six J, Merckx R et al (2002) Short-term effects of biological and physical forces on aggregate formation in soils with different clay mineralogy. Plant Soil 246:185–200
Derenne S, Largeau C (2001) A review of some important families of refractory macromolecules: composition, origin, and fate in soils and sediments. Soil Sci 166:833–847
Dilkes NB, Jones DL, Farrar J (2004) Temporal dynamics of carbon partitioning and rhizodeposition in wheat. Plant Physiol 134:706–715
Domanski G, Kuzyakov Y, Siniakina SV et al (2001) Carbon flows in the rhizosphere of ryegrass (*Lolium perenne*). J Plant Nut Soil Sci 164:381–387
Drake JE, Gallet-Budynek A, Hofmockel KS et al (2011) Increases in the flux of carbon below-ground stimulate nitrogen uptake and sustain the long-term enhancement of forest productivity under elevated CO_2. Ecol Lett 14:349–357
Ellerbrock RH, Hohn A, Gerke HH (1999) Characterization of soil organic matter from a sandy soil in relation to management practice using FT-IR spectroscopy. Plant Soil 213:55–61
Ellert BH, Janzen HH, VandenBygaart AJ et al (2008) Measuring change in soil organic carbon storage. In: Carter MR, Gregorich EG (eds) Soil sampling and methods of analysis. Taylor & Francis, Boca Raton
Elliott ET (1986) Aggregate structure and carbon, nitrogen, and phosphorus in native and cultivated soils. SSSA J 50:627–633
Eusterhues K, Rumpel C, Kleber M et al (2003) Stabilisation of soil organic matter by interactions with minerals as revealed by mineral dissolution and oxidative degradation. Org Geochem 34:1591–1600
Feng XJ, Simpson MJ (2008) Temperature responses of individual soil organic matter components. J Geophys Res Biogeosci 113

Fierer N, Schimel JP (2002) Effects of drying-rewetting frequency on soil carbon and nitrogen transformations. Soil Biol Biochem 34:777–787
Fierer N, Schimel JP (2003) A proposed mechanism for the pulse in carbon dioxide production commonly observed following the rapid rewetting of a dry soil. SSSA J 67:798–805
Fierer N, Grandy AS, Six J et al (2009) Searching for unifying principles in soil ecology. Soil Biol Biochem 41:2249–2256
Fischer H, Meyer A, Fischer K et al (2007) Carbohydrate and amino acid composition of dissolved organic matter leached from soil. Soil Biol Biochem 39:2926–2935
Flessa H, Ludwig B, Heil B et al (2000) The origin of soil organic C, dissolved organic C and respiration in a long-term maize experiment in Halle, Germany, determined by C-13 natural abundance. J Plant Nut Soil Sci 163:157–163
Fontaine S, Bardoux G, Abbadie L et al (2004) Carbon input to soil may decrease soil carbon content. Ecol Lett 7:314–320
Fontaine S, Barot S, Barre P et al (2007) Stability of organic carbon in deep soil layers controlled by fresh carbon supply. Nature 450:277–U210
Franko U (1997) Modellierung des Umsatzes der organischen Bodensubstanz. Arch Acker Pfl Boden 41:527–547
Friedlingstein P, Cox P, Betts R et al (2006) Climate-carbon cycle feedback analysis: results from the C[4]MIP model intercomparison. J Climate 19:3337–3353
Gholz HL, Wedin DA, Smitherman SM et al (2000) Long-term dynamics of pine and hardwood litter in contrasting environments: toward a global model of decomposition. Glob Chang Biol 6:751–765
Gillabel J, Cebrian-Lopez B, Six J et al (2010) Experimental evidence for the attenuating effect of SOM protection on temperature sensitivity of SOM decomposition. Glob Chang Biol 16:2789–2798
Golchin A, Oades J, Skjemstad J et al (1994) Study of free and occluded particulate organic matter in soils by solid state 13C Cp/MAS NMR spectroscopy and scanning electron microscopy. Aust J Soil Res 32:285–309
Golchin A, Oades JM, Skjemstad JO et al (1995) Structural and dynamic properties of soil organic-matter as reflected by C-13 natural-abundance, pyrolysis mass-spectrometry and solid-state C-13 NMR-spectroscopy in density fractions of an Oxisol under forest and pasture. Aust J Soil Res 33:59–76
Gu L, Post WM, King AW (2004) Fast labile carbon turnover obscures sensitivity of heterotrophic respiration from soil to temperature: a model analysis. Glob Biogeochem Cycle 18:GB1022
Haile-Mariam S, Collins HP, Wright S et al (2008) Fractionation and long-term laboratory incubation to measure soil organic matter dynamics. SSSA J 72:370–378
Hammes K, Schmidt MWI, Smernik RJ et al (2007) Comparison of quantification methods to measure fire-derived (black/elemental) carbon in soils and sediments using reference materials from soil, water, sediment and the atmosphere. Glob Biogeochem Cycle 21:GB3016
Hanson PJ, Edwards NT, Garten CT et al (2000) Separating root and soil microbial contributions to soil respiration: a review of methods and observations. Biogeochem 48:115–146
Harmon ME, Silver WL, Fasth B et al (2009) Long-term patterns of mass loss during the decomposition of leaf and fine root litter: an intersite comparison. Glob Chang Biol 15:1320–1338
Hege U, Offenberger K (2006) Effect of differentiated mineral fertilization and organic manuring on yield, product quality and N balances in the International Permanent Organic Nitrogen Fertilization Experiment (IOSDV). Puch Arch Agron Soil Sci 52:535–550
Heimann M, Reichstein M (2008) Terrestrial ecosystem carbon dynamics and climate feedbacks. Nature 451:289–292
Heinze S, Raupp J, Joergensen RG (2010) Effects of fertilizer and spatial heterogeneity in soil pH on microbial biomass indices in a long-term field trial of organic agriculture. Plant Soil 328:203–215
Heitkamp F, Raupp J, Ludwig B (2009) Impact of fertilizer type and rate on carbon and nitrogen pools in a sandy Cambisol. Plant Soil 319:259–275
Heitkamp F, Raupp J, Ludwig B (2011) Effects of fertilizer type and rate on labile soil fractions of a sandy Cambisol – long term and short-term dynamics. J Plant Nut Soil Sci 174:121–127

Heitkamp F, Jäger N, Flessa H et al (2012a) Effect of fertilization on respiration from different sources in a sandy Cambisol of an agricultural long-term experiment. Arch Agron Soil Sci. doi:10.1080/03650340.2011.555762
Heitkamp F, Wendland M, Offenberger K et al (2012b) Implications of input estimation, residue quality and carbon saturation on the predictive power of the Rothamsted Carbon Model. Geoderma 170:168–175
Helfrich M, Ludwig B, Buurman P et al (2006) Effect of land use on the composition of soil organic matter in density and aggregate fractions as revealed by solid-state C-13 NMR spectroscopy. Geoderma 136:331–341
Hickler T, Smith B, Prentice IC et al (2008) CO_2 fertilization in temperate FACE experiments not representative of boreal and tropical forests. Glob Chang Biol 14:1531–1542
Hugh HAL (2007) Soil freeze-thaw cycle experiments: trends, methodological weaknesses and suggested improvements. Soil Biol Biochem 39:977–986
Jacob M, Viedenz K, Polle A et al (2010) Leaf litter decomposition in temperate deciduous forest stands with a decreasing fraction of beech (*Fagus sylvatica*). Oecologia 164:1083–1094
Jacobs A, Helfrich M, Dyckmans J et al (2011) Effects of residue location on soil organic matter turnover: results from an incubation experiment with 15N-maize. J Plant Nut Soil Sci 174:634–643
Jäger N, Stange CF, Ludwig B et al (2011) Emission rates of N_2O and CO_2 from soils with different organic matter content from three long-term fertilization experiments-a laboratory study. Biol Fert Soils 47:483–494
Jensen LS, Salo T, Palmason F et al (2005) Influence of biochemical quality on C and N mineralisation from a broad variety of plant materials in soil. Plant Soil 273:307–326
Joergensen RG, Wichern F (2008) Quantitative assessment of the fungal contribution to microbial tissue in soil. Soil Biol Biochem 40:2977–2991
Jones DL, Hodge A, Kuzyakov Y (2004) Plant and mycorrhizal regulation of rhizodeposition. New Phytol 163:459–480
Kieft TL, Soroker E, Firestone MK (1987) Microbial biomass response to a rapid increase in water potential when dry soil is wetted. Soil Biol Biochem 19:119–126
Kiem R, Kögel-Knabner I (2002) Refractory organic carbon in particle-size fractions of arable soils II: organic carbon in relation to mineral surface area and iron oxides in fractions <6 μm. Org Geochem 33:1699–1713
Kim D-G, Vargas R, Bond-Lamberty B et al (2011) Effects of soil rewetting and thawing on soil gas fluxes: a review of current literature and suggestions for future research. Biogeosci Discuss 8:9847–9899
Kirschbaum MUF (1995) The temperature dependence of soil organic matter decomposition, and the effect of global warming in soil organic C storage. Soil Biol Biochem 27:753–760
Kirschbaum MUF (2006) The temperature dependence of organic matter decomposition – still a topic of debate. Soil Biol Biochem 38:2510–2518
Knorr W, Prentice IC, House JI et al (2005) Long-term sensitivity of soil carbon turnover to warming. Nature 433:298–301
Kögel-Knabner I, Guggenberger G, Kleber M et al (2008) Organo-mineral associations in temperate soils: integrating biology, mineralogy, and organic matter chemistry. J Plant Nut Soil Sci 171:61–82
Kuzyakov Y (2006) Sources of CO_2 efflux from soil and review of partitioning methods. Soil Biol Biochem 38:425–448
Kuzyakov Y (2011) How to link soil C pools with CO_2 fluxes? Biogeosciences 8:1523–1537
Kuzyakov Y, Gavrichkova O (2010) Time lag between photosynthesis and carbon dioxide efflux from soil: a review of mechanisms and controls. Glob Chang Biol 16:3386–3406
Kuzyakov Y, Larionova AA (2005) Root and rhizomicrobial respiration: a review of approaches to estimate respiration by autotrophic and heterotrophic organisms in soil. J Plant Nut Soil Sci 168:503–520
Kuzyakov Y, Friedel JK, Stahr K (2000) Review of mechanisms and quantification of priming effects. Soil Biol Biochem 32:1485–1498
Kuzyakov Y, Subbotina I, Chen H et al (2009) Black carbon decomposition and incorporation into soil microbial biomass estimated by 14C labeling. Soil Biol Biochem 41:210–219

Lamparter A, Bachmann J, Goebel M-O et al (2009) Carbon mineralization in soil: Impact of wetting-drying, aggregation and water repellency. Geoderma 150:324–333

Leakey ADB, Ainsworth EA, Bernacchi CJ et al (2009) Elevated CO_2 effects on plant carbon, nitrogen, and water relations: six important lessons from FACE. J Exp Bot 60:2859–2876

Ledford H (2008) Forestry carbon dioxide projects to close down. Nature 456:289–289

Leifeld J, Fuhrer J (2005) The temperature response of CO_2 production from bulk soils and soil fractions is related to soil organic matter quality. Biogeochemistry 75:433–453

Leuschner C, Jungkunst HF, Fleck S (2009) Functional role of forest diversity: pros and cons of synthetic stands and across-site comparisons in established forests. Basic Appl Ecol 10:1–9

Ludwig B, Schulz E, Rethemeyer J et al (2007) Predictive modelling of C dynamics in the long-term fertilization experiment at Bad Lauchstädt with the Rothamsted Carbon model. Eur J Soil Sci 58:1155–1163

Lundquist EJ, Scow KM, Jackson LE et al (1999) Rapid response of soil microbial communities from conventional, low input, and organic farming systems to a wet/dry cycle. Soil Biol Biochem 31:1661–1675

MacCarthy P (2001) The principles of humic substances. Soil Sci 166:738–751

Manderscheid R, Pacholski A, Weigel H-J (2010) Effect of free air carbon dioxide enrichment combined with two nitrogen levels on growth, yield and yield quality of sugar beet: Evidence for a sink limitation of beet growth under elevated CO(2). Eur J Agric 32:228–239

Martens DA (2000) Plant residue biochemistry regulates soil carbon cycling and carbon sequestration. Soil Biol Biochem 32:361–369

Matzner E, Borken W (2008) Do freeze-thaw events enhance C and N losses from soils of different ecosystems? A review. Eur J Soil Sci 59:274–284

Melin E (1930) Biological decomposition of some types of litter from north American forests. Ecology 11:72–101

Muhr J, Goldberg SD, Borken W et al (2008) Repeated drying-rewetting cycles and their effects on the emission of CO_2, N_2O, NO, and CH_4 in a forest soil. J Plant Nutr Soil Sci 171:719–728

Navarro-García F, Casermeiro MÁ, Schimel JP (2012) When structure means conservation: Effect of aggregate structure in controlling microbial responses to rewetting events. Soil Biol Biochem 44:1–8

Oades JM (1984) Soil organic-matter and structural stability – mechanisms and implications for management. Plant Soil 76:319–337

Oades JM (1993) The role of biology in the formation, stabilization and degradation of soil structure. Geoderma 56:377–400

Paul EA, Morris SJ, Conant RT et al (2006) Does the acid hydrolysis-incubation method measure meaningful soil organic carbon pools? SSSA J 70:1023–1035

Piccolo A, Spaccini R, Haberhauer G et al (1999) Increased sequestration of organic carbon in soil by hydrophobic protection. Naturwissenschaften 86:496–499

Potthoff M, Dyckmans J, Flessa H et al (2005) Dynamics of maize (Zea mays L.) leaf straw mineralization as affected by the presence of soil and the availability of nitrogen. Soil Biol Biochem 37:1259–1266

Powers JS, Montgomery RA, Adair EC et al (2009) Decomposition in tropical forests: a pan-tropical study of the effects of litter type, litter placement and mesofaunal exclusion across a precipitation gradient. J Ecol 97:801–811

Prescott CE (2010) Litter decomposition: what controls it and how can we alter it to sequester more carbon in forest soils? Biogeochemistry 101:133–149

Rasse DP, Rumpel C, Dignac MF (2005) Is soil carbon mostly root carbon? Mechanisms for a specific stabilization. Plant Soil 269:341–356

Reichstein M, Bednorz F, Broll G et al (2000) Temperature dependence of carbon mineralisation: conclusions from a long-term incubation of subalpine soil samples. Soil Biol Biochem 32:947–958

Rillig MC, Allen MF (1999) What is the role of arbuscular mycorrhizal fungi in plant-to-ecosystem responses to elevated atmospheric CO_2? Mycorrhiza 9:1–8

Rochette P, Angers DA, Flanagan LB (1999) Maize residue decomposition measurement using soil surface carbon dioxide fluxes and natural abundance of carbon-13. SSSA J 63:1385–1396

Rosier CL, Hoye AT, Rillig MC (2006) Glomalin-related soil protein: assessment of current detection and quantification tools. Soil Biol Biochem 38:2205–2211
Rovira P, Kurz-Besson C, Couteaux MM et al (2008) Changes in litter properties during decomposition: a study by differential thermogravimetry and scanning calorimetry. Soil Biol Biochem 40:172–185
Rühlmann J (1999) A new approach to estimating the pool of stable organic matter in soil using data from long-term field experiment. Plant Soil 213:149–160
Rumpel C, Koegel-Knabner I (2011) Deep soil organic matter-a key but poorly understood component of terrestrial C cycle. Plant Soil 338:143–158
Ryan MG, Law BE (2005) Interpreting, measuring, and modeling soil respiration. Biogeochemistry 73:3–27
Schenck zu Schweinsberg-Mickan M, Joergensen RG, Mueller T (2010) Fate of ^{13}C- and ^{15}N-labelled rhizodeposition of Lolium perenne as function of the distance to the root surface. Soil Biol Biochem 42:910–918
Schmidt MWI, Torn MS, Abiven S et al (2011) Persistence of soil organic matter as an ecosystem property. Nat Geosci 478:49–56
Schmitt A, Glaser B, Borken W et al (2008) Repeated freeze-thaw cycles changed organic matter quality in a temperate forest soil. J Plant Nut Soil Sci 171:707–718
Schulze E-D, Luyssaert S, Ciais P et al (2009) Importance of methane and nitrous oxide for Europe's terrestrial greenhouse-gas balance. CarboEuropeTeam 2:842–850
Seifert A-G, Trumbore S, Xu X et al (2011) Variable effects of labile carbon on the carbon use of different microbial groups in black slate degradation. Geochim Cosmochim Acta 75:2557–2570
Sierra J (1990) Analysis of soil nitrogen mineralization as estimated by exponential models. Soil Biol Biochem 22:1151–1153
Siewert C (2004) Rapid screening of soil properties using thermogravimetry. SSSA J 68:1656–1661
Silver WL, Miya RK (2001) Global patterns in root decomposition: comparisons of climate and litter quality effects. Oecologia 129:407–419
Smith P, Smith JU, Powlson DS et al (1997) A comparison of the performance of nine soil organic matter models using datasets from seven long-term experiments. Geoderma 81:153–225
Sollins P, Homann P, Caldwell BA (1996) Stabilization and destabilization of soil organic matter: mechanisms and controls. Geoderma 74:65–105
Solomon S, Qin D, Manning DAC et al (2007) Climate change 2007: the physical science basis – contribution of Working Group I to the fourth assessment report of the Intergovernmental Panel on climate change. Cambridge University Press, Cambridge
Swift MJ, Heal OW, Anderson JM (1979) Decomposition in terrestrial ecosystems. Blackwell Scientific Publications, Oxford
Teepe R, Brumme R, Beese F (2001) Nitrous oxide emissions from soil during freezing and thawing periods. Soil Biol Biochem 33:1269–1275
Thevenot M, Dignac M-F, Rumpel C (2010) Fate of lignins in soils: a review. Soil Biol Biochem 42:1200–1211
Tisdall JM, Oades JM (1982) Organic-matter and water-stable aggregates in soils. J Soil Sci 33:141–163
Treseder KK, Allen MF (2000) Mycorrhizal fungi have a potential role in soil carbon storage under elevated CO_2 and nitrogen deposition. New Phytol 147:189–200
Treseder KK, Turner KM (2007) Glomalin in ecosystems. Soil Sci Soc Am J 71:1257–1266
Trofymow JA, Moore TR, Titus B et al (2002) Rates of litter decomposition over 6 years in Canadian forests: influence of litter quality and climate. Can J For Res-Revue Canadienne De Recherche Forestiere 32:789–804
Trumbore S (2006) Carbon respired by terrestrial ecosystems – recent progress and challenges. Glob Chang Biol 12:141–153
Trumbore SE, Czimczik CI (2008) An uncertain future for soil carbon. Science 321:1455–1456
Valentini R, Matteucci G, Dolman AJ et al (2000) Respiration as the main determinant of carbon balance in European forests. Nature 404:861–865

Vanhala P, Karhu K, Tuomi M et al (2007) Old soil carbon is more temperature sensitive than the young in an agricultural field. Soil Biol Biochem 39:2967–2970

Vicca S, Janssens IA, Flessa H et al (2009) Temperature dependence of greenhouse gas emissions from three hydromorphic soils at different groundwater levels. Geobiology 7:465–476

von Lützow M, Kögel-Knabner I (2009) Temperature sensitivity of soil organic matter decomposition – what do we know? Biol Fert Soils 46:1–15

von Lützow M, Kogel-Knabner I, Ekschmitt K et al (2006) Stabilization of organic matter in temperate soils: mechanisms and their relevance under different soil conditions – a review. Eur J Soil Sci 57:426–445

Waldrop MP, Firestone MK (2004) Altered utilization patterns of young and old soil C by microorganisms caused by temperature shifts and N additions. Biogeochemistry 67:235–248

Wall DH, Bradford MA, St John MG et al (2008) Global decomposition experiment shows soil animal impacts on decomposition are climate-dependent. Glob Chang Biol 14:2661–2677

Wang Y, Hsieh Y-P (2002) Uncertainties and novel prospects in the study of the soil carbon dynamics. Chemosphere 49:791–804

Werth M, Kuzyakov Y (2008) Root-derived carbon in soil respiration and microbial biomass determined by ^{14}C and ^{13}C. Soil Biol Biochem 40:625–637

West TO, Post WM (2002) Soil organic carbon sequestration rates by tillage and crop rotation: a global data analysis. SSSA J 66:1930–1946

Wetterstedt JAM, Persson T, Agren GI (2010) Temperature sensitivity and substrate quality in soil organic matter decomposition: results of an incubation study with three substrates. Glob Chang Biol 16:1806–1819

Wiseman CLS, Puttmann W (2005) Soil organic carbon and its sorptive preservation in central Germany. Eur J Soil Sci 56:65–76

Wutzler T, Reichstein M (2008) Colimitation of decomposition by substrate and decomposers – a comparison of model formulations. Biogeosciences 5:749–759

Yoo G, Yang XM, Wander MM (2011) Influence of soil aggregation on SOC sequestration: a preliminary model of SOC protection by aggregate dynamics. Ecol Eng 37:487–495

Yuste JC, Nagy M, Janssens IA et al (2005) Soil respiration in a mixed temperate forest and its contribution to total ecosystem respiration. Tree Physiol 25:609–619

Zhang D, Hui D, Luo Y et al (2008) Rates of litter decomposition in terrestrial ecosystems: global patterns and controlling factors. J Plant Ecol-UK 1:85–93

Chapter 19
Soil Carbon and Nitrogen Interactions and Biosphere-Atmosphere Exchange of Nitrous Oxide and Methane

Klaus Butterbach-Bahl and Michael Dannenmann

Abstract Carbon (C) sequestration in terrestrial ecosystems and here specifically in soils is currently discussed as a potential strategy to contribute to reducing atmospheric carbon dioxide (CO_2) concentrations. However, increases in soil C stocks may also have adverse effects on the exchange of greenhouse gases (GHGs) between terrestrial ecosystems and the atmosphere. In view of the unprecedented perturbation of the global nitrogen (N) cycle, increases in soil C stocks and the ongoing saturation of terrestrial ecosystems with reactive forms of N (Nr) may result in a stimulation of soil nitrous oxide (N_2O) emissions. These largely unexplored ecosystem C-N interactions and their importance for biosphere-atmosphere GHG exchange need to better understood to finally assess the climate benefits of C sequestration in soils.

Keywords CN interactions • N_2O • CH_4 • Nitrification • Denitrification

---

K. Butterbach-Bahl (✉)
Karlsruhe Institute of Technology, Institute of Meteorology and Climate Research, Atmospheric Environmental Research (IMK-IFU), Kreuzeckbahnstr. 19, 82467 Garmisch-Partenkirchen, Germany
e-mail: klaus.butterbach-bahl@kit.edu

M. Dannenmann
Institute of Forest Botany and Tree Physiology, University of Freiburg, Georges-Koehler-Allee 53/54, 79110 Freiburg, Germany
e-mail: Michael.Dannenmann@kit.edu

R. Lal et al. (eds.), *Recarbonization of the Biosphere: Ecosystems and the Global Carbon Cycle*, DOI 10.1007/978-94-007-4159-1_19,

Abbreviations

NH_3	ammonia
NH_4^+	ammonium
C	Carbon
CO_2	carbon dioxide
N_2	dinitrogen
GHGs	greenhouse gases
CH_4	methane
NO_3^-	nitrate
NO	nitric oxide
N	nitrogen
N_2O	nitrous oxide
Nr	reactive forms of N
SOC	soil organic carbon pools
SOM	soil organic matter

19.1 Introduction

Decomposition of soil carbon (C) stocks following conversion of natural soils to agricultural soils and intensification of agricultural management has significantly contributed to a depletion of soil organic carbon pools (SOC) worldwide. Lal (2004) pointed out that following cultivation of natural soils 60% or up to 75% of the SOC pools may be lost in temperate or tropical regions, respectively. It has been estimated that since 1850 changes in land use may have resulted in a loss of carbon (C) from terrestrial ecosystems of about 156 Pg C (Houghton 2007), with losses of SOC due to mineralization amounting to about one-third or 52 ± 8 Pg C and losses due to erosion to 26 ± 9 Pg C (Lal 2004). In the last years the reversion of historical SOC losses via the intentional implementation of improved land management practices – e.g., higher inputs of residue C, adoption of less intensive cropping systems or temporary vegetative cover between agricultural crops – is discussed as a potential strategy to contribute to reducing atmospheric carbon dioxide (CO_2) concentrations (Smith et al. 2008; Conant et al. 2011). However, increasing SOC stocks will not only affect ecosystem C cycling, but also ecosystem nitrogen (N) cycling and soil microbial processes involved in the production and oxidation of other non-CO_2 greenhouse gases (GHGs), namely nitrous oxide (N_2O) and methane (CH_4). In addition, changes in global N cycling as driven by human activities need also to be considered while implementing strategies to increasing soil C sequestration, since additional N inputs and increased availability of soil C may alter the net balance of the biosphere atmosphere exchange of GHGs at the ecosystem as well as at the landscape scale. This chapter will therefore provide examples of how C-N interactions and changes in N trace gas exchange between terrestrial ecosystems and the atmosphere may potentially affect the climate benefits of C sequestration in soils.

19.1.1 Pertubation of the Global Nitrogen Cycling and Soil Carbon-to-Nitrogen Ratios

Human activities have not only resulted in a perturbation of the global C cycle, but even more significantly in an unprecedented change of global N cycling. Following the introduction of industrial ammonia (NH_3) production via the Haber-Bosch process N fertilizers have increasingly been used to meet the demand of the growing world population for food and feed (Erisman et al. 2009). Furthermore, N fixing crops are increasingly cultivated also adding N to the biosphere. Finally, reactive N (Nr) compounds are also created incidentally during combustion processes. The human perturbation of the natural N cycle has led to an unprecedented accumulation of Nr in the biosphere (Erisman et al. 2009). For 2005 it is estimated that due to human activities, 187 Tg of N have been added to the global biosphere, which approximately relates to a doubling of global N cycling (Galloway et al. 2008). Not all of this N is ultimately denitrified, but a substantial, though hardly to quantify amount of Nr may remain in terrestrial ecosystems and here mainly in the soil. Galloway et al. (2004) estimated that from the 0.268 Pg N which were added annually during the 1990s to terrestrial ecosystems approximately 0.060 Pg N year^{-1} may be stored in terrestrial ecosystems, i.e., 20–25% of the total input of Nr to terrestrial ecosystems. In contrast, it is assumed that in pre-industrial times Nr inputs and outputs were balanced. In conclusion, it is very much likely to assume that Nr is accumulating in the global biosphere and that Nr stocks of all terrestrial ecosystems are increasing.

What is the consequence of this? Batjes (1996) estimated that global soil C stocks for 0–30 cm are in the range of 906–969 Pg C and that to 100 cm soil depth 2,150–2,300 Pg C may be stored. For soil N stocks the respective numbers are: =63–67 Pg N for the 0–30 cm soil layer and 133–140 Pg N for 0–100 cm soil layer, respectively. Following Galloway et al. (2004) that from the 0.268 Pg N year^{-1} added to terrestrial ecosystems 0.060 Pg N year^{-1} are remaining in ecosystems, and assuming that the major fate is its storage in soils this will lead to a narrowing of soil C:N ratios (Fig. 19.1). If soil C stocks are not increasing and 10% of the additional Nr is stored in the topsoil (0–30 cm) the global average soil C:N ratio may narrow for the first 30 cm of soil from 14.4 to 13.8 in the coming 100 years (Fig. 19.1). However, soil C:N ratios may further widen if soils get re-carbonized, whereas if soil C loss due to cultivation of soils continues the narrowing of soil C:N ratio may accelerate (Fig. 19.1).

Significant changes in forest floor and mineral topsoil soil C:N ratios following simulated increased atmospheric N deposition has been observed e.g., in a study by Andersson et al. (2002) for a series of *Picea abies* (L.) stands in Sweden or by Lovett and Goodale (2011) for a mixed oak wood stand in Northeastern US. In a report for the German Environmental Agency, Kiese et al. (2009) provided data for a pine forest in the Northeastern German Lowlands where due to increased atmospheric N deposition forest floor C:N ratios decreased within 20 years from 31 to 27 and in the top 10 cm of the mineral soil from 26 to 23,

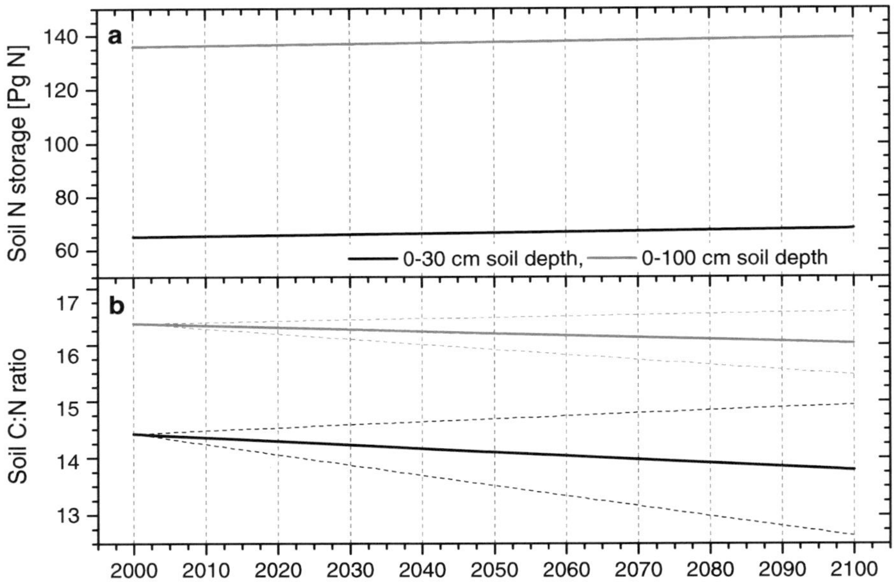

Fig. 19.1 Changes in global soil N stocks from 2000 to 2100 assuming that soil C stocks are remaining unchanged, anthropogenic N inputs are remaining at the same level as in the 1990s and that total N additions to terrestrial ecosystems are equaling 0.268 Pg N $year^{-1}$ (Galloway et al. 2004) and that about 10% or 0.030 Pg $year^{-1}$ of the N input to terrestrial ecosystems are either stored in the topsoil (0–30 cm) or in soils to 100 cm depth. (**a**) Shows changes in total soil N stocks, whereas (**b**) displays resulting changes in the soil C:N ratio either for 0–30 cm or 0–100 cm soil depth, respectively. Data for global soil N and soil C stocks were taken from Batjes (1996). *Dashed lines* in the (**b**) are indicating changes in the soil C:N ratio if C loss due to cultivation continues or if soils are re-carbonized. For both scenarios changes in soil C stocks (either 0–30 cm or 0–100 cm) at a rate of 78 Pg C 100 $year^{-1}$ were assumed

respectively (Fig. 19.2). These observed decreases could be simulated well by a biogeochemical model based on reported data of atmospheric N deposition at the forest site (Kiese et al. 2009).

19.1.2 Coupling of Carbon and Nitrogen Turnover Processes

As pointed out earlier the re-carbonization of the biosphere has been highlighted recently as an important approach to mitigate the increase and to consolidate atmospheric CO_2 concentrations by converting the atmospheric CO_2 into biotic or abiotic C sequestered in vegetation or soil pools. Due to the tight-coupling of C and N cycles in soils and ecosystems (Fig. 19.3) – e.g., mineralization of organic matter does not only lead to break down of C substrates and the release of CO_2 but also to the liberation of inorganic N – anthropogenic induced changes in rates of biotic C-sequestration in terrestrial ecosystems will directly affect N turnover processes

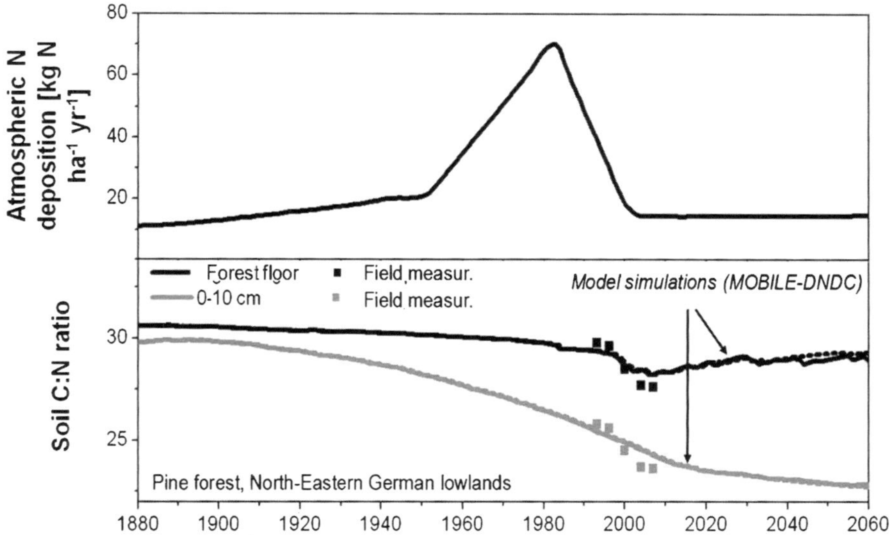

Fig. 19.2 Observed changes in forest floor and soil (0–10 cm) C:N ratios (*points lower graph*) of a pine stand in the Northeastern German Lowlands as a response to increased atmospheric N deposition (*upper graph*). The *lower graph* also displays simulated changes in soil C:N ratios using the biogeochemical model MOBILE-DNDC, and historic and predicted climate and N deposition data as drivers. For further details see Kiese et al. (2009)

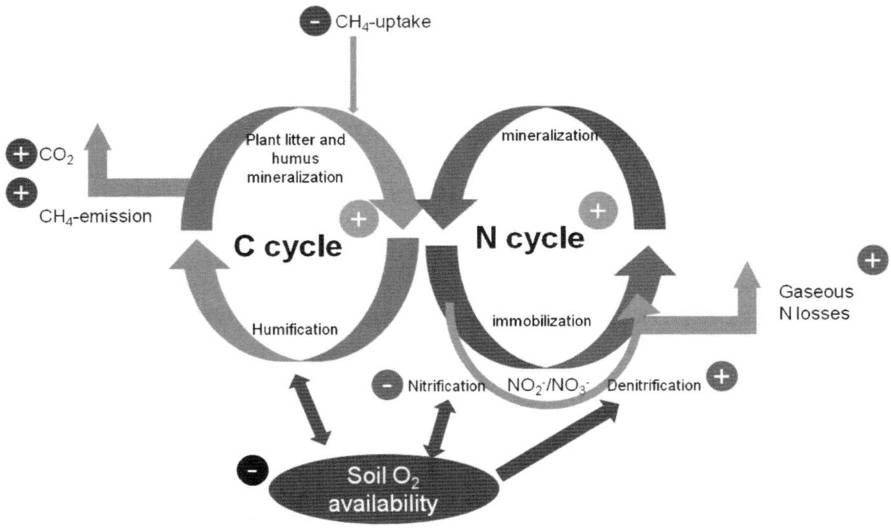

Fig. 19.3 Coupling of ecosystem C and N cycling and possible consequences of an acceleration of both cycles with regard to soil-atmosphere C and N exchange. +: enhancement –: attenuation (Figure modified following Blagodatsky et al. 2011)

in soils, and thus, also the biosphere–atmosphere exchange of gaseous N and C compounds (Li et al. 2005).

Consequently, increased rates of C sequestration in soils are likely to accelerate not only C but also N turnover, in particular when agricultural N fertilizer use and atmospheric N deposition in many regions globally remain at the present high levels. Increased C and N cycling may result in increased soil anaerobiosis, due to increased microbial oxygen consumption during mineralization of organic matter. This may lead to increases in soil anaerobiosis, which may result in elevated soil N_2O emissions and a weakening of the sink strengths of upland soils for atmospheric CH_4 (Butterbach-Bahl et al. 2011), due to a stimulation of denitrification and methanogenesis versus plant and microbial N immobilization or CH_4 oxidation in soils.

19.2 Soil Organic Carbon and Magnitude of Nitrous Oxide Emissions

In their literature review on C:N interactions, Li et al. (2005) revealed a very significant positive relationship between SOC content of soils and the magnitude of N_2O emissions (Fig. 19.4). There are several reasons why SOC contents are a major control of soil N_2O emissions. First, organic C is the basis for any biotic mineralization process and with increasing SOC contents – except for conditions of strict

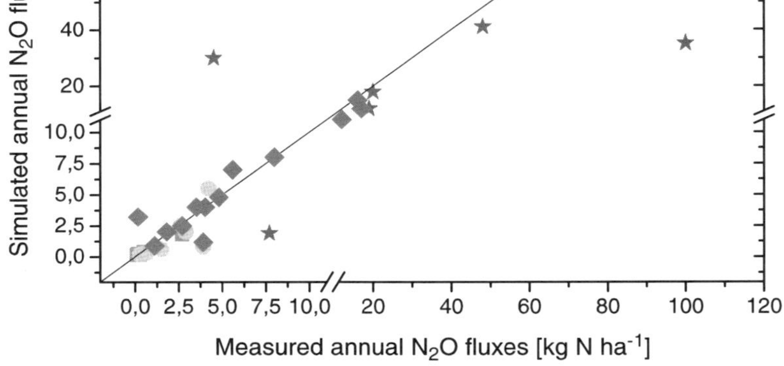

Fig. 19.4 Comparison of observed and simulated annual N_2O emissions from agricultural soils. Symbols indicate soil organic carbon (*SOC*) content in soil in kg SOC kg^{-1} soil. A general trend to higher N_2O emissions for higher SOC is apparent in both field and simulated results (Figure adapted from Li et al. 2005)

anaerobiosis like in peatlands or if soil moisture/temperature is hampering microbial activity like in arid and semi-arid regions – mineralization activities are increasing too. The oxygen consumption by microbial mineralization may result in an increased frequency and increased spatial extend of anaerobic microsites, where mineralized and oxidized inorganic Nr may be transformed by nitrification and denitrification processes into N_2O. Second, the availability of labile C substrates is a prerequisite for denitrification besides the availability of oxidized Nr substrates. Simultaneous availability of sufficient concentrations of readily available C and Nr substrates for denitrification is more likely to occur in soils rich in organic matter, while in soils with low organic C content denitrification activity and, thus, also N_2O formation, may even be hampered due to missing C substrates. Such a situation is e.g. occurring in many arable soils in the North China Plain, one of the key regions of food production in China. For this region, several reports indicate, that soil N_2O emissions are rather low at mineral fertilizer application of up to 600 kg N ha^{-1} $year^{-1}$ with annual cumulative emission rates of <4 kg N_2O-N ha^{-1} $year^{-1}$ (Ju et al. 2011; Liu et al. 2010). Also the N fertilizer N_2O emission factor of 0.5–0.7% of added fertilizer (Ding et al. 2007) for wheat-maize (*Triticum spp.-Zea mays* L.) rotations in the North China Plain is significantly below the global average of 1% as suggested by IPCC (2006). The main reason for low N_2O emissions seems to be the low C content of the soils and not the availability of nitrate (NO_3^-), which accumulates in the soil. Most likely C substrate supply limits denitrification and, thus, also N_2O production by denitrification. If these soils are re-carbonized, the large stocks of inorganic Nr in surface and subsurface soils may largely get denitrified, thereby potentially releasing significant amounts of N_2O as by-products of denitrification. Comparable situations, i.e., that the availability of readily available C is limiting denitrification and N_2O formation has been described for agro-ecosystems as well as for natural ecosystems (Morley and Baggs 2010; Kammann et al. 2008; Butterbach-Bahl and Dannenmann 2011).

19.3 Soil C:N Ratio and Microbial N Turnover Processes

The major long-term sink of atmospheric Nr input into ecosystems is soil organic matter (SOM), as was shown in many ^{15}N-tracing experiments (Morier Jaquet et al. 2008; Nadelhoffer et al. 1999; Tietema et al. 1998). Since Nr deposition via atmospheric pathways is often exceeding plant N demand in natural and semi-natural ecosystems, the atmospheric N input has led to N saturation of natural and semi-natural ecosystems, resulting in a decrease in the soil C:N ratio both in the organic layer and in the mineral soil of natural and semi-natural ecosystems within large regions of Europe, Asia and North America (Galloway et al. 2004; Corré et al. 2007; Velthoff et al. 2011; Butterbach-Bahl et al. 2011). Even if atmospheric Nr input in terrestrial ecosystems may not further increase in future, the C:N ratio in mineral topsoils will further decrease (Fig. 19.1). The N saturation and associated decline in soil C:N ratios involves detrimental effects such as soil eutrophication, -acidification,

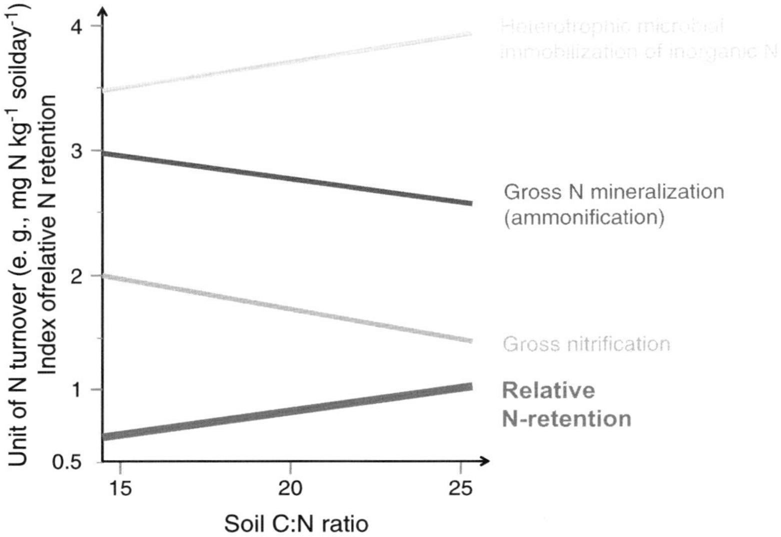

Fig. 19.5 Schematic representation of the relationships between soil C:N ratio and gross rates of inorganic N production (ammonification, nitrification) and consumption (microbial immobilization). When C:N ratios narrow below a value of approximately 25, e.g. as a consequence of atmospheric Nr deposition, gross ammonification as well as nitrification will increase, while heterotrophic microbial immobilization of inorganic Nr will decrease. Thus, the index of microbial Nr retention in the soil (i.e., the ratio between gross microbial immobilization and total gross inorganic N production) is decreasing, indicating an opening of the N cycle and increased N_2O emissions. Mechanisms of impacts of the C:N ratio on single N cycle processes are provided in the text

reduced plant biodiversity and reduction in tree population, while it has a large potential to increase C sequestration in plant biomass and eventually also in soil (Snyder et al. 2009; Velthoff et al. 2011).

Besides these effects, the C:N ratio is – as well as total C and N content and -availability – a major driver of soil microbial N turnover processes. This is not only the case for processes such as ammonification, nitrification, denitrification and microbial immobilization of inorganic N, but for the relative importance or even dominance of single processes such as nitrification over competing processes such as microbial ammonium (NH_4^+) immobilization. In particular, the relative importance of nitrification and denitrification over the processes of microbial immobilization of inorganic Nr (and mycorrhizal/plant uptake) determine whether there are large or small losses of N_2O from soil (Tietema and Wessel 1992; Stockdale et al. 2002; Butterbach-Bahl et al. 2011). By regulating this delicate balance, the C:N ratio plays a crucial role in the characterization of the N cycle, i.e., if it is open (high N loss along gaseous and hydrological pathways) or closed (internal N cycling facilitates ecosystem N retention). In the following section, the effect of the C:N ratio on ecosystem Nr retention and loss is explained at the level of the single N cycle processes (Fig. 19.5).

The lower the soil C:N ratio, the larger the N yield during depolymerization and mineralization of a given amount of SOM. Thus, ammonification will increase with decreasing C:N ratio (Frankenberger and Abdelmagid 1985; Hart et al. 1994). Subsequently, the balance of partitioning of NH_4^+ to the potentially competing microbial processes of heterotrophic immobilization into biomass of free living microorganisms and autotrophic nitrification (i.e. conversion to NO_3), as well as plant or mycorrhizal uptake of NH_4^+, determines whether N is retained in the ecosystem or subject to increased risk of N loss via gaseous or hydrological pathways.

Due to the long residence time of Nr in plant biomass, plant Nr uptake results in persistent ecosystem N retention until disturbances such as harvest or fire. However, plants are in general poor competitors for Nr against microbes at the process level (Rennenberg et al. 2009). Even in fast growing spruce (*Picea* spp) forests, plant N uptake may at least be one order of magnitude lower than gross N mineralization, thus consuming only for a minor part of inorganic N produced in soil (Kreutzer et al. 2009).

Heterotrophic NH_4^+ assimilating microorganisms generally tend to outcompete autotrophic nitrifiers (Verhagen and Laanbroek 1991; Hart et al. 1994; Verhagen et al. 1995; Booth et al. 2005). However, this may not necessarily be the case in agricultural soils, when there is no NH_4^+ limitation (Burger and Jackson 2003). The availability of labile C compounds is the major controller of heterotrophic microbial immobilization of NH_4^+-N (Woodmansee and Duncan 1980; Compton and Boone 2002; Booth et al. 2005; Accoe et al. 2004, 2005). At decreasing C:N ratios, microbial N uptake may be C-limited, leaving more NH_4^+ available to the process of autotrophic nitrification, thus opening pathways to potential N loss such as N_2O emission to the atmosphere.

In contrast to nitrification, the incorporation of NH_4^+-N into microbial cell walls, characterized by short residence times, leads to retention of N in the ecosystem. This may be of short-term, when there is dieback and remineralization of microbial residues leading to a rapid and repeated re-allocation of mineral N to the soil NH_4^+ pool. However, despite the rapid turnover of microbial biomass in soil, microbial immobilization can also lead to long-term stabilization of N. Among the processes are microbially mediated long-term N stabilization by the accumulation of bacterial-derived residues such as proteins, amino acids and amino sugars on the surface of clay minerals (Sollins et al. 2006). Decreasing soil C:N ratios could also reduce the long-term N stabilization in soil via reducing microbial immobilization, which could contribute to increased risk of N loss. Hence, the C:N ratio is of crucial importance for the balance of NH_4^+ partitioning to processes promoting either N retention or N loss. Beside its effect on nitrification at the substrate level, the C:N ratio is also affecting the balance of consumption of the end product of nitrification. Specifically, heterotrophic microbial NO_3^- immobilization will also be dependent on C availability, i.e., decreasing with decreasing C:N ratio, promoting accumulation of soil NO_3^-, which would be available for denitrification and associated N_2O production and –loss.

Via its effects on ammonification, nitrification as well as on microbial immobilization, the soil C:N ratio is a crucial determinant of the microbial Nr retention capacity (Booth et al. 2005). The effect of the soil C:N ratio on microbial N retention was shown in some field studies, e.g., for a Rendzic Leptosol in a mountainous beech (*Fagus sylvatica* L.) forest in Southern Germany. In this study, variations in soil C:N ratio triggered by different microclimate and management were found to be correlated with microbial N retention, i.e., decreased C:N ratios lead to decreased relative importance of microbial N retention (Dannenmann et al. 2006, 2007).

By promoting nitrification, the decreasing C:N ratio may also promote the formation of N_2O as a by-product. Besides its effect on nitrification, also the formation of N_2O as a facultative end-product of denitrification may be increased, when a narrow C:N ratio is leading to NO_3^- accumulation in soil, promoting denitrification in general, while impairing the last step of denitrification, i.e. the conversion of N_2O to the end product dinitrogen (N_2) catalyzed by the enzyme N_2O reductase (e.g., Groffman et al. 2006; Wang et al. 2011). Thus, narrowing C:N ratios favour both N_2O production by nitrification and denitrification. However, denitrification requires also labile C substrates. In this context, increased root exudation of monomeric C compounds, promoted by generally improved plant growth as a consequence of atmospheric Nr deposition and, in particular, due to increased atmospheric CO_2 levels, may further promote denitrification (Kammann et al. 2008; Butterbach-Bahl and Dannenmann 2011). Denitrification may be further promoted by increased soil respirations and plant water use efficiency, both increasing the soil anaerobic volume in a changing climate (Butterbach-Bahl and Dannenmann 2011). Hence, in a changing climate, NO_3^- accumulation in soil as a consequence of Nr deposition and narrowed C:N ratios may increasingly meet environmental conditions favourable for denitrification, bearing the potential for high N_2O emissions from soil.

19.4 Soil Carbon-to-Nitrogen Ratios and Soil Nitrogen Trace Gas Emissions

The most important soil microbial processes involved in the formation of N_2O and nitric oxide (NO) are nitrification and denitrification. The magnitude of both processes will strongly depend on the availability of Nr substrates. For autotrophic nitrification NH_3 is the relevant Nr substrate, while for denitrification inorganic Nr oxides as well as labile C substrates are needed. Heterotrophic nitrifiers may use also organic N compounds instead of NH_3 to produce NO_3^- and produce as side products as well NO and N_2O. Therefore, a close link between Nr and C availability and N_2O emissions can be expected. Specifically, close relationship between soil C:N ratios and magnitude of N trace gas emissions have been shown for forest soils in various climate zones (Fig. 19.6). Based on soil emission measurements at various tropical rain forest sites in Queensland, Australia, and a detailed analysis of soil properties, Breuer et al. (2000) showed that topsoil C:N ratios are a significant predictor for understanding the spatial variability of soil N_2O emissions. Comparable results were also reported by Pilegaard et al. (2006) while analyzing environmental controlling variables of soil N_2O and NO emissions for 15 different forest ecosystem sites across Europe (Fig. 19.6). At least for N_2O a significant relationship was found

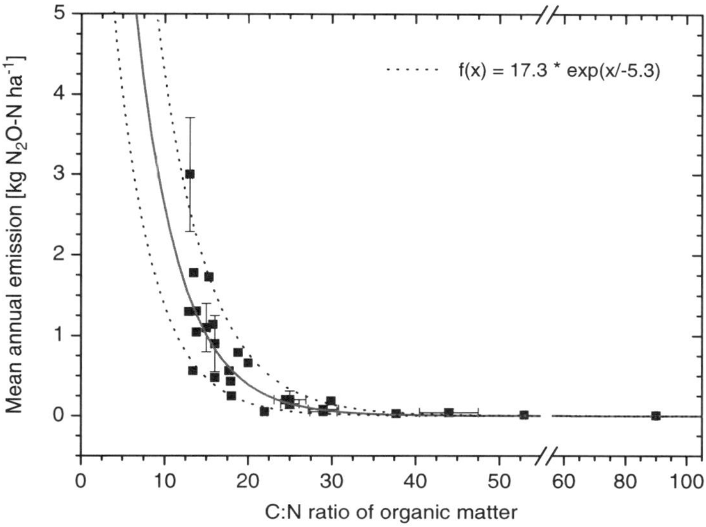

Fig. 19.6 Correlation between topsoil C:N ratio and mean annual nitrous oxide emissions for different forest sites across Europe (Data were taken from Table 1 of Klemedtsson et al. (2005) and Tables 2 and 5 from Pilegaard et al. (2006))

with exponential increasing soil N_2O emission rates with decreasing soil C:N ratios in the topsoil. The study of Klemedtsson et al. (2005) on N_2O emissions from drained forested histosols in Sweden shows that a strong negative relationship between N_2O emissions and soil C:N ratios exist. This relationship has been successfully used to estimate emissions at other sites in Finland and Germany.

The exponential increase in forest soil N_2O emissions with decreasing C:N ratios is in good agreement with the observation that the risk for elevated NO_3^- leaching from forest soils is increasing exponentially if soil C:N ratios are smaller than 25 (Gundersen et al. 2006; Butterbach-Bahl et al. 2011). In both cases the ecosystem N retention capacities are likely to be exceeded and losses of N to the hydrosphere and atmosphere are increasing. With regard to the re-carbonization of soils this clearly indicates, that the N status of the ecosystem needs to be explored, e.g., by recording soil C:N ratios as indicators – and it will be necessary to consider total ecosystem N inputs either by atmospheric N-input and/or by atmospheric-N-input plus organic or inorganic fertilizers or even weathering of bedrock N (Morford et al. 2011).

19.5 Nitrogen Availability and Ecosystem Carbon Sequestration

Input of Nr to terrestrial ecosystems is also a measure to increase ecosystem C sequestration. For agricultural systems Nr has been reported to play an important role in soil C storage either by promoting crop dry matter production and/ or by chemically stabilizing C in the soil (Snyder et al. 2009). However, though several studies have shown that nitrogen fertilization results in higher levels of soil C over time (Paustian et al. 1992; Wilts et al. 2004) the stimulating effect may be rather small (about 2%) as shown in the meta-data analysis by Liu and Greaver (2009). Besides agricultural ecosystems, several recent studies have shown that Nr deposition to European forest ecosystems has resulted in increased forest growth and C sequestration (De Vries et al. 2006, 2011; Schulze et al. 2010). With regard to the net climate balance of atmospheric Nr deposition, C sequestration was thereby largely outweighing reductions in soil CH_4 uptake or stimulations in soil N_2O emissions (De Vries et al. 2011). However, long-term assessments are still missing and the stimulation of forest growth may cease following a few decades while stimulative effects on soil N_2O emissions may persist over longer periods of time.

19.6 Conclusions

Re-carbonization of terrestrial ecosystems is likely to be a sustainable strategy to stabilize or even lowering of atmospheric CO_2 concentrations. Nevertheless, it still needs to be assessed if specifically an increase in soil C stocks will indeed lead to a net reduction of GHG emissions from terrestrial ecosystems, i.e., if benefits due to C sequestration are not specifically outweighed by the stimulation of N_2O emissions from soils. The reason for this is the unprecedented perturbation of the global N cycle. If current rates of direct (fertilization) and indirect reactive N inputs (mainly atmospheric N deposition) continues, C:N ratios in soils may further narrow even if additional C is sequestered in soils. Both, C accumulation as well as the narrowing of soil C:N ratios may result in a stimulation of microbial N turnover and associated emissions of N_2O, which has as a global warming potential of about 300 over a 100-year period as compared to CO_2. Therefore, a thorough quantification of the biogeochemical interaction of C and N needs to be incorporated into assessment frameworks to accurately evaluate the climate benefits of a re-carbonization of the terrestrial environment.

Acknowledgments We thank the German Science Foundation (Deutsche Forschungsgemeinschaft) for generously funding the work of Michael Dannenmann through contract number DA 1217/2-1. This work is a contribution to the EU funded integrated project NitroEurope.

References

Accoe F, Boeckx P, Busschaert J et al (2004) Gross N transformation rates and net N mineralization rates related to the C and N contents of soil organic matter fractions in grassland soils of different age. Soil Biol Biochem 36:2075–2087

Accoe F, Boeckx P, Videla X et al (2005) Estimation of gross nitrogen transformations and nitrogen retention in grassland soils using FLUAZ. Soil Sci Soc Am J 69:1967–1976

Andersson P, Berggren D, Nilsson I (2002) Indices for nitrogen status and nitrate leaching from Norway spruce (*Picea abies* (L.) Karst.) stands in Sweden. For Ecol Manage 157:39–53
Batjes NH (1996) Total carbon and nitrogen in the soils of the world. Eur J Soil Sci 47:151–163
Blagodatsky S, Grote R, Kiese R et al (2011) Modelling of microbial carbon and nitrogen turnover in soil with special emphasis on N-trace gases emission. Plant Soil 346:297–330
Booth MS, Stark JM, Rastetter E (2005) Controls on nitrogen cycling in terrestrial ecosystems: a synthetic analysis of literature data. Ecol Monogr 75:139–157
Breuer L, Papen H, Butterbach-Bahl K (2000) N_2O-emission from tropical forest soils of Australia. J Geophys Res 105:26353–26378
Burger M, Jackson LE (2003) Microbial immobilization of ammonium and nitrate in relation to ammonification and nitrification rates in organic and conventional cropping systems. Soil Biol Biochem 35:29–36
Butterbach-Bahl K, Dannenmann M (2011) Denitrification and associated soil N_2O emissions due to agricultural activities in a changing climate. Curr Opin Environ Sci 3:389–395
Butterbach-Bahl K, Gundersen P, Ambus P et al (2011) Nitrogen processes in terrestrial ecosystems. In: Sutton MA, Howard CM, Erisman JW et al (eds) The European nitrogen assessment: sources effects, and policy perspectives. Cambridge University Press, Cambridge
Compton JE, Boone RD (2002) Soil nitrogen transformations and the role of light fraction organic matter in forest soils. Soil Biol Biochem 34:933–943
Conant RT, Ogle SM, Paul EA et al (2011) Measuring and monitoring soil organic carbon stocks in agricultural lands for climate mitigation. Front Ecol Environ 9:169–173
Corré M, Brumme R, Veldkamp E et al (2007) Changes in nitrogen cycling and retention processes in soils under spruce forests along a nitrogen enrichment gradient in Germany. Glob Change Biol 13:1509–1527
Dannenmann M, Gasche R, Ledebuhr A et al (2006) Effects of forest management on soil N cycling in beech forests stocking on calcareous soils. Plant Soil 287:279–300
Dannenmann M, Gasche R, Papen H (2007) Nitrogen turnover and N_2O production in the forest floor of beech stands as influenced by forest management. J Plant Nutr Soil Sci 170:134–144
De Vries W, Reinds GJ, Gundersen P et al (2006) The impact of nitrogen deposition on carbon sequestration in European forests and forest soils. Glob Change Biol 12:1151–1173
De Vries W, Kros J, Reinds GJ et al (2011) Quantifying impacts of nitrogen use in European agriculture on global warming potential. Curr Opin Environ Sci 3:291–302
Ding WX, Meng L, Cai ZC, Han FX (2007) Effects of long-term amendment of organic manure and nitrogen fertilizer on nitrous oxide emission in a sandy loam soil. J Environ Sci China 19:185–193
Erisman JW, Sutton MA, Galloway J et al (2009) How a century of ammonia synthesis changed the world. Nat Geosci 1:636–639
Frankenberger WT, Abdelmagid HM (1985) Kinetic parameters of nitrogen mineralization rate of leguminous crops incorporated into soil. Plant Soil 87:257–271
Galloway JN, Dentener FJ, Capone DG et al (2004) Nitrogen cycles: past, present, future. Biogeochemistry 70:153–226
Galloway JN, Townsend AR, Erisman JW et al (2008) Transformation of the nitrogen cycle: recent trends, questions, and potential solutions. Science 320:889–892
Groffman PM, Altabet MA, Böhlke JK et al (2006) Methods for measuring denitrification: diverse approaches to a difficult problem. Ecol Appl 16:2091–2122
Gundersen P, Schmidt IK, Raulund-Rasmussen K (2006) Leaching of nitrate from temperate forests – effects of air pollution and forest management. Environ Rev 14:1–57
Hart S, Nason GE, Myrold DD, Perry DA (1994) Dynamics of gross nitrogen transformations in an old-growth forest: the carbon connection. Ecology 75:880–891
Houghton RA (2007) Balancing the global carbon budget. Annu Rev Earth Planet Sci 35:313–347
IPCC (2006) IPCC Guidelines for National Greenhouse Gas Inventories, prepared by the National Greenhouse Gas Inventories Programme, IGES, Hayama, Japan
Ju X, Lu X, Gao Z et al (2011) Processes and factors controlling N_2O production in an intensively managed low carbon calcareous soil under sub-humid monsoon conditions. Environ Pollut 159:1007–1016

Kammann C, Müller C, Grünhage L et al (2008) Elevated CO_2 stimulates N_2O emissions in permanent grassland. Soil Biol Biochem 40:2194–2205
Kiese R, Wochele S, Grote R et al (2009) Modellierung und kartierung räumlich differenzierter wirkungen von stickstoffeinträgen in Ökosysteme im rahmen der UNECE-Luftreinhaltekonvention. Final report for the "FuE-Vorhaben FKZ 205 85 239", German Environmental Protection Agency (Umweltbundesamt), Dessau, Germany
Klemedtsson L, Von Arnold K, Weslien P et al (2005) Soil C:N ratio as a scalar parameter to predict nitrous oxide emissions. Glob Change Biol 11:1142–1147
Kreutzer K, Butterbach-Bahl K, Rennenberg H et al (2009) The complete nitrogen cycle of a N-saturated spruce forest ecosystem. Plant Biol 11:694–700
Lal R (2004) Soil carbon sequestration impacts on global climate change and food security. Science 304:1623–1627
Li C, Frolking S, Butterbach-Bahl K (2005) Carbon sequestration can increase nitrous oxide emissions. Clim Change 72:321–338
Liu L, Greaver TL (2009) Nitrogen addition stimulates emissions of biogenic greenhouse gases in terrestrial and wetland ecosystems. Ecol Lett 12:1103–1117
Liu C, Zheng X, Zhou Z et al (2010) Nitrous oxide and nitric oxide emissions from an irrigated cotton field in Northern China. Plant Soil 1–2:123–134
Lovett GM, Goodale CL (2011) A new conceptual model of nitrogen saturation based on experimental nitrogen addition to an oak forest. Ecosystems 14:615–631
Morford SL, Houlton BZ, Dahlgren RA (2011) Increased forest ecosystem carbon and nitrogen storage from nitrogen rich bedrock. Nature 477:78–81
Morier Jaquet I, Guenat C, Siegwolf R et al (2008) Dynamics of atmospheric nitrogen deposition in a temperate calcareous forest soil. J Environ Qual 37:2012–2021
Morley N, Baggs EM (2010) Carbon and oxygen controls on N_2O and N_2 production during nitrate reduction. Soil Biol Biochem 42:1864–1871
Nadelhoffer KJ, Downs MR, Fry B (1999) Sinks for ^{15}N-enriched additions to an oak forest and a red pine plantation. Ecol Appl 9:72–86
Paustian K, Parton WJ, Persson J (1992) Modeling soil organic matter in organic amended and N-fertilized long-term plots. Soil Sci Soc Am J 56:476–488
Pilegaard K, Skiba U, Ambus P et al (2006) Factors controlling regional differences in forest soil emission of nitrogen oxides (NO and N_2O). Biogeosciences 3:615–661
Rennenberg H, Dannenmann M, Gessler A et al (2009) Nitrogen balance in forest soils: nutritional limitation of plants under climate change stresses. Plant Biol 11:4–23
Schulze ED, Ciais P, Luyssaert S et al (2010) The European carbon balance, part 4: integration of carbon and other trace-gas fluxes. Glob Change Biol 16:1451–1469
Smith P, Martino D, Cai Z et al (2008) Greenhouse gas mitigation in agriculture. Philos Trans R Soc Lond B Biol Sci 363:789–813
Snyder CS, Bruulsema TW, Jensen TL et al (2009) Review of greenhouse gas emissions from crop production systems and fertilizer management effects. Agric Ecosyst Environ 133:247–266
Sollins P, Swanston C, Kleber M et al (2006) Organic C and N stabilization in a forest soil: evidence from sequential density fractionation. Soil Biol Biochem 38:3313–3324
Stockdale E, Hatch DJ, Murphy DV et al (2002) Verifying the nitrification to immobilization ratio (N/I) as a key determinant of potential nitrate loss in grassland and arable soils. Agronomie 22:831–838
Tietema A, Wessel WW (1992) Gross nitrogen transformations in the organic layer of acid forest ecosystems subjected to increased atmospheric nitrogen input. Soil Biol Biochem 24:943–950
Tietema A, Emmett B, Gundersen P et al (1998) The fate of ^{15}N labelled nitrogen deposition in coniferous forest ecosystems. For Ecol Manage 101:19–27
Velthoff G, Barot S, Bloem J et al (2011) Nitrogen as a threat to European soil quality. In: Sutton MA, Howard CM, Erisman JW et al (eds) The European nitrogen assessment: sources effects, and policy perspectives. Cambridge University Press, Cambridge

Verhagen FJM, Laanbroek HJ (1991) Competition for ammonium between nitrifying and heterotrophic bacteria in dual energy-limited chemostats. Appl Environ Microbiol 57:3255–3263

Verhagen FJM, Laanbroek HJ, Woldendorp JW (1995) Competition for ammonium between plant roots and nitrifying and heterotrophic bacteria and the effects of protozoan grazing. Plant Soil 170:241–250

Wang R, Willibald G, Feng Q et al (2011) Measurement of N_2, N_2O, NO, and CO_2 emissions from soil with the gas-flow-soil-core technique. Environ Sci Technol 45:6066–6072

Wilts AR, Reicosky DC, Allmaras RR et al (2004) Long-term corn residue effects: harvest alternatives, soil carbon turnover, and root-derived carbon. Soil Sci Soc Am J 68:1342–1351

Woodmansee RG, Duncan DA (1980) Nitrogen and phosphorus dynamics and budgets in annual grasslands. Ecology 61:893–904

Chapter 20
Carbon Storage and Sequestration in Subsoil Horizons: Knowledge, Gaps and Potentials

Cornelia Rumpel, Abad Chabbi, and Bernd Marschner

Abstract Carbon (C) sequestered in subsoils generally contributes to more than half of the total stocks within a soil profile. In contrast to topsoil, organic matter (OM) stored in subsoil horizons is characterised by high mean residence times. While the mechanisms and controls of OM stabilisation in topsoils are fairly well understood, processes and dynamics of subsoil OM stabilisation appear to be controlled by other factors. We summarize the available data on C quantities, chemical composition as well as decomposition and stabilisation dynamics in subsoils and discuss the relevance of these processes for longterm carbon storage. Moreover, the importance of spatial dist ribution of SOM and its degraders is addressed. The objective of this chapter was to discuss the controls of carbon sequestration in subsoil horizons in order to explore the possibility to increase soil carbon stocks by carbonising subsoil horizons. Subsoil C is characterised by much higher solubility after destruction of the mineral phase compared to topsoil C suggesting that much of it consists of small molecules stabilised by interaction with the mineral phase.

C. Rumpel
CNRS, Laboratoire de Biogéochimie et Ecologie des Milieux Continentaux, UMR 7618 (UPMC-CNRS-UPEC-IRD-AgroParisTech)
e-mail: cornelia.rumpel@grignon.inra.fr

A. Chabbi (✉)
INRA, Laboratoire de Biogéochimie et Ecologie des Milieux Continentaux, UMR 7618 (UPMC-CNRS-UPEC-IRD-AgroParisTech)

INRA Unité de Recherche Pluridisciplinaire Prairies et Plantes Fourragères UR P3F, Centre de Recherche Poitou- Charentes, INRA, Le Chêne – RD 150, BP 6, 86600 Lusignan, France
e-mail: Abad.chabbi@lusignan.inra.fr

B. Marschner
Geographisches Institut Bodenkunde/Bodenoekologie, Ruhr-Universitaet Bochum, Universitätsstr. 150, D-44801 Bochum, Germany
e-mail: bernd.marschner@rub.de

R. Lal et al. (eds.), *Recarbonization of the Biosphere: Ecosystems and the Global Carbon Cycle*, DOI 10.1007/978-94-007-4159-1_20,

The chemical composition of OM stabilised by mineral interactions in subsoil horizons is different from those of topsoils. Precursors of this C may be root-derived and microbially processed organic matter as well as organic matter transported in dissolved form or by bioturbation. The reports about the environmental controls of subsoil OM degradation are conflicting, most likely due to site-specificity. The lower dynamics of temperature and soil moisture in subsoils may thus enhance or reduce OM mineralization, while low nutrient availability is a more common limiting factor. Spatial distribution of OM may determine the likelihood of its stabilisation at long time scales, which may be most related to absence of energy-rich material needed for decomposition. Therefore a suitable strategy for increasing C stocks in deep horizons may be the addition of highly stable OM such as biochar or highly aliphatic material.

Keywords Sub-soil • Organic matter stabilization • Long-term carbon storage • Root-deprived carbon • Biochar • Aliphatic materials • Dissolved organic matter • Radiocarbon age • C:N ratio • ^{14}C • Carbon sequestration potential • Mineral interactions • Spatial heterogeneity • Vertical tongues

Abbreviations

qCO_2	biomass-specific respiration rates
C	carbon
DOC	dissolved organic carbon
DOM	dissolved organic matter
MRTs	mean residence time
SOC	soil organic carbon
SOM	oil organic matter
OM	organic matter

20.1 Introduction

Subsoils may play an important role in the re-carbonisation of the biosphere. They store much higher amounts of organic matter than the topsoil horizons (Batjes 1996). Moreover, the carbon (C) stored in the subsoil horizons generally has residence times as high as several thousand years (Paul et al. 1997), suggesting that soil organic matter (SOM) is characterised by a high degree of stabilisation. Another prominent feature of subsoil horizons is much lower C contents compared to topsoil horizons. Therefore it was hypothesised, that there is a huge possibility to increase this stock.

Since the last 10 years, much research was devoted to understand the reasons for the high stability and influencing factors of carbon stabilisation in subsoil horizons (Rumpel and Kögel-Knabner 2011). These studies showed that main C sources of subsoil OM are dissolved organic matter (Kaiser and Guggenberger 2000), root biomass (Rasse et al. 2005) and physically or biologically transported particulate organic matter (Don et al. 2008). Organic matter (OM) in subsoil horizons is characterized by a high radiocarbon age but the reasons for this are not entirely clear. Decreasing C/N ratio, increasing $\delta^{13}C$ and analysis of single OM compounds suggest, that microbial-derived OM may contribute with higher proportions to OM in subsoil horizons than plant-derived OM (Liang et al. 2008; Rumpel et al. 2010). However, it was shown that the number of organisms is low and that the diversity of the microbial biomass in top and subsoil horizons may be quite different (Taylor et al. 2002).

Most studies of carbon sequestration in subsoil horizons concerned the stabilization processes, which have been elucidated for several soil types. The chemical composition of the stabilized carbon is still a matter of debates, because of methodological problems due to the low carbon contents. Another open question concerns the fate of SOM precursors within different soil depths.

In this chapter we present pool sizes and stability of SOM in subsoils as well as the state of knowledge concerning its chemistry and stabilization mechanisms. The aim of this review was to evaluate the relevance of factors controlling litter decomposition and stabilization in subsoil for long-term carbon storage and suggest management options that might be implemented to increase these stocks.

20.2 Subsoil Organic Matter: Pool Sizes and Stability

In the global carbon cycle, the soil contains the largest active terrestrial reservoir. Total soil carbon stock estimates for the top 1 m range between 1,500 and 2,000 Pg ($1.5–2.0 \times 10^{18}$ g) (Jobbagy and Jackson 2000; Janzen 2005) and are even higher if the recently corrected estimates for boreal soils with permafrost (Tarnocai et al. 2009) are considered. With annual C-fluxes into and out of soils estimated to be around 75 Pg (Schlesinger and Andrews 2000) about 3–5% of the global soil carbon pool is highly dynamic. In the past, carbon flux measurements and modelling have mostly considered the top 20 or 30 cm of the soil where C-concentrations, root densities and microbial activities are generally highest. However, depending on climate zone and land use, this soil compartment contains only 30–50% of the C-stocks of the first meter (Jobbagy and Jackson 2000). If the deeper subsoil down to 3 m is also considered, the contribution of topsoil carbon stocks to total soil C-pools is only 20–40% while an equivalent amount of the C-stocks may be found below 1 m depth (Fig. 20.1). If this is taken into account, global soil carbon stocks may be up to 840 Pg higher than stock estimates based on the top meter only (Jobbagy and Jackson 2000).

Another distinct property of subsoil OM is its high apparent ^{14}C age. As shown in Fig. 20.2, the ^{14}C age of bulk SOM below 30 cm depth generally increases continuously indicating mean residence times of several 10^3–10^4 years. Large pool size and high radiocarbon age suggest that subsoil OM has accumulated at very low rates over very long time periods and therefore appears to be very stable. Consequently,

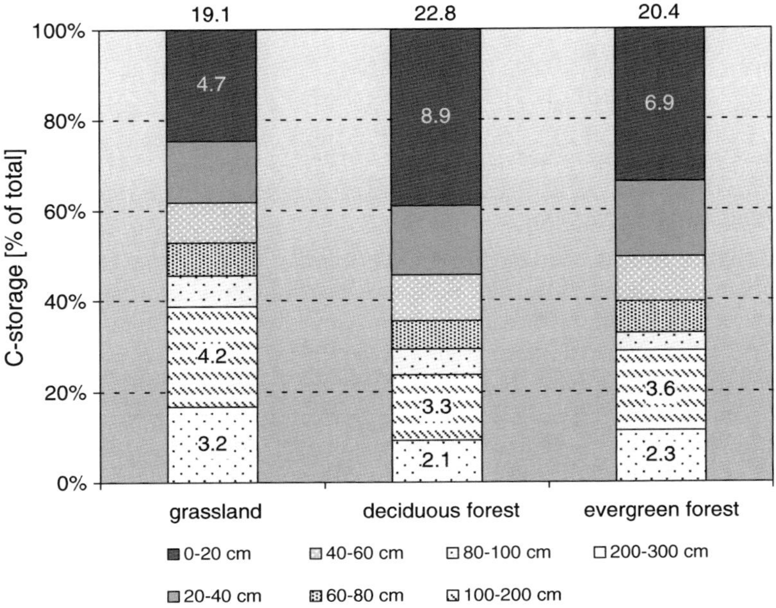

Fig. 20.1 Vertical distribution of mean organic carbon stocks in soils of the temperate zone. The numbers in and above the *columns* give C-stocks in kg m^{-2} (Based on data from Jobbagy and Jackson 2000)

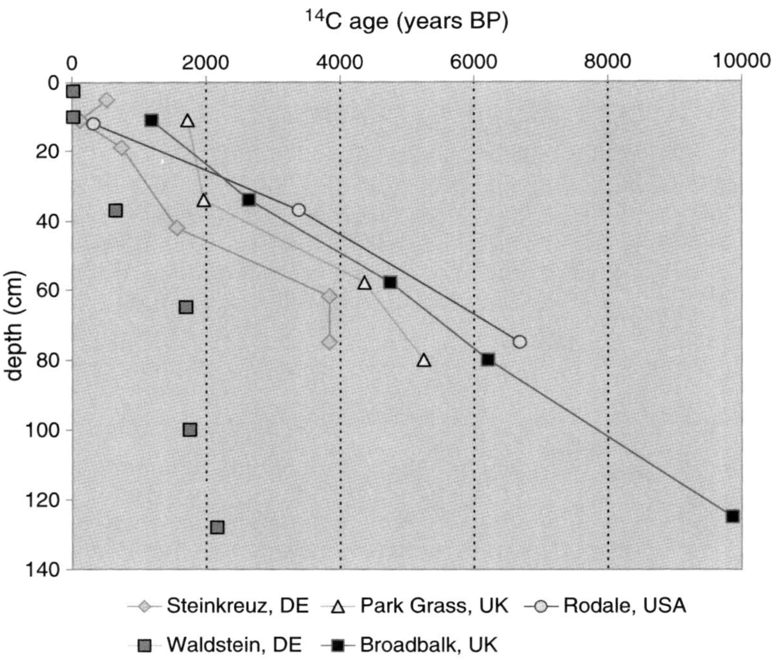

Fig. 20.2 ^{14}C age depth profiles for bulk soil organic matter at different sites in Germany, UK and USA (Data from Kögel-Knabner et al. 2008; Jenkinson and Coleman 2008; Paul et al. 2001)

subsoil OM was considered as not relevant for global C-cycle due to a low risk for destabilization. However, it was shown by a distinct bomb-derived ^{14}C enrichment in pasture subsoils from California and New Zealand that 10–40% of the subsoil SOM pool has annual to decadal turnover times (Baisden and Parfitt 2007).

This might be explained by fresh C-inputs derived from roots, dissolved organic matter (DOM) and bioturbation, which constantly deliver young carbon to the subsoil microorganisms that utilize C for growth and catabolism. Consequently, some SOM is decomposed in subsoils and contributes to the global C cycle. While very little is known about carbon fluxes or turnover rates in the subsoil several explanations have been proposed why subsoil SOM is so seemingly old and inert:

1. Subsoil SOM largely consists of recalcitrant compounds that are the least degradable end-products of numerous metabolic and biochemical processes and which thus accumulate during the degradation of fresh biomass inputs.
2. The annual inputs of fresh OM into the subsoil are so low that their sequestration in the subsoil does not significantly contribute to pool size changes or ^{14}C activity over centuries.
3. Old SOM in subsoils is stabilized by sorption to pedogenic minerals, leaving only limited sorption potential for fresh OM inputs that are thus either mineralized or leached.
4. The high radiocarbon age of bulk subsoil OM is caused by a small fraction of fossil or geogenic C. The mean residence time of SOM is thus grossly overestimated.
5. Microbial densities in subsoils are low and activities are limited by environmental factors like low O_2 and low temperatures so that SOM is mineralized at extremely low rates.
6. Microbial communities and activities are limited to very few hot spots where fresh OM is supplied with DOM, roots, exudates or burrowing animals. Large parts of the bulk soil are so sparsely populated that even potentially degradable OM can persist.

20.3 Subsoil Organic Matter: Chemistry

Concomitantly with decreasing carbon content, enrichment in stable C and N isotopes of SOM with increasing soil depth was observed in most studies and often related to the fact, that SOM in subsoil horizons is in general more transformed and enriched in strongly degraded plant material and/or microbial-derived compounds (exp. Högberg 1997; Boström et al. 2007). This is corroborated by the low C/N ratios. However, the changes of the stable carbon isotope signature with depth have also been related to chemical composition of SOM, which depends on soil inherent stabilisation mechanisms (Krull and Skjemstad 2003). Due to N limitation in most natural ecosystems, the depth trends of ^{15}N in forest soils may be strongly controlled by ectomycorrhizal fungi activity, which was found to lead to significant enrichment in ^{15}N (Lindahl et al. 2007) and accretion of ^{15}N enriched microbial compounds (Huygens et al. 2008).

As far as the bulk chemical composition of SOM in deep soil horizons is concerned few studies have tackled this issue, due to low carbon concentrations and high losses upon removal of the mineral phase. Rumpel et al. (2004) examined the chemical composition of bulk SOM in fine particle size fractions of B and C horizons of two soil profiles with contrasting pedogenic processes by solid-state NMR spectroscopy. Small particle size fractions showed increasing OM enrichment with depth. The results indicated that soil may act like a chromatographic system preserving specific compounds in well defined soil horizons. This is in line with a pyrolysis study of bulk soil sampled from top- and subsoil of an Andisol containing high amounts of stabilised carbon. Statistical assessment of the pyrolysis signature showed evidence for clear chemical differences between top and subsoil horizons (Rumpel et al. 2012).

In the traditional concept of SOM turnover models, fresh biomass inputs are metabolized and mineralized by soil organisms and further altered by biochemical polymerisation and condensation reactions to such a degree that only a small fraction of biochemically stable, i.e. recalcitrant compounds remain in the soil and thus form the SOM fraction with the highest ^{14}C age. Consequently, the increasing ^{14}C age with soil depth was attributed to the slow downward movement and accumulation of such recalcitrant compounds. But recently, little evidence was found for recalcitrance being an important mechanism of SOM stabilisation (Marschner et al. 2008) and numerous studies show that subsoil SOM contains compounds of potentially high degradability such as simple microbial carbohydrates and amino sugars (Krull and Skjemstad 2003; Liang and Balser 2008).

The only plant-derived organic matter compound that seems to have a higher intrinsic stability than others is "black carbon" formed from incomplete combustion during vegetation fires and, in urban areas, from coal and related products (Marschner et al. 2008). It is quite common in many soils (Preston and Schmidt 2008; Rodionov et al. 2010) and its contribution to the total SOM pool has been found to increase in subsoils where it reaches up to 35% of SOM (Brodowski et al. 2007; Rodionov et al. 2010). Due to the heterogeneous composition and different sources it remains unclear if black carbon has persisted for centuries or millennia or if it is only associated with other old SOM stabilized by other mechanisms. However, even black carbon can be degraded microbially as shown by Rethemeyer et al. (2005) who identified the incorporation of coal-derived black carbon into microbial PLFAs by ^{14}C analysis. Since some black carbon in subsoils carries a distinct signal from recent fossil fuel burning (Brodowski et al. 2007), translocation to greater depths may be a rather rapid process.

This shows that a high ^{14}C age of bulk SOM does not necessarily indicate a high stability. Trumbore (2009) has pointed out that bulk ^{14}C data biases towards longer turnover times due to the high proportion of old SOM in subsoils. Still, ^{14}C analyses of CO_2 and dissolved organic carbon (DOC) in subsoils provide evidence for a substantial turnover of old SOM in greater depths (Trumbore 2009).

Assuming that much of the SOM inputs into subsoils occur via roots (Rasse et al. 2005), it also needs to be taken into account that in forest ecosystems roots may live for decades so that fresh root litter may already carry an older ^{14}C signal (Trumbore 2009).

Some studies have suggested that the high ^{14}C-age of subsoil OM is due to an increasing contribution of geogenic or fossil organic matter with depth. Sedimentary rocks such as shales, carbonates or loess generally contain geogenic carbon from 0.1% to 2.0% (Tissot and Welte 1984) that may be 10^4–10^8 years old and therefore displays very little to no detectable ^{14}C. In subsoils with very low concentrations of recent carbon, this geogenic carbon can dilute the ^{14}C signal strongly and lead to gross overestimation of turnover times (Trumbore 2009). In loess-derived soils, 10,000–13,000 year old SOC fractions have even been isolated in topsoils (Paul et al. 2001; Helfrich et al. 2007) and had obviously been deposited with the loess at the end of the last glaciation. Recent inputs of ^{14}C-free fossil fuel-derived black carbon may also be relevant in certain urban and industrial areas where it mostly accumulates in the topsoil but is also transferred to greater depth (Brodowski et al. 2007; Flessa et al. 2008), where it can strongly affect the apparent ^{14}C age of SOM. However, high ^{14}C ages were also recorded in soils developed on carbon-fee parent rock like Granite, which were not influenced by BC deposition (Rumpel et al. 2002), indicating that most probably soil inherent stabilisation mechanisms lead to long residence times of SOM in subsoils.

20.4 Fluxes and Fate of Fresh OM Inputs into the Subsoil

Inputs of organic matter into subsoils may occur via three pathways: (i) leaching of DOM, (ii) root dieback and root exudates, and (iii) translocation of particulate organic matter through large pores or through bioturbation. In temperate forests, annual DOC fluxes into the mineral soil amount to 115–500 kg ha^{-1} and to 10–200 kg ha^{-1} in 20–30 cm depth (Michalzik et al. 2001; Kalbitz and Kaiser 2008). Based on annual DOC retention rates of 30–400 kg ha^{-1} and SOC pool sizes, Guggenberger and Kaiser (2003) calculated that the mineral-associated SOM pool of such forest sites could have built up within 186–1,730 years through sorption and stabilisation by soil minerals. However, if steady state is assumed and the limited sorption sites are taken into account Guggenberger and Kaiser (2003) estimate mean residence times of sorbed SOM to be <50 years, which of course stands in sharp contrast to the high mean ^{14}C ages of subsoil SOM. Similarly, Kalbitz and Kaiser (2008) and Michalzik et al. (2003) calculated that DOM fluxes can supply 25–89% to the mineral soil SOM pool.

Root derived carbon inputs into the subsoil are more difficult to estimate. It is generally assumed that 40–60% of the net-fixed carbon is allocated to the roots (Rasse et al. 2005; Jones et al. 2009) which amounts to ca. 3–6 tC ha^{-1} $year^{-1}$ for cereals and grasses (estimated from above- and belowground productivity data compiled Ellenberg and Leuschner (2010)) and to about 5 tC ha^{-1} $year^{-1}$ for temperate broad-leaved forests (Ellenberg and Leuschner 2010). Much of this carbon is rapidly mineralized, either by root respiration or in the rhizosphere (Jones et al. 2009). In a Central European beech forest, Gries (Ellenberg and Leuschner 2010) measured a turnover of fine and large roots of 1.3–1.6 tC ha^{-1} $year^{-1}$, while

3.4–3.7 tC ha^{-1} $year^{-1}$ were consumed by root respiration. Root tissue is generally degraded more slowly than shoot tissue (Rasse et al. 2005; Hansson et al. 2010) so that root derived biomass inputs may have longer residence times in the mineral soil. However, only little is known about root growth and turnover in subsoils or about the influence of vegetation type and soil properties.

A root origin of OM in subsoil horizons was evidenced by the relation between deep soil carbon stocks and plant functional types on a global scale (Jobbagy and Jackson 2000). Compound specific analysis of root and shoot biomarkers have further supported a high input of root carbon into subsoil horizons (Mendez-Millan et al. 2012).

A field incubation with ^{13}C and ^{15}N labeled root litter in three different depths of an agricultural soil showed similar amounts of C and N remaining after 3 years (Sanaullah et al. 2011). The decomposition dynamics, microbial biomass and potential activity was different in top- and subsoil, while similar SOM protection mechanisms were recorded. In the long run, the better abiotic decay conditions prevailing in subsoil horizons may compensate for their poorer physico-chemical characteristics and lead to rapid decomposition of root litter in subsoil horizons (Sanaullah et al. 2011).

Native SOM behaved quite differently compared to. Much higher amounts were remaining after 3 years of incubation and if any carbon seemed to be lost at a higher rate compared to nitrogen (Sanaullah et al. 2011). The C/N ratio of particulate organic matter, occluded OM and OM <50 μm was lower for native SOM compared to the labeled counterpart. This be related to the fact, that most native SOM may be highly decomposed and protected by association with soil minerals as well as micropores (Eusterhues et al. 2003; Guggenberger and Kaiser 2003).

Similarly, results from C3/C4 vegetation change studies and from FACE experiments indicate that most of the fresh carbon input into the subsoil is rapidly mineralized, as subsoil CO_2 largely consists of fresh sources and only little fresh carbon is incorporated into subsoil C pools (Rasse et al. 2005; Bernhardt et al. 2006; Taneva et al. 2006). However, if longer time periods are considered based on ^{14}C analyses, an enrichment of subsoil SOC with modern bomb-derived ^{14}C has been detected at Rothamsted (Jenkinson and Coleman 2008) and in pasture soils from New Zealand and California since the early 1950s (Baisden and Parfitt 2007).

Migration of particulate organic matter into subsoils via colloidal transport, peloturbation, bioturbation or cryoturbation may also substantially contribute to subsoil C pools in some soils, but these processes have not been yet quantified.

Overall, the relative contribution of the different C-fluxes into subsoils is largely unknown as well as their controlling factors in different climatic zones, vegetation types or soil units.

20.5 Relevance of Stabilisation Mechanisms for C Sequestration Potential in Subsoil Horizons

Due to their low carbon content, subsoils have been suspected to be potentially large carbon sinks. The relevance of the four stabilisation mechanisms for increasing C-sequestration in subsoils is shown in Tab. Although chemical recalcitrance may

Table 20.1 Stabilisation mechanisms of SOM and their relevance as well as C-sequestration potential in subsoil horizons

Mechanism	+ or – relevance shown by	C-sequestration potential for more than decades
Chemical recalcitrance		
Root tissue	+ Rasse et al. (2005)	No
Microbial products	– Gleixner et al. (2002)	No
Aliphatic compounds	+ Lorenz et al. (2007)	Yes
Black carbon	+ Brodowski et al. (2007)	Yes
	+ Rodionov et al. (2010)	Yes
Incorporation into soil aggregates	+ Rasmussen et al. (2005)	yes
	+ Sanaullah et al. (2011)	Not determined
	+ Moni et al. (2010a)	Yes
Sorption to minerals	+ Kaiser and Guggenberger (2003)	Yes
	+ Kalbitz and Kaiser (2008)	Yes
	+ Kleber et al. (2007)	Yes
	+ Eusterhues et al. (2007)	No
	+ Kögel-Knabner et al. (2008)	Yes
	+ Spielvogel et al. (2008)	Yes

no longer be regarded as a stabilisation mechanism able to increase the residence time of OM in general, it was suggested that specific compounds, such as black carbon and aliphatic compounds may remain for long time in subsoils because of their high intrinsic stability against microbial decomposition (Table 20.1). The importance of incorporation into soil aggregates as stabilisation mechanism for SOM was found to be of similar importance in top and subsoil horizons (Sanaullah et al. 2011). It was shown to be able to greatly increase the mean residence time of SOM (Rasmussen et al. 2005; Moni et al. 2010a). The most important SOM stabilisation mechanism in subsoils, able to greatly increase their C sequestration potential for more than decades was cited by many authors to be the interaction with soil minerals (Table 20.1).

20.6 Mineral Interactions, the Main Stabilisation Mechanism for OM in Subsoils?

Organic matter stored in subsoil horizons of many soil types regardless the land-use or climatic region was found to be stabilized mainly by interactions with soil minerals, in particular iron or aluminum oxides (Eusterhues et al. 2003; Kleber et al. 2005; Moni et al. 2010b), although also direct interaction with Al cations in acid soil (Matus et al. 2006; Bardy et al. 2007) and Ca in alkaline soil (Muneer and Oades 1989; Peinemann et al. 2005) have been cited. The stabilization of OM through intimate interaction with the mineral phase was first deducted from high carbon losses after demineralization of soil. Such a treatment leads to the destruction of the mineral phase and loss of soluble OM compounds. Soluble compounds are many small molecules like organic acids or soluble sugars. Evidence for stabilization of polysaccharides as well as amino acids in subsoil horizons was provided in several studies (Rumpel et al. 2010; Liang et al. 2008; Spielvogel et al. 2008). The stabilization of these labile compounds in subsoil horizons may be related to high contribution of dissolved OM, which might be the most important precursor for subsoil OM (Kaiser and Guggenberger 2002). Moreover, root exudates, which are excreted in close proximity, may be a likely origin of small highly soluble molecules stabilized by interactions with the mineral phase in subsoil horizons (Rasse et al. 2006).

The close correlation between the proportion of HF-soluble SOC and ^{14}C age of bulk SOC in subsoils shows that mineral interactions are important SOC stabilisation mechanisms (Spielvogel et al. 2008). However, in two acid forest soils, the HF-soluble SOC fraction in the C horizons was up to 3,200 years younger than the HF-resistant SOC fraction (Eusterhues et al. 2007). This may be due to: (i) a continuous exchange of old sorbed SOC by young soluble compounds, (ii) a higher stability of sorbed particulate SOC or (iii) the formation of insoluble SOC compounds as a prerequisite for long-term stabilisation. This means that even with a limited sorption capacity, soluble organic compounds can be immobilized through exchange processes until they are transformed into truly insoluble and thus stable compounds.

On the other hand, mineral weathering and the formation of pedogenic minerals continuously provide reactive species and fresh surfaces for sorptive stabilisation and co-precipitation of soluble SOC (Spielvogel et al. 2008; Kalbitz and Kaiser 2008). Still, the fixation rates of fresh SOC inputs must be much lower than those derived from mass-balance estimates of DOC fluxes, according to which the mineral-associated pool should be completely replaced within 300–1,700 years, if the whole HF-soluble fraction would be subject to these exchange processes (Guggenberger and Kaiser 2003; Kalbitz and Kaiser 2008). This clearly is not the case as the HF-soluble fraction of these subsoils has ^{14}C ages of 2,000–3,500 years (Eusterhues et al. 2007). This discrepancy may be due to the fact that freshly sorbed OM is still more easily degradable than older mineral-associated SOC (Kalbitz et al. 2005), also indicated by the relatively young age of CO_2 from subsoil respiration (Fierer et al. 2003). On the other hand, the DOC available for "exchange reactions" in the subsoil may not be young at all but originate from a cascade of exchange reactions during profile passage (Kalbitz and Kaiser 2008; Sanderman and Amundson 2008) or from microbial degradation of old SOC (Gleixner et al. 2002) and would thus not greatly contribute to the "rejuvenation" of sorbed SOC.

Evidence for this is found in the increasing ^{14}C age of DOC with soil depth (Don and Schulze 2008) and in the very high ^{14}C ages of DOC leached from agricultural drainages and found in rivers of small catchments (Sickman et al. 2008). The mineral associated SOC pool in subsoils may therefore be more dynamic than suggested by mean residence time (MRTs) calculated from ^{14}C data.

20.7 Biological Parameters at Depth

The abundance of microbial biomass, the most important factor affecting SOM content, is generally declining with soil depth (Taylor et al. 2002; Anderson and Domsch 1989; Ekelund et al. 2001) as well as its activity (Fang and Moncrieff 2005). The depth gradient is most pronounced for soil fauna and fungi so that bacterial biomass or cell numbers become dominant below 30–50 cm depth (Taylor et al. 2002, Ekschmitt et al. 2008) and can extend at similar biomass levels down to 8 m depth in prairie soils (Dodds et al. 1996). The opposite was noted for the Cmic/Corg ratio, suggesting that in deeper horizons a higher proportion of organic C is present as microbial biomass (Agnelli et al. 2004). However, Lavahun et al. (1996) and Agnelli et al. (2004) both determined strong increases in biomass-specific respiration rates (qCO_2) with depth and attributed this to a less efficient substrate utilization due to the higher recalcitrance of subsoil organic matter. Similarly, von Lützow and Kögel-Knabner (2009), Fierer et al. (2003), and Karhu et al. (2010) explained the higher temperature sensitivity of OM mineralisation in subsoils with its decreasing substrate quality. However, in view of the above-mentioned processes, the decreasing substrate use efficiency may be explained by the increasing degree of SOM stabilisation.

Despite low biomass contents, subsoils are metabolically active and contain substantial numbers of microorganisms (Taylor et al. 2002). Studies on pesticide degradation showed that microbial activity was still significant below the rooting zone down to a depth of 2.5 m (Ajwa et al. 1988; Charnay et al. 2005). Similar observations were made concerning amino acid biodegradation (Kemmitt et al. 2008). Microbial activity in subsurface horizons was found to be similar to that measured in surface soil when normalized to biomass size (Blume et al. 2002). The absence of a decreasing trend is interesting, as several authors observed differences in microbial community structure between surface and subsurface horizons.

In general, the genetic and metabolic diversity was reduced in subsoil horizons (Goberna et al. 2005; Agnelli et al. 2004). Carbon sources for soil bacteria with increasing depth increasingly consisted of older SOM derived C (Kramer and Gleixner 2008), suggesting that these are used as C sources by the soil organisms present. The use of subsoil C may further be enhanced by bioturbation, which was shown to accelerate decomposition of old carbon below the soil surface (Briones et al. 2011). However, the activity of earthworms below the soil surface and their effect on the input of fresh litter as well as the decomposition of ancient C have not yet been quantified yet.

20.8 Environmental Factors at Depth

When incubated in the laboratory, samples from subsoils often show a similar or even higher SOC mineralisation than topsoil samples, indicating that physical parameters like temperature, aeration or soil structure limit SOM degradation in

Table 20.2 Environmental factors, which can limit microbial decomposition at depth

Factor	Influence on decomposition	Experimental setup	References
Temperature	Low	Laboratory	Gillabel et al. (2010)
	Low	Field	Sanaullah et al. (2011)
	High	Laboratory	Fierer et al. (2003)
Water	Low	Field	Sanaullah et al. (2011)
	High	Field	Rovira et al. (2003)
Nutrients	Low	Laboratory	Salomé et al. (2010)
	High	Laboratory	Fierer et al. (2003)
		Field	Garcia-Pausas et al. (2008)
Energy	Low	Laboratory	Salomé et al. (2010)
	Low	Laboratory	
	High	Laboratory	Fontaine et al. (2007)
	Low	Field	Sanaullah et al. (2011)
Low microbial density and diversity	High	Field	Agnelli et al. (2004)
	High	Field	Ekschmitt et al. (2008)

subsoils (Lavahun et al. 1996; Ajwa et al. 1988; Fierer et al. 2003; Agnelli et al. 2004; Fang and Moncrieff 2005; Salomé et al. 2010, see Table 20.2). However, when OM decomposition is assessed under field conditions, influence of these factors is much more variable and does in some cases not confirm the laboratory results (Table 20.2).

Low O_2 and high CO_2 partial pressures *in situ* have been suspected to reduce microbial activity in subsoils. This may be true for certain fungi and oxidative exocellular enzymes (Ekschmitt et al. 2008), but Salomé et al. (2010) found no reductions in microbial activity in undisturbed subsoil samples incubated at low O_2 partial pressure. While in their experiments, microbial activity was not stimulated by nutrient additions, Fierer et al. (2003) found subsoil microbial activity to be clearly N- and P-limited. This result is supported by a field study, which showed that in subsurface horizons of mountain grassland soils the microbial activity was limited by the availability of N (Garcia-Pausas et al. 2008).

There is currently a strong debate about the question of substrate limitations in subsoils. Fontaine et al. (2007) observed a strong increase in mineralisation of old subsoil SOM after addition of ^{14}C-labelled cellulose, while Salomé et al. (2010) found such priming effects to be much more pronounced in topsoil samples. During a 3 year field incubation experiment with ^{13}C labelled litter at depth no substantial loss of native SOM could be observed (Sanaullah et al. 2011). Xiang et al. (2008) assume that microorganisms in subsoils are C-starved but attribute this not to the absence of degradable substrates but to the inaccessibility due to long diffusion pathways in the sparsely populated subsoil. The high SOC mineralisation rates observed in laboratory incubation studies with disturbed subsoil samples (Fierer et al. 2003; Salomé et al. 2010) may therefore simply result from the improved contact between substrate and degraders.

20.9 Spatial Heterogeneity of Old and Young Carbon

The spatial heterogeneity of subsoil C distribution and fluxes is much higher than in topsoils (Syswerda et al. 2011) due to the greater relevance of preferential flow paths, roots and animal burrows for C-inputs. Compared to the bulk soil, such hot spots contain more SOC and C_{mic} (Bundt et al. 2001), more degraded lignin monomers and younger SOC with wider C:N ratio (Chabbi et al. 2009) and higher enzyme activities (Grierson and Comerford 2000; Dong et al. 2007). Nunan et al. (2003) have identified steep gradients in microbial densities down to the μm-scale in the vicinity of pores and Young et al. (2008) point out that even in biologically very active topsoils less than 1% of the surface area is colonized by microorganisms which can affect the soil environment only up to 10 μm from the cell surface. The spatial segregation of consumers and substrate therefore is more a rule than an exception in soils. Xiang et al. (2008) point out that substrate availability in subsoils may largely be diffusion controlled and therefore can only operate at the μm scale while in topsoils convective transport of solutes due to wetting/drying and root activity also take place at larger scales in the pore system.

The hot spots of microbial activity may include the rhizosphere (Joergensen 2000), and/or preferential flow pathways (Bundt et al. 2001). Chabbi et al. (2009) showed, that vertical tongues, which are a few centimetres wide and more than a meter long are occurring in soil (Fig. 20.3). These tongues may represent 14–20% of the soil volume and have a significantly higher carbon contents compared to the soil matrix. Carbon within the soil matrix was found to be characterised by ^{14}C ages

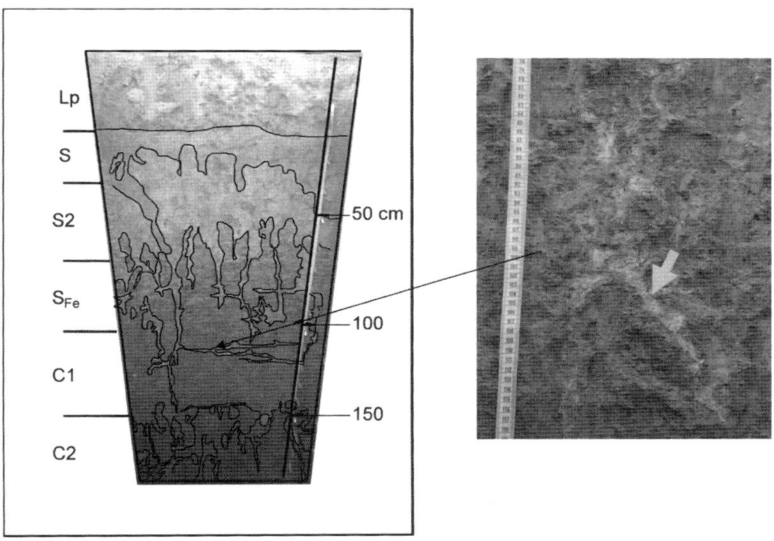

Fig. 20.3 Occurrence of vertical tongues within a soil profile

of several thousand years, whereas modern carbon was found in tongues (Chabbi et al. 2009). The tongues may be ancient root channels, related to oak forest, which was growing at the site in former times.

The occurrence of zonation within a soil profile is a common phenomenon and generally relation to preferential flow pathways, root channels and earthworm casts. Preferential flow pathways were found to persist for decades leading to chemical gradients between flowpaths and matrix (Hagedorn and Bundt 2002). Ancient root channels may also be the favourite location for colonisation of new roots from plants growing at the moment on the site (Rasse and Smucker 1998). Therefore both processes may be contributing to the input of young carbon into the subsoil horizons. The input of fresh plant material into subsoil horizons could further accelerate the turnover of stabilised carbon (Fontaine et al. 2007), which may explain the very different ^{14}C ages in adjacent parts of the soil profile.

20.10 Management Options to Increase SOM Storage in Subsoil Horizons

Carbon storage potential of subsoil below 1 m depth was estimated to be 760 and 1,520 Pg C (Lorenz and Lal 2005). Management options to increase SOM storage in subsoil horizons are essentially focused on increasing carbon input into deep soil. Input may be manipulated by the use of plants with particular root systems or by irrigation of soil with water rich in dissolved organic matter (Table 20.3). Rooting depth of plants is partly dependent on genetics and therefore breeding of crop plants with deep long-lived roots has been suggested to increase carbon sequestration in deep soil horizons (Kell 2011). The author claims that doubling the steady state depth of roots from 1 to 2 m could have a significant beneficial impact on lowering atmospheric CO_2 levels. However, such a C sequestration may only be possible if input of easily decomposable compounds does not lead to loss of old stabilized C compounds by priming effects. Such a loss was observed, when soil was irrigated

Table 20.3 Management options to increase deep soil C storage

Management option	References	Carbon sequestration potential
Breeding of plants with deep roots	Kell (2011)	Yes/to be tested
Selection of plants with high suberin/ phenol content in roots	Lorenz et al. (2007)	Yes
Irrigation with C-rich water	Jueschke et al. (2008)	No
Input of colloidal biochar		Yes/ to be tested
Input of lipid precursors for abiotic polymerization processes	Lorenz et al. (2007)	To be tested
Colonisation of soil with anecic earthworm species		To be tested
Mycorrhization of plant roots		To be developed

for several decades with wastewaters containing high amounts of dissolved organic matter (Jueschke et al. 2008). This practice may therefore not be recommended to increase C storage in subsoils.

More promising for increasing subsoil carbon stocks may be the input of recalcitrant compounds, such as colloidal biochar or aliphatic molecules. Possible plant breeding strategies to increase aliphatic compound input may concern increase of the plants suberin content (Lorenz et al. 2007). Addition of colloidal biochar to soil may lead to its rapid migration into deeper soil layers (Brodowski et al. 2007; Rumpel et al. 2009). The addition of such material may also ameliorate the soils physical and chemical properties (Lehmann 2007).

Other possibilities to increase subsoil carbon stocks include the colonization of soil with anecic earthworm species, which would transfer litter deposited on the soil surface down to a depth of up to 4 m. Mycorrhization of plant roots could be an option, as plants transfer up to 20% of the carbon they fix to their fungal partner. However, such manipulations of natural processes need to be carefully tested and we also need to keep in mind that all plant carbon that is allocated belowground is not available for aboveground production.

20.11 Conclusions and Open Question

This review shows that a common explanation for the apparently slow turnover and high ^{14}C age of subsoil SOM is highly unlikely or are only relevant at a few sites. It is quite evident that old stabilized SOM compounds in any soil depth are very diverse and not characterized by an inherently low degradability. Subsoil SOM therefore appears not to be the accumulated recalcitrant end product of numerous biochemical degradation or condensation reactions. The concept of the subsoil as a passive receptor of non-degradable biochemical end products must be rejected for other reasons, too. Firstly, the current estimates of C-fluxes with DOC and roots into the subsoil are high enough to substantially increase the subsoil C-pools within decades. Therefore, most of this OM must be mineralized within months to decades, showing that there is considerable microbial activity in this seemingly inert soil compartment. Secondly, there is growing evidence that the subsoil C-pool itself is much more dynamic than suggested by its mean ^{14}C age. This has not been considered in soil C-modelling and can have important implications for the susceptibility of C-pools to land use and climate change and for management options aiming at sequestering carbon in subsoils.

Still, the high mean ^{14}C age of subsoil SOM indicates that only a small fraction of this pool takes part in such exchange and turnover reactions. To date it is unclear to what extent the old subsoil carbon pool derives from fossil or geogenic sources that would bias the mean ^{14}C age towards higher values or from old plant-derived SOM that is effectively stabilized against biochemical degradation through interactions with (pedogenic) soil minerals. This can only be resolved by the analysis of specific soil fraction and individual compounds that can be related to well defined sources.

The other major open question regarding subsoil SOM dynamics concerns the activity of soil organisms as well as their spatial distribution in relation to their substrates. Based on the few studies that have addressed this issue, it seems quite conceivable that active and passive SOM pools are largely controlled by the presence or absence of the sparse and patchily distributed microbial population. The environmental controls for the size, distribution and activity of such "hot spots" are not known, but are likely related to root growth, preferential flow paths and earthworm burrows.

The most promising strategy for increasing subsoil carbon stocks may be increasing the input of compounds stable against microbial decomposition, such like aliphatic compounds or black carbon. Another possibility is the colonisation of soil with aneic earthworms and/or the manipulation of plants either to increase root growth, exudation or mycorrhization. However, such practices still need to be either developed or tested.

References

Agnelli A, Ascher J, Corti G et al (2004) Distribution of microbial communities in a forest soil profile investigated by microbial biomass, soil respiration and DGGE of total and extracellular DNA. Soil Biol Biochem 36:859–868

Ajwa HA, Rice CW, Sotomayor D (1988) Carbon and nitrogen mineralization in tallgrass prairie and agricultural soil profiles. Soil Sci Soc Am J 62:942–951

Anderson T, Domsch KH (1989) Ratios of microbial biomass carbon to total organic-carbon in arable soils. Soil Biol Biochem 21:471–479

Baisden WT, Parfitt RL (2007) Bomb C-14 enrichment indicates decadal C pool in deep soil? Biogeochemistry 85:59–68

Bardy M, Bonhomme C, Fritsch E et al (2007) Al speciation in tropical podzols of the upper Amazon Basin: a solid-state 27Al MAS and MQMAS NMR study. Geochim Cosmochim Acta 71:3211–3222

Batjes NH (1996) Total carbon and nitrogen in the soils of the world. Eur J Soil Sci 47:151–163

Bernhardt ES, Barber JJ, Pippen JS, Taneva L, Andrews JA, Schlesinger WH (2006) Long-term effects of free air CO_2 enrichment (FACE) on soil respiration. Biogeochemistry 77:91–116

Blume E, Bischoff M, Reichert J et al (2002) Surface and subsurface microbial biomass, community structure and metabolic activity as a function of soil depth and season. Appl Soil Ecol 20:171–181

Boström B, Comstedt C, Ekblad A (2007) Isotope fractionation and 13C enrichment in soil profiles during the decomposition of soil organic matter. Oecologia 153:89–98

Briones MJI, Garnett MH, Ineson P (2011) Soil biology and warming play a key role in the release of 'old C' from organic soils. Soil Biol Biochem 42:960–967

Brodowski S, Amelung W, Haumaier L et al (2007) Black carbon contribution to stable humus in German arable soils. Geoderma 139:220–228

Bundt M, Widmer F, Pesaro M, Zeyer J, Blaser P (2001) Preferential flow paths: biological 'hot spots' in soils. Soil Biol Biochem 33:729–738

Chabbi A, Kögel-Knabner I, Rumpel C (2009) Stabilised carbon in subsoil horizons is located in spatially distinct parts of the soil profile. Soil Biol Biochem 41:256–271

Charnay MP, Tuis S, Coquet Y (2005) Spatial variability in C-14-herbicide degradation in surface and subsurface soils. Pest Manage Sci 61:845–855

Dodds WK, Banks MK, Clenan CS et al (1996) Biological properties of soil and subsurface sediments under abandoned pasture and cropland. Soil Biol Biochem 28:837–846
Don A, Schulze E (2008) Controls on fluxes and export of dissolved organic carbon in grasslands with contrasting soil types. Biogeochemistry 91:117–131
Don A, Steinberg B, Schoening I et al (2008) Organic carbon sequestration in earthworm burrows. Soil Biol Biochem 40:1803–1812
Dong S, Brooks D, Jones MD et al (2007) A method for linking in situ activities of hydrolytic enzymes to associated organisms in forest soils. Soil Biol Biochem 39:2414–2419
Ekelund F, Ronn R, Christensen S (2001). Distribution with depth of protozoa, bacteria and fungi in soil profiles from three Danish forest sites. Soil Biol Biochem 33:475–481
Ekschmitt K, Kandeler E, Poll C, Brune A, Buscot F, Friedrich M, Gleixner G, Hartmann A, Kastner M, Marhan S, Miltner A, Scheu S, Wolters V (2008) Soil-carbon preservation through habitat constraints and biological limitations on decomposer activity. J Plant Nutri and Soil Sci 171:27–35
Ellenberg H, Leuschner CH (2010) Vegetation mitteleuropas mit den alpen. Ulmer Verlag, Stuttgart
Eusterhues K, Rumpel C, Kleber M et al (2003) Stabilisation of soil organic matter by interactions with minerals as revealed by mineral dissolution and oxidative degradation. Org Geochem 34:1591–1600
Eusterhues K, Rumpel C, Kögel-Knabner I (2007) Composition and radiocarbon age of HF-resistant soil organic matter in a Podzol and a Cambisol. Org Geochem 38:1356–1372
Fang C, Moncrieff JB (2005) The variation of soil microbial respiration with depth in relation to soil carbon composition. Plant Soil 268:243–253
Fierer N, Allen AS, Schimel JP et al (2003) Controls on microbial CO_2 production: a comparison of surface and subsurface soil horizons. Glob Change Biol 9:1322–1332
Flessa H, Amelung W, Helfrich M et al (2008) Storage and stability of organic matter in a Luvisol and Phaeozem with continuous maize cropping: a synthesis. J Plant Nutr Soil Sci 171:36–51
Fontaine S, Barot S, Barré P et al (2007) Stability of organic carbon in deep soil layers controlled by fresh carbon supply. Nature 450:277–281
Garcia-Pausas J, Casals P, Camarero L et al (2008) Factors regulating carbon mineralization in the surface and subsurface soils of Pyrenean mountain grasslands. Soil Biol Biochem 40:2803–2810
Gillabel J, Cebrian-Lopez B, Six J et al (2010) Experimental evidence for the attenuating effect of SOM protection on temperature sensitivity of SOM decomposition. Glob Change Biol 16:2789–2798
Gleixner G, Poirier N, Bol R et al (2002) Molecular dynamics of organic matter in a cultivated soil. Org Geochem 33:357–366
Goberna M, Insam H, Klammer S et al (2005) Microbial community structure at different depths in disturbed and undisturbed semiarid Mediterranean forest soils. Microb Ecol 50:315–326
Grierson PF, Comerford NB (2000) Non-destructive measurement of acid phosphatase activity in the rhizosphere using nitrocellulose membranes and image analysis. Plant Soil 218:49–57
Guggenberger G, Kaiser K (2003) Dissolved organic matter in soil: challenging the paradigm of sorptive preservation. Geoderma 113:3–4
Hagedorn F, Bundt M (2002) The age of preferential flow paths. Geoderma 108:119–132
Hansson K, Kleja DB, Kalbitz K et al (2010) Amounts of carbon mineralised and leached as DOC during decomposition of Norway spruce needles and fine roots. Soil Biol Biochem 42:178–185
Helfrich M, Flessa H, Mikutta R et al (2007) Comparison of chemical fractionation methods for isolating stable soil organic carbon pools. Eur J Soil Sci 58:1316–1329
Högberg P (1997) Transley review no. 95–15 N natural abundance in soil-plant systems. New Phytol 137:179–203
Huygens D, Denef K, Vandeweyer R et al (2008) Do nitrogen isotope patterns reflect microbial colonization of soil organic matter fractions? Biol Fertil Soil 44:955–964
Janzen HH (2005) Soil carbon: a measure of ecosystem response in a changing world? Can J Soil Sci 85:467–480

Jenkinson DS, Coleman K (2008) The turnover of organic carbon in subsoils. Part 2. Modelling carbon turnover. Eur J Soil Sci 59:400–413

Jobbagy EG, Jackson RB (2000) The vertical distribution of soil organic carbon and its relation to climate and vegetation. Ecol Appl 10:423–436

Joergensen RG (2000) Ergosterol and microbial biomass in the rhizosphere of grassland soils. Soil Biol Biochem 32:647–652

Jones DL, Nguyen C, Finlay RD (2009) Carbon flow in the rhizosphere: carbon trading at the soil-root interface. Plant Soil 321:5–33

Jueschke E, Marscner B, Tarchitzky J et al (2008) Effects of treated wastewater irrigation on the dissolved and soil organic carbon in Israeli soils. Water Sci Technol 57:727–733

Kaiser K, Guggenberger G (2000) The role of DOM sorption to mineral surfaces in the preservation of organic matter in soils. Org Geochem 31:711–725

Kaiser K, Guggenberger G (2002) Accumulation of organic matter in soils related to sub-micron porosity-Evaluation by microscopy and microspectroscopy. Abstracts of Papers of American Chemical Society 223:U604–U605

Kaiser K, Guggenberger G (2003) Dissolved organic matter in soil: changing the paradigm of sorptive preservation. Geoderma 113:293–310

Kalbitz K, Kaiser K (2008) Contribution of dissolved organic matter to carbon storage in forest mineral soils. J Plant Nutr Soil Sci 171:52–60

Kalbitz K, Schwesig D, Rethemeyer J et al (2005) Stabilization of dissolved organic matter by sorption to the mineral soil. Soil Biol Biochem 37:1319–1331

Karhu K, Fritze H, Hamalainen K et al (2010) Temperature sensitivity of soil carbon fractions in boreal forest soil. Ecology 91:370–376

Kell DB (2011) Breeding crop plants with deep roots: their role in sustainable carbon, nutrient and water sequestration. Ann Bot 108:407–418

Kemmitt SJ, Wright D, Murphy DV et al (2008) Regulation of amino acid biodegradation in soil as affected by depth. Biol Fertil Soil 44:933–941

Kleber M, Mikutta R, Torn MS et al (2005) Poorly crystalline mineral phases protect organic matter in acid subsoil horizons. Eur J Soil Sci 56:717–725

Kleber M, Sollins P, Sutton R (2007) A conceptual model of organo-mineral interactions in soils:self-assembly of organic molecular fragments into zonal structures on mineral surfaces. Biogeochemistry 85. doi:10.1007/s10533-007-9103-5

Kögel-Knabner I, Guggenberger G, Kleber M et al (2008) Organo-mineral associaltions in temperate soils: integrating biology, mineralogy and organic matter chemistry. J Plant Nutr Soil Sci 171:61–82

Kramer C, Gleixner G (2008) Soil organic matter in soil depth profiles: distinct carbon preferences of microbial groups during carbon transformation. Soil Biol Biochem 40:425–433

Krull ES, Skjemstad JO (2003) d13C and d15N profiles in 14C-dated Oxisol and Vertisols as a function of soil chemistry and mineralogy. Geoderma 112:1–29

Lavahun MFE, Joergensen RG, Mayer B (1996) Activity and biomass of soil microorganisms at different depths. Biol Fertil Soil 23:38–42

Lehmann J (2007) Bioenergy in the black. Front Ecol Environ 5:381–387

Liang C, Balser TC (2008) Preferential sequestration of microbial carbon in subsoils of a glacial-landscape toposequence, Dane County, WI, USA. Geoderma 148:113–119

Liang B, Lehmann J, Solomon D, Sohi S, et al (2008) Stability of biomass -derived black carbon in soils. Geochimica Et Cosmochimica Acta 72:6069–6078

Lindahl BD, Ihrmark K, Boberg J et al (2007) Spatial separation of litter decomposition and mycorrhizal nitrogen uptake in a boreal forest. New Phytol 173:611–620

Lorenz K, Lal R (2005) The depth distribution of soil organic carbon in relation to land use and management and the potential of carbon sequestration in subsoil horizons. Adv Agron 88:35–66

Lorenz K, Lal R, Preston CM, Nierop KGJ (2007) Strengthening the soil organic carbon pool by increasing contributions from recalcitrant aliphatic bio(macro)molecules. Geoderma 142:1–10

Marschner B, Brodowski X, Dreves A et al (2008) How relevant is recalcitrance for the stabilization of organic matter in soils? J Plant Nutr Soil Sci 171:91–132

Matus F, Amigo X, Kristiansen SM (2006) Aluminium stabilization controls organic carbon levels in Chilean volcanic soils. Geoderma 132:158–168

Mendez-Millan M, Dignac MF, Rumpel C et al (2012) High contribution of maize root derived-C to soil organic carbon throughout an agricultural soil profile assessed by compound-specific 13C analysis. Org Geochem 42(12):1502–1511

Michalzik B, Kalbitz K, Park JH et al (2001) Fluxes and concentrations of dissolved organic carbon and nitrogen: a synthesis for temperate forests. Biogeochemistry 52:173–205

Michalzik B, Tipping E, Mulder J, Lancho JFG, Matzner E, Bryant CL, Clarke N, Lofts S, Esteban MAV (2003) Modelling the production and transport of dissolved organic carbon in forest soils. Biogeochemistry 66:241–264

Moni C, Rumpel C, Virto I et al (2010a) Relative importance of adsorption versus aggregation for organic matter storage in subsoil horizons of two contrasting soils. Eur J Soil Sci 61:958–969

Moni C, Chabbi A, Nunan N et al (2010b) Spatial dependence of organic carbon-metal relationships. A multi-scale statistical analysis, from horizon to field. Geoderma 158:120–127

Muneer M, Oades JM (1989) The role of Ca-organic interactions in soil aggregate stability. 2. Field studies with C-14 labelled straw, CaCO3, and CaSO42H2O. Aust J Soil Res 27:401–409

Nunan N, Wu KJ, Young IM et al (2003) Spatial distribution of bacterial communities and their relationships with the micro-architecture of soil. FEMS Microbiol Ecol 44:203–215

Paul EA, Follett RF, Leavitt SW et al (1997) Radiocarbon dating for determination of soil organic matter pool sizes and dynamics. Soil Sci Soc Am J 61:1058–1067

Paul EA, Collins HP, Leavitt SW (2001) Dynamics of resistant soil carbon of Midwestern agricultural soils measured by naturally occurring 14C abundance. Geoderma 104:239–256

Peinemann N, Guggenberger G, Zech W (2005) Soil organic matter and its lignin component in surface horizons of salt-affected soils of the Argentinean Pampa. Catena 60:113–128

Preston CM, Schmidt MWI (2008) Black (pyrogenic) carbon: a synthesis of current knowledge and uncertainties with special consideration of boreal regions. Biogeoscience 3:397–420

Rasmussen C, Torn MS, Southard RJ (2005) Mineral assemblage and aggregates control carbon dynamics in a California conifer forest. Soil Sci Soc Am J 69:1711–1721

Rasse DP, Smucker AJM (1998) Root recolonization of previous root channels in corn and alfalfa rotations. Plant Soil 204:203–212

Rasse DP, Rumpel C, Dignac MF (2005) Is soil carbon mostly root carbon? Mechanisms for a specific stabilisation. Plant Soil 269:341–356

Rasse DP, Mulder J, Moni C, Chenu C (2006) Carbon turnover kinetics with depth in a French loamy soil. Soil Sci. Soc. Am. J. 70:2097–2105

Rethemeyer J, Kramer C, Gleixner G et al (2005) Radiocarbon analysis of functional-defined and molecular organic matter fractions from agricultural soil profiles. Geoderma 128:94–105

Rodionov A, Amelung W, Peinemann N et al (2010) Black carbon in grassland ecosystems of the world. Glob Biogeochem Cycle 24. doi:10.1029/2009GB003669

Rovira P, Vallejo VR (2002) Labile and recalcitrant pools of carbon and nitrogen in organic matter decomposing at different depths in soil:an acid hydrolysis approach. Geoderma 107:109–141

Rumpel C, Kögel-Knabner I (2011) Deep soil organic matter – a key but poorly understood component of terrestrial C cycle. Plant Soil 338:143–158

Rumpel C, Kögel-Knabner I, Bruhn F (2002) Vertical distribution, age, and chemical composition of organic carbon in two forest soils of different pedogenesis. Org Geochem 33:1131–1142

Rumpel C, Eusterhues K, Kögel-Knabner I (2004) Location and chemical composition of stabilized organic carbon in topsoil and subsoil horizons of two acid forest soils. Soil Biol Biochem 36:177–190

Rumpel C, Ba A, Darboux F et al (2009) Erosion budget of pyrogenic carbon at meter scale and process selectivity. Geoderma 154:131–137

Rumpel C, Eusterhues K, Kögel-Knabner I (2010) Non-cellulosic neutral sugar contribution to mineral associated organic matter in top-and subsoil horizons of two acid forest soils. Soil Biol Biochem 42:379–382

Rumpel C, Rodríguez-Rodríguez A, González-Pérez JA et al (2012) Contrasting composition of free and mineral-bound organic matter in top and subsoil horizons of Andosols. Biol Fertil Soil (in press)

Salomé C, Nunan N, Pouteau V et al (2010) Carbon dynamics in topsoil and in subsoil may be controlled by different regulatory mechanisms. Glob Change Biol 16:416–426
Sanaullah M, Chabbi A, Leifeld J et al (2011) Decomposition and stabilization of root litter in top- and subsoil horizons: what is the difference? Plant Soil 338:127–141
Sanderman J, Amundson R (2008) A comparative study of dissolved organic carbon transport and stabilization in California forest and grassland soils. Biogeochemistry 89:309–327
Schlesinger WH, Andrews JA (2000) Soil respiration and the global carbon cycle. Biogeochemistry 48:7–20
Sickman JO, DiGiorgio CL, Davisson ML et al (2008) Identifying sources of dissolved organic carbon in agriculturally dominated rivers using radiocarbon age dating: Sacramento-San Joaquin River Basin, California. Biogeochemistry 99:79–96
Spielvogel S, Prietzel J, Kögel-Knabner I (2008) Soil organic matter stabilisation in acidic forest soils is preferential and soil type-specific. Eur J Soil Sci 59:674–692
Syswerda SP, Corbin AT, Mokma DL et al (2011) Agricultural management and soil carbon storage in surface vs. deep layers. Soil Sci Soc Am J 75:92–101
Taneva L Pippen JS, Schlesinger WH, Gonzalez-Meler MA (2006) The turnover of carbon pools contributing to soil CO(2) and soil respiration in a temperate forest exposed to elevated CO(2) concentration. Glob Change Biol 12:983–994
Tarnocai C, Canadell JG, Schuur EAG et al (2009) Soil organic carbon pools in the northern circumpolar permafrost region. Glob Biogeochem Cycle 23:GB2023. doi:10.1029/2008GB003327
Taylor JP, Wilson B, Mills MS et al (2002) Comparison of microbial numbers and enzymatic activities in surface soils and subsoils using various techniques. Soil Biol Biochem 34:387–401
Tissot BP, Welte DH (1984) Petroleum formation and occurrence, 2nd edn. Springer, Berlin
Trumbore S (2009) Radiocarbon and soil carbon dynamics. Annu Rev Earth Planet Sci 37:47–66
Von Lützow M, Kögel-Knabner I (2009) Temperature sensitivity of soil organic matter decomposition-what do we know? Biol Fertil Soil 46:1–15
Xiang SR, Doyle A, Holden PA et al (2008) Drying and rewetting effects on C and N mineralization and microbial activity in surface and subsurface California grassland soils. Soil Biol Biochem 40:2281–2289
Young IM, Crawford JW, Nunan N et al (2008) Microbial distribution in soil: physics and scaling. In: Sparks D (ed) Advances in agronomy, vol 100. Academic Press, Burlington, pp 81–121

Chapter 21
Transforming Carbon Dioxide from a Liability into an Asset

Carlo Rubbia

Abstract Over the past decades it has become clear that carbon dioxide (CO_2) emissions into the Earth's atmosphere, resulting from the combustion of fossil fuels, are the main causes of global warming. Several approaches to limit the release of CO_2 have been investigated and are gradually being implemented. One approach is the introduction of renewable energy, such as solar or wind power. However, as fossil fuels will still remain the mainstay of energy production over the next few decades, carbon capture and storage (CCS), the capture and sequestration of CO_2 under the sea or in disused gas and oil wells, is viewed by many as an important way to limit man-made climate change. Notwithstanding, the permanence of stored CO_2 in the Earth's crust for many thousand years is problematic, as the whole process is very energy consuming and the risk of deadly accidental leaks is very high. An alternative to sequestration is presently being investigated at the IASS (Institute for Advanced Sustainability Studies) within the framework of the Earth, Energy and Environment – E^3 Cluster, i.e., the combination of CO_2 with hydrogen, obtained from natural gas (methane (CH_4)) by pyrolysis. Pyrolysis is a procedure that allows the capture of carbon (C) in pure form as solid carbon black, which can subsequently either be used industrially or safely stored. This concept is aimed at circumventing greenhouse gas (GHG) emissions when utilizing CH_4 resources either as natural gas or in the form of clathrates, of which huge reserves, exceeding natural gas reserves are expected to exist. The hydrogen (H_2) and the already spent CO_2 from fossil burning can be combined to form methanol. Methanol as a liquid fuel, either used directly or converted into a higher hydrocarbon, can replace oil in applications such as fuel for transport or as the starting material in the manufacture of many chemicals now derived from oil.

C. Rubbia (✉)
IASS Institute for Advanced Sustainability Studies, e.V., Berliner Strasse 130, 14467 Potsdam, Germany
e-mail: carlo.rubbia@iass-potsdam.de

R. Lal et al. (eds.), *Recarbonization of the Biosphere: Ecosystems and the Global Carbon Cycle*, DOI 10.1007/978-94-007-4159-1_21,

Keywords Carbon dioxide emission • CCS • Clathrates • Methane pyrolysis • Methanol Economy

Abbreviations

C	carbon
CCS	carbon capture and storage ()
CERN	Conseil Européen pour la Recherche Nucléaire (the European Organization for Nuclear Research)
CO_2	carbon dioxide
CO	carbon monoxide
E^3 Cluster	Earth Energy and Environment
GHG	greenhouse gas
H_2	hydrogen
IASS	Institute for Advanced Sustainability Studies
IPCC	Intergovernmental Panel on Climate Change
CH_4	methane
NOAA	National Oceanic and Atmospheric Administration
SRM	Steam Reforming of Methane
STP	Standard Temperature and Pressure
TDM	thermal decomposition of methane

21.1 Introduction

Man-made climate change caused by the combustion of fossil fuels is threatening both the environment as well as continuing economic growth. Also, the finiteness of the natural reserves of fossil fuels will affect the economy in the future. The Intergovernmental Panel on Climate Change (IPCC) has reported in 2007 that the emission of greenhouse gases (GHGs) is the most likely the cause of global warming (IPCC 2007). It is clear that economic growth is the greatest factor contributing to the yearly increase in carbon dioxide (CO_2) emissions. For example, during 2009 the production of CO_2 decreased by 1.3% because of the reigning economic crisis (Friedlingstein et al. 2010). However, during 2010, with the gradual return to normality, CO_2 emissions have continued to grow again at the rate of about 3% per year (NOAA 2011).

An important factor that has to be taken into consideration in the search for alternative energy sources is that natural fossil fuel reserves are limited. Oil and natural gas (about 500 Pg [1 Pg = 10^{15} g] each) are expected to reach their limits sooner rather than later, due both to increases in consumption, especially in the developing countries, and the progressive reduction of easily available resources. There is plenty of cheap and readily available coal (5,000 Pg), which however has a higher carbon (C) footprint on burning compared to natural gas.

The rising trend in the use of fossil fuels is unsustainable because it causes progressive global warming. A future based on a continued expansion of "business as usual" is inevitably increasing the vulnerability to severe energy supply disruptions and resulting price shocks. There is a large gap between political promises and the reality of budgetary and other priorities. We are unfortunately continuing to develop a fossil-fuel dominated society, which will inevitably cause major environmental disruptions for many millennia ahead of us.

Is there a way of reconciling the use of fossil fuels and the effects on climate? Politicians' hopes are high. But the job is daunting, since already today over 30 Pg of CO_2 are produced every year (The International Energy Agency [IEA] 2010a). The lifetime of CO_2 in the atmosphere is incredibly long. The mean atmospheric CO_2 lifetime, in the order of 10^4 years, is in stark contrast with the "popular" perception of only several hundred years (Caldeira and Wickett 2005; Archer 2005). The 300-year simplification misses the immense longevity of the tail on the CO_2 lifetime, and hence its interaction with major ice sheets, ocean methane (CH_4) clathrate deposits, and future glacial/interglacial cycles. For the "best guess" case the mean lifetime of fossil CO_2 is about 30–35 thousand years, to be associated with the today's public perception of nuclear energy, with the lifetime of Pu-239 of 26 thousand years.

21.2 Political Action

In response to these threats, new technologies for energy sources, such as solar energy, wind energy, biomass and geothermal energy, have made huge progress over the last three decades, and now supply an increasingly important part of the energy production in many countries (IEA 2010b). These renewable sources do not produce CO_2 and therefore limit the impact of human activity on climate change. The research at IASS includes also new technologies for thermal solar power generation by DESERTEC, a European solar power generation project in North Africa (Knies et al. 2009). Transporting renewable energy to Europe via long-distance superconducting lines, with the aim of reducing transmission losses and leaving smaller environmental footprints, is also part of the IASS research programs.

Decarbonisation policies will have to take into account what goals need to be achieved by 2050. For example, if we want to keep the global temperature rise below 2°C, emission levels in 2050 will have to be 20–50% lower than those of 2000 (IPCC 2007). Among the most important consequences are rising sea levels. According to several projections, the sea level will rise more than 1 m by the end of the twenty-first century (Weiss et al. 2011; Overpeck and Weiss 2009).

21.3 Public Funding

Since the oil price shocks in the 1970s and 1980s, we have enjoyed relatively inexpensive and plentiful energy supplies. The easy availability of resources, the absence of C constraints and the commercial imperatives of market forces have not only left us highly dependent on fossil fuels, but have also tempered the interest for innovation and investment in new technologies for alternative energy.

Several factors contribute to resistance to innovation in sustainable energy. There is little interest in funding new technologies and infrastructures for replacing existing and entrenched C-based infrastructures. Dominant actors, imposed price caps, changing regulatory frameworks and network connection challenges are other obstacles to innovation. New technologies are at least initially more expensive than those they replace, while not providing a better service in terms of revenue.

The immediate benefits of innovation in sustainable energy tend to accrue to society rather than to the markets. Some new technologies, such as carbon capture and sequestration (CCS), face problems in social acceptance and often require additional up-front integration expenditures to fit into the existing energy infrastructures. Also, legal and administrative barriers render innovation in the energy sector more difficult. In addition to the scale of investments, these factors result in long lead times in energy innovation. It may often take decades to achieve substantial market investments.

In short, there is neither a natural market appetite nor a short-term business benefit for such technologies. Compared to other emerging technologies, the number of so-called *green patents* is low. Of all the patents filed yearly (1.91 million in 2008), 3,424 patent applications in four energy-related technology fields – fuel cells, solar, wind and geothermal energy – were filed in 2009. However, this is an increasing figure compared to the 584 applications in the same fields in 2000. Solar energy-related patent applications accounted for about 60% of this total increase (World Intellectual Property Indicators 2010).

Public intervention to support energy innovation is, thus, both necessary and justified. According to a report by the Bruegel Policy Contribution (Brussels), government intervention should consist of carbon pricing and the increased funding of research and development (Aghion et al. 2009).

In order to reconcile sustainable development and economic growth with the threat of environmental degradation, coherent strategic choices have to be made relying on truly innovative scientific and technological developments. These do not only include new ways for exploiting renewable energy resources, but also the development of new technologies for transporting, transforming and storing energy.

21.4 Carbon Capture and Sequestration

It is clear that fossil fuels will still be used on a large scale for the next few decades, and it will also take a comparable time before a sustainable carbon-free energy becomes a dominant reality. As a transitional technology for controlling the CO_2 concentration in the atmosphere, CCS is now increasingly being considered for capturing the CO_2 before it can reach the atmosphere (Florin and Fennell 2010). The CO_2 is extracted from the flue gases of electric power plants burning fossil fuels. The recovered CO_2 is then sequestrated in underground geological storage, such as disused oil and natural gas wells. An alternative approach is storage at the bottom of the sea. Currently, CCS is already used by the oil industry at a level of a few million Mg (1 Mg = 10^6 g) per year. About \$3.4 billion have been spent by the United States, and similar incentives exist in Europe and elsewhere.

However, CCS cannot be viewed as a definitive technology for achieving global carbon-free energy production. The capture, liquefaction and transport of CO_2 consume about 20% of the fuel burned in a power plant (Florin and Fennell 2010). The amounts of CO_2 to be stored underground are huge. Burning of 1 Mg of carbon coal produces 3.67 Mg of CO_2. At today's rate of 3.23 Pg of coal mined per year this would require the sequestration of 10 Pg CO_2 per year (Schrag 2007). It is questionable whether enough suitable structures underground could be found to store 1,000 Pg of CO_2 during the twenty-first century.

There also remains the question of the permanence of the underground storage because of possible degradation of the sealed wells as well as the risk of accidental leaks. Although not toxic (as for example carbon monoxide (CO)), higher concentrations of CO_2 can cause death when reaching the Earth's surface, because it is a powerful brain vein dilator. In a few minutes, unconsciousness occurs at 15% of relative concentration and immediate death at 30%. Deaths start being observed already at 9%.

Injection of CO_2 into the ocean also increases acidity. Over the last 150 years the ocean pH has dropped from 8.2 to 8.1 because of the increased concentration of CO_2 in sea water, with dramatic consequences to marine biota (Schiermeier 2011). It reduces calcification, reproduction growth, circulatory oxygen supply, and increases mortality. Immediate mortality is expected close to injection points, or near CO_2 lakes on the sea floor.

A safer method of C sequestration is mineral carbonization, also known as mineral C sequestration. Captured CO_2 is allowed to react with minerals present in geologic formations, such as olivine (magnesium iron silicate) or serpentine (hydrous magnesium iron phyllosilicate). The CO_2, either liquid or gaseous, reacts with the mineral and becomes a solid part of it and is thus safely and permanently stored. At least 40,000 times more C is contained in carbonate rocks than in the atmosphere, showing that the risk of release in the atmosphere is minimal (Oelkers et al. 2008). In principle, some energy can be recovered with mineral carbonation, and it can be the source of construction material and industrial reagents. A 500 MW power plant would need slightly more than 30,000 Mg of magnesium silicate ore per day to sequestrate the 10,000 Mg of CO_2 it produces per day (Goldberg et al. 2001). However, to make the process economically viable, faster reaction pathways would be required. Also, it is difficult to recover the low-grade heat, and the mining of the required minerals is viewed as an environmental drawback. Nonetheless, rock formations that react spontaneously with CO_2 are found in many parts of the world.

21.5 Carbon Dioxide Increase in the Atmosphere

The combustion of fossil fuels on a large scale started with the Industrial Revolution more than 200 years ago. The combustion of coal was first used for powering steam engines and is still now the major conventional energy source for electricity. More recently, burning of oil and gas has steadily increased the concentration of CO_2 in the atmosphere. Over the last 200 years, the concentration of CO_2 has increased from about 280 parts per million (ppm) to over 390 ppm (Neftel et al. 1994). Data taken at Mauna Loa Observatory by the National Oceanic and Atmospheric Administration (NOAA), starting in 1960, show that the CO_2 concentration increased from 1960 to 2010 by more than 60 ppm (NOAA 2011). The measurements also reveal an acceleration of the growth rate from the year 2007 to the beginning of 2011, when the CO_2 concentration increased from 383 to 391 ppm. A further increase, peaking around 410–425 ppm, is expected during the second decade of the twenty-first century (MacKay 2008). After this peak, the CO_2 concentration is expected to gradually return to 355 ppm by 2100.

However, the CO_2 added to the atmosphere is highly persistent. Several models have been elaborated showing that CO_2 remains in the atmosphere for thousands of years (Archer et al. 2009). For example, following a 5,000 Pg "pulse" of anthropogenic C emission, 17–33% will still be present after 1,000 years, 10–15% after 10,000 years, and 7% after 100,000 years, as presented in Fig. 21.1 (Archer 2005). This shows that the increase of CO_2 will affect the Earth's climate for centuries to come, yet the public and the scientific community have not fully appreciated the gravity of this situation. As a comparison, the lifetime of Pu-239 is 26,000 years and is in contrast quite relevant to today's public perception of nuclear energy decisions.

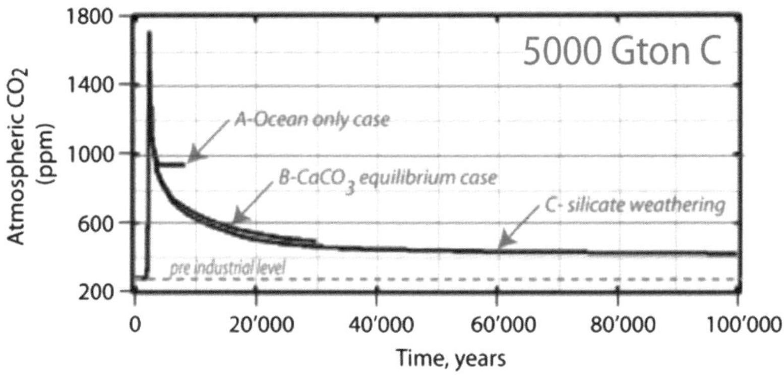

Fig. 21.1 Residual atmospheric CO_2 as a function of time (Archer 2005)

21.6 Towards New Technologies

We believe that the time has come to develop new technologies that will add to, and even replace, some of these existing technologies and are investigating energy production from fossil fuels without CO_2 emissions, as an alternative to CCS. We are also investigating a technology to use the already generated CO_2 for the production of a liquid substitute to oil, i.e., methanol. When used directly or converted into a higher hydrocarbon, methanol can replace oil in applications such as fuel for transport or as the starting material in the manufacture of many chemicals now derived from oil.

The realisation that huge amounts of CH_4 are trapped in clathrates in permafrost regions and below the sea floor has led us to investigate how the total amount of CH_4 stored in clathrates can be assessed and how methods for recovery of this CH_4 can be developed, since this could represent an important new energy source in the distant future.

Although uranium fuel is an energy source without CO_2 emissions, nuclear energy has largely fallen out of favour because of safety problems and the still largely unresolved problem of the long-term storage of nuclear waste. Further, uranium reserves are limited as well and the production of plutonium in fission reactors is a remaining possible cause of the proliferation of nuclear weapons. As an alternative, thorium represents a novel, almost inexhaustible nuclear fuel. The research in thorium reactors, started at Conseil Européen pour la Recherche Nucléaire (the European Organization for Nuclear Research (CERN) and elsewhere, is important because thorium reactors are inherently safer and more environmentally friendly than the current nuclear reactors burning uranium. The lifetime of the nuclear waste is much shorter, of the order of few hundred years. The link to military applications is severed.

21.7 An Alternative – Burning Natural Gas Without Carbon Dioxide Emissions

Natural gas, which consists mainly of CH_4, is widely used for heating and the generation of electric power. The combustion produces vast amounts of CO_2 that are released into the atmosphere. An important step in circumventing CO_2 emissions is the transformation of the large amounts of fossil natural gas into H_2 through the spontaneous thermal decomposition of methane (TDM): $CH_4 \rightarrow 2H_2 + C$. When heated, CH_4 dissociates spontaneously into hydrogen (H_2) gas and solid carbon black. Several experimental approaches have been reported in the literature (Muradov 2001; Muradov et al. 2006). The CH_4 is passed through a heated tube and H_2 is recuperated from the top of the tube, while carbon black can be removed mechanically from the bottom of the tube, as schematically depicted in Fig. 21.2.

The direct combustion of CH_4 will produce 890 kJ mol^{-1}, while the combustion of the resulting 2 mols of H_2 is 2×286 kJ mol^{-1}. The product theoretical efficiency is then 64%. The carbon black represents 394 kJ mol^{-1} C and its mass is 75% of the

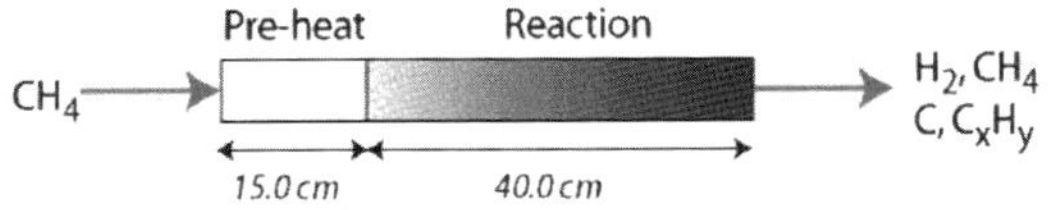

Fig. 21.2 Experimental analysis of direct thermal methane cracking (Courtesy of Alberto Abánades, IASS and Universidad Politécnica de Madrid)

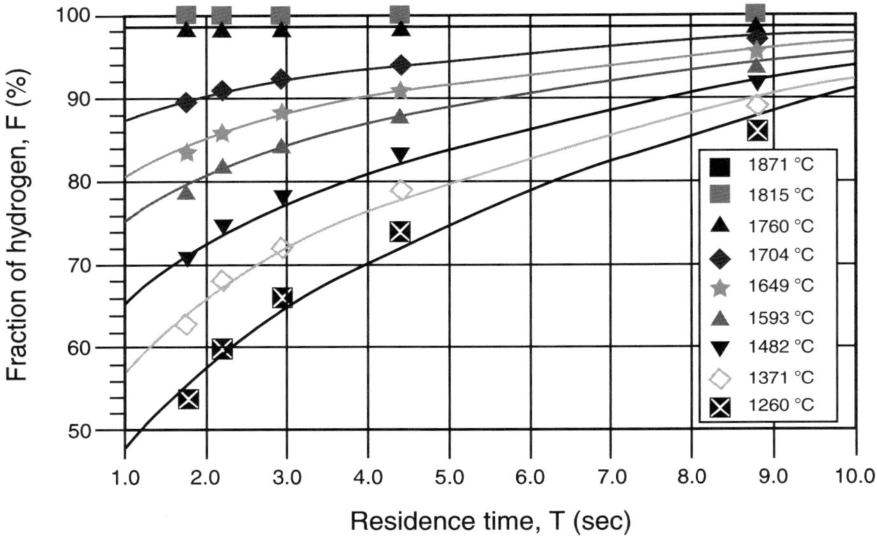

Fig. 21.3 Dissociation of methane at different temperatures and residence times (Courtesy of Alberto Abánades, IASS and Universidad Politécnica de Madrid)

CH_4. The relatively small endothermic contribution of 76 kJ mol^{-1} (Muradov 2001) is 13% of the H_2 produced. The total energy balance is then:

$$\begin{array}{cccc} CH_4 & \Delta_{heat} & C & 2\times H_2 \\ 890\ \text{kJmol}^{-1} & +76\ \text{kJmol}^{-1} \rightarrow & 394\ \text{kJmol}^{-1} & +572\ \text{kJmol}^{-1} \end{array} \tag{21.1}$$

If the CH_4 pyrolysis takes place at a temperature higher than 1,600°C, it is experimentally observed that the dissociation rate reaches almost 100 % after a residence time of a few seconds, as estimated in Fig. 21.3. The dissociation temperature can be lowered by using a catalyst. Instead of a heated tube, CH_4 can be bubbled through a molten salt or metal (*Fe, Sn, NaCl, NaF)*, whereby the metal could as well serve as a catalyst. A decrease of common pressure in the system shifts the reaction equilibrium in favour of H_2 generation. At 700°C and under a common pressure of 0.1 MPa, the H_2 concentration reaches 90 vol% at equilibrium. The recuperated carbon black can be safely sequestered or used in a number of commercial applications including use as a filler or construction material and in the production of carbon compounds, such as ethanol, and many compounds now derived from oil.

Table 21.1 A comparison of reforming and pyrolysis of NG for H_2 production (Steinberg 1999)

Item	SRM–reforming	TDM–pyrolysis
Reaction chemistry	$CH_4 + 2H_2O = CO_2 + 4H_2$	$CH_4 = C + 2H_2$
Mols H_2 per mol CH_4	4	2
Endothermic Ht of reaction kcal/mol CH_4	60	18
At 80% thermal eff. process heat in kcal/mol H_2	18.8	11.3
Process thermal efficiency for H_2 production–%	75	58
CO_2 emission mols CO_2/mol H_2	0.43	0.0
Lbs CO_2 gas/MMBTU	155	
Lbs C solid/MMBTU	0	49
Process unit operations	1. Reformer 2. Shift 3. CO_2 separation	1. Pyrolyzer 2. CH_2 separation if needed
Sequestration % net energy reduction	Liquid CO_2, in ocean, gas wells, aquifers = ~15%	Solid C, in land fill, mines o market = ~0%
Net energy efficiency %	75–15 = 60% Energy lost = 40%	58% Energy stored = 42%
By-product value	Low	High materials potential
Uncertainties	Possible hazardous environmental Effects	Minimal
Process development	Well developed	Needs development

Currently, the standard industrial process for producing H_2 from CH_4, called Steam Reforming of Methane (SRM), consists of the reaction of CH_4 with water, i.e., $CH_4 + 2H_2O \leftrightarrow CO_2 + 4H_2$. Although SRM also liberates the H_2 from water and produces 4 mols H_2 per mol CH_4 instead of the 2 mols of H_2 in pyrolysis, the disadvantage of SRM is the production of CO_2. Although the net energy efficiencies are comparable (60% and 58%), pyrolysis may turn out to be a cheaper production method for H_2 than SRM. Furthermore, SRM loses energy (40%) while pyrolysis stores energy in the form of carbon black, as shown in Table 21.1.

Another approach towards the pyrolysis of CH_4 has been developed by Kvaerner Engineering (Norway), known as the CB and H process (Gaudernack and Lynum 1998). Using cooling water and electricity to produce a plasma arc, CH_4 is separated into pure C and H_2 at a temperature of 1,600°C. Disadvantages of this technique are the high cost of the electricity and the inefficiency of the process. A pilot plant produces about 500 kg of pure carbon (carbon black) per hour and 2,000 Nm^3/h of H_2 from 1,000 Nm^3/h of natural gas (37.08 GJ h^{-1}) and 2,100 kWel (7.56 GJ h^{-1}) and 1,000 kW of high-temperature steam.

The next step is the planned construction of a plant capable of producing 120,000 Nm^3/h of H_2 with a cost estimated to be somewhere in the vicinity of $ 200 Mil.

21.8 The Methanol Economy

Although the natural reserves of oil still represent many years' supply, it is expected that prices will be driven up when demand surpasses supply, creating a lasting oil crisis with dire economic consequences. Today, the economy is still largely dependent on an uninterrupted supply of oil; the most vulnerable sectors are transportation with 97% reliance on oil and industry with 42%. The use of vehicles will increase enormously in emerging economies. For example, an estimate by researchers in China projects the total number of vehicles in their country to increase to over 860 million by 2030 (Zhang et al. 2010). Today the total number of cars in the US is 260 million (National Transportation Statistics 2011).

In order to avoid the massive extraction of oil from shale and tar sands and the extraction of heavy oils, we have to envisage a substitute that can replace oil in transport. The most promising approach is the methanol economy, which combines the use of H_2 obtained from CH_4 with the reuse of CO_2 as an alternative to the sequestration of CO_2 (Olah 2005; Olah et al. 2009a). The main concepts proposed by George Olah are schematically represented in Fig. 21.4. CO_2 used in the production of methanol is ultimately released back into the atmosphere, but since it has been taken out of the atmosphere in the first place, we can say that the combustion of the obtained methanol leaves no carbon footprint. The operation of a car with

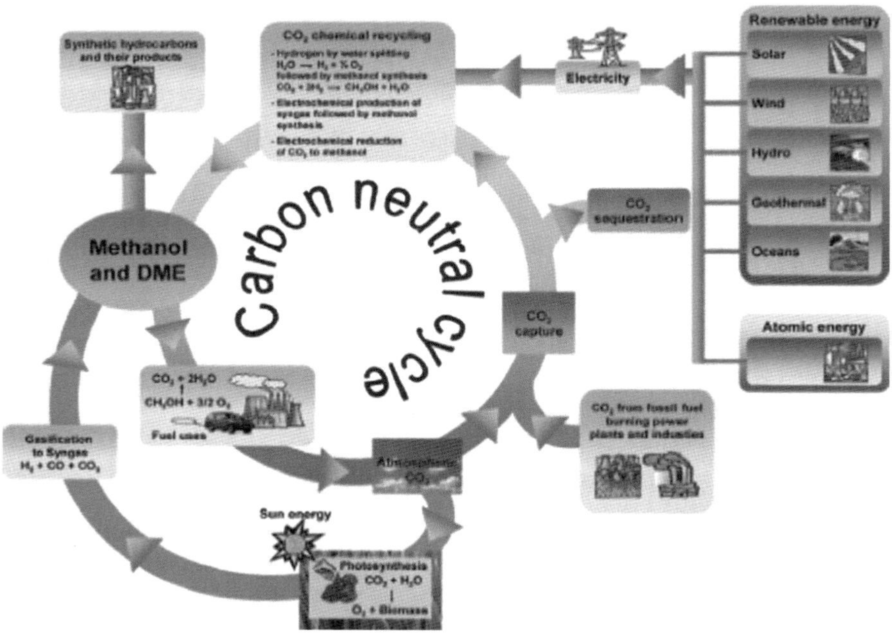

Fig. 21.4 Carbon dioxide recycling in the methanol economy (Olah et al. 2009a)

methanol would be completely without a carbon footprint, if the energy obtained from the dissociation of CH_4 for the creation of H_2 and the energy required for combining hydrogen with CO_2 to produce methanol would be of renewable or nuclear energy origin. Methanol is also a convenient starter material for many chemicals that now are petroleum derivatives, such as ethanol, ethylene, propylene and many industrial polymers. Compared to H_2 as a possible fuel, methanol is much better suited for powering vehicles because its use does not demand a new distribution infrastructure or major changes in engine technology.

21.9 Clathrates – New, Unconventional Source of Natural Gas

The progressive decarbonisation of fossil fuels will necessarily be followed by an increase in the use of CH_4. CH_4 has a low C footprint as compared to coal, which releases 1.8 times more C upon burning than CH_4 per unit of released energy (Olah et al. 2009b).

Over the last few decades, large amounts of CH_4 have been discovered in the Earth's crust in the form of methane hydrates. Gas hydrates such as methane hydrates are inclusion compounds (also called clathrates) that consist of a water framework. The hydrogen bonded water molecules form cavities that enclose small molecules such as CH_4. Although many gases may form gas hydrates, CH_4 is the predominant species in naturally occurring clathrates. Figure 21.5 shows dissociating methane hydrates with burning methane.

Fig. 21.5 Methane, released from the clathrate burns, a phenomenon known as "burning ice" (Courtesy of Manja Luzi, IASS)

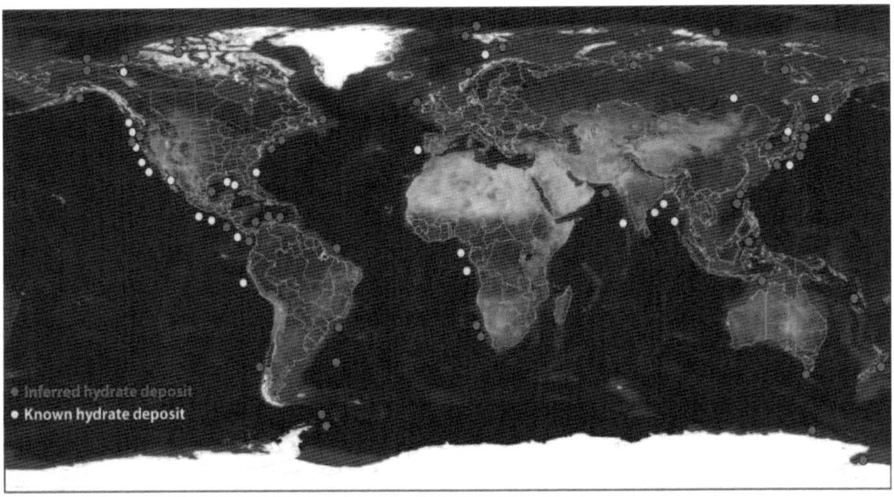

AR Hester KC, Brewer PG. 2009.
Annu. Rev. Mar. Sci. 1:303–27

Fig. 21.6 Distribution of observed and inferred clathrate deposits

In general, methane hydrates form from liquid water or ice and in presence of sufficient amounts of gas under elevated pressures and at low temperatures. These conditions are given in permafrost regions, for instance in the Siberian tundra, in marine sediments of the continental shelves at up to 2,000 m from the sea floor, in deep lake sediments such as Lake Baikal, but also in gas and oil pipelines. Methane hydrates are very efficient at storing the gas – 1 l of hydrate can store 164 l of CH_4 (at standard temperature and pressure (STP)).

Natural gas hydrate deposits have been investigated in several areas onshore and offshore, for example, in the Nankai Trough off Japan, the Canadian Arctic, and the Pacific Ocean. These findings allow us to infer that methane hydrate deposits are distributed worldwide. Figure 21.6 shows a map presenting the worldwide locations of known and inferred gas hydrate reservoirs. Figure 21.6 indicates that the majority of the hydrate deposits is situated in ocean continental margins, especially at tectonic plate junctions. Hydrate accumulations in permafrost regions are much less reported (Sloan and Koh 2008).

The actual global gas hydrate inventory is subject to uncertainties and controversy. Estimates range from 3,000 Pg (marine hydrate deposits only), to 10,000 Pg, or even to as much as 63,400 Pg (marine hydrate deposits only) of C bound in hydrates (Buffett and Archer 2004; Kvenvolden 1988; Klauda and Sandler 2005). By comparison, known natural gas reserves are about 500 Pg C.

Because the conventional reserves of natural gas may reach a limit later this century, the CH_4 trapped in clathrates is viewed as a new, climate-friendly energy resource that could be exploited in the future.

On the other hand, since the stability of gas hydrate deposits is strongly related to temperature and pressure variability and because CH_4 itself is a potent GHG, it is expected that a climate change with temperatures rising above 2°C might cause non-linear effects such as a run-away condition with an increased emission of CH_4 from the Earth's crust. The most severe extinction that occurred 251 million years ago, where about 96% of all living species became extinct, was probably caused by a massive degassing of methane clathrates caused by a pre-existing temperature rise of 6°C. Recently, researchers reported an increase of CH_4 venting from the sea floor near West Spitsbergen caused by northward flowing currents attributed to global warming.

Investigations on physical and chemical properties of gas hydrates, the development of exploration and production techniques, as well as the role of gas hydrates in the global C cycle will continue to develop as important research areas.

21.10 Conclusions

The problems concerning energy in a future sustainable world and the related potential threats are central to the scope of modern science: addressing these challenges can be carefully evaluated and solutions can be found. But the question remains if mankind will have the common political will to act in the face of these dangers and if it can reach a wide understanding enabling it to assess the probability of avoiding these threats. These are global problems that humanity is poorly equipped to handle.

In a world in which numerical population growth may become decoupled from scientific development it is unclear if humanity will take up the challenge of qualitative growth. It may instead fall into a pattern of slower development that could become stagnant and lead to decay.

The human capital of the most educated section of society is a decisive factor in establishing norms of social conduct, determined by higher education and attitudes and values propagated, for example, by the media.

Choices will have to be taken over the next 10–15 years that will have profound consequences for energy security, for climate change, for growth and for jobs. The cost of action may be high, but the price of inactivity much higher.

Acknowledgements The contributions of Alexander Hellemans, Manja Luzi, Adela Marian, Delia Salmieri and Alberto Varone from IASS are gratefully acknowledged.

References

Aghion P, Veugelers R, Serre C (2009) Cold start for the green innovation machine. Bruegel policy contribution, issue 12. http://aei.pitt.edu/12219/1/pc_climateparvcs_231109.pdf

Archer D (2005) Fate of fossil fuel CO_2 in geologic time. J Geophys Res 110, C09S05. doi:10.1029/2004JC002625

Archer D, Eby M, Brovkin V et al (2009) Atmospheric lifetime of fossil fuel carbon dioxide. Annu Rev Earth Planet Sci 37:117–134. doi:10.1146/annurev.earth.031208.100206
Buffett B, Archer D (2004) Global inventory of methane clathrate: sensitivity to changes in the deep ocean. Earth Planet Sci Lett 227:185–199
Caldeira K, Wickett ME (2005) Ocean model predictions of chemistry changes from carbon dioxide emissions to the atmosphere and ocean. J Geophys Res 110, C09S04
Florin N, Fennell P (2010) Carbon capture technology: future fossil fuel use and mitigating climate change. Grantham Institute for Climate Change, briefing paper no. 3. Imperial College, London. http://workspace.imperial.ac.uk/climatechange/Public/pdfs/Briefing%20Papers/Grantham%20Briefing%20paper_Carbon%20Capture%20Technology_November%202010.pdf
Friedlingstein P, Houghton RA, Marland G et al (2010) Update on CO_2 emissions, letter to the editor. Nat Geosci 3:81–812. doi:10.1038/ngeo1022
Gaudernack B, Lynum S (1998) Hydrogen from natural gas without release of CO_2 to the atmosphere. Int J Hydrog Energy 23:1087–1093. doi:10.1016/S0360-3199(98)00004-4
Goldberg P, Chen ZY, O'Connor W et al (2001) CO_2 mineral sequestration studies in US. In: Proceedings of the first national conference on carbon sequestration, Washington, DC, 14–17 May 2001, session. 6c. http://www.netl.doe.gov/publications/proceedings/01/carbon_seq/6c1.pdf
IEA (2010a) CO_2 emissions from fuel combustion: highlights. IEA, Paris. http://www.iea.org/co2highlights/CO2highlights.pdf
IEA (2010b) Key world energy statistics. IEA, Paris. http://www.iea.org/textbase/nppdf/free/2010/key_stats_2010.pdf
IPCC (2007) Summary for policymakers. In: Climate change 2007: the physical science. IPCC. Geneva. http://www.ipcc.ch/publications_and_data/publications_ipcc_fourth_assessment_report_wg1_report_the_physical_science_basis.htm
Klauda JB, Sandler SI (2005) Global distribution of methane hydrate in ocean sediment. Energy Fuel 19:459–470
Knies G, Moeller U, Straub M (2009) Clean power from the deserts. In: Knies G, Moeller U, Straub M (eds) The DESERTEC concept for energy, water and climate security. White paper, Desertec Foundation, Hamburg
Kvenvolden KA (1988) Methane hydrate – a major reserve of carbon in the shallow geosphere? Chem Geol 71:41–51
MacKay DJC (2008) Sustainable energy – without the hot air. UIT, Cambridge. http://withouthotair.com/
Muradov N (2001) Hydrogen via methane decomposition: an application for decarbonization of fossil fuels. Int J Hydrog Energy 26:1165–1175. doi:10.1016/S0360-3199(01)00073-8
Muradov N, Smith F, Huang C et al (2006) Autothermal catalytic pyrolysis of methane as a new route to hydrogen production with reduced CO_2 emissions. Catal Today 116:281–288. doi:10.1016/j.cattod.2006.05.070
National Transportation Statistics (2011) U.S. Department of Transportation, Table 1–11. http://www.bts.gov/publications/national_transportation_statistics/
Neftel A, Friedli H, Moor E et al (1994) Historical carbon dioxide record from the Siple Station ice core. In: Trends: a compendium of data on global ghange. Carbon Dioxide Information Analysis Center, Oak Ridge National Laboratory, U.S. Department of Energy, Oak Ridge. http://cdiac.ornl.gov/trends/co2/siple.html
NOAA (2011) Trends in atmospheric carbon dioxide. NOAA, http://www.esrl.noaa.gov/gmd/ccgg/trends/
Oelkers EH, Gislason SR, Matter J (2008) Mineral carbonation of CO_2. Elements 4:333–337
Olah GA (2005) Beyond oil and gas: the methanol economy. Angew Chem Int 44:2636–2639
Olah GA, Goeppert A, Prakash GKS (2009a) Chemical recycling of carbon dioxide to methanol and dimethyl ether: from greenhouse gas to renewable, environmentally carbon neutral fuels and synthetic hydrocarbons. J Org Chem 74:487–498
Olah GA, Goeppert A, Prakash GKS (2009b) Beyond oil and gas: the methanol economy, 2nd edn. Wiley-VCH, Weinheim

Overpeck JT, Weiss JL (2009) Projections of future sea level becoming more dire. Proc Natl Acad Sci USA 106:21461–21462

Schiermeier Q (2011) Earth's acid test. Nature 47:154–156

Schrag DP (2007) Preparing to capture carbon. Science 315:812–813

Sloan ED, Koh CA (2008) Clathrate hydrates of natural gases, 3rd edn. CRC Press, Boca Raton

Steinberg M (1999) Fossil fuel decarbonization technology for mitigating global warming. Int J Hydrog Energy 24:771–777. doi:10.1016/S0360-3199(98)00128-1

Weiss JL, Overpeck JT, Strauss B (2011) Implications of recent sea level rise science for low-elevation areas in coastal cities of the conterminous U.S.A. Clim Change 105:635–645. doi:10.1007/s10584-011-0024-x

World Intellectual Property Indicators (2010) World Intellectual Property Organization. http://www.wipo.int/freepublications/en/intproperty/941/wipo_pub_941_2010.pdf

Zhang Q, Tian W, Zheng Y et al (2010) Fuel consumption from vehicles of China until 2030 in energy scenarios. Energy Policy 38:6860–6867. doi:10.1016/j.enpol.2010.06.058

Chapter 22
Bioenergy and Biospheric Carbon

Tim Beringer and Wolfgang Lucht

Abstract Land use change is responsible for about 15% of global greenhouse gas (GHG) emissions. Major efforts are underway to reduce deforestation and expansion of agriculture for food production which are responsible for the ongoing depletion of terrestrial carbon (C) stocks. At the same time, the global demand for biomass as an energy source is rising. Recent analyses of the global energy system suggest that large amounts of bioenergy are needed to achieve ambitious climate protection targets at reasonable economic costs. Efforts to cut GHG emissions and increase energy security with bioenergy may, thus, stimulate future land use changes leading to decarbonization of the terrestrial biosphere. Various sustainability standards are discussed in order to reduce the social and environmental impacts of large-scale bioenergy production. In this study, a dynamic global vegetation model was used to simulate the cultivation of modern lignocellulosic energy crops under different scenarios of land availability that consider competing land use objectives for food security, nature conservation and C pool protection. The analyses indicate that global bioenergy potentials are lower than previously thought, because many regions with potentially high productivity do not qualify for the establishment of dedicated biomass plantations. Human encroachment into these areas generates large C emissions that would delay any climate-benefits for centuries. However, biomass may still play a significant role in future energy production because large areas remain where

T. Beringer (✉)
Earth System Analysis, PIK, Potsdam Institute for Climate Impact Research, Telegrafenberg A 31, Potsdam 14473, Germany
e-mail: tim.beringer@pik-potsdam.de

W. Lucht
Potsdam Institute for Climate Impact Research, Potsdam, Germany

Department of Geography, Humboldt Universität zu Berlin, Berlin, Germany
e-mail: lucht@pik-potsdam.de

R. Lal et al. (eds.), *Recarbonization of the Biosphere: Ecosystems and the Global Carbon Cycle*, DOI 10.1007/978-94-007-4159-1_22,

simulated biomass yields offset emissions from land use change within a short period of time. Exploiting these potentials, however, will inevitably affect a large number of ecosystems that are already under pressure from human activities.

Keywords Biofuel • Bioenergy • Greenhouse gas emissions • Deforestation • Terrestrial carbon stocks • Biomass • Climate protection • Decarbonization • Lignocellulosic energy crops • Bioenergy plantations • Land use change • Human activities • Renewable energy • Carbon capture and storage • Land use • Policy framework • Biodiversity • Contemporary carbon cycle • Degradation of natural ecosystems • Environmental constraints • Agronomic productivity • Crop residues • Soil fertility • Poplar • Switchgrass • REDD

Abbreviations

C	carbon
CCS	carbon capture and storage
CO_2	carbon dioxide
GHG	greenhouse gas
IPCC	Intergovernmental Panel on Climate Change
CH_4	methane
N_2O	nitrous oxide
REDD	Reducing Emissions from Deforestation and Forest Degradation
RE	renewable energy
SOC	soil organic carbon

22.1 Introduction

In a recent comprehensive assessment of renewable energy (RE) potentials for climate change mitigation, the Intergovernmental Panel on Climate Change (IPCC) concluded that bioenergy, wind, and solar technologies will be the most important renewable sources in future energy systems (IPCC 2011). Over a wide range of scenarios, bioenergy is projected to become the single largest RE source in the coming decades. Biomass can be used across all sectors for the production of electricity, heat, and transportation fuels. Biomass plays a central role in ambitious climate protection targets, because bioenergy production with carbon capture and storage (CCS) generates negative emissions required for fast reductions of GHG emissions at comparatively low mitigation costs (Edenhofer et al. 2010; Krey and Clarke 2011).

At the same time, serious concerns are raised about the massive expansion of global bioenergy production. Biomass has a low energy density meaning that large quantities of feedstock are required to produce significant amounts of bioenergy.

It is therefore expected that a growing demand for biomass will lead to a further intensification and expansion of human land use that is already responsible for serious environmental degradation and social changes worldwide (Foley et al. 2005). Available sources of biomass include residues from agriculture and forestry as well as dedicated energy crops. Additional appropriation of biomass for bioenergy may thus contribute to the human interference with the global biogeochemical cycles leading to further depletion of terrestrial C stocks through deforestation and energetic use of residues from agriculture and forestry (Blanco-Canqui and Lal 2009; Fargione et al. 2008).

C is at the center of the growing influence of humanity on the global environment that manifests itself in climate and land use change. C emissions from the combustion of fossil fuels and large scale deforestation are mainly responsible for increasing atmospheric concentrations of carbon dioxide (CO_2) and other GHGs. There is growing evidence that unabated climate change and the progressive degradation of natural resources threaten the planetary life support systems and may ultimately overwhelm the adaptive capacities of societies and natural ecosystems with large negative consequences for human well-being (WBGU 2007). At the center of the problem lies the universal dependency of modern societies on fossil energy sources. Ambitious climate protection efforts based on the massive expansion of renewable energies to reduce GHG emissions and the development of modern grid infrastructures are a prerequisite for a transformation towards increased global sustainability based on low-C societies (WBGU 2011).

Bioenergy is a prime example for the complex challenges involved in managing such a great transformation. While supporters of increased biomass use highlight the prospects of fossil fuel substitution and rural development especially in the developing countries (IEA Bioenergy 2009), there are growing concerns about the consequences of massive expansion of human land use into natural ecosystems (Fargione et al. 2008). At the same time, economic models of the energy system project significantly higher mitigation costs if not enough biomass will become available in the future for electricity production in combination with CCS (Leimbach et al. 2010). Bioenergy as an element of both the energy and the land use system interacts with sinks and sources of C in terrestrial ecosystems and changes in anthropogenic GHG emissions in multiple ways.

Comprehensive policy frameworks will be required to ensure that further development of bioenergy and biofuels use contributes to climate protection, energy security and rural development (Tilman et al. 2009). Without an extensive set of regulations, uncontrolled expansion of biomass cultivation into native ecosystems can lead to the depletion of C stocks that will delay any GHG benefits of bioenergy for decades to centuries (Gibbs et al. 2008) and accelerate biodiversity loss (Secretariat of the Convention on Biological Diversity 2010). Concerns about deteriorating food security and rising prices for staple crops due to the imminent competition between food and energy production have also been raised. Market incentives that promote the cultivation of crops for the bioenergy market could result in reduced food output, leading to reduced self-sufficiency and greater need for food imports. Prices of agricultural feedstocks suitable for biofuel production such as maize (*Zea mays* L.) and soybean are also becoming increasingly dependent on the development of global energy markets. A tighter link between oil and food commodity prices involves also the risk for higher price volatility in agricultural input and food markets (FAO 2011).

22.2 The Contemporary Carbon Cycle

Human activities are responsible for fundamental perturbations of the natural C cycle. Since the onset of the industrial revolution, atmospheric CO_2 concentrations increased by 40% (Global Carbon Project 2010) while methane (CH_4) concentrations are now about 150% higher than in 1,750 (Richardson et al. 2009). In order to satisfy the growing worldwide demand for food and shelter, mankind has modified more than half of the Earth' ice-free land surface (Ellis and Ramankutty 2008) for agriculture and settlements, and harvests nearly a quarter of the biosphere's primary production for food and feed annually (Haberl et al. 2007). Changes in land use account for about 15% of global CO_2 emissions (Friedlingstein et al. 2010). Large scale deforestation for agriculture occurs predominantly in tropical regions where it threatens terrestrial C stocks and biodiversity (Gibbs et al. 2010; Koh et al. 2011). Expansion of human land use may lead to further increase in CO_2 and nitrous oxide (N_2O) emissions from deforestation, soil cultivation and fertilization (Crutzen et al. 2008). Enhanced C flows from deforestation and land management reduced soil C storage by 10% and biomass C stocks by 25%, respectively (Bondeau et al. 2007).

Aside from its role in the GHG balance, moving C from the biosphere into the atmosphere and subsequently the oceans is also of great concern. Specifically, C losses from the soil hamper crucial ecosystem properties such as nutrient- and water-holding capacities which affect soil fertility and agronomic yield potentials (Lal 2009). The ongoing degradation and destruction of natural ecosystems poses also a threat to global biodiversity. Habitat fragmentation and loss in combination with accelerating climate change may drive a large number of animal and plant species into extinction (Barnosky et al. 2011).

In addition, natural terrestrial ecosystems are C sinks and absorb roughly one quarter of anthropogenic CO_2 emissions. Thus, the observed annual increase in atmospheric CO_2 concentration is substantially smaller than the amount of CO_2 released each year (Global Carbon Project 2010). The currently observed enhanced CO_2 uptake by terrestrial ecosystems is explained by longer growing season, increasing precipitation and the positive effects of CO_2 fertilization which enhance plant growth and C storage in growing biomass (Fischlin et al. 2007).

Expanding bioenergy production can affect terrestrial C storage in several ways. The establishment of dedicated biomass plantations may replace natural ecosystems altering the exchange of C between vegetation, soils and atmosphere. In most cases, this will result in short and long term C losses from the ecosystem and, thus, increased GHG emissions (Fargione et al. 2008, Fig. 22.1). Current practices of

Fig. 22.1 Simulated changes in C storage (vegetation and soil) based on the LPJmL model when pristine ecosystems are converted into biomass plantations with lignocellulosic energy crops under current climate. This figure does not consider current agricultural areas and their effect on terrestrial C stocks, but illustrates the consequences of expanding human land use for bioenergy production into natural vegetation. An equal share of tree and grass energy crops is assumed within all grid cells where bioclimatic conditions allow cultivation

biofuel production based on the use of food and oil crops from existing croplands may cause an expansion of food production elsewhere causing indirect land use changes (Searchinger et al. 2008). The altered natural vegetation may have continued to function as C sinks for many decades. Such foregone C sinks are externalities of bioenergy production generally not accounted for in the analysis of GHG mitigation opportunities from energy crops (Searchinger 2010). Where biomass plantations, in particular using lignocellulosic crops, are established on current or abandoned croplands, improvements in soil quality, including soil organic carbon (SOC) content have been observed (Tolbert et al. 2002).

Residues from agriculture and forestry are another potential source of biomass that involves fewer risks for large scale land use change than energy crop cultivation. The use of non-edible plant materials also avoids competition with food production. However, indiscriminate removal of crop and wood residues from fields and forests carries risks for land quality and productivity because nutrients and C are removed with the residues (Blanco-Canqui and Lal 2009).

22.3 Bioenergy and Terrestrial Carbon

Firewood and charcoal were humanities primary source of energy over most historic times, and have only been replaced recently by coal, oil and natural gas during the last 150 years. Today, biomass is the largest renewable source of energy and provides about 10% of the global primary energy demand (IPCC 2011). However, most of this biomass is used traditionally, and combusted directly for cooking and heating in simple devices with low conversion efficiencies. As a consequence, indoor air pollution causes serious damages to human health and is a wide-spread problem in many developing countries (Bailis et al. 2005).

In contrast, modern bioenergy systems can produce various end-products using different organic feedstock such as residues from agriculture and forestry, organic wastes as well as oil, starch and lignocellulosic energy crops. A number of technologies are available or under development to convert biogenic materials into electricity and heat and gaseous, liquid or solid fuels (Faaij 2006). Because the transportation sector is almost exclusively depending on vehicles with internal combustion engines powered by liquid hydrocarbons, the potential to substitute petroleum for fuels produced from bioresources has received particular interest from governments and industry (Ragauskas et al. 2006).

Bioenergy is generally (IPCC 2006) and may therefore contribute to the reduction of GHG emissions from energy production and transport. Blending mandates for biofuels are an example for political incentives that aim to increase the market share of biofuels. Biomass is an exceptionally versatile energy carrier and, in particular, the possibility to produce wood pellets, biogas or biofuels allows transportation and storage on existing infrastructures. In modern integrated energy systems, bioenergy has the ability to balance intermittent power flows from solar power plants and wind turbines (WBGU 2009). Another goal of increased bioenergy use as an alternative to fossil energy carriers is improved energy security and reduced dependence on imports of oil and natural gas from politically unstable regions (IEA Bioenergy 2009). Biomass can also substitute petroleum-based raw materials in the chemical industry (Ragauskas et al. 2006). It can be assumed that rising crude oil prices will also raise the demand for biomaterials in this sector which may lead to growing competition between energetic and raw material uses of biomass (SRU 2007).

Global demand for biomass increased steadily during recent decades and this trend is expected to continue (Haberl et al. 2007; WBGU 2009). According to the FAO (2009), the world population will rise to more than nine billion by 2050, increasing the demand for food and feed by 70%. The future demand for biomass will not only reflect increasing consumption of vegetable and animal food by a growing and more affluent world population but also the expanding use of bioenergy and renewable raw materials (IEA 2009, 2011; WBGU 2011). Because the conversion of solar energy into new biomass during photosynthesis operates at low efficiencies, large amounts of land and water will be required to produce sizable additional quantities of feedstock. Continuing intensification and expansion of agriculture is therefore expected to further increase the pressure on global land use.

22.4 Global Bioenergy Potentials Under Environmental Constraints

A number of recent studies indicated that most current practices to convert food-derived carbohydrates or plant oils into ethanol and biodiesel have only limited capabilities to reduce GHG emissions from the energy and transport sector. However, they compete directly with food production on the most fertile lands. C emissions related to direct and indirect land-use change may be larger than the amount of CO_2

saved by replacing fossil fuels. For example, it takes up to 400 years to pay back this so-called C debt for tropical forests (Fargione et al. 2008; Gibbs et al. 2008; Searchinger et al. 2008). Nitrogen leaching is another far-reaching consequence from the dependency of modern food crop production N fertilizers (Donner and Kucharik 2008). Residues from agriculture and forestry as well as lignocellulosic energy crops are generally regarded as more sustainable sources for biomass (Tilman et al. 2009).

Crop residues are left in the field after harvest and their use for bioenergy production may be a sustainable source for biomass. Their role in the recycling of C and nutrients, however, warrants a more specific evaluation of sustainable residue removal rates. In most cases, the utilization of residues from agriculture and forestry avoids indirect land use changes. On the field, non-harvested residues serve as carriers for C and nutrients cycled between vegetation and soils and, thus, perform important functions within the ecosystem. Crop residues replenish nutrients and SOC pools, provide food for soil biota, protect against erosion, and filter runoff (Lal 2005). These ecosystem functions contribute to maintaining agronomic productivity and, consequently, removing too much crop residues may undermine soil fertility of the land or lead to increased fertilizer use to maintain yields. Site specific environmental conditions determine the amount of residues that may be collected without long-term negative effects for soil physical, chemical, and biological properties. For corn stover, annual removal rates beyond 25% have been found to reduce SOC pools and grain yields (Blanco-Canqui and Lal 2007). Use of harvest residues from forests for bioenergy presents a number of threats to local ecosystems. Branches and leaves contain more nutrients than stem wood and their removal may lead to impoverished soil fertility and reduced tree growth in the long-term (Åkselsson et al. 2007; Chapin et al. 2002). Coarse woody debris is also an important substrate for invertebrates and fungi that perform essential functions in soils including maintenance of soil structure, regulation of hydrological processes, and soil detoxification (FAO 2005).

Annual bioenergy potentials from agricultural and forestry residues therefore remain uncertain. Recent analysis indicate that between 50 and 100 EJ year^{-1} (1 EJ=10^{18} J) may become available from residues, municipal solid wastes and animal manures (Haberl et al. 2010; WBGU 2009). In the face of further increasing food demand for a population that may surpass ten billion at the end of the century (UN 2011), using crop residues to maintain and sustain soil fertility is the first priority. According to the most recent assessments, producing large amounts of bioenergy will thus require a large contribution from dedicated biomass plantations (IPCC 2011).

Most global assessments of future bioenergy production have not considered the negative effects of establishing biomass plantations such as C losses from land use change. Thus, the assessments likely overestimate the amount of bioenergy potentially available without negative consequences for climate protection. In a recent study, simulations of large scale lignocellulosic energy crop cultivation from a global biosphere model were combined with a range of sustainability constraints for food security, nature conservation, and C protection restrict the availability of land for additional agricultural areas in the future (Beringer et al. 2011).

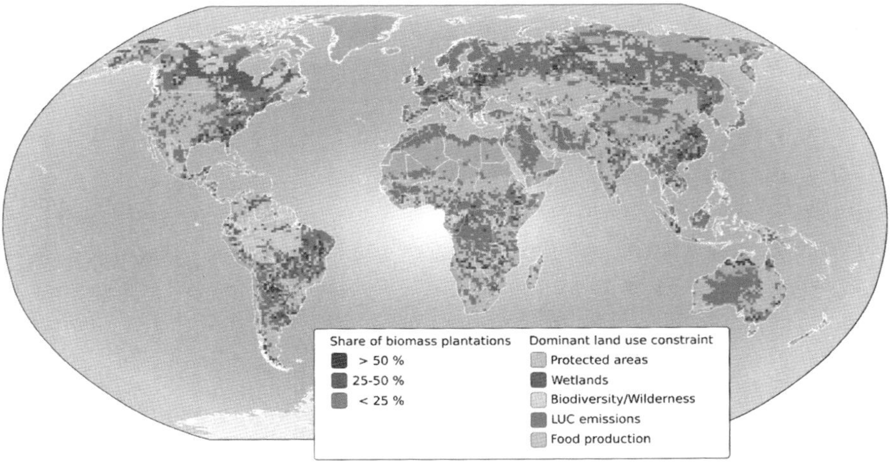

Fig. 22.2 Spatial distribution of potential bioenergy plantations and their relative share within grid cells for a scenario without cropland expansion (Beringer et al. 2011). Dominant land use constraints are shown where the cultivation of energy crops is not allowed in this scenario. These areas are excluded because they are already protected (protected areas), belong to the remaining wetlands (wetlands), contain large contiguous areas of undisturbed ecosystems with high nature conservation value due to the provision of important ecosystem services or high concentrations of biodiversity (wilderness/biodiversity), where C losses would result from land conversion that cannot be compensated by subsequent C uptake within a compensation period of 10 years (land use change (*LUC*) emissions), or because they are already agricultural areas required for food production (food production). Note that in most cases several constraints occur simultaneously

Large scale cultivation and yield potentials of different lignocellulosic energy crops like poplar (*Populus* spp.) or switchgrass (*Panicum virgatum* L.) were simulated with the dynamic global vegetation LPJmL (Bondeau et al. 2007; Sitch et al. 2003; Gerten et al. 2004). LPJmL integrates major biogeochemical and biophysical ecosystem processes to calculate C and water fluxes in vegetation and soils at the global scale. The model was driven by a range of climate scenarios from the fourth IPCC assessment report (Meehl et al. 2007) to include the effects of changes in temperature and precipitation on biomass yields. In order to quantify the potential role of biomass plantations for energy in a future energy system, land availability for dedicated energy crops was determined following the guardrail concept of the German Advisory Council on Global Change (WBGU 2009). In this approach, different categories of land are systematically excluded from future human use because they are required for future food production, nature conservation or GHG mitigation. These areas include current croplands and pastures, pristine forests, wetlands, and protected areas. The remaining land resources are assumed to be available for the cultivation of energy crops (Fig. 22.2).

According to these scenarios, the global bioenergy potential from dedicated biomass plantations ranges between 30 and 120 EJ annually, depending on the different scenario assumptions about competing land requirements for food production and nature conservation. Using all renewable water resources not required for food production and natural ecosystems for irrigation, the total potential increases to 50–170 EJ $year^{-1}$. Expanding modern bioenergy production in the coming decades may, thus, provide more than 15% of anthropogenic energy demand which is expected to rise to 600–1,000 EJ $year^{-1}$ (IEA Bioenergy 2009). Producing large amounts of biomass, however, will come at the expense the World's remaining natural ecosystems, because the cultivation of energy crops is very land intensive. Depending on the scenario, agricultural areas for food, feed and bioenergy production may expand up to 30% and agricultural water demand for irrigation may double compared to current consumption. Although the most valuable ecosystems in terms of biodiversity and C storage are preserved in the scenarios, the remaining areas may still have considerable regional importance for wildlife and local communities. Examples include the semiarid scrub forests of the Caatinga, the savannas of the Cerrado and the Chaco, as well as the grassland of the Humid Pampas. All these regions have experienced manifold effects of human activities in the past so that natural ecosystems are often degraded and fragmented remnants of the original habitat. At the same time, they represent valuable and unique ecosystems that also have cultural value.

22.5 Outlook and Conclusions

Human land use may deliver co-benefits for energy security, rural development, and climate protection, but only if it is internationally coordinated and regionally adapted. The contribution of modern bioenergy technologies to the reduction of GHG emissions depends on the effective protection and management of terrestrial C stocks, because the large scale establishment of biomass plantations in addition to further areas for future food production introduces artificial C flows that alter the exchange of C between biosphere and atmosphere (Melillo et al. 2009; Searchinger 2010). If C-rich forests continue to be cleared for the cultivation of oil-palms (*Elaeis* spp.), large emissions of CO_2 occur that outweigh any gains from energy crop C sequestration for decades. Today, around 15% of anthropogenic CO_2 emissions are related to deforestation, forest degradation, and peatland destruction (van der Werf et al. 2009). Forest protection through biomass certification schemes or the Reducing Emissions from Deforestation and Forest Degradation (REDD) initiative are therefore important elements of sustainable future bioenergy strategies. REDD could become part of a post-2012 climate agreement to provide financial incentives for developing countries to better protect their forest resources. However, increasing demand for biomass in addition to rising food production requirements can lead to increasing land scarcity and, hence, greater opportunity costs of the land (Lambin and Meyfroidt 2011; Smith et al. 2010). Higher expenditures required to compensate landowners for their foregone profits from deforestations and agriculture may erode the economic viability of REDD schemes (Butler et al. 2009).

Promoting the increased use of agricultural residues and biogenic wastes instead of expanding energy crop cultivation is also reasonable from a climate mitigation and nature conservation perspective (WBGU 2009). Residue removal rates on croplands, however, need to be adapted to local site conditions to prevent losses of SOM and nutrients that may decrease soil fertility and reduce crop yields in the long term (Blanco-Canqui and Lal 2009).

Even though long term climate benefits of bioenergy may justify a C prize for additional land use change that will be compensated over time, any further alteration of the Earth' land surface will increase the pressure on water resources (Berndes 2002) and adds to the ongoing loss and fragmentation of habitats that is already a serious threat for biodiversity (Barnosky et al. 2011). A recent study showed that grassland ecosystems can host similar levels of animal species diversity although they store significantly less biomass C (Strassburg et al. 2010). A narrow focus of sustainability criteria and guidelines for biomass cultivation on climate protection and C stock may, thus, ensure the positive impact of bioenergy use on GHG emissions, but still entail negative effects for biodiversity. Comprehensive policy instruments need to be established that integrate a range of standards for bioenergy production in relation to GHG emissions, competition with food production, conservation of natural resources including soils, water, and biodiversity.

References

Åkselsson C, Westling O, Sverdrup H et al (2007) Impact of harvest intensity on long-term base cation budgets in Swedish forest soils. Water Air Soil Pollut Focus 7:201–210

Bailis R, Ezzati M, Kammen DM (2005) Mortality and greenhouse gas impacts of biomass and petroleum energy futures in Africa. Science 308:98–103

Barnosky AD, Matzke N, Tomiya S et al (2011) Has the Earth's sixth mass extinction already arrived? Nature 470:51–57

Beringer T, Lucht W, Schaphoff S (2011) Bioenergy production potential of global biomass plantations under environmental and agricultural constraints. Glob Change Biol Bioenergy 3:299–312

Berndes G (2002) Bioenergy and water – the implications of large-scale bioenergy production for water use and supply. Glob Environ Change 12:253–271

Blanco-Canqui H, Lal R (2007) Soil and crop response to harvesting corn residues for biofuel production. Geoderma 141:355–362

Blanco-Canqui H, Lal R (2009) Crop residue removal impacts on soil productivity and environmental quality. Crit Rev Plant Sci 28:139–163

Bondeau A, Smith P, Zaehle S et al (2007) Modelling the role of agriculture for the 20th century global terrestrial carbon balance. Glob Change Biol 13:679–706

Butler RA, Koh LP, Ghazoul J (2009) REDD in the red: palm oil could undermine carbon payment schemes. Conserv Lett 2:67–73

Chapin FS, Matson PA, Mooney HA (2002) Principles of terrestrial ecosystem ecology. Springer, Berlin

Crutzen PJ, Mosier AR, Smith KA et al (2008) N_2O release from agro-biofuel production negates global warming reduction by replacing fossil fuels. Atmos Chem Phys 8:389–395

Donner SD, Kucharik CJ (2008) Corn-based ethanol production compromises goal of reducing nitrogen export by the Mississippi River. Proc Natl Acad Sci USA 105:4513–4518

Edenhofer O, Knopf B, Barker T et al (2010) The economics of low stabilization: model comparison of mitigation strategies and costs. Energy J 31:11–48
Ellis EC, Ramankutty N (2008) Putting people in the map: anthropogenic biomes of the world. Front Ecol Environ 6:439–447
Faaij APC (2006) Modern biomass conversion technologies. Mitig Adapt Strateg Glob Change 11:343–375
FAO (2005) The importance of soil organic matter: key to drought-resistant soil and sustained food and production. Food and Agriculture Organization of the United Nations (FAO), Rome
FAO (2009) How to feed the world in 2050. Food and Agriculture Organization of the United Nations (FAO), Rome
FAO (2011) The state of food and agriculture 2010-11. Food and Agriculture Organization of the United Nations (FAO), Rome
Fargione J, Hill J, Tilman D et al (2008) Land clearing and the biofuel carbon debt. Science 319:1235–1238
Fischlin A, Midgley GF, Price JT et al (2007) Climate change 2007: impacts, adaptation and vulnerability. In: Parry ML, Canziani OF, Palutikof JP et al (eds) Contribution of Working Group II to the fourth assessment report of the Intergovernmental Panel on Climate Change. Cambridge University Press, Cambridge
Foley JA, DeFries RS, Asner GP et al (2005) Global consequences of land use. Science 309:570–574
Friedlingstein P, Houghton RA, Marland G et al (2010) Update on CO_2 emissions. Nat Geosci 3:811–812
Gerten D, Schaphoff S, Haberlandt U et al (2004) Terrestrial vegetation and water balance-hydrological evaluation of a dynamic global vegetation model. J Hydrol 286:249–270
Gibbs HK, Johnston M, Foley JA et al (2008) Carbon payback times for crop-based biofuel expansion in the tropics: the effects of changing yield and technology. Environ Res Lett 3:34001
Gibbs HK, Ruesch AS, Achard F et al (2010) Tropical forests were the primary sources of new agricultural land in the 1980s and 1990s. Proc Natl Acad Sci USA 107:16732–16737
Global Carbon Project (2010) Carbon budget 2009. http://www.globalcarbonproject.org/carbonbudget/
Haberl H, Erb K-H, Krausmann F et al (2007) Quantifying and mapping the human appropriation of net primary production in earth's terrestrial ecosystems. Proc Natl Acad Sci USA 104:12942–12947
Haberl H, Beringer T, Bhattacharya SC et al (2010) The global technical potential of bio-energy in 2050 considering sustainability constraints. Curr Opin Environ Sustain 2:394–403
IEA (2009) World energy outlook 2009. OECD Publishing, Paris
IEA (2011) Technology roadmap biofuels for transport. http://www.iea.org/papers/2011/biofuels_roadmap.pdf
IEA Bioenergy (2009) Bioenergy – a sustainable and reliable energy source. http://www.ieabioenergy.com/LibItem.aspx?id=6479
IPCC (2006) Guidelines for national greenhouse gas inventories. IPCC, Geneva
IPCC (2011) IPCC special report on renewable energy sources and climate change mitigation. Cambridge University Press, Cambridge/New York
Koh LP, Miettinen J, Liew SC et al (2011) Remotely sensed evidence of tropical peatland conversion to oil palm. Proc Natl Acad Sci USA 108:5127–5132
Krey V, Clarke L (2011) Role of renewable energy in climate mitigation: a synthesis of recent scenarios. Clim Policy 11:1131–1158
Lal R (2005) World crop residues production and implications of its use as a biofuel. Environ Int 31:575–584
Lal R (2009) Soil degradation as a reason for inadequate human nutrition. Food Secur 1:45–57
Lambin EF, Meyfroidt P (2011) Global land use change, economic globalization, and the looming land scarcity. Proc Natl Acad Sci USA 108:3465–3472
Leimbach M, Bauer N, Baumstark L et al (2010) Technological change and international trade – insights from REMIND. Energy J 31:1786–1793

Meehl GA, Stocker TF, Collins WD et al (2007) Global climate projections. In: Solomon S, Qin D, Manning M et al (eds) Climate change 2007: the physical science basis. Contribution of Working Group I to the fourth assessment report of the Intergovernmental Panel on Climate Change. Cambridge University Press, Cambridge/New York

Melillo JM, Reilly JM, Kicklighter DW et al (2009) Indirect emissions from biofuels: how important? Science 326:1397–1399

Ragauskas AJ, Williams CK, Davison BH et al (2006) The path forward for biofuels and biomaterials. Science 311:484–489

Richardson K, Steffen W, Schellnhuber H-J et al (2009) Synthesis report from climate change. Global Risks, Challenges & Decisions. http://climatecongress.ku.dk

Searchinger TD (2010) Biofuels and the need for additional carbon. Environ Res Lett 5:024007

Searchinger TD, Heimlich R, Dong F et al (2008) Use of U.S. croplands for biofuels increases greenhouse gases through emissions from land-use change. Science 319:1238–1240

Secretariat of the Convention on Biological Diversity (2010) Global Biodiversity Outlook 3. http://www.cbd.int/gbo3/

Sitch S, Smith B, Prentice IC et al (2003) Evaluation of ecosystem dynamics, plant geography and terrestrial carbon cycling in the LPJ dynamic global vegetation model. Glob Change Biol 9:161–185

Smith P, Gregory PJ, van Vuuren DP et al (2010) Competition for land. Philos Trans R Soc Lond B Biol Sci 365:2941–2957

SRU (2007) Klimaschutz durch Biomasse. Erich Schmidt Verlag, Berlin

Strassburg BBN, Kelly A, Balmford A et al (2010) Global congruence of carbon storage and biodiversity in terrestrial ecosystems. Conserv Lett 3:98–105

Tilman D, Socolow R, Foley JA et al (2009) Energy. Beneficial biofuels – the food, energy, and environment trilemma. Science 325:270–271

Tolbert VR, Todd DE, Mann LK et al (2002) Changes in soil quality and below-ground carbon storage with conversion of traditional agricultural crop lands to bioenergy crop production. Environ Pollut 116:S97–S106

UN (2011) World population prospects: the 2010 revision. United Nations, New York

van der Werf GR, Morton DC, DeFries RS et al (2009) CO_2 emissions from forest loss. Nat Geosci 2:737–738

WBGU (2007) Climate change as a security risk. Earthscan, London/Sterling

WBGU (2009) Future bioenergy and sustainable land use. Earthscan, London/Sterling

WBGU (2011) Welt im Wandel: Gesellschaftsvertrag für eine große Transformation. WBGU, Berlin

Chapter 23
The Economics of Land and Soil Degradation-Toward an Assessment of the Costs of Inaction

Joachim von Braun and Nicolas Gerber

Abstract Recarbonization of the biosphere is a desirable objective in view of climate change and greenhouse gas (GHG) emission problems. Yet it is confronted with at least two challenges. First, there are increased trade-offs between biomass uses in the emerging bio-economy (e.g., food-fuel competition). This impacts the role of soils for carbon (C) sequestration. These trade-offs are to be reconciled by accelerated knowledge and innovation intensive approaches in a "Green Growth" strategy. Secondly, the degradation of the earth's lands and soils is increasingly recognized as a global problem as extent and impacts are increasingly affecting and affected by environmental and social vulnerability as well as climate change. Both of these challenges cannot be met without a comprehensive assessment of the land and soil degradation issue. A review of the state of the art on the quantification and mapping of degradation, its effects and driving forces, and its economic valuation is provided here. Further, a framework for a global assessment of costs of Action versus Inaction against degradation is proposed.

Keywords Land and soil degradation • Economic valuation • State of knowledge • Methodology

J. von Braun (✉)
Zentrum für Entwicklungsforschung der Universität Bonn, Walter-Flex-Strasse 3, Bonn 53113, Germany
e-mail: jvonbraun@uni-bonn.de

N. Gerber
Center for Development Research, University of Bonn, Walter-Flex-Strasse 3, 53113 Bonn, Germany
e-mail: ngerber@uni-bonn.de

R. Lal et al. (eds.), *Recarbonization of the Biosphere: Ecosystems and the Global Carbon Cycle*, DOI 10.1007/978-94-007-4159-1_23,

Abbreviations

C	carbon
CO_2	carbon dioxide
the CAVI approach	costs induced by continued degradation
BöR	German Bio-economy Council
GLADA	Global Land Degradation Assessment
GLADIS	Global Land Degradation Information System
GHG	greenhouse gas
LSD	land and soil degradation
LADA	Land Degradation Assessment in the Drylands
MAC	marginal abatement cost curve
MA	Millennium Ecosystem Assessment
NPP	net primary production
NDVI	Normalized Difference Vegetation Index
RUE	rainfall use efficiency
RESTREND	residual trend analysis
SOC	soil organic carbon
SOM	soil organic matter
SA	South Asia
SSA	Sub-Saharan Africa
SLWM	sustainable land and water management
TEEB	The Economics of Ecosystems and Biodiversity
UNCCD	United Nations Convention to Combat Desertification

23.1 Introduction

The bio-economy is defined as the cross-sectoral element of the economy which produces, processes or uses any type of biological resources, as well as the services related to this element (BöR 2011). The bio-economy, thus, covers very different types of activities, spanning from the traditional production of raw biological material (e.g., traditional forestry or agriculture) to frontier industries in the field of biotechnology. The existence of the bio-economy is a fact. Human development has been based on the use of biological resources from the beginning, and humankind has continuously developed and benefited from innovations in the way that it can consume these biological resources (BöR 2011).

As the demand for biological resources increases – due to factors such as population growth, changes in consumption behavior, the globalization of the world economy, the development of new bio-based products, or the need to find substitutes to decrease the world's dependence on inorganic raw materials and fossil fuels in the energy, construction and industrial sectors – competition over the same biological resources will intensify. Further, the value chains of the different sectors of the (bio-) economy become increasingly intertwined and complex.

An example of this trend is the intensification of the link between agricultural and energy markets (von Braun 2008). Fuel prices drive food prices not only as input factors but also because of the competition between biofuels and food products over land resources and other inputs. This, compounded with a growing world consumption of food and a reduction in the growth rate of agricultural production, recently resulted in increased volatility of food prices, with devastating effects for the poorest (von Braun 2008).

As value chains for biomass are generally becoming longer and competition over biological intensifies, it is crucial that biomass-use efficiency along the value chains increases. Priorities must be set in the use of biomass. The German Bio-economy Council (BöR) notes that the material use of biomass creates higher values than its use in the energy sector (BöR 2011). Thus, the argument of the BöR is that value chains must be organized so that biomass can be used first in the material chain, and then in the energy chain. These priorities should also include by-products. For example, by-products of the agricultural production are increasingly used for and developed into feed, fertilizer and energy value chains (e.g., grain – oil mills, sugar – bio-ethanol). Even higher value product chains will be available for these by-products in the future as biotechnology will enable to break down and extract specific substances from biological material.

In several instances, the increase in competition over biological resources across the sectors of the bio-economy is aggravated by the fact that stocks are already being depleted at alarming rates. Such is the case for productive land and soils around the world. Arable land per capita worldwide decreased from 0.45 ha in 1950, to 0.35 ha in 1970 and 0.22 ha in 2000, and is projected to decrease to 0.15 ha by 2050 (Lal and Stewart 2010), with alarming trends between 1970 and 2005 in Sub-Saharan Africa (SSA) and South Asia (SA) (from 0.50 to 0.25 ha and 0.28–0.14 ha, respectively, Nkonya et al. 2011). As the competition over land resources intensifies, land prices are set to rise worldwide. The phenomenon of "land grabs", i.e., the (trans)national commercial acquisition of large tracks of land, has strongly increased in recent years and is further evidence of the increased competition over land resources.

Land degradation, in its most recent and inclusive definition, refers to the change in productivity and change in the provision of ecosystem services and the human benefits derived from them (Nkonya et al. 2011). This definition is very inclusive and covers issues such as desertification (i.e., land degradation in the drylands), soil degradation, as well as the degradation of what is produced from the soil (i.e., decrease in biomass production and cover, for example due to deforestation or other land cover changes, or decreased capacity of soil to sustain biomass growth). Soil degradation in this chapter is defined as a decrease in soil quality, e.g., a decrease in the amount of soil nutrients and in the concentration of soil micro-organisms, that is associated with a decrease in the ability of the soil to sustain agricultural production in particular, or to sustain other terrestrial ecosystem services and benefits. The role of soil micro-organisms for soil organic carbon (SOC) sequestration is one of the scientific frontiers, with scientists studying the impact of increased temperatures on their activity, amongst other things (Reichelt 2009).

The recarbonization of the biosphere is understood in this chapter as the process of drawing carbon (C) back out of the atmosphere and sequestering it on land, either in the vegetation cover or in soil. The assimilation of atmospheric carbon dioxide (CO_2) in plants is a result of the photosynthesis process. The capture of atmospheric CO_2 in soils is the result of several processes, both "natural" and "man-made". The natural process is the result of decomposition of dead plant biomass, which in turn enriches the soil with SOC and is an important source of nutrient for plant growth. Thus soils naturally store C. The combined total of C stored in soils worldwide is several times higher than that stored in biomass or in the atmosphere (Blaustein 2010). In that respect, soils hold the potential to contribute to the decrease in atmospheric CO_2 concentration if managed wisely, but also to aggravate atmospheric CO_2 levels if managed unwisely. As expressed by Lal (2004), SOC is an important resource in itself, the main component of soil organic matter (SOM), a major source of nutrient, and SOC levels ought to be restored as a mean to improve the productivity of soil and land resources. Careful management of the C cycle in agriculture may increase soil productivity and have the added benefit of decreasing atmospheric CO_2 concentrations. The conversion of grass lands into crop land has for example a negative impact on the atmospheric CO_2 balance, as grasslands trap much C into their root system, are less disturbed by management practices (i.e. tillage) and receive less fertilizer than croplands. The conversion of forests into crop land releases also C into the atmosphere due to enhanced oxidation of C-rich soil organic matter (SOM). Similarly, tillage farming practices or the burning of agricultural residues release SOC through oxidation. Man-made processes of C sequestration also exist and are currently developed, as alternative and supplementary solutions to reductions in greenhouse gas (GHG) emissions. They include the injection of CO_2 into the oceans or into underground geological formations, as well as trapping C in the form of various carbonated minerals.

Land and soil degradation are caused by two types of processes, i.e., processes which are typically man-made, and natural processes over which humans have little or no control. Their effects, however, can in many cases be mitigated by human interventions. Causes of land and soil degradation are reviewed in more detail in a later section. The emphasis at this stage of the paper is on the existence of a link between land and soil degradation (LSD) and human actions. These actions, or lack thereof, are the enactments of choices made by land users (defined broadly, from farmers to national institutions), individually or collectively. These choices are usually the results of decision-making based on the information at the disposal of the land users and on the set of rules administering land use and management. These conditions which frame the decision-making process are referred to as "institutions" and are broadly defined as "the rules of the game" under which actors form their decisions. Similarly to other cases of depletive natural resource uses, it is then assumed that land use decisions leading to LSD are the consequences of a rational decision from a single land user's perspective given the set of institutions.

If the degradation of land and soil resources is sub-optimal from a societal point of view (i.e., not the single user's view but society as whole including future land users), it is likely that the land user's decisions are ignoring impacts of his/her decisions on other (potentially future) members of the society. Economists refer to this as the externalities caused by the present exploitation of the land resources. A typical externality mentioned in this context is the siltation of water ways caused by the erosion of agricultural land. The externality can also be imposed on land users of the future. The current land user is depleting the productive capacity of the land in a way that is rational from his/her perspective, but not from an intergenerational one. In this sense, land and soil quality are public goods from a societal and multigenerational perspective, and there are no incentives for the current land user to conserve the productive capacity of the land for the future. Institutional arrangements, such as national laws or market interactions, can influence and modify the incentive structures affecting land use and management. The role of policy makers is then to ensure that these structures deliver the socially desirable set of land use decisions, including soil and land conservation actions. From an economic standpoint, it is crucial that these actions meet the requirements of effectiveness and cost efficiency, in order to guarantee their economic sustainability. The recarbonization of the soils is one solution envisioned to the issues of soil degradation and climate change. Thus, it is important that recarbonization is evaluated along other options, technical as well as institutional (i.e., influencing the way land users form their decisions, such as policies aiming at decreasing CO_2 emissions), according to the criteria of effectiveness (i.e., what results can be achieved in terms of soil improvement and climate control) and efficiency (i.e., at what costs can these achievements be attained).

23.2 Land and Soil Degradation in Economics

A change in LSD matters because of its impacts on the benefits that people derive from the use of the land, soils and services they can provide. Many of these benefits are not directly observable or quantifiable through market interactions, as markets for many environmental goods and services do not exist. Hence, as a first step for the evaluation of LSD it is important to catalogue and evaluate all the ecosystem services that un-degraded lands and soils provide. All types of terrestrial ecosystems (and their services and the benefits they provide to humans) should be covered in a comprehensive global economic assessment of LSD. This includes anthropogenic ecosystems – i.e., ecosystems which are heavily influenced by people (Ellis and Ramankutty 2008) such as agro-ecosystems, planted forests, rangelands, urbanized zones. Meanwhile, a vast majority of the literature investigating the impacts of LSD, its costs and mitigation deals with agro-ecosystems. Yet, agro-ecosystems (defined as spatially and functionally coherent units of agricultural activity) are strongly linked to other ecosystems, for instance through the provision of ecosystem functions such as supporting and regulating services (e.g., climate regulation, water purification).

UNCCD (1996) defines land as “the terrestrial bio-productive system that comprises soil, vegetation, other biota, and the ecological and hydrological processes that operate within the system". As such, land provides ecosystem services categorized in the Millennium Ecosystem Assessment (MA 2005) as supporting services (e.g., soil development, primary production, nutrient cycling), provisioning services (e.g., the delivery of food, fibre, forage, fuelwood, biochemicals, fresh water), regulating services (e.g., water regulation, pollination/seeds, climate regulation) and cultural services (e.g., recreation, landscapes, heritage, aesthetic).

Whilst MA (2005) was instrumental in illustrating the importance of ecosystem services to human well-being, the concept of ecosystem services is not always perfectly suitable for framing the economic valuation of land resources. This point is clear in Balmford et al. (2008), i.e., what economists seek to value are the benefits people derive from the ecosystem services, not the services themselves. The concept of ecosystem services as proposed in the MA (2005) lends itself to a problem of double-counting. Although the separation between a service and the actual benefits people draw from it is not necessarily straightforward, clearly the valuation of nutrient cycling as a supporting ecosystem service and of food production as a provisioning service illustrates the double-counting issue. At this point in time, not enough is known about the relationships among the various types of services and between the different services and benefits for an economic valuation of LSD to focus solely on the benefits derived from terrestrial ecosystems. Focusing on the measurement of benefits whenever possible, and keeping in mind the issue of double-counting when falling back on literature estimates of the value of services when needed, is ,thus, crucial under current knowledge. Yet, the challenges posed by the dynamic interactions between LSD and the opportunity costs of land use change, altering the nature of ecosystems and of the benefits and services they provide, demand a global coverage of the latter. Unfortunately and despite considerable advances in the economic valuation of ecosystem services in recent years, there are still many gaps in this field in terms of the coverage of specific ecosystem services and gaps in the geographical coverage of such valuation.

23.3 The Relationship Between Climate Change and Land and Soil Degradation

Climate change and LSD are related through the interactions between the land surface, the soils and the atmosphere. These interactions involve multiple processes, with impact flows running from the land to the atmosphere and vice versa. The feedback effects between climate change and land degradation are not yet fully understood.

Climate change impacts land degradation because of its longer-term trends and because of its impacts on the occurrence of extreme events and increased climate variability. Climate change trends include the increase in temperature and change of rainfall patterns. These are two determinants in the creation and evolution of soils, notably through their impact on the distribution of vegetation. Further, climate variability, a predicted consequence of climate change, holds the potential for some of the most severe human impacts. For instance, the occurrence and severity of droughts has been related to actual declines in economic activity. In SSA in particular, climate variability will affect growing periods and yields, and is expected to intensify land degradation and affect the ability of land management practices to maintain land and water resources in the future (Pender et al. 2009). However, it must also be noted that climate change is not only a negative influence on land degradation. For example, agro-climatic conditions are expected to improve in some areas.

Simultaneously, land and soil degradation impacts on climate change through the direct effects of degradation processes on the land surface, which then affects for instance atmospheric circulation patterns, as well as through the effects of LSD on land use, land use changes in turn affecting the climate.

In those complex interrelationships between climate change and LSD, sustainable land and water management (SLWM) can play a crucial mitigating role. Notably, research has shown the links between soil C sequestration and its impact on climate change and food security (Lal 2004). Soil C sequestration transfers atmospheric CO_2 into the soils, hence mitigating its climate change impacts. Increasing soil C stocks in turn has a positive impact on crop productivity, at least past a certain minimum threshold (World Bank 2010). Thus, SLWM practices that sequester large amounts of soil C can provide a win-win-win solution to the issues of climate change, land degradation and some of its human dimensions, such as food security (Lal 2006). Examples of such practices, which aim at the recarbonization of the biosphere, include no-till farming, cover crops, manuring and agroforestry (Lal 2004).

The extent of the win-win-win situations mentioned above and the conditions under which they can be realized are areas where more systematic research is required. As climate change and variability will impact different regions in different ways, so too will their consequences vary in terms of LSD in general. Further, the linkages between land and climate systems hold important keys to the valuation of the costs of LSD and of land conservation or restoration.

23.4 Why Is a Global Assessment of Land and Soil Degradation Necessary?

LSD is a widespread environmental issue affecting all climatic zones. Desertification is a term often used to describe LSD in arid, semi-arid and dry sub humid areas (as illustrated by the definition of the UNCCD) while studies focusing on the observation of land degradation as a loss of vegetation cover clearly identify that land degradation is very important in humid areas which account for almost 80% of the world's total degraded land area (Bai et al. 2008). Naturally, the results of studies on the extent of land degradation are dependent on the definition of LSD that they use and on the indicators of degradation they rely on. One clear fact remains nonetheless, i.e., although the type of degradation and/or the processes through which degradation takes place can be highly local specific, the issue of LSD is global in terms of its extent. LSD is also a global issue in terms of its impacts on human well-being. It affects land users (e.g., farmers), the consumers of land products (e.g., food), the users of other ecosystem services which are affected by land degradation (referring to the notion of externalities mentioned earlier, e.g., the users of sedimentated water ways), as well as consumers and producers of goods which are affected by land degradation through market mechanisms (e.g., producers and consumers of goods using land products as inputs, such as bread, or competing with land products for inputs such as labor, water, capital). All these elements of the general issue of LSD, and the fact that land quality is a global public good, point towards the need for a global assessment of the costs of LSD.

The economic assessment of environmental and climatic problems has received increased international attention in recent years. For example, the Stern Review on the Economics of Climate Change was released for the British government in October 2006 (Stern 2006). The Economics of Ecosystems and Biodiversity (TEEB) initiative was launched as a consequence of the G8+5 Environmental Ministers meeting in Potsdam, Germany, in March 2007, and has since produced several publications aimed at emphasizing the value of ecosystems and biodiversity as well as on ways how they can be managed. We propose to follow a framework similar to that put forward by those reports, i.e., an economic evaluation of the costs of action (i.e., the costs of mitigating land degradation) versus the costs of inaction (i.e., the costs induced by continued degradation) – the CAVI approach.

The Stern Review (2006) demonstrated the appeal of the CAVI approach lying in the immediate comparability of the results which matter to policy and decision-makers. It allows to answer the following question: what is the discounted value of the economic impact of letting land degradation taking place, which carries no immediate costs but future costs in terms of decreased terrestrial ecosystem services and benefits, compared to the discounted economic impacts of undertaking actions against land degradation now carrying immediate costs in terms of actions and future benefits in terms of sustained (relative to the inaction scenario) terrestrial ecosystem services and benefits? Implementing the CAVI approach is not an easy task. For the assessment to be unbiased, a wide range of effects and outcomes must be integrated in the analysis. They should include direct and indirect effects of LSD on the biosphere and their direct and indirect economic outcomes across the value chain of terrestrial ecosystem services and benefits and all their links to human well-being.

Land use decisions affect terrestrial ecosystem services and the stream of benefits they provide and, thus, impacting human well-being. The costs and benefits of these impacts can take several forms:

- On- and off-site costs and benefits, i.e., occurring either at the location where land degradation takes place (e.g., increased fertilizer used to compensate for depleted soil nutrients), or in other locations – the idea of externality (e.g., sediments resulting from erosion are carried in water ways, with potential costs – maintenance of dams, or benefits – increased fertility for downstream agricultural land).

- Indirect costs and benefits, which refer to the socio-economic impacts of LSD which go beyond the socio-economic group in which they are created, e.g., the food security impacts beyond the land-users, or the economy-wide effects of increased input prices for non-agricultural sectors due to the increase in agricultural prices and increased demand for inputs in agriculture. Indirect costs and benefits could be broadly defined as externalities, but giving them a specific name and differentiating them from the other externalities described above should highlight their importance and how little is known about them. They are a crucial component of a global assessment of the human dimensions of LSD.
- Current and future costs and benefits, i.e., LSD economic impacts felt instantly, during the life-cycle of the current land users, or in future generations – the time dimension is particularly important as certain forms of LSD are slow processes and/or processes whose consequences might not be fully observable for the current land users.

All categories need to be considered in order to cover not only private costs and benefits (i.e., those of the land user), but (more crucially) social costs and benefits. Thus, reflecting the overall impacts of LSD on human-well-being from the society's (and the planner's) perspective.

The CAVI approach and its implementation in the case of LSD is discussed in a later section.

23.5 The Causes of Land and Soil Degradation

In order to assess land degradation, it is important to identify its driving forces. They are generally classified as proximate or underlying causes of land degradation. The former can be split into biophysical causes and in unsustainable land management practices. The latter are those causes which indirectly affect the proximate causes, for instance the reasons why unsustainable land management occurs. Examples of the two categories are presented below.

23.5.1 Proximate Causes

The **biophysical proximate causes** of land degradation include topography (e.g., slope, exposure to sea water flux in costal zones, exposure to volcanic activity), land cover, climatic conditions (in particular extreme and variable weather patterns, such as heavy rainfalls, storms, cyclones, wild fires), soil erodibility, pests, and diseases.

The relationship between climate (change) and land degradation is particularly interesting and complex. In some cases, the impacts of climate are of sufficient intensity to induce ecological land degradation without human interference. Examples include dry and hot climate and natural bush fires, or monsoon rain and erosion. However, very often human activities trigger or exacerbate ecological land degradation (Barrow 1991). In an era of climate change, such human-induced triggers of or contributions to ecological land degradation might increase. The dynamics of the climate change – land degradation, through the interactions of the land (surface and below surface) with the atmosphere, involve multiple, simultaneous and changing processes and feed-back effects between the two systems. Not all interactions are fully understood yet.

Amongst **unsustainable land management practices**, the most frequently cited examples include land clearing, overgrazing, cultivation on steep slopes, bush burning, pollution of land and water sources and soil nutrient.

23.5.2 Underlying Causes

Institutions (including policies and incentive structures for land users) and other socio-economic factors affect the proximate causes of land degradation. In particular, they largely determine why the unsustainable management of land resources takes place. Underlying causes of LSD are particular relevant to study in a global assessment of the costs of action versus inaction against LSD, as they are the most directly influenced by land management (and other) policies. Some of the most commonly cited examples of underlying causes are presented below. As many examples show, the impact of the underlying cause on LSD is often ambiguous and context specific. Such complex interactions point towards the need for tailored and well-researched policy interventions against LSD. Such interventions require more research efforts for a better understanding of the key economic drivers of land management decisions and of the trade-offs that land users face between economic and ecological goals. These issues are likely to become more complex as the bio-economy grows.

National level policies can have a strong influence on land management practices, for instance through regulation or incentive structures. This influence can be direct, i.e., the policy directly targets land management (e.g., the payment for ecosystem services scheme to conserve forest and biodiversity in Costa Rica, Pagiola 2008). This influence can also be indirect, for example a policy targeting poverty reduction can indirectly impact on land management (e.g., a reform of land tenure systems). Indirect impacts can be positive or negative with regard to LSD. For example, agricultural policies, farm or input subsidies in particular, naturally have an enormous influence on land management. Subsidies have contributed to higher adoption of fertilizer use in several developing countries (Heffer and Prud'homme 2009), a mostly positive result from the LSD perspective. Yet subsidies can also contribute to environmental pollution through the excessive application of fertilizer (Mulvaney et al. 2009).

Over the past four decades, **international policies** and initiatives have increasingly been oriented towards sustainable development (Sanwal 2004), with strong impacts downstream the whole way to community-level land management. Examples of general initiatives include the Rio Summit of 1992, the Millennium Summit of 2000, the 2002 Johannesburg Summit on Sustainable Development. The Millennium

Ecosystem Assessment (MA) and TEEB illustrate global research efforts. An international initiative that directly addresses land degradation is the United Nations Convention to Combat Desertification (UNCCD) which went into effect 1996. The Asian Green Revolution was initiated as an international strategy aimed at increasing agricultural productivity to meet the increasing demand for food and had strong local impacts in terms of LSD, contributing to a reduction in the conversion of land to agricultural production (Borlaug 2000; Hazell 2010).

On the other hand, international policies can also have negative (indirect) impacts on LSD. International trade agreements are often mentioned in this respect. For example, some argue trade liberalization, coupled with financial and macroeconomic national policies and as well as growing world demand impacts the Brazilian agriculture, especially soybean production in the Brazilian Center-West – a Savanna area called Cerrado and a notoriously fragile ecosystem with naturally low levels of nutrients, supporting an important biodiversity (Mayrand et al. 2005). Erosion is a major issue with estimates of 6 kg of soil erosion for each kilo of crop produced (Mayrand et al. 2005). A double impact due to intensification and expansion can be observed with further indirect impacts through the marginalization of small farmers. The latter can often take the form of exclusion of the small farmers from "1st choice" input markets, including good land, capital, machinery, fertilizers, trapping them into the 2nd class food production chain relying on unsustainable and suboptimal production techniques and levels (yields below the production possibility frontier and agriculture extension on deforested land). Then land degradation further increases the level of marginalization closing the LSD – poverty trap.

Market access for agricultural producers can impact on LSD either positively or negatively, through complex incentives based on agricultural output and input markets. For example, good market access can lead to higher producer prices and cheaper agricultural inputs prices for the land users. Both can serve as incentives for investment in land management (Pender et al. 2006). On the other hand, easy market access can provide alternative livelihood opportunities. This can increase the opportunity cost of labour in agriculture and, thus, provide negative incentives to invest in labour intensive land management practices (Scherr and Hazell 1994). The application of other land and soil-preserving inputs such as manure can also be affected by (lack of) market access, for instance, through interactions with alternative fuel sources such as coal. Many other interactions between agricultural inputs, their substitutes and complements, through market transactions partly determine the adoption of sustainable land management practices and the level of LSD.

Boserup (1965) showed how agriculture can intensify under high **population density**. In support of this theory, empirical studies have illustrated the positive relationship between population density and land improvement (Bai et al. 2008; Tiffen et al. 1994; von Braun et al. 1991). However, evidence of the opposite also exists (Grepperud 1996). Indeed, the link between a decrease in per capita cropland availability (Katyal and Vlek 2000) and increasing food demand per capita worldwide (due to higher income levels) on one hand, and increasing global levels of LSD (e.g., conversion of forests into cropland, expansion of agriculture into more fragile land, unsustainable nutrient outflows) on the other hand, is intuitive. The differences between evidence found in various studies can be explained among other things by a wide range of conditioning factors, such as agricultural marketing, whose impacts have been discussed above, and by the scale of the analysis.

Secure **land** (user) **rights** are intuitively important to incentivize land users to invest in soil and water conservation measures. The reason is that such measures typically have high up-front costs and delayed benefits. Indeed, several studies have illustrated the positive impact of land titles on land management in different countries (Deininger and Chamorro 2004; Lopez 1997; Alston et al. 1995). However, Brasselle et al. (2002) showed that the literature on the topic provides rather inconclusive evidence on the sign of the impacts. Land titling efforts in particular (i.e., efforts to establish safer land right systems) have often failed to improve land management, to increase agricultural productivity, or to reduce poverty (Place and Hazell 1993; Deininger 2003) and decreased rangeland productivity through the permanent settlement of nomadic populations (Thomas and Middleton 1994). Further, in some cases insecure land rights can act as an incentive for increased investment in land improving strategies as the land user applies them to enhance his/her tenure security (Besley 1995; Place and Otsuka 2002; Brasselle et al. 2002). From these examples, it is evident that the impacts of land tenure policies are often context-dependent. This calls for local-specific policies to improve land management, reduce poverty, and achieve other objectives. In particular, the gender implications of land tenure systems are important to consider. For instance, if women cannot own land, they often cannot have access to agricultural inputs either leading them to adopt unsustainable land management practices.

The literature on the links between **poverty** and LSD reveals two diverging paradigms. In the first one, the idea that poverty leads to LSD which leads to yet more poverty is central, forming a vicious circle (Way 2006; Cleaver and Schreiber 1994; Scherr 2000). The reasons why the vicious circle is formed vary among authors, but a commonly cited cause is the lack of investment capability or "investment poverty" (Reardon and Vosti 1995), preventing poor farmers to invest in land conservation measures for lack of labour and/or capital means. As Safriel and Adeel (2005) pointed out, "*poverty is not only a result but also a cause of degradation*". Poverty is also often associated with the isolation of farmers on marginal, low productivity land (Rockström et al. 2003). In the second paradigm, different authors argue that the key factor is the strong dependence of the livelihood of poor farmers on the state of their land. This gives them equally strong incentives to utilize their limited labour and capital resources to maintain their soils and mitigate LSD, provided that the conditions under which they operate (i.e., legal system, market access) allows them to allocate their resources efficiently (de Janvry et al. 1991). As an illustration, Nkonya et al. (2008) found a negative relationship between soil erosion and livestock endowment. Thus, in the case of poverty and LSD, arises again a situation where the sign of the impact of the underlying cause on LSD cannot be generally determined and needs to be investigated on a case by case basis, along with many confounding factors. Other underlying causes of LSD listed in Nkonya et al. (2011) include infrastructure development, access to agricultural extension services, and local institutions.

23.6 The Extent of Land and Soil Degradation: What Is Known?

Although we have chosen the terminology of LSD in this paper, the definitions of land and its degradation widely cited in the literature include the state of soils as part of the "inputs" in the production land ecosystem goods and services (UNCCD 1996, Part 1, Article 1e, FAO 1979). The reason for choosing this terminology comes from our perception that the recent global studies of land degradation have shifted the focus away from the need to understand the processes which govern the productivity of soils and how these are affected and by what, when a decrease in the productivity of the soils is observed. Rather, most recent studies focus on the results of a decrease in soil productivity as these are more readily observable. For example changes in vegetation cover (type and/or intensity) can indicate land (and soil) degradation and are "easily" observed globally and across time via satellite imaging and remote sensing technologies. It should not be forgotten that indicators of changes of vegetation cover (a visible outcome) are proxies of land and soil degradation (a less visible process).

A commonly used index of vegetation cover is the Normalized Difference Vegetation Index (NDVI). It is computed based on the reflection of red and near-infrared light by the vegetation surface. Taking into account factors such as climate, soils, terrain and land use, deviations from the norm can be interpreted as land degradation or improvement. However, increases in the value of the NDVI can hide actual land degradation, for example, because of atmospheric CO_2 fertilization (Vlek et al. 2010). Further, it cannot differentiate between types of vegetation and cannot identify land use change. The NDVI, thus, seems to provide crucial information when coupled with additional local-level data and information and ground-truthing. Yet it remains the most used indicator of LSD at the global scale due to its availability and easy computation. There are other global indicators of land degradation or improvement which are computed from satellite imagery (e.g., net primary production, NPP; rainfall use efficiency, RUE; or residual trend analysis, RESTREND). Yet they are related to the NDVI and thus suffer from the same drawbacks. As a result, the few global assessments of land degradation have simultaneously been recognized as important achievements and criticized for their inaccuracy.

From a human perspective, global land degradation particularly matters because the availability of "good" or fertile land at the global scale can be considered as a public good. Nonetheless, not all land degradation is equally "humanly" concerning in the short to medium term. What policy makers need to take action against LSD are to be based upon assessments of where it occurs and impact human the most. Such areas can conceptually be viewed as areas where the cost of inaction against LSD is the highest, or areas where the net benefits of action against LSD is the highest. For example, areas with high population density, a high incidence of poverty and a strong reliance on the land resources in the livelihood strategies of the poor are clear priorities for action. Thus, global assessments of land degradation should aim to combine ecological indicators with socio-economic indicators. The Land Degradation Assessment in the Drylands (LADA), the Global Land Degradation Assessment (GLADA), the Global Land Degradation Information System (GLADIS) and Vlek et al. (2010) have all made advances in this direction. Population density is the most commonly applied information related to human impacts with relationship to land degradation. Vlek et al. (2010) added land use information in their analysis whereas GLADIS further offers the possibility to combine NDVI maps with information on agricultural and forest values and production trends, accessibility, tourism, human development index. The results of numerous studies are discussed in a review of global assessments of land degradation provided in Nkonya et al. (2011). They highlighted the shortcomings in the current state of knowledge about the global state of land and soil resources and their human impacts, as well as in methodological issues (choice of indicators of degradation and confounding factors). For the purpose of the discussion here, it is important to keep in mind that although the NDVI and related indicators currently provide the only empirical tools for global assessments of LSD, they have clear shortcomings. In particular, their ground-truthing reveals many (and large) errors, their relationship with actual LSD is still debated (see for example, Vlek et al. 2010), and their application and treatment in parallel with socio-economic indicators and models is still hampered by a lack of compatibility in data format and nature. Further, a comprehensive methodology to overcome these issues, such as that outlined in Nkonya et al. (2011), has not yet been applied. It will require concerted efforts by many parties to produce a global and integrated assessment of LSD. As a positive sign for the future, several independent research groups and other stakeholder groups have come together around the issue of LSD and are already making contacts in that respect. Simultaneously, the economic perspective on land degradation is increasingly recognized as a federative perspective to gather knowledge and energy around the global LSD issue.

Figure 23.1 below is a map showing the evolution of the average NDVI between the baseline period 1982–1986 and the end period 2002–2006. The average is computed based on the sum of bi-weekly observations collected throughout the two periods, at pixel level and resolution of 8 × 8 km. Areas of land improvement are colored in blue, areas of land degradation in red. In Nkonya et al. (2011), this map is presented in combination with different socio-economic indictors. The results presented in that report and their discussion exemplify the complexity of the relationship between socio-economic indicators and land degradation.

23.7 Land and Soil Degradation: The Costs of Action Versus Inaction Approach

Nkonya et al. (2011) presented a methodology for the cost of action versus inaction approach in the case of land degradation. The main points of their arguments are presented here. Degradation affects the economic value of land because this value is based on the capacity of the land to provide services. These services include

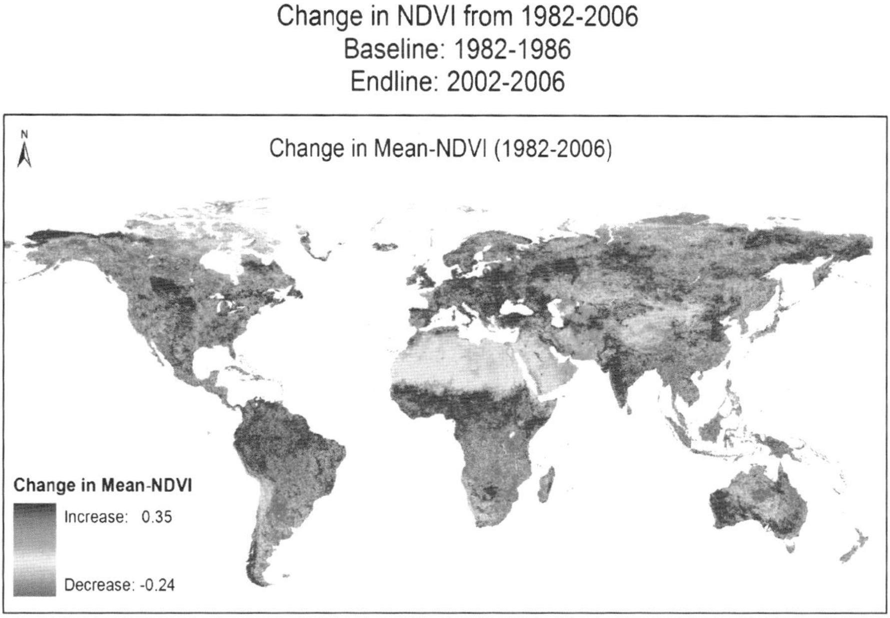

Fig. 23.1 The evolution of vegetation cover worldwide between 1982 and 2006 (Cartography: V. Graw, ZEF, University of Bonn, April 2011, Data: Global Inventory Modeling and Mapping Studies (GIMMS))

physical output (e.g., food and resource production) and other services impacting on human well-being (e.g., recreational parks, water filtration, flood control). Ethical, philosophical, and cultural considerations that give ecosystems a value, irrespective of their benefits to humans, are undeniably important but are not part of this assessment. It can be assumed that they would increase the incentives to take actions to protect land resources. No studies provide an economic valuation of the impacts of LSD on ecosystem services in general. Rather, they focus on specific types of ecosystem services and on selected benefits that human draw from these services. The most commonly investigated are the so-called provisioning services (MA 2005), and the particular benefits are food or feed production. Nkonya et al. (2011) give an account of the type of ecosystem benefits and geographical cover of many LSD studies. Clearly, more work is needed to get to a comprehensive global and integrated assessment of LSD. The important message is that all impacts of LSD must be represented in a global assessment. That includes the impacts on:

- the environment;
- the economy, through market interactions between land ecosystem "products", their inputs, and other up- and down-stream goods and services;
- humans, i.e., land users and their dependents as well as society as a whole.

The impact of environmental factors on human well-being is not straightforward. Economists are still debating what a good measure of well-being might be, and how this is linked to economic performance beyond measures of gross domestic product (Stiglitz et al. 2009). Surely, LSD has an impact on the amount of agricultural production and on its price (captured in the GDP measure), but its impact on human well-being extend beyond this. So it is not only the variety of ecosystem services and benefits impacted by LSD that must be extended for a global assessment, but also the ways in which these are valued, taking into account all economic, social and environmental costs and benefits.

Following influential reports on the valuation of the impacts of climate change (Stern 2006) and of biodiversity loss (TEEB 2010) on human well-being, an economic assessment the costs of action (i.e., costs of mitigating land degradation) versus the costs of inaction (i.e., costs induced by continued degradation) against land degradation must be undertaken to guide policy makers and inform all stakeholders of the LSD debate and the public in general. This approach aims to answer the following questions:

- What is it worth taking action against LSD (benchmarked against the costs of inaction)?
- Where and when should action take place to yield the most beneficial impact from society's point of view? To answer this question, one should assess and compare:
  - Where the costs of action are the lowest,
  - Where the costs of inaction are the highest,
  - Where the impact on human well-being is the highest.

Since land degradation is a process that occurs over time, intertemporal considerations will characterize land users' decisions. This means that the benefits derived from land use (and the value of the land) need to be maximized over time and that land users are assumed to continuously choose between land-degrading and land-conserving practices. In many cases, it turns out that the costs of prevention of LSD are much smaller than the costs of rehabilitation (Schwilch et al. 2009), favoring early action.

From an economic perspective, the current profits of adopting land-degrading practices are continuously compared against the future benefits deriving from the adoption of land-conservation practices. A rational land user will let degradation take place until the benefits from adopting a conservation practice equal the costs of letting additional degradation occur. Each land user determines his own optimal 'private' rate of land degradation. This optimal 'private' rate of land degradation depends mainly on the costs and benefits that the land user directly experiences – such as yield declines due to degradation. Typically, productivity losses are referred to as on-site costs (i.e., taking place on the land user's area of land). Hence, those ecosystem services that result in lower production levels are considered in his decisions but those that do not become measurable in terms of lost production are neglected.

In fact, many of the costs related to land degradation do not directly impact an individual land user. As a consequence, the private rate of degradation is likely to exceed the optimal rate of degradation from society's viewpoint. From a society's point of view, all costs and benefits (including externalities) that occur due to ongoing land degradation need to be considered to result in the optimal 'social' rate of land degradation. To achieve this, a global assessment must go beyond the consideration of on-site and direct costs that land users experience in terms of lower yields for instance. It must also account for changes in the value of the benefits derived from all ecosystem services that may be affected on and off-site, (e.g. off-site costs caused by the sedimentation of waterways), as well as for all indirect effects (e.g. economy wide impacts through market interactions, threats to food security and the social impacts of food insecurity, poverty and other outcomes affecting the society).

Institutional factors can prevent individual land users to reach a socially optimal decision in terms of land management. For example, imperfect or unenforced land rights, distorted and volatile market prices, lack of information about future damages related to degradation, and imperfect or missing credit markets, are among the factors that prevent land users from investing in potentially profitable soil conservation measures. Anything that creates uncertainty about the future benefits of conservation measures reduces farmers' incentives to adopt them. A global and integrated economic assessment of land degradation must account for all factors that influence the costs of action, including the institutional factors that partly determine the costs of soil and water conservation measures, as well as the costs involved in changing the institutional structures and incentives governing land management.

23.8 Implementing the CAVI Approach in the Case of Land and Soil Degradation

This approach can be made operational by comparing marginal costs and benefits (i.e., costs and benefits of a very small change in the level of degradation) related to degradation. For the application of this method it is paramount that information about the marginal social cost related to continued degradation (marginal costs of non-action) and the marginal social cost related to conservation (cost of action) can be gathered. Below we provide a generic example of how this can be carried out in the case of agriculture.

First, to construct the marginal cost curves production functions that link the extent of degradation to the maximum agricultural output associated with a technology (non-conserving or conserving) must be developed. This allows capturing the on-site productivity loss as the most direct impact of LSD on farmers. Besides direct costs and benefits of land degradation, off-site costs and benefits as well as indirect effects need to be taken into account as well. In order to come up with a socially optimal level of degradation, a mix of economic methods has to be identified to address the various on- and off-site, direct and indirect costs and benefits (Nkonya et al. 2011).

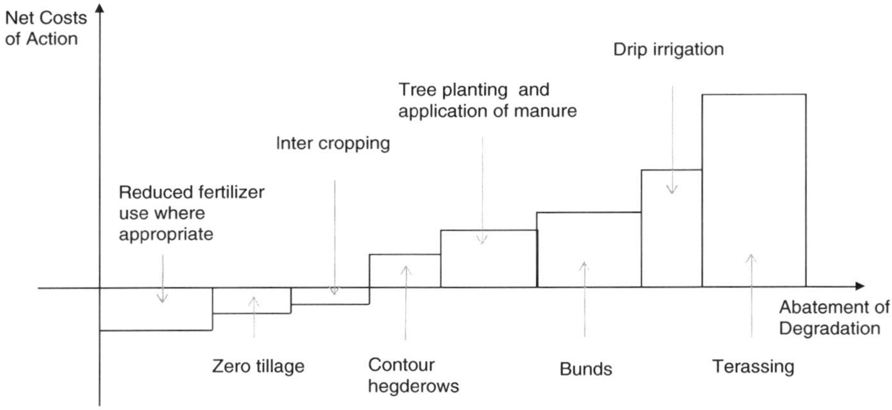

Fig. 23.2 Marginal abatement cost curve, a stylized example (Nkonyå et al. 2011)

As time plays a vital role and the impact of land degradation may aggravate over time, time considerations must be incorporated as well. Costs and benefits that arise over time have to be discounted in order to be comparable. We support the use of a low discount factor as in Stern (2006) for a global, intergenerational issue such as LSD. Due to the current lack of knowledge on the long-term impacts of agricultural practices on degradation rates (and potential price fluctuations), uncertainty has to be incorporated in the analysis as well.

The marginal cost of Action curve (often referred to as marginal abatement cost curve (MAC)) consists of various measures (soil and water conservation techniques, institutions, policies) and their cost to abate degradation by one unit. On the MAC, each point along the curve shows the cost of a combination of action(s) to abate degradation by one additional (called marginal) unit, given the existing level of degradation. Marginal changes refer in this case to changes in LSD caused by a single (combination of) measure(s). The rising MAC curve (positive slope) indicates that as a higher level of abatement is achieved, the cost of the next unit of abatement increases: the MAC is an increasing function of the level of abatement.

In practice the MAC curve is difficult to observe or estimate. One way to approximate it based on conservation measures is illustrated in Fig. 23.2 below. Such a construction of the MAC curve is only an approximation of the real MAC curve. Nonetheless, similar techniques have been successfully applied in other contexts of natural resource conservation to guide policy choices (McKinsey and Company 2009 for the case of water). As (combinations of) abatement strategies are applied (independently to each other but within a given study region), their impacts on specific processes of degradation (e.g., levels of soil nutrient, water retention, or erosion) are measured, controlling for other factors affecting degradation (weather/climate, slope, working practices of the farmers). Given the number of "units" of degradation which are abated by these measures and given their total cost, an average cost of abatement over the range of abated degradation is computed, albeit in abstraction of how much abatement had already been achieved before the implementation of this specific (combination of) strategy. The horizontal aggregation of average costs over given (small) ranges of degradation abatement can be viewed as an approximation to the MAC curve.

In this example, strategies with negative costs are included. These illustrate cases where correcting current production practices strategies simultaneously decrease land degradation and production costs for a given level of output. A typical case of such win-win situation is when fertilizers have been over-used, leading to strongly decreasing marginal returns in yields per unit of fertilizer and causing degradation issues such as salinity and other chemical degradation. The most expensive conservation strategies to the right are typically labor and/or equipment intensive.

The marginal cost of inaction curve represents the continued impact of non-conserving/land mining agricultural (and other land use) practices on costs. In extreme cases, it can come to land being abandoned by the farmer (at least its original or most profitable land use is abandoned), in which case the cost of inaction is equal to the value of the fore-gone production, net of the costs of conservation measures. A crucial feature of the costs of inaction is that they will tend to rise with time and with increasing levels of land degradation. The short term strategy of delaying action can initially pay-off, as the early costs of inaction are lower than the up-front costs of action. However, past a certain threshold, one can expect that the cost of delayed abatement rises sharply above the cost of earlier action. For example, this can happen if delayed action makes it more difficult to restore already lost productivity and mitigate continued negative off-site effects: action now is cheaper than action later. This effect is compounded by the fact that delayed action increases the price of productive land, as non-degraded land has become scarcer.

Such analytical framework should be undertaken in several representative areas, thus bringing the site specificities of LSD into its global economic assessment. So far, valuation studies of the costs and benefits of land degradation/land improvement have focused on agro-ecosystems and their provisioning services. The framework needs to be developed, in combination with knowledge built in projects such as TEEB, to cover more terrestrial ecosystem services and their benefits. The approach should be as comprehensive as current science and knowledge allows, including all the services affected directly or indirectly by LSD. This can be achieved similarly to the analysis of agro-ecosystem services, relying on representative case studies. The case studies have to be representative of different ecologies, livelihoods and institutional settings. Thus, and in order to have statistically valid results, the case studies need to be drawn from a global sampling frame. As a second component of the global coverage of the land degradation issues, a global assessment must go beyond case studies and incorporate the transboundary dimensions of land degradation. These dimensions can be punctual (e.g., erosion in country A causing sedimentation of dams in country B) or at larger scale (e.g., land degradation in a specific area having impacts on global climate or on at larger scale food prices). Such transboundary effects of LSD must first be observed, recorded and then accounted for through integrated (i.e., geographically and sector-wise connected) and dynamic (i.e., accounting for the time dimension) modeling approaches.

23.9 The Institutional Set Up for a Global Integrated Assessment of Land and Soil Degradation

Clearly, the realization of a global assessment of the costs of action versus inaction of LSD requires much cooperation across regions – to gather the representative case studies – and disciplines – to better integrate the natural science aspects of LSD and its indicators with its human impacts. Lal (2010) also points to the need for soil scientists to cooperate with several other disciplines in order to globally address LSD issues. Such a concerted effort cannot materialize and gain global credibility without the appropriate institutional set-up including review processes. The key lessons learnt from previous large scale environmental assessments (Stern 2006; TEEB 2010, Intergovernmental Panel on Climate Change) is that scientific evidence and results must be developed independently and without intervention of policy stakeholders. Further, the scientific work must be submitted to strict peer review processes. Only if these two conditions are fulfilled can the scientific evidence on and assessment of the global costs of LSD gain global acceptance. However, the process needs to lead to policy-results, and thus should not take place without consultation with the policy stakeholders of the LSD issues. Continued dialogue between the scientific sphere and the policy sphere is required to ensure that the science-based evidence becomes policy-relevant. Nkonya et al. (2011) proposed an institutional set-up fulfilling these three conditions of independence, quality and relevance. It is given in the diagram below (Fig. 23.3).

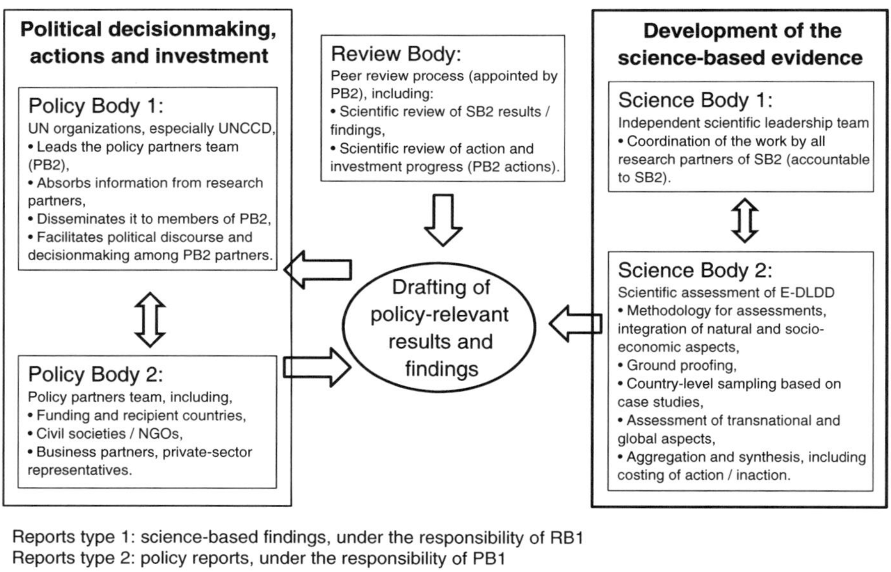

Fig. 23.3 Institutional set-up for a global assessment of LSD in a cost of action versus inaction framework (Nkonya et al. 2011)

23.10 Harmonizing Land and Soil Management in a Growing Bio-economy

Based on the global economic assessment proposed for LSD, a general framework for policy action (including science policy) for land and soil management can be laid down. Crucially, as we have discussed throughout this paper, this assessment must be time dependent. As such, it must make reference to the global trends of the bio-economy, broadly characterized by an increased competition over biological resources, and land in particular. Priority setting among sectors and value chains of the bio-economy is required, but needs to be based on objective, science-based and economic evidence. In this context, the interactions between climate change and LSD depicted earlier need to be factored in. Solutions to the LSD issue presenting potential win-wins, such as the recarbonization of the biosphere, should play a prominent role in resolving the issue of LSD under the likely scenario of the development of the bio-economy. Nonetheless, all solutions need to be appraised objectively in terms of their opportunity costs as well as their benefits. In the current world context of fast growth in demand for food and biomass, slow growth in agricultural and generally biomass production, fast growth of output prices (e.g., food) and resources prices (e.g., land, water) and increased undernutrition, the recarbonization of the biosphere (through improved land management practices) has the potential to drive up the price of biomass and food, thus increasing the human costs of LSD. Solutions to such contra-productive effects take time to implement such as enhanced agricultural (and generally biomass) productivity and/or changes in consumption behavior.

23.11 Conclusion

The main messages of this paper can be summarized as follows:

- Preventing the degradation of or restoring the land and soil resources is essential to ensure the sustainable production of increasing amounts of biomass;
- The recarbonization of the biosphere is one strategy with great potential to restore/conserve soil resources under these circumstances, with added benefits in terms of climate change mitigation;
- Nonetheless, the economics of recarbonizing the biosphere, and more generally of LSD and its mitigation, must be clarified in order to take the step of policy implementation;
- Awareness of the importance of the LSD issue and of the importance of soil and land resources in the climate change debate have risen in recent time and time is now ripe for an global and integrated assessment of land and soil degradation, based on the concept of the costs of action versus inaction;
- The LSD issue is a global issue and thus needs global action and mechanisms to guide such action;

- The mechanisms and institutional arrangements for the implementation of the global integrated assessment of LSD must follow best practices exemplified (or not) in other global economic assessments of nature's resources.

Acknowledgments This chapter partly draws on a recent study on the economics of land degradation by ZEF and IFPRI (Nkonya et.al. 2011) of which von Braun and Gerber are co-authors. The authors are also grateful to two anonymous reviewers, whose comments helped to improve this chapter.

References

Alston LJ, Libecap GD, Schneider R (1995) Property rights and the preconditions for markets: the case of the Amazon frontier. J Inst Theor Econ 151:89–107
Bai ZG, Dent DL, Olsson L, Schaepman ME (2008) Proxy global assessment of land degradation. Soil Use Manage 24:223–234
Balmford A, Rodrigues A, Walpole M, ten Brink P, Kettunen M, Braat L (2008) The economics of biodiversity and ecosystems: scoping the science. Final report. European Commission, Cambridge, UK (contract: ENV/070307/2007/486089/ETU/B2)
Barrow CJ (1991) Land degradation: development and breakdown of terrestrial environments. Cambridge University Press, Cambridge
Besley T (1995) Property rights and investment incentives: theory and evidence from Ghana. J Polit Econ 103:903–937
Blaustein RJ (2010) Recarbonizing the earth. World Watch Mag 23:24–28
BöR (The BioEconomy Council) (2011) Bio-economy innovations: research and technological development to ensure food security, the sustainable use of resources and competitiveness. BöR, Berlin
Borlaug NE (2000) The green revolution revisited and the road ahead. Special 30th anniversary lecture. Norwegian Nobel Institute, Oslo
Boserup E (1965) The conditions of agricultural growth: the economics of agrarian change under population pressure. Aldine Press, New York
Brasselle F, Brasselle A, Gaspart F, Platteau J-P (2002) Land tenure security and investment incentives: puzzling evidence from Burkina Faso. J Dev Econ 67:373–418
Cleaver KM, Schreiber GA (1994) Reversing the spiral: the population, agriculture, and environment nexus in sub-Saharan Africa. World Bank, Washington, DC
De Janvry A, Fafchamps M, Sadoulet E (1991) Peasant household behavior with missing markets: some paradoxes explained. Econ J 101:1400–1417
Deininger K (2003) Land policies for growth and poverty reduction. World Bank policy research report. World Bank/Oxford University Press, Washington, DC/Oxford, UK
Deininger K, Chamorro J (2004) Investment and equity effects of land regularization: the case of Nicaragua. Agric Econ 30:101–116
Ellis EC, Ramankutty N (2008) Putting people in the map: anthropogenic biomes of the world. Front Ecol Environ 6:439–447
FAO (Food and Agricultural Organization of the United Nations) (1979) A provisional methodology for soil degradation assessment. FAO, Rome
Grepperud S (1996) Population pressure and land degradation: the case of Ethiopia. J Environ Econ Manage 30:18–33
Hazell P (2010) An assessment of the impact of agricultural research in South Asia since the green revolution. In: Pingali P, Evenson R (eds) Handbook of agricultural economics, vol 4. Elsevier, Amsterdam, pp 3469–3530
Heffer P, Prud'homme M (2009) Fertilizer outlook 2009–2013. 77th International Fertilizer Industry Association annual conference, Shanghai, 25–27 May

Katyal JC, Vlek PL (2000) Desertification: concept, causes, and amelioration. ZEF discussion papers on development policy no 33. Center for Development Research, Bonn

Lal R (2004) Soil carbon sequestration impacts on global climate change and food security. Science 304:1623–1627

Lal R (2006) Enhancing crop yields in the developing countries through restoration of the soil organic carbon pool in agricultural lands. Land Degrad Dev 17:197–209

Lal R (2010) Managing soils to address global issues of the twenty-first century. In: Lal R, Stewart BA (eds) Food security and soil quality. CRC Press, Boca Raton, pp 5–22

Lal R, Stewart BA (2010) Introduction: food security and soil quality. In: Lal R, Stewart BA (eds) Food security and soil quality. CRC Press, Boca Raton, pp 1–4

Lopez R (1997) Land titles and farm productivity in Honduras. World Bank, Washington, DC, mimeo

MA (Millenium Ecosystem Assessment) (2005) Dryland systems. In: Hassan R, Scholes R, Ash N (eds) Ecosystem and well-being: current state and trends. Island Press, Washington, DC, pp 623–662

Mayrand K, Paquin M, Dionne S (2005) From boom to dust? Agricultural trade liberalization, poverty, and desertification in rural drylands: the role of UNCCD. UNISFÉRA, Montreal

McKinsey and Company (2009) Charting our future. Economic frameworks to inform decision-making. 2030 Water Resources Group. http://www.mckinsey.com/App_Media/Reports/Water/Charting_Our_Water_Future_Full_Report_001.pdf. Visited Nov 2011

Mulvaney RL, Khan SA, Ellsworth TR (2009) Synthetic nitrogen fertilizers deplete soil nitrogen: a global dilemma for sustainable cereal production. J Environ Qual 38:2295–2314

Nkonya E, Pender J, Kaizzi K, Kato E, Mugarura S, Ssali H, Muwonge J (2008) Linkages between land management, land degradation, and poverty in Sub-Saharan Africa: the case of Uganda. IFPRI research report #159, Washington, DC

Nkonya E, Gerber N, Baumgartner P, von Braun J, De Pinto A, Graw V, Kato E, Kloos J, Walter T (2011) The economics of land degradation: toward an integrated global assessment, vol 66, Development economics and policy series. Peter Lang GmbH, Frankfurt a.M

Pagiola S (2008) Payments for environmental services in Costa Rica. Ecol Econ 65:712–724

Pender J, Place F, Ehui S (2006) Strategies for sustainable land management in the East African highlands. International Food Policy Research Institute, Washington, DC

Pender JL, Ringler C, Magalhaes M, Place F (2009) The role of sustainable land management for climate change adaptation and mitigation in sub-Saharan Africa. TerrAfrica. http://startinternational.org/library/items/show/5. Visited Nov 2011

Place F, Hazell T (1993) Productivity effects on indigenous land tenure systems in sub-Saharan Africa. Am J Agric Econ 75:10–19

Place F, Otsuka K (2002) Land tenure systems and their impacts on agricultural investments and productivity in Uganda. J Dev Stud 38:105–128

Reardon T, Vosti S (1995) Links between rural poverty and the environment in developing countries: asset categories and investment poverty. World Dev 23:1495–1506

Reichelt C (2009) Climate buried in the soil. Max Planck Research 4/2009, Berlin, pp 80–87

Rockström J, Barron J, Fox P (2003) Water productivity in rain-fed agriculture: challenges and opportunities for smallholder farmers in drought-prone tropical agroecosystems. In: Kijne JW, Barker R, Molden D (eds) Water productivity in agriculture: limits and opportunities for improvement. CAB International, Wallingford, pp 145–162

Safriel UN, Adeel Z (2005) Dryland systems. In: Hassan R, Scholes R, Ash N (eds) Ecosystems and human well-being: current state and trends, vol 1. Island Press, Washington, DC, pp 623–662

Sanwal M (2004) Trends in global environmental governance: the emergence of a mutual supportiveness approach to achieve sustainable development. Glob Environ Polit 4:16–22

Scherr S (2000) Downward spiral? Research evidence on the relationship between poverty and natural resource degradation. Food Policy 25:479–498

Scherr S, Hazell P (1994) Sustainable agricultural development strategies in fragile lands. Environment and production technology division discussion paper, no. 1. International Food Policy Research Institute, Washington, DC

Schwilch G, Bestelmeyer B, Bunning S, Critchley W, Kellner K, Liniger HP et al (2009) Lessons from experiences in Monitoring and Assessment of Sustainable Land Management (MASLM). Paper presented at the first scientific UNCCD conference on Monitoring and Assessment of Sustainable Land Management, Buenos Aires, Argentina
Stern N (2006) Stern review: the economics of climate change. UK Treasury, London
Stiglitz JE, Sen A, Fitoussi J-P (2009) Report by the Commission on the measurement of economic performance and social progress. http://www.stiglitz-sen-fitoussi.fr/documents/rapport_anglais.pdf. Visited Nov 2011
TEEB (The Economics of Ecosystems and Biodiversity) (2010) Mainstreaming the economics of nature: a synthesis of the approach, conclusions and recommendations of TEEB, Malta
Thomas D, Middleton N (1994) Desertification: exploding the myth. British Library, West Sussex
Tiffen M, Mortimore M, Gichuki F (1994) More people, less erosion: environmental recovery in Kenya. Wiley, London
UNCCD (United Nations Convention to Combat Desertification) (1996) United Nations Convention to Combat Desertification in countries experiencing serious drought and/or desertification, particularly in Africa. http://www.unccd.int/convention/text/convention.php. Visited Nov 2011
Vlek PL, Le QB, Tamene L (2010) Assessment of land degradation, its possible causes, and threat to food security in sub-Saharan Africa. In: Lal R, Stewart BA (eds) Food security and soil quality. CRC Press, Boca Raton, pp 57–86
von Braun J (2008) Rising food prices: what should be done? IFPRI policy brief, April 2008, Washington, DC
von Braun J, de Haen H, Blanken J (1991) Commercialization of agriculture under population pressure: effects on production, consumption, and nutrition in Rwanda. IFPRI research report 85. International Food Policy Research Institute, Washington, DC
Way SA (2006) Examining the links between poverty and land degradation: from blaming the poor toward recognizing the rights of the poor. In: Johnson P, Mayrand K, Paquin M (eds) Governing global desertification: linking environmental degradation, poverty, and participation. Ashgate, Burlington, pp 27–41
World Bank (2010) Sub-Saharan Africa—managing land in a changing climate: an operational perspective for sub-Saharan Africa. Sustainable Development Department, Africa Region, report 54134-AFR. World Bank, Washington, DC

Chapter 24
Assessment of Carbon Sequestration Potential in Coastal Wetlands

James T. Morris, James Edwards, Stephen Crooks, and Enrique Reyes

Abstract This paper describes model (Marsh Equilibrium Model) simulations of the unit area carbon sequestration potential of contemporary coastal wetlands before and following a projected 1 m rise in sea level over the next century. Unit rates ranged typically from 0.2 to 0.3 Mg C ha^{-1} $year^{-1}$ depending primarily on the rate of sea-level rise, tidal amplitude, and the concentration of suspended sediment (TSS). Rising sea level will have a significant effect on the carbon sequestration of existing wetlands, and there is an optimum tide range and TSS that maximize sequestration. In general, the results show that carbon sequestration and inventories are greatest in mesotidal estuaries. Marshes with tidal amplitudes <50 cm and TSS < 20 mg l^{-1} are unlikely to survive a 1 m rise in sea level during the next century. The majority of the United States coastline is dominated by tidal amplitude less than 1 m. The areal extent of coastal wetlands will decrease following a 1 m rise in sea level if existing wetland surfaces <1 m fail to maintain elevation relative to mean sea level, i.e. expansion by transgression will be limited by topography. On the other hand, if the

J.T. Morris (✉)
The Belle Baruch Institute for Marine & Coastal Sciences, University of South Carolina, Columbia, SC 29208, USA
e-mail: morris@inlet.geol.sc.edu

J. Edwards
Marine Sciences Program, University of South Carolina, Columbia, SC 29208, USA
e-mail: Edwar257@email.sc.edu

S. Crooks
Climate Change Services, ESA PWA | Environmental Hydrology, 550 Kearny Street, Suite 900, San Francisco, CA 94108, USA
e-mail: SCrooks@esassoc.com

E. Reyes
Department of Biology, East Carolina University, Greenville, NC 27858, USA
e-mail: REYESE@ecu.edu

R. Lal et al. (eds.), *Recarbonization of the Biosphere: Ecosystems and the Global Carbon Cycle*, DOI 10.1007/978-94-007-4159-1_24,

existing vegetated surfaces survive, coastal wetland area could expand by 71%, provided there are no anthropogenic barriers to migration. The model-derived contemporary rate of carbon sequestration for the conterminous United States was estimated to be 0.44 Tg C $year^{-1}$, which is at the low end of earlier accounts. Following a 1 m rise in sea level, with 100% survival of existing wetland surfaces, rates of carbon sequestration rise to 0.58 and 0.73 Tg C $year^{-1}$ at TSS = 20 and 80 mg l^{-1}, respectively, or 32–66% higher than the contemporary rate. Globally, carbon sequestration by coastal wetlands accounts for probably less than 0.2% of the annual fossil fuel emission. Thus, coastal wetlands sequester a small fraction of global carbon fluxes, though they take on more significance over long time scales. The deposits of carbon in wetland soils are large. There have been large losses of coastal wetlands due to their conversion to other land uses, which creates opportunities for restoration that are locally significant.

Keywords Marsh equilibrium model • Suspended solids • Carbon sequestration • Coastal ecosystems • Coastal wetlands • Tidal marshes • Mangroves • Carbon stocks • Autochthonous • Sea level rise • Anthropogenic disturbance • Holocene • Organic rich soil • Subsidence • Diking • Drainage • Digital elevation model • Tide range • Primary productivity • Tidal amplitude

Abbreviations

B_s	standing biomass density
C	carbon
D	depth of the marsh surface below MHW
DEM	Digital Elevation Model
EOS	end-of-season
GHG	greenhouse gas
MEM	Marsh Equilibrium Model
MHW	mean high water
MSL	mean sea level
Mg	megagram
NWI	National Wetlands Inventory
OM	organic matter
RSLR	rate of sea-level rise
k_r	refractory fraction of root and rhizome production
B_r	root and rhizome production
φ	root:shoot quotient
ρ	sediment dry bulk density
q	settling velocity
SRTM	Shuttle Radar Topography Mission
SOC	soil organic carbon
m	suspended solids

Tg	Teragram
T	tide range
TSS	total suspended solids
k_s	trapping coefficient

24.1 Introduction

Coastal wetlands, including tidal marshes and mangroves, hold significant stocks of carbon (C), mostly buried in long-term soils storage (Donato et al. 2011). The source of organic compounds is primarily, but not exclusively, autochthonous, derived from continuous production of root biomass. The in-situ decayed products together with mineral sediments contribute to soil as the marsh builds with sea level rise. Storage in wet conditions with low oxygen availability protects these organics from microbial degradation, effectively removing sequestered C from circulation (Hedges et al. 1999; Freeman et al. 2001).

The global extent of coastal wetlands prior to major anthropogenic disturbance represented the long-term accumulation of organic-bearing coastal alluvium throughout the mid to late Holocene; a relatively quiescent period of gradual eu-static sea-level rise (typically 1 mm or less per year) (Gehrels et al. 2011). Gradual sea-level rise over this time fostered conditions favoring the accumulation of deep sequences of organic–rich soil, commonly of 3–5 m in depth and in some places deeper (Redfield and Rubin 1962). In locations subject to subsidence, either through soft sediment compaction in deltaic areas or tectonic crustal movement, the contribution of mineral sediments is a critical component of the marsh building process in the face of enhanced relative sea-level rise. Under conditions of low or even negative rates of sea-level rise, marshes with soils consisting predominantly of organic material may be found. Their existence is unbuffered by mineral sediment supply and is potentially sensitive to accelerated rates of sea-level rise if space is not available for landward migration.

Outside Europe, North America and Australia, the extent of tidal wetlands is poorly documented, and the extent of drained wetlands less so (Armentano and Menges 1986). It is now known that Europeans diked and drained most of their coastal wetland areas beginning around the Roman Era and continued with real enthusiasm during the seventeenth and eighteenth centuries. This practice spread to the New World. Between 1850 and the 1960s (when protective legislation was put in to place), extensive areas of coastal wetlands along the east and west coasts of the United States were diked. In states such as California more than 95% of all coastal wetland areas were converted to other land uses.

China also has long history of coastal wetland diking and drainage, beginning during the late Han Dynasty (BC 202 to AD 220). For example, of the 30,000 km^2 of wetland that built up over the past 4,000 years at the mouth of large rivers in Jiangsu Province only 900 km^2 remained undeveloped by the turn of the twenty-first century (An et al. 2007). Of the 43,000 km^2 of coastal wetlands that existed across China in 1950, 51% were converted to other land uses by the end of the century (He and Zhang 2001; An et al. 2007). It is estimated that a total of 133,500 km^2 of croplands, fishponds, saltponds and residential land was created from conversion of coastal wetlands in China (Yang and Chen 1995, reported in An et al. 2007).

This chapter presents results of simulations of the unit area C sequestration potential of coastal wetlands under current conditions before and after a projected 1 m rise in sea level. A methodology is described that arrives at an estimate of wetland area based on a digital elevation model of the coastline coupled with a sampling of tide ranges. Satellite altimetry-derived estimates of current wetland area and future wetland area in the United States following a 1 m rise in sea level are shown, allowing for transgression, with and without survival of existing wetland area. Applying model-derived unit area C sequestration to estimates of wetland areas yields the spatially integrated potential C sequestration in wetlands with different sea-level rise scenarios. Finally, the data are extrapolated to a global scale and the rates of C sequestration are placed in the context of the global C cycle.

24.2 Development of the Marsh Equilibrium Model (MEM)

Fundamental to the analysis presented are the predictions of a theoretical model that describe feedbacks among the plant community, sediments and tides and that explain the dependency of the relative elevation of a salt marsh on rising sea level (Morris et al. 2002). The model assumes that the sedimentation of suspended solids carried by tides over the marsh surface increases with the concentration of suspended solids (m), duration of flooding (Krone 1985; Friedrichs and Perry 2001), and standing biomass density (B_s) (Morris et al. 2002). Flood duration is proportional to the depth (D) of the marsh surface below mean high water (MHW) divided by the tide range (T).

In addition to surface deposition, production of organic matter (OM), primarily of roots and rhizomes, contributes to the total accumulation rate (Reed 1995; Turner et al. 2001). These surface and subsurface processes can be expressed as:

$$dS / dt = m(q + k_s B_s) D^2 / T + k_r B_r \tag{24.1}$$

Parameter q is the settling velocity, k_s is a trapping coefficient, and k_r is the refractory fraction of annual root and rhizome production (B_r). The production of roots and rhizomes (B_r) is proportional to the end-of-season (EOS) standing biomass density (B_s) by way of a belowground turnover rate and a root:shoot quotient (φ). Biomass density (B_s) is a function of the depth of the marsh surface below mean high water (MHW) (Morris et al. 2002):

$$B_s = aD + bD^2 + c \tag{24.2}$$

Coefficients a, b, and c determine the growth range and optimum depth below MHW. Their values were determined by bioassay (Morris 2007). Substituting for B_s into Eq. 24.1 gives:

$$dS/dt = c\varphi k_r + aD\varphi k_r + D^2(ck_s m + mq + b\varphi k_r T)/T + aD^3 k_s m/T + bD^4 k_s m/T \tag{24.3}$$

For the purposes of this paper, a root:shoot quotient of 2.2, a root and rhizome turnover of 1.5 $year^{-1}$ and $k_r = 0.02$ were applied, which gives an effective sequestration rate of 0.066 $year^{-1}$ of the EOS standing biomass. This combination of parameter values best described the vertical profile of sediment OM at North Inlet, South Carolina, USA. These and other assumptions can be explored using an interactive version of the model that can be found on the world-wide-web at http://jellyfish.geol.sc.edu/model/marsh/mem.asp. From Eq. 24.3 it can be seen that the depth of the marsh surface (D) below MHW and the rate of sea-level rise (RSLR) are proportional when the marsh surface and mean sea level are in equilibrium:

$$RSLR \propto dS/dt \propto D \tag{24.4}$$

Thus, as the rate of sea-level rise increases, the depth of the marsh surface must increase, or in other words its elevation relative to mean sea level (MSL) must decrease.

The rate of change of the marsh elevation is obtained by dividing the sedimentation rate dS/dt by bulk density (ρ), which is commonly expressed as a function of the OM content of the sediment (Jeffrey 1970; Harrison and Bocock 1981):

$$\rho = (\varphi k_r B_s)/(m(\varphi + k_s B_s)D^2/T + \varphi k_r B_s) \tag{24.5}$$

24.3 Digital Elevation Model

A coastal Digital Elevation Model (DEM) for the United States was constructed, and when combined with tide data, allowed estimates of potential wetland area. For the United States, the elevation data from the Shuttle Radar Topography Mission (SRTM) were applied (Farr et al. 2007). The SRTM-derived wetland estimates were compared with classified imagery from the National Wetlands Inventory data base to determine if the SRTM data, which has near global coverage, could be used as a reasonable proxy for classified imagery as a means of estimating the coverage of coastal wetlands. The SRTM data are available at a 3 arc-sec (about 90 m) resolution. Mean vertical accuracy was reported to range from –0.7 to 1.8 m with standard deviation of as great as 5.9 m, depending on location (Rodriguez et al. 2006).

The SRTM data were summarized spatially according to the tidal amplitude along the adjacent coast using bin sizes of 0–1, 1–2 and 2–3 m. The assumption was made that MSL has a SRTM elevation of 0 and that any surface within the 0–1 m contour and that lies within a region with tidal amplitude ≤1 m is potentially, intertidal emergent wetland. Similarly, surfaces within the 0–2 m contours in areas with tides of 1–2 m tidal amplitude were assumed to be wetland. Combining the topographic data with the spatial distribution of tidal amplitudes gives a rough picture of the frequency distribution of intertidal area.

Limited by the resolution of the SRTM, only the land area that lies between 1 m contours can be quantified. In areas where the tidal amplitude may be only a small fraction of a meter, the use of SRTM data should overestimate wetland area. For example, areas on coastlines with tidal amplitudes of 0.1 m are unlikely to support marsh at elevations that exceed perhaps 0.5 m, though by definition a potential wetland was operationally defined as a surface with an SRTM elevation of 0 m, i.e., 0 to <1 m. On the other hand, SRTM data should underestimate wetland area in a region with tidal amplitude of about 0.7–0.99 m, because vegetated wetland surface could exist at elevations above 1 m, exceeding the SRTM 0 m contour. Furthermore, the −1 m SRTM contour was assumed to be below the zone of vegetation, though wetland plants will grow at elevations lower than MSL, perhaps to −0.5 m in meso- and macrotidal regions (McKee and Patrick 1988).

Contemporary estimates of coastal wetland area in the United States are based on the National Wetlands Inventory (NWI) data base, which is derived from classified imagery (U.S. Department of the Interior, Fish and Wildlife Service, Washington, D.C. http://www.fws.gov/wetlands/). From the NWI, the areas classified as estuarine, intertidal emergent or E2EM were extracted and summarized, and these were compared with the SRTM-derived estimates of potential and current wetland areas.

24.4 Tidal Data

In this chapter, the potential, vegetated intertidal area of the coastline is defined as land that falls within 1 tidal amplitude of MSL. Tidal data for 1,723 NOS/CO-OPS stations that have currently accepted datum were obtained from the U.S. National Oceanographic and Atmospheric Administration and mapped onto the coastal DEM (Fig. 24.1). The tide gage stations cover the United States and its territories only. Station data included the mean tide range (MN = mean high water level – mean low water level), the latitude, and longitude. Tide data were reported for the 1983–2001 epoch. Averaged over all stations, the grand mean tide range was 1.23 ± 1.08 m (± 1 SD).

24.5 Unit Area Carbon Sequestration

The MEM was used to compute the contemporary and future rates of unit area C sequestration (Fig. 24.2). The forecast was based on a future scenario in which sea level rise accelerates, rising ultimately to a level that is 1 m above present day MSL. Because the relative elevation of a coastal wetland will change with the rate of sea-level rise, and affects primary productivity, the rates of C sequestration over a century were integrated. The computations were made for combinations of two variables that affect sediment accretion and wetland response to rising sea level, namely the concentration of total suspended solids (TSS) and tidal amplitude

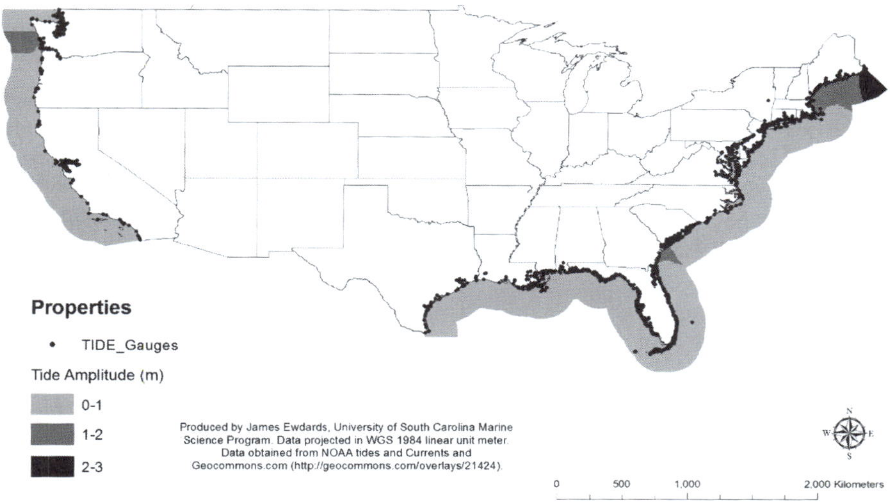

Fig. 24.1 Tidal amplitudes from the network of tide gages (gages are plotted as *dots*) mapped around the U.S. coastline

Fig. 24.2 Computed rate of carbon sequestration for coastal wetlands that are equilibrated to a contemporary rate of sea-level rise of 0.24 cm year^{-1} (*grey surface*), and following an accelerating rate of sea-level rise that reaches the 1 m elevation in a century (*color surface*)

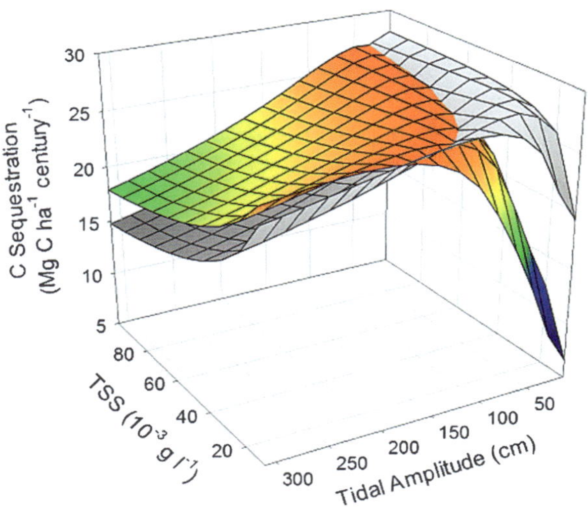

(Fig. 24.2). Figures 24.2 and 24.3 each show two surfaces; the one in grey represents the equilibrium solution to different combinations of tidal amplitude and TSS for a constant rate of sea-level rise of 0.24 cm year^{-1}. The color surfaces show the non-equilibrium solutions when sea level was increased by 100 cm by the end of the twenty-first century.

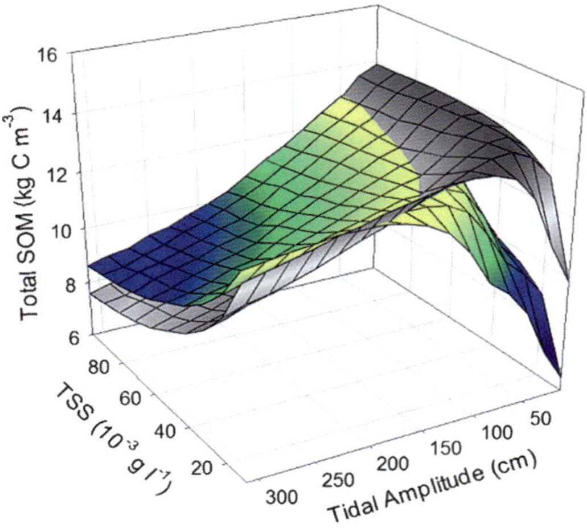

Fig. 24.3 Computed inventory of sediment organic matter in coastal wetlands that are equilibrated to a contemporary rate of sea-level rise of 0.24 cm year^{-1} (*grey surface*), and following an accelerating rate of sea-level rise that reaches the 1 m elevation in a century (*color surface*)

The response surfaces are complex and show that there are optimum combinations of TSS and tidal amplitude that maximize C sequestration and C stocks. Rising sea level will have a significant effect on C sequestration of existing wetlands. In some cases the effect of rising sea level will be positive while in others the effect will be negative. This occurs because in some cases (e.g., macrotidal estuaries) the marshes are initially in equilibrium at an elevation that is considerably greater than optimal for the vegetation, and a rapid rise in sea level will at first increase productivity. Other scenarios (e.g., microtidal estuaries) start with marshes equilibrated much closer to the optimum elevation. In general, the results indicate that C sequestration is greatest in mesotidal estuaries (Fig. 24.2).

The MEM predictions of C sequestration tend to fall around the lower end of empirical estimates that range from 18 to 1,713 g C m^{-2} year^{-1} or Mg C ha^{-1} century^{-1} (Chmura et al. 2003). Based on the C inventory of sediment cores and ^{14}C measurements, Choi and Wang (2004) reported C burial rates ranging from 18 to 193 g C m^{-2} year^{-1} in mid- to low-elevation salt marshes, but they also reported that these rates were considerably higher than long-term rates of C sequestration due to the slow but continuous decomposition of OM over time.

The permanent burial of C is equivalent to the product of the production of roots and rhizomes and the OM fraction resistant to decay. Figures 24.2 and 24.3 were generated using a 7% sequestration rate (the product of the resistant fraction and production), which is needed to fit the model to typical sediment SOM profiles from minerotrophic marshes typical of the southeast United States. Neither the quantity of labile C in sediment nor its addition to sediment contributes to C sequestration, unless the site is one that is being restored or newly colonized, in which case the increase in the inventory of labile C does contribute to sequestration. Labile C by definition will decay, but on a restoration site the addition of living root and rhizome biomass represents the *de novo* addition of C that remains approximately constant as long as the vegetation survives.

24.6 Unit Area Carbon Inventory

The current inventory or density of C in wetland sediments varies greatly. The C density varies primarily with the rate of relative sea-level rise, but tide range also has an effect (Fig. 24.3). This occurs because biomass production is a function of relative elevation, and the elevation is a function of the rate of sea-level rise. Tidal amplitude is also significant because it determines the vertical range over which with vegetation can grow, and the feasible growth range increases with increasing tidal amplitude. The C density in wetland sediments is largely a function of dilution by mineral sediment, which, like primary production, is a function of relative elevation. Peatlands widely occur in northern latitudes in the eastern United States. Interestingly, these are the regions with the greatest (2–3 m) tidal amplitude (Fig. 24.1), but they are also occur where isostatic rebound is important (Shipp et al. 1991). Isostatic rebound would have lowered the relative rate of sea-level rise for thousands of years, allowing the marshes to equilibrate at the top of the tidal frame. Thus, peat forms when the relative RSLR is very low and when a marsh equilibrates at the top of the tidal frame where the opportunity for mineral inputs is limited and where, at very low RSLR, the accretion of OM alone is sufficient to keep the marsh surface in equilibrium with MSL. Furthermore, there is probably a positive feedback between the concentration of soil organic carbon (SOC) and its preservation. As the C density rises, the concentration of phenolic compounds rises and helps to preserve OM (Freeman et al. 2004), further increasing the C density.

The depth of accreted sediment and C sequestered in a given wetland is dependent on time and rate of relative sea-level rise. Sediment depth for a given wetland in equilibrium with sea level is by definition equal to the rate of sea-level rise multiplied by its age. Assuming that existing wetlands have been in equilibrium with sea level during the late Holocene when rates of sea-level rise averaged <1 mm $year^{-1}$, then marshes must have sediments that range in depth from near zero, for those that have recently transgressed, to perhaps as much as 7 m for the oldest wetlands or those that are 7,000 years in age.

24.7 Potential, Future and Contemporary Coastal Wetland Area

Comparisons of NWI and SRTM-derived estimates of wetland areas were made for the eastern United States coastline known to support coastal wetlands and differing in tidal influence. The states of North Carolina, South Carolina, and Georgia have tidal amplitudes, averaged over all stations by state of 0.4, 0.77 and 1.03 m, respectively. Potential, topographically-derived wetland area in North Carolina (SRTM = 0 m) was 433 km^2, while the NWI E2EM area was 725 km^2 (Table 24.1). Here the topographic method underestimated NWI-wetland area by 40%. At the other end of the tidal spectrum, Georgia has a potential vegetated wetland area of 608 km^2 by the topographic method (SRTM = 0 + SRTM = 1), which was 54% lower than the NWI area of 1,337 km^2. In South Carolina the topographic method also underestimated NWI-wetland area (Table 24.1).

Table 24.1 Comparison of potential wetland areas computed using SRTM data with intertidal emergent marsh area derived from classified imagery (NWI data)

	Intertidal wetland area (km²)[a]				
	North Carolina	South Carolina	Georgia	Subtotal	Total U.S.
NWI (E2EM)	725	1,398	1,337	3,460	18,372
SRTM −1 to 0 m zone	177	346	76	599	3,052
SRTM 0 to <1 m zone	433	1,078	278	1,789	14,722
SRTM 1 to <2 m zone	0	188	330	518	1,057
SRTM 2 to <3 m zone	0	0	0	0	4
SRTM subtotal 0 to <3 m	433	1,266	608	2,307	15,783

[a]Areas within each SRTM zone are assumed to be potential wetlands only if they are adjacent to a coast with tidal amplitude great enough to flood that contour. NWI wetland area is that defined as estuarine intertidal, vegetated area (Dahl 2011)

Based on this small sample size it is estimated that the topographic method of delineating wetlands using SRTM data on a regional level can be approximately 45–90% of the actual, contemporary coastal wetland area. There are several possible interpretations that explain the discrepancy, including that wetlands extend to elevations lower than the SRTM = 0 m contour. Indeed, inclusion of the SRTM −1 class in the national data results in a total wetland area of 18,834 km^2, which is in good agreement with the NWI data (Table 24.1). Another source of error occurs when tidal amplitudes approach the upper limit of the maximum SRTM bin size and wetlands extend to a higher elevation, as they do in much of South Carolina and Georgia. A higher resolution DEM should improve the results, but for the purposes of constructing a global budget, the SRTM data give a reasonable approximation of potential wetland area.

At a national level the current spatial distribution of intertidal areas and the potential for marsh transgression following a 1 m rise in MSL were addressed. The coastline of the conterminous United States is dominated by tides less than 1 m (Fig. 24.1) and the great majority of potentially vegetated intertidal area lies at elevations between 0 and 1 m. Specifically, 73% of the U.S. coastline had tidal amplitudes of 1 m or less, and 24% had tidal amplitudes of 1–2 m. Of the total intertidal area, 93% lies within 0–1 m (Table 24.2). When broken down by tidal amplitude, intertidal areas along coastlines with 1–2 m tidal amplitudes were 43% and 57% within the 0–1 and 1–2 m elevations, respectively (Table 24.4). Coastlines with 2–3 m tidal amplitudes were distributed between 27% and 45% across elevation bins (Table 24.5).

If MSL increases by 1 m and no contemporary vegetated surface lower than 1 m survives, then there will be a 22% loss of total potential wetland area (SRTM-derived wetland area) from 15,783 to 12,264 km^2 (Table 24.2). That is, the topography of the landscape over which the marsh will transgress is limited in area even when all anthropogenic barriers are removed or prevented. Alternatively, assuming 100% survival of surfaces in the 0 m elevation class (i.e., 1 cm year^{-1} accretion), as much as a 71% increase in intertidal area is possible. Note that both of these projections assume that transgression inland is limited only by elevation and not by barriers

Table 24.2 Contemporary and projected, cumulative intertidal areas by topographic zone (intertidal elevation) along the coastline of the conterminous United States

Current intertidal elevation (m)	Current area (km^2) within each zone	Area (km^2) following a 1 m rise, no survival	Area (km^2) following a 1 m rise, 100% survival[a]
0–1 (SRTM contour)	14,722 (93%)	0	0
1–2	1,057 (7%)	11,224 (92%)	25,946 (96%)
2–3	4 (<0.5%)	1,035 (8%)	1,035 (4%)
3–4	0	5 (<0.5%)	5 (<0.5%)
Totals	15,783 (100%)	12,264 (100%)	26,986 (100%)

[a]Assumes all contemporary marshes survive and gain 1 m in elevation

Table 24.3 Contemporary and projected, intertidal areas by topographic zone (intertidal elevation) along coastlines with a 0–1 m tidal amplitude

Current intertidal elevation (m)	Current area (km^2) within each zone	Area (km^2) following a 1 m rise, no survival	Area (km^2) following a 1 m rise, 100% survival
0–1 (SRTM contour)	13,911	0	0
1–2	0	10,168	24,079
Totals	13,911	10,168	24,079

Table 24.4 Contemporary and projected, intertidal areas by topographic zone (intertidal elevation) along coastlines with a 1–2 m tidal amplitude

Current intertidal elevation (m)	Current area (km^2) within each contour	Area (km^2) following a 1 m rise, no survival	Area (km^2) following a 1 m rise, 100% survival
0–1 (SRTM contour)	804 (43%)	0	0
1–2	1,053 (57%)	1,053 (51%)	1,857 (64%)
2–3	0	1,031 (49%)	1,031 (36%)
Totals	1,857 (100%)	2,084 (100%)	2,888 (100%)

Table 24.5 Contemporary and projected, intertidal areas by topographic zone (intertidal elevation) along coastlines with a 2–3 m tidal amplitude

Current intertidal elevation relative to MSL (m)	Current area (km^2) within each zone	Area (km^2) following a 1 m rise, no survival	Area (km^2) following a 1 m rise, 100% survival
0–1 (SRTM contour)	6 (45%)	0	0
1–2	4 (27%)	4 (29%)	10 (53%)
2–3	4 (28%)	4 (31%)	4 (21%)
3–4	0	5 (40%)	5 (27%)
Totals	14 (100%)	13 (100%)	19 (100%)

such as sea walls. Thus, the future C sequestration of coastal wetlands depends on barriers to upland migration and accretion rates sufficient to maintain elevation relative to MSL (Table 24.2).

24.8 Spatially Integrated Rates of Carbon Sequestration

Salt marsh coverture at either the national or global level is subject to variable accuracy (Mitra et al. 2005). In the United States, the estimates of coastal wetland area range from the present SRTM-derived estimate of 15,783 km^2 (this chapter) to the current NWI estimate of 18,372 km^2 (Dahl 2011). For this range of wetland areas the contemporary rates of C sequestration weighted by tide range, and the effect of a 1 m rise in sea-level over the next century was computed. One of the variables affecting unit area C sequestration is the TSS concentration, for which there are no national or global data bases. However, a majority of estuaries reported by (Kirwan et al. 2010) had TSS concentrations ranging from 20 to 80 mg l^{-1}. For marshes in equilibrium with a constant rate of sea-level rise of 0.24 cm $year^{-1}$, TSS concentration of 20×10^{-3} g l^{-1}, and tidal amplitudes of 50–250 cm, the MEM-derived, unit C sequestration (Fig. 24.2) ranged from 28 to 22 Mg C ha^{-1} $century^{-1}$. At a TSS of 80×10^{-3} g l^{-1} the MEM-derived, unit C sequestration ranged from 26 to 17 Mg C ha^{-1} $century^{-1}$ for tides ranging from 50 to 250 cm.

The MEM-derived contemporary rate of C sequestration for the conterminous United States is 0.44 Tg C $year^{-1}$ at a TSS of 20×10^{-3} g l^{-1}. At a TSS of 80×10^{-3} g l^{-1} the rate is only 10% lower. The arithmetic mean C sequestration of 210 g m^{-2} $year^{-1}$ from the meta-analysis of Chmura et al. (2003) gave an integrated rate of 5 Tg C $year^{-1}$ for the conterminous U.S. Using their data, a geometric mean rate of 115 gC m^{-2} $year^{-1}$ was calculated, which would halve their integrated rate, but that is still greater than the MEM estimate. Thus, this discrepancy may be a function of the inclusion of labile C in the empirical measurements of total organic C density in sediment cores. The estimate in this study is more consistent with the radiocarbon technique used by Choi and Wang (2004).

As sea-level rise accelerates this century, the change in C sequestration by coastal wetlands will depend on how successfully the existing marsh surfaces maintain a relative elevation that continues to support the vegetation. For marshes in equilibrium with an initial rate of sea-level rise of 0.24 cm $year^{-1}$, and with sea level rising 1 m in 100 years, the MEM-derived, unit C sequestration (Fig. 24.2) ranged from 22 to 25 Mg C ha^{-1} $century^{-1}$ for tidal amplitudes of 50 and 250 cm, with a maximum of 27 Mg C ha^{-1} $century^{-1}$ for the 150 cm tidal amplitude at TSS of 20×10^{-3} g l^{-1}. Note that the rate of C sequestration in the case of accelerating sea-level rise is not constant, which is the rationale for computing a rate that is integrated over a century. The corresponding, spatially integrated rates for the conterminous United States, with no survival of marshes at current elevations less than 1 m, are 0.27 and 0.33 Tg C $year^{-1}$ at TSS of 20 and 80×10^{-3} g l^{-1}, respectively. This is a significant reduction below the contemporary rate (0.44 Tg C $year^{-1}$), even when transgression was assumed to be unimpeded by anthropogenic barriers. However, with 100% survival of existing wetland surfaces less than 1 m in elevation, rates of C sequestration rise to 0.58 and 0.73 Tg C $year^{-1}$ at TSS of 20 and 80×10^{-3} g l^{-1}, respectively, or 32–66% higher than the contemporary rate.

MEM simulations predict that most of these 1 m sea-level rise scenarios are survivable for 100 years when the *in silico* marshes were started at elevations equilibrated with a rate of SLR of 0.24 cm $year^{-1}$. The exceptions are those marshes in regions of low tidal amplitude. Marshes with tidal amplitude of ≤50 cm and TSS ≤ 20 × 10^{-3} g l^{-1} will not survive. Nevertheless, before those marshes drown, toward the end of the century, they will continue to sequester C. Not simulated here was the fate of marshes with starting elevations of 50, 150 and 250 cm, which are the elevations corresponding to the SRTM bin sizes.

On a global scale, MEM estimates of coastal wetland C sequestration range from 4.5 to 15.4 Tg C $year^{-1}$. These estimates are small in comparison to global fossil fuel emissions of 8 Pg C $year^{-1}$ (Canadell et al. 2007), i.e., 0.05–0.2%, but unlike other natural ecosystems that approach a C equilibrium at maturity, coastal wetlands will continue to sequester C, because they are depositional environments.

Another large uncertainty in the global C sequestration of coastal wetlands is the estimated areal extent of mangrove and salt marsh wetland types. It ranges from 159,760 to 552,361 km^2 (Mcleod et al. 2011), a range that is unacceptably large and points to the need for an alternative means of quantification, such as satellite altimetry. The global coverage provided by SRTM is a good starting point.

The prospect for recarbonization through wetland restoration is significant on the basis of percentage change, but not in terms of global fluxes. The 51% conversion of coastal wetlands to other land uses in China (He and Zhang 2001; An et al. 2007) is probably typical of global patterns and is consistent with figures reported earlier (Armentano and Menges 1986). There are, therefore, opportunities for mitigation of greenhouse gas (GHG) emissions on a local scale. Moreover, saline wetlands have an advantage over freshwater wetlands in terms of net sequestration of short-lived GHGs in that they emit little or no methane (Bridgham et al. 2006).

24.9 Conclusions

Simulations were made of the unit area carbon sequestration potential of contemporary coastal wetlands at different rates of sea-level rise, for various concentrations of TSS and tidal amplitudes. Salt marsh vegetation occupies approximately the upper half of the tidal frame, and biomass density varies within this range from minima at the extremes to a maximum where the elevation is optimal for growth. Understanding this is key to forecasting how carbon sequestration will change with sea-level. Computed rates of sequestration typically fell between 0.2 and 0.3 Mg C ha^{-1} $year^{-1}$. Rates generally decline when TSS is high, because marshes equilibrate at higher elevations within the tidal frame where productivity is lower. When sea level was accelerated to 1 m over the next century, sequestration rates declined at low tidal amplitudes, but increased at high tidal amplitudes. At low tidal amplitude, the marsh surface transitions quickly to a suboptimal elevation for primary production, while at high tidal amplitude marshes are able to maintain an elevation favorable to primary production for a longer time than that possible in microtidal estuaries.

Spatially integrated rates of carbon sequestration depend largely on tidal amplitude and the projected rate of sea-level rise. Tidal amplitude and local topography determine the areal extent of intertidal habitat suitable for marsh development. These variables also interact to determine how the areal extent of marsh habitat will change following a rise in sea level. The great majority of the coastline of the United States is dominated by tidal amplitudes less than 1 m. The area of coastal wetlands will decrease following a 1 m rise in sea level if existing wetland surfaces <1 m fail to maintain elevation relative to mean sea level, i.e. expansion by transgression will be limited by topography. On the other hand, if the existing vegetated surfaces survive, coastal wetland area could expand by 71%, provided there are no anthropogenic barriers to migration. The model-derived contemporary rate of carbon sequestration for the conterminous United States was estimated to be 0.44 Tg C $year^{-1}$, which is at the low end of earlier accounts. With 100% survival of existing wetland surfaces following a 1 m rise in sea level, rates of carbon sequestration rise to 0.58 and 0.73 Tg C $year^{-1}$ at TSS = 20 and 80 mg l^{-1}, respectively, or 32–66% higher than the contemporary rate. Globally, carbon sequestration by coastal wetlands is a small fraction of other global fluxes. It represents probably less than 0.2% of the annual fossil fuel emission. However, the opportunity for restoration is significant locally considering large losses of wetland habitat that have been caused by wetland conversion to other land uses.

Acknowledgements This work was supported by grants from the NSF, SERDP, NOAA and USGS. No endorsement of the conclusions by these agencies is implied. Contribution no. 1644 of the Belle W. Baruch Institute for Marine & Coastal Sciences, University of South Carolina.

References

An S, Li H, Guan B et al (2007) China's natural wetlands: past problems, current status, and future challenges. Ambio 36:335–342

Armentano TV, Menges ES (1986) Patterns of changes in the carbon balance of organic soil-wetlands of the temperate zone. J Ecol 74:755–774

Bridgham SD, Megonigal JP, Keller JK et al (2006) The carbon balance of North American wetlands. Wetlands 26:889–916

Canadell JG, Pataki D, Gifford R et al (2007) Saturation of the terrestrial C sinks, Chapter 6. In: Canadell JG, Pataki D, Pitelka L (eds) Terrestrial ecosystems in a changing world, The IGBP series. Springer, Berlin

Chmura GL, Anisfeld SC, Cahoon DR et al (2003) Global carbon sequestration in tidal, saline wetland soils. Glob Biogeochem Cycle 17. doi:10.1029/2002GB00197

Choi Y, Wang Y (2004) Dynamics of carbon sequestration in a coastal wetland using radiocarbon measurements. Glob Biogeochem Cycle 18, GB4016. doi:10.1029/2004GB002261

Dahl TE (2011) Status and trends of wetlands in the conterminous United States 2004 to 2009. Fish and Wildlife Service, Washington, DC

Donato DC, Kauffman JB, Murdiyarso D et al (2011) Mangroves among the most carbon-rich forests in the tropics. Nat Geosci 4:293–297

Farr TG, Rosen PA, Caro E et al (2007) The shuttle radar topography mission. Rev Geophys 45, RG2004 1:33

Freeman C, Ostle N, Hojeong K (2001) An enzymic 'latch' on a global carbon store. Nature 409:149

Freeman C, Ostle NJ, Fenner N et al (2004) A regulatory role for phenol oxidase during decomposition in peatlands. Soil Biol Biochem 36:1663–1667

Friedrichs C, Perry J (2001) Tidal salt marsh morphodynamics. J Coast Res Special Issue 27:7–37

Gehrels WR, Horton BP, Kemp AC et al (2011) Sivan Sea-level records of the last 2000 years hold the key to understanding contemporary and future sea-level change. EOS Trans Am Geophys Union 92:289–29

Harrison AF, Bocock KL (1981) Estimation of soil bulk-density from loss-on-ignition value. J Appl Ecol 18:919–927

He Y, Zhang MX (2001) Study on wetland loss and its reasons in China. Chin Geogr Sci 11:241–245

Hedges JI, Hu FS, Devol AH et al (1999) Sedimentary organic matter preservation: a test for selective degradation under oxic conditions. Am J Sci 299:529–555

Jeffrey DW (1970) A note on the use of ignition loss as a means for the approximate estimation of soil bulk density. J Ecol 58:297–299

Kirwan ML, Guntenspergen GR, D'Alpaos A et al (2010) Limits on the adaptability of coastal marshes to rising sea level. Geophys Res Lett 37:L23401. doi:10.1029/2010GL045489

Krone RB (1985) Simulation of marsh growth under rising sea levels. In: Waldrop WR (ed) Hydraulics and hydrology in the small computer age. Hydraulics Division of the ASCE, Lake Buena Vista

McKee KL, Patrick WH Jr (1988) The relationship of smooth cordgrass (*Spartina alterniflora*) to tidal datums: a review. Estuaries 11:143–151

Mcleod E, Chmura GL, Bouillon S et al (2011) A blueprint for blue carbon: toward an improved understanding of the role of vegetated coastal habitats in sequestering CO_2. Front Ecol Environ 9:552–560

Mitra S, Wassmann R, Vlek PLG (2005) An appraisal of global wetland area and its organic carbon stock. Curr Sci 88:25–35

Morris JT (2007) Estimating net primary production of salt-marsh macrophytes, Chapter 7. In: Fahey TJ, Knapp AK (eds) Principles and standards for measuring primary production. Oxford University Press

Morris JT, Sundareshwar PV, Nietch CT et al (2002) Responses of coastal wetlands to rising sea level. Ecology 83:2869–2877

Redfield AC, Rubin M (1962) The age of salt marsh peat and its relation to recent changes in sea level at Barnstable Massachusetts. Proc Natl Acad Sci USA 48:1728–1734

Reed DJ (1995) The response of coastal marshes to sea-level rise: survival or submergence? Earth Surf Process Landf 20:39–48

Rodriguez E, Morris CS, Belz JE (2006) Global assessment of the SRTM performance. Photogramm Eng Remote Sens 72:249–260

Shipp RC, Belknap DF, Kelley JT (1991) Seismic-stratigraphic and geomorphic evidence for a post-glacial sea-level lowstand in the northern Gulf of Maine. J Coast Res 7:341–364

Turner RE, Swenson EM, Milan CS (2001) Organic and inorganic contributions to vertical accretion in salt marsh sediments. In: Weinstein MP, Kreeger DA (eds) Concepts and controversies in tidal marsh ecology. Springer, New York

Yang S, Chen J (1995) Coastal salt marsh and mangrove swamps in China. Chin J Oceanogr Limnol 13:318–324

Chapter 25
Research and Development Priorities Towards Recarbonization of the Biosphere

Rattan Lal, Klaus Lorenz, Reinhard F. Hüttl, Bernd Uwe Schneider, and Joachim von Braun

Abstract Despite the importance of the terrestrial biosphere for the global carbon (C) cycle and its potential to reduce the rate of enrichment of atmospheric carbon dioxide (CO_2) by anthropogenic emissions, there is incomplete and insufficient scientific knowledge to identify sources and sink of C, risks of biomes to climate change, and site-specific practices to recarbonizing the biosphere. Two options of mitigating climate change through management of biomes are (i) to enhance, manage and sustain biomass production and prolong the residence time of biomass C, and (ii) to improve the C balance within the biosphere. In addition, there is a lack of modus operandi on developing science-policy, nexus to identify and implement appropriate policy interventions to promote adoption of land use and management practices leading to recarbonization of the biosphere. In addition to reducing the magnitude of anthropogenic sources (e.g., deforestation, peatland cultivation, drainage of wetlands, excessive tillage), it is also important to identify and enhance the capacity of land-based C sinks. Further, C sequestration in the terrestrial biosphere

R. Lal (✉)
Carbon Management and Sequestration Center, The Ohio State University, 2021 Coffey Road, Columbus, OH 43210, USA
e-mail: lal.1@osu.edu

K. Lorenz
Global Soil Forum, IASS Institute for Advanced Sustainability Studies e.V., Berliner Strasse 130, 14467 Potsdam, Germany
e-mail: klaus.lorenz@iass-potsdam.de

R.F. Hüttl • B.U. Schneider
Deutsches GeoForschungszentrum GFZ, Telegrafenberg, 14473 Potsdam, Germany
e-mail: reinhard.huettl@gfz-potsdam.de; bernd_uwe.schneider@gfz-potsdam.de

J. von Braun
Zentrum für Entwicklungsforschung, Universität Bonn, Walter-Flex-Str. 3, 53113 Bonn, Germany
e-mail: jvonbraun@uni-bonn.de; a.vdveen@uni-bonn.de

R. Lal et al. (eds.), *Recarbonization of the Biosphere: Ecosystems and the Global Carbon Cycle*, DOI 10.1007/978-94-007-4159-1_25,

must compliment and not threaten or compete with other functions such as food production, water resources, nutrients and biodiversity. Priority biomes for recarbonization are peatlands, wetlands, degraded/desertified lands, and agroecosystems. Biomes with risks of positive feedback to climate change are permafrost and peatlands, and the soil organic carbon (SOC) pool. A global platform/instrument is needed to enhance soil-policy nexus, promote synergism and complimentary among organizations, addressing this issue of global significance.

Keywords Anthropogenic emissions • Greenhouse gases • Anthromes • Anthropocene • Global warming • Priority biomes • Peatlands • Permafrost • Positive feedback • Global carbon budget • Unknown carbon sink • Unknown carbon source • Cryosols • Carbon density • Rice paddy • Deforestation • Climate change risk analysis • CO_2 fertilization effect • Carbon burial in lakes • Dissolved organic carbon • Wild fires • The human dimensions

Abbreviations

C	carbon
CCS	Carbon Capture and Storage
CO_2	carbon dioxide
DOC	dissolved organic carbon
GCC	global carbon cycle
GHGs	greenhouse gases
LULCC	land use/land cover change
CH_4	methane
NPP	net primary production
POC	particulate organic carbon
SOC	soil organic carbon

25.1 Introduction

It is widely recognized that the response of the terrestrial biosphere to natural and anthropogenic perturbation strongly impacts the global carbon cycle (GCC), and can also affect the magnitude of the present and future climate change (Rayner 2010). While the terrestrial biosphere has been a major source of atmospheric carbon dioxide (CO_2) and other greenhouse gases (GHGs) since major transformation of the biosphere began by deforestation (Williams 2000) and the onset of settled agriculture millennia ago (Ruddiman 2003, 2006), there exist a C sink capacity in world's soil and vegetation that can be filled by judicious management of terrestrial biomes. Realization of the high potential drawdown of atmospheric CO_2 through carbon (C) sequestration in the biosphere estimated at ~50–70 ppm of CO_2 by 2150

(Hansen et al. 2008; House et al. 2003) poses numerous challenges which must be overcome. The basic principles of enhancing the terrestrial C pool in soils and vegetation are known (Lenton 2010). However, techniques of soil/plant/animal systems must be identified, validated and fine-tuned under site-specific conditions. Once identified, these techniques must be promoted for local adoption with due consideration of the human dimensions (i.e., economic, social, political, gender and cultural issues). Trying to keep global warming within 2°C limits as proposed by the Copenhagen Accord (Macintosh 2010) necessitates additional research about options and feasibility. Thus, the objective of this chapter is to outline research, development and outreach priories toward the goal of recarbonization of the terrestrial biosphere.

25.2 Researchable Priorities

25.2.1 Priority Biomes

The first step is to identify priority biomes with a high C sink capacity. Chapter 2 in this volume listed that priority biomes include peatland, degraded lands, and agricultural lands. In the event of projected global warming, soils under permafrost may thaw and contribute immensely to CO_2 and methane (CH_4) emissions with a positive feedback to climate change.

Permafrost is defined as subsurface earth material (i.e., soil, sediment, rocks) which remains frozen for at least two consecutive years (Schuur et al. 2008). Permafrost regions occupy 22% of the exposed land area (Zhang et al. 1999), and also occur in the highlands. The environmental and ecological characteristics of the permafrost (e.g., temperature, thickness, geographic continuity, soil C content) depend on the surface energy balance which varies widely due to latitude and altitude. The thickness of the permafrost layer ranges also widely in the glaciated areas and unglaciated regions (Schuur et al. 2008). Therefore, there are several types of permafrost: continuous, discontinuous, sporadic and isolated (Tarnocai et al. 2009). Similar to non-permafrost pedons, permafrost soils also comprise of a wide range of materials such as organic and inorganic components, rocks and gravels as skeletal materials, and of course, the ice. The ice may be clearly visible as wedges in layers or lenses or included in soil pores. Consequently, soils have a patterned ground caused by cryoturbation, and permafrost soils do not make a stable foundation for roads or buildings.

The permafrost region is an important reservoir of soil C. It has a high soil organic C (SOC) content, which can be a major climate-induced positive feedback to the anthropogenic climate change by decomposition and emission of CO_2 and CH_4 (Schuur et al. 2008). Yet, there are numerous uncertainties regarding the area, total soil C pool, and processes and properties which affect the dynamic of the large but highly sensitive region. Consequently, estimates of the SOC pool vary widely ranging from 1,220 to 1,576 Pg C (1 Pg = 10^{15} g) for 1-m depth, 2,376–2,456 Pg C

Table 25.1 Global carbon budget in 2010 (Adapted from Peters et al. 2011)

Parameter	Annual flux (Pg C $year^{-1}$)
Sources	
Fossil fuel + cement	9.1 ± 0.5
Land use	0.9 ± 0.7
Sinks	
Atmosphere	5.0 ± 0.2
Ocean	2.4 ± 0.5
Land	2.6 ± 1.0

Table 25.2 Deficiency of the data in Table 25.1

Sources	Flux (Pg C $year^{-1}$)	References
1. Human respiration	0.16	Oelkers and Cole (2008)
2. Animal respiration	0.86	Steinfeld et al. (2006)
3. Fire	3.94	Seiler and Crutzen (1980) and Levine et al. (1999)
4. Erosion	1.14	Lal (2003)
5. Atmospheric brown clouds	?	Iyngarasan and Ramanathan (2011)
6. Urban ecosystems	?	Pataki et al. (2011)
7. Abiotic carbon dioxide fluxes	?	Serrano-Ortiz et al. (2010)

to 2-m depth, and 2,727–2,807 Pg C for 3-m depth (Tarnocai et al. 2009). Schuur et al. (2008) estimated the SOC pool in permafrost soils at 1,872 Pg C, of which 277 Pg C is contained in peat soils. Total SOC pool to 3-m depth in all soils is estimated at 2,344 Pg C (Jobbágy and Jackson 2000). Thus, estimates of the total SOC pool need to be improved.

25.2.2 Uncertainties in the Global Carbon Budget

Establishing the C cycle and quantifying different components at the biome and global scales is important to understanding the "missing" C sources and sinks. The issue of "missing sink" or residual sink has been raised since 1990 (Tans et al. 1990), but there are also new sources which may become major contributors with the projected global warming (Woodwell et al. 1998). The magnitude and location of terrestrial C sources and sinks, existing and the new with the projected warming, are contributors to major uncertainties (House et al. 2003) in preparing credible and quantitative components of the GCC. It is difficult to balance the GCC without knowledge of the current and future C sources and sinks. The data in Table 25.1 for 2010 show the residual of the global CO_2 budget attributed to a land sink of 2.6 ± 1.0 Pg C (Peters et al. 2011). This was more than 1 Pg C below the previous 2 years but the land C sink is highly variable and also varies naturally (Le Quéré et al. 2009; Friedlingstein et al. 2010). Thus, there may also be an unknown C source. A compilation of some unknown sources are listed in Table 25.2. These include human respiration (Oelkers and Cole 2008), wild fires (Levine et al. 1999; Seiler and

Crutzen 1980), soil erosion (Lal 2003; Van Oost et al. 2007; Stallard 1998; Smith et al. 2005; Wang et al. 2010), Atmospheric Brown Clouds (Iyngarasan and Ramanathan 2011), urban ecosystems (Rayner 2010), and abiotic CO_2 flows due to weathering processes and subterranean cavity ventilation (Serrano-Ortiz et al. 2010). Several biomes which are presently sink of CO_2 (Cryosols, permafrost) may become major C sources with the projected warming and positive feedback. Thus, reconciling inconsistences in sources and sinks (House et al. 2003) remains a high priority. How terrestrial ecosystems could affect earth's climate depends on how alterations in their structure and functioning may occur due to land use and management (Foley et al. 2003), and on their effects on biophysical factors such as albedo, evaporation and surface roughness affecting climate on various temporal and spatial scales (Pielke et al. 2002).

25.2.3 Saturation of the Terrestrial Carbon Sinks

It is widely argued that the terrestrial biosphere has been a net C sink since 1980s to the magnitude of 0.3–0.5 Pg C $year^{-1}$ (Canadell et al. 2007). Its large year-to-year variability is primarily driven by variability in precipitation, surface temperature and radiation (Le Quéré et al. 2009). However, the current terrestrial sinks may not be permanent features of the terrestrial biosphere. The analyses presented in Chap. 2 of this book (Table 2.8) also showed that the capacity of the natural sinks (ocean, land) may be declining. Furthermore, the C sink capacity cannot be sustained indefinitely although reaching an equilibrium in soil profile SOC levels may be a long-term process (Wutzler and Reichstein 2007). The longevity of land sinks depends on a wide range of factors, including historic and past land use change, the magnitude of CO_2 fertilization effect, dynamics of high-C density biomes (i.e., peatlands, permafrost) (Canadell et al. 2007), among others.

25.2.4 The Anthromes and Anthropocene

The concept of "the Anthropocene" (Crutzen 2002) and of anthromes is extremely pertinent in the view of sustainable management of the Anthropocene. These concepts highlight the importance of human activity in altering the biosphere. However, there is a discussion about the on-set of the Anthropocene. Did it start with the beginning of the Industrial Revolution ~1,750 with the combustion of fossil fuel (Crutzen and Stoermer 2000) or did the anthropogenic greenhouse era begin thousands of years ago with the start of deforestation, cultivation or rice paddies and domestication of ruminants (Ruddiman 2003, 2006; Williams 2000)? Any consensus on this issue may help in prioritizing the strategies of biosequestration vs. Carbon Capture and Storage (CCS). Any focus on recarbonization of the biosphere will necessitate credible data on the land uses and rates of land use change since the dawn of human civilization (Goldewijk et al. 2011; Pielke et al. 2011).

25.2.5 Response of the Soil and Terrestrial Carbon Pools to Climate Change

The rate of mineralization of the SOC pool strongly depends on temperature and moisture regimes (Eglin et al. 2010). Further, the strong relation between SOC pool and soil quality also determines the net primary production (NPP) and the above- and below-ground biomass C. The climate-induced changes in soil structure, available water capacity, and nutrient retention and availability may strongly impact the SOC pool, its composition, turnover, and the attendant pedospheric processes. Yet, the response of the terrestrial C pool to the projected climate change is not known.

25.2.6 Climate Change Risk Analysis of the Terrestrial Ecosystems

There is a strong need to conduct climate-change risk analysis, both biophysical and socio-economical, of the terrestrial ecosystems. Identification and implementation of strategies to recarbonize the biosphere require credible risk analysis. Scholze et al. (2006) observed that the climate warming may induce high risks of forest loss in Amazonia but also in Eurasia China, Canada and Central America. The extent of areas affected by the projected global warming may depend on the magnitude of change in temperature (Scholze et al. 2006).

25.2.7 Management of the Anthromes

Judicious management of the terrestrial biosphere is essential to its recarbonization. The importance of aiming at forest and soils as potential sinks to reduce atmospheric CO_2 has been recognized by the scientific community (Hansen et al. 2008). Improved vegetation and soil management practices can recarbonize the biosphere by recovering some of the SOC released in the past as result of land use/land cover change (LULCC). SOC-accreting land use and management practices include phyto-engineering towards plants with deeper and bushy root ecosystems, increasing photosynthetic efficiency, optimized C_3/C_4 plant species mixtures, replacing annual with perennial crops, and soil addition of black carbon compounds. Yet, numerous questions remain about the terrestrial C stocks and biodiversity (Midgley et al. 2010), climate control of terrestrial C exchange on continental scale (Yi et al. 2010), the terrestrial biochemical feedbacks in the climate system (Arneth et al. 2010), among others. The pros and cons of technological options, the role of markets and pricing, and the vast diversity of site-specific conditions, need to be assessed. Karlen et al. (2009) responded to the suggestion about ocean burial of crop residues. Similarly, there are even fewer data especially for more fertile soils and soils of boreal and temperate regions but stronger claims about the potential of using biochar on soil C sequestration (Lal 2010). Comprehensive analyses are needed to assess the interaction of many terrestrial mitigation strategies and their competitive interactions (Obersteiner et al. 2010). For example, the role of biofuel production on delivery of C-neutral biomass for production of energy is mostly assessed on the assumption that the overall terrestrial sink remains constant overtime (Obersteiner et al. 2010). The competing uses of producing biofuel feedstock must be objectively assessed.

25.2.8 The Carbon Dioxide Fertilization Effect

There exists a lot of uncertainty about the magnitude (potential) and constraints to realizing the potential of CO_2 fertilization (Pacala and Socolow 2004; Lobell and Field 2008). For example, forest NPP is increased by elevated CO_2 but the response can diminish over time (Norby and Zak 2011). Further, the C accumulation at elevated CO_2 is driven by interactions between the C and N cycles, and probably also between the C and P cycles. Forest plant community structure may change but elevated CO_2 may have only minor effects on microbial community structure (Norby and Zak 2011). However, the CO_2 fertilization effect in boreal and tropical forest ecosystems is even less well known compared to temperate forest ecosystems (Hickler et al. 2008). Tree ring evidence indicates that the total magnitude of actual CO_2 fertilization in global forests over the twentieth century is relatively small, and that any CO_2-induced growth increases do not result in C allocations to persistent C pools (Gedalof and Berg 2010). In agroecosystems elevated CO_2 moderately stimulated photosynthesis of C_3 crops associated with limited gains in grain yield (Ainsworth et al. 2008). However, no significant yield stimulation was reported under well-watered conditions for C_4 crops but both C_3 and C_4 crops potentially profit from reduced demand for water at elevated CO_2. Also, modest photosynthetic responses of grassland species to long-term elevated CO_2 has been observed representing strong acclimation (i.e., down-regulation) of photosynthesis (Lee et al. 2011). Thus, long-lived strong stimulation of photosynthesis in terrestrial ecosystems by elevated CO_2 may be unlikely.

25.2.9 Carbon Burial in Lakes, Coastal Wetlands and Depositional Sites

The identification of unknown sinks needs credible information on fate and magnitude of C transported into aquatic ecosystems, and depositional sites. Globally, about 1 Pg C $year^{-1}$ is delivered to inland waters from the landscape, and 0.9 Pg C $year^{-1}$ to the ocean (Cole et al. 2007). Specifically, about 0.3 Pg of dissolved organic carbon (DOC) and 0.2 Pg of particulate organic carbon (POC) are discharged annually by rivers to the global ocean representing 50% of total C export by rivers (Junk et al. 1989). However, very little terrestrial derived organic C is in the global ocean, and only about 30% of terrestrial derived organic C buried in marine sediments is of terrestrial origin in muddy deltaic regions with high sedimentation rates (Bianchi 2011). Dean (1999) estimated that if all lakes bury organic C at the same rate as Lake Michigan (5 $g\,m^{-2}$ $year^{-1}$), then the annual rate of C burial in 1.18×10^{12} m^2 area of freshwater lakes is 6 Tg $year^{-1}$. Coastal wetlands are also important to C dynamics (Aufdenkampe et al. 2011). Global freshwater systems including lakes and wetlands are estimated to release between 0.7 and 3.3 Pg C annually as CO_2 (summarized in Butman and Raymond 2011). However, this flux may be underestimated as temperate rivers between 25°N and 50°N alone may release about 0.5 Pg C $year^{-1}$ (Butman and Raymond 2011). Despite their importance, the information on all pieces of jigsaw to integrate CO_2 fluxes in the coastal ecosystems and inland waters is not available (Borges 2005; Cole et al. 2007; McLaughlin et al. 2003).

25.2.10 Wild Fires

Wild fires have a strong impact on the GCC with the global annual area burned by fire for the years 1997–2008 ranging from 330 to 431 Mha (Giglio et al. 2010). Average global fire C emissions for the 1997–2009 period were 2.0 Pg C $year^{-1}$ and ranged from 1.5 to 2.8 Pg C $year^{-1}$ for individual years (van der Werf et al. 2010). Projected climate change may increase the intensity and frequency of wild fires (Flannigan et al. 2009). In this regards, improved fire data are needed on C dynamics, and the production and retention of black carbon compounds such as char, charcoal, biochar or soot.

25.2.11 The Human Dimensions

Recarbonization of the biosphere shall contribute to improved livelihoods of people. It can be part of a sound green growth strategy. Recarbonization is a public goods production. However, it must not be seen in isolation as recarbonization – if overextended – may compete with current use of biomass, incl. for food and feed. Thus, key policy instruments must minimize risks and enhance opportunities for the optimal management of C in the terrestrial biosphere. Research and development for enhanced biomass production and improved efficiency of its utilization, including for recarbonization, is part of a strategy. Given that most of world land users are small farmers and many of them are women farmers, inclusion of these main actors in a recarbonization strategy is needed from the beginning. The cost of inaction may be much higher than that of judicious action such as payments for ecosystem services including that for C sequestration in cropland soils (Robbins 2011). Realizing important ancillary benefits (e.g., agronomic production, food security, water security) also have strong policy implications.

25.3 Conclusions

Research focus is needed to understanding the role of natural and managed ecosystems on the C cycles at biome and global scales. This requires quantitative assessment of the impact of anthropogenic activities/perturbations on C pool and fluxes in present and future climatic conditions. Data from long-term field studies are needed to assess the impact of land use and management on C pools and fluxes for ecologically-sensitive biomes (i.e., permafrost, peatlands, wetlands, tropical rainforest), intensively-managed ecosystems (e.g., agroecosystems, urban lands), and those which can be managed to enhance the terrestrial C pool (e.g., desert biomes, degraded soils and desertified ecosystems). Further, new long-term field experiments must be established to identify land use and management strategies to enhance biospheric C pools, in particular, for systems depleted in C by previous LULCC.

The rapidly changing social and bioclimatic conditions necessitate a continual reassessment of researchable priorities, identification of science-policy nexus, developing and strengthening of the mechanisms and ways by which the scientific knowledge is produced and delivered to key stakeholder and policy makers. Policy, practices, and behavior of people vary widely across the world and over time. Thus, the role of people and communities must be integrated in all initiatives. This requires a trans-disciplinary understanding of the inter-relationship of the biophysical processes of enhancing the C pool in terrestrial biosphere with the human dimensions pertaining to economics, social and political factors. Carbon markets need to be developed to reward land managers through payments for ecosystem services. There is a strong need to initiate a global platform/instrument which can prioritize the issues concerning science-policy nexus, synthesize existing scientific knowledge into a language that policy makers can use, create awareness among policy makers about issues of global significance, and enhance synergisms and complimentary among institutions involved.

References

Ainsworth EA, Beier C, Calfapietra C et al (2008) Next generation of elevated [CO_2] experiments with crops: a critical investment for feeding the future world. Plant Cell Environ 31:1317–1324

Arneth A, Harrision SP, Zaehle S et al (2010) Terrestrial biogeochemical feedbacks in the climate system. Nat Geosci 3:525–532

Aufdenkampe AK, Mayorga E, Raymand PA et al (2011) Riverine coupling of biogeochemical cycles between land, oceans, and atmosphere. Front Ecol Enivorn 9(1):53–60. doi:10.1890/100014

Bianchi TS (2011) The role of terrestrially derived organic carbon in the coastal ocean: a changing paradigm and the priming effect. Proc Natl Acad Sci USA 108:19473–19481

Borges AV (2005) Do we have enough pieces of the jigsaw to integrate CO(2) fluxes in the coastal ocean? Estuaries 28(1):3–27. doi:10.1007/BF02732750

Butman D, Raymond PA (2011) Significant efflux of carbon dioxide from streams and rivers in the United States. Nat Geosci 4:839–842
Canadell JG, Pataki D, Gifford R et al (2007) Saturation of the terrestrial C sinks, Chapter 6. In: Canadell JG, Pataki D, Pitelka L (eds) Terrestrial ecosystems in a changing world, The IGBP series. Springer, Berlin
Cole JJ, Prairie YT, Caraco NF et al (2007) Plumbing the global carbon cycle: integrating inland waters into the terrestrial carbon budget. Ecosystems 10(1):171–184. doi:10.1007/s10021-006-9013-8
Crutzen PJ (2002) The Anthropocene. J Phys IV France 12: Pr 10–1 to Pr 10–5. doi:10.1051/jp4:20020447
Crutzen PI, Stoermer EF (2000) The "Anthropocene". IGBP Newsl 41:12
Dean WE (1999) Magnitude and significance of carbon burial in lakes, reservoirs and northern peatlands. USGA Fact Sheet FS-058-99, Washington, DC
Eglin T, Ciais P, Pias SL et al (2010) Historical and future perspectives of global soil carbon response to climate and land-use changes. Tellus 62B:700–718
Flannigan MD, Krawchuk MA, de Groost WJ et al (2009) Implications of changing climate for global wild fire. Int J Wildland Fire 18:483–507
Foley JA, Costa MC, Deline C et al (2003) Green surprise? How terrestrial ecosystems could affect earth's climate. Front Ecol Environ 1:38–44
Friedlingstein P, Houghton RA, Marland G et al (2010) Update on CO_2 emissions. Nat Geosci 3:811–812
Gedalof Z, Berg AA (2010) Tree ring evidence for limited direct CO_2 fertilization of forests over the 20th century. Glob Biogeochem Cycle 24, GB3027. doi:10.1029/2009GB003699
Giglio L, Randerson JT, van der Werf GR et al (2010) Assessing variability and long-term trends in burned area by merging multiple satellite fire products. Biogeosciences 7:1171–1186
Goldewijk KK, Neusen A, Van Drecht G et al (2011) The HYDE 3.1 spatially explicit database of human0induced global land-use change over the past 12,000 years. Glob Ecol Biogeogr 20:73–86
Hansen J, Sato M, Kharecha P et al (2008) Target atmospheric CO_2: where should humanity aim? Open Atmos Sci 2:217–231
Hickler T, Smith B, Prentice IC et al (2008) CO_2 fertilization in temperate FACE experiments not representative of boreal and tropical forests. Glob Change Biol 14:1531–1542
House JI, Prentice IC, Ramakutty N et al (2003) Reconciling apparent inconsistencies in estimates of terrestrial CO_2 sources and sinks. Tellus 55B:345–363
Iyngarasan M, Ramanathan V (2011) Atmospheric brown clouds: and integrated approach for understanding and mitigating climate change and other environmental problems resulting from atmospheric pollution. ICAC News 45:20–23
Jobbágy E, Jackson R (2000) The vertical distribution of soil organic carbon and its relation to climate and vegetation. Ecol Appl 10:423–436
Junk WJ, Bayley PB, Sparks RE (1989) The flood-pulse concept in river-floodplain system. In: Dodge DP (ed) Proceedings of the international large river symposium. Special publication. Can J Fisheries and Aquatic Science 106:110–127
Karlen D, Lal R, Follett RF et al (2009) Crop residues: the rest of the story. Environ Sci Technol 43:8011–8015
Lal R (2003) Soil erosion and the global C budget. Environ Int 29:437–450
Lal R (2010) Managing soils and ecosystems for mitigating anthropogenic carbon emissions and advancing global food security. Bioscience 60:708–721
Le Quéré C, Raupach MR, Canadell JG et al (2009) Trends in the sources and sinks of carbon dioxide. Nat Geosci 2:831–836
Lee TD, Barrott SH, Reich PB (2011) Photosynthetic responses of 13 grassland species across 11 years of free-air CO_2 enrichment is modest, consistent and independent of N supply. Glob Change Biol 17:2893–2904
Lenton TM (2010) The potential of land-based biological CO_2 removal to lower future atmospheric CO_2 concentration. Carbon Manage 1:145–160

Levine JS, Bobbe T, Ray N et al (1999) Wildland fires and environment. UNEP, Nairobi, Kenya. p 46 (ISBN: 92-807-1742-1)
Lobell DB, Field CB (2008) Estimation of the CO_2 fertilization effect using growth rate anomalies of CO_2 and crop yields since 1961. Glob Change Biol 14:39–45
Macintosh A (2010) Keeping warming within the 2°C limits after Copenhagen. Energy Policy 38:2964–2975
McLaughlin CJ, Smith CA, Buddemeier RW et al (2003) Rivers, runoff, and reefs. Glob Planet Change 39(1–2):191–199. doi:10.1016/SO921-8181(03)00024-9
Midgley GF, Bknd WJ, Kapos V (2010) Terrestrial carbon stocks and biodiversity: key knowledge gaps and some policy implications. Curr Opin Environ Sustain 2:264–270
Norby RJ, Zak DR (2011) Ecological lessons from free-air CO_2 enrichment (FACE) experiments. Annu Rev Ecol Evol Syst 42:181–203
Obersteiner M, Böttcher H, Yamagata Y (2010) Terrestrial ecosystem management for climate change mitigation. Curr Opin Environ Sustain 2:271–276
Oelkers EH, Cole DR (2008) Carbon dioxide sequestration: a solution to a global problem. Elements 4:305–310
Pacala S, Socolow R (2004) Stabilization wedges: solving the climate problem for the next 50 years with the current technologies. Science 305:968–972
Pataki DE, Carreiro MM, Cherrier J, Grulke NE, Jennings V, Pincete S, Pouyat RV, Whitlow TH, Zipperer WC (2011) Coupling biogeochemical cycles in urban environments: ecosystem services, green solutions, and misconceptions. Front Ecol Enivorn 9(1):27–36
Peters GP, Marland G, Le Quéré C et al (2011) Rapid growth in CO_2 emissions after the 2008–2009 global financial crisis. Nat Clim Change. doi:10.1038/nclimate1332
Pielke RA, Marland G, Betts RA et al (2002) The influence of land-use change and landscape dynamics on the climate system: relevance to climate-change policy beyond the radiative effect of greenhouse gases. Philos Trans R Soc Lond A 360:1705–1719
Pielke RA Sr, Potman A, Niyosi D et al (2011) Landuse/land cover changes and climate: modeling analysis and observational evidence. WIRE Clim Change 2:828–850
Rayner PJ (2010) The current state of carbon-cycle data assimilation. Curr Opin Environ Sustain 2:2890296
Robbins M (2011) Crops and carbon: paying farmers to combat climate change. Earthscan, New York
Ruddiman WF (2003) The anthropogenic greenhouse era began thousands of years ago. Clim Change 61:261–293
Ruddiman WF (2006) On the Holocene CO_2 rise: anthropogenic or natural? EOS Trans Am Geophys Union 87:352–353
Scholze M, Knorr W, Arnell NW (2006) A climate-change risks analysis for world ecosystems. Proc Natl Acad Sci USA 35:13116–13120
Schuur EAG, Bockheim J, Canadell JG et al (2008) Vulnerability of permafrost carbon to climate change: implications for the global carbon cycle. Bioscience 58:701–714. doi:10.1641/B580807
Seiler W, Crutzen PJ (1980) Estimates of gross and net fluxes of carbon between the biosphere and the atmosphere from biomass burning. Clim Change 2:207–247
Serrano-Ortiz P, Roland M, Sanchez-Moral S et al (2010) Hidden, abiotic CO_2 flows and gaseous reservoirs in the terrestrial carbon cycle: review and perspectives. Agric For Meteorol 150:321–329
Smith SV, Sleezer RO, Renwick WH et al (2005) Fates of eroded soil organic carbon. Ecol Appl 15:1929–1940
Stallard RF (1998) Terrestrial sedimentation and the carbon cycle: coupling weathering and erosion to carbon burial. Glob Biogeochem Cycle 12:231–257
Steinfeld H, Gerber P, Wassenaan T et al (2006) Livestock's long show. FAO, Rome
Tans PD, Fund IY, Takahasi T (1990) Observational constraints on the global atmospheric CO_2 budget. Science 247:1431–1438
Tarnocai C, Canadell JG, Schuur EAG et al (2009) Soil organic carbon pools in the northern circumpolar permafrost region. Glob Biogeochem Cycle 23, GB2023. doi:10.1029/2008GB003327

van der Werf GR, Randerson JT, Giglio L et al (2010) Global fire emissions and the contribution of deforestation, savanna, forest, agricultural, and peat fires (1997–2009). Atmos Chem Phys 10:11707–11735

Van Oost K, Wuine TA, Glovers G et al (2007) The impact of agricultural soil erosion in the global C cycle. Science 318:626–629

Wang Z, Govers G, Steegen A et al (2010) Catchment-scale carbon redistribution and delivery by water erosion in an intensively cultivated area. Geomorphology 124:65–74

Williams M (2000) Dark ages and dark areas: global deforestation in the deep past. J Hist Geogr 26:28–46

Woodwell GM, Mackenzie FT, Houghton RA et al (1998) Basic feedbacks in the warming of the earth. Clim Change 40:495–518

Wutzler T, Reichstein M (2007) Soils apart from equilibrium – consequences for soil carbon balance modelling. Biogeosciences 4:125–136

Yi C, Ricciuto D, Li R et al (2010) Climate control of terrestrial carbon exchange across biomes and continents. Environ Res Lett 5(034007)

Zhang T, Barry RG, Knowles K et al (1999) Statistics and characteristics of permafrost and ground-ice distribution in the northern hemisphere. Polar Geogr 23:132–154

Index

R. Lal et al. (eds.), *Recarbonization of the Biosphere: Ecosystems and the Global Carbon Cycle*, DOI 10.1007/978-94-007-4159-1,

C

T